ENVIRONMENTAL ENGINEERING
A Design Approach

Arcadio P. Sincero Sr., D.Sc., P.E.
Assistant Professor of Civil Engineering
Morgan State University

Gregoria A. Sincero, M. Eng., P.E.
Water Resources Engineer IV
Maryland Department of the Environment

Prentice Hall, Upper Saddle River, New Jersey 07458

Library of Congress Cataloging-in-Publication Data

Sincero, Arcadio Pacquiao.
    Environmental engineering: a design approach / Arcadio Pacquiao
Sincero, Gregoria Alivio Sincero.
        p.   cm.

    Includes bibliographical references and index.
    ISBN 0-02-410564-3
    1. Environmental engineering.   I. Sincero, Gregoria Alivio.
II. Title.
    TD146.S54   1996
    628--dc20   95-19010
                                                          CIP

Acquisitions editor: *Bill Stenquist*
Editorial/production supervision: *Barbara Marttine Cappuccio*
Editorial assistant: *Meg Weist*
Copyeditor: *Barbara Zeiders*
Cover design: *Jayne Conte*
Manufacturing buyer: *Donna Sullivan*

© 1996 by Arcadio P. Sincero and Gregoria A. Sincero
Prentice-Hall, Inc.
Simon & Schuster/A Viacom Company
Upper Saddle River, New Jersey 07458

The author and publisher of this book have used their best efforts in preparing this book. These efforts include the development, research, and testing of the theories and programs to determine their effectiveness. The author and publisher make no warranty of any kind, expressed or implied, with regard to these programs or the documentation contained in this book. The author and publisher shall not be liable in any event for incidental or consequential damages in connection with, or arising out of, the furnishing performance, or use of these programs.

Printed in the United States of America

10 9 8 7 6 5 4 3 2

ISBN 0-02-410564-3

Prentice-Hall International (UK) Limited, *London*
Prentice-Hall of Australia Pty. Limited, *Sydney*
Prentice-Hall Canada Inc., *Toronto*
Prentice-Hall Hispanoamericana, S.A., *Mexico*
Prentice-Hall of India Private Limited, *New Delhi*
Prentice-Hall of Japan, Inc., *Tokyo*
Simon & Schuster Asia Pte. Ltd., *Singapore*
Editora Prentice-Hall do Brasil, Ltda., *Rio de Janeiro*

Contents

5   INTRODUCTION TO ENVIRONMENTAL QUALITY MODELING   167

6   CONVENTIONAL WATER TREATMENT   225

8   SLUDGE TREATMENT AND DISPOSAL   358

9   ADVANCED WASTEWATER AND WATER TREATMENT AND LAND TREATMENT SYSTEMS   400

10 POLLUTION FROM COMBUSTION AND ATMOSPHERIC POLLUTION — 467

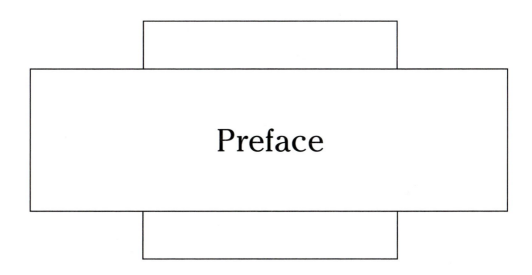

Preface

This book is intended for junior and senior students in undergraduate engineering who are required to take the environmental engineering course. Practicing engineers as well as students and professors in graduate schools may also find this book useful as a reference or as a textbook. The book could be covered in three semesters at three credits per semester. For schools offering only one or two courses in environmental engineering this book gives the instructor the liberty of picking the particular topics required in a given curriculum design. At Morgan State University, for example, this book is used in two environmental engineering courses. One is in the junior year when the student has already passed fluid mechanics and general chemistry, and the other in the senior year. Only relevant topics in appropriate chapters as required by the curriculum in the civil engineering department of this institution are picked for classroom discussions. It is a requirement that students have passed a fluid mechanics course before using this book.

Essentially all aspects of environmental engineering—water and wastewater, environmental hydrology, environmental hydraulics and pneumatics, air, solid waste, noise, environmental modeling, hazardous waste—are covered. It is therefore ideal for courses in environmental engineering, where a grounding in the principles of the overall discipline of environmental engineering as they relate to design is desired. Both principles and design are taught in this book. Principles are enunciated in the simplest way possible. Equations presented are first derived, except those that are obtained empirically. Statements such as "It can be shown ..." are not used. The intent is to impart the principles and concept of the subject matter, which may not be done by using it-can-be-shown statements. In addition, nobody can guarantee remembering how "it can be be

xv

shown …," and when a book includes so many "it can be showns," reading it becomes very frustrating, especially when the reader is in doubt and has no way to know whether or not the equations presented are correct. After principles are presented, example problems follow that illustrate the techniques of design using the concept presented. Each chapter includes numerous problems that can be worked out by students and assigned as homework by the instructor. Some of these problems may be assigned as computer homework.

In the environmental profession, environmental modeling is a fragmented discipline. People in the water quality area have their own water quality models; people in the groundwater quality area have their own groundwater models; and people in the air quality area have their own air quality models. Modelers in one area are normally not comfortable to do modeling in another area simply because they have not been exposed in these other areas. In Chapter 5 we show that these areas are all related and that modeling is based simply on the conservation principle. The unifying concept behind the conservation principle is the Reynolds transport theorem.

The unifying concept approach is also applied to settling and cake filtration. The theory of settling is derived in Chapter 6, the result also being applied in Chapter 7. However, since settling is simply the falling down of particles in response to a force field, the theory is also applied to the "settling" of particles in cyclones and electrostatic precipitators, as well as of particulates in airflow settling chambers. The theory of cake filtration, derived in Chapter 8, is also applied to determine the pressure drop across reverse osmosis membranes and pressure drops across dust cakes in baghouse filters.

Equivalents and equivalent weights are two troublesome and confusing subjects. If the chemistry or environmental engineering literature were reviewed, these subjects will be found not to be well explained. Equivalents and equivalent weights in a unified fashion are explained herein using the concept of reference species.

There are two prerequisites for this book: a course in general chemistry and a course in fluid mechanics. *The fluid mechanics requirement is heavily stressed, since this book may not be used by students who have not had the course.* To study Chapter 14, the reader must have completed the sound component of physics. In general, no chapter of the book is a prerequisite for subsequent chapters.

Portions of Chapters 5, 6, 7, and 12 require background knowledge of the Reynolds transport theorem. It is strongly suggested that students be thoroughly grounded in this theorem before attempting to discuss the material covered in these chapters. Understanding would be very difficult without this background knowledge. This theorem is discussed in any good book on fluid mechanics and is also discussed in Appendix 21.

As with any other textbook, some errors may have slipped through. Readers are encouraged to bring any errors or omissions to the attention of the author or the publisher.

Acknowledgments

No author can take entire credit for publishing a book. In one way or another, many people have helped me in seeing this book to completion.

I acknowledge Bill Stenquist, executive editor at Prentice Hall, who put his trust in my abilities and whose criticisms helped shape this book. I also acknowledge the panel of reviewers; their criticisms helped to make the book better. Meg Weist, editorial assistant at Prentice Hall, was always very helpful in every way during the course of this project. I also acknowledge the extraordinary effort made by the executive editor and the production editors, Barbara M. Cappuccio and Jennifer Wenzel, in pushing to meet the publishing deadline. My thanks also go to members of the editorial, production, marketing, and manufacturing departments who helped to meet the publishing schedule. The copy editor, Barbara Zeiders, has made an extraordinary effort in editing the manuscript.

The book was written during my tenure at Morgan State University. I acknowledge the administrators of this institution, in particular, Dr. Earl S. Richardson, president; Dr. Clara I. Adams, vice president for academic affairs; Dr. Eugene M. DeLoatch, dean of the School of Engineering; and Dr. Lewis P. Clopton, chairman of the Department of Civil Engineering. I make special mention of a colleague, Dr. Robert Johnson, who was Acting Chairman of the Department of Civil Engineering when I came on board.

I would like to thank the following people who reviewed this book: Rodney A. James, Montana College of Mineral Science and Technology; Howard M. Liljestrand, University of Texas, Austin; Tim Keener, University of Cincinatti; Stephan J. Nix, Syracuse University; Neil J. Hutzler, Michigan Technological University; C. David Cooper,

University of Central Florida; Donald E. Modesitt, University of Missouri-Rolla; and Gene F. Parkin, University of Iowa.

This acknowledgment would not be complete if I did not mention three of my former professors who helped me in a large number of ways: Dr. Roscoe F. Ward, Dr. Rolf T. Skrinde, and Prof. Mainwaring B. Pescod. Dr. Ward was formerly dean and is now professor in the School of Applied Science at Miami University. Dr. Skrinde is a professor of civil/environmental engineering at Seattle University. Prof. Pescod was (and probably still is) a professor at the University of New Castle upon Tyne, U.K. They were my professors at the Asian Institute of Technology, Bangkok, where I earned my master's degree in environmental engineering.

I acknowledge my wife, Gregoria, for writing Chapter 11 on solid waste management, and my students, especially two former students, Al-Nisa Montague and Della Morrison, for their helpful suggestions and proofreading of parts of the chapters on water and wastewater treatment.

Appendix 22 is a guide to the literature on and instructions for the use of the water quality model computer program WAT25. WAT25 is an improved version of FREE1P5, a water quality model for discharge into free-flowing streams written by this author. (The original of FREE1P5 was TI-59, the model used to set discharge limits for all sewage treatment plants in the state of Maryland.) My sons, Roscoe and Arcadio, wrote WAT25, which can now model discharge into both free-flowing streams and one-dimensional tidal waters (such as one-dimensional estuaries). Whereas inputting and revision of input into FREE1P5 were difficult, these tasks are now very simple and easy using WAT25. Designed to be interactive, the model is very easy to run and can handle any number of discharges and stations, depending only on the amount of disk space on the diskette or hard drive. The diskette provided with this book contains the source code in QuickBASIC v. 4.5 and the executable version (WAT25.EXE) of the model.

I dedicate this book to members of my family: Gregoria, my wife; Roscoe and Arcadio Jr., my sons; the late Gaudiosa Pacquiao Sincero, my mother; Santiago, my father; the late Aguido Alivio and the late Teodora Managase Alivio, my father-in-law and mother-in-law, respectively; Meliton, my brother; Anelda and Feliza, my sisters; and Col. Miguel M. Alivio, M.D., of the armed forces of the Philippines, my brother-in-law.

Arcadio P. Sincero
Morgan State University

About the Authors

Gregoria A. Sincero is a senior level Water Resources Engineer in the Maryland Department of the Environment. She was also a former professor at the Cebu Institute of Technology, Philippines. She holds a Bachelor's degree in Chemical Engineering from the Cebu Institute of Technology and a Master's degree in Environmental Engineering from the Asian Institute of Technology, Bangkok, Thailand. She is a registered Professional Engineer in the Commonwealth of Pennsylvania. She is a member of the American Water Works Association.

Mrs. Sincero has practical experience both in engineering and governmental regulations. She was a Senior Chemist/Microbiologist in the Ashburton Filters of Baltimore City. In the state of Maryland, she was a Water Resource Engineer in the Water Resources Administration, Department of Natural Resources and a Water Resources Engineer in the Office of Environmental Programs, Department of Health and Medical Hygiene, before joining in her present position.

Arcadio P. Sincero is Assistant Professor of Civil Engineering at Morgan State University, Baltimore, Maryland, where he is coordinator for courses in environmental and hydraulic engineering. He was also a former professor at the Cebu Institute of Technology, Philippines. He has a Bachelor's degree in Chemical Engineering from the Cebu Institute of Technology, a Master's degree in Environmental Engineering from the Asian Institute of Technology, Bangkok, and a Doctor of Science degree in Environmental Engineering from the George Washington University. He is a registered professional engineer in the Commonwealth of Pennsylvania and in the state of Maryland. He is a

Member of the American Society of Civil Engineers, a Member of the American Institute of Chemical Engineers, a member of the Water Environment Federation, a member of the Asian Society of Environmental Protection, and a member of the American Association of University Professors.

Dr. Sincero has a wide variety of practical experiences. He was a shift supervisor in a copper processing plant and a production foreman in a corn starch processing plant in the Philippines. He was a CPM (Critical Path Method) Planner in a construction management firm and a Public Works Engineer in the City of Baltimore. In the State of Maryland, he was a Public Health Engineer in the Bureau of Air Quality and Noise Control, Department of Health and Mental Hygiene; a Water Resources Engineer in the Water Resources Administration, Department of Natural Resources; a Water Resources Engineer in the Office of Environmental Programs, Department of Health and Mental Hygiene; and a Water Resources Engineer in the Water Management Admininstration, Maryland Department of the Environment (MDE). His last position for the State of Maryland was as a Chief of Permits Division of the Construction Grants and Permits Program of the Water Management Administration at MDE. These practical experiences have allowed Dr. Sincero to gain a wide range of environmental engineering and regulatory experiences: air, water, solid waste, and environmental quality modeling.

CHAPTER 1

Introduction

This chapter is a brief introduction to the book. Environmental engineering is defined and the important topics of environmental engineering covered in the book are discussed. Environmental legislation that has been passed in the United States and some environmental tragedies are also mentioned briefly.

ENVIRONMENTAL ENGINEERING

What is environmental engineering? The answer to this question must be tied to the definition of engineering itself. Engineering may be defined as the application, under constraints of scientific principles to the planning, design, construction, and operation of structures, equipment, and systems for the benefit of society. If the tasks performed by environmental engineers were examined, it would be found that the engineers deal with structures, equipment, and systems that are designed to *protect and enhance the quality of the environment and to protect and enhance public health and welfare.* For example, the environmental engineer plans, designs, constructs, and operates sewage treatment plants to prevent the pollution of receiving streams. In other words, these structures are built to *protect and enhance water quality.* The environmental engineer also builds and operates water treatment plants. Clean, bacteriologically safe drinking water *protects and enhances public health.* Environmental engineers plan, design, construct, and operate air pollution control equipment. The resulting cleaner air is conducive to people's good health and prevents the deterioration of materials through the harmful effects of air pollution. Such equipment thus *protects and enhances public health and welfare*.

All these activities are performed by environmental engineers under constraints, which may involve shortage of funds, political pressure, inadequate space in which to build the structures, social and racial considerations, and similar considerations that limit the freedom of design. *Environmental engineering will now be defined as the application of engineering principles, under constraint, to the protection and enhancement of the quality of the environment and to the enhancement and protection of public health and welfare*.

Traditionally, environmental engineering has been the province of the civil engineering profession. Before 1968, this branch of engineering was called the *sanitary engineering* option of civil engineering in the United States and in countries with colonial ties to the United States such as the Philippines, where it was a specialized study after graduation as a civil engineer. In Great Britain and in countries with colonial ties to Great Britain, such as India, this profession was called *public health engineering*. Around 1968, the name *sanitary engineering* or *public health engineering* was changed to *environmental engineering*.

COVERAGE

In this book we address the following important topics of environmental engineering: environmental chemistry and biology, environmental hydrology, environmental hydraulics and pneumatics, water treatment, wastewater treatment, solid waste management, air pollution control, hazardous waste management and risk assessment, noise pollution and control, and environmental quality modeling.

Environmental Chemistry and Biology

Chemistry and biology are huge disciplines in themselves. Topics discussed in this book include those that are of direct relevance to environmental engineering. These include discussions on pathogens and indicator organisms, tests for the coliform group of organisms, drinking water standards, equivalents and equivalent weights, microbial thermodynamics, solubility product constants, biological combustion, chemistry of the anaerobic process, and eutrophication chemistry.

Environmental Hydrology

Environmental engineers need to design reservoirs and they need to control groundwater contamination. They may also be needed to design the drainage system of a parking lot, shopping centers, developments, and the like. In this section we therefore discuss surface water hydrology, reservoirs, groundwater hydrology, and groundwater contamination.

Environmental Hydraulics and Pneumatics

Treated water must be distributed to consumers and the resulting spent water must be collected for conveyance to a sewage treatment plant. Also, pumping stations are needed for the collection of sewage and the distribution of treated water. In addition, polluted air from various processes in industry needs to be captured and conveyed to air pollution control devices. We discuss water supply and distribution, collection and transmission of water, collection of wastewater, pumping stations, and pneumatic systems.

Environmental Quality Modeling

Mathematical models are developed to determine the effect of discharge on the quality of the environment. These models include models for surface water quality, air quality, and groundwater quality.

Water Treatment

Topics include the normal unit processes and operations essential for treating a raw water source to produce drinkable water. Topics included are settling, filtration, coagulation and flocculation, water softening, and disinfection. Advanced water treatment is also discussed.

Wastewater Treatment

Treatment topics discussed include primary, secondary, and advanced treatments. Principles of aeration and sludge treatment and disposal are also discussed. Advanced treatment includes nitrification and denitrification and such physical processes as reverse osmosis and carbon adsorption. Wetland treatment systems are also discussed. Sludge treatment includes such topics as anaerobic and aerobic sludge treatment as well as cake filtration and composting.

Pollution from Combustion and Atmospheric Pollution

Topics include discussion of pollutants that result from combustion, such NO_x, SO_x, and CO. Automotive emission control, photochemical smog, the NO_2 photolytic cycle, acid rain, tropospheric ozone destruction due to CFCs, and greenhouse effects are all discussed.

Solid Waste Management

The three basic functional elements of solid waste management—collection, processing, and disposal—are discussed. Five basic aspects in the implementation of sanitary landfilling are also discussed: site selection, leachate control, gas control, operation plan, and permit application.

Air Pollution Control

In general, the types of control discussed correspond to the type of air pollutants emitted: gas and particulate. Particulate controls discussed include cyclones, gravitational settling, electrostatic precipitators, scrubbers, and bagfilters. Gaseous controls include absorption in towers, adsorption to activated carbon, and incineration. Indoor air pollution is also discussed.

Hazardous Waste Management and Risk Management

Topics include RCRA (Resource Conservation and Recovery Act) wastes, radioactive wastes, the hazardous waste transportation and manifest, methods of final disposal of nuclear wastes, and risk assessment.

Noise Pollution and Control

Methods of noise measurement, noise criteria, and control are discussed. Noise is also discussed in three types of settings: industrial, community, and airport. Indoor and outdoor noise propagation is discussed.

ENVIRONMENTAL TRAGEDIES AND LEGISLATIONS

In 1948, Donora, a small town at the bend of the Monongahela River in Pennsylvania (population 14,000), had three main industrial plants: a steel mill, a wire mill, and a zinc plating plant. During the last week in October 1948, a weather condition that prevented mixing of air prevailed in the area, resulting in an air pollution condition called *smog*. The smog contained primarily very fine particles admixed with sulfur dioxide. By Wednesday, the pollution became so intense that a streak of carbon appeared to hang motionless in the air, and the visibility was so poor that even the natives were lost.[1] By Friday, doctors' offices and hospitals were inundated with calls for medical attention. By the end of the episode, 21 people had died and about 6000 people became ill.

A second notable episode occurred in London, England, in 1952. On Thursday, December 4 of that year, a high-temperature air mass that hovered over southern England created adverse, nonmixing air conditions. A fog settled over London and as particulate and sulfur dioxide rose to higher levels, the atmosphere started to blacken. Visibility was reduced to zero and "a white collar became almost black within 20 minutes."[2] Smog is especially irritating to the respiratory system, and many people developed red eyes, burning throats, and nagging coughs. The elderly and people with chronic respiratory problems began to die, as well as otherwise healthy people whose jobs kept them near

---

[1]H. H. Schrenk and others (1949). "Air Pollution in Donora, Pa." *U.S. Public Health Service Bulletin 306.*

[2]C. Woolf (1969). Letter to the editor. *Archives of Environmental Health,* 18(715).

the smog. On December 9 when the smog lifted, 4000 Londoners had died. The episode was enough to move the British to pass the Clean Air Act of 1956. The United States had passed its Clean Air Act of 1955 a year earlier. From this brief account, control of air pollution could no longer be ignored. Even so, however, more deaths followed in London in 1956, 1957, 1959, 1962, and 1963. Even in New York, 200 to 400 people died of air pollution in 1963.

A third episode worthy of note reflects the effect of hazardous wastes. In 1892, William T. Love built a canal in Niagara Falls, New York. He intended to use the power generated from this canal to transform the area into an industrial complex. The proposed power to be generated by the canal could not compete with the new technology of the alternating-current electricity; Love's vision did not materialize. Nevertheless, the canal was built, although it was never finished. In the 1930s, the Hooker Chemical and Plastic Corporation, whose products included pesticides, plasticizers, and caustic, began filling the north end of the canal. Subsequently, other chemical companies began using the canal as a chemical waste dump. In 1952, Hooker Chemical closed and capped the site. Residences then began to be built around and on the site, and a school was built on top of the canal. Part of the site was converted into a park where children amused themselves with the sparks and explosions from chunks of phosphorus they picked from the ground. Not long afterward, residents discovered black liquid seeping into their basements. Later, puddles of chemicals surfaced in backyards. In 1974, one resident found his pool raised 2 feet above ground, pushed by a yellow-, blue-, and orchid-colored groundwater. Soon thereafter, a woman gave birth to a child with a damaged heart, bone blockages of the nose, deformed ears, and a cleft palate. Later, the child developed a double row of teeth on the bottom gumline. Health surveys in the neighborhood revealed that spontaneous abortions were 250 times above normal, blood tests showed increases in liver damage, and birth defects were high, including clubfeet, deafness, and mental retardation. High levels of chemical wastes were found in the groundwater, including benzene, chloroform, trichloroethylene, trichlorophenol, and seven additional carcinogens. All in all, 22,000 U.S. tons of chemical wastes were deposited in the Love canal over the years. Love could not have envisioned this tragedy that befell Niagara Falls almost a century after his time. Clearly, an activity considered beneficial or even heroic today could easily lead to a disaster tomorrow.

A fourth episode, which focuses on the importance of microbiology, relates to water. In 1854, Dr. John Snow, a public health worker, discovered that a cholera epidemic was rampant around the vicinity of a pump on Broad Street in London, England. Even cases some distance from the pump could be traced back to Broad Street. Because people of his time did not have any knowledge of the application of microbiology to drinking water, Snow's contemporaries did not accept his evidence. It is alleged that he actually removed the handle of the pump to bring the epidemic to a halt.

Additional environmental tragedies are described in the appendixes, and as this book is being written many more are occurring in various parts of the world. In an attempt to prevent the recurrence of such tragedies, laws and regulations have been passed worldwide. Environmental legislation passed by the U.S. Congress is listed in Appendix 13, and various offices and branches of the U.S. Environmental Protection

Agency are listed in Appendix 14. Passage of such laws and regulations has produced three types of persons in our society: the *environmentalist,* who often does not care about the economy; the *industrialist,* who often does not care about the environment; and the *lawyers* who are hired by environmentalists and industrialists to help settle disputes. Better understanding among these groups is another aim of this book.

GLOSSARY

Engineering. The application, under constraints, of scientific principles to the planning, design, construction, and operation of structures, equipment, and systems for the benefit of society.

Environmental engineering. The application of engineering principles, under constraints, to the protection and enhancement of the quality of the environment and to the protection and enhancement of public health and welfare.

Public health engineering. An older term for *environmental engineering* in the United Kingdom and countries with historical ties to the United Kingdom.

Sanitary engineering. An older term for *environmental engineering* in the United States and countries with historical ties to the United States.

BIBLIOGRAPHY

SCHRENK, H. H., and others (1949). "Air Pollution in Donora, Pa." *U.S. Public Health Service Bulletin 306.*

WOOLF, C. (1969). Letter to the editor. *Archives of Environmental Health*, 18(715).

CHAPTER 2

Environmental Chemistry and Biology

In this chapter we discuss the basic chemistry and biology required in the practice of environmental engineering. The topics include pathogens and indicator organisms, tests for the coliform group, drinking water standards, biological combustion, chemistry of the anaerobic process, equivalents and equivalent weights, wastewater characteristics, kinetics, and others.

PATHOGENS AND INDICATOR ORGANISMS

It is the job of environmental engineers to provide consumers with clear, pathogen-free water. Treated water must be analyzed to ensure that this is so. Throughout human history, water has been a source of life as well as a source of death. The body is mostly H_2O, and a large amount of water must be taken in daily, but if the water contains pathogens, a person could get ill. In the case of the Broad Street pump, when John Snow removed the handle, the 1854 epidemic of Asiatic cholera in London subsided. Table 2-1 shows some common waterborne pathogens and the diseases they produce.

It is impractical to analyze all the organisms in Table 2-1. In addition, determining the concentration of pathogenic organisms on a routine basis is a hazard to the analyst. Moreover, the job would be most time consuming and costly. For these reasons, a surrogate organism is used to indicate the presence of pathogens. Organisms used for this purpose are called *indicator organisms*. An ideal indicator organism must satisfy at least four requirements: The organism must be present when the pathogen is present, the

TABLE 2-1   COMMON WATER-BORNE PATHOGENS

Organisms	Disease
Viruses	
Enteric cytopathogenic human orphan (ECHO)	Aseptic meningitis, infantile diarrhea
Poliomyelitis	Acute anterior poliomyelitis, infantile paralysis
Unknown viruses	Infectious hepatitis
Bacteria	
Francisella tularensis	Tularemia
Leptospirae	Leptospirosis
Salmonella paratyphi	Paratyphoid fever
Salmonella typhi	Typhoid fever
Shigella	Shigellosis (bacillary dysentery)
Vibrio comma (*Vibrio cholerae*)	Cholera
Protozoa	
Entamoeba histolytica	Amebiasis
Giardia lamblia	Giardiasis
Helminths (parasitic worms)	
Dracunculus medinensis	Dracontiasis
Echinococcus	Echinococcosis
Schistosoma	Schistosomiasis

organism must be absent when the pathogen is absent, the organism must be cheap to analyze, and the organism must not itself be a pathogen.

For a person to get diseased, the pathogen must live inside the human body so the indicator organism itself must come from inside the body. Most of the coliform group of organisms reside in the alimentary tract. Members of this group have therefore been used as indicator organisms. The coliform group, which is present in the intestinal tract, is composed of the genera *Escherichia, Aerobacter, Klebsiella*, and *Paracolobacterium*. When the pathogen is present in a sample of drinking water, the coliform group is thus also present. This satisfies the first requirement. However, the coliform group may be present even when the pathogen is absent. At first this may not seem to satisfy the second requirement; however, close scrutiny would reveal that the presence of the indicator even when the pathogen is absent is an added factor of safety. In practice, a criterion is set that if a certain amount of the indicator organisms is present, it will be assumed that pathogens are also present. If the amount of indicator organisms present is less than the criterion, pathogens are presumed absent. For the third requirement, it is relatively cheaper to analyze coliforms than to analyze the pathogens themselves. For one thing, only one type of organism is analyzed rather than the entire range of pathogens. For the fourth requirement, although some coliforms are infectious, they are generally regarded as nonpathogenic.

TEST FOR THE COLIFORM GROUP

The test for the coliform group may be qualitative or quantitative. As the name implies, the qualitative test simply attempts to identify the presence or absence of the coliform group. The quantitative test, on the other hand, quantifies the presence of organisms. In all tests, whether qualitative or quantitative, all media, utensils, and the like must be sterilized. This is to preclude unwanted organisms that can affect the results of the analysis.

Qualitative Tests

There are three types of qualitative test: presumptive, confirmed, and completed. These tests rely on the property of the coliform group to produce a gas during fermentation of lactose. An inverted small vial is put at the bottom of a fermentation tube. The vial traps the gas, forming bubbles that indicate positive gas production.

In a *presumptive test,* a portion of the sample is inoculated into a number of test tubes containing lactose broths and other ingredients necessary for growth. The tubes are then incubated at $35 \pm 0.5°C$. Liberation of gases within 24 to 48 ± 3 hours indicates a positive presumptive test. Since organisms other than coliforms can also liberate gases at this fermentation temperature, a positive presumptive test simply signifies the possible presence of the coliform group.

An additional test, called a *confirmed test,* is carried out to confirm the presumptive test result. The theory behind this test uses the property of the coliforms to ferment lactose even in the presence of a green dye. Noncoliform organisms cannot ferment lactose in the presence of this dye. The growth medium is called *brilliant green lactose bile broth.* A wire loop from the positive presumptive test is inoculated into fermentation tubes containing the broth. As in the presumptive test, the tubes are incubated at $35 \pm 0.5°C$. Liberation of gases within 24 to 48 ± 3 hours indicates a positive confirmed test.

The coliform group of organisms comes not only from the intestinal tracts of human beings but also from the outside surroundings. There are times when the fecal variety needs to be differentiated from nonfecal forms, as in pollution surveys of a water supply raw water source. For this and similar situations, the procedure is modified by raising the incubation temperature to $44.5 \pm 0.2°C$. The broth used is the EC medium, which still contains lactose. The high incubation temperature precludes the nonfecal forms from metabolizing the EC medium. Positive test results are heralded by the evolution of fermentation gases within 24 ± 2 hours.

Normally, qualitative tests are conducted up to the confirmed test only. There may be situations, however, when organisms must actually be seen to be identified. Both presumptive and confirmed tests are based on circumstantial evidence only, not on actually seeing the organisms. If desired, a *completed test* is performed.

In this test, a wire loop from the positive confirmed test fermentation tube is streaked across either an EMB (eosine methylene blue) or Endo's agar in prepared petri dishes. The dishes are then incubated at $35 \pm 0.5°C$ for 24 ± 2 hours. At the end of this period the dishes are examined for growth. *Typical colonies* are growths that exhibit a *green sheen*; *atypical colonies* are light colored and without the sheen. The typical colonies confirm the presence of the coliform group, while the atypical colonies neither confirm nor deny their presence. (The EMB or Endo agar method can also be used as an alternative to the brilliant green lactose bile broth method.)

The typical or atypical colony from the EMB or Endo plate is inoculated into an agar slant in a test tube and to a fermentation tube containing lactose. If no gas evolves within 24 to 48 ± 3 hours of incubation at $35 \pm 0.5°C$, the completed test is negative. If gases evolve from the tubes, the growth from the agar slant is smeared into a microscopic slide and the Gram-stain technique performed. If *spores* or *gram-positive rods* are present, the completed test is negative. If gram-negative rods are present, the completed test is positive.

Quantitative Tests

In some situations, actual enumerations of the number of organisms present may be necessary. This is the case, for example, for limitations imposed on a discharge permit. In general, there are two ways to enumerate coliforms: by either the *membrane-filter* or *multiple-tube* technique. The *membrane-filter technique* consists of filtering under vacuum a volume of water through a membrane filter having pore openings of 0.45-μm, placing the filter in a petri dish and incubating, and counting the colonies that develop after incubation. The 0.45-μm opening retains the bacteria on the filter. The volume of sample to be filtered depends on the anticipated bacterial density. Sample volumes that yield plate counts of 20 to 80 colonies on the petri dish are considered most valid. The standard volume filtered in potable water analysis is 100 mL. If the anticipated concentration is high, requiring a smaller volume of sample (such as 20 mL), the sample must be diluted to disperse the bacteria uniformly. The dispersion will ensure a good spread of the colonies on the plate for easier counting.

M-Endo medium is used for the coliform group and M-FC medium for fecal coliforms. The culture medium is prepared by putting an absorbent pad on a petri dish and pipetting enough M-Endo or M-FC medium into the pad. After filtration, the membrane is removed with forceps and placed directly on the pad. The culture is then incubated at $35 \pm 0.5°C$ in the case of the coliform group and at $44.5 \pm 0.2°C$ in the case of fecal coliforms. At the end of the incubation period of 24 ± 2 hours, the number of colonies that developed is counted and reported as number of organisms per 100 mL.

The other method of enumerating coliforms is through use of the multiple-tube technique. This method is statistical in nature and the result is reported as the most probable number of organisms, MPN. Hence the other name of this method is the *MPN technique*. This technique is an extension of the qualitative techniques of presumptive, confirmed, and completed tests. In other words, MPN results can be presumptive, confirmed, and completed MPNs.

The procedure is similar to that of the qualitative method in that the liberation of gases from the consumption of lactose is used as a criterion. The difference is that the number of tubes liberating gases is used to calculate the MPN. A serial dilution such as 10-, 1-, 0.1-, and 0.01-mL portions of the sample is prepared. Each of these portions is inoculated into a set of five tubes. For example, for the 10-mL portion, each of five fermentation tubes is inoculated with 10-mL portions of the sample. For the 1-mL portion, each of five fermentation tubes is also inoculated with 1-mL portions of the sample, and so on with the rest of the serial dilutions. In this example, since there are four serial dilutions (10, 1, 0.1, and 0.01) there are four sets of five tubes each, making a total of 20 fermentation tubes.

The number of tubes liberating gases is counted from each of the set of five tubes. This information is then used to compute statistically the most probable number of organisms in the sample per 100 mL. A statistical table prepared by the American Public Health Association is reproduced as Table 2-2. The first three columns are "the combination of positives" in only three dilutions. Hence if the analysis had been done on more than three dilutions, only three can be adopted for use of this table. The 0.01 is said to be the highest dilution in the serial dilution of 1, 0.1, and 0.01. To use Table 2-2 for a number of dilutions greater than three, pick the highest three dilutions showing evolution of gases and correct the result obtained, as exemplified in Example 2-2.

TABLE 2-2   MPN INDEX AND 95% CONFIDENCE LIMITS FOR COLIFORM COUNTS BY THE MULTIPLE-TUBE FERMENTATION TECHNIQUE

Number of tubes giving positive reaction				95% Confidence level	
5 of 10-mL	5 of 1-mL	5 of 0.1-mL	MPN per 100 mL	Lower	Upper
0	0	0	<2	—	—
0	0	1	2	1	10
0	1	0	2	1	10
0	2	0	4	1	13
1	0	0	2	1	11
1	0	1	4	1	15
1	1	0	4	1	15
1	1	1	6	2	18
1	2	0	6	2	18
2	0	0	4	1	17
2	0	1	7	2	20
2	1	0	7	2	21
2	1	1	9	3	24
2	2	0	9	3	25
2	3	0	12	5	29
3	0	0	8	3	24
3	0	1	11	4	29
3	1	0	11	4	29
3	1	1	14	6	35
3	2	0	14	6	35
3	2	1	17	7	40
4	0	0	13	5	38
4	0	1	17	7	45
4	1	0	17	7	46
4	1	1	21	9	55
4	1	2	26	12	63
4	2	0	22	9	56
4	2	1	26	12	65
4	3	0	27	12	67
4	3	1	33	15	77
4	4	0	34	16	80
5	0	0	23	9	86
5	0	1	30	10	110

TABLE 2-2 MPN INDEX AND 95% CONFIDENCE LIMITS FOR COLIFORM COUNTS BY THE MULTIPLE-TUBE FERMENTATION TECHNIQUE (CONTINUED)

Number of tubes giving positive reaction				95% Confidence level	
5 of 10-mL	5 of 1-mL	5 of 0.1-mL	MPN per 100 mL	Lower	Upper
5	0	2	40	20	140
5	1	0	30	10	120
5	1	1	50	20	150
5	1	2	60	30	180
5	2	0	50	20	170
5	2	1	70	30	210
5	2	2	90	40	250
5	3	0	80	30	250
5	3	1	110	40	300
5	3	2	140	60	360
5	3	3	170	80	410
5	4	0	130	50	390
5	4	1	170	70	480
5	4	2	220	100	580
5	4	3	280	120	690
5	4	4	350	160	820
5	5	0	240	100	940
5	5	1	300	100	1300
5	5	2	500	200	2000
5	5	3	900	300	1900
5	5	4	1600	600	5300
5	5	5	≥ 1600	—	—

ALPHA, AWWA, and WEF (1992). *Standard Methods for the Examination of Water and Wastewater.* Reprinted by permission of the American Public Health Association.

Example 2-1

A sample of water is analyzed for the coliform group using three sample portions: 10 mL, 60 mL, and 600 mL. Each of these portions is filtered through five filter membranes using the membrane-filter technique. The results of the colony counts are as follows: 10-mL portions: 6, 7, 5, 8, 6; 60-mL portions: 30, 32, 33, 31, 25; and 600-mL portions: 350, 340, 360, 370, 340. What is the number of coliforms per 100 mL of the sample?

Solution Using the criterion of 20 to 80 as the most valid count in a membrane-filter technique, the results of the 10- and 600-mL portions can be excluded in the calculation.

$$\text{Average count for the 60-mL portions} = \frac{30 + 32 + 33 + 31 + 25}{5}$$

$$= 30.2 \text{ coliforms per 60 mL}$$

Therefore,

$$\text{coliforms per 100 mL} = \frac{30.2}{60}(100) = 50.3 \text{ average}   \textbf{Answer}$$

Example 2-2

A river water is analyzed for total and fecal coliforms using the MPN technique. The results are as follows:

Serial dilution	Number of positive tubes out of five		
	Lactose	BGB	EC
10	5	5	5
1.0	3	3	2
0.1	2	2	2
0.01	1	0	0
0.001	0	0	0

What are the presumptive and confirmed MPNs and the fecal coliform MPN? Also give the ranges of these values at the 95% confidence level. BGB stands for "brilliant green bile broth" and EC stands for the medium for fecal coliforms. Use Table 2-2.

Solution For the presumptive total coliform MPN, the three highest dilutions are 1.0, 0.1, and 0.01, corresponding to the positive readings of 3, 2, and 1. From Table 2-2, for a serial dilution of 10, 1.0, and 0.1 and positive readings of 3, 2, and 1, the MPN = 17. Therefore, for a serial dilution of 1.0, 0.1, and 0.01,

$$MPN = 17(10) = 170 \text{ organisms/100 mL} \quad \textbf{Answer}$$

$$\text{At 95\% confidence, range: } 70 \leq MPN \leq 400 \quad \textbf{Answer}$$

For the confirmed test, the three highest dilutions are 10, 1.0, and 0.1, corresponding to positive readings of 5, 3, and 2. From Table 2-2,

$$MPN = 140 \text{ organisms/100 mL} \quad \textbf{Answer}$$

$$\text{At 95\% confidence, range: } 60 \leq MPN \leq 360 \quad \textbf{Answer}$$

For the fecal test, the three highest dilutions are 10, 1.0, and 0.1, corresponding to the positive readings of 5, 2, and 2. From the table,

$$MPN = 90 \text{ fecal coliforms/100 mL} \quad \textbf{Answer}$$

$$\text{At 95\% confidence, range: } 40 \leq MPN \leq 250 \quad \textbf{Answer}$$

It is clear from Example 2-1 that MPN is not a very accurate technique. For one thing, the numbers are simply based on circumstantial evidence—the organisms are not actually "seen" and counted but simply estimated based on the evolution of gases. As the membrane-filter technique is perfected, it may replace this cumbersome MPN technique entirely as a method of enumerating the number of organisms contained in a sample.

DRINKING WATER STANDARDS

In the past, since waters were pristine and naturally clean, water standards were not necessary. In modern times, however, a host of toxic and hazardous chemicals have already been discharged into the environment such that some raw water supplies are not immune from contamination. The setting of standards is therefore necessary. In addition, only a few of these chemicals can be removed by standard processes. This means that if present in the raw water, these chemicals will probably also be present in the drinking water tap. *This situation calls for a very careful selection of the raw water supply source and its vigilant protection in daily operation of the system.*

The first environmental act in the United States relating to the protection of drinking water was the Public Health Service Act of 1912 (see Appendix 13). Not many activities by Congress relating to drinking water transpired until several decades later, when the President of the United States signed the Safe Drinking Water Act (SDWA) of 1974 on December 16, 1974. Thereafter, standard setting has become the norm. For example, on May 7, 1991, the U.S. Environmental Protection Agency (USEPA) set final regulations, which require a treatment technique for lead and copper. This requirement is triggered by a lead action level of 0.015 mg/L or a copper action level of 1.3 mg/L, determined as the 90th percentile values of the concentrations measured at the taps of customers.

The major provisions of the SDWA of 1974 are the establishment of *primary regulations* for the protection of public health and the establishment of *secondary regulations* for aesthetic considerations such as taste, appearance, and odor. The primary regulations result in the establishment of *primary standards,* and the secondary regulations result in the establishment of *secondary standards.* The SWDA was amended in 1986.

One of the regulatory functions of the USEPA (or simply, EPA) is the setting of standards [maximum contaminant levels (MCLs)] or goals [maximum containment level goals (MCLGs)] for classes of substances that can be present in water supplies in harmful concentrations. These substances include volatile organic chemicals (VOCs), synthetic organic chemicals (SOCs), inorganic chemicals (IOCs), microbiological contaminants, disinfection by-products (DBPs), and radionuclides. The appendixes provide tables of drinking water parameters.

The quality characteristics of water fall under three categories: chemical, biological, and physical. VOCs, SOCs, IOCs, and DBPs are chemical characteristics. Microbiological contaminants are biological characteristics. Radionuclide contaminants can be either physical or chemical. Gross alpha-particle, beta-particle, and photon radioactivities are physical characteristics, while concentrations of uranium, radium 226 and radium 228, and radon are chemical characteristics. The other physical characteristics of concern in drinking water include turbidity, color, taste and odor, and temperature. Chlorides, fluorides, iron, manganese, lead, copper, nitrate, sodium, sulfate, and zinc are the other chemical characteristics that are also of concern in drinking water. Figure 2-1 is a photograph of atomic absorption apparatus that may be used to analyze very low concentrations of metals in drinking water.

Figure 2-1   Atomic absorption apparatus. (Used by permission of the Perkin-Elmer Corporation)

Turbidity

Done photometrically, *turbidity* is a measure of the extent to which suspended matter in water either absorbs or scatters radiant light energy impinging on the suspension. The original measuring apparatus, the *Jackson turbidimeter,* was based on the absorption principle. A standardized candle was placed under a graduated glass tube housed in a black metal sheet so that the light from the candle can only be seen from above the tube. The water sample was then poured slowly into the tube until the candle flame was no longer visible. The turbidity was then read on the graduation etched on the tube. The unit of turbidity is the *turbidity unit* (TU), which is equivalent to the turbidity produced by 1 mg/L of silica (SiO_2). SiO_2 was used as the reference standard. Turbidities in excess of 5 TU are easily detected in a glass of water and are objectionable not necessarily for health but for aesthetic reasons.

At present, turbidity measurements are done conveniently through the use of photometers. A beam of light from a source produced by a standardized electric bulb is passed through a sample vial. The light that emerges from the sample is then directed to a photometer that measures the light absorbed. The readout is calibrated in terms of turbidity. A chemical, formazin, that provides a more reproducible result has now replaced silica as the standard. Accordingly, the unit of turbidity is now also expressed as *formazin turbidity units* (FTUs).

A second method of measurement is by light scattering. The sample "scatters" the light that impinges on it. The scattered light is then measured by putting the photometer at right angles to the original direction of the light generated by the light source. This measurement of light scattered at a 90° angle is called *nephelometry*. The unit of turbidity in nephelometry is the *nephelometric turbidity unit* (NTU).

Color

Color is the perception registered as radiation of various wavelengths strikes the retina of the eye. Materials decayed from vegetation and inorganic matter impart color to water, which is objectionable, again, not for health reasons but for aesthetics. Natural colors give a yellow-brownish appearance to water; the natural tendency to associate this color with urine. The unit of measurement of color is the platinum in potassium chloroplatinate (K_2PtCl_6). One milligram per liter of Pt in K_2PtCl_6 is 1 unit of color. The secondary MCL for color is 15 color units.

Taste and Odor

Taste is the perception registered by the taste buds, while odor is the perception registered by the olfactory nerves. There should be no noticeable taste and odor at the point of use of the water. The numerical value of odor or taste is determined quantitatively by measuring a volume of sample A (in milliliters) and diluting it with a volume of sample B (in milliliters) of an odor-free distilled water so that the odor of the resulting mixture is just barely detectable at a total mixture volume of 200 mL. The unit of odor or taste is then expressed in terms of a threshold number as follows:

$$\text{TON or TTN} = \frac{A + B}{A} \tag{2-1}$$

where TON is the threshold odor number and TTN is the threshold taste number. The secondary standard for odor is 3.

Example 2-3

A sample of water requires 113 mL of distilled water to render the odor barely detectable. What is the volume of the sample, and what is the TON?

Solution

$$A + 113 = 200 \qquad A = 87 \text{ mL} \quad \textbf{Answer}$$

$$\text{TON} = \frac{A + B}{A} = \frac{87 + 113}{87} = 2.3 \quad \textbf{Answer}$$

Temperature

Most people find water at temperatures of 10 to 15°C most palatable. Groundwaters and waters from mountainous areas are normally within this range.

Chlorides

Chlorides in concentrations of 250 mg/L or greater are objectionable to most people. Hence the secondary standard for chlorides is 250 mg/L. Whether concentrations of 250 mg/L are objectionable or not, however, depends on the degree of acclimation of the user to the water. In Antipolo, a barrio of Cebu in the Philippines, the normal source of water of the residents is a spring that emerges along the shoreline between a cliff and the sea. As such, the fresh water is contaminated by salt water before being retrieved by the people. The salt imparts to the water a high concentration of chlorides. Chloride contaminants could go as high as 2000 mg/L; however, even with concentrations this high, the people continue to use the source and are accustomed to the taste.

Fluorides

The absence of fluorides in drinking water encourages dental caries or tooth decay; excessive concentrations of the chemical produce mottling of the teeth or dental fluorosis. Hence managers and operators of water treatment plants must be careful that the exact concentrations of the fluorides are administered to the drinking water. Optimum concentrations of 0.7 to 1.2 mg/L are recommended, although the actual amount in specific circumstances depends on the air temperature, since air temperature influences the amount of water that people drink. For example, the USEPA published the fluoride primary standards shown in Table 2-3, indicating that the MCL values depend on the ambient air temperature.

TABLE 2-3 EPA PRIMARY DRINKING WATER MCL FOR FLUORIDE

Annual average of maximum daily air temperatures of community in which water system is situated (°C)	Maximum contaminant levels (mg/L)
12.0 and below	2.4
12.1–14.6	2.2
14.7–17.6	2.0
17.7–21.4	1.8
21.5–26.2	1.6
26.3–32.5	1.4

Environmental Protection Agency (1975). "National Interim Primary Drinking Water Regulations," *Federal Register, Part IV*.

Iron and Manganese

Iron and manganese are objectionable in water supplies because they impart brownish colors to laundered goods. Fe also affects the taste of beverages such as tea and coffee.

Mn flavors tea and coffee with medicinal taste. The SMCLs (secondary MCLs) for Fe and Mn are, respectively, 0.3 and 0.05 mg/L.

Lead and Copper

Clinical, epidemiological, and toxicological studies have demonstrated that lead exposure can adversely affect human health. The three systems in the human body most sensitive to lead are the blood-forming system, the nervous system, and the renal system. In children, blood levels from 0.8 to 1.0 μg/L can inhibit enzymatic actions. Also, in children, lead can alter physical and mental development, interfere with growing, decrease attention span and hearing, and interfere with heme synthesis. In older men and women, lead can increase blood pressure.[1] Lead is emitted into the atmosphere as Pb, PbO, PbO_2, $PbSO_4$, PbS, $Pb(CH_3)_4$, $Pb(C_2H_5)_4$, and lead halides. In drinking water, it can be emitted from pipe solders.

The source of copper in drinking water is the plumbing used to convey water in the house distribution system. In small amounts it is not detrimental to health, but it will impart an undesirable taste to the water. In appropriate concentrations, copper can cause stomach and intestinal distress. It also causes Wilson's disease. A type of polyvinyl chloride (PVC) pipe called CPVC can be used to replace copper for household plumbing.

Nitrate

Nitrate is objectionable because it causes *methemoglobinemia* (infant cyanosis or blue babies) in infants. The MCL is 10 mg/L.

Sodium

The presence of sodium in drinking water can affect persons suffering from heart, kidney, or circulatory ailments. Sodium may elevate blood pressures of susceptible persons.

Sulfate

The sulfate ion is one of the major anions occurring naturally in water. It produces a cathartic or laxative effect on people when present in excessive amounts in drinking water. Its SMCL is 250 mg/L.

Zinc

Zinc is not considered detrimental to health, but it will impart undesirable taste to drinking water. Its SMCL is 5 mg/L.

---

[1]F. W. Pontius (1991). "How to Comply with the Lead and Copper Rule," *Opflow*, pp. 1–5.

SOME SWDA REGULATORY REQUIREMENTS

The SWTR promulgated by the EPA on June 29, 1989 sets forth primary drinking water regulations for treatment of surface waters and groundwaters that may be under the influence of surface waters. The regulations require filtration and/or disinfection in lieu of establishing MCLs for turbidity, *Giardia lamblia, Legionella,* viruses, heterotrophic bacteria, and many other pathogenic organisms that may be removed by these treatment techniques. The regulations also call for an MCLG of zero for *Giardia liamlia, Legionella,* and the viruses. No MCLG was set for heterotrophic plate count or turbidity. The turbidity goal of the American Water Works Association (AWWA) is less than 0.1 TU (turbidity unit).

The coliform rule promulgated by the EPA on June 19, 1989 requires primary drinking water MCL values for total coliforms, including fecal coliforms and *Escherichia coli.* Larger systems are required to collect each month at least 40 samples and must not show positive results for 5% of the samples. Smaller systems are allowed to collect samples fewer than 40 per month but must not show positive results in more than one sample. The accepted methods include the multiple-tube fermentation technique (MTF), membrane-filter technique (MF), minimal media ONPG-MUG test (Colilert system, MMO-MUG), or presence–absence coliform test (P-A). Regardless of the method used, the standard sample volume for total coliform testing is 100 mL.

EQUIVALENTS AND EQUIVALENT WEIGHTS

In the literature, the definition of equivalents and equivalent weights is confused, and no universal definition is attempted. In this section we unify the definition of these terms by utilizing the concept of *reference species.* The mass of any substance participating in a reaction per unit of this reference species is the *equivalent weight* of the substance, and the mass of the substance divided by this it equivalent weight is the *number of equivalents* of the substance.

The reference species is one of only two possibilities: the *electrons* involved in an oxidation-reduction reaction or the *positive* (or, alternatively, the *negative*) charges in all the other reactions. The concept of the reference species is explained in the following example. Take the following two reactions of phosphoric acid with sodium hydroxide:

$$H_3PO_4 + 2NaOH \rightarrow 2Na^+ + HPO_4^{2-} + 2H_2O$$

$$H_3PO_4 + 3NaOH \rightarrow 3Na^+ + PO_4^{3-} + 3H_2O$$

In these reactions the reference species is the positive electric charge or the negative electric charge of the reactants on the left. (*Take note that the reference species is taken from the reactant, not from the product.*) The positive charge comes from the Na^+ of NaOH or the H^+ of H_3PO_4. The negative charge comes from the OH^- of NaOH or the HPO_4^{2-} of H_3PO_4 in the first reaction or the PO_4^{3-} of H_3PO_4 in the second reaction.

Let us consider the positive electric charge and the first reaction. Since Na^+ has a charge of $+1$ and the coefficient of the term of NaOH is 2, two positive charges

$(2 \times 1 = 2)$ are involved. In the case of H_3PO_4, the equation shows that the acid breaks up into HPO_4^{2-} and other substances, with one H still "clinging" to the PO_4 on the right-hand side of the equation. This indicates that two H^+'s are involved in the breakup. Since the charge of H^+ is $+1$, two positive charges are involved. In the cases of both Na^+ and H^+, reference species are the two positive charges and the number of reference species is two. Therefore, in the first reaction, the equivalent weight of a substance participating is obtained by dividing the term (including the coefficient) by 2. Hence for the acid, the equivalent weight is $H_3PO_4/2$; for the base, the equivalent weight is $2NaOH/2$; and so on.

In the second reaction, based on the positive electric charges, the number of reference species is three. Hence in this reaction, for the acid, the equivalent weight is $H_3PO_4/3$; for the base, the equivalent weight is $3NaOH/3$ and so on. These observations show that the equivalent weight of a substance depends on the chemical reaction it participates in.

Following are the third and fourth reactions:

$$Fe(HCO_3)_2 + 2Ca(OH)_2 \rightarrow Fe(OH)_2 + 2CaCO_3 + 2H_2O$$

$$4Fe(OH)_2 + O_2 + 2H_2O \rightarrow 4Fe(OH)_3$$

In the third reaction, since the ferrous iron has a charge of $+2$, ferrous bicarbonate has two positive charges of references species. In the case of the calcium hydroxide, however, since calcium has a charge of $+2$ and the coefficient of the term is 2, the number of reference species is four. In this situation there are two alternatives: two or four positive charges as the number of reference species. *Either* can be used provided that when one is chosen, all subsequent calculations must be based on the one particular choice. However, adopt the convention that *the number of reference species is the one that is the lowest*. Hence the number of reference species in the third reaction is two—not four—and all the equivalent weights of the participating substances are obtained by dividing each term of the reaction by 2.

In the fourth reaction, ferrous iron has been oxidized to the ferric form from an oxidation state of $+2$ to $+3$. Hence in this reaction, electrons are involved. Therefore, adopt the second convention that *whenever oxidation-reduction is involved in a given chemical reaction, electron transfer is used as the reference species rather than the positive or negative charges*. For ferrous hydroxide, the number of electrons involved is four. For oxygen, the atom has been reduced from zero to -2 per atom. Since there are two oxygen atoms, the total number of electrons involved is also equal to four $(2 \times 2 = 4)$. Hence to obtain the equivalent weights of all the participating substances, each term of the equation must be divided by 4.

In the discussion above, the unit of the reference species was not established. A convenient unit would be the mole (i.e., mole of electrons or mole of positive or negative charges). The mole can be a milligram-mole, gram-mole, and so on. The mass unit of measurement of the equivalent weight would then correspond to the type of mole used for the reference species. For example, if the mole used is the gram-mole, the mass of the equivalent weight would be expressed in grams of the substance per gram-mole of the reference species; and if the mole used is the milligram-mole, the equivalent weight would be expressed in milligrams of the substance per milligram-mole of the reference species; and so on.

Since the reference species is used as the the standard of reference, its unit, the mole, can be said to have a unit of 1 equivalent. From this, the equivalent weight of a participating substance may be expressed as the mass of the substance per equivalent of the reference species. But since the substance is equivalent to the reference species itself, the equivalent weight may also be expressed as the mass of the substance per equivalent of the substance.

The general formula for finding the equivalent weight of a substance is

$$\text{equivalent weight} = \frac{\text{term in balanced equation}}{\text{number of moles of reference species}} \tag{2-2}$$

The formula for obtaining the mass of a substance given its number of equivalents is

$$\text{mass of a substance} = \text{number of equivalents} \times \text{equivalent weight} \tag{2-3}$$

Example 2-4

Water containing 2.5 mol of calcium bicarbonate and 1.5 mol of calcium sulfate is softened using lime and soda ash. How many grams of calcium carbonate solids are produced using (a) the method of equivalent weights, and (b) the balanced chemical reaction?

Solution   (a) The chemical reactions are

$$Ca(HCO_3)_2 + CaO \rightarrow 2CaCO_3 + H_2O$$

$$CaSO_4 + Na_2CO_3 \rightarrow CaCO_3 + Na_2SO_4$$

Molecular weights:

$$Ca(HCO_3)_2 = 40 + 2(1 + 12 + 48) = 162$$

$$CaSO_4 = 40 + 32 + 4(16) = 136$$

$$CaCO_3 = 40 + 12 + 3(16) = 100$$

For the first reaction, the number of reference species is 2 mol of positive charges. Hence equivalent weights are

$$Ca(HCO_3)_2 = \frac{162}{2} = 81$$

$$CaCO_3 = \frac{2(100)}{2} = 100$$

Thus

$$\text{equivalents of } Ca(HCO_3)_2 = \frac{2.5(162)}{81} = 5 = \text{equivalents of } CaCO_3$$

Therefore,

$$\text{carbonate solids from the bicarbonate} = 5(100) = 500 \text{ g}$$

For the second reaction, the number of reference species is 2 mol of positive charges. Hence equivalent weights are

$$CaSO_4 = \frac{136}{2} = 68$$

$$CaCO_3 = \frac{100}{2} = 50$$

Thus

$$\text{equivalents of } CaSO_4 = \frac{1.5(136)}{68} = 3 = \text{equivalents of } CaCO_3$$

and

$$\text{carbonate solids from the sulfate} = 3(50) = 150 \text{ g}$$

$$\text{total grams of carbonate solids produced} = 500 + 150 = 650 \text{ g} \quad \textbf{Answer}$$

(b) From the first reaction,

$$\text{carbonate solids produced} = \frac{2(100)}{162}(2.5)(162) = 500 \text{ g}$$

From the second reaction,

$$\text{carbonate solids produced} = \frac{100}{136}(1.5)(136) = 150 \text{ g}$$

Therefore,

$$\text{total carbonate solids produced} = 500 + 150 = 650 \text{ g} \quad \textbf{Answer}$$

MICROBIAL THERMODYNAMICS

The study of the relationships between heat and other forms of energy is called *thermodynamics.* Since all living things utilize heat, the science of thermodynamics may be used to evaluate life processes. An example of a life process is the growth of bacteria when wastewater is fed to them to treat the waste. Knowledge of microbial thermodynamics is therefore very important to environmental engineers involved in cleaning up wastewaters. Variables involved in the interchange of heat and energy are called *thermodynamic variables.* Examples of these variables are temperature, pressure, volume and free energy.

In the discussion of microbial thermodynamics, one form of Gibbs free energy is used. Its use is based on the concept that free energy is the net or maximum energy available for work. This form of Gibbs free energy is $\Delta G = -nF\Delta E$, where n represents the number of faradays involved in a reacion, F the number of coulombs per faraday, and ΔE the voltage difference. This is discussed further below.

Life processes involve electron transport. Specifically, the mitochondrion and the chloroplast are the sites of this electron movement in the eucaryotes. In the procaryotes, this function is embedded in the sites of the cytoplasmic membrane. As far as electron movement is concerned, life processes are similar to a battery cell, where electrons move because of electrical pressure, the voltage difference. By the same token, electrons move in the organism because of the same electrical pressure, the voltage difference.

Microbial thermodynamics deals with processes involving oxidation-reduction reactions. In an oxidation-reduction reaction, 1 mol of electrons involved is called the faraday, which is equal to 96,494 coulombs (c). A mole of electrons is equal to 1 equivalent of any substance. Therefore, a faraday is equal to 1 equivalent.

Let n be the number of faradays of charge per mole of a substance participating in a reaction. Let the general reaction be represented by the following half-cell reaction of Zn:

$$Zn \rightarrow Zn^{2+} + 2e \tag{2-4}$$

In the couple Zn/Zn^{2+}, as represented by reaction (2-4), there is electric pressure between two cells. Now take another couple, such as Mg/Mg^{2+}, whose half-cell reaction would be similar to that of reaction (2-4). The couples Mg/Mg^{2+} and Zn/Zn^{2+} do not possess the same voltage potential. If the two couples are connected, they form a cell. Since their voltage potentials are not the same, a voltage difference would be developed between their electrodes.

Let the voltage between the electrodes of the cell above be measured by a potentiometer. Designate this voltage difference by ΔE. In potentiometric measurements, no electrons are allowed to flow and thus only the voltage tendency of the electrons to flow is measured. Since no electrons are allowed to flow, no energy is dissipated due to friction of electrons "rubbing along the wire." Hence any energy associated with this no-electron-flow process represents the maximum energy available. Since voltage is energy in joules per coulomb of charge, the energy associated can be calculated from the voltage difference. This associated energy corresponds to a no-friction-loss process; it is therefore a maximum energy.

Let n, the number of faradays involved in a reaction, be multiplied by F, the number of coulombs per faraday. The result, nF, is the number of coulombs involved in the reaction. If nF is multiplied by ΔE, the total associated energy in the potentiometric experiment above is obtained. Since the voltage measurement was done with no energy loss, by definition, this associated energy represents the free-energy change of the cell (i.e., the maximum energy change in the cell). In symbols,

$$\Delta G = -nF\Delta E \tag{2-5}$$

where a negative sign is prefixed to show that the reaction represented by the equation is spontaneous.

As mentioned, the battery cell process is analogous to the living cell process of the mitochondrion, the chloroplast, and the electron transport system in the cytoplasmic membrane of the procaryotes. Hence equation (2-5) can represent the basic thermodynamics of a microbial system. In the living cell, organic materials are utilized for both energy (oxidation) and synthesis (reduction). Microorganisms that utilize organic materials for energy are called *heterotrophs.* Those that utilize inorganics for energy are called *autotrophs.* Autotrophs utilize CO_2 and HCO_3^- for the carbon needed for cell synthesis; heterotrophs utilize organic materials for their carbon source. Autotrophs that use inorganic chemicals for energy are called *chemotrophs;* those that use sunlight are *phototrophs.*

Somehow, in life processes, the production of energy from release of electrons does not occur automatically but through a series of steps that produce a high-energy-containing compound. This high-energy-containing compound is adenosine triphosphate (ATP). Although ATP is not the only high-energy-containing compound, it is by far the major one that fuels synthesis in the cell. ATP is the energy currency that the cell relies on for energy supply.

The energy function of ATP is explained as follows: ATP contains two high-energy bonds. To form these bonds, energy must be released from an energy source through electron transfer. As soon as the energy is released, it is captured and stored in the two high-energy bonds of ATP. On demand, hydrolysis of the bond releases the stored energy, which the cell can then use for synthesis and cell maintenance. ATP is produced from adenosine diphosphate (ADP) by coupling the release of electrons to the reaction of organic phosphates and ADP to produce ATP. There are two modes of production of ATP: *substrate-level phosphorylation* and *oxidative phosphorylation.* In the former, the electrons released by the energy source are simply absorbed by an intermediate in the immediate surroundings within the system. This electron absorption is accompanied by an energy release. The transference of the electrons is simple and does not require an elaborate electron transport system. Fermentation is a substrate-level phosphorylation process that uses intermediate absorbers such as formaldehyde. Substrate-level phosphorylation is inefficient and produces only a few molecules of ATP.

In the oxidative phosphorylation mode, the electron moves from one electron carrier to another in a series of reduction and oxidation steps. For a hydrogen-containing energy source, the series starts with the initial removal of a hydrogen atom from the molecule of the source. Since the hydrogen is carrying the electron in its valence shell, the electron is removed from the molecule. This hydrogen reduces a series of intermediate carriers such as nicotinamide adenine dinucleotide (NAD). The reduction of NAD produces $NADH_2$. The series continues on with further reduction and oxidation steps. The entire line of reduction and oxidation constitutes the electron transport system. At strategic points of the transport system, ATP is produced from ADP and inorganic phosphates.

The other version of oxidative phosphorylation used by autotrophs involves the release of electrons directly from an inorganic energy source. An example of this is the release of electrons from NO_2^-, oxidizing NO_2^- to NO_3^-.

The electrons, both from the hydrogen and the one directly from the energy source, emerge from the system to reduce a final external electron acceptor. The acceptors may be one of the following: for aerobic processes, the acceptor is O_2; for anaerobic processes, the acceptors are NO_3^-, SO_4^{2-}, and CO_2. When the acceptor is NO_3^-, the system is said to be *anoxic.*

As mentioned before, free-energy changes are normally reported for values at standard conditions. In biochemistry, in addition to the requirement of unit activity for concentrations of reactants and products, the hydrogen ion concentration is set arbitrarily at pH 7.0. Following this convention, equation (2-5) may be written as

$$\Delta G_0' = -nF \ \Delta E_0' \qquad (2\text{-}6)$$

The primes emphasize the fact that the standard condition now requires the $\{H^+\}$ to be 10^{-7} mol/L.

In environmental engineering it is customary to call the substance oxidized the *electron donor* and the substance reduced the *electron acceptor*. The electron donor is normally considered as "food." Equation (2-4) is an example of an electron donor reaction, Zn being the donor. The reverse of equation (2-4) is an example of an electron acceptor reaction, Zn^{2+} being the electron acceptor. McCarty[2] derived values for free energy of half-reactions for various electron donors and acceptors utilized in a bacterial system. These are shown in Table 2-4.

[2]P. L. McCarty (1975). "Stoichiometry of Biological Reactions." *Progress in Water Technology*. Pergamon Press, London.

TABLE 2-4 HALF-REACTIONS FOR BACTERIAL SYSTEMS[a]

	$\Delta G_o'$ (kcal/electron-mol)
Reactions for bacterial cell synthesis	
Ammonia as nitrogen source:	
$\frac{1}{5}CO_2 + \frac{1}{20}HCO_3^- + \frac{1}{20}NH_4^- H^+ + e \longrightarrow \frac{1}{20}C_5H_7O_2N + \frac{9}{20}H_2O$	—
Nitrate as nitrogen source:	
$\frac{1}{28}NO_3^- + \frac{5}{28}CO_2 + \frac{29}{28}H^+ + e \longrightarrow \frac{1}{28}C_5H_7O_2N + \frac{11}{28}H_2O$	—
Reactions for electron acceptors	
Oxygen:	
$\frac{1}{4}O_2 + H^+ + e \longrightarrow \frac{1}{2}H_2O$	−18.675
Nitrate:	
$\frac{1}{5}NO_3^- + \frac{6}{5}H^+ + e \longrightarrow \frac{1}{10}N_2 + \frac{3}{5}H_2O$	−17.128
Sulfate:	
$\frac{1}{8}SO_4^{2-} + \frac{19}{16}H^+ + e \longrightarrow \frac{1}{16}H_2S + \frac{1}{16}HS^- + \frac{1}{2}H_2O$	5.085
Carbon dioxide:	
$\frac{1}{8}CO_2 + H^+ + e \longrightarrow \frac{1}{8}CH_4 + \frac{1}{4}H_2O$	5.763
Heterotrophic reactions for organic electron donors	
Domestic wastewater:	
$\frac{1}{50}C_{10}H_{19}O_3N + \frac{9}{25}H_2O \longrightarrow \frac{9}{50}CO_2 + \frac{1}{50}NH_4^+ + \frac{1}{50}HCO_3^- + H^+ + e$	−7.6
Protein (amino acids, proteins, nitrogenous organics)	
$\frac{1}{66}C_{16}H_{24}C_5N_4 + \frac{27}{66}H_2O \longrightarrow \frac{8}{33}CO_2 + \frac{2}{33}NH_4^+ + \frac{31}{33}H^+ + e$	−7.7
Carbohydrates (cellulose, starch, sugars):	
$\frac{1}{4}CH_2O + \frac{1}{4}H_2O \longrightarrow \frac{1}{4}CO_2 + H^+ + e$	−10.0
Grease (fats and oils)	
$\frac{1}{46}C_8H_{16}O + \frac{15}{46}H_2O \longrightarrow \frac{4}{23}CO_2 + H^+ + e$	−6.6
Acetate:	
$\frac{1}{8}CH_3COO^- + \frac{3}{8}H_2O \longrightarrow \frac{1}{8}CO_2 + \frac{1}{8}HCO_3^- + H^+ + e$	−6.609
Propionate:	
$\frac{1}{14}CH_3CH_2COO^- + \frac{5}{14}H_2O \longrightarrow \frac{1}{7}CO_2 + \frac{1}{14}HCO_3^- + H^+ + e$	−6.664

TABLE 2-4 HALF-REACTIONS FOR BACTERIAL SYSTEMS[a] (CONTINUED)

	$\Delta G'_o$ (kcal/electron-mol)
Benzoate: $\frac{1}{30}C_6H_5COO^- + \frac{13}{20}H_2O \longrightarrow \frac{1}{5}CO_2 + \frac{1}{30}HCO_3^- + H^+ + e$	-6.892
Ethanol: $\frac{1}{12}CH_3CH_2OH + \frac{1}{4}H_2O \longrightarrow \frac{1}{6}CO_2 + H^+ + e$	-7.592
Lactate: $\frac{1}{12}CH_3CHOHCOO^- + \frac{1}{3}H_2O \longrightarrow \frac{1}{6}CO_2 + \frac{1}{12}HCO_3^- + H^+ + e$	-7.873
Pyruvate: $\frac{1}{10}CH_3COCCO^- + \frac{2}{5}H_2O \longrightarrow \frac{1}{5}CO_2 + \frac{1}{10}HCO_3^- + H^+ + e$	-8.545
Methanol: $\frac{1}{6}CH_3OH + \frac{1}{6}H_2O \longrightarrow \frac{1}{6}CO_2 + H^+ + e$	-8.965
Autotrophic reactions for inorganic electron donors	
$Fe^{2+} \longrightarrow Fe^{3+} + e$	17.780
$\frac{1}{2}NO_2^- + \frac{1}{2}H_2O \longrightarrow \frac{1}{2}NO_3^- + H^+ + e$	9.43
$\frac{1}{8}NH_4^+ + \frac{3}{8}H_2O \longrightarrow \frac{1}{8}NO_3^- + \frac{5}{4}H^+ + e$	8.245
$\frac{1}{6}NH_4^+ + \frac{1}{3}H_2O \longrightarrow \frac{1}{6}NO_2^- + \frac{4}{3}H^+ + e$	7.852
$\frac{1}{6}S + \frac{2}{3}H_2O \longrightarrow \frac{1}{6}SO_4^{2-} + \frac{4}{3}H^+ + e$	-4.657
$\frac{1}{16}H_2S + \frac{1}{16}HS^- + \frac{1}{2}H_2O \longrightarrow \frac{1}{8}SO_4^{2-} + \frac{19}{16}H^+ + e$	-5.085
$\frac{1}{8}S_2O_{2/3}^- + \frac{5}{8}H_2O \longrightarrow \frac{1}{4}SO_4^{2-} + \frac{5}{4}H^+ + e$	-5.091
$\frac{1}{2}H_2 \longrightarrow H^+ + e$	-9.670
$\frac{1}{2}SO_{2/3}^- + \frac{1}{2}H_2O \longrightarrow \frac{1}{2}SO_4^{2-} + H^+ + e$	-10.595

[a]$\{H^+\} = 10^{-7}$ mol/L, e = number of electron-moles.

Example 2-5

At standard conditions, (a) what is the oxidation potential for the oxidation of sewage and (b) how many moles of electrons are involved per mole of sewage? (c) Calculate the equivalent weight.

Solution (a) The oxidation reaction is

$$\frac{1}{50}C_{10}H_{19}O_3N + \frac{9}{25}H_2O \rightarrow \frac{9}{50}CO_2 + \frac{1}{50}NH_4^+ + \frac{1}{50}HCO_3^- + H^+ + e$$

$$\Delta G'_0 = -7.6\frac{kcal}{electron\text{-}mol} = -7.6(4.18)\frac{kJ}{electron\text{-}mol} = -31.77\frac{kJ}{electron\text{-}mol}$$

Therefore,

$$\Delta E'_0 = -\frac{\Delta G'_0}{nF} = -\frac{-31.77(1000)}{1(92,494)} = 0.343 \text{ V} \quad \textbf{Answer}$$

(b) From the half-reaction, the coefficient of $C_{10}H_{19}O_3N$ (sewage) is $\frac{1}{50}$. Hence multiplying both sides of the equation by 50 produces the moles of electrons per mole of sewage:

$$\text{moles of electrons per mole of sewage} = 50 \quad \textbf{Answer}$$

(c) $C_{10}H_{19}O_3N = 12(10) + 1.0(19) + 16(3) + 14 = 201$

Therefore,

$$\text{equivalent weight} = \frac{201}{50} = 4.02 \text{ g/Eq} \quad \textbf{Answer}$$

As shown in Example 2-5, the oxidation of sewage is favorable thermodynamically. Also, the voltage potential for a favorable reaction is positive.

Example 2-6

In the utilization of NH_3 by *Nitrosomonas* under aerobic conditions: (a) show the donor and acceptor reactions, (b) calculate the free-energy change of the reaction without synthesis, and (c) repeat part (b) considering cell synthesis. Assume 0.15 mg of *Nitrosomonas* synthesized per milligram of NH_3-N utilized.

Solution (a) From Table 2-4 the donor reaction is

$$\tfrac{1}{6}NH_4^+ + \tfrac{1}{3}H_2O \rightarrow \tfrac{1}{6}NO_2^- + \tfrac{4}{3}H^+ + e \quad \textbf{Answer}$$

Since the process is aerobic, the electron acceptor is O_2. Hence from Table 2-4, the acceptor reaction is

$$\tfrac{1}{4}O_2 + H^+ + e \rightarrow \tfrac{1}{2}H_2O \quad \textbf{Answer}$$

(b) The donor and acceptor reactions will be manipulated algebraically.

$$\tfrac{1}{6}NH_4^+ + \tfrac{1}{3}H_2O \rightarrow \tfrac{1}{6}NO_2^- + \tfrac{4}{3}H^+ + e \qquad \Delta G_0' = 7.852 \qquad \text{(a)}$$

$$\tfrac{1}{4}O_2 + H^+ + e \rightarrow \tfrac{1}{2}H_2O \qquad\qquad \Delta G_0' = -18.675 \qquad \text{(b)}$$

Adding reactions (a) and (b) yields

$$\tfrac{1}{6}NH_4^+ + \tfrac{1}{4}O_2 \rightarrow \tfrac{1}{6}NO_2^- + \tfrac{1}{3}H^+ + \tfrac{1}{6}H_2O$$

$$\Delta G_{0(\text{reaction})}' = -10.823 \frac{\text{kcal}}{\text{electron-mole}} \quad \textbf{Answer}$$

(c) Basis: 1 mg of initial NH_3-N. From Table 2-4 the synthesis reaction is

$$\tfrac{1}{5}CO_2 + \tfrac{1}{20}HCO_3^- + \tfrac{1}{20}NH_4^+ + H^+ + e \rightarrow \tfrac{1}{20}C_5H_7O_2N + \tfrac{9}{20}H_2O \qquad \text{(c)}$$

Therefore, by reaction (c),

$$\text{equivalent weight of } NH_3\text{-N} = \tfrac{14}{20} = 0.7 \text{ mg/mEq}$$

and

$$1 \text{ mg } NH_3\text{-N (the basis of calculations)} = \frac{1}{0.7} = 1.43 \text{ mEq } NH_3\text{-N}$$

$$C_5H_7O_2N = 12(5) + 1(7) + 16(2) + 14 = 113$$

Therefore, using the reaction above, we obtain

$$\text{equivalent weight of } C_5H_7O_2N = \frac{113}{20} = 5.65 \text{ mg/mEq}$$

$$0.15 \text{ mg } C_5H_7O_2N \Rightarrow \frac{0.15}{5.65} = 0.0265 \text{ mEq } C_5H_7O_2N = 0.0265 \text{ mEq of N in } C_5H_7O_2N$$

$$\Rightarrow \frac{0.15}{113} = 0.00133 \text{ mol}$$

Therefore,

NH_3-N remaining after production of *Nitrosomonas* $= 1.43 - 0.0265 = 1.40$ mEq

$$= 1.4(0.7) = 0.98 \text{ mg} = \frac{0.98}{14} = 0.07 \text{ mmol}$$

Therefore, on the basis of 1 mg of NH_3-N, the actual reactions are

Donor: $0.07NH_4^+ + 0.07\left(\dfrac{1/3}{1/6}\right)H_2O \rightarrow 0.07NO_2^- + 0.07\left(\dfrac{4/3}{1/6}\right)H^+ + \dfrac{0.07}{1/6}e$

$$0.07NH_4^+ + 0.14H_2O \rightarrow 0.07NO_2^- + 0.56H^+ + 0.42e \qquad (d)$$

Synthesis: $0.0053CO_2 + 0.00133HCO_3^- + 0.00133NH_4^+ + 0.027H^+ + 0.027e \rightarrow$

$$0.00133C_5H_7O_2N + 0.012H_2O \qquad (e)$$

From reaction (d), 0.42 electron mole has been supplied by 1 mg of NH_3-N. In reaction (e), 0.027 electron mole has been used for synthesis. Therefore,

$$\text{electron-moles supplied for energy production} = 0.42 - 0.027 = 0.393$$

and

$$\text{actual energy reaction: } 0.098O_2 + 0.393H^+ + 0.393e \rightarrow 0.197H_2O \qquad (f)$$

From reaction (d) and Table 2-4,

$$\Delta G'_{0(donor)} = 0.42(7.852) = 3.30 \text{ kcal/mg } NH_3\text{-N}$$

From reaction (f) and Table 2-4,

$$\Delta G'_{0(energy)} = 0.393(-18.675) = -7.34 \text{ kcal/}NH_3\text{-N}$$

Adding reactions (d), (e), and (f), gives

$$\Delta G'_0 = -7.34 + 3.30 = -4.04 \text{ kcal} = -\frac{4.04}{0.42}$$

$$= -9.62 \frac{\text{kcal}}{\text{electron-mole donated}} \qquad \textbf{Answer}$$

Note: Comparing this answer with the answer in part (b), energy used for synthesis $= 10.823 - 9.62 = 1.20$ kcal per electron-mole donated.

BIOLOGICAL COMBUSTION

Biological combustion is a form of combustion in which an organism utilizes molecular O_2 to oxidize a substrate producing energy. As discussed in the section on the thermodynamics of a microbial process, electrons are moved through an elaborate electron-transport system before O_2 is utilized. Using $C_xH_yO_zN_aS_b$ to represent any biological matter and using the half-reaction for sewage in Table 2-4 as the model, the electron donor reaction may be written as

$$\frac{1}{y + 4x - 3a - 2z - 2b}C_xH_yO_zN_aS_b + \frac{2x + a - z}{y + 4x - 3a - 2z - 2b}H_2O \rightarrow$$

$$\frac{x - a}{y + 4x - 3a - 2z - 2b}CO_2 + \frac{a}{y + 4x - 3a - 2z - 2b}NH_4^+ \qquad (2\text{-}7)$$

$$+ \frac{a}{y + 4x - 3a - 2z - 2b}HCO_3^- + \frac{b}{y + 4x - 3a - 2z - 2b}H_2S + H^+ + e$$

For a complete biological combustion, reaction (2-7) may be combined with the oxygen acceptor reaction of Table 2-4, producing

$$\frac{1}{y + 4x - 3a - 2z - 2b}C_xH_yO_zN_aS_b + \frac{1}{4}O_2 \rightarrow \frac{x - a}{y + 4x - 3a - 2z - 2b}CO_2$$

$$+ \frac{a}{y + 4x - 3a - 2z - 2by}NH_4^+ + \frac{a}{y + 4x - 3a - 2z - 2b}HCO_3^- \qquad (2\text{-}8)$$

$$+ \frac{b}{y + 4x - 3a - 2z - 2b}H_2S + \frac{y - 5a - 2b}{2(y + 4x - 3a - 2z - 2b)}H_2O$$

As shown, C has been converted to CO_2. Living things always produce carbon dioxide during metabolism. Also, nitrogen is bonded to hydrogen in the organic substance as is S. Hence the fate of nitrogen and sulfur is the production of ammonia and hydrogen sulfide, respectively. Conversion to NO or oxides of sulfur needs an extra step and more energy; it is therefore not favored.

Reaction (2-8) represents a complete biological combustion of organic matter. In more realistic situations, however, some form of residue from synthesis always remains after the process. From Table 2-4, using the synthesis reaction involving ammonia as the model for synthesis and letting $C_dH_eO_fN_gS_h$ represent the residue, the synthesis reaction may be written as

$$Y_c(d - g)CO_2 + gY_cHCO_3^- + gY_cNH_4^+ + Y_chH_2S$$

$$+ Y_c(e + 4d - 3g - 2f - 2h)H^+ + Y_c(e + 4d - 3g - 2f - 2h)e \longrightarrow \qquad (2\text{-}9)$$

$$Y_cC_dH_eO_fN_gS_h + Y_c(2d + g - f)H_2O$$

where Y_c is the cell yield in moles of $C_dH_eO_fN_gS_h$ per mole of $C_xH_yO_zN_aS_b$, the electron donor.

From reaction (2-7), the number of electron moles available per mole of donor is $y + 4x - 3a - 2z - 2b$. Hence, after synthesis [reaction (2-9)], the number of electron moles available for the acceptor reaction is $(y + 4x - 3a - 2z - 2b) - Y_c(e + 4d - 3g - 2f - 2h)$. The acceptor reaction of Table 2-4 may now be written as

$$
\frac{(y + 4x - 3a - 2z - 2b) - Y_c(e + 4d - 3g - 2f - 2h)}{4}\,O_2
$$

$$
+ [(y + 4x - 3a - 2z - 2b) - Y_c(e + 4d - 3g - 2f - 2h)]\,H^+
$$

$$
+ [(y + 4x - 3a - 2z - 2b) - Y_c(e + 4d - 3g - 2f - 2h)]\,e \longrightarrow
$$

$$
\frac{(y + 4x - 3a - 2z - 2b) - Y_c(e + 4d - 3g - 2f - 2h)}{2}\,H_2O
$$

(2-10)

Multiplying reaction (2-7) throughout by $y + 4x - 3a - 2z - 2b$ and combining the result with reactions (2-9) and (2-10), the overall biological oxidation that contains residues of organic matter is obtained:

$$
C_xH_yO_zN_aS_b + \frac{y + 4x - 3a - 2z - 2b - Y_c(e + 4d - 3g - 2f - 2h)}{4}\,O_2 \rightarrow
$$

$$
Y_c\,C_dH_eO_fN_gS_h + [(x - a) - Y_c(d - g)]CO_2 + (a - gY_c)NH_4^+
$$

$$
+ (a - gY_c)HCO_3^- + (b - Y_ch)H_2S
$$

(2-11)

$$
+ \frac{y + 2Y_ch - 5a - 2b - Y_c(e - 5g)}{2}\,H_2O
$$

To use reaction (2-11), Y_c and $C_dH_eO_fN_gS_h$ must first be determined by a bench- or pilot-plant study. Once this is done, all the terms of the equation can be calculated.

As shown in reaction (2-11), NH_4^+ is produced. In solid waste composting, this ammonium may stay with the compost; however, in a wastewater treatment plant, the aeration may be extended long enough. If this is done, *Nitrosomonas* and *Nitrobacter* may develop, which can convert nitrogen to nitrite and then to nitrate, respectively. This important topic is discussed in Chapter 11.

Example 2-7

Ultimate analysis of a mixture of sewage sludge and solid waste subjected to composting shows the following results: $x = 50$, $y = 100$, $z = 40$, and $a = 1$. The plant composts 100 metric tons (tonnes)/day of the mixture. Assuming that 20% excess air is used, (a) calculate the cubic meters per minute of air a blower must force into the pile measured at a standard temperature and pressure (STP) of 0°C and 1 atm pressure for complete biological combustion, and (b) calculate the air requirement (20% excess air) if 30% of the original C is converted to carbon dioxide, and the average empirical formula of the residue is $C_{70}H_{98}O_{28}N$. Assume an incoming air humidity of 0.016 mol of H_2O per mole of dry air.

Solution (a) $\dfrac{1}{y + 4x - 3a - 2z - 2b} C_x H_y O_z N_a S_b + \dfrac{1}{4}O_2 \rightarrow$

$$\dfrac{x - a}{y + 4x - 3a - 2z - 2b} CO_2 + \dfrac{a}{y + 4x - 3a - 2z - 2b} NH_4{}^+$$

$$+ \dfrac{a}{y + 4x - 3a - 2z - 2b} HCO_3{}^-$$

$$+ b\,H_2S + \dfrac{y - 5a - 2b}{2(y + 4x - 3a - 2z - 2b)} H_2O$$

$$C_x H_y O_z N_a S_b = C_{50}H_{100}O_{40}N$$

Therefore,

$$\dfrac{1}{y + 4x - 3a - 2z - 2b} = \dfrac{1}{100 + 4(50) - 3(1) - 2(40) - 2(0)} = \dfrac{1}{217} = 0.0046$$

$$\dfrac{y - 5a - 2b}{2(y + 4x - 3a - 2z - 2b)} = \dfrac{100 - 5(1) - 2(0)}{2(217)} = 0.218$$

Therefore,

$$0.0046 C_{50}H_{100}O_{40}N + 0.25 O_2 \rightarrow 0.226 CO_2 + 0.0046 NH_4{}^+ + 0.0046 HCO_3{}^- + 0.218 H_2O$$

$$\text{number of reference species} = 0.25(4) = 1 \text{ mol of electron}$$

$$\text{no. equiv. } C_{50}H_{100}O_{40}N = \text{no. equiv. } O_2 = \dfrac{100}{0.0046 C_{50}H_{100}O_{40}N/1}$$

$$= \dfrac{100}{0.0046(1354.8)/1} = \dfrac{100}{6.23} = 16.04 \text{ tonne-Eq/day}$$

$$\text{moles wet air used} = 1.2 \left\{ \dfrac{100}{21}(16.04)\left(\dfrac{0.25 O_2}{1}\right)\left(\dfrac{1}{O_2}\right) \right.$$

$$\left. + (0.016)\left[\dfrac{100}{21}(16.04)\left(\dfrac{0.25 O_2}{1}\right)\left(\dfrac{1}{O_2}\right)\right] \right\}$$

$$= 1.2 \left\{ \dfrac{100}{21}(16.04)\left(\dfrac{0.25}{1}\right) + (0.016)\left[\dfrac{100}{21}(16.04)\left(\dfrac{0.25}{1}\right)\right] \right\}$$

$$= 1.2(19.09 + 0.306) = 23.28 \text{ tonne-mol/day}$$

Therefore,

$$\text{cubic meters of air at STP} = 23.28(10^6)(22.4)(10^{-3})\left[\frac{1}{24(60)}\right]$$

$$= 362.13 \text{ per minute} \quad \textbf{Answer}$$

(b) $C_xH_yO_zN_aS_b + \dfrac{y + 4x - 3a - 2z - 2b - Y_c(e + 4d - 3g - 2f - 2h)}{4}O_2$

$$\rightarrow Y_cC_dH_eO_fN_gS_h$$

$$+ [(x - a) - Y_c(d - g)]CO_2 + (a - gY_c)NH_4^+$$

$$+ (a - gY_c)HCO_3^- + (b - Y_ch)H_2S$$

$$+ \frac{y + 2Y_ch - 5a - 2b - Y_c(e - 5g)}{2}H_2O$$

$$C_dH_eO_fN_gS_h = C_{70}H_{98}O_{28}N$$

$$\text{carbon gasified} = \frac{C_{50}}{C_{50}H_{100}O_{40}N}(100)(0.30) = \frac{12(50)}{1354.8}(100)(0.30)$$

$$= 13.39 \text{ tonnes/day}$$

$$\text{carbon converted to } C_{70}H_{98}O_{28}N = \frac{12(50)}{1354.8}(100) - 13.39 = 30.90 \text{ tonnes/day}$$

Therefore,

$$Y_c = \frac{30.90/[12(70)]}{100/1354.8} = 0.50$$

and

$$\frac{y + 4x - 3a - 2z - 2b - Y_c(e + 4d - 3g - 2f - 2h)}{4}$$

$$= \frac{100 + 4(50) - 3(1) - 2(40) - 2(0) - 0.50[98 + 4(70) - 3(1) - 2(28) - 2(0)]}{4}$$

$$= 14.375$$

$$\text{no. reference species} = 14.375(4) = 57.5 \text{ mol of electron}$$

$$\text{no. equiv. } C_{50}H_{100}O_{40}N \text{ gasified} = \text{no. equiv. } O_2 = \frac{100(0.3)}{C_{50}H_{100}O_{40}N/57.5}$$

$$= \frac{100(0.3)}{1354.8/57.5} = 1.27 \text{ tonne-Eq/day}$$

$$\text{moles wet air used} = 1.2 \left\{ \frac{100}{21}(1.27)\left(\frac{14.375O_2}{57.5}\right)\left(\frac{1}{O_2}\right) \right.$$

$$+ (0.016)\left[\frac{100}{21}(1.27)\left(\frac{14.375O_2}{57.5}\right)\left(\frac{1}{O_2}\right)\right]\right\}$$

$$= 1.2\left\{\frac{100}{21}(1.27)(8)\left(\frac{1}{O_2}\right) + (0.016)\left[\frac{100}{21}(1.27)(8)\left(\frac{1}{O_2}\right)\right]\right\}$$

$$= 1.2(1.51 + 0.024) = 1.84 \text{ tonne-mol/day}$$

Therefore,

$$\text{cubic meters of air at STP} = 1.84(10^6)(22.4)(10^{-3})\left[\frac{1}{24(60)}\right]$$

$$= 28.62 \text{ per minute} \quad \textbf{Answer}$$

CHEMISTRY OF THE ANAEROBIC PROCESS

A biological decomposition of organic waste done in the absence of air is called an *anaerobic process.* When the electron acceptor is the nitrate ion, the process is called *anoxic*; otherwise, the process is simply a normal anaerobic process. As in the case of the biological combustion process, the reaction may or may not go to completion. In fact, incomplete process is the rule rather than the exception. The decomposition of solid wastes in a sanitary landfill and the digestion of sludge in a sewage treatment plant are two processes that use the chemistry of the anaerobic process.

When $C_xH_yO_zN_aS_b$ is decomposed or digested by life processes, CO_2 is produced. This is true whether the process is aerobic or anaerobic. In the anaerobic process, there is the question of what the electron acceptor is. When nitrates and sulfates are added intentionally, the electron acceptors are these ions. However, when nothing is added intentionally, what would be the acceptor? As mentioned, when $C_xH_yO_zN_aS_b$ is digested, CO_2 is produced. Under anaerobic condition during the digestion of $C_xH_yO_zN_aS_b$, no nitrates and sulfates will be formed, since anaerobic processes are done in the absence of oxygen. Therefore, when there is no intentional addition of acceptors, the logical acceptor would be carbon dioxide. Hence, from Table 2-4, use the CO_2 acceptor reaction in the following derivation of the anaerobic chemical reaction.

As in biological combustion, anaerobic reactions can be complete or incomplete. First, the complete anaerobic reaction will be derived. The donor reaction will still be [reaction (2-7)]

$$\frac{1}{y+4x-3a-2z-2b}C_xH_yO_zN_aS_b + \frac{2x+a-z}{y+4x-3a-2z-2b}H_2O \rightarrow$$

$$\frac{x-a}{y+4x-3a-2z-2b}CO_2 + \frac{a}{y+4x-3a-2z-2b}NH_4^+$$

$$+ \frac{a}{y+4x-3a-2z-2b}HCO_3^-$$

$$+ \frac{b}{y+4x-3a-2z-2b}H_2S + H^+ + e \qquad (2\text{-}12)$$

The acceptor reaction is

$$\tfrac{1}{8}CO_2 + H^+ + e^- \rightarrow \tfrac{1}{8}CH_4 + \tfrac{1}{4}H_2O \qquad (2\text{-}13)$$

Adding reactions (2-12) and (2-13) produces the overall reaction for the complete-reaction anaerobic process:

$$\frac{1}{y+4x-3a-2z-2b}C_xH_yO_zN_aS_b + \frac{4x+7a+2b-2z-y}{4(y+4x-3a-2z-2b)}H_2O \rightarrow$$

$$\frac{4x+2z+2b-5a-y}{8(y+4x-3a-2z-2b)}CO_2 + \frac{1}{8}CH_4 + \frac{a}{y+4x-3a-2z-2b}NH_4^+ \qquad (2\text{-}14)$$

$$+ \frac{a}{y+4x-3a-2z-2b}HCO_3^- + \frac{b}{y+4x-3a-2z-2b}H_2S$$

Now, derive the reaction that produces a residue. The half-cell reactions for the donor and synthesis reactions are the same as those of biological oxidation, reactions (2-7) and (2-9), respectively. The difference is the acceptor reaction; for aerobic, the acceptor is oxygen, while for anaerobic, the acceptor is carbon dioxide. From reactions (2-7) and (2-9), the number of electron moles left for the electron acceptor per mole of $C_xH_yO_zN_aS_b$ is $(y+4x-3a-2z-2b) - Y_{dc}(e+4d-3g-2f-2h)$, where Y_{dc} is the Y_c for anaerobic conditions. This modifies the CO_2 electron acceptor reaction, reaction (2-13), as follows:

$$\frac{(y+4x-3a-2z-2b)-Y_{dc}(e+4d-3g-2f-2h)}{8}CO_2$$

$$+ [(y+4x-3a-2z-2b)-Y_{dc}(e+4d-3g-2f-2h)]H^+$$

$$+ [(y+4x-3a-2z-2b)-Y_{dc}(e+4d-3g-2f-2h)]e^- \rightarrow$$

$$\frac{(y+4x-3a-2z-2b)-Y_{dc}(e+4d-3g-2f-2h)}{8}CH_4 \qquad (2\text{-}15)$$

$$+ \frac{(y+4x-3a-2z-2b)-Y_{dc}(e+4d-3g-2f-2h)}{4}H_2O$$

Next, we reproduce the synthesis reaction, (2-9), but change Y_c to Y_{dc}.

$$Y_{dc}(d - g)CO_2 + gY_{dc}HCO_3^- + gY_{dc}NH_4^+ + Y_{dc}hH_2S$$

$$+ Y_{dc}(e + 4d - 3g - 2f - 2h)H^+ + Y_{dc}(e + 4d - 3g - 2f - 2h)e^- \rightarrow \qquad (2\text{-}16)$$

$$Y_{dc}C_dH_eO_fN_gS_h + Y_{dc}(2d + g - f)H_2O$$

Multiplying reaction (2-7) throughout by $y + 4x - 3a - 2z - 2b$ and combining the result with reactions (2-15) and (2-16), the overall anaerobic reaction that contains residues of organic matter, $C_dH_eO_fN_gS_h$, is

$$C_xH_yO_zN_aS_b + \frac{4x + 7a + 2b - 2z - y - Y_{dc}(4d + 7g + 2h - 2f - e)}{4}H_2O \rightarrow$$

$$Y_{dc}C_dH_eO_fN_gS_h$$

$$+ \frac{4x + 2z + 2b + Y_{dc}(5g + e - 4d - 2f - 2h) - 5a - y}{8}CO_2 \qquad (2\text{-}17)$$

$$+ (a - gY_{dc})NH_4^+ + (a - gY_{dc})HCO_3^- + (b - hY_{dc})H_2S$$

$$+ \frac{(y + 4x - 3a - 2z - 2b) - Y_{dc}(e + 4d - 3g - 2f - 2h)}{8}CH_4$$

From the stoichiometry of the reactions, it is possible to calculate the various masses involved. For example, the masses of H_2O and CH_4 per unit mass of $C_xH_yO_zN_aS_b$ from reaction (2-17) are, respectively,

$$\frac{\text{mass } H_2O}{\text{mass } C_xH_yO_zN_aS_b} = \frac{\frac{4x+7a+2b-2z-y-Y_{dc}(4d+7g+2h-2f-e)}{4}H_2O}{C_xH_yO_zN_aS_b}$$

$$\frac{\text{mass } CH_4}{\text{mass } C_xH_yO_zN_aS_b} = \frac{\frac{(y+4x-3a-2z-2b)-Y_{dc}(e+4d-3g-2f-2h)}{8}CH_4}{C_xH_yO_zN_aS_b} \qquad (2\text{-}18)$$

Similar equations can be derived for the rest.

Example 2-8

Ultimate analysis of a mixture of sewage sludge and solid waste deposited in a sanitary landfill produces the following results: $x = 50$, $y = 100$, $z = 40$, and $a = 1$. The landfill processes 100 tonnes/day of combined sludge and solid waste ($C_{50}H_{100}O_{40}N$). What is the water requirement for a complete decomposition to take place?

Solution

$$\frac{1}{y + 4x - 3a - 2z - 2b}C_xH_yO_zN_aS_b + \frac{4x + 7a + 2b - 2z - y}{4(y + 4x - 3a - 2z - 2b)}H_2O \rightarrow$$

$$\frac{4x + 2z + 2b - 5a - y}{8(y + 4x - 3a - 2z - 2b)}CO_2 + \frac{1}{8}CH_4$$

$$+ \frac{a}{y + 4x - 3a - 2z - 2b}NH_4^+ + \frac{a}{y + 4x - 3a - 2z - 2b}HCO_3^-$$

$$+ \frac{b}{y + 4x - 3a - 2z - 2b}H_2S$$

$$C_xH_yO_zN_aS_b = C_{50}H_{100}O_{40}N = 12(50) + 100 + 16(40) + 14 = 1354$$

$$\frac{\text{mass } H_2O}{\text{mass } C_xH_yO_zN_aS_b} = \frac{\frac{4x+7a+2b-2z-y}{4(y+4x-3a-2z-2b)}H_2O}{\frac{1}{y+4x-3a-2z-2b}C_xH_yO_zN_aS_b}$$

$$= \frac{[4(50) + 7(1) + 2(0) - 2(40) - 100]18}{4(1354)}$$

$$= 0.09$$

Therefore,

$$\text{water requirement} = 0.09(100) = 9\frac{\text{tonnes}}{\text{day}} \quad \textbf{Answer}$$

Example 2-9

In Example 2-8, if 30% of the original C is converted to carbon dioxide and the average empirical formula of the residue is $C_{70}H_{98}O_{28}N$, calculate the water requirement.

Solution

$$\frac{\text{mass } H_2O}{\text{mass } C_xH_yO_zN_aS_b} = \frac{\frac{4x+7a+2b-2z-y-Y_{dc}(4d+7g+2h-2f-e)}{4}H_2O}{C_xH_yO_zN_aS_b}$$

$$C_dH_eO_fN_gS_h = C_{70}H_{98}O_{28}N$$

$$\text{Carbon gasified} = \frac{C_{50}}{C_{50}H_{100}O_{40}N}(100)(0.30) = \frac{12(50)}{1354}(100)(0.30) = 13.3 \text{ tonnes/day}$$

$$\text{Carbon converted to } C_{70}H_{98}O_{28}N = \frac{12(50)}{1354}(100) - 13.30 = 31.0 \text{ tonnes/day}$$

Therefore,

$$Y_{dc}C_{70} = \frac{31.0/[12(70)]}{100/1354} = 0.50$$

$$C_xH_yO_zN_aS_b = C_{50}H_{100}O_{40}N$$

Thus,

$$\frac{\text{mass } H_2O}{\text{mass } C_xH_yO_zN_aS_b}$$

$$= \frac{\frac{4(50)+7(1)+2(0)-2(40)-100-0.50[4(70)+7(1)+2(0)-2(28)-98]}{4}(18)}{1354} = -0.12$$

Therefore,

$$\text{water requirement} = -0.12(100) = -12\frac{\text{tonnes}}{\text{day}} = 0 \quad \textbf{Answer}$$

WASTEWATER CHARACTERISTICS

Important parameters used to characterize an organic waste are given in the constituents of municipal waste shown in Table 2-5.

TABLE 2-5 TYPICAL COMPOSITION OF UNTREATED DOMESTIC WASTEWATER[a]

Constituent	Concentration		
	Strong	Medium	Weak
Biochemical oxygen demand (BOD$_5$ at 20°C)	420	200	100
Total organic carbon (TOC)	280	150	80
Chemical oxygen demand (COD)	1000	500	250
Total solids	1250	700	300
Dissolved	800	500	230
Fixed	500	300	140
Volatile	300	200	90
Suspended	450	200	70
Fixed	75	55	20
Volatile	375	145	50
Settleable solids (mL/L)	20	10	5
Total nitrogen	90	50	20
Organic	35	15	10
Free ammonia	55	35	10
Nitrites	0	0	0
Nitrates	0	0	0
Total phosphorus as P	18	10	5
Organic	5	3	1
Inorganic	13	7	4
Chlorides	110	45	30
Alkalinity as CaCO$_3$	220	110	50
Grease	160	100	50

[a]All values except settleable solids are expressed in mg/L.

The *total solids* content of a wastewater are the materials left after water has been evaporated from the sample. The evaporation is normally done at 103 to 105°C. Total solids, the residue after evaporation, may be classified as *dissolved* or *suspended*. The dissolved solids fraction includes, by definition, the colloidal particles. If the solids are not dissolved, they are, by definition, suspended. The *Standard Methods* also calls suspended solids *nonfilterable residues* and the dissolved solids *filterable residues*. The *settleable* fraction is included in the suspended solids and is the volume of the solids after settling for 30 minutes in a cone-shaped container called an *Imhoff cone* [*sludge volume index* (SVI)]. Settleable solids are an approximate measure of the volume of sludge that will settle by sedimentation. The *fixed* portions of the solids (dissolved and suspended) are those that remain as a residue when the sample is decomposed at 600°C. Those that disappear are called *volatile solids*. Figure 2-2 shows Imhoff cones.

Figure 2-2   Imhoff cones.

The *biochemical oxygen demand* (BOD) of wastewater is a measure of the oxygen-consuming property of the waste expressed in terms of oxygen that is consumed biologically; it is a measure of waste strength. BOD_5 is a waste strength that represents the amount of oxygen that will be consumed in 5 days at a given temperature. The normal temperature used for BOD analysis is 20°C. This temperature standardizes all BOD_5 results from one laboratory to another, other parameters considered constant. Intrinsic in the biochemical oxygen demand is the reaction of the wastewater carbon with oxygen in the biochemical process. Hence the strength of the waste to consume oxygen can also be expressed in terms of the carbon it contains. *Total organic carbon* is the carbon content of an organic waste. *Chemical oxygen demand* (COD) is a measure of the total ability of a wastewater to consume oxygen. The method of analysis normally uses potassium dichromate as the oxidant. Because of its nonbiological portion, COD is greater than BOD.

An organic matter may contain protein. The protein, in turn, contains nitrogen, which is about 16% of the protein mass. The nitrogen in protein is an organic nitrogen. *Organic nitrogen*, therefore, is a measure of the protein content of an organic waste. When organic matter is attacked by microorganisms, its protein hydrolyzes into ammonia. *Free ammonia* is the hydrolysis product of organic nitrogen. The *nitrites* and *nitrates* are the results of the oxidation of ammonia to nitrites by *Nitrosomonas* and the oxidation of nitrites to nitrates by *Nitrobacter*, respectively. The sum of the organic, free ammonia, nitrite, and nitrate nitrogens is called *total nitrogen*. The sum of ammonia and organic nitrogens is called *Kjeldahl nitrogen*.

As shown in Table 2-5, the other constituent of an organic waste is phosphorus, which is classified into organic and inorganic. The species of nitrogen and phosphorus can cause eutrophication of receiving bodies of water if they are discharged in large amounts. Only one of the elements nitrogen and phosphorus needs to be controlled. Phosphorus control is recommended. The last three constituents shown in Table 2-5 are chlorides, alkalinity, and grease. Grease is very troublesome in sewers, as it can accumulate in bends, causing sewage to back up into basements of homes. Other contaminants found in sewage are pathogens, refractory organics, and heavy metals.

Since BOD_5 is very important in the design of wastewater treatment plants, this parameter will discussed at length. In general, there are two types of BOD laboratory methods of analysis: one where dilution is necessary and one where dilution is not necessary. When the BOD of a sample is small, such as in river waters, dilution is not necessary. Otherwise, the sample would have to be diluted. Table 2-6 sets the criteria for determining the dilution required. This table shows that there are two ways that dilution can be made: as a percent mixture or by direct pipetting into 300-mL BOD bottles. Normally, BOD analysis is done using 300-mL incubation bottles.

TABLE 2-6   RANGES OF BOD MEASURABLE WITH VARIOUS DILUTIONS OF SAMPLES

Using percent mixtures		By direct pipetting into 300-mL bottles	
% mixture	Range of BOD_5	mL	Range of BOD_5
0.01	35,000–70,000	0.01	40,000–100,000
0.03	10,000–35,000	0.05	20,000–40,000
0.05	7,000–10,000	0.10	10,000–20,000
0.1	3,500–7,000	0.30	4,000–10,000
0.3	1,400–3,500	0.50	2,000–4,000
0.5	700–1,400	1.0	1,000–2,000
1.0	350–700	3.0	400–1,000
3.0	150–350	5.0	200–400
5.0	70–150	10.0	100–200
10.0	35–70	30.0	40–100
30.0	10–35	50.0	20–40
50.0	5–10	100	10–20
100	0–5	300	0–10

Since BOD analysis attempts to measure the oxygen equivalent of a given waste, the environment inside the BOD bottle must be conducive to uninhibited bacterial growth. The parameters of importance for maintaining this type of environment are freedom from toxic materials, favorable pH and osmotic pressure conditions, optimal amount of nutrients, and the presence of significant amount of population of mixed organisms of soil origin. Through long years of experience, it has been found that synthetic dilution water prepared from distilled water or demineralized water is best for BOD work, since the presence of such toxic substances as chloramine, chlorine, and copper can easily be controlled. The maintenance of favorable pH can be assured by buffering the dilution water at about pH 7.0 using potassium and sodium phosphates. The potassium and sodium ions, along with the addition of calcium and magnesium ions, can also maintain the proper osmotic pressure, as well as provide the necessary nutrients in terms of these elements. The phosphates, of course, provide the necessary phosphorus nutrient requirement. Ferric chloride, magnesium sulfate, and ammonium chloride supply the requirements for iron, sulfur, and nitrogen.

Since the sample submitted for analysis may contain different types of "BOD constituents," a population of mixed organisms is accordingly required. The sample, however, may not contain any organisms at all. Such is the case, for example, of an industrial waste, which can be completely sterile. For this situation, the dilution water must be seeded with organisms from an appropriate source. In domestic wastewaters, all the organisms needed are already there; hence these wastewaters can serve as good sources of seed organisms. Experience has shown that a seed volume of 2.0 mL per liter of dilution water is all that is needed.

Calculation of BOD

In the subsequent development, the formulation will be based on the assumption that the dilution method is used. If, in fact, the method used is direct, that is, no dilution, the dilution factor that appears in the formulation will simply be equated to zero. The technique of determining the BOD of a sample is to find the difference in dissolved oxygen (DO) concentration between the final and initial time after a period of incubation at some controlled temperature. This difference, converted to mass of oxygen per unit volume of sample (such as mg/L), is the BOD. Let I be the initial DO of the sample, which has been diluted with seeded dilution water, and F be the final DO of the same sample after the incubation period. The difference would then represent a BOD, but since the sample is seeded, a correction must be made for the BOD of the seed. This necessitates running a blank.

Let I' represent the initial DO of a volume Y of the blank composed of only the seeded dilution water; also, let F' be the final DO after incubating this blank at the same time and temperature as the sample. If X is the volume of the seeded dilution water mixed with the sample, the DO correction would be $(I' - F')(X/Y)$. Letting D be the fractional dilution, the BOD of the sample is

$$\text{BOD} = \frac{(I - F) - (I' - F')(X/Y)}{D} \qquad (2\text{-}19)$$

In this equation, if the incubation period is 5 days, the BOD is called the *5-day biochemical oxygen demand*, BOD_5. It is understood that unless it is specified, BOD_5 is a BOD measured at the standard temperature of incubation of 20°C. If incubation is done for a long period of time, such as 20 to 30 days, it is assumed that all the BOD has been consumed. The BOD under this situation is the ultimate; hence it is called *ultimate* BOD, or BOD_u.

BOD_u, in turn, can have two fractions: one due to carbon and the other due to nitrogen. As mentioned before, in the biochemical reaction, carbon reacts with oxygen; also, nitrogen, in the form of ammonia, reacts with oxygen. If the BOD reaction is allowed to go to completion with the ammonia reaction inhibited, the resulting ultimate BOD is called *ultimate carbonaceous* BOD, or CBOD. Since *Nitrosomonas* and *Nitrobacter*, the organisms for the ammonia reaction, cannot compete very well with carbonaceous bacteria (the organisms for the carbon reaction) the reaction during the first few days of incubation up to approximately 5 or 6 days is mainly carbonaceous. Hence BOD_5 is mainly carbonaceous. If the reaction is uninhibited, after 5 or 6 days of incubation the BOD also contains the nitrogenous BOD. BOD is normally reported in units of mg/L.

Experience has demonstrated that a dissolved oxygen concentration of 0.5 mg/L does not affect BOD. Also, it has been learned that a depletion of less than 2.0 mg/L produces erroneous BOD results. Hence it is important that in BOD work, the concentration of DO in the incubation bottle should not fall below 0.5 mg/L and that the depletion after the incubation period should not be less than 2.0 mg/L.

Example 2-10

Ten milliliters of sample is pipeted directly into a 300-mL incubation bottle. The initial DO of the diluted sample is 9.0 mg/L and its final DO is 2.0 mg/L. The initial DO of the dilution water is also 9.0 mg/L, but the final DO is 8.0 mg/L. The temperature of incubation is 20°C. If the sample is incubated for 5 days, what is the BOD_5 of the sample?

Solution

$$BOD_5 = \frac{(I - F) - (I' - F')(X/Y)}{D}$$

$I = 9.0$ mg/L; $F = 2.0$ mg/L; $I' = 9.0$ mg/L; and $F' = 8.0$ mg/L. $X = 300 - 10 = 290$ mL and $Y = 300$ mL, since the dilution water would have to be incubated in a 300-mL bottle.

$$D = \frac{10}{300} = 0.033$$

Therefore,

$$BOD_5 = \frac{(9.0 - 2.0) - (9.0 - 8.0)(290/300)}{0.033} = 183 \text{ mg/L} \quad \textbf{Answer}$$

Example 2-11

A suspended solids analysis is run on a sample. The tare mass of the crucible and filter is 55.3520 g. A sample of 260 mL is then filtered and the residue dried to constant mass at 103°C. If the constant mass of the crucible, filter, and the residue is 55.3890 g, what is the suspended solids content (SS) of the sample?

Solution

$$SS = \frac{55.3890 - 55.3520}{260(10^{-3})}(1000) = 142 \text{ mg/L} \quad \textbf{Answer}$$

SOLUBILITY PRODUCT CONSTANT

Let the reactant be represented by a solid molecule A_aB_b. As this reactant is mixed with water, it dissolves into its constituent solute ions. From general chemistry, the equilibrium dissolution reaction is

$$A_aB_b \rightleftharpoons aA + bB \qquad (2\text{-}20)$$

The equilibrium constant is

$$\frac{K = \{A\}^a\{B\}^b}{\{A_aB_b\}} \qquad (2\text{-}21)$$

Since A_aB_b is a solid, the concentrations of A and B are small compared to that of A_aB_b. For this reason, by convention, the product $K\{A_aB_b\}$ is considered a constant and is designated as K_{sp}. K_{sp} is called the *solubility product constant* of the equilibrium dissolution reaction. Equation (2-21) now transforms to

$$K_{sp} = \{A\}^a\{B\}^b \qquad (2\text{-}22)$$

Table 2-7 is a table of solubility product constants of representative compounds.

TABLE 2-7 SOLUBILITY PRODUCT CONSTANTS OF REPRESENTATIVE COMPOUNDS

Equilibrium reaction	K_{sp} at 25°C	Significance
$AgCl \rightleftharpoons Ag^+ + Cl^-$	$1.8(10^{-10})$	Chloride analysis
$Al(OH)_3 \rightleftharpoons Al^{3+} + 3\ OH^-$	$1.9(10^{-33})$	Coagulation
$BaSO_4 \rightleftharpoons Ba^{2+} + SO_4{}^{2-}$	10^{-10}	Sulfate analysis
$CaCO_3 \rightleftharpoons Ca^{2+} + CO_3{}^{2-}$	$4.8(10^{-9})$	Hardness removal, scales
$Ca(OH)_2 \rightleftharpoons Ca^{2+} + 2\ OH^-$	$7.9(10^{-6})$	Hardness removal
$CaSO_4 \rightleftharpoons Ca^{2+} + SO_4{}^{2-}$	$2.4(10^{-5})$	Flue gas desulfurization
$Ca_3(PO_4)_2 \rightleftharpoons 3\ Ca^{2+} + 2\ PO_4{}^{3-}$	10^{-25}	Phosphate removal
$CaHPO_4 \rightleftharpoons Ca^{2+} + HPO_4{}^{2-}$	$5(10^{-6})$	Phosphate removal
$CaF_2 \rightleftharpoons Ca^{2+} + 2\ F^-$	$3.9(10^{-11})$	Fluoridation
$Cr(OH)_3 \rightleftharpoons Cr^{3+} + 3\ OH^-$	$6.7(10^{-31})$	Heavy metal removal
$Cu(OH)_2 \rightleftharpoons Cu^{2+} + 2\ OH^-$	$5.6(10^{-20})$	Heavy metal removal
$Fe(OH)_3 \rightleftharpoons Fe^{3+} + 3\ OH^-$	$1.1(10^{-36})$	Coagulation, iron removal, corrosion
$Fe(OH)_2 \rightleftharpoons Fe^{2+} + 2\ OH^-$	$7.9(10^{-15})$	Coagulation, iron removal, corrosion
$MgCO_3 \rightleftharpoons Mg^{2+} + CO_3{}^{2-}$	$ca(10^{-5})$	Hardness removal, scales
$Mg(OH)_2 \rightleftharpoons Mg^{2+} + 2\ OH^-$	$1.5(10^{-11})$	Hardness removal, scales
$Mn(OH)_2 \rightleftharpoons Mn^{2+} + 2\ OH^-$	$4.5(10^{-14})$	Manganese removal
$Ni(OH)_2 \rightleftharpoons Ni^{2+} + 2\ OH^-$	$1.6(10^{-14})$	Heavy metal removal
$Zn(OH)_2 \rightleftharpoons Zn^{2+} + 2\ OH^-$	$4.5(10^{-17})$	Heavy metal removal

Example 2-12

The practical limit of technology in the removal of calcium carbonate hardness at normal plant operating conditions is 40 mg/L of $CaCO_3$. What are the concentrations of the calcium and carbonate ions at 25°C?

Solution   From Table 2-7, $K_{sp}(CaCO_3) = 5 \times 10^{-9}$ and

$$CaCO_3 \rightleftharpoons Ca^{2+} + CO_3^-$$

Let x be the concentration of Ca^{2+}. From the reaction the concentration of the carbonate ion is also x. Hence substituting into the equilibrium equation yields

$$x^1(x^1) = 5(10^{-9}) = x^2$$

$$x = 7.1(10^{-5}) \text{ mol/L} = \{Ca^{2+}\} = \{CO_3^{2-}\} \quad \textbf{Answer}$$

EUTROPHICATION

Eutrophication is a natural process of aging of a body of water. It is a result of a very slow process of natural sedimentation of microscopic organisms which takes geologic times to complete. The completion of the process results in the extinction of the water body.

The process of eutrophication is propelled by increasing concentrations of nutrients necessary for biological activity. First, envision a clean, clear water. In this condition, since the nutrients available are minimal, there is no significant biological activity in the water column that can support sedimentation. The water body is healthy and the condition is called *oligotrophic*. As time progresses, however, nutrients can build up. A water body with nutrient concentration supporting biological activity that is not objectionable but above that of oligotrophic conditions is considered *mesotrophic*. In the next stage of the life cycle the water becomes *eutrophic*. This is characterized by murky water with an accelerated rate of sedimentation of microorganisms. The final life stage before extinction is a pond, marsh, or swamp.

The life cycle depicted above takes geologic times to complete. Human activity, however, encourages the production of nutrients and shortens the eutrophication cycle. People who have a green lawn are prolific producers of nutrients. Farmlands are an excellent source of nutrients. The Chesapeake Bay in Maryland, for example, is loaded with nutrients coming from farmlands as far away as New York.

Microorganisms are composed of carbon, hydrogen, oxygen, nitrogen, phosphorus, and trace elements. For microorganisms to survive, they must be provided with these elements in the form of nutrients. It is therefore possible to limit the growth of these microorganisms by controlling the input of these elements into a body of water. Theoretically, introduction of carbon, hydrogen, and oxygen may be limited were it not for the fact that these elements are already in abundance in the environment. Carbon is already introduced in the form of CO_2 from the air, and hydrogen and oxygen are components of H_2O.

As far as eutrophication control is concerned, the nutrients of utmost importance are nitrogen and phosphorus. Nitrogen is utilized by organisms in the form of NH_4^+, NO_2^-, and NO_3^-; phosphorus is utilized in the form of orthophosphorus. Algae, the prime cause of eutrophication, are produced by photosynthesis. In the chloroplast, using the energy from sunlight, the water molecule breaks down, releasing electrons for synthesis:

$$276H_2O \rightarrow 138O_2 + 552H^+ + 552e \quad\quad\quad (2\text{-}23)$$

CO_2 from the air dissolved in water, together with other nutrients and trace elements, then takes the electrons released from water to form algae as follows:

$$106CO_2 + 16NO_3^- + HPO_4^{2-} + 570H^+ + 552e + \text{trace elements} \rightarrow$$
$$(CH_2O)_{106}(NH_3)_{16}H_3PO_4 \cdot \text{trace elements (the algae)} + 154H_2O \quad (2\text{-}24)$$

Adding reactions (2-23) and (2-24) produces the total reaction,

$$106CO_2 + 16NO_3^- + HPO_4^{2-} + 122H_2O + 18H^+ + \text{trace elements} \rightarrow$$
$$(CH_2O)_{106}(NH_3)_{16}H_3PO_4 \cdot \text{trace elements}^3 + 138O_2 \quad (2\text{-}25)$$

In water quality modeling, the concentration of algae is normally expressed in terms of chlorophyll a, $C_{55}H_{72}MgN_4O_5$.

As evidenced from reaction (2-25), control of only one element is all that is needed to limit the growth of algae. Of the elements nitrogen and phosphorus, which should be controlled? It has been found that the blue greens can fix nitrogen from the air. Blue greens can survive almost anywhere and can easily outgrow other algae under adverse conditions. For this and other reasons, it is not prudent to control nitrogen. Phosphorus control is recommended.[4]

Example 2-13

The Back River sewage treatment plant in the city of Baltimore discharges to the Back River estuary. The state of Maryland imposes on its permit a total phosphorus limitation of 2.0 mg/L. Assuming that all the total phosphorus converts to orthophosphorus, what are the expected (a) algae concentration, and (b) chlorophyll a concentration, assuming that chlorophyll a is 1.5% of algae? Assume the dilution effect of the estuary to be negligible (of course, this is not true).

Solution   (a) $106CO_2 + 16NO_3^- + HPO_4^{2-} + 122H_2O + 18H^+ + \text{trace elements} \rightarrow$

$(CH_2O)_{106}(NH_3)_{16}H_3PO_4 \cdot \text{trace elements} + 138O_2$

$(CH_2O)_{106}(NH_3)_{16}H_3PO_4 = (12 + 18)(106) + 17(16) + 3 + 31 + 64 = 3550$

---

[3]W. Stumm and J. J. Morgan (1981). *Aquatic Chemistry*, 2nd ed., Wiley-Interscience, New York.

[4]A. P. Sincero (1984). "Eutrophication and the Fallacy of Nitrogen Removal." *Pollution Engineering*. Pudvan Publishing Co., Northbrook, Ill.

Therefore,

$$\text{theoretically expected algae concentration} = \frac{3550}{31}(2) = 229.0 \text{ mg/L} \quad \textbf{Answer}$$

(b) Theoretically expected chlorophyll a concentration $= 0.015(229)(1000)$

$$= 3435 \ \mu g/L \quad \textbf{Answer}$$

The computed concentrations in Example 2-13 are unrealistic; the dilution effect of the receiving stream should be considered. This will result in a smaller orthophosphorus concentration. Also, settling of total phosphorus not considered in the solution to the problem. In a real estuary, a high concentration of 300 $\mu g/L$ of chlorophyll a is not uncommon.

Example 2-14

(a) From reactions (2-23) to (2-25), find the equivalent weight of algae. (b) A sample of estuarine water containing 300 $\mu g/L$ of chlorophyll a is saturated with oxygen and incubated under dark conditions. Assuming that sufficient oxygen is present in the incubation bottle to satisfy the reaction, how many mg/L of oxygen is consumed? Assume that algae contains 1.5% chlorophyll a.

Solution

(a) $\hspace{3cm} 276H_2O \rightarrow 138O_2 + 552H^+ + 552e$

$$106CO_2 + 16NO_3^- + HPO_4^{2-} + 570H^+ + 552e + \text{trace elements} \rightarrow$$

$$(CH_2O)_{106}(NH_3)_{16}H_3PO_4 \cdot \text{trace elements} + 154H_2O$$

algae

$$106CO_2 + 16NO_3^- + HPO_4^{2-} + 122H_2O + 18H^+ + \text{trace elements} \rightarrow$$

$$(CH_2O)_{106}(NH_3)_{16}H_3PO_4 \cdot \text{trace elements} + 138O_2$$

From the reactions, 552 mol of electrons is consumed per mole of algae. Therefore,

$$\text{equivalent weight of algae} = \frac{3550}{552} = 6.43 \text{ mg/mEq} \quad \textbf{Answer}$$

(b) $(CH_2O)_{106}(NH_3)_{16}H_3PO_4 \cdot \text{trace elements} + 138O_2 \rightarrow$

$$106CO_2 + 16NO_3^- + HPO_4^{2-} + 122H_2O + 18H^+ + \text{trace elements}$$

$$\text{mg algae} = \frac{300}{0.015(1000)} = 20 \text{ per liter} = \frac{20}{6.43} = 3.11 \text{ mEq/L}$$

$$\text{equivalent weight } O_2 = \frac{138(32)}{552} = 8.0 \text{ mg/mEq}$$

Therefore,

$$\text{mg/L } O_2 \text{ consumed} = 3.11(8) = 24.88 \text{ mg/L} \quad \textbf{Answer}$$

The Kinetics of Eutrophication

The condition of excessive growth of plants, both attached and planktonic, in a water body is called *eutrophication.* The major cause of eutrophication is the excessive discharge of the nutrients nitrogen and phosphorus. In the context of chemical or biological reactions, *kinetics* means the rate of change of concentration with respect to time (i.e., a derivative with respect to time). Designate this derivative by R. (Knowledge of eutrophication kinetics gained in this chapter will be used in the discussion on environmental quality modeling in a subsequent chapter.)

Five kinetic processes are involved in eutrophication: phytoplankton (floating algae), phosphorus, nitrogen, CBOD, and DO kinetics. The schematic is depicted in Figure 2-3, where the five processes can be identified. *State variable* means the variable whose R derivative is to be formulated. There are a total of eight state variables indicated in the schematic: PHYT, ON, NO_3, NH_3, OP, OPO_4, CBOD, and DO.

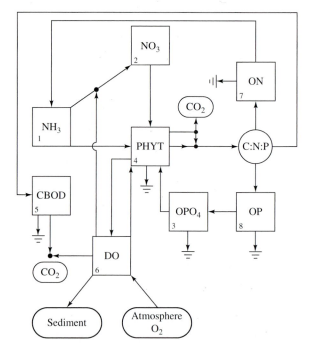

Figure 2-3 Schematic diagram of eutrophication kinetics.

Phytoplankton (PHYT) require nutrients for growth. This is indicated by the input arrows of nitrates (NO_3), ammonia (NH_3), and orthophosphorus (OPO_4). During photosynthesis, oxygen is released, and during respiration, oxygen is needed by the phytoplankton. [Algae (i.e., phytoplankton) also respire.] During respiration, CO_2 is released. The phytoplankton body contains nitrogen, carbon, and phosphorus. As the phytoplankton die, these elements go to the elements pool, designated as C:N:P. In the two-arrow pathway for CO_2, the arrow going downward simply indicates that the carbon in CO_2 goes to the elements pool.

Upon death of the phytoplankton, the phosphorus goes to the elements pool as organic phosphorus, OP. There are, subsequently, two kinetic processes that OP undergoes: settling, as indicated by the arrow going downward, and hydrolysis to OPO_4. The orthophosphorus, OPO_4, may be sorbed onto particles; hence it is indicated as being settled. The other portion of OPO_4 is then consumed by the phytoplankton.

Also, upon death of the phytoplankton, nitrogen goes to the pool as organic nitrogen and ammonia. Some of the organic nitrogen settles and some is hydrolyzed to NH_3. By the action of *Nitrosomonas* and *Nitrobacter*, ammonia is oxidized to NO_3. The nitrate may then be used by the phytoplankton, or be denitrified to gaseous nitrogen, N_2, if the surrounding becomes anoxic. The carbon portion of the elements pool is designated as the carbonaceous oxygen demand (CBOD). The solid portion of the CBOD settles and the other portion is utilized by carbonaceous microorganisms, utilizing oxygen in the process. The CBOD that settles becomes sediment oxygen demand (SOD).

The last kinetic process is the dissolved oxygen (DO). The sources of oxygen are the atmosphere and phytoplankton photosynthesis. The dissolved oxygen is used by *Nitrosomonas* and *Nitrobacter* in oxidizing the NH_3 to nitrate. It is also used by the CBOD kinetics and to oxidize the SOD.

Phytoplankton kinetics. As shown in the figure, phytoplankton is the fourth state variable; call its concentration c_4. The kinetic coefficients involved are growth (μ_p), death (k_d), and settling (k_{s4}). Therefore, the R derivative, R_4, is

$$R_4 = (\mu_p - k_d - k_{s4})c_4 \qquad (2\text{-}26)$$

The concentration of phytoplankton is normally measured in terms of active chlorophyll a, $C_{55}H_{72}MgN_4O_5$. *Active chlorophyll a* is to be differentiated from *inactive* or *dead chlorophyll a*, which is also called *pheophytin.*

The growth coefficient μ_p may be deduced as a fraction of the maximum growth rate coefficient μ_{pm}, modified by the effect of light intensity (f_I), temperature (f_T), and limiting nutrient (f_N). There appear to be two extremes in the effect of light intensity on growth kinetics: at very low light intensity, growth is inhibited, and at very intense light intensity, growth is photoinhibited. Between these two extremes is the intensity where growth rate is a maximum. The intensity corresponding to this point is called the *saturating growth rate light intensity, I_s.*

Call μ_{pm} at 20°C μ_{pm20}. Considering the modifying factors f_I, f_T, and f_T, μ_p may be written as

$$\mu_p = \mu_{pm20} f_I f_T f_N \qquad (2\text{-}27)$$

The bottom three plots in Figure 2-4 show the effect of changing the light intensity I from the optimum on the concentration of phytoplankton. P/P_s is the ratio of the

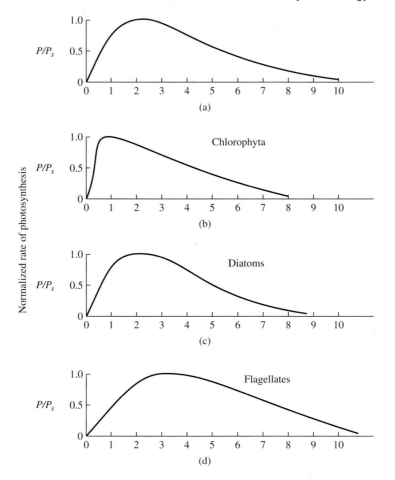

Normalized rate of photosynthesis

(a)

Chlorophyta

(b)

Diatoms

(c)

Flagellates

(d)

Figure 2-4   Normalized rate of photosynthesis versus incident light intensity.

concentration of phytoplankton P produced at I to the concentration of the phytoplankton P_s produced at I_s. The top part of the figure is a plot of the equation proposed by Steele:[5]

$$\frac{P}{P_s} = f_I = \frac{I}{I_s}\exp\left(-\frac{I}{I_s}+1\right) \tag{2-28}$$

where I_s is around 300 langleys (Ly) per day and f_I is the fractional attenuation of the maximum growth due to a different light intensity (mentioned before).

From the Lambert–Beers law, $I = I_0 \exp(-K_e z)$, where I_0 is the intensity at the surface of the water body, K_e is an extinction coefficient, and z is the water depth.

[5]J. H. Steele (1965). "Primary Production in Aquatic Environments," in C. R. Goldman (ed.), *Notes on Some Theoretical Problems in Production Ecology.* Memorial Institute for Idrobiology, University of California Press, Berkeley, Calif., 18 Suppl., pp. 383–398.

Substituting in equation (2-28), f_I at any depth is

$$f_I = \frac{I_0 \exp(-K_e z)}{I_s} \exp\left[-\frac{I_0 \exp(-K_e z)}{I_s} + 1\right] \qquad (2\text{-}29)$$

A factor may also have to be incorporated to consider the effect of cloud cover.

An empirical relationship relating K_e to the Secchi depth z_s has been developed.[6] The Secchi depth is the depth at which a white disk about 1 ft in diameter will disappear from view when immersed in a body of water. The relationship is

$$K_e = \frac{1.8}{z_s} \qquad (2\text{-}30)$$

The Arrhenius factor f_T used to correct for temperature is normally written in the form θ_g^{T-20}, where the initial temperature is 20°C. Hence the μ_m to be used is μ_{m20} at 20°C. $\theta_g = 1.068$ is generally used.

The effect of nutrient limitation is based on Monod's concept. *Nutrient limitation* means that the rate of growth of the organism is retarded because of the small concentration of the single limiting nutrient. Normally, the nutrient that limits growth is either nitrogen or phosphorus. To ascertain which is limiting, the Monod expression for nitrogen and phosphorus should be evaluated and the smaller of the two chosen. In other words, with f_N as the correction factor,

$$f_N = \min\left(\frac{[N]}{K_N + [N]}, \frac{[P]}{K_P + [P]}\right) \qquad (2\text{-}31)$$

where [N] and [P] are the concentrations of dissolved nitrogen and phosphorus, respectively, and K_N and K_P are the respective half-saturation constants. The comma separating the ratios inside the parentheses means that only the minimum of the two will be chosen. These ratios are the Monod expression for nitrogen and phosphorus, respectively. If nutrients other than nitrogen and phosphorus are the limiting terms, instead, their Monod expression should be evaluated for the minimum and then chosen.

The dissolved forms of nitrogen are ammonia, nitrite, and nitrate, and the dissolved form of phosphorus is orthophosphorus. Substituting the expressions of the modifying factors in equation (2-27) yields

$$\mu_p = \mu_{pm20}\left\{\frac{I_0 \exp(-K_e z)}{I_s} \exp\left[\frac{-I_0 \exp(-K_e z)}{I_s} + 1\right]\right\}$$
$$\theta_g^{T-20} \min\left(\frac{[N]}{K_N + [N]}, \frac{[P]}{K_P + [P]}\right) \qquad (2\text{-}32)$$

The value of μ_{pm20} should be determined in the laboratory under simulated light conditions. A value of 2.0 per day has been reported.[7]

---

[6] A. M. Beeton (1958). "Relationship between Secchi Disk Readings and Light Penetration in Lake Huron." *American Fisheries Society Transactions*, 87, pp. 73–79.

[7] USEPA (1988). *WASP4, A Hydrodynamic and Water Quality Model: Model Theory, User's Manual, and Programmers Guide.* EPA/600/3-87/039. Environmental Research Laboratory, U.S. Environmental Protection Agency, Athens, Ga. p. 71.

Example 2-15

The concentration of active chlorophyll a at the 4-m depth is 10 μg/L . The Secchi depth is 9.2 m and the surface light intensity I_o is equal to 2200 Ly/day. K_N and K_P are 25 μg/L and 1.0 μg/L, respectively. If [N] and [P] are 50 μg/L and 10 μg/L, respectively, calculate R_4. Assume that $k_{s4} = 0.14$ per day, $k_d = 0.05$ per day, $z_s = 9.2$ m, and temperature = 25°C.

Solution

$$R_4 = (\mu_p - k_d - k_{s4})c_4$$

$$\mu_p = \mu_{pm20} \left\{ \frac{I_0 \exp(-K_e z)}{I_s} \exp\left[\frac{I_0 \exp(-K_e z)}{I_s} + 1\right]\right\} \theta_g^{T-20} \min\left(\frac{[N]}{K_N + [N]}, \frac{[P]}{K_P + [P]}\right)$$

Assume that $I_s = 300$ Ly/day, $\mu_{pm20} = 2$ per day, and $\theta_g = 1.068$ per day.

$$K_e = \frac{1.8}{z_s} \qquad z_s = 9.2 \text{ m}$$

Therefore,

$$K_e = \frac{1.8}{9.2} = -0.2 \text{ m}$$

$$\mu_p = 2\left(\frac{2200 \exp[(-0.2)(4)]}{300} \exp\left\{-\frac{2200 \exp[(-0.2)(4)]}{300} + 1\right\}\right)$$

$$(1.068^{25-20})\left[\min\left(\frac{50}{25 + 50}, \frac{10}{1 + 10}\right)\right]$$

$$= 2(3.3)(0.1)(1.39)[\min(0.67, 0.91)] = 0.61 \text{ per day}$$

$$R_4 = (0.61 - 0.05 - 0.14)(10) = 4.2\frac{\mu g}{\text{L-day}} \qquad \textbf{Answer}$$

Phosphorus kinetics.   From Figure 2-3 the state variables involved in phosphorus kinetics are phytoplankton c_4, organic phosphorus (OP) c_8, and orthophosphorus (OPO$_4$) c_3. The pool of organic phosphorus is increased by the death of phytoplankton and is decreased by mineralization to orthophosphorus and by settling. Mathematically, this phenomenon is expressed by the R derivative R_8 as

$$R_8 = k_d a_{pc} c_4 - k_{h8} c_8 - k_{s8} c_8 \qquad (2\text{-}33)$$

Here k_{h8} is the mineralization rate coefficient mainly caused by enzymatic hydrolysis, and k_{s8} the settling rate of organic phosphorus, and a_{pc} the ratio of phosphorus to active chlorophyll a in phytoplankton.

There is sorption between orthophosphorus and suspended solids. This interaction is expressed by the partition coefficient. The sorbed portion of orthophosphorus causes its reduction by settling. As indicated by Figure 2-3, the concentration of orthophosphorus is also reduced by utilization of phytoplankton and increased by the mineralization of organic phosphorus. Let c_{3p} be the concentration of orthophosphorus sorbed onto the solids. $c_{3p} = K_{sp3} c_3$, where K_{sp3} is the K_{sp} (partition coefficient) for orthophosphorus onto the solids m_s. The R derivative of orthophosphorus R_3 is then given by

$$R_3 = k_{h8}c_8 - \mu_p a_{pc}c_4 - k_s m_s c_{3p} = k_{h8}c_8 - \mu_p a_{pc}c_4 - k_s m_s K_{sp3}c_3 \qquad (2\text{-}34)$$

where μ_p is as defined before and k_s is the settling velocity coefficient of the solids. $\mu_p a_{pc}c_4$ is equal to $k_{mp3}c_3$, where k_{mp3} is the microbial decay coefficient for orthophosphorus due to phytoplankton. Note that c_{3p} is expressed as the mass of orthophosphorus per unit mass of solids.

Nitrogen kinetics. From Figure 2-3, the state variables involved in nitrogen kinetics are ammonia (NH_3) c_1, nitrite-nitrate (NO_3) c_2, phytoplankton (PHYT) c_4, and organic nitrogen (ON) c_7. Organic nitrogen is formed during respiration and upon death of the phytoplankton. Also, during respiration and upon death, ammonia is formed. Settling and enzymatic hydrolysis to ammonia also reduce the concentration of organic nitrogen; the R derivative R_7 is then

$$R_7 = f_{ON}k_d a_{nc}c_4 - k_{h7}c_7 - k_{s7}c_7 \qquad (2\text{-}35)$$

Here f_{ON} is the fraction of cellular nitrogen converted to organic nitrogen upon death and during respiration, k_{h7} the enzymatic hydrolysis coefficient, k_{s7} the settling velocity coefficient, and a_{nc} the ratio of nitrogen to active chlorophyll a in phytoplankton.

When organic nitrogen is hydrolyzed, the product is ammonia. Ammonia is used directly by phytoplankton and by *Nitrosomonas* and *Nitrobacter*. Phytoplankton incorporate ammonia into cells during synthesis. *Nitrosomonas* produces nitrites, and *Nitrobacter* produces nitrates, indicated collectively in Figure 2-3 as NO_3. During algal respiration and death, ammonia is also directly produced. Taking all these processes into consideration, the R derivative for ammonia R_1 is

$$R_1 = k_{h7}c_7 - \mu_p a_{nc}c_4 - k_{mb1}c_1 + (1 - f_{ON})k_d a_{nc}c_4 \qquad (2\text{-}36)$$

$\mu_p a_{nc}c_4$ is equal to $k_{mp1}c_1$, where k_{mp1} is the microbial decay coefficient for ammonia due to phytoplankton and k_{mb1} is the microbial decay coefficient for ammonia due to the bacteria *Nitrosomonas* and *Nitrobacter* (the nitrifiers).

The NO_3 (nitrite and nitrate) produced from ammonia is either utilized by phytoplankton for the synthesis of cells or denitrified to nitrogen gas and other gases, such as N_2O. In utilization of the nitrogen nutrients, however, phytoplankton have the preference for ammonia. The nitrites and nitrates will only be utilized in accordance with a factor called the *ammonia preference factor*, P_{NH3}; the affected part of the R derivative R_2 will therefore be multiplied by the correction factor, $1 - P_{NH3}$. The expression for P_{NH3} is[8]

$$P_{NH3} = [NH_3]\frac{[NO_3]}{(K_N + [NH_3])(K_N + [NO_3])}$$
$$+ [NH_3]\frac{K_N}{([NH_3] + [NO_3])(K_N + [NO_3])} \qquad (2\text{-}37)$$

K_N is the half-saturation constant for nitrogen.

---

[8] Ibid., p. 81.

The R derivative R_2 for NO_3 is then

$$R_2 = k_{mb1}c_1 - (1 - P_{NH3})\mu_p a_{nc} c_4 - k_{md2} c_2 \qquad (2\text{-}38)$$

where k_{md2} is the microbial decay coefficient of NO_3 due to denitrification. The first term of equation (2-38) represents the conversion of ammonia to NO_3, the second represents the utilization by phytoplankton, and the third represents the process of denitrification.

Example 2-16

What is the ammonia preference factor in example 2-15 if [N] is broken down as $[NH_3] = 30\ \mu g/L$ and $[NO_3^-] = 20\ \mu g/L$?

Solution

$$P_{NH3} = [NH_3]\frac{[NO_3]}{(K_N + [NH_3])(K_N + [NO_3])} + [NH_3]\frac{K_N}{([NH_3] + [NO_3])(K_N + [NO_3])}$$

$$= 30\frac{20}{(25 + 30)(25 + 20)} + 30\frac{25}{(30 + 20)(25 + 20)} = 0.58 \quad \textbf{Answer}$$

CBOD kinetics. The state variables involved in CBOD kinetics are phytoplankton c_4 and carbonaceous oxygen demand (CBOD) c_5. c_5 is increased by the death of phytoplankton; it is decreased by the biochemical reaction with oxygen and by settling. R_5 is then

$$R_5 = k_d a_{cc} c_4 - k_{mb5} c_5 - k_{s5} c_5 \qquad (2\text{-}39)$$

a_{cc} is the CBOD/active chlorophyll a ratio in phytoplankton, k_{mb5} is the microbial decay coefficient of CBOD due to bacteria, and k_{s5} is the settling velocity coefficient for CBOD. It is to be noted that the concentration c_5 does not distinguish between dissolved and sorbed CBOD but is the total CBOD per unit volume of water.

DO kinetics. In DO kinetics, the following state variables are involved: c_6 (DO), c_5 (CBOD), c_1 (NH_3), and c_4 (PHYT). The concentration of the DO is increased by aeration from the atmosphere and by photosynthesis. It is consumed by CBOD, microbial degradation (oxidation) of ammonia, plant respiration, and sediment oxygen demand (SOD). The SOD is a demand for oxygen exerted by bottom sediments and is incorporated in that part of c_1 and CBOD released from bottom sediments.

The difference between the dissolved oxygen that the water body can hold at the given temperature and pressure and the actual dissolved oxygen concentration is called the *oxygen deficit D*. The driving force for the aeration is this deficit, $O_s - c_6 = D$, where O_s is the saturated DO at the prevailing temperature and pressure. In DO kinetics, $k_{mb1}c_1$ is expressed in terms of its oxygen equivalent, $k_n L_n$, where k_n is the microbial decay coefficient of ammonia when c_1 is expressed in terms of its oxygen equivalent, L_n. $k_{mb5}c_5$ is already expressed in terms of its oxygen equivalent, and by analogy with $k_{mb1}c_1$, is expressed as $k_c L_c$, where k_c is k_{mb5} and L_c is c_5. Calling the effects of photosynthesis and respiration P and R, respectively, the R derivative R_6 is

$$R_6 = k_2(O_s - c_6) + P - R - k_c L_c - k_n L_n \tag{2-40}$$

where k_2 is the overall mass transfer coefficient for aeration.

The photosynthetic reaction which produces oxygen was written as

$$106CO_2 + 16NO_3^- + HPO_4^{2-} + 122H_2O + 18H^+ + \text{trace elements} \rightarrow$$

$$(CH_2O)_{106}(NH_3)_{16}H_3PO_4 \cdot \text{trace elements} + 138O_2 \tag{2-41}$$

Photosynthetic reactions are related to phytoplankton growth. From reaction (2-41) the ratio of oxygen produced to phytoplankton a_{op} is $[138O_2]/[(CH_2O)_{106}(NH_3)_{16}H_3PO_4] = [138(32)]/3550 = 1.24$, mass of oxygen per unit mass of phytoplankton. If the mass ratio of phytoplankton to active chlorophyll a is a_{pc}, then

$$P = a_{op}a_{pc}\mu_p c_4 = 1.24 a_{pc}\mu_p c_4$$

$$= 1.24 a_{pc}\mu_{pm20} \left\{ \frac{I_0 \exp(-K_e z)}{I_s} \exp\left[-\frac{I_0 \exp(-K_e z)}{I_s} + 1 \right] \right\} \tag{2-42}$$

$$\theta_g^{T-20} \left\{ \min\left(\frac{[N]}{K_N + [N]}, \frac{[P]}{K_P + [P]} \right) \right\} c_4$$

The reverse of reaction (2-41) can be considered as the reaction for respiration. Hence by a similar argument, R is

$$R = 1.24 a_{pc}k_d c_4 \tag{2-43}$$

GLOSSARY

Adenosine triphosphate (ATP). The energy currency of the cell.

Algae. Eucaryotic protists (which may be unicelled, multicelled, or colonial) capable of carrying out photosynthesis.

Anaerobic process. Biological decomposition of organic waste in the absence of air.

Anoxic process. A process of decomposition in which the electron acceptor is the nitrate ion.

Biochemical oxygen demand. A measure of the strength of an organic substance indicated by the amount of equivalent oxygen consumed in its biological decomposition.

Biological combustion. A form of combustion in which an organism utilizes molecular oxygen to oxidize a substrate producing energy.

Carbonaceous biochemical oxygen demand (CBOD). The biological oxygen demand corresponding to the carbon content of a substance.

Chemical oxygen demand. A measure of the total ability of a substance to consume oxygen.

Chloroplast. An organelle that forms the site of the electron-transport system of photosynthetic eucaryotes.

Combustion. Reaction of a substance with molecular oxygen, producing energy.

Completed test for coliforms. Proceeding from the confirmed test, a wire loop is streaked into a dish containing nutrients and organisms identified using the Gram stain technique.

Composting. An enhanced process of rapidly oxidizing a solid material using atmospheric oxygen.

Confirmed test for coliforms. Proceeding from the presumptive test and incubating at $35 \pm 0.5°C$ using the brilliant green lactose bile broth, this test is positive if gases are liberated within 24 to 48 ± 3 hours.

Cytoplasmic membrane. Found directly beneath the cell wall, forms as a protective covering of the cytoplasm.

Equivalent weight (or equivalent mass). The mass of a substance participating in a chemical reaction per unit mole of the reference species.

Eucaryote. A protist whose DNA are arranged in an easily recognizable chromosome structure.

Eutrophication. The natural process of aging of a body of water.

Five-day biochemical oxygen demand (BOD_5). A biochemical oxygen demand exerted in 5 days.

Free ammonia. Hydrolysis product of organic nitrogen.

Kinetics. The rate of change of the concentration of a substance with respect to time.

Kjeldahl nitrogen. The sum of the concentrations of ammonia and organic nitrogen.

Nitrogenous biochemical oxygen demand (NBOD). The biochemical oxygen demand corresponding to the nitrogen content of a substance.

Organic nitrogen. Nitrogen content of an organic substance.

Oxygen deficit. The difference between the saturation dissolved oxygen (DO_{sat}) and the actual dissolved oxygen (DO).

Pheophytin. Dead chlorophyll a.

Presumptive test for coliforms. Incubated at $35 \pm 0.5°C$, with liberation of gases within 24 to 48 ± 3 hours.

Procaryote. A protist whose DNA are not arranged in an easily recognizable structure as the chromosomes are.

Reference species. The moles of electrons or the moles of positive (or alternatively, negative) oxidation states participating in a given chemical reaction.

Settleable solids. The volume of solids that settles after a predetermined period of time.

Suspended or nonfiltrable solids. The fraction of total solids that does not pass through a filter.

Thermodynamics. Study of the interrelationships of heat and other forms of energy.

Thermodynamic variables. Variables involved in the study of the interrelationship of heat and other forms of energy.

Total nitrogen. The sum of the concentrations of free ammonia, nitrite, and nitrate nitrogens.

Total organic carbon. The total carbon content of a substance.

Total solids. The material left after water has been evaporated from a sample.

Ultimate biochemical oxygen demand. Biochemical oxygen demand exerted after an infinite time.

Volatile solids. The solids that disappear after a sample is decomposed at 600°C.

SYMBOLS

a_{cc}	CBOD/active chlorophyll a ratio
a_{nc}	nitrogen/active chlorophyll a ratio
a_{op}	oxygen/phytoplankton ratio
a_{pc}	phytoplankton/active chlorophyll a ratio
ADP	adenosine diphosphate
ATP	adenosine triphosphate
c_1	concentration of NH_3-N
c_2	concentration of NO_2-NO_3-N
c_3	concentration of orthophosphorus
c_4	concentration of phytoplankton chlorophyll a
c_5	concentration of CBOD
c_6	concentration dissolved oxygen
c_7	concentration of organic nitrogen
c_8	concentration of organic phosphorus
f_{ON}	fraction of cellular nitrogen converted to organic nitrogen
F	96,494 coulombs per faraday
G	Gibbs free energy
k_c	CBOD decay coefficient
k_n	nitrogenous decay coefficient
K_e	extinction coefficient
K_N	nitrogen half-saturation constant
K_P	phosphorus half-saturation constant
K_{sp}	solubility product constant
L_c	CBOD
L_n	nitrogenous oxygen demand (NBOD)
MPN	most probable number

PROBLEMS

2-1. What are the three steps in the qualitative test for coliforms? At what temperature is the normal coliform test conducted? At what temperature is the fecal coliform test conducted? Why?

2-2. At what intervals of time are the fermentation tubes checked for evolution of gases? At what intervals of time are the plates inspected for growth?

2-3. A presumptive multiple-tube fermentation test of a river sample yielded the following results:

Serial dilution	Number of positives
1.0	5
0.1	5
0.01	3
0.001	2
0.0001	0

What is the MPN?

2-4. Solve Problem 2-3 assuming that 0.01 is the highest dilution.

2-5. Solve Problem 2-3 assuming that 1.0 is the highest dilution.

2-6. Samples of 500-mL portions analyzed using the filter-membrane technique produce the following results: 350, 340, 370, and 340 of coliforms. What is the concentration of coliforms per 100 mL?

2-7. Assuming that the density of coliforms in Problem 2-6 is the same as that of water, what is the concentration in mg/L?

2-8. In a taste test, 67 mL of the sample is used. What are the volume of distilled used and the TTN?

2-9. Develop a method of taste and odor testing that will not restrict the final volume of distilled water plus sample to 200 mL.

2-10. Discuss the significance of gram-positive and gram-negative bacteria.

2-11. $Mg(OH)_2$ reacts with CO_2 according to the reaction below. If 100 g of $Mg(OH)_2$ is reacted, calculate the amount of CO_2 used using (**a**) equivalent weights considering the charges in CO_2 as the reference species and considering the charges in $Mg(OH)_2$ as the reference species, and (**b**) the balanced chemical reaction.

$$Mg(OH)_2 + 2CO_2 \rightarrow Mg^{2+} + 2(HCO_3)^-$$

2-12. Repeat part (a) of Problem 2-11 using the method developed in this book.

2-13. At standard conditions: (**a**) What is the oxidation potential for the oxidation of ethanol? (**b**) What is the oxidation potential for its reduction? (**c**) How many moles of electrons are involved per mole of ethanol? (**d**) Calculate the equivalent weight.

2-14. In the utilization of NO_2^- by *Nitrobacter* under aerobic conditions, (**a**) show the donor and acceptor reactions, (**b**) calculate the free-energy change of the reaction without synthesis, and (**c**) repeat part (b) considering cell synthesis. Assume that 0.02 mg of *Nitrobacter* is produced per milligram of NO_2-N.

2-15. Account for the difference between parts (b) and (c) of Problem 2-14.

2-16. Ultimate analysis of a mixture of sewage sludge and solid waste subjected to composting yields the following results: H = 6.0%, C = 45.0%, N = 2.0%, O = 33.0%, S = 0.5%, ash = 13.5%, and moisture = 4.8%. The plant composts 100 tonnes/day. Assuming that 23% excess air is used, calculate (**a**) the cubic meters per minute of air a blower must force into the pile measured at STP for a complete biological combustion, and (**b**) the air requirement (20% excess air) if 30% of the original C is converted to carbon dioxide and the average empirical formula of the residual organics is $C_{30}H_{51}O_{26}N$. Assume that the incoming air humidity is 0.016 mol of H_2O per mole of dry air.

2-17. The residual organics in Problem 2-16 contains 5% aluminum. Derive the corrected formula.

2-18. Ultimate analysis of a mixture of sewage sludge and solid waste deposited in a sanitary landfill yields the following results: H = 6.0%, C = 45.0%, N = 2.0%, O = 33.0%, S = 0.5%, ash = 13.5%, and moisture = 4.8%. The landfill processes 100 tonnes/day of the mixture of sludge and solid waste. (**a**) What is the water requirement for complete decomposition to take place? (**b**) If it is estimated that 70% of the mixture ends up as residue and the average empirical formula of the residual organics is $C_{30}H_{51}O_{26}N$, calculate the water requirement.

2-19. Derive

$$\frac{1}{y + 4x - 3a - 2z - 2b}C_xH_yO_zN_aS_b + \frac{1}{4}O_2 \rightarrow \frac{x - a}{y + 4x - 3a - 2z - 2b}CO_2$$

$$+ \frac{a}{y + 4x - 3a - 2z - 2by}NH_4^+ + \frac{a}{y + 4x - 3a - 2z - 2b}HCO_3^-$$

$$+ \frac{b}{y + 4x - 3a - 2z - 2b}H_2S + \frac{y - 5a - 2b}{2(y + 4x - 3a - 2z - 2b)}H_2O$$

2-20. An organic waste has the following composition: H = 6.0%, C = 45.0%, N = 2.0%, O = 33.0%, S = 0.5%, ash = 13.5%, and moisture = 4.8%. Using the reaction derived in Problem 2-19, how much O_2 is needed to decompose this waste?

2-21. Derive

$$C_xH_yO_zN_aS_b + \frac{y + 4x - 3a - 2z - 2b - Y_c(e + 4d - 3g - 2f - 2h)}{4}O_2 \rightarrow$$

$$Y_cC_dH_eO_fN_gS_h + [(x - a) - Y_c(d - g)]CO_2 + (a - gY_c)NH_4^+ + (a - gY_c)HCO_3^-$$

$$+ (b - Y_ch)H_2S + \frac{y + 2Y_ch - 5a - 2b - Y_c(e - 5g)}{2}H_2O$$

2-22. Derive

$$\frac{1}{y + 4x - 3a - 2z - 2b}C_xH_yO_zN_aS_b + \frac{4x + 7a + 2b - 2z - y}{4(y + 4x - 3a - 2z - 2b)}H_2O \rightarrow$$

$$\frac{4x + 2z + 2b - 5a - y}{8(y + 4x - 3a - 2z - 2b)}CO_2 + \frac{1}{8}CH_4 + \frac{a}{y + 4x - 3a - 2z - 2b}NH_4^+$$

$$+ \frac{a}{y + 4x - 3a - 2z - 2b}HCO_3^- + \frac{b}{y + 4x - 3a - 2z - 2b}H_2S$$

2-23. An organic waste has the following composition: H = 6.0%, C = 45.0%, N = 2.0%, O = 33.0%, S = 0.5%, ash = 13.5%, and moisture = 4.8%. Using the reaction derived in Problem 2-22, how much H_2O is needed to decompose this waste? How much H_2S is produced?

2-24. Derive

$$C_x H_y O_z N_a S_b + \frac{4x + 7a + 2b - 2z - y - Y_{dc}(4d + 7g + 2h - 2f - e)}{4} H_2O \rightarrow$$

$$Y_{dc} C_d H_e O_f N_g S_h + \frac{4x + 2z + 2b + Y_{dc}(5g + e - 4d - 2f - 2h) - 5a - y}{8} CO_2$$

$$+ (a - gY_{dc})NH_4{}^+ + (a - gY_{dc})HCO_3{}^-$$

$$+ (b - hY_{dc})H_2S + \frac{(y + 4x - 3a - 2z - 2b) - Y_{dc}(e + 4d - 3g - 2f - 2h)}{8} CH_4$$

2-25. A 10-mL sample is pipeted directly into a 300-mL incubation bottle. The initial DO of the diluted sample is 9.0 mg/L and its final DO is 2.0 mg/L. The dilution water is incubated in a 200-mL bottle, and the initial and final DOs are, respectively, 9.0 and 8.0 mg/L. If the sample and the dilution water are incubated at 20°C for 5 days, what is the BOD_5 of the sample at this temperature?

2-26. Solve Problem 2-25 if the volume of the incubation bottle is 200 mL.

2-27. Volatile suspended solids analysis is run on a sample. The tare mass of the crucible and filter is 55.3520 g. A sample of 260 mL is then filtered and the residue dried and decomposed at 600°C to drive off the volatile matter. Assume that the filter does not decompose at this temperature. After decomposition, the constant weight of the crucible and the residue was determined and found to be equal to 30.3415 g. What is the volatile suspended solids if the total suspended solids is 142 mg/L?

2-28. Assuming that the temperature is 25°C, calculate the concentration of the other ion indicated by a question mark.

$$[OH^-] = 0.6(10^{-5})M, [H^+] = ?$$

$$[H^+] = 4(10^{-4})M, [OH] = ?$$

2-29. Alum is added to a sample of water with sufficient alkalinity. The quantity of the alum is enough to produce 40 mg/L $Al(OH)_3$. Assuming that the temperature is 25°C, what will be the equilibrium concentration of Al^+ and OH^-?

2-30. Analysis of algae yields the following formula: $(CH_2O)_{106}(NH_3)_{16}H_3PO_4$. From the knowledge of the breakdown of water molecules during photosynthesis, derive

$$106CO_2 + 16NO_3{}^- + HPO_4{}^{2-} + 122H_2O + 18H^+ + \text{trace elements} \rightarrow$$

$$(CH_2O)_{106}(NH_3)_{16}H_3PO_4 \cdot \text{trace elements} + 138O_2$$

2-31. The Back River sewage treatment plant in Baltimore discharges to the Back River estuary. The state of Maryland imposes on its permit a total phosphorus limitation of 2.0 mg/L. Assuming that all the total phosphorus converts to orthophosphorus and that the dilution effect of the estuary cannot be neglected, calculate the concentration of algae expected.

2-32. A sample of estuarine water containing 300 μg/L of chlorophyll a is saturated with oxygen and incubated under dark conditions. Assuming that sufficient oxygen is present in the incubation bottle to satisfy the reaction, how many mg/L of oxygen is consumed? Do not use the method of equivalents but use the balanced chemical equation directly. Assume that the percent chlorophyll a in algae is 1.5%.

2-33. A wastewater treatment plant discharges treated sewage into a stream. If the concentration of BOD_5 1 mile downstream from the point of discharge is 5 mg/L and k_m is assumed to be 0.2 per day (in terms of BOD_5), what is the value of R_m at this point?

2-34. The deoxygenation coefficient to the base 10 for a certain waste is 0.1 per day. What fraction is BOD_5 to CBOD of this waste?

2-35. What are the fractions of H, C, O, N, and P in algae?

2-36. The concentration of active chlorophyll a at a 4-m depth is 10 μg/L. The Secchi depth is 9.2 m and the surface light intensity, I_o, is equal to 2200 Ly/day. K_N and K_P are 25 μg/L and 1.0 μg/L, respectively. If [N] and [P] are 40 μg/L and 10 μg/L, respectively, calculate R_4. Assume that $k_{s4} = 0.14$ per day, $k_d = 0.05$ per day, and temperature = 25°C.

2-37. Algae is grown artificially in a laboratory with all the necessary nutrients supplied. Subsequently, the supply of phosphorus is withdrawn. By considering its formula, what will happen to the concentration of algae?

2-38. The state of Maryland requires that all sewage discharge to the Upper Chesapeake Bay not exceed 2.0 mg/L of total phosphorus. Comment on this requirement.

2-39. Reviewing Problem 2-38, why do you think that the state of Maryland did not impose nitrogen limitations?

2-40. What is the ammonia preference factor if [N] is broken down as $[NH_3] = 45$ μg/L and $[NO_3^-] = 10$ μg/L. K_N and K_P are 25 μg/L and 1.0 μg/L, respectively.

BIBLIOGRAPHY

APHA, AWWA, and WEF (1992). *Standard Methods for the Examination of Water and Wastewater*, 15th ed. American Public Health Association, Washington, D.C.

BEETON, A. M. (1958). "Relationship between Secchi Disk Readings and Light Penetration in Lake Huron." *American Fisheries Society Transactions*, 87, pp. 73–79.

McCARTY, P. L. (1975). "Stoichiometry of Biological Reactions." *Progress in Water Technology*. Pergamon Press, London.

PEAVY, H. S., D. R. ROWE, and G. TCHOBANOGLOUS (1985). *Environmental Engineering*. McGraw-Hill, New York.

SAWYER, C. N., and P. L. McCARTY (1978). *Chemistry for Environmental Engineers*. McGraw-Hill, New York.

SINCERO, A. P. (1984). "Eutrophication and the Fallacy of Nitrogen Removal." *Pollution Engineering*. Pudvan Publishing Co., Northbrook, Ill.

STEELE, J. H. (1965). "Primary Production in Aquatic Environments," in C. R. Goldman (ed.), *Notes on Some Theoretical Problems in Production Ecology*. Memorial Institute for Idrobiology, University of California Press, Berkeley, Calif., 18 Suppl., pp. 383–398.

STUMM, W., and J. J. MORGAN (1981). *Aquatic Chemistry*, 2nd ed. Wiley-Interscience, New York.

USEPA (1988). *WASP4, A Hydrodynamic and Water Quality Model: Model Theory, User's Manual, and Programmers Guide*. EPA/600/3-87/039. Environmental Research Laboratory, United States Environmental Protection Agency, Athens, Ga.

CHAPTER 3

Environmental Engineering Hydrology

Environmental engineering hydrology refers to that part of the huge discipline of hydrology which engineers use in the practice of environmental engineering. In this chapter we discuss surface water hydrology, reservoirs, groundwater hydrology, and control of groundwater contamination.

SURFACE WATER HYDROLOGY

Hydrology may be defined as the science that deals with the study of the *properties, distribution,* and *behavior* of water in nature. The key word in this definition is *nature.* To be a part of this definition, water must be "in nature," not in homes and other artificial places.

The backbone in the study of hydrology is the hydrologic or water cycle, depicted in Figure 3-1. The major parts of the hydrologic cycle are precipitation, infiltration (and percolation), surface runoff, and evaporation and transpiration. Since the

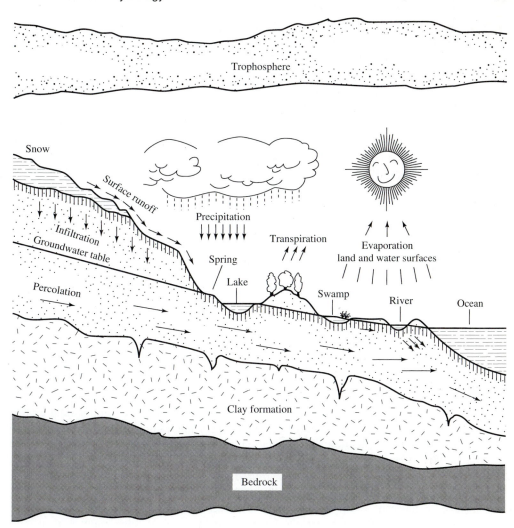

Figure 3-1 Hydrologic cycle.

hydrologic cycle is a cycle, in its discussion we can start anywhere, but it is most appropriate to start with precipitation. *Precipitation* results when moisture in the clouds falls as rain, snow, sleet, hail, or ice, depending on conditions in the atmosphere. As precipitation reaches the ground, the first processes that transpire are the filling of depressions, called *depression storage,* and the *interception* of precipitation by leaves and branches of trees and similar structures. Soon infiltration begins. *Infiltration* is a process in which precipitation enters the underground through the surface. After satisfaction of the infiltration capacity, as well as depression storage and interception, *surface runoff* begins. As shown in the figure, runoff eventually joins lakes, swamps, rivers,

and oceans. Some of the infiltrates continue to follow the pull of gravity and percolate downward through the unsaturated zone of the earth's crust. The water that remains in the unsaturated zone becomes the vadose water. The extra water that cannot
be held in the unsaturated or vadose zone percolates farther and joins the saturated
zone below, forming the huge reservoir of water underneath the earth. The water in
this reservoir is called *groundwater*. The upper part of the groundwater is called the
water table. Due to attraction by the surrounding soils upon the water molecules, a
water formation exists above the water table. This formation is called the *capillary
fringe.*

Some groundwaters surface out from the ground to form springs and to become part
of lakes, rivers, swamps, and finally, oceans. From these surfaces, water is evaporated.
Water is also evaporated from intercepted precipitation on wet leaves of plants and trees,
buildings, grounds, and similar structures. Another way of evaporating water into the air
is through transpiration, a process carried out by plants. These waters finally evaporate
to form clouds. To complete the cycle, convective, orographic, and cyclonic processes
operate on clouds, again forming precipitation.

Of the portions of the hydrologic cycle, precipitation and runoff are parts of surface
water hydrology. These subjects are discussed next.

Precipitation

It is important for environmental engineers to know how to measure precipitation and
how to make calculations of precipitation. Calculation of runoff is an important function
in urban drainage. The engineer may be hired to design a storm drainage system, and for
this, must calculate precipitation to obtain runoff. The design of an impounding reservoir
is another reason why environmental engineers must know about precipitation. Water
quality is also a function of precipitation, and this subject can have an important bearing
on how permit limits are set on discharge permits.

In general, precipitation is measured by recording and nonrecording raingages. The
standard raingage is the 8-in. cylindrical can. An 8-in. funnel is inserted at the open
end of the can. With the assembly standing upright, the funnel catches the rain and
directs it to a small tube inside the can. The amount of rain caught is then measured
as depth of rainfall, corrected for the 8-in. projected area of the catching funnel. The
standard raingage is the nonrecording type. The U.S. National Weather Service maintains
approximately 14,000 standard raingages in the United States. Examples of recording
gages are the *tipping bucket* and the *weighing raingage*. Their principles of operation
are similar in that the amount of rainfall is recorded on a chart. As opposed to the
standard raingage, the data on the recording rain gages are continuous. This type of data
recording allows one to compute rainfall intensity, which is important for calculating
runoff (discussed later).

Rainfall data on a particular station are random events. Storm durations vary from
one storm to the next. Rainfall intensity varies within a storm and from storm to storm.
Because of the random nature, they must be analyzed statistically. If the gage is of the
recording type, the whole spectrum of the storm event would have been recorded. In

design, what storm occurrence would be chosen? Of course, there are many storm events that the design can be based on, but which should be used? Because of the random nature of storm events, no event on the record should be chosen arbitrarily, but instead, all of them must be taken into consideration at once. Our object, then, is to synthesize a single storm to be used for design based on the several storms on record. In other words, basing the choice on all the records in the station, a single spectral distribution of rainfall intensity against time will be obtained. This spectral distribution corresponds to the return period to be used in the design. *Return period* or *interval* is the number of years, on the average, that a particular storm event is equaled or exceeded. The steps in the procedure for determining the distribution, called *probability distribution analysis,* are as follows:

Step 1. Scan all the years of record (such as N years of record) and determine the longest storm duration that ever occurred. Divide this duration into suitable subdurations. (Note that subdurations are always referred to the beginning of the storm. For example, a 50-min subduration is a 50-min interval of the storm measured from the beginning.) After deciding on the subdurations, pick the smallest one, such as 5 min, and proceed to step 2.

Step 2. Scan the N years of record again and pick the highest depth of rainfall in each of the N years corresponding to the subduration just chosen. Convert these depths to intensity of rainfall by dividing the respective depth by the subduration. For example, let the subduration chosen be 5 minutes and the maximum 5-min depth of rainfall in a certain year be 1.4 cm. The intensity of rainfall i for this "certain year" is then

$$i = \frac{1.4}{5}(60) = 16.8 \text{ cm/h}$$

Step 3. Arrange the intensities of rainfall obtained in step 2 in an array from the highest to the lowest (i.e., in decreasing order) and calculate the probability p that each intensity value is equaled or exceeded using the formula

$$p = \frac{m}{N+1} \tag{3-1}$$

where m is the order number. The distribution resulting from the use of equation (3-1) is called the *probability* or *frequency distribution* of the set.

Before giving an example of the use of this formula, it will be discussed. Let $p = m/N$. When m equals N, the array element corresponding to m is the smallest in the set. The probability p is 1. This means that the smallest value will be exceeded. This is, of course, true but only when applied to the current set of array. However, when applied in general, there is no certainty that the smallest will be exceeded, since even the smallest may not even occur. What is certain is that the smallest has the highest probability of being exceeded. To incorporate this fact, the denominator in equation (3-1) has therefore been modified by adding 1 to N.

Example 3-1

Analysis of 10 years of precipitation data yields the following results for the first to the tenth years, respectively (in cm): 1.4, 1.8, 1.6, 1.0, 1.3, 1.7, 1.9, 1.0, 0.9, and 1.2. Assuming that these data are for 5-min subduration, calculate the probability distribution.

Solution   Arrange the data in descending order as in the first column below and proceed with the calculations as indicated.

Precipitation depth (cm)	Precipitation intensity (cm/h)	m	$\dfrac{m}{N+1} = \dfrac{m}{11}$	
1.9	22.8[a]	1	0.09	
1.8	21.6	2	0.18	
1.7	20.4	3	0.27	
1.6	19.2	4	0.36	
1.4	16.9	5	0.45	**Answer**
1.3	15.6	6	0.55	
1.2	14.4	7	0.64	
1.0	12.0	8	0.73	
1.0	12.0	9	0.82	
0.9	10.8	10	0.91	

[a] $22.8 = (1.9/5)(60)$.

Step 4. Repeat steps 2 and 3 for the next subduration, and so on.

Step 5. After completion of step 4, a number of probability distributions equal to the number of subdurations would have been established. Now, establish the return intervals desired, such as once in every 2 years, once in every 5 years, once in every 20 years, ... , once in every 100 years, and so on. From the definition of return interval, the corresponding probability of occurrence can be determined. From the statement "once in every so many years," the corresponding probability is 1 divided by the "so many years." The "so many years" is the return interval T. Therefore, the probability p is

$$p = \frac{1}{T} \tag{3-2}$$

Step 6. Choose the first return interval (or the corresponding probability) from the several return intervals established in step 5. From the probability distributions obtained in step 4, determine the magnitudes of the precipitation intensity corresponding to this return interval for all subdurations previously chosen. For example, if the first return interval picked is once in every 2 years, the corresponding probability is $\frac{1}{2} = 0.5$. Then, from the probability distributions obtained in step 4, determine the corresponding precipitations for all the subdurations using the probability of 0.5. There were several subdurations originally chosen, but take the 5-min subduration as an illustration. In this case the result of Example 3-1 is handy. From this result, the probability of 0.5 is sandwiched between 0.45 and 0.55, as shown below.

Precipitation (cm/h)	m	$\dfrac{m}{N+1} = \dfrac{m}{11}$
16.9	5	0.45
x	—	0.50
15.6	6	0.55

By interpolation,

$$
\begin{aligned}
16.9 &\Rightarrow 0.45 \\
x &\Rightarrow 0.50 \\
15.6 &\Rightarrow 0.55
\end{aligned}
\qquad
\frac{x - 16.9}{15.6 - 16.9} = \frac{0.50 - 0.45}{0.55 - 0.45}
$$

$$x = 16.25 \text{ cm/h}$$

Therefore, the magnitude of the precipitation intensity corresponding to the probability of 0.5 for a 5-min subduration is 16.25 cm/h. The same procedure is applied to the probability distributions of all the other subdurations.

Step 7. Pick the second return interval as established in step 5. Repeat step 6 for this second interval. After completing, pick the third return interval as established in step 5 and repeat step 6. Do this cycle of repetition for the fourth, fifth, . . . , and the last return interval.

Step 8. After completing step 7, the following table of values will have been obtained:

First return interval		Second return interval		$\cdots$	Final return interval	
Duration	Precipitation	Duration	Precipitation		Duration	Precipitation
X_{11}	Y_{11}	X_{21}	Y_{21}		X_{l1}	Y_{l1}
X_{12}	Y_{12}	X_{22}	Y_{22}		X_{l2}	Y_{l2}
$\vdots$						

This table still needs to be rearranged to conform to real storms. Normally, in actual storms the peak occurs at about one-third its duration. Therefore, the precipitation columns should be rearranged to conform to this criterion. It is to be noted that the table records intensity (precipitation), duration, and frequency and thus is called an *intensity–duration–frequency* (IDF) *table.* Technically, if the return interval is T years, the storm is called a T-year storm (e.g., a 2-year storm, a 10-year storm, a 25-year storm).

Example 3-2

To use the probability distribution analysis, records for a sufficient number of years should be used to produce a good estimate. Since the computation is very long and tedious, this is normally done using a computer. For illustrative purposes, however, use only 10 years of record. A weather service station is scanned for precipitation to perform a probability distribution analysis. The results are as follows (in cm): 5-min subduration: 1.4, 1.8, 1.6, 1.0, 1.3, 1.7, 1.9, 1.0, 0.9, and 1.2; 20-min subduration: 2.96, 3.0, 2.86, 2.75, 3.10, 3.20, 2.84, 2.69, 2.78, and 2.91; 50-min subduration: 3.98, 3.92, 3.70, 3.89, 4.30, 4.00, 4.10, 3.99, 4.08, and 3.87; and 100-min duration: 4.70, 4.60, 4.76, 4.89, 4.30, 4.98, 4.85, 4.67, 4.79, and 4.77. Derive the intensity–duration–frequency table.

Solution

Step 1. The subdurations chosen are the 5-, 20-, 50-, and 100-min durations. The 5-min duration is the smallest, so we choose it first.

Step 2. The highest precipitations in each year for the 5-min subduration have been chosen, and they are 1.4, 1.8, 1.6, 1.0, 1.3, 1.7, 1.9, 1.0, 0.9, and 1.2.

Precipitation depth (cm)	Precipitation intensity (cm/h)
1.4	16.8
1.8	21.6
1.6	19.2
1.0	12.0
1.3	15.6
1.7	20.4
1.9	22.8
1.0	12.0
0.9	10.8
1.2	14.4

Step 3.

Precipitation depth (cm)	Precipitation intensity (cm/h)	m	$\dfrac{m}{N+1} = \dfrac{m}{11}$
1.9	22.8	1	0.09
1.8	21.6	2	0.18
1.7	20.4	3	0.27
1.6	19.2	4	0.36
1.4	16.9	5	0.45
1.3	15.6	6	0.55
1.2	14.4	7	0.64
1.0	12.0	8	0.73
1.0	12.0	9	0.82
0.9	10.8	10	0.91

Step 4. For the 20-min subduration:

Precipitation depth (cm)	Precipitation intensity (cm/h)	m	$\frac{m}{N+1} = \frac{m}{11}$
3.20	9.60	1	0.09
3.10	9.30	2	0.18
3.0	9.00	3	0.27
2.96	8.88	4	0.36
2.91	8.73	5	0.45
2.86	8.58	6	0.55
2.84	8.52	7	0.64
2.78	8.34	8	0.73
2.75	8.25	9	0.82
2.69	8.07	10	0.91

For the 50-min subduration:

Precipitation depth (cm)	Precipitation intensity (cm/h)	m	$\frac{m}{N+1} = \frac{m}{11}$
4.30	5.16	1	0.09
4.1	4.92	2	0.18
4.08	4.90	3	0.27
4.00	4.80	4	0.36
3.99	4.79	5	0.45
3.98	4.78	6	0.55
3.92	4.70	7	0.64
3.89	4.67	8	0.73
3.87	4.64	9	0.82
3.70	4.44	10	0.91

For the 100-min subduration:

Precipitation depth (cm)	Precipitation intensity (cm/h)	m	$\frac{m}{N+1} = \frac{m}{11}$
4.98	2.99	1	0.09
4.89	2.93	2	0.18
4.85	2.91	3	0.27
4.79	2.87	4	0.36
4.77	2.86	5	0.45
4.76	2.86	6	0.55
4.70	2.82	7	0.64
4.67	2.80	8	0.73
4.60	2.76	9	0.82
4.3	2.58	10	0.91

Step 5. Establish as desired the 2-year and 100-year return intervals.
Step 6. The first return interval is the 2-year interval; $p = 0.5$.
For the 5-min subduration:

Precipitation depth (cm)	Precipitation intensity (cm/h)	m	$\dfrac{m}{N+1} = \dfrac{m}{11}$
1.4	16.9	5	0.45
	x		0.50
1.3	15.6	6	0.55

$$\frac{x - 16.9}{15.6 - 16.9} = \frac{0.50 - 0.45}{0.55 - 0.45}$$

$$x = 16.25 \text{ cm/h}$$

For the 20-min subduration:

2.91	8.73	5	0.45
	x		0.50
2.86	8.58	6	0.55

$$\frac{x - 8.73}{8.58 - 8.73} = \frac{0.50 - 0.45}{0.55 - 0.45} = 0.5$$

$$x = 8.66 \text{ cm/h}$$

For the 50-min subduration:

3.99	4.79	5	0.45
	x		0.50
3.98	4.78	6	0.55

$$\frac{x - 4.79}{4.78 - 4.79} = \frac{0.50 - 0.45}{0.55 - 0.45}$$

$$x = 4.78 \text{ cm/h}$$

For the 100-min subduration:

4.77	2.86	5	0.45
	x		0.50
4.76	2.86	6	0.55

$$x = 2.86 \text{ cm/h}$$

Step 7. The second return interval is the 100-year interval; $p = 1/100 = 0.01$.

For the 5-min subduration:

	x		0.01
1.9	22.8	1	0.09
1.8	21.6	2	0.18

$$\frac{22.8 - x}{21.6 - x} = \frac{0.09 - 0.01}{0.18 - 0.01} = 0.47$$

$$x = 23.86 \text{ cm/h}$$

For the 20-min subduration:

	x		0.01
3.20	9.60	1	0.09
3.10	9.30	2	0.18

$$\frac{9.60 - x}{9.30 - x} = \frac{0.09 - 0.01}{0.18 - 0.01} = 0.47$$

$$x = 9.87 \text{ cm/h}$$

For the 50-min subduration:

	x		0.01
4.30	5.16	1	0.09
4.1	4.92	2	0.18

$$\frac{5.16 - x}{4.92 - x} = \frac{0.09 - 0.01}{0.18 - 0.01} = 0.47$$

$$x = 5.37 \text{ cm/h}$$

For the 100-min subduration:

	x		0.01
4.98	2.99	1	0.09
4.89	2.93	2	0.18

$$\frac{2.99 - x}{2.93 - x} = \frac{0.09 - 0.01}{0.18 - 0.01} = 0.47$$

$$x = 3.04 \text{ cm/h}$$

Step 8.

2-year interval		100-year interval	
Duration	Precipitation (intensity)	Duration	Precipitation (intensity)
5	16.25	5	23.86
20	8.66	20	9.87
50	4.78	50	5.37
100	2.86	100	3.04

Rearranging yields

2-year interval		100-year interval	
Duration	Precipitation (intensity)	Duration	Precipitation (intensity)
5	8.66	5	9.87
20	16.25	20	23.86
50	4.78	50	5.37
100	2.86	100	3.04

Surface Runoff: Rational Method

There are three methods of estimating runoff: empirical, statistical, and the use of rainfall–runoff relations. For example, the Maryland Geological Survey divides the state into several regions. For each region, empirical formulas are derived from data obtained from the U.S. Geological Survey (USGS) stream gaging stations. Needless to say, these formulas are not transportable anywhere, since they apply only in the regions for which they have been derived. The statistical method of estimating runoff needs a fair amount of data. It is true that there are now a fair number of records of stream gaging in various

USGS stations; however, they apply only to the stream where the station is located. Often, however, runoff data are needed in locations where no gaging station can be found. For those places fortunate enough to have stations, a statistical method such as probability distribution analysis can be used accurately. In other places, rainfall–runoff relations have been used.

The *curve-number method* developed by the Soil Conservation Service, the *unit hydrograph,* and the *rational method* are methods of rainfall–runoff relations. The rational method is very simple to use and is very popular. In parking lot design, the rational method is normally used to determine the peak storm. Figure 3-2 shows a grating of a storm inlet at a parking lot. This inlet catches runoff above it.

Figure 3-2 Grating of inlet in parking lot.

The rational formula is very simple:

$$Q = CiA \tag{3-3}$$

where Q is the runoff rate, C the dimensionless coefficient of imperviousness of the ground surface, i the intensity of rainfall, and A the drainage area. A can be easily planimetered from a USGS map. The units of Q depend on the units of i and A used. Values of C that have been used are shown in Table 3-1.

Equation (3-3) assumes a uniform drainage area. In the real world, however, the area is heterogeneous. To account for this heterogeneity, the areas are weighted by the coefficient of imperviousness and summed; thus equation (3-3) is revised as follows:

$$Q = i \sum CA \tag{3-4}$$

In general, there are two ways of obtaining the value of i. One is, again, through the use of empirical formulas but is applicable only in places for which they were derived. The other is to get i from the IDF table. The return or recurrence interval of the runoff corresponds to the return or recurrence interval of the intensity of storm rainfall being used. For example, if the storm is a 2-year storm, the runoff is correspondingly a 2-year runoff or flood. To obtain i, the duration t_c is needed. In the rational method t_c is called the *time of concentration.*

Consider a point in the drainage area where a runoff rate is needed. Assume that the rain has just started. To avoid flooding, the maximum rate of flow is required in design. To obtain this rate of flow, i is needed to be substituted in CiA. In turn, to get i, t_c is needed. Now, which t_c is to be used, the t_c in the beginning of the rainfall event or the t_c some time later?

In the beginning of a rainfall event, A is the smallest, since in the beginning not all areas in the drainage basin are contributing flows to the point under consideration. Accordingly, the runoff Q may be the smallest. Here comes the concept of the time of concentration. All flows in the basin must concentrate at the point considered at some particular time if the flow at the point were to be the maximum. The question is: At what time will all flows be concentrated at the point? The answer is that all flows will be concentrated at the point when all areas are contributing to the flow. The corresponding time, if found, is clearly the time to be used to obtain i. This time is called the time of concentration, and the point under consideration is, appropriately, called the *point of concentration.*

To have a valid definition of the time of concentration, the following assumptions must be made: (1) All slopes at every point in the basin toward the point of concentration must be equal, and (2) the contributing areas of the basin must be homogeneous. From these assumptions it follows that the runoff in the basin will converge uniformly toward the point of concentration. Having agreed on the uniformity of convergence toward the point, consider another point in the basin located the greatest distance away. As long as the water particles at this point have not arrived, we cannot consider all flows to be concentrated, and therefore the rate of flow transpiring at the point cannot be assured to be the maximum. Therefore, we define the time of concentration as the time it takes for the particles at the farthest point to reach the point of concentration, *assuming that the flow is converging uniformly toward the point.*

In the real world, assumptions 1 and 2 are never true; hence the assumption of uniform convergence is, also, never true. The actual time of concentration would have to be greater than the ideal one where uniformity of convergence toward the point of concentration is assumed. In practice, the longest time is used as the measure of the time of concentration.

Two formulas that may be used for computing the time of concentration are[1,2]

$$t_c = \left[\frac{3.35(10^{-6})L^3}{h} \right]^{0.385} \tag{3-5}$$

[1] P. A. Kirpich (1940). "Time Concentration of Small Agricultural Watersheds." *Civil Engineering,* 10(6).

[2] G. A. Hathaway (1945). "Design of Drainage Facilities." *Transactions of ASCE,* 110.

where L is the channel length in feet and h is the difference in elevation in feet between the upper and lower limits of the drainage basin; and

$$t_c = \left(\frac{2Ln}{3\sqrt{S}}\right)^{0.47} \tag{3-6}$$

where L is the channel length in feet, S the slope of the basin, and n is Manning's roughness coefficient. Values of n to be used in equation (3-6) are listed in Table 3-1.

TABLE 3-1 COEFFICIENTS OF IMPERVIOUSNESS AND MANNING'S n FOR VARIOUS TYPES OF SURFACES

Type of surface	C
Asphaltic cement streets	0.80–0.95
Gravel driveways and walks	0.15–0.25
Lawns, sandy soil	
2% slope	0.05–0.15
2–7% slope	0.15–0.20
> 7% slope	0.20–0.30
Lawns, heavy soil	
2% slope	0.10–0.15
2–7% slope	0.15–0.25
> 7% slope	0.25–0.35
Portland cement streets	0.80–0.95
Paved driveways and walks	0.75–0.85
Watertight roofs	0.80–0.95

	n
Average grass cover	0.45
Bare packed soil, free of stones	0.15
Dense grass cover	0.85
Poor grass cover or moderately rough surface	0.20
Smooth pavements	0.02

Example 3-3

The diagram that follows is a schematic of a drainage area. The runoff from area A drains to manhole 4, the runoff from area B drains to manhole 3, and the runoff from area C drains to manhole 2. The time of flow between manholes is 5 min. A is equal to 0.05 acre, B is equal to 15 acres, and C is equal to 30 acres. Assume all areas to be squares. The average channel slope is 0.0006. For the value of Manning's n, assume smooth-pavement surfaces. Using the IDF obtained in Example 3-2, calculate the maximum flow from manhole 2 to 1 for a 2-year storm. The coefficients of imperviousness are as follows: for $A = 0.80$, for $B = 0.90$, and for $C = 0.95$.

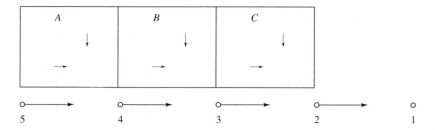

Solution First compute t_c for each area using equation (3-6).

$$t_c = \left(\frac{2Ln}{3\sqrt{S}} \right)^{0.47}$$

For area A: From Table 3-1, $n = 0.02$.

$$\text{Area} = 0.05 \text{ acre} = 0.05(43{,}560) = 2178 \text{ ft}^2$$

Therefore,

$$\text{length of one side} = \sqrt{2178} = 46.67$$

Let L be the length of the diagonal of the square area. Then $L = 46.67\sqrt{2} = 66$ ft. Thus

$$t_c = \left[\frac{2(66)(0.02)}{3\sqrt{0.0006}} \right]^{0.47} = 5.38 \text{ min, overland flow time to manhole 4}$$

The point of concentration is point 2. Total $t_c =$ overland t_c (called the inlet time) plus time of travel in the sewer from manhole 4 to 2, or

$$\text{total } t_c = 5.38 + 2(5) = 10.38 \text{ min}$$

For area B:

$$\text{length of one side} = \sqrt{43{,}560(15)} = 808.33 \text{ ft} \qquad L = 808.33\sqrt{2} = 1143.15 \text{ ft}$$

Thus

$$t_c = \left[\frac{2(1143.15)(0.02)}{3\sqrt{0.0006}} \right]^{0.47} = 21 \text{ min, overland flow time to manhole 3}$$

$$\text{total } t_c = 21 + 5 = 26 \text{ min}$$

For area C:

$$L = \sqrt{43{,}560(30)}\sqrt{2} = 1616.67 \text{ ft}$$

Thus

$$t_c = \left[\frac{2(1{,}616.67)(0.02)}{3\sqrt{0.0006}} \right]^{0.47} = 24.21 \text{ min, overland flow time to manhole 2}$$

$$\text{total } t_c = 24.21 \text{ min}$$

The highest $t_c = 26$ min, so $t_c = 26$ min.

Part of step 8 of Example 3-2 is reproduced below, where the 26 min is being inserted between the 20- and the 50-min durations.

2-year interval		100-year interval	
Duration	Precipitation (intensity)	Duration	Precipitation (intensity)
5	8.66	5	9.87
20	16.25	20	23.86
26	x		
50	4.78	50	5.37

By interpolation,

$$\frac{x - 16.25}{4.78 - 16.25} = \frac{26 - 20}{50 - 20}$$

$$x = 13.96 \text{ cm/h}$$

$$Q = i \sum CA = 13.96\,[0.80(0.05) + 0.90(15) + 0.95(30)]$$

$$= 586.88 \text{ cm-acre/h}$$

RESERVOIRS

In general, there are two types of reservoirs that an environmental engineer must know: impounding reservoir and distribution reservoir. An *impounding reservoir* is a basin constructed in the valley of a stream or river for the purpose of holding streamflow so that the stored water may be released to satisfy demand when supply is insufficient. The *distribution reservoir,* on the other hand, is a reservoir constructed to equalize the supply and demand of the community or user for treated water and to provide supplies in times of emergencies. The main difference between an impounding reservoir and a distribution reservoir is that the former holds untreated water while the latter holds treated water. The water held by an impounding reservoir may not all end up as treated water for community use but may be used for other purposes, such as irrigation.

The most important function of both the impounding and the distribution reservoirs is to ensure that water be available at all times. To be able to supply water at all times does not, however, mean that the largest reservoir must be constructed. It means that the reservoir must be adequately sized. Adequate sizing of a reservoir means that it must not be too large and it must not be too small. To ensure this, the rate of water consumption of the community or users should be known; in addition, for impounding reservoirs, the streamflow during drought conditions should be calculated.

Sizing of Impounding Reservoirs and Low-Flow Analysis

To size an impounding reservoir, one of the analysis to be done is that of predicting streamflow. In this instance, a probability distribution analysis for drought conditions must be performed, which is the reverse of the prediction of maximum precipitation that has already been discussed. Since drought conditions are being addressed, in low-streamflow analysis, instead of starting from the largest flow, the ordering starts from the lowest flow.

In the analysis for maximum or peak storms, the subdurations used are in terms of minutes. The reason is that storms last only for minutes, hours, or a few days, which can conveniently be subdivided into subdurations of several minutes. On the other hand, droughts do not just last for a few minutes, hours, or days but for several days, months, or even years. The subdurations used in low-flow analysis is therefore normally in terms of days. In synthesizing drought flows, steps 1 to 8 for the probability distribution analysis of storms is used, with two main differences: (1) the ordering is started from the lowest flow, and (2) the subdurations are expressed in terms of days rather than in terms of minutes.

In peak runoff determinations, the rate is usually expressed as an instantaneous rate. This is so since the instantaneous rate is the one that is used to size such structures as storm sewers and spillways. In low-flow analysis, on the other hand, the instantaneous value is not the critical factor. The rate is therefore expressed as the average over the subduration (i.e., the equivalent or *sustained flow* during the subduration).

Because the ordering is started from the lowest, a flow magnitude represents the highest equivalent sustained flow for the stated subduration that can be equaled but not exceeded in a given return interval. Thus a 7Q10 flow represents the highest equivalent sustained flow for a 7-day subduration that can be equaled but not exceeded once in 10 years. The 2Q10 has a similar meaning. It is to be noted that the ordering is in terms of years. If there are only 10 years of record, the highest order is 10, and if there are 50 years of record, the highest order is 50. Thus the return interval derived from this ordering is always in terms of years.

Sizing of reservoirs is done by applying the principle of material balance, where the material being referred to is the mass. Since the density of water is practically constant, the material or mass balance is essentially a volume balance, which states that

$$\text{required storage} = \text{inflow} - \text{losses} - \text{demand} \qquad (3\text{-}7)$$

The inflow to the reservoir is a function of streamflow. It is also, theoretically, a function of precipitation, but for drought conditions, there will hardly be any precipitation that this parameter is neglected. Losses are the loss due to evaporation and the loss due to seepage from the reservoir, and demand is the water consumption by the community. Demand may also include such quantities of water that are needed by downstream users not benefited by the construction of the reservoir dam and for a required release of water for resident fish and anadromous fish such as salmon and striped bass.

Example 3-4

A community has decided to construct an impounding reservoir to supply water for a 5-year drought. The raw water is to be processed in a water treatment plant. Low-streamflow analysis of the stream before construction of the dam yielded the following subduration-average flow pairs: 7, 50; 15, 54; 30, 61; 60, 72; 120, 98; 180, 228; 365, 710. The subdurations are in terms of days and the average flows are in terms of cubic feet per second (cfs). The other analysis results are as follows: (1) subdurations and combined evaporation–seepage loss pairs: 7, 1.2; 15, 2.5; 30, 5.0; 60, 10.0; 120, 20.0; 180, 30; 365, 60.8 (subdurations in days, evaporation–seepage in inches); and (2) subdurations and community demand pairs: 7, 2800; 15, 6000; 30, 12,000; 60, 24,000; 120, 48,000; 180, 73,500; 365, 140,000 (subdurations in days, demand in cfs-days). Assume that the reservoir pool area increases to 1000 acres after construction. Neglecting downstream releases required, determine the volume of the reservoir.

Solution   First, convert all units to acre-ft:

$$\text{cfs-days} = \frac{\text{ft}^3}{\text{sec}} \left(\frac{86,400 \text{ sec}}{\text{day}} \right) (\text{day}) \left(\frac{\text{acre-ft}}{43,560 \text{ ft}^3} \right) = 1.98 \text{ acre-ft}$$

For evaporation since the area of reservoir is 1000 acres,

$$1 \text{ in.} = \frac{1}{12}(1000) = 83.3 \text{ acre-ft}$$

In the computation below, the material balance equation (3-7) is applied in every subduration.

(1)	(2)	(3)	(4)	(5)	(6)	(7)
			Evapo-ration			
Subdura-tion (days)	Inflow (cfs)	Inflow volume (acre-ft)	plus seepage (acre-ft)	Consumer demand (acre-ft)	Col. (3) minus col. (4) (acre-ft)	Col. (5) minus col. (6) (acre-ft)
7	50	693	100	5,544	593	4,951
15	54	1,604	208	11,880	1,396	10,484
30	61	3,623	417	23,760	3,206	20,554
60	72	8,554	833	47,520	7,721	39,799
120	98	23,285	1,666	95,040	21,619	73,421
180	228	81,260	2,499	145,530	78,761	66,769
365	710	513,117	5,065	277,200	508,052	−230, 852

Col. (3): 693 = 50(7)(1.98); 50 is the inflow for the 7-day subduration
Col. (4): 100 = 1.2(83.3); 1.2 is the evaporation-seepage for the 7-day duration
Col. (5): 5544 = 2800(1.98); 2800 is the consumer demand for the 7-day duration
Col. (6): 593 = 693 − 100, net inflow after correction for losses
Col. (7): 4451 = 5544 − 593

The entries in column (7) represent the storage requirement for a given duration. Of all the storages indicated, the largest will hold the other storage requirements. Therefore, choose the largest volume, which is 73,421 acre-ft, corresponding to the 120-day drought. Thus

$$\text{volume of reservoir} = 73,421 \text{ acre-ft} \quad \textbf{Answer}$$

If desired, the answer may be refined further by revising the calculations to consider smaller subduration intervals around the "120-day" drought.

GROUNDWATER HYDROLOGY

Hydrology is defined as the science that deals with the study of the properties, distribution, and behavior of water in nature. In conformance with this definition, *groundwater hydrology* may be defined as the science that deals with the study of the properties, distribution, and behavior of water in nature as it occurs underneath the surface of the earth. The subject of groundwater is a broad field. Some books on this subject deal with well-drilling technology; others deal with the movement of groundwater. In this section we deal with the environmental aspect of groundwater hydrology.

Underground Waters

Figure 3-3 is a schematic of the various underground waters. In general, with respect to underground waters, there are two zones: the vadose and groundwater zones. Not all underground waters are groundwater. Underground water whose pressure is less than atmospheric is *vadose water,* while underground water whose pressure is greater than atmospheric is *groundwater.*

The vadose region has an upper and a lower zone. The upper zone is where infiltration from precipitation enters. That portion of the upper zone into which roots of the plants penetrate is called the *root zone.* Vadose water in this zone is held by molecular attraction to soil grains, thus making it available to the plants. Some impervious layer may exist in the vadose zone. As such, percolating water from the infiltrate may be trapped in this layer, forming a *perched water.* The top of the perched water is called a *perched water table.* As shown, the perched water rises to form a fringe. This fringe is called a *perched water capillary fringe.*

The groundwater finally comes to rest at the very bottom of the figure. It is shown resting on the top of an aquitard, aquiclude, or aquifuge (defined later). It can no longer move downward but rests like water resting in a tank. In fact, the groundwater zone is a huge underground reservoir, which, of course, has to be leaking. There is no other direction in which the groundwater can go but must follow the contour of the bottom, restraining formation; hence groundwater flows in a general horizontal direction. The assumption of a purely horizontal direction is called the *Dupuit–Forcheimer assumption.* The reservoir holding the groundwater is called an *aquifer.*

Groundwater aquifers may be confined or unconfined. When impervious layers bound the upper and the lower parts, the aquifer is called a *confined aquifer;* otherwise,

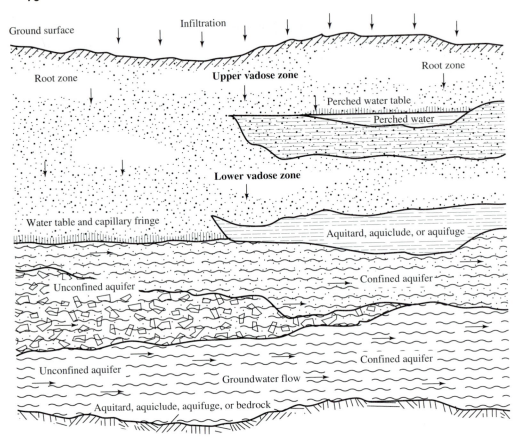

Figure 3-3   Underground waters.

it is called an *unconfined aquifer.* These confining layers are called aquitard, aquiclude, or aquifuge. An *aquitard* is a formation that holds water but cannot transmit it in such a quantity to be productive regionally. An *aquiclude* is also a formation that holds water but cannot transmit it, and an *aquifuge* neither holds nor transmit water. These formations are generally called *confining layers.*

     When the head of the water entering a confining layer is greater than the elevation of the formation, the groundwater is under pressure. Imagine in Figure 3-3 that a well penetrates the layer. If the top portion of this well is open to the atmosphere, water will gush out under the heavy pressure. This type of well is called an *artesian well,* and the aquifer is called an *artesian aquifer.* The level to which the water will rise above or below the ground surface, when not confined, is called the *piezometric head.* The piezometric head thus forms a piezometric surface over the aquifer.

     In Figure 3-3, next to the confined aquifer, is the unconfined aquifer. By hydrostatics, the top of this aquifer, called the *water table,* is at atmospheric pressure. Any point below the water table is greater than atmospheric and the pressure is equal to the atmospheric pressure plus the product of the specific weight and the depth from the surface. Above the water table surface, water molecules rise due to the attraction between the molecules and the surface of the soil grains and due to the surface tension of water. This band of water is called the *groundwater capillary fringe*. Hence the water molecules at the capillary fringe are said to be under surface tension. Since the pressure at the top of the water table is atmospheric and since the capillary fringe is above the water table, the pressure at this level is below atmospheric. Thus the pressure is negative gage. The piezometric head corresponds to the water-table level of an unconfined aquifer.

Well Hydraulics

To control or, otherwise, mitigate the impact of pollution into a groundwater resource, it is important that the physics of groundwater hydraulics be understood. From fluid mechanics, the head form of the Hagen–Poiseuille equation may be solved for the velocity V:

$$V = \frac{\gamma D^2}{32\mu} \frac{h}{l} = K \frac{h}{l} \tag{3-8}$$

The ratio h/l is nothing but the loss of head h per unit length, which is the average energy slope throughout the length l. In groundwater flow, the velocity is very small. The contribution of the velocity head to the total energy is therefore neglected. Thus the total energy is simply equal to the piezometric head. To get a more accurate energy or piezometric slope, $\partial h/\partial l$ is used to represent the slope at a particular point along l but not across the whole length of l. (Note that the partial notation is used, indicating that h could be also a function of variables other than l.) Using this ratio for the slope in equation (3-8) will give the corresponding velocity at the point. In addition, since h may increase or decrease with l, put a plus or minus sign before the derivative to make the equation more general. In other words, when h increases with l, the derivative is positive; and when it decreases with l, the derivative is negative. Therefore,

$$V = \pm K \frac{\partial h}{\partial l} \tag{3-9}$$

Equation (3-9) is called *Darcy's law,* and K is called *Darcy's coefficient, hydraulic conductivity*, or *coefficient of permeability*. V is called the *Darcy velocity*. Representative values of K are shown in Table 3-2.

Figures 3-4 and 3-5 are schematics of wells penetrating an unconfined and a confined aquifer, respectively. At equilibrium, the head h is only a function of distance l and not of time t. Therefore, $\partial h/\partial l = dh/dl$. Applying equation (3-9), assuming equilibrium flow, the groundwater flow Q to the well in Figure 3-4 is

$$Q = AV = AK \frac{dh}{dl} = AK \frac{dh}{dr} \tag{3-10}$$

TABLE 3-2 REPRESENTATIVE VALUES OF DARCY'S COEFFICIENT (K), AND OTHER MATERIAL PROPERTIES

Granular material	K (m/day)	Consolidated material	K (m/day)
Gravel	100–1,000	Rock	
Sand and gravel mixes	25–100	Aquifers	0.001–10
Fine sand	1–10	Core samples	0–400
Medium sand	5–30	Shale	10^{-7}
Coarse sand	25–100	Dense, solid rock	$<10^{-5}$
Clay, sand, and gravel mixes (till)	0.001–0.1	Fractured or weathered volcanic rock	0–1050
Clay soils (surface)	0.01–0.15	Sandstone	0.001–0.1
Deep clay beds	10^{-8}–10^{-2}	Carbonate rock with secondary porosity	0.01–01
Loam soils (surface)	0.1–1		

Material	Porosity	Specific yield	Specific retention	Hydraulic conductivity (m/day)
Basalt	0.10	0.07	0.03	$10^{-8} - 1000$
Clay	0.50	0.05	0.45	$10^{-7} - 10^{-4}$
Granite	0.001	0.0008	0.0002	$10^{-8} - 1$
Gravel	0.20	0.18	0.02	$100 - 7000$
Limestone	0.20	0.18	0.02	$10^{-4} - 6000$
Sand	0.25	0.20	0.05	$0.05 - 130$
Soil	0.60	0.40	0.20	$10^{-3} - 10$
Sandstone	0.10	0.05	0.05	$10^{-5} - 1$

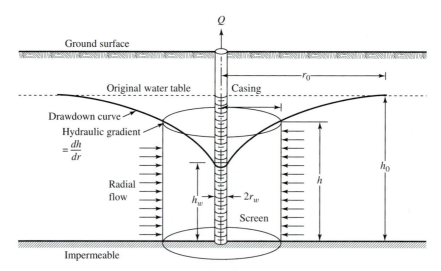

Figure 3-4 Radial flow to well penetrating unconfined aquifer. (From M. J. Hammer (1986). *Waste and Wastewater Technology*, John Wiley and Sons, New York.)

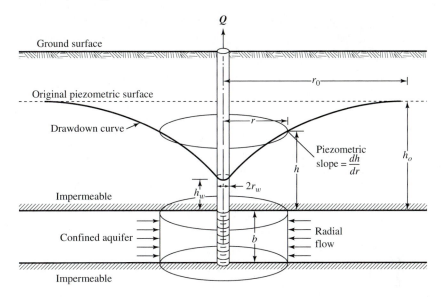

Figure 3-5   Radial flow to well penetrating confined aquifer. (From M. J. Hammer (1986). *Waste and Wastewater Technology*, John Wiley and Sons, New York.)

Note that l is r in the figure, which is any distance circumscribing a cylinder about the well. Also, the slope dh/dr is positive since r increases with h. A is the lateral area of the circumscribing cylinder, which is equal to $2\pi rh$. Substituting this value of A into equation (3-10) and integrating between the limits of h to h_0 and r to r_0,

$$Q = \pi K \frac{h_0^2 - h^2}{\ln(r_0/r)} \qquad (3\text{-}11)$$

The radius r_0 is called the *radius of influence*, and the locus of r and h is called the *drawdown curve*. The distance from any point along the drawdown curve to the original piezometric level is called the *drawdown*.

The area A of the circumscribing cylinder for the confined aquifer is $2\pi rb$, where b is the thickness of the aquifer. Using this term for A and similarly applying equation (3-9) to the well in Figure 3-5, the formula for the groundwater flow to the well is

$$Q = 2\pi K b \frac{h_0 - h}{\ln(r_0/r)} \qquad (3\text{-}12)$$

Example 3-5

A 0.60-m well is constructed in a confined aquifer of $b = 20$ m. Assume the aquifer to be composed of a mixture of sand and gravel. The upper confining impermeable layer extends from the top of the aquifer to the ground surface and has a thickness of 40 m.

The original piezometric surface is 10 m below the ground surface. Pumping has stabilized at the following drawdowns: 2.5 m at the observation well 25 m away from the pumped well and 4.0 m at the observation well 3 m away from the pumped well. Calculate the flow Q.

Solution   From equation (3-12),

$$Q = 2\pi K b \frac{h_0 - h}{\ln(r_0/r)}$$

From Table 3-2, use $K = 50$ m/day. Use the top of the aquifer as the reference datum. $h_0 = 40 - 10 - 2.5 = 27.5$ m, $r_0 = 25$ m, $h = 40 - 10 - 4 = 26$ m, and $r = 3$ m. Thus

$$Q = 2\pi 50(20) \left[\frac{27.5 - 26}{\ln(25/3)} \right] = 4445 \text{ m}^3/\text{day} = 51.4 \text{ L/s} \textbf{Answer}$$

The three other terms in Table 3-2 will now be defined: porosity, specific yield, and specific retention. *Porosity* is the fraction of space in the bulk volume of any substance. Hence in the case of an aquifer, it is the fraction of the bulk volume of the aquifer occupied by water. *Specific yield* is the fraction of water that drains freely under the influence of gravity in a given bulk volume of an unconfined aquifer. The fraction of water that remains after the rest has drained out is called the *specific retention* of the bulk volume. The term *specific yield* cannot be applied to a confined aquifer since a confined aquifer is acted upon not only by gravity but also by a positive pressure head. Hence the aquifer cannot be drained freely under gravity. A more general term, *storage coefficient*, is therefore coined to apply to both the confined and unconfined states of the aquifer. The *storage coefficient* is defined as the fraction of water that is drained from an aquifer when the piezometric head is lowered a unit distance. Another term used in groundwater hydraulics is transmissibility T. It is defined as the product of the Darcy coefficient K and the depth of the aquifer.

In 1935, Theis[3] presented an unsteady-state analysis assuming a homogeneous and isotropic aquifer. The formula is

$$Z_r = \frac{Q}{4\pi T} \int_u^\infty \frac{e^{-u}}{u} du \tag{3-13}$$

where Z_r is the drawdown in an observation well r distance from a pumped well. T is the transmissibility coefficient and u is defined as

$$u = \frac{r^2 s_c}{4 T t} \tag{3-14}$$

where s_c is the storage coefficient and t is the time of pumping from the start. Although

---

[3]C. V. Theis (1935). "The Relation between the Lowering of the Piezometric Surface and the Rate and Duration of a Well Using Groundwater Storage." *Transactions of the American Geophysical Union*, 16, pp. 519–524.

Theis originally developed equation (3-13) for a confined aquifer, it can also be applied to an unconfined aquifer, provided that the drawdown is relatively small.[4]

The integral of equation (3-13) is called the *well function of u, W(u)*. e^{-u} is always dimensionless, and since du divides u, $W(u)$ is therefore dimensionless. The integral may be solved by expressing it in a series:

$$\int_u^\infty \frac{e^{-u}}{u} du = W(u) = -0.5772 - \ln u + u - \frac{u^2}{2(2!)} + \frac{u^3}{3(3!)} - \frac{4^4}{4(4!)} + \cdots \qquad (3\text{-}15)$$

where ! is the factorial notation.

Example 3-6

A 0.60-m well is constructed in an unconfined aquifer of $b = 20$ m. Assume the aquifer to be composed of sand. The original piezometric surface is 10 m below ground surface. After pumping the well for 1 day, the drawdown in the observation well 25 m away is 2.5 m. Calculate the pumping rate Q that effected this drawdown. Maintaining this pumping rate, what will the drawdown in the observation well be after 1 year?

Solution Since the aquifer is unconfined, s_c is equal to the specific yield. From Table 3-2, use $s_c = 0.22$. Also, $K = 50$ m/day.

$$u = \frac{r^2 s_c}{4Tt}$$

Thus

$$u = \frac{25^2(0.22)}{4[50(20)(1)]} = 0.034 \text{ (dimensionless)}$$

$$W(u) = -0.5772 - \ln u + u - \frac{u^2}{2(2!)} + \frac{u^3}{3(3!)} - \frac{u^4}{4(4!)} + \cdots$$

$$= -0.5772 - \ln 0.034 + 0.034 - \frac{0.034^2}{4} + \frac{0.034^3}{18} - \frac{0.034^4}{96}$$

$$= 2.836$$

Therefore,

$$Q = \frac{Z_r(4\pi T)}{W(u)} = \frac{2.5(4\pi)(50)(20)}{2.836} = 11,077.55 \text{ m}^3/\text{day} \quad \textbf{Answer}$$

After 1 year of pumping,

$$u = \frac{25^2(0.22)}{4[50(20)(365)]} = 0.0000942 \quad \text{(dimensionless)}$$

$$W(u) = -0.5772 - \ln 0.0000942 + 0.0000942 - \frac{0.0000942^2}{4} + \frac{0.0000942^3}{18}$$

$$= 8.692$$

---

[4] A. L. Prasuhn (1987). *Fundamentals of Hydraulic Engineering*. Holt, Rinehart and Winston, New York, p. 95.

Thus

$$Z_r = \frac{QW(u)}{4\pi T} = \frac{11,077.55(8.692)}{4(\pi)(50)(20)} = 7.66 \text{ m} \quad \textbf{Answer}$$

CONTAMINATION OF GROUNDWATER

Groundwater is a very important source of drinking water. About one-third of U.S. water needs for drinking water is supplied from this groundwater resource. Historically, groundwater was considered to be so safe that it has been consumed directly from water wells without further treatment. In underdeveloped countries, practically all villages rely on groundwater from shallow-dug wells. Cebu City, the second largest city in the Philippines, has relied solely on groundwater, after its impounding reservoir was silted up due to poor management. Water pumps are distributed strategically throughout the city to supply water directly to consumers.

In the United States, however, along with the benefits of comfortable living, lies the specter of groundwater contamination. In New Jersey, for example, a community was forced to rely on bottled water because of contamination of its water supply. Incidence of high cancer rates suspected to be the result of the contamination of water supply is often reported in the media. In one instance, the water supply was reported to be close to a chemical plant. But unfortunately, the persons affected happened to be making their livelihood in the very chemical factory that was polluting their drinking water.

Once a groundwater formation is contaminated, it is very difficult for it to be restored. Several drinking wells across the United States have been shut down. There are several ways that an aquifer can be contaminated: through a legally or an illegally constructed hazardous waste landfill, old sanitary landfill as well as new, injection wells, impoundments, and illegal dumping grounds.

Figure 3-6 is a schematic of the various ways that contaminants enter the groundwater system. Starting from the left-hand side of the figure, contaminated water from a land spreading operation percolates through the aquifer. Also shown is the injection well. Injection wells are normally well designed. Those used for disposing highly toxic wastes are usually designed for a depth of not less than 1000 ft. The majority are in the range 2000 to 6000 ft. Quite a few are beyond 7000 ft, with some reaching 8000 ft below the ground surface. These depths are well below the location of productive potable aquifers. Nonetheless, potable aquifers can still be contaminated by leakage through the casing, as indicated in the figure. In addition, as we discuss later, injection-well disposal has a grave potential threat for environmental damage.

Next is the leakage through septic tanks and cesspools. Septic tanks have been found to fail and to leak profusely, causing environmental damage. For example, studies of the Rhode River estuary in Maryland suggest that the cause of its eutrophication is septic tank leakage from a community on the Mayo Beach peninsula bordering the water body. All the septic tanks have now been abandoned in favor of a centralized treatment

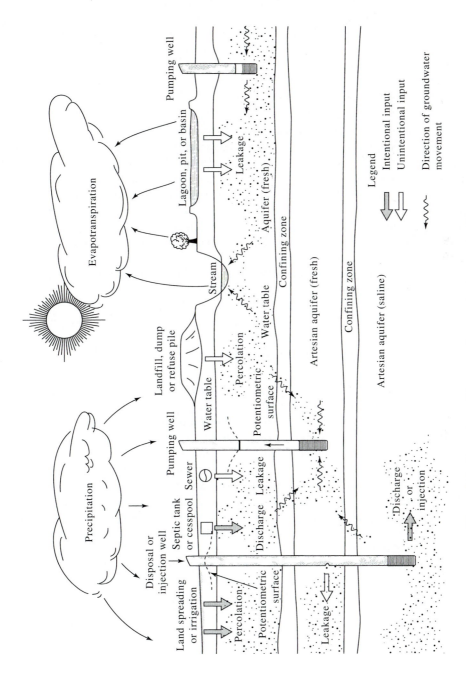

Figure 3-6 Several possible ways of contaminating a groundwater system.

system, a part of which is an artificial wetland. The other possible sources of contamination shown in the figure are the landfill in the middle and the lagoon on the right-hand side. After the contamination, the aquifer is pumped, one of the purposes of which could be for drinking water supply purposes. The two pumps are shown on the right side and near the middle of the figure.

Injection-Well Disposal Contamination

More may be said of the potential hazards of injection wells in addition to the one depicted in Figure 3-6. Hazardous wastes deposited by injection wells underground are like time bombs waiting to erupt on future generations. For example, wastes are deposited in anticlines and synclines of folds[5] as waste reservoirs. But there is no guarantee that conditions stay this way. Synclines and anticlines, faults and fractures are a manifestation of a changing earth crust. Hence in future generations they may be transformed into a productive aquifer. By injection disposal, the present generation is depositing time bombs for future generations to discover sitting on productive aquifers that could be developed for water supply. Just as past generations were indiscriminately dumping hazardous wastes to haunt the present generation, the present generation could be doing the same things to haunt future generations. A sketch of an injection well is shown in Figure 3-7; will it hold?

CONTROL OF GROUNDWATER POLLUTION

The control of groundwater pollution may include physical containment and extraction of contaminated water for subsequent treatment. The *physical containment method* involves literally stopping the plume of an advancing pollution by a barrier. A trench about 1 to 2 m wide is dug from the surface down to the depth of the impermeable layer to stop the plume. The trench is then backfilled with an impermeable material. Backfill materials that have been used are a mixture of soil and bentonite.[6] Of course, this method of control is only temporary while other measures are applied to clean up the aquifer.

Figure 3-8 shows sectional views of contaminant plumes. Although the figure indicates the spills as sources of contamination, it can generally represent other sources as well. To remove the contaminants by extraction, the whole width of the plume must be captured by the pumping operation. The capture zone may be delineated, approximately, using the superposition of a plane sink and a plane uniform flow.

A sink is indicated on the top left of Figure 3-9. The sink is the effect of the pumping operation. On the right-hand side is the uniform flow, which represents the flow of the groundwater conveying the contaminants. The objective is to derive the coordinates

[5]D. L. Warner and J. H. Lehr (1981). *Subsurface Wastewater Injection.* Premier Press, Berkeley, Calif. p. 31.

[6]G. M. Masters (1991). *Introduction to Environmental Engineering and Science.* Prentice Hall, Englewood Cliffs, N.J., p. 165.

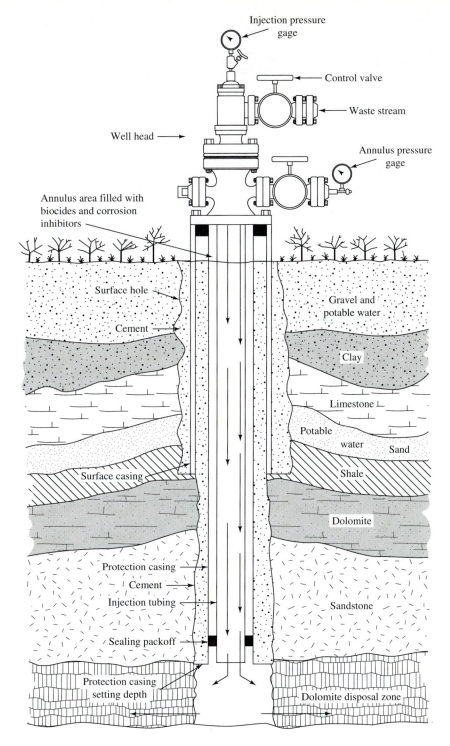

Figure 3-7 Design of hazardous waste injection well.

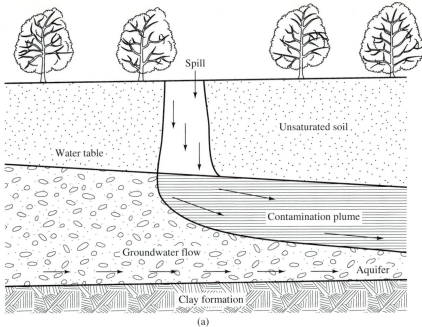

(a)

Dissolved contamination plume.

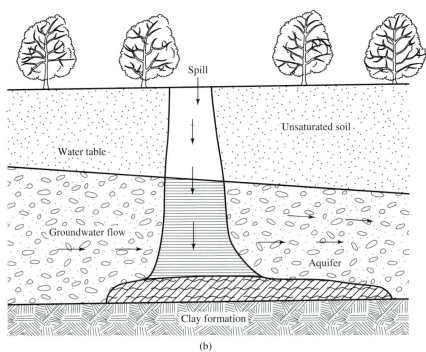

(b)

Immiscible plume denser than water.

Figure 3-8 Sectional views of contaminant plumes: (a) dissolved contamination plume; (b) immiscible plume denser than water.

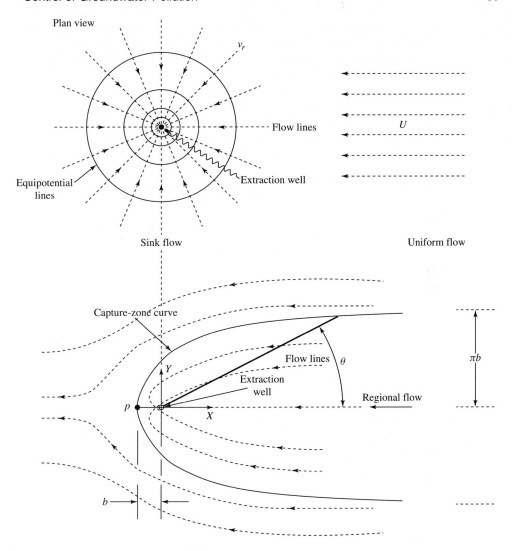

Figure 3-9   Derivation of capture zone.

of the capture zone delineated at the bottom portion of the figure. Once delineated, proper location of the extraction pump and its size can be specified.

For the proper perspective and to appreciate the limitations of the formula to be derived, we define the term *potential flow* as inviscid, incompressible, and irrotational flow of fluids. The term *inviscid* means that the flow is frictionless. Despite the obvious presence of friction in groundwater flow, since the motion is very, very slow, the flow may be assumed inviscid. That is, since the motion is very slow, the effect of friction is not felt; hence the flow is equivalent to being inviscid. Water is, of course, incompressible.

The term *rotation* refers to the fact that when a fluid parcel moves, it is dragged by neighboring parcels because of friction. This dragging cause the parcel to rotate. The term *irrotational* then connotes the absence of friction; hence in potential flow the parcel does not rotate. Again, since the motion is very, very slow, the effect of friction is negligible, such that groundwater flow may be assumed irrotational as well. To conclude, it is assumed that groundwater flow is a potential flow. In addition, it will be assumed that the flow is a plane flow (i.e., two-dimensional in the horizontal plane).

The technique of derivation is to superimpose the sink flow due to pumping and the groundwater flow, assumed to be uniform. This superposition will result in a point p called the *stagnation point*, where the sink and the uniform flows separate. At the stagnation point, the velocity of the groundwater U and that of the sink v_r due to the extraction will be equal. Assume that the sink is extracting m cubic units of flow per unit depth of the aquifer. Therefore,

$$m = (2\pi r)v_r\eta \tag{3-16}$$

where r is the radius of a "cylinder" of flow, v_r the radial velocity of flow, and η the porosity. Since $v_r = U$ at the stagnation point p,

$$m = 2\pi b U \eta \tag{3-17}$$

U is not the same as the Darcy velocity V, an apparent velocity. U is the actual velocity of the particles of water that flow through the interstices of the formation. Since the interstitial cross-sectional area is much smaller than the gross superficial cross-sectional area from which V is derived, U is much larger and is related to V by the porosity as follows: $U = V/\eta$, where η is the porosity.

To proceed further, find the equation of the superimposed streamline that passes through the stagnation point. First, solve for the stream function of the sink flow. From fluid mechanics, $(1/r)(\partial\Psi/\partial\theta) = -v_r$, where Ψ is the sink-flow stream function and θ is the polar cylindrical coordinate. Substituting this equation in equation (3-16) and integrating for Ψ gives

$$\Psi = -\frac{m}{2\pi\eta}\theta \tag{3-18}$$

The stream function differential equation for the uniform groundwater flow is $(\partial\Psi/\partial y) = -U$, where y is the familiar vertical y coordinate. Integrating this for Ψ, noting that $y = r\sin\theta$, we obtain

$$\Psi = -Ur\sin\theta \tag{3-19}$$

Superimposing equations (3-18) and (3-19) produces

$$\Psi = -\frac{m}{2\pi\eta}\theta - Ur\sin\theta \tag{3-20}$$

Equation (3-20) is the equation of the superimposed streamlines of the sink and groundwater flows of the lower figure of Figure 3-9. At the stagnation point, $\theta = \pi$ and the equation becomes

$$\Psi_{stagnation} = -\frac{m}{(2\eta)}$$
(3-21)

From equation (3-17), $m/(2\eta) = \pi bU$ and $m/(2\pi\eta) = bU$. Substituting equation (3-21), with $\Psi_{stagnation} = -m/(2\eta) = -\pi bU$ and $m/(2\pi\eta) = bU$ in equation (3-20), the equation of the streamline passing through the stagnation point is

$$-\pi bU = -bU\theta - Ur\sin\theta$$
(3-22)

Solving for $r\sin\theta$ gives

$$r\sin\theta = b(\pi - \theta) = y$$
(3-23)

Equation (3-23) gives the coordinates of the envelope of the capture-zone curve. Referring to Figure 3-9, as $\theta \to 0$, y equals the half-width of the capture zone. Hence the width of the capture zone w_c, twice the half-width, is

$$w_c = 2\pi b$$
(3-24)

Equation (3-24) gives the width of the capture zone of one extraction pump. Assuming n extraction pumps, the total width W_c of the combined capture zones is simply the sum of the individual widths. If all the pumps have the same capacity,

$$W_c = nw_c = 2n\pi b$$
(3-25)

To extract all the contaminants in the flow, W_c must be equal to the width of the plume. Therefore, to design an extraction pumping system, the width of the plume should first be determined and made equal to W_c. Having obtained W_c, b is calculated from equation (3-25). The value of m is then solved using equation (3-17). This value gives the size of each pump. The exact locations of the pumps in the field may be determined by plotting equation (3-23) separately for each pump so that the overall capture curve will include or envelop the boundary of the plume. This will involve trial and error until the desired envelope is obtained.

Coefficient of Retardation

The velocity U used in the equations above is not the velocity of the contaminant plume but, as defined, the velocity of the groundwater. In general, when the contaminant travels in the formation, it is retarded. Hence define the coefficient of retardation C_r as the ratio of the velocity U_p of the centroid of the contaminant plume divided by the groundwater velocity U, as follows:

$$C_r = \frac{U_p}{U}$$
(3-26)

Some values of coefficient of retardation are shown in Table 3-3, which indicates that C_r is a function of the time of travel.

Example 3-7

A contaminant plume of carbon tetrachloride from a spill is discovered. The largest width of the plume is approximately 80 m. The depth of the contaminated aquifer is 30 m. Field investigation indicates that the aquifer is mainly sandstone and that the groundwater elevation is 30 m above the river 3300 m downstream. (a) How long will it take for the spill to reach the river? (b) If management decides to extract the compound using two pumps, determine their sizes and the spacing between them in the field.

Solution (a) $V = K \frac{dh}{d\ell}$; assume that $K = 10^{-2}$ m/day; $\eta = 0.11$. Then

$$V = 10^{-2} \left(\frac{30 - 0}{3300} \right) = 9.1(10^{-5}) \text{ m/day} \qquad U = \frac{9.1(10^{-5})}{0.11} = 8.3(10^{-4}) \text{ m/day}$$

From Table 3-3, for carbon tetrachloride, $C_r = 0.40$. Thus

$$\text{time to reach the river} = \frac{3300}{0.4(8.3)(10^{-4})} = 9.9(10^6) \text{ days} = 27{,}349 \text{ years} \quad \textbf{Answer}$$

(b) $W_c = nw_c = 2n\pi b$

$$b = \frac{W_c}{2n\pi} = \frac{80}{2(2)\pi} = 6.4 \text{ m}$$

$$m = 2\pi bU\eta = 2\pi (6.4)(8.3)(10^{-4})(0.11) = 0.0037 \ \frac{\text{m}^3}{\text{day-m}}$$

Thus

$$\text{size of one pump, } Q = 0.0037(30) = 0.11 \text{m}^3/\text{day} \quad \textbf{Answer}$$

$$\text{Spacing} = 2\pi b = 2\pi (6.4) = 40.0 \text{ m} \quad \textbf{Answer}$$

Note: Management does not need to install a pump.

TABLE 3-3 COEFFICIENTS OF
RETARDATION OF SOME COMPOUNDS

Compound	Travel time (days)	C_r
Hexachloroethane	100	0.12
Dichlorobenzene	100	0.17
	400	0.12
	640	0.11
Tetrachloroethylene	100	0.25
	400	0.19
	640	0.17
Bromoform	100	0.43
	400	0.36
	640	0.34
Carbon tetrachloride	100	0.48
	400	0.43
	640	0.40

GLOSSARY

Aquiclude. An underground formation containing water that cannot be transmitted.

Aquifer. An underground reservoir or formation where water flows at rate that can productively be retrieved regionally.

Aquifuge. An underground formation that does not hold water.

Aquitard. An underground formation containing water that cannot be transmitted at such quantity to be productive regionally.

Capillary fringe. A curtain of water that forms above the water table due to the attraction of the soil formation on the water.

Capture zone curve. The limit that defines the extent of groundwater contamination that can be extracted by a pump or combination of pumps.

Coefficient of retardation. The ratio of the velocity of the contaminant plume to the groundwater velocity.

Confining layer. A layer of aquitard, aquiclude, or aquifuge that restricts the vertical movement of the water in the aquifer.

Depression storage. Fraction of rainfall that fills puddles and depressions on the surface of the earth.

Distribution reservoir. A reservoir constructed to equalize the supply and demand for treated water in a community and to provide supply of water during emergency situations.

Drawdown. The vertical distance from the original surface of the groundwater to the existing surface of the pumped groundwater.

Drawdown curve. The locus of the drawdown.

Groundwater. The precipitation that gravitates toward the lowest possible point under the surface of the earth; its pressure is greater than atmospheric.

Groundwater hydrology. The study of the properties, distribution, and behavior of water in nature as it occurs under the surface of the earth.

Hydraulic conductivity, permeability coefficient, or Darcy's coefficient. The proportionality coefficient in Darcy's law.

Hydrology. The science that deals with the study of the properties, distribution, and behavior of water in nature.

Impounding reservoir. A reservoir in the valley of a stream or river constructed for the purpose of holding streamflow; this flow may be released to satisfy demand during times of low stream flows.

Infiltration. The process of entering of precipitation from the air into the subsurface of the ground.

Inviscid flow. Frictionless flow.

Irrotational flow. A flow where the fluid has no tendency to form vortices.

Point of concentration. The point in a drainage area where a maximum rate of flow is to be determined.

Porosity. The fraction of space in the bulk volume of any substance.

Potential flow. The inviscid, incompressible, and irrotational flow of fluids.

Precipitation. Any of the forms of H_2O that result when clouds become heavily laden with moisture.

Probability distribution analysis. A method of statistical analysis where the probability of occurrence of events are arrayed either in descending or ascending order.

Radius of influence. The horizontal distance from the center of a well to the point where the drawdown is zero.

Return period or return interval. The average number of years that a particular event is equaled or exceeded.

Runoff. That part of precipitation that flows on the surface of the earth or just immediately beneath the surface of the earth.

Specific yield. The fraction of water that drains freely under the effect of gravity in a given volume of unconfined aquifer.

Spillway. A channel built at a point along a dam or reservoir used to convey excess water from the reservoir.

Storage coefficient. The fraction of water that will drain from an aquifer under the effect of gravity per unit lowering of the piezometric head; in an unconfined aquifer, it is equal to the specific yield.

Time of concentration. In a given drainage area, the time it takes for the area to contribute at a particular point the maximum rate of flow; it is the time it takes a particle from the farthest point in the drainage area to the point under consideration assuming that the flow converges uniformly toward the point.

Vadose water. Water in the vadose zone whose pressure is less than atmospheric.

Vadose zone. The zone in the underground formation from the surface to the groundwater.

Watershed. The same as catchment area.

Water table. The surface of a groundwater.

SYMBOLS

C	coefficient of runoff in the rational method
i	rainfall intensity
IDF	intensity–duration–frequency
m	order number in probability distribution calculation; flow per unit depth of aquifer
mgd	million gallons per day
n	the Manning coefficient of roughness; number of extraction pumps
p	probability that a certain event is equaled or exceeded
r	radius of concentric cylinder about a well
r_o	radius of influence of a well

s_c	storage coefficient
t	time
t_c	time of concentration
T	transmissivity
u	$r^2 s_c / 4T t$
U	velocity of groundwater
U_p	velocity of centroid of contaminant plume of a single dump of the contaminant
v_r	radial velocity of flow to an extraction well
w_c	half-width of a capture zone corresponding to one extraction pump
W_c	largest width of contaminant plume
Z_r	drawdown of well
η	porosity
θ	polar coordinate
Ψ	stream function

PROBLEMS

3-1. Define *hydrology* and *environmental engineeering hydrology*.

3-2. Define *surface hydrology* and *groundwater hydrology*.

3-3. Differentiate between *groundwater hydrology* and *hydrogeology*.

3-4. A total of 13 in. of precipitations falls on a 45-mi^2 drainage area in 24 h. Calculate the total volume of runoff during the period.

3-5. What is the average rainfall intensity in Problem 3-4?

3-6. Express the volume of runoff in Problem 3-4 in cubic meters.

3-7. For the weather station nearest your hometown, record the average precipitation for each month of the year and construct a monthly bar graph. The information may be obtained from the *Climatological Data*, U.S. National Weather Service.

3-8. What is the highest rainfall intensity in Problem 3-7?

3-9. Analysis of 10 years of precipitation data for a 20-min subduration yields the following results for the first up to the tenth year, respectively (in cm): 2.96, 3.0, 2.86, 2.75, 3.10, 3.20, 2.84, 2.69, 2.78, and 2.91. Calculate the probability distribution.

3-10. In Problem 3-9, what is the rainfall intensity having a return interval of 20 years?

3-11. A weather service station is scanned for precipitation to perform a probability distribution analysis. The results are as follows (in cm): 5-min subduration: 1.4, 1.8, 1.6, 1.0, 1.3, 1.7, 1.9, 1.0, 0.9, and 1.2; 20-min subduration: 2.96, 3.0, 2.86, 2.75, 3.10, 3.20, 2.84, 2.69, 2.78, and 2.91; 50-min subduration: 3.98, 3.92, 3.87, 3.89, 4.30, 4.00, 4.10, 3.99, 4.08, and 3.87; and 100-min subduration: 4.70, 4.60, 4.76, 4.89, 4.30, 4.98, 4.85, 4.67, 4.79, and 4.77. Synthesize the 10-year storm.

3-12. For the 10-year storm of Problem 3-11, what is the average intensity for the 20-min-duration storm?

3-13. A method of estimating a missing precipitation record is by weighting observations of nearby stations by the inverse distance to the point of the missing record. The expression is

$$P_0 = \frac{\sum_{i=1}^{N}(1/r_{0i}^2)P_i}{\sum_{i=1}(1/r_{01}^2)}$$

where r_{0i} is the distance between station i and station 0. In analyzing the long-term precipitation records of four stations, one station was found to have missing records for the months of June, July, and August in a particular year. The following are the records in mm depth of rainfall: First station: June, 54; July, 48; and August, 44. Second station: June, 67; July, 56; and August, 49. Third station: June, 75; July, 44; and August, 57. The long-term average precipitations for all four stations are as follows: First station: June, 60; July, 48; and August, 45. Second station: June, 65; July, 56; and August, 47. Third station: June, 70; July, 65; and August, 60. Fourth and missing station: June, 67; July, 60; and August, 55. Estimate the missing precipitations using $r_{41} = 5$ km, $r_{42} = 1.5$ km, and $r_{43} = 10$ km.

3-14. The diagram below is a schematic of a drainage area. What are the maximum inlet flows to manholes 4, 3, and 2 from the respective areas A, B, and C? At what respective times are these occurring? With rainfall intensity $i = 13.96$ cm/h, calculate the 10-year storm from manhole 2 to 1. The runoff from area A drains to manhole 4, the runoff from area B drains to manhole 3, and the runoff from area C drains to manhole 2. The time of flow between manholes is 5 minutes. A is equal to 0.05 acre, B is equal to 15 acres, and C is equal to 30 acres. Assume all areas to be squares. The average channel slope is 0.0006. For the value of Manning's n, assume smooth-pavement type of surfaces. The coefficients of imperviousness are as follows: for $A = 0.80$, for $B = 0.90$, and for $C = 0.95$.

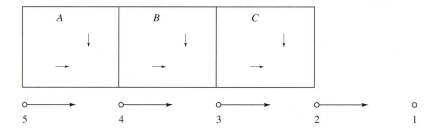

3-15. For the figure in Problem 3-14, a 20,000-gal holding tank is placed at manhole 2 to reduce the peak flow from manhole 2 to 1. The runoff will be diverted to the tank only when the flow to manhole 2 exceeds 20 cfs, which, thereafter, the diverted flow accepted will be two-thirds of the difference between the peak flow from the drainage area and 20 cfs. After the tank is full, all flow goes down the sewer, and when the total flow from the drainage area goes below 20 cfs, a pump starts to empty the tank at a rate of 745 gpm. What is the maximum rate of flow from manhole 2 to manhole 1?

3-16. What size sewer should be provided between manhole 2 and manhole 1 in Problem 3-15? Use a self-cleaning velocity of 2.5 fps in the sewer.

3-17. In Problem 3-15, how many minutes after the storm starts will the pump have emptied the tank?

3-18. In Problem 3-15, what percentage of the peak flow from the drainge area does the peak flow in the sewer represent?

3-19. In Problem 3-15, at what time after the storm starts does the peak flow occur in the sewer?

3-20. Given the design rainfall intensity as $i = 3.5/(3 + t_c)$ for a concrete area of an airport, calculate the peak rate of runoff. The area is 30 acres and $t_c = 15$ min.

3-21. If a rectangular channel is used to convey the drainage from the airport in Problem 3-20 and if a self-cleaning velocity of 2.5 fps is to be maintained, what will be the depth of flow in the channel?

3-22. Solve Problem 3-21 using a circular channel.

3-23. One of the two following 7-day drought figures is a 10-year drought and the other is a 5-year drought: 50 cfs and 45 cfs. Which is the 10-year drought and which is the 5-year drought? Why?

3-24. Calculate the diameter of the circular pipe used to convey the 10-year flow in Problem 3-23.

3-25. A community has decided to construct an impounding reservoir to supply water for a 10-year drought. Analysis of the stream before construction of the dam yielded the following subduration-average flow pairs: 7, 45; 15, 49; 30, 56; 60, 67; 120, 93; 180, 220; 365, 702. The subdurations are in terms of days and the average flows are in terms of cfs. The other analysis results are as follows: (1) subdurations and combined evaporation–seepage loss pairs: 7, 1.0; 15, 2.3; 30, 3.5; 60, 7.0; 120, 17.0; 180, 25; 365, 57 (subdurations in days, evaporation–seepage in inches); and (2) subdurations and community demands pairs: 7, 2600; 15, 5000; 30, 11,000; 60, 22,000; 120, 46,000; 180, 72,000; 365, 138,000 (subdurations in days, demand in cfs-days). Assume that the reservoir pool area increases to 1000 acres after construction. Assuming a downstream release requirement of 50 cfs, determine the volume of the reservoir.

3-26. What critical rate of consumer demand will be satisfied by the reservoir of Problem 3-25?

3-27. For the downstream release requirement of Problem 3-25, what is the head over the spillway, assuming it to be in the form of a broad-crested weir?

3-28. A dam has been constructed to hold a volume of 90,000 acre-ft. The average annual runoff from the drainage area is approximately 40 cfs. How many years of average annual runoff will the reservoir be able to hold?

3-29. Neglecting evaporation and seepage, a spillway of what flow capacity should be provided in Problem 3-28?

3-30. A 0.60-m well is constructed in a confined aquifer of $b = 20$ m. Assume the aquifer to be composed of fine sand. The upper confining impermeable layer extends from the top of the aquifer to the ground surface and has a thickness of 40 m. The original piezometric surface is 10 m below ground surface. Pumping has stabilized at the following drawdowns: 2.5 m at the observation well 25 m away from the pumped well, and 4.0 m at the observation well 2 m away from the pumped well. Calculate the flow Q.

3-31. Two wells are dug 430 m apart. Both fully penetrate a 200-m-thick confined aquifer. They are both pumped at a rate of 0.1 m^3/min. If the hydraulic conductivity is 2 m/day, and the piezometric head was originally 120 m above the top of the aquifer, calculate the drawdown to be expected halfway between the wells. Assume a radius of influence of 300 m.

3-32. A 0.60-m well is constructed in an unconfined aquifer of $b = 20$ m. Assume the aquifer to be composed of fine sand. The original piezometric surface is 10 m below ground surface.

After pumping the well for 1 day, the drawdown in the observation well 25 m away was 2.5 m. Calculate the pumping rate Q that effected this drawdown.

3-33. Maintaining the pumping rate in Problem 3-32, what will be the drawdown in the observation well after 1 year?

3-34. What is the volume above the original water surface if injection of 0.1 m³/min of hazardous waste resulted in a condition of steady state. The water level in the injection well is observed to be 10 m above the original groundwater surface. Assume that the mounded hazardous waste surface coincides with the original water level at a distance of 180 m from the well.

3-35. A well starts pumping at a rate of 990 m³/day. After 1 day the rate was increased to 1500 m³/day. What is the drawdown at an observation well 1000 m from the pumped well 3 days after the start of pumping? The aquifer is confined with the following characteristics: $T = 1400$ m²/day and $s_c = 10^{-4}$.

3-36. A contaminant plume of dichlorobenzene from a spill is discovered. The largest width of the plume is 80 m and the depth of the aquifer is 30 m. Field investigation indicates that the aquifer is mainly limestone with a K value of 20 m/day, a porosity of 0.20, and a groundwater elevation of 30 m above the river 300 m downstream. How long will it take for the spill to reach the river?

3-37. Repeat Problem 3-36 if the plume is bromoform.

3-38. In Problem 3-36, if management decides to extract the compound using three pumps, determine their sizes and the spacing between them in the field.

3-39. Repeat Problem 3-38 if the plume is bromoform.

3-40. It has been determined that an extraction rate of 0.006 m³/s is sufficient to decontaminate a contaminate aquifer. The aquifer has an average depth of 30 m, a hydraulic conductivity of 10^{-2} m/s, and an average gradient of 0.002. If the plume is 100 m wide, locate a single extraction pump to remove the contaminant totally.

3-41. Solve Problem 3-40 if three extraction pumps are to be used.

BIBLIOGRAPHY

HATHAWAY, G. A. (1945). "Design of Drainage Facilities." *Transactions of ASCE*, 110.

KIRPICH, P. A. (1940). "Time Concentration of Small Agricultural Watersheds." *Civil Engineering*, 10(6).

MASTERS, G. M. (1991). *Introduction to Environmental Engineering and Science*. Prentice Hall, Englewood Cliffs, N.J.

McGHEE, T. J. (1991). *Water Supply and Sewerage*. McGraw-Hill, New York.

PRASUHN, A. L. (1987). *Fundamentals of Hydraulic Engineering*. Holt, Rinehart and Winston, New York.

ROBERSON, J. A., J. J. CASSIDY, and M. F. CHAUDHRY (1988). *Hydraulic Engineering*. Houghton Mifflin, Boston.

THEIS, C. V. (1935). "The Relation between the Lowering of the Piezometric Surface and the Rate and Duration of a Well Using Groundwater Storage." *Transactions of the American Geophysical Union*, 16, pp. 519–524.

WARNER, D. L., and J. H. LEHR (1981). *Subsurface Wastewater Injection*. Premier Press, Berkeley, Calif.

CHAPTER 4

Environmental Engineering Hydraulics and Pneumatics

It is the purpose of this chapter to discuss various hydraulic and pneumatic engineering tools needed by environmental engineers in practice. The sections are broadly categorized under the headings of water supply and distribution systems, collection of wastewater, pumping stations, and pneumatic systems.

WATER SUPPLY AND DISTRIBUTION OF WATER

Distribution of water refers to the actual delivery of treated water to homes, businesses, and industries. (It is axiomatic that no water should be delivered without treatment.) Before water can be treated and distributed, it has to be collected from a water supply source and transmitted to a water treatment plant.

Collection of Water

In general, there are two types of sources of raw water: surface and ground. In searching for a surface water supply, there are two things to be considered: the catchment area and the reservoir site. The *catchment area* (also called *watershed* or *drainage area*) is that area of ground that intercepts the precipitation that forms into runoff toward a point of concentration. Since along with the legitimate water runoff, sediments could also be carried, the catchment area must be chosen carefully if a reservoir is to be constructed downstream. Areas that yield too much sediment should be avoided, as this will shorten the life of the reservoir. A catchment area must also be of satisfactory sanitary quality. The area may be patrolled to ensure that the water quality does not become unsafe.

Where there are no natural lakes available, a reservoir may be constructed. The siting of the reservoir should meet some criteria, among which could be the following:

1. The surface topography should create a high ratio of reservoir to dam volume. This means, for example, that a surface topography converging into a gorge is desirable. Construction of a dam across the gorge would entail a smaller volume of dam materials, thus saving the community while providing a reservoir of relatively larger storage volume for water.

2. The subsurface geology at the dam site should provide a safe foundation for the dam and spillway. This is a very important consideration as thousands of lives have been lost due to dam failures.

3. The catchment area for the reservoir should be sparsely populated. When the dam is filled up, large areas will be inundated: this means an entire community, with houses, churches, playgrounds, schools, and cemeteries. Therefore, it is better for the water utility to deal only with a sparsely populated area, if it is available.

4. If possible, the reservoir should be located at a higher elevation so that water can be transmitted to the water treatment plant by gravity. This saves having to pump the water.

Raw surface water supplies can also come from large rivers and lakes in lowland areas, which, with additional impounding, provide an abundant supply of water. These sources are normally under multijurisdictional control (i.e., the management of the catchment area is a concern not only of a single community but of several). It may even transcend beyond more than one state or even country. Such is the case, for example, of the Nile River, the longest river in the world, which traverses 4000 miles through central Africa to the Mediterranean Sea.

As mentioned, the other source of water is the groundwater. Groundwaters are drawn (1) from the pores of alluvial (waterborne), glacial, or aeolian (windblown) deposits of granular materials such as sand, gravel, and sandstone; (2) from solution passages, cleavage planes, caverns, fissures, or fractures of rocks including limestone, slate, shale, and igneous rocks; and (3) from combinations of (1) and (2). *Limestone* is an impure calcium carbonate. *Sandstone* is sand grains cemented together and may be very pervious. *Shale* is a result of the consolidation of clays and is generally impervious acting as aquiclude. *Slate* is a further transformation of shales and is also relatively impervious.

Transmission of Water

The word *transmission* is the term used for the conveyance of raw water from the collection source to the water treatment plant and for the conveyance of the treated water from the treatment plant to the distribution mains. Transmission conduits before the treatment plant may be under pressure or under a free-flow condition. Transmission conduits after the treatment plant for the treated water is always under pressure. Examples of free-flow conditions are the flows in canals, flumes, grade aqueducts, and grade tunnels. Examples of pressure-flow conditions are the flows in pressure aqueducts, pressure tunnels, pipelines and force mains, and depressed pipes or inverted siphons. *Canals* are open channels formed by a balanced cut-and-fill on the ground surface. *Flumes* are also open channels formed of wood, masonry, or metal that are supported on or above the ground. *Aqua* means water and *duct* means conduit; therefore, *aqueduct* means a conduit used to convey water. The term *aqueduct* refers only to large conduits, however. Tunnels are openings constructed through rocks or other formations. When the pressure of the flowing water inside the aqueduct or tunnel is at atmospheric, the conduit is called a *grade aqueduct* or *grade tunnel*; when the pressure of the flowing water is above atmospheric, the conduit is called a *pressure aqueduct* or *pressure tunnel*. A *force main* is that part of a main where the water is pumped (forced) to a higher elevation. An *inverted siphon* is a pressure conduit laid underground to convey water through an obstruction such as crossing a river or conveying water under a road.

Distribution System

Treated waters are distributed to the city by gravity, by pumps alone, by pumps and reservoirs storage, or by a combination of gravity, pumping, and storage. The most desirable method is gravity distribution, and the least desirable method is pumped distribution alone.

All pipes in the distribution system conveying water that ultimately leads to consumers' premises are called *mains.* Starting from the house or building, water is supplied here through the *house, building*, or *service pipe*. This service pipe is connected to a small main along the street. The small mains are the actual distributors of water and are called *service, distributor*, or *street mains*. The service mains are connected to a second, larger system of mains that serves larger areas and at spacings of two to four blocks. These second and larger mains are called *secondary feeders*. The word *feeder* is exactly what it is; it is a feeder that "feeds" water to the service main. The secondary feeders are, again, connected to the third and largest mains in the system. These largest mains are called *primary feeders*. Primary feeders are normally spaced 1 to 2 km apart; hence they can circumscribe an entire community. Finally, the primary feeders are connected to the main, which is connected, in turn, to the transmission system conduit. This last main is the *main supply* or *trunk feeder*. Since the primary feeder resembles the artery that carries blood from the heart to all parts of the human body, they are also called *arterial mains*. Arterial mains form the basic structure of the system and carry flows from trunk feeders and pumps to distribution reservoirs and to various districts of the city. The secondary feeders, on the other hand, aside from supplying the service mains, carry the large volumes of water needed for firefighting purposes. The service or distributor mains also resemble the capillaries that ultimately supply blood to the tissues. They are also, therefore, called *capillary mains*, to distinguish them from arterial mains. The smallest capillary main should not be less than 150 mm. If dead ends are to be allowed, the smallest capillary main should not be less than 200 mm.

All pipes in the system should form a grid that prevents the establishment of dead ends. Dead ends cause sediments to build up, since the pipe at these places is not subjected to a continuous flow of water. When the system is disinfected after repairs, disinfectants cannot go into dead ends, since there is no continuous flow there. On the other hand, water flows continuously in the grid system form of distribution. Hence sediments cannot build up and the system can be easily disinfected after repairs.

Figure 4-1 shows a typical residential service connection (a) and distribution system (b). Notice that the water in the grid system of pipes can flow continuously in a loop. In the grid distribution system (3 of b), the biggest conduit introduced at the middle of the system and to the left of it is the transmission conduit. This then joins with the trunk feeder. The three conduits perpendicular to the trunk feeder running up and down the figure are the primary feeders. The primary feeders form into loops at their extremities at the uppermost and bottommost parts of the figure. Bigger rectangles appear at the bottom and right-hand side of the grid. These represent the secondary feeders which, as shown, connect to the primaries. At the top of the figure are smaller rectangles, representing the street mains. In 4 of b, the mains are at dead ends. This type of a distribution system should be avoided.

The final components making up the distribution system are the valves necessary to control the flow, the fire hydrants, the distribution reservoirs and standpipes needed to equalize supply and pressures in the system, and the meters to measure the flows that go to the various consumers. A *standpipe* is a vertical, cylindrical pipe of height considerably larger than its diameter designed to hold water and to maintain the needed pressure. When the height is equal to or less or only a little bit larger than the diameter, the cylinder is simply a *tank—a storage tank.*

Water Quantity Requirements

In the design of distribution systems and other parts of a waterworks system, there are three flow terms that an environmental engineer needs to know: average flow, maximum daily flow, and maximum hourly flow. The *average flow* is the average daily flow, computed over a year as follows:

$$x = \frac{y}{365(p)} \tag{4-1}$$

where x is the consumption in units of volume per capita per day, y the total cubic units of water delivered to the distribution system, and p the midyear population served by the distribution system.

From this definition it is evident that there would be a number of values, depending on the number of years that the averages are computed. For purposes of design, the engineer must decide which value to use. The highest value may not be used arbitrarily since this may result in overdesign; on the other hand, the lowest value may also not be used for a similar but reverse reasoning. This deduction implies the use of probability distribution analysis to pick the proper average flow value. Depending on the source of information, average values may be very different as the following table of average

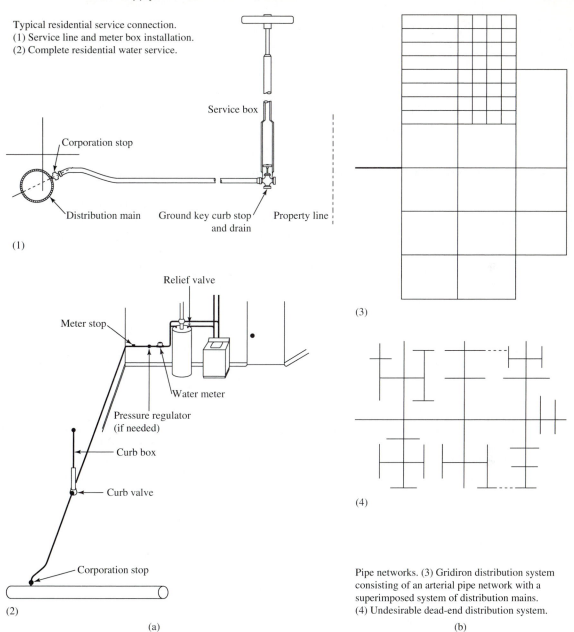

Typical residential service connection.
(1) Service line and meter box installation.
(2) Complete residential water service.

Service box

Corporation stop

Distribution main   Ground key curb stop   Property line
                          and drain

(1)

Relief valve

Meter stop

Water meter

Pressure regulator
(if needed)

Curb box

Curb valve

Corporation stop

(2)

(a)

(3)

(4)

Pipe networks. (3) Gridiron distribution system
consisting of an arterial pipe network with a
superimposed system of distribution mains.
(4) Undesirable dead-end distribution system.

(b)

Figure 4-1   (a) Typical residential service connection. (b) A grid distribution system.
(M. J. Hammer (1986); *Water and Wastewater Technology*. Reprinted by permission of
John Wiley & Sons, New York, New York.)

usages (in gallons per capita per day) indicates:

User	Data from source A, gal per cap per day	Data from source B, gal per cap per day
Domestic	35	64
Commercial and industrial	30	65
Public and losses	35	20

The table shows that you are likely to obtain widely differing data from different sources and therefore have to gather your own.

The design of a distribution system must satisfy two requirements: (1) the water demands of regular domestic, industrial, and commercial consumers, and (2) the water demands of the other principal consumer—water for firefighting. Ideally, to meet these requirements, the system must be designed to satisfy all peak instantaneous consumer demands. In practice, however, for two reasons, the satisfaction of instantaneous demand is not considered: (1) the probability that the peak demands of *all* normal domestic, industrial, and commercial consumers and firefighting requirements will happen coincidentally is low; and (2) the system design is sufficiently far reaching in that as long as the design period is not exceeded, the system has excess capacity. The design of the distribution system is therefore based on either one of the following: maximum day plus firefighting requirements, or maximum hour if this exceeds the former combination of maximum day and fire flow. To get the maximum day and maximum hour, the *proper* average daily flow is multiplied by the ratios of the maximum day and maximum hour to the average day, respectively.

The question of what maximum ratios to use is not easy to answer. There are, literally, hundreds of maxima in the record. Philosophically, there should be only one maximum. However, this maximum may be so high that its use will result in overdesign. Again, as in the case of the choice of the proper average, this deduction implies the use of probability distribution analysis.

The record may be scanned for the occurrence of the daily maximum hourly and divided by the proper daily average. The results obtained are then arranged serially in decreasing order and frequency distribution calculated. The probability obtained from this calculation is a daily probability, as distinguished from the probabilities of occurrence of storms that are based on years and hence are yearly probabilities. A weekly, monthly, or yearly probability of the maximum hourly may also be calculated, but the daily probability is more accurate. The maximum hourly ratio that is equaled or exceeded, corresponding to the desired probability of occurrence, is then interpolated from the calculated probability distribution. The same procedure is followed for the ratio of the maximum day to the average day. Weekly or yearly probabilities may also be calculated.

Taking into account the probabilistic nature of the maximum hourly and the maximum daily ratios, the following definitions are made: The *maximum hourly flow* is the highest of the hourly flows of the appropriate probability of occurrence; the *maximum daily flow* is the highest of the daily flows of the appropriate probability of occurrence.

In addition, the *average daily flow* is the yearly daily average flow of the appropriate probability of occurrence.

Example 4-1

As in the probability distribution analysis of precipitation and runoff, a sufficient number of years of record should be available to have accurate results. But in this example, for illustration purposes, only 11 figures (not necessarily years) will be used. A review of the water consumption of a small American town revealed the statistics in the table below (the dividing line between a small and a large community is taken as 50,000 people). (a) What are the ratios of the maximum day and maximum hour to the average day if the 10% probability of occurrence is the criterion for choosing the ratios? (b) If the average daily flow is 0.49 m³/capita-day, what are the maximum hourly and daily flows, respectively?

Item	Ratio to average	
	Max. day	Max. hour
1	1.48	2.25
2	1.53	2.30
3	1.50	2.31
4	1.44	2.21
5	1.51	2.35
6	1.49	2.27
7	1.42	2.15
8	1.53	2.39
9	1.54	2.36
10	1.52	2.28
11	1.42	2.21

Solution Arranging the data serially from highest to lowest and computing the probabilities give us the following table:

(a)

Serial number	Ratio to average		$\dfrac{m}{N+1}$
	Max. day	Max. hour	
1	1.54	2.39	0.08
2	1.53	2.36	0.17
3	1.53	2.35	0.25
4	1.52	2.31	0.33
5	1.51	2.30	0.42
6	1.50	2.28	0.50
7	1.49	2.27	0.58
8	1.48	2.25	0.67
9	1.44	2.21	0.75
10	1.42	2.21	0.83
11	1.42	2.15	0.92

Interpolating, we find

1	1.54	2.39	0.08
2	x	y	0.10
3	1.53	2.35	0.25

$$\frac{x - 1.54}{1.53 - 1.54} = \frac{0.10 - 0.08}{0.25 - 0.08} \qquad x = 1.54 \quad \textbf{Answer}$$

$$\frac{y - 2.39}{2.35 - 2.39} = \frac{0.10 - 0.08}{0.25 - 0.08} \qquad y = 2.39 \quad \textbf{Answer}$$

(b) Maximum hourly flow $= 2.39(0.49) = 1.17$ m^3/capita-day   **Answer**

Maximum daily flow $= 1.54(0.49) = 0.75$ m^3/capita-day   **Answer**

For expansion to an existing system, the records that already exist may be analyzed to obtain the design values. For an entirely new system, the records of a similar community nearby may be utilized to obtain the design values. Of course, in selecting the final design value, there are other factors to be considered. For example, the practice of metering consumption encourages the consumer not to waste water. Therefore, if the record analyzed contains a period when a meter was used and one when it was not, the conclusion drawn from statistical analysis must take this fact into consideration. The record might also contain years when use of water was heavily curtailed because of drought and years when use was unrestricted. This condition tends to make the data inhomogeneous. Considerable engineering judgment must therefore be exercised to arrive at the final value to use.

Concurrent with estimation of the various maxima is the forecast of the population values to be used in design. A number of methods that should be reviewed exist in the literature. Water intakes, wells, treatment plants, pumping, and transmission lines are normally designed using the maximum daily flow with hourly variations handled by storage. For emergency purposes, standby units should also be installed.

Water Pressure Requirements

The distribution system should satisfy pressure requirements. First, the pressure must enable the water to come out of the faucet, showerheads, and similar outlets at desirable speeds and flow rates. Second, as residential houses are normally constructed one to three stories high, the pressure must also enable the water from fire hoses to reach the roofs of three-story residential homes during conflagrations. Third, the pressure must enable the water to reach the upper stories of tall buildings in the downtown areas of cities, and have sufficient residual pressure at faucets and similar outlets and for firefighting purposes. For residential areas, to meet the foregoing requirements, the recommended residual gage pressures in the street mains are at least 278 kilopascal gage (kPag) [40 pounds per square inch gage (psig)]. Pressure in excess of

690 kPag (100 psig) is undesirable. For high-value districts, the recommended pressures normally range from 417 to 520 kPag (60 to 75 psig) in the streets. By using this pressure range, however, water reaches only several stories high. To provide water to the upper stories, booster pumps are provided to store water in rooftop tanks or water towers.

The distribution system is not designed for higher pressures for two reasons:

1. The rate of discharge Q is directly proportional to the square root of the head loss (or pressure drop). Increasing pressures would mean excessive pressure drops. Hence increasing pressure will increase leakage in the system.

2. High pressures cause excessive stresses in the plumbing systems, which can result in pipes bursting. This will require installing pressure-reducing valves. Pipes and fittings used for plumbing systems are designed for a maximum pressure of 1030 kPag (48 psig).

Engine pumpers can deliver high rates of flow (1500 gpm) at adequate pressures at the nozzle to the scene of fire. Therefore, it is not necessary to build the pressures to a much higher level at the fire hydrants to combat fire. However, if the water is overdrawn from the mains, a vacuum could be created inside the pipes, sucking in pollution from outside. Theoretically, a gage pressure of zero inside the pipe will not allow the pollutants to be sucked in; however, as a factor of safety, the pressures inside the mains may be only allowed to drop to values of 69 to 139 kPag (10 to 20 psig) during the time of fighting fires.

Example 4-2

Hydrants are normally spaced to cover a radius of 60 m. Assuming a rubber-lined hose of $2\frac{1}{2}$-in. diameter is used, calculate how high the water will rise if the hose is aimed at an angle of 60° with the horizontal and (a) the hose is connected to a hydrant in a high-value district; (b) the hose is connected to a hydrant in a residential area. Assume a *standard fire stream* of 250 gpm at a temperature of 20°C.

Solution   (a) Cross-sectional area $A = \pi(2.5/12)^2/4 = 0.034$ ft²; assume a hose length, $\ell = 60$ m $= 196.86$ ft. As 250 gpm $= 0.56$ cfs, cross-section velocity $V = 0.56/0.034 = 16.47$ fps inside hose. From Appendix 18,

$$\text{density } \rho = 998 \text{ kg/m}^3 = \frac{998(2.2)}{32.17(3.281^3)} = 1.93 \, \frac{\text{slug}}{\text{ft}^3}$$

$$\text{viscosity } \mu = 10\left(10^{-4}\right) \text{ kg/m-s} = 10(10^{-4})\left[\frac{2.2}{32.17(3.281)}\right] = 2.08(10^{-5})\frac{\text{slug}}{\text{ft-s}}$$

$$\text{specific weight } \gamma = 998(9.81) \, \frac{N}{\text{m}^3} = 62.09 \, \frac{\text{lb}_f}{\text{ft}^3}$$

$$\text{Reynolds number, Re} = \frac{(2.5/12)(16.47)(1.93)}{2.08(10^{-5})} = 3.18(10^5)$$

Using the Moody diagram (Appendix 17), assuming a smooth hose, $f = 0.0145$. Therefore,

$$\text{head loss } h_f = f\frac{\ell V^2}{D \cdot 2g} = 0.0145\left(\frac{196.86}{2.5/12}\right)\frac{16.47^2}{2(32.2)} = 57.71 \text{ ft}$$

neglecting entrance loss from hydrant to hose. D is the diameter of the hose and g is the gravitational constant. If entrance loss is not to be neglected, $h_f = h + K(V^2/2g)$. Applying the energy equation from the base of the hydrant to the tip of the nozzle assuming a hydrant pressure of 75 psig $=$ 10,800 psfg (pounds per square foot gage),

$$\frac{V_1^2}{2g} + z_1 + \frac{P_1}{\gamma} - h_f = \frac{V_2^2}{2g} + z_2 + \frac{P_2}{\gamma}$$

where P_1 is the pressure at point 1, P_2 the pressure at point 2, z_1 the elevation of point 1 from datum, and z_2 the elevation of point 2 from datum. Assume that $z_1 = z_2$; and $P_2 = 0$.

$$\frac{16.47^2}{2g} + \frac{10,800}{62.09} - 57.71 = \frac{V_2^2}{2g}$$

$$V_2 = 88.07 \text{ fps}$$

From kinematics,

$$V^2 = V_0^2 + 2gs \Rightarrow 0 = (88.07\sin 60°)^2 + 2(-32.2)s$$

$$s = 90.35 \text{ ft} \quad \textbf{Answer}$$

(b) Assume that $P = 40$ psig $= 5760$ psfg and $P = 20$ psig $= 2880$ psfg.

$$\frac{16.47^2}{2g} + \frac{5760}{62.09} - 57.71 = \frac{V_2^2}{2g} \qquad \frac{16.47^2}{2g} + \frac{2880}{62.09} - 57.71 = \frac{V_2^2}{2g}$$

$$V_2 = 50.28 \text{ fps} \qquad\qquad V_2 = \text{negative}$$

$$V^2 = V_0^2 + 2gs \Rightarrow 0 = (50.28\sin 60°)^2 + 2(-32.2)s$$

$$s = 29.44 \text{ for 40 psig at hydrant} \quad \textbf{Answer}$$

Distribution Reservoirs

As soon as raw water has been treated, it is ready for distribution to the community. To equalize the supply and demand for treated water, to provide more uniform pressure throughout the distribution system, and to provide for emergency supply such as fire-fighting, distribution reservoirs are normally constructed. Figure 4-2 shows a picture of a distribution reservoir at Bonita Springs, Florida, and Figure 4-3 shows two possible locations around a distribution system.

C in Figure 4-3 represents the community. As shown, the distribution reservoirs can be placed either between the pumping station and the community or beyond the

Figure 4-2 Distribution reservoir at Bonita Springs, Florida.

community. The uppermost part of the figure shows that without the reservoir during a maximum day with fire supply, it needs 329 ft of energy to pump the water through the distribution system, whereas in the middle and bottom parts, with the installation of the distribution reservoir, this energy requirement has been reduced to 227 to 250 ft. Thus distribution reservoirs save energy. Also, by comparing the slopes of the energy lines, the system with the reservoir constructed has a more uniform pressure.

Distribution reservoirs are sized using the material balance equation. As discussed later, the demand for the maximum day is the one used for the calculation. An example of reservoir sizing is illustrated in the next example.

Example 4-3

The estimated hourly requirement for the maximum day in a small city is tabulated below. Fire flow has been determined to be 65 L/s for a conflagration of 2 h. If pumping is to start at hour 0800 and end at hour 1600, calculate the storage requirement.

Solution Length of pumping $1600 - 0800 = 8.0$ h

Hour ending	Demand (m^3)	Water pumped (m^3)	Required storage (m^3)
0100	160	0	160
0200	129	0	129
0300	146	0	146
0400	138	0	138
0500	159	0	159
0600	196	0	196
0700	274	0	274
0800	390	0	390
0900	480	961.25	—
1000	500	961.25	—
1100	473	961.25	—
1200	451	961.25	—
1300	444	961.25	—
1400	447	961.25	—
1500	434	961.25	—
1600	405	961.25	—
1700	417	0	417
1800	421	0	421
1900	406	0	406
2000	343	0	343
2100	270	0	270
2200	223	0	223
2300	199	0	199
2400	185	0	185
$\sum$	7690	7690	4056

Therefore,

$$\text{storage requirement} = 4056 \text{ m}^3 + \text{fire-flow requirement}$$

$$= 4056 + \frac{65}{1000}(60)(60)(2) = 4524 \text{ m}^3 \quad \textbf{Answer}$$

Fire-Flow Requirements

Although the actual yearly consumption of water for firefighting is very small, the rate of use during a conflagration is very high and can even govern the sizing of distribution system components. In rating cities and municipalities for insurance purposes, the Insurance Service Office (ISO) considers the following components of a water supply for evaluation of adequacy and reliability: source of supply, pumping capacity, power supply, mains, spacing of valves, and location of fire hydrants. These are all the essential components of the firefighting facilities of a municipality. The ISO[1] calculates fire-flow

---

[1]T. J. McGhee (1991). *Water Supply and Sewerage*. McGraw-Hill, New York. p. 15.

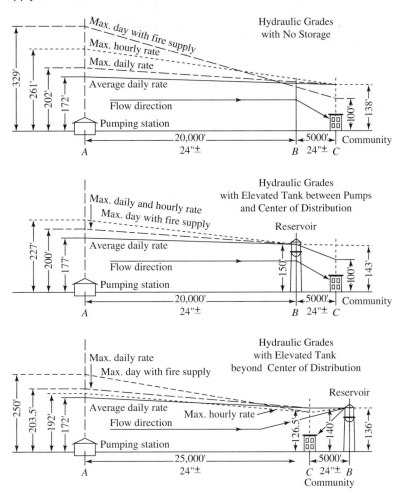

Figure 4-3 Two possible locations of distribution reservoirs in a distribution system.

requirement F in m³/s using the following formula:

$$F = 3.7(10^{-3})C(A^{0.5}) \tag{4-2}$$

where C is a dimensionless coefficient related to the type of construction: 1.5 for wood-frame construction, 1.0 for ordinary construction, 0.8 for noncombustible construction, and 0.6 for fire-resistive construction. A is the total floor area, excluding the basement, in m². For fire-resistive constructions, the six successive floor areas are used if the vertical openings are not protected, while if the openings are protected, only the three successive floors are used.

Regardless of calculation results, the ISO recommends that the maximum values of F cannot exceed the following: 0.54 m³/s for wood-frame or ordinary constructions and 0.38 m³/s for noncombustible, fire-resistive, or one-story building of any type of construction. Also, regardless of calculation results, the minimum value of F should not be less than 0.32 m³/s. For high-fire-hazard loadings such as chemical works, explosives, oil refineries, paint shops, and solvent extraction, the value of F may be increased up to 25%. However, the maximum value after the adjustment should not exceed 0.76 m³/s. The ISO also recommends that the required fire flow be available for the expected time of duration of a fire as shown in Table 4-1.[2]

TABLE 4-1   FIRE-FLOW DURATIONS

Fire flow (m³/s)	Duration (h)	Fire flow (m³/s)	Duration (h)	Fire flow (m³/s)	Duration (h)
≥ 0.61	10	0.54–0.61	9	0.48–0.54	8
0.42–0.48	7	0.36–0.42	6	0.30–0.36	5
0.24–0.30	4	0.16–0.24	3	≤ 0.16	2

Synthesis of the Maximum Day

Just as the progress of a peak storm over its duration was synthesized in Chapter 3, the progress of the use of water over a maximum day may also be synthesized. As mentioned before, no one maximum day may be chosen arbitrarily to represent the design maximum, but rather, the maximum day should be derived statistically. The statistical derivation is similar to that used for the synthesis of the maximum storm; it uses the probability distribution analysis. To illustrate, the data for water consumption used in Example 4-3 may be used.

At the end of hour 0100, the consumption was 160 m³. This means that the data were analyzed for the interval 2400–0100. Throughout the duration of the record, the largest consumption values for this interval of time in each day, week, or year, depending on the type of probability desired, were chosen and arranged in descending order and the probability distribution analysis performed as was done for the synthesis of the peak storm. Then, depending on the criterion of probability chosen (5% chance, 10% chance of being equaled or exceeded, and so on), the consumption for the interval 2400–0100 was interpolated from the calculated distribution. In the example, the consumption for the interval was 160 m³. The same calculation procedure was used for the interval 0100–0200, the interval 0200–0300, the interval 0300–0400, and so on. According to the table, the maximum consumption for the interval of 0100–0200 was 129 m³. For the interval of 0200–0300, it was 146 m³, and so on. Table 4-1 therefore represents the synthetic consumption for the maximum day using a certain probability criterion, such as 10%.

[2]M. J. Hammer (1986). *Water and Wastewater Technology.* John Wiley & Sons, New York, New York. p. 185.

Example 4-4

If the city in Example 4-3 is assumed to be built of buildings of ordinary construction as a general characterization with a total combined floor area of 2000 m^2 per story, compute the storage requirement. Assume that the average number of stories per building is three.

Solution

$$F = 3.7(10^{-3})C(A^{0.5})$$

$$= 3.7(10^{-3})(1.0)[2000(3)]^{0.5} = 0.29 \text{ m}^3/\text{s}$$

From Table 4-1, the expected duration is 4 h. From Example 4-3, the required storage for normal demand is 4056 m^3. Therefore,

$$\text{total storage requirement} = 4056 + 0.29(60)(60)(4) = 8232 \text{ m}^3 \quad \textbf{Answer}$$

Hydraulics of the Distribution System

The purpose of a hydraulic analysis of a distribution system is to determine the pressure contour and flow patterns of the system. Information on the pressure contour and flow patterns will enable the environmental engineer to determine if the system can meet the demands for which it was designed. Of the methods available, the most practical that lends itself easily to computer programming is the *Hardy Cross method*. In this method the distribution system is divided into interconnected loops. To make it simpler to visualize, Figure 4-4 shows only three loops out of a multitude of loops in a distribution system. To derive the method, consider loop *IHBAI* or *IABHI*. The principle used in the derivation is that given any two points A and I, the head loss in going from A to I, h_{fAI} through the direct path A to I is equal to the head loss in going from A to I, h_{fABHI}, through the indirect path A to B to H to I. This is true because if the head loss h_{fAI} were less than the head loss h_{fABHI}, no flow would pass through AB, BH, and HI. The flow would simply bypass the loop and go directly from A to I through the direct path AI. Hence, the sum of the clockwise head losses $\sum h_{fc}$ between any two points in a loop is equal to the sum of the counterclockwise head losses $\sum h_{fcc}$. In symbols,

$$\sum h_{fc} = \sum h_{fcc} \tag{4-3}$$

From fluid mechanics,

$$h_f = f\frac{\ell V^2}{D \cdot 2g} = f\frac{\ell}{2gDA^2}Q^2 = kQ^2 \tag{4-4}$$

where h_f is the head loss, f the Fanning friction factor, V the velocity, D the diameter of the pipe, Q the rate of flow, A the cross-sectional area of flow, and k is a proportionality constant. Therefore, substituting in equation (4-3) gives us

$$\sum kQ_c^2 = \sum kQ_{cc}^2 \tag{4-5}$$

where Q_c is the clockwise flow and Q_{cc} is the counterclockwise flow.

In Figure 4-4 there is an inflow at A and three outflows at B, H, and I. According to equation (4-5), these flows must apportion themselves so that the equation will be satisfied. In the derivation of the Hardy Cross method, this apportioning is done by

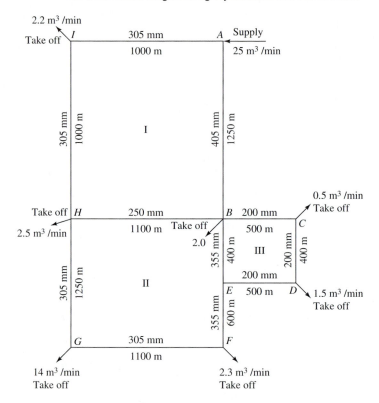

Figure 4-4   Derivation of the Hardy Cross method.

trial-and-error. Assume that in the course of the trial-and-error procedure, the clockwise Q was assumed larger by ΔQ. Accordingly, the counterclockwise Q is smaller by ΔQ. To restore the balance, the flows are corrected as

$$\sum k(Q_c - \Delta Q)^2 = \sum k(Q_{cc} + \Delta Q)^2 \qquad (4\text{-}6)$$

A very important point must now be noted. The assumption has been made that the clockwise flow is larger than the counterclockwise flow. This assumption sets the convention that the correction or error ΔQ, whether positive or negative, must be sub-tracted from Q_c and added to Q_{cc}. This is indicated in equation (4-6). If the assumption of the magnitudes of the flows had been the other way around, these statements would have to be reversed, accordingly, and equation (4-6) revised to interchange the plus and minus signs before ΔQ to reflect the other assumption.

Expanding equation (4-6) by the binomial theorem, neglecting the second power of ΔQ after the expansion, and solving for ΔQ yields

$$\Delta Q = \frac{\sum k Q_c^2 - \sum k Q_{cc}^2}{\sum 2k Q_c + \sum 2k Q_{cc}} = \frac{\sum k Q_c^2 + \sum(-k Q_{cc}^2)}{\sum 2k Q_c + \sum 2k Q_{cc}} = \frac{\sum k Q^2}{2 \sum k Q} \qquad (4\text{-}7)$$

In the numerator, by putting the minus sign inside the summation sign, a second convention has just been set: The product of the square of the counterclockwise flow Q_{cc} and k must always be negative; the product of the square of the clockwise flow Q_c and k must always be positive. Also, the denominator of equation (4-7) sets the third convention: The respective products of Q_c and Q_{cc} with k must always be positive.

Example 4-5

For the given source and load in Figure 4-4: (a) how will the flow be distributed in the network, and (b) what will be the pressure at load points B, H, and I if the pressure at the source is 60 psig? Assume that all pipes are horizontal and $f = 0.012$.

Solution   Name a pipe by the letters at its endpoints, such as pipe AI or IA in Figure 4-4. Also, a suffix c or cc will be affixed to the name, depending on whether the direction of flow in the pipe is clockwise or counterclockwise, respectively; that is, if the flow is clockwise, AIc or IAc, or if the flow is counterclockwise, $AIcc$ or $IAcc$.

$$k = f\frac{\ell}{2gDA^2} = 0.012\left[\frac{\ell}{2gD(\pi D^2/4)^2}\right] = 9.92(10^{-4})\frac{\ell}{D^5}$$

(a) First pass:

Loop I (starting point less $= A$, ending point $= H$):

(1)	(2)	(3)	(4)	(5)	(6)	(7)	(8)
Pipe	D (m)	ℓ (m)	k	Q (m³/ min)	Q (m³/s)	kQ^2	kQ
ABc	0.41	1250	107.0	13	0.22	5.18	23.54
BHc	0.25	1100	1117.4	3	0.05	2.79	55.87
$AIcc$	0.31	1000	346.5	12	0.2	−13.86	69.3
$IHcc$	0.31	1000	346.5	9.8	0.16	−8.87	55.44

$$\Delta Q = \frac{\sum kQ^2}{2\sum kQ} = \frac{-14.76}{2(204.15)} = -0.036 \qquad \sum \qquad -14.76 \qquad 204.15$$

Col. (1): With respect to starting point A, AB is clockwise, BH is clockwise, AI is counterclockwise, and IH is counterclockwise.

Col. (2): The diameters are given.

Col. (3): The lengths of pipes are given.

Col. (4): $k = f\dfrac{\ell}{2gDA^2} = 0.012\left[\dfrac{\ell}{2gD(\pi D^2/4)^2}\right] = 9.92(10^{-4})\dfrac{\ell}{D^5}$.

Col. (5): Given or assumed.

Col. (6): $\dfrac{\text{Col. (5)}}{60}$.

Col. (7): From $\dfrac{\sum kQ^2}{2\sum kQ}$, kQ^2 must be calculated for each pipe length. Note that AI and IH have negative signs.

Col. (8): From $\dfrac{\sum kQ^2}{2\sum kQ}$, kQ must be calculated for each pipe length. Note that all entries in the column are positive.

Loop II (starting point = B, ending point = G):

(9)	(10)	(11)	(12)	(13)	(14)	(15)	(16)
Pipe	D (m)	ℓ (m)	k	Q (m³/min)	Q (m³/s)	kQ^2	kQ
BEc	0.36	400	65.62	6.5	0.11	0.77	7.22
EFc	0.36	600	98.44	6.0	0.10	0.98	9.84
FGc	0.31	1100	381.15	3.7	0.06	1.45	22.87
$BHcc$	0.25	1100	1117.39	3	0.05	−2.79	55.87
$HGcc$	0.31	1250	433.12	10.3	0.17	−12.76	73.63

$$\Delta Q = \frac{\sum kQ^2}{2\sum kQ} = \frac{-12.35}{2(169.43)} = -0.036 \qquad \sum \qquad -12.35 \qquad 169.43$$

Note: The same explanations hold as in loop I.

Loop III (starting point = B, ending point = D):

(17)	(18)	(19)	(20)	(21)	(22)	(23)	(24)
Pipe	D (m)	ℓ (m)	k	Q (m³/ min)	Q (m³s)	kQ^2	kQ
BCc	0.20	500	1550	1.5	0.025	0.97	38.75
CDc	0.20	400	1240	1.0	0.02	0.50	24.8
$BEcc$	0.36	400	65.62	6.5	0.11	−0.79	7.22
$EDcc$	0.20	500	1550	0.5	0.008	−0.11	12.4

$$\Delta Q = \frac{\sum kQ^2}{2\sum kQ} = \frac{0.57}{2(83.17)} = 0.0034 \qquad \sum \qquad 0.57 \qquad 83.17$$

First correction:

| (25) | (26) | | | (27) | (28) |
| | Correction from ΔQ | | | | |
Pipe	Loop I	Loop II	Loop III	Q previous	Q corrected
AB	−0.036c	0	0	0.22	0.26+
BH	−0.036c	−0.036cc	0	0.05	0.05+
AI	−0.036cc	0	0	0.20	0.16+
IH	−0.036cc	0	0	0.16	0.12+
BE	0	−0.036c	0.0034cc	0.11	0.15+
EF	0	−0.036c	0	0.10	0.14+
FG	0	−0.036c	0	0.06	0.10+
HG	0	−0.036cc	0	0.17	0.13+
BC	0	0	0.0034c	0.025	0.022+
CD	0	0	0.0034c	0.02	0.020+
ED	0	0	0.0034cc	0.008	0.011+

Notes: In col. (26), suffix c denotes clockwise and suffix cc denotes counterclockwise for the direction of flow in the pipe where correction is to be made. In col. (28), "+" means that the direction of flow in the pipe remains the same as it was before correction, and "−" means that direction is now reversed after correction.

To get correction for BE: From convention, ΔQ (= −0.036) for loop II must be subtracted from the previous Q (= 0.11) since the direction of flow in BE is clockwise in loop II; and ΔQ (= 0.0034) for loop III must be added to the previous Q (= 0.11) since BE is counterclockwise in loop III. Therefore,

$$\text{corrected } Q \text{ for } BE = 0.11 - (-0.036) + (0.0034) = 0.1494 \approx 0.15$$

(Also, the direction of flow remains the same as it was before corrections were applied.) Corrections for the other conduits are similar.

Second Pass:

Loop I (starting point = A, ending point = H):

| (1) | (2) | (3) | (4) | (5) | (6) | (7) | (8) |
Pipe	D (m)	ℓ (m)	k	Q (m³/ min)	Q (m³/s)	kQ^2	kQ
ABc	0.41	1250	107.0		0.26	7.23	27.82
BHc	0.25	1100	1117.4		0.05	2.79	55.87
$AIcc$	0.31	1000	346.5		0.16	−8.87	55.44
$IHcc$	0.31	1000	346.5		0.12	−4.99	41.58

$$\Delta Q = \frac{\sum kQ^2}{2\sum kQ} = \frac{-3.84}{2(180.71)} = -0.010 \qquad \sum \qquad -3.84 \qquad 180.71$$

Loop II (starting point = B, ending point = G):

(9) Pipe	(10) D (m)	(11) ℓ (m)	(12) k	(13) Q (m³/ min)	(14) Q (m³/s)	(15) kQ^2	(16) kQ
BEc	0.36	400	65.62		0.15	1.48	9.84
EFc	0.36	600	98.44		0.14	1.93	13.78
FGc	0.31	1100	381.15		0.10	3.81	38.12
$BHcc$	0.25	1100	1117.39		0.05	−2.79	55.87
$HGcc$	0.31	1250	433.12		0.13	−7.32	56.31

$$\Delta Q = \frac{\sum kQ^2}{2\sum kQ} = \frac{-2.89}{2(173.92)} = -0.0083 \qquad \sum \qquad -2.89 \qquad 173.92$$

Loop III (starting point = B, ending point = D):

(17) Pipe	(18) D (m)	(19) ℓ (m)	(20) k	(21) Q (m³/ min)	(22) Q (m³/s)	(23) kQ^2	(24) kQ
BCc	0.20	500	1550		0.026	1.048	40.3
CDc	0.20	400	1240		0.022	0.60	27.28
$BEcc$	0.36	400	65.62		0.15	−1.48	9.84
$EDcc$	0.20	500	1550		0.0064	−0.063	9.92

$$\Delta Q = \frac{\sum kQ^2}{2\sum kQ} = \frac{-0.105}{2(87.34)} = -0.00060 \qquad \sum \qquad -0.105 \qquad 87.34$$

Second correction:

(25) Pipe	(26) Correction from ΔQ			(27) Q previous	(28) Q corrected
	Loop I	Loop II	Loop III		
AB	−0.010c	0	0	0.26	0.27+
BH	−0.010c	−0.0083cc	0	0.05	0.052+
AI	−0.010cc	0	0	0.16	0.15+
IH	−0.010cc	0	0	0.12	0.11+
BE	0	−0.0083c	−0.00060cc	0.15	0.158+
EF	0	−0.0083c	0	0.14	0.148+
FG	0	−0.0083c	0	0.10	0.108+
HG	0	−0.0083cc	0	0.13	0.122+
BC	0	0	−0.00060c	0.022	0.023+
CD	0	0	−0.00060c	0.020	0.021+
ED	0	0	−0.00060c	0.011	0.012+

The summations under cols. (7), (15), and (23) should all be zero to have the correct balance of heads and flows. However, since the procedures for succeeding passes are the same, we will stop at this point. The distribution of flows are, therefore, shown in col. (28) of the second correction.   **Answer**

(b) Since the pipes are given to be horizontal, the only loss of energy is due to the friction, $h_f = \sum k Q^2$. If the pipes were not assumed horizontal, the energy equation must be applied in each segment of the loops. The source is at point A with an inflow of 25 m^3/min. The loads are at points B, H, I, G, F, C, and D.

For pressure at point A:

$$1 \text{ lb}_f = 1 \text{ slug} \left(1\frac{\text{ft}}{\text{s}^2}\right)\left(\frac{32.2 \text{ lb}_m}{\text{slug}}\right)\left(\frac{454 \text{ g}}{\text{lb}_m}\right)\left(\frac{\text{kg}}{1000 \text{ g}}\right)\left(\frac{1 \text{ m}}{3.281 \text{ ft}}\right) = 4.46 \frac{\text{kg-m}}{\text{s}^2}$$

$$1\frac{\text{lb}_f}{\text{in}^2} = 1 \text{ psi} = \frac{4.46\frac{\text{kg-m}}{\text{s}^2}}{\text{in}^2}\left(\frac{12^2 \text{ in}^2}{\text{ft}^2}\right)\left(\frac{3.281^2 \text{ ft}^2}{\text{m}^2}\right) = 6913.69 \frac{\text{kg}}{\text{m-s}^2}$$

$$1 \text{ N} = 1 \text{ kg}\left(1\frac{\text{m}}{\text{s}^2}\right) = \frac{\text{kg-m}}{\text{s}^2} \qquad \text{thus} \quad \frac{\text{kg}}{\text{m-s}^2} = \frac{\text{N}}{\text{m}^2}$$

Therefore,

$$1 \text{ psi} = 6913.69 \frac{\text{N}}{\text{m}^2} = 6.91 \frac{\text{kPa}}{\text{m}^2}$$

$$P_A = 60(6.91) = 414.60 \text{ kPag}$$

For pressure at point I:

$$h_{AI} = \sum k Q^2 \text{ or } \Delta P_{AI} = \left(\sum k Q^2\right)(\gamma) = \frac{346.5(0.15)^2(998)(9.81)}{1000} = 76.33 \text{ kPa}$$

assuming a temperature of 20°C. Therefore,

$$P_I = 414.60 - 76.33 = 338.27 \text{ kPag} \quad \textbf{Answer}$$

For pressure at point B:

$$\Delta P_{AB} = \frac{107(0.27^2)(998)(9.81)}{1000} = 76.37 \text{ kPa}$$

Therefore,

$$P_B = 414.60 - 76.37 = 338.23 \text{ kPa} \quad \textbf{Answer}$$

Note: The pipes should also be checked for excessive velocities. Normal velocities for design range from 0.90 to 1.50 m/s.

For pressure at point H:

$$\Delta P_{BH} = \frac{346.5(0.11^2)(998)(9.81)}{1000} = 41.05 \text{ kPa}$$

Therefore,

$$P_H = 338.27 - 41.05 = 297.22 \text{ kPa} \quad \textbf{Answer}$$

Design Period

In designing the water supply system, a period of design must be decided. The information below for the period of design of various water supply components may be useful to arrive at this decision.

> Large dams and conduits: 25 to 50 years
>
> Wells, distribution systems, and filter plants: 10 to 25 years
>
> Pipes more than 300 mm (12 in.) in diameter: 20 to 25 years
>
> Secondary mains less than 300 mm (12 in.) in diameter: full development

COLLECTION OF WASTEWATER

Wastewater is the spent water after homes, commercial establishments, industries, public institutions, and other entities have used their waters for various purposes. It is synonymous with *sewage*, although sewage is a more general term that refers to the liquid waste that is conveyed in a sewer, and hence may contain, in addition to wastewater, stormwater, infiltration, and inflow. A *sewer* is a channel or conduit intended for the conveyance of liquid wastes. Wastewater that comes from the sanitary conveniences of homes, institutions, industries, commercial establishments, and other entities is called *sanitary sewage*. Sanitary sewage from homes is called *domestic sewage*. Stormwater: is also called *storm sewage*, since it is actually a contaminated liquid. Infiltration is that portion of precipitation that enters sewers through cracks in the pipe and faulty joints. *Inflow* refers to the precipitation that enters the manhole through holes in its cover and from roof leaders through illegal connections. *Manhole* is an opening that allows a person access to the underground sewer system for maintenance and other purposes.

   The act of gathering all the sewage is called *collection*. The system used to collect the sewage is called the *collection system*. Strictly, a *wastewater collection system* collects wastewater only. In practice, however, it also collects infiltration and inflow; that is, it collects all the sewage. *Sewerage* is a term that refers not only to the collection of wastewater or sewage but also to its treatment and disposal. *Sewerage works* or *sewage works* include all the physical structures required for the collection, treatment, and disposal of sewage.

   In the past, three types of collection system were normally used: sanitary, stormwater, and combined. At present, because of the stricter regulations of the Clean Water Act, the combined system is no longer allowed. As the names imply, the sanitary system conveys only sanitary sewage, and the stormwater system conveys only stormwater. The combined system conveys both sanitary sewage and stormwater in the same conduit. At present, combined systems have already reached their design capacity and are frequently overflowing. These overflows are called *combined sewer overflows* (CSOs). Washington, D.C. and Cumberland and Cambridge in Maryland have combined systems. They all have CSO problems.

Figure 4-5 is a schematic drawing showing the various parts of a collection system. Starting from a building or house, sewage from various sources inside (sinks, bathtubs, etc.) is collected into the *building* or *house sewer,* shown at the bottom of the figure. This sewer is just outside the house and within the property boundary of the owner's premises. The sewage from the house sewer then goes to the *branch* or *lateral sewer.* Lateral sewers are the first element of the public collection system, and they are used to collect sewage from one or more houses or buildings. The laterals then join the

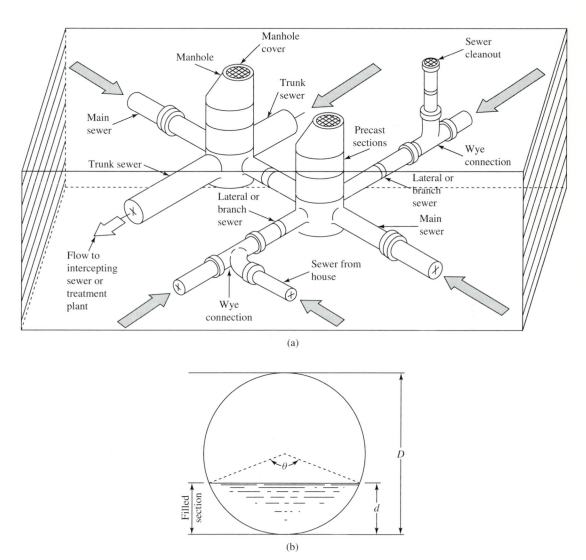

(a)

(b)

Figure 4-5   (a) Definition sketch of wastewater collection system. (b) Geometric elements of circular sewer.

main sewer, which is a sewer that collects all sewage from one or more laterals and conveys it to the *trunk sewer*. The trunk sewer conveys the sewage either to an *intercepting sewer* or directly to a sewage treatment plant through an outfall sewer. The trunk sewer is analogous to a tree trunk. The trunk is the largest part of a tree, so is the trunk sewer in the collection system. The trunk sewer is the largest part of the branching mains, which resemble the branches of a tree. As in the water distribution system, laterals, mains, and trunk sewers are all called mains. Any main smaller than a given main is a submain of the main. As the name says, an intercepting sewer is a sewer that intercepts flows. It is a large conduit that intercepts flows from a number of trunk or main sewers. The sewer that conducts the sewage to a treatment plant is called an *outfall sewer*. An outfall sewer is also the sewer that conducts the treated effluent from a treatment plant or the sewage from the collection system to a point of final discharge.

Quantity of Sewage

To design the wastewater collection system, the quantities to be collected must be known. The amount of wastewater produced is related directly to the amount of water consumed, although not all of the water supply delivered to a community ends up as sewage. The discrepancies may be accounted for by the volume used in sprinkling lawns, car washing, leakage in the distribution system, and other losses. Values of 60 to 130% of the water consumption have been cited in the literature.[3] Many designers assume for design purposes that the value of wastewater produced equals water consumed, including the allowance for infiltration and inflow.

Data abound in the literature on the amount of water usage from which wastewater production can be judged. For example, low-income, single-family homes consume 270 liters per capita per day, while the rich homes consume 380 liters per capita per day. For restaurants, the consumption is 30 liters per customer per day; and for department stores, the consumption is 40 liters per employee per day. If you were the consultant, would you really trust these figures? They may be used as reference, but should not be trusted when designing facilities that cost thousands of dollars.

Since places are, literally, dotted with collection systems, it may be more suitable to measure the flow in the sewer. For an expansion to an existing system, the flow measurement can be done easily, since the system already exists. For brand-new collection systems, the metering may be done in a nearby town or village, where the characteristics of the system and the habit of the people are similar to those of the proposed. The requirement of similarity is very important. To have accurate predictions of sewage flows, it is obvious that the reference town should have characteristics similar to those proposed. This method of obtaining data is much better than getting them from the literature, where the data may have been obtained from faraway and dissimilar places such as Iraq and Kuwait or even Kuala Lumpur.

---

[3]S. R. Qasim (1985). *Wastewater Treatment Plants: Planning, Design, and Operation.* Holt, Rinehart and Winston, New York. p. 19.

The purpose of sewer metering is twofold: to measure the dry-weather flow and to measure the combined infiltration inflow. The information on the dry-weather flow will enable the designer to determine the minimum flow. Sewer systems are designed as open channels. As such, in order to hold in check the deposition of waste matters onto sewer bottoms (called *inverts*), a minimum self-cleaning velocity of 0.60 to 0.75 m/s (2.0 to 2.5 fps) should be provided—and this means providing this velocity at the minimum flow. When added to the dry-weather maximum flow obtained during dry-weather flow measurement, the information on infiltration inflow will determine the ultimate maximum flow expected in the sewer. Again, this ultimate maximum flow is used to size the sewers.

To determine what time of year the metering of both dry-weather and infiltration-inflow (wet weather) surveys are to be done, precipitation records should be consulted. The records will show at what months dry-weather and wet-weather conditions are expected. In Maryland, for example, dry weather is normally June to August. From the results of the dry-weather survey, the ratio of minimum and maximum dry flows to the average can be determined. The flow records in the treatment plant corresponding to the dry-weather months may constitute the data for the dry-weather survey.

The dry-weather flow survey will establish a typical pattern of dry-weather flow. The design pattern may then be derived by probability distribution analysis, similar to the method of synthesizing the maximum day for water use. Or, if accuracy is not much of a concern, the pattern may be derived by simply picking a particular pattern from the several patterns obtained during the survey. Having established the pattern, the typical wet-weather flow pattern is superimposed over the dry-weather flow pattern. Once superimposed, the contribution of the infiltration can be determined. The determination of the wet-weather flow pattern is derived similarly as the dry-weather flow pattern.

Inflow to the sewer system appears as a burst of increased flow to the system. The engineer must therefore be on the watch of the progress of the precipitation record to ascertain when the heavy rain has occurred. The flow monitoring record is then inspected to determine the time of the burst. This occurs later from the time the heavy rain occurs because of time lag in the drainage basin and in the sewers. The flow monitoring pattern that contains the burst is then superimposed over the two previous patterns. From the superimposition, the inflow can be determined.

Above the maximum dry weather flow, the value on infiltration–inflow may then be added to obtain the ultimate maximum flow needed to size the sewers. The ultimate minimum flow may, similarly, be obtained. This flow is also used to size the sewers to ensure that they are self-cleaning at low-flow conditions.

Example 4-6

A representative section of the Town of Middletown collection system was studied for its dry-weather and wet-weather flow characteristics. The study, which lasted for three months, showed the average peak dry-weather flow to be 205 L/s, the average minimum dry-weather flow to be 46 L/s, and the average flow to be 96 L/s. The combined infiltration–inflow was found to be 70 L/s. The sewage inflow records of the Middletown Sewage Treatment Plant were also analyzed for the average flow of the driest months of each year (average of the flows during the successive months of June, July, and August of each year of record) yielding the following probability distribution.

Average dry-weather flow (L/capita-day)	Probability p that flow is equaled or exceeded
310	0.019
307	0.080
304	0.23
300	0.38
⋮	⋮
160	0.84
147	0.99

(a) Find the ratios of the minimum and the maximum flows to the average flow. (b) If the midyear population of Middletown is 58,600, what are the present minimum, maximum, and average flows of the town assuming 10% probability of occurrence as the criterion for determining flows?

Solution

(a)
$$\frac{\text{Minimum}}{\text{Average}} = \frac{46}{96} = 0.48 \quad \textbf{Answer}$$

$$\frac{\text{Maximum}}{\text{Average}} = \frac{205}{96} = 2.14 \quad \textbf{Answer}$$

(b)

307	0.08
x	0.10
304	0.23

$$\frac{x - 307}{304 - 307} = \frac{0.10 - 0.08}{0.23 - 0.08}$$

$$x = 306.6 \text{ L/capita-day}$$

Therefore,

$$\text{average flow at 10\% probability} = 306.6(58,600) = 17,966,760 \text{ L/day} = 0.21 \text{ m}^3/\text{s}$$

$$\text{infiltration–inflow allowance} = \frac{70}{96} = 0.73 \text{ per average flow}$$

$$\text{minimum flow} = (0.48)(0.21) = 0.10 \text{ m}^3/\text{s} \quad \textbf{Answer}$$

$$\text{maximum flow} = (2.14 + 0.73)(0.21) = 0.60 \text{ m}^3/\text{s} \quad \textbf{Answer}$$

Self-Cleaning Hydraulics of Sewers

In the hydraulic design of sewers, one of the most important consideration is assuring that the flow will be self-cleaning at all depths of flow. Sewers normally flow partially full, especially at the laterals. At the shallower depths of flow, velocities tend to slow down, causing debris to settle at the invert. This leads, ultimately, to clogging of the sewers. Hydraulic design techniques must therefore be developed to assure that flows are self-cleaning at all times. Normally, a velocity of 0.60 to 0.75 m/s will be self-cleaning; hence if this velocity can be maintained from full depth at maximum flow to the shallowest depth at the minimum flow, freedom from undue maintenance of the sewers due to premature clogging may be assured. Formulas for self-cleaning properties will now be developed.

The formula used to compute velocities in sewers V and v for full and partial flows, respectively, is normally the Manning equation,

$$V = \frac{1}{N} R^{2/3} S^{1/2} \qquad \text{full flow} \qquad (4\text{-}8)$$

$$v = \frac{1}{n} r^{2/3} s^{1/2} \qquad \text{partial flow} \qquad (4\text{-}9)$$

where N and n are roughness coefficients for full and partial flows, respectively; R and r are hydraulic radii for full and partial flows, respectively; and S and s are energy slope for full and partial flows, respectively.

Both sides of equations (4-8) and (4-9) may each be multiplied by cross-sectional areas A and a, for full and partial flows, respectively, producing the corresponding equations below. Q and q are the full and partially full rates of flow, respectively.

$$Q = \frac{A}{N} R^{2/3} S^{1/2} \qquad \text{full flow} \qquad (4\text{-}10)$$

$$q = \frac{a}{n} r^{2/3} s^{1/2} \qquad \text{partial flow} \qquad (4\text{-}11)$$

Dividing equation (4-9) by equation (4-8) and equation (4-11) by equation (4-10) produce the corresponding equations.

$$\frac{v}{V} = \frac{N}{n} \left(\frac{r}{R}\right)^{2/3} \left(\frac{s}{S}\right)^{1/2} \qquad (4\text{-}12)$$

$$\frac{q}{Q} = \frac{N}{n} \frac{a}{A} \left(\frac{r}{R}\right)^{2/3} \left(\frac{s}{S}\right)^{1/2} \qquad (4\text{-}13)$$

To dislodge the debris at the invert for conditions of all depths of flow, the tractive stress on the debris at partial flow must be equal to the tractive stress that dislodges the same debris at full flow. This stress is equal to the component of the weight of water parallel to the invert and is given by $\gamma V \sin\theta$, (where γ is the specific weight of water, V the volume, and θ the angle of slope of the invert) divided by the area A_s upon which the debris is resting. V/A_s equals R for full flow and equals r for partial flow. For small angles $\sin\theta$ is equal to $\tan\theta$, which, in turn, is the slope of the invert. In addition, for uniform flows, the slope of the invert is equal to the energy slope. Hence

$$\gamma r s = \gamma R S \qquad (4\text{-}14)$$

$$\frac{r}{R} = \frac{S}{s} \qquad (4\text{-}15)$$

Substituting equation (4-15) in equations (4-12) and (4-13) gives

$$\frac{v}{V} = \frac{v_s}{V} = \frac{N}{n}\left(\frac{r}{R}\right)^{1/6} \tag{4-16}$$

$$\frac{q}{Q} = \frac{N}{n}\frac{a}{A}\left(\frac{S}{s}\right)^{1/6} \tag{4-17}$$

v_s is the self-cleaning v at partial flow. If V is a self-cleaning velocity at full flow, then to be self-cleaning at partial flow, v_s must be equal to V; hence $v_s/V = 1$. Solving equation (4-16) for N/n yields

$$\frac{N}{n} = \left(\frac{R}{r}\right)^{1/6} \tag{4-18}$$

Substituting in equation (4-17) produces

$$\frac{q}{Q} = \frac{a}{A}\left(\frac{R}{r}\right)^{1/6}\left(\frac{S}{s}\right)^{1/6} \tag{4-19}$$

In the hydraulic design of sewers, q and Q are known. Assuming a self-cleaning velocity, a, A, S, and R can then subsequently be found. If the wetted perimeter is known, r can be computed from a. Once r is known, the desired slope s can then be computed.

In particular, for a circular sewer, the wetted perimeter may be derived by referring to Figure 4-5. From the figure at the bottom, the wetted perimeter p is

$$p = \frac{\theta}{360}\pi D \tag{4-20}$$

Also, from the figure, the area a is

$$a = \text{area of sector} - \text{area of triangle} \tag{4-21}$$

$$\text{area sector} = \frac{\theta}{360}\left(\frac{\pi D^2}{4}\right) \tag{4-22}$$

$$\text{area of triangle} = \frac{1}{2}\left[2\left(\frac{D}{2}\right)\left(\sin\frac{\theta}{2}\right)\right]\left(\frac{D}{2}\cos\frac{\theta}{2}\right) \tag{4-23}$$

Substituting equations (4-22) and (4-23) in equation (4-21) and simplifying, we have

$$a = \frac{D^2}{4}\left(\frac{\pi\theta}{360} - \frac{\sin\theta}{2}\right) \tag{4-24}$$

From knowledge of the value of a, θ can be solved from equation (4-24). Once known, p can be computed, and it follows that r can also be computed. Once r is found, the value of s can be solved.

The hydraulic radius r may also be solved by dividing equation (4-24) by equation (4-20). The result is

$$r = \frac{D}{4}\left(1 - \frac{360\sin\theta}{2\pi\theta}\right) \tag{4-25}$$

Also, from the figure, the half-angle of θ may also be found as follows:

$$\cos\frac{\theta}{2} = 1 - 2\left(\frac{d}{D}\right) \tag{4-26}$$

The value of n varies with the depth of flow. Table 4-2 shows the variation of N/n with the depth of flow in circular conduits as given by the ratio of partial-flow depth d to full depth D. Values of N for different conduit materials are given in Table 4-3.

TABLE 4-2 VARIATION OF N/n WITH THE DEPTH OF FLOW IN CIRCULAR CONDUITS

d/D	N/n	d/D	N/n	d/D	N/n	d/D	N/n
1.00	1.00	0.90	0.92	0.80	0.89	0.70	0.84
0.60	0.81	0.50	0.80	0.40	0.79	0.30	0.79
0.20	0.79	0.10	0.80	0.00	—	—	—

TABLE 4-3 VALUES OF N FOR DIFFERENT CONDUIT MATERIALS

Conduit material	Condition of interior surface			
	Best	Good	Fair	Bad
Tile pipe				
Vitrified (glazed)	0.010	0.012	0.014	0.017
Unglazed	0.011	0.013	0.015	0.017
Concrete pipe	0.012	0.013	0.015	0.016
Cast-iron pipe, coated	0.011	0.012	0.013	—
Brick sewers				
Glazed	0.011	0.012	0.013	0.015
Unglazed	0.012	0.013	0.015	0.017
Steel pipe				
Welded	0.010	0.011	0.013	—
Riveted	0.013	0.015	0.017	—
Concrete-lined channels	0.012	0.014	0.016	0.018

Example 4-7

The facility plan for a development recommended that the minimum and maximum flows to be considered for the design of a proposed collection system are 5037 and 15,319 m³/day, respectively. (a) What size of trunk sewer should be used to carry these flows? (b) To ensure self-cleaning at all times, at what slope should the sewer be laid? (c) What is the minimum depth of flow? Assume a concrete pipe sewer.

Solution (a) Assume a self-cleaning velocity of 0.60 m/s. Thus

$$A = \frac{15,319/[24(60)(60)]}{0.60} = \frac{0.177}{0.60} = 0.30 \text{ m}^2 = \frac{\pi D^2}{4}$$

$D = 0.62$ m; if not available, use the next-larger pipe.   **Answer**

(b) $a = \dfrac{5,037/[24(60)(60)]}{0.60} = 0.097 \text{ m}^2 = \dfrac{D^2}{4}\left(\dfrac{\pi\theta}{360} - \dfrac{\sin\theta}{2}\right)$

$$= \dfrac{0.62^2}{4}\left(\dfrac{\pi\theta}{360} - \dfrac{\sin\theta}{2}\right)$$

$$1.01 = 0.00873\theta - \dfrac{\sin\theta}{2} = Y$$

θ	0.00873θ	$\dfrac{\sin\theta}{2}$	Y
110	0.96	0.47	0.49
x			1.01
150	1.31	0.25	1.06

$\dfrac{x - 110}{150 - 110} = \dfrac{1.01 - 0.49}{1.06 - 0.49}$

$x = 146.49°$

$$p = \dfrac{146.49}{360}\pi(0.62) = 0.79$$

$$r = \dfrac{0.097}{0.79} = 0.12 \text{ m}$$

$$V = \dfrac{1}{N}R^{2/3}S^{1/2}$$

From Table 4-3, $N = 0.012$; $R = (\pi 0.62^2/4)/(\pi\, 0.62) = 0.155$ m. Therefore,

$$S^{1/2} = \dfrac{0.60(0.012)}{0.155^{2/3}} = 0.0249 \qquad S = 0.00062$$

$$\dfrac{q}{Q} = \dfrac{a}{A}\left(\dfrac{R}{r}\right)^{1/6}\left(\dfrac{S}{s}\right)^{1/6}$$

$$\dfrac{5037}{15,319} = \left(\dfrac{0.097}{0.3}\right)\left(\dfrac{0.155}{0.12}\right)^{1/6}\left(\dfrac{0.00062}{s}\right)^{1/6}$$

$$\left(\dfrac{0.00062}{s}\right)^{1/6} = 0.97$$

$$s = 0.00074 \quad \textbf{Answer}$$

(c) $\cos\dfrac{\theta}{2} = 1 - 2\dfrac{d}{D}$ $\cos\dfrac{146.49}{2} = 1 - 2\left(\dfrac{d}{0.62}\right)$

$$d = 0.22 \text{ m} \quad \textbf{Answer}$$

Design Period

In designing the wastewater collection system, a period of design must be decided. The following information for the period of design of various collection system components may be useful to arrive at this decision.

Laterals and submains less than 380 mm (15 in) in diameter: full development

Main sewers, outfalls, and interceptors: 40 to 50 years

Treatment works: 10 to 25 years

PNEUMATIC SYSTEMS

The word *pneumatic* refers to air and other gases. Therefore, a *pneumatic system* is one in which air or other gases are conveyed around a structure to serve some purpose. Examples of pneumatic systems are the air-conditioning and heating systems of a residential home. In air conditioning, air is drawn from outside and forced to contact the surfaces of a unit that has been cooled, cooling the incoming air. The cooled air is then distributed to various points around the house. In heating, the same process operates, except that instead of a cooled surface, a heated surface is used. In environmental pollution control, the pneumatic system is a system composed of the collection, conveyance, and treatment of polluted air, which by analogy to the wastewater collection system may be called a *waste air collection system.*

The components of a waste-air collection system are the hood, the fan, the duct, and the air pollution control device that treats the polluted air before discharge to the ambient atmosphere. By analogy to *sewerage works* (*sewage works* or *wastewater works*), the term *waste air works* will be used to refer to the physical structures required for the collection, treatment, and disposal of the waste air system. The *hood* is a device or structure that captures the emissions from a process. To capture these emissions, a suction force is necessary. This is provided by the *fan*. The captured emissions are then conducted by means of a *duct* to the air pollution control device. The duct is just a conduit analogous to the conduits or pipes in the wastewater collection or water distribution system. It is simply called a *duct* when dealing with pneumatics.

Figure 4-6 shows a schematic of the waste-air collection system for a particular industrial operation. Point 1 is a *dump hopper*, where materials are "dumped" for subsequent conveyance by the bucket elevator 2. A *bucket elevator* is one in which materials are scooped at the bottom of the structure by a series of buckets attached to a continuous chain that travels up and down the elevator structure. Unit 2 is indicated simply as a rectangular box, but inside it are these continuous buckets. The bucket elevator then conveys (elevates) the materials to the ribbon blender 3. A *ribbon blender* is one in which a ribbon-shaped blade is used to blend the materials. The blended materials are then dropped to drums at 4.

The waste air collection system of the above-mentioned industrial operation is indicated by the lines A, B, C, D, E, F, G, and H. Lines A, B, D, and F are called *branch*

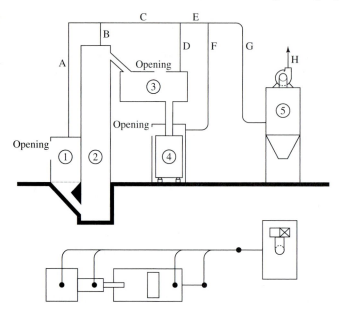

Figure 4-6   Waste air collection system.

ducts or *laterals*, just as the lines that collect sewage from individual homes are called *branches* or *laterals*. Since dusts are created from the dumped materials, branch duct A collects the waste air from hopper 1 through a hood over the hopper. For the same reason, dust-laden waste air is also collected at the elevator, ribbon blender, and drum booth. The branches then convey their contents into the *main duct*, labeled at the top of the sketch as CE. The main duct conveys its contents to the *outfall* or *trunk duct* G, which, in turn, introduces the waste air into the air pollution control device 5. In the case of dusts, a bagfilter is the control device normally used to clean the air (5 in the sketch). A *bagfilter* is the same as the vacuum cleaner used in homes, but much, much larger. H is the exhaust to the ambient air. A more simplified schematic is shown at the bottom.

Hood

Even how elaborate the duct system and the pollution control devices are, the treatment of the whole waste airs will not be as efficient without a good catch of the pollutants emitted from the process. This makes the catching hood a very important element in the whole waste air collection works systems design. Hoods must be carefully designed to catch all emissions, especially toxic emissions.

In general, there are three types of hoods: capturing hoods, canopy hoods, and enclosures (Figure 4-7). *Capturing hoods* are those designed to create conveying velocities at points away from the direct influence of the hood. As shown in part (c) of the figure, the pollutants emitted from the tank are not under the direct influence of the hood. To catch them, sufficient conveying velocities must be created on the tank surface to draw the air into the hood. *Canopy hoods* are those constructed above an emitting surface to

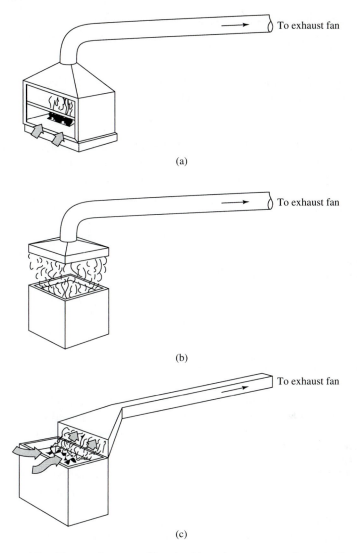

Figure 4-7   Three main types of hoods: (a) enclosures—contain contaminants released inside the hood; (b) canopy hoods—catch contaminants that rise into them; (c) capturing hoods—reach out to draw in contaminants.

catch the emissions. As indicated in part (b), the emissions are under the direct influence of the hood above them. If the emitting surface is hot, the air flows into the canopy by mere convection. *Enclosures*, on the other hand, completely encloses the emission source, except for an access opening, as shown in part (a).

   Capturing hoods.   Denote Y as the fraction of the centerline velocity at the entrance to a circular hood of area A. At a distance x from the hood entrance measured parallel to the centerline, Y is given by Della Valle as

$$\frac{Y}{1-Y} = \frac{0.1A}{x^2} \tag{4-27}$$

A and x can be expressed in any consistent set of units. Combining equation (4-27) with the equation of continuity, $Q = AV$,

$$Q = v_x(10x^2 + A) \tag{4-28}$$

where Q is the rate of airflow into the hood, called *ventilation volume*, and v_x is the velocity at a given point a distance x measured parallel to the centerline from the entrance to the hood. If half of the hood is constructed against a wall, the other half, called the *virtual hood*, may be imagined to exist. Calling the area of the virtual hood plus the actual "half-hood" A_v and the flow as Q_v, and using equation (4-28) yields

$$Q_v = v_x(10x^2 + A_v) \tag{4-29}$$

But $A_v = 2A$ and $Q_v = 2Q$. A and Q are the actual area and flow to the half-hood, respectively. Substituting in equation (4-29), we obtain

$$Q = \frac{v_x(10x^2 + 2A)}{2} \tag{4-30}$$

To use the equations above, v_x must hold at the null point of the emission. The *null point* is defined as the maximum extent that a pollutant particle will wander in the absence of a capturing influence. To capture this particle at the null point, v_x must catch it at this point and convey it toward the hood. A minimum design velocity at the null point of 15 to 30 m/min (50 to 100 fpm) is usually provided. To get the null point for a specific application process, an actual experiment should be conducted before design. The formulas above hold for hoods of circular cross sections and for hoods of rectangular cross sections up to a length/width ratio of 3:1.

Example 4-8

A paint spray booth has a 3-m × 2-m entrance. The null point was determined by an experiment to be 1.5 m measured parallel to the centerline from the face of the booth. If a velocity of 100 fpm is to be provided at the null point, (a) what is the ventilation volume required? and (b) what is the face velocity at the entrance to the booth?

Solution   Since the bottom part of the opening to the booth is set against the floor, the formula to be used is the one involving the virtual hood.

(a)
$$Q = \frac{v_x(10x^2 + 2A)}{2}$$

$$v_x = \frac{100}{3.281} = 30.48 \text{ m/min}$$

$$Q = \frac{30.48[10(1.5^2) + 2(2)(3)]}{2} = 526.78 \text{ m}^3/\text{min} \quad \textbf{Answer}$$

(b) $$V = \frac{526.78}{2(3)} = 87.63 \text{ m/min} \quad \textbf{Answer}$$

A special type of capturing hood is the *slot hood*, which is used to capture emissions from an open tank as shown in Figure 4-8. The hood of width b_s is constructed on the periphery on the tip of the tank. Slot hoods and capturing hoods, in general, should only be used to capture emissions from cold processes as opposed to hot processes. The formulas developed for hoods so far are not applicable to this type of hood, since those formulas apply only up to a maximum length/width ratio of the opening of 3:1. To design the slot hood an actual experiment should be conducted on a similar installation and ventilation rate determined from there; or a pilot study may be conducted. For the ventilation rate determined to be appropriate, the slot velocity should not be below 610 m/min (2000 fpm).

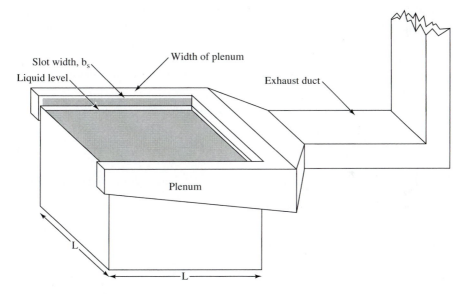

Figure 4-8 Slot hood for capture of emissions from open-surface tank.

Example 4-9

A slot hood is to be used to capture emissions from a chrome-plating tank. The tank measures 2 m by 3 m on the surface. The slot is to be put on one of the 2-m sides and on both of the 3-m sides. From an actual experiment, the ventilation rate was determined to be 200 cfm/ft^2 of tank surface. Determine the slot-hood dimensions.

Solution

$$200 \ \frac{\text{cfm}}{\text{ft}^2} = \frac{200(3.281^2)}{3.281^3} = 60.96 \ \frac{\text{m}^3/\text{min}}{\text{m}^2}$$

Therefore,

$$Q = 60.96(2)(3) = 365.74 \text{ m}^3/\text{min}$$

$$\text{length of hood} = 2(3) + 2 = 8 \text{ m} \quad \textbf{Answer}$$

Assume a slot velocity of 610 m/min. Then

$$\text{width of slot} = \frac{365.74}{610(8)} = 0.075 \text{ m} = 7.5 \text{ cm} \quad \textbf{Answer}$$

Canopy Hoods

Canopy hoods may be high or low. Low canopy hoods may go as low as convenience in operation permits. A very high canopy hood results in excessive ventilation requirements. When heat is added to air, such as over a heated process tank, the density of the surrounding air becomes relatively greater than that of the heated air. This condition causes a convective motion as the higher-density surrounding air rushes toward the lower-density heated air. The heated air can rise at a velocity as high as 122 m/min (400 ft/m). The diameter D_c in meters of the rising air at a distance x_f in meters above a hypothetical point source, as shown in Figure 4-9, is given empirically by[4]

$$D_c = 0.43x_f^{0.88} \tag{4-31}$$

From the figure,

$$x_f = z + y \tag{4-32}$$

where z is the distance from the hot-point source to the hypothetical point source in meters and is given empirically by

$$z = 2.59(D_s)^{1.138} \tag{4-33}$$

D_s is the diameter of the hot point source in meters. As shown in the figure, y is also in meters. D_f is the diameter of the hood.

According to Hemeon,[4] the velocity of the rising heated air v_f in m/min is given in terms of x_f and the heat added to the air q_c in kJ/min as follows:

$$v_f = \frac{7.85}{x_f^{0.29}}(q_c)^{1/3} \tag{4-34}$$

$$q_c = 0.15A_s(\Delta T)^{1.25} \tag{4-35}$$

where A_s is the area of hot source in m^2 and ΔT the temperature difference between the hot source and the ambient air in kelvin. The following design criteria have been

---

[4]D. A. Danielson (1973). *Air Pollution Engineering Manual*. Office of Air Quality Planning and Standards, U.S. Environmental Protection Agency, Research Triangle Park, NC p. 35.
[4]Ibid.

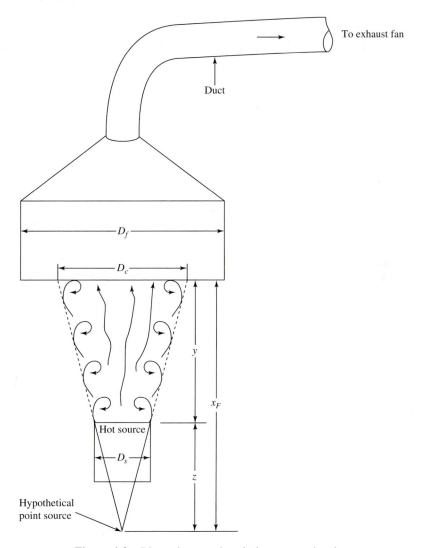

Figure 4-9 Dimensions used to design canopy hoods.

recommended: Increase D_c by $0.8y$ to obtain D_f and increase the ventilation rate obtained by the use of equation (4-34) by 20%.

The formulas above may be used for rectangular sources by considering the two pairs of opposite sides of the rectangle as two circular sources and superposing the results of the calculation. This is illustrated in the next example.

Example 4-10

A batch-galvanizing operation uses a batch kettle measuring 2 m by 3 m. Because of interference of the crane used to load the kettle, the hood must be located 3.5 m above the molten zinc surface. The metal temperature is 885°F and the room temperature is approximately 80°F. Determine (a) the dimensions of the hood and (b) the ventilation rate.

Solution

(a)
$$z = 2.59(D_s)^{1.138}$$

For the 2-m side of the kettle,

$$z_2 = 2.59(2)^{1.138} = 5.70 \text{ m}$$

Thus,

$$x_{f2} = 5.70 + 3.5 = 9.2 \text{ m}$$

$$D_{c2} = 0.43x_{f2}^{0.88} = 0.43(9.2)^{0.88} = 3.03 \text{ m}$$

and

$$D_{f2} = D_{c2} + 0.8y = 3.03 + 0.8(3.5) = 5.83 \text{ m} \quad \text{for one side of hood}$$

For the 3-m side of the kettle,

$$z_3 = 2.59(3)^{1.138} = 9.04 \text{ m}$$

Thus,

$$x_{f3} = 9.04 + 3.5 = 12.54 \text{ m}$$

$$D_{c3} = 0.43x_{f3}^{0.88} = 0.43(12.54)^{0.88} = 3.98 \text{ m}$$

and

$$D_{f3} = D_{c3} + 0.8y = 3.98 + 0.8(3.5) = 6.78 \text{ m} \quad \text{for the other side of hood}$$

Thus,

dimension of hood: 5.83 m by 6.78 m   **Answer**

(b)
$$v_f = \frac{7.85}{x_f^{0.29}}(q_c)^{1/3}$$

$$q_c = 0.15A_s(\Delta T)^{1.25} = 0.15(2)(3)\left(\frac{885 - 80}{1.8}\right)^{1.25} = 1851 \text{ kJ/min}$$

$$x_{f\text{ave}} = \frac{9.2 + 12.54}{2} = 10.87 \text{ m}$$

Thus,

$$v_f = \frac{7.85}{10.87^{0.29}}(1851)^{1/3} = 48.23 \text{ m/min}$$

and

$$Q = 1.20(48.23)[5.83(6.78)] = 2288 \text{ m}^3/\text{min} \quad \textbf{Answer}$$

Enclosures

An enclosure hood completely encloses the source of emissions except for an access opening. In the case of large enclosures, the access opening provides room for workers to move about. In its design, three things must be considered: the leakage through openings above the level of the hot source due to chimney effects, indraft velocity, and ventilation requirement for thermal draft. *Indraft velocity* refers to the velocity of the inflow of air from outside the enclosure to the inside replacing the volume of air exhausted from inside the enclosure to satisfy ventilation requirement. *Thermal draft* refers to the convection flow setup inside the enclosure as the emission rises. Sufficient ventilation must be provided for this draft to exhaust the pollutants to the ducting system; otherwise, the pollutants would simply bounce back. The ventilation requirement may be computed using the same method for the canopy calculation. Indraft velocities that have been used range from 30 to 240 m/min (100 to 800 fpm). The higher ranges are used for air contaminants released with extremely great force, as in the direct-arc steel-melting furnace. The enclosure should be designed to prevent leakage. However, some situations may prevent the enclosure to be leakage-proof. Hemeon developed an equation for leakage velocities v_e in meters per minute by assuming the openings to be orifices:

$$v_e = 33.13 \left(\frac{\ell_0 q_c}{A_0 T} \right)^{1/3} \tag{4-36}$$

where ℓ_0 is the vertical distance from the upper boundary of the access opening to the leakage orifice, q_c is as defined before, A_0 is the area of the orifice, and T is the average temperature of the air inside the hood in kelvin.

Example 4-11

If the emissions from the galvanizing kettle of Example (4-10) are to be captured by an enclosure, determine the total ventilation requirement. The enclosure is 8 m wide by 10 m long by 6 m high. The access opening is 3 m high by 10 m long and runs along the length of the enclosure. Total leakage area is approximately 0.093 m^2 and is located at the top of the enclosure. Assume an indraft velocity of 60 m/min. The average temperature of the air inside the hood is 66°C.

Solution Assume thermal draft requirements the same as in Example (4-10). Hence $D_{c2} = 3.03$ m, $D_{c3} = 3.98$ m, and $v_f = 48.23$ m/min. Thus,

$$\text{thermal draft requirement} = 48.23(3.03)(3.98) = 581.62 \ \text{m}^3/\text{min}$$

$$v_e = 33.13 \left(\frac{\ell_0 q_c}{A_0 T} \right)^{1/3}$$

$$q_c = 0.15 A_s (\Delta T)^{1.25} = 1851 \ \text{kJ/min} \quad \text{from Example (4-10)}$$

$$v_e = 33.13 \left(\frac{\ell_0 q_c}{A_0 T} \right)^{1/3} = 33.13 \left[\frac{(6-3)(1851)}{0.093(66+273)} \right]^{1/3} = 185.71 \ \text{m/min}$$

$$q_0 = A_0 v_e = 185.71(0.093) = 17.27 \ \text{m}^3/\text{min}$$

$$\text{indraft requirement} = 60(3)(10) = 1800 \text{ m}^3/\text{min}$$

Thus, total indraft plus leakage requirements $= 1800 + 17.27 = 1817.27 \text{ m}^3/\text{min} > 581.62 \text{ m}^3/\text{min}$ for the thermal draft requirement, and

$$\text{total ventilation requirement} = 1817.27 \text{ m}^3/\text{min} \quad \textbf{Answer}$$

Duct System

Important calculations to be performed in the design of a duct system are those for the head losses. As in water and wastewater systems, the head losses involved are those for the straight runs of pipe or duct, transitions, and fittings losses. The method of calculations has already been discussed. An important consideration is that the velocities inside the duct must not fall below the minimum conveying velocities found to work in practice, especially for dust emissions. (In the jargon of the air pollution specialist, dust emissions are also called *fugitive dusts* or *particulates*.) The following values have been recommended: (1) for gases, vapors, smokes, fumes, and very light dusts, minimum velocity = 610 m/min; (2) for medium-density dry dusts, minimum velocity = 915 m/min; (3) for average industrial dusts, minimum velocity = 1220 m/min; and (4) for heavy dusts, minimum velocity = 1525 m/min. Examples of the first category are all vapors, gases, and smokes, zinc and aluminum oxide fumes, wood, flour, and cotton lint; for the second category: buffing lint, sawdust, grain, rubber, and plastic dusts; for the third category: sandblasting and grinding dusts, wood-shavings dust, and cement dust; and for the fourth category: lead and foundry shakeout dusts, and metal turnings.

Tables 4-4 and 4-5 show various transition losses. In particular, Table 4-4 expresses the loss in terms of the equivalent feet of straight duct. For example, a 6-in. duct of throat radius of $1.5D(1.5 \times 6 = 9 \text{ in.})$, the equivalent length is 7 ft. The definition of throat radius is indicated in the table. In Table 4-5 the loss factor coefficients for expansion and contraction are multiplied by the respective velocity head differences between the sections at D_1 and D_2. The results are always considered positive. For abrupt contractions of angles greater than $60°$, the loss factor K is multiplied by the velocity head at D_2.

It will be recalled that in the Hardy Cross method, the head loss in a loop from a starting point to an endpoint going in a clockwise direction is made equal to the head loss from the same starting point to the same endpoint but going in the counterclockwise direction. In the pneumatics of ducts, the same principle is applied. Refer to Figure 4-6. The pressure at 1 is the same as the pressure at 2. Since the points are of the same pressure, to form one loop, point 1 or 2 may be considered as the starting point and another point, the upper intersection of B, as the ending point. From Hardy Cross, $h_{1A} = h_{12B}$. This loop is the loop of the dump hopper and the bucket elevator, loop $1AB21$. The second loop of the system is the one formed between the bucket elevator and the ribbon mixer. As in the first loop, the pressures at 2 and 3 are all at atmospheric. This loop may be called $2BCD32$. The Hardy Cross head loss equation for the loop can be $h_{2BC} = h_{23D}$. The third and last loop is that formed between the ribbon mixer and

TABLE 4-4 AIR FLOW RESISTANCE EXPRESSED AS EQUIVALENT FEET OF STRAIGHT DUCT

Diameter of duct (in.)	90° Elbow throat radius (R)			60° Elbow throat radius (R)			45° Elbow throat radius (R)			Branch entry angle of entry (θ)		
	1.0D	1.5D	2.0D	1.0D	1.5D	2.0D	1.0D	1.5D	2.0D	45°	30°	15°
3	5	4	3	4	3	2	2	1	1	3	2	1
4	7	5	4	5	4	3	4	3	2	5	3	1
5	9	6	5	7	5	4	5	4	3	6	4	2
6	11	7	6	8	5	4	6	4	3	7	5	2
7	12	9	7	9	6	5	7	5	4	9	6	3
8	14	10	8	11	7	6	8	5	4	11	7	3
9	17	12	10	12	9	7	9	6	5	12	8	4
10	20	13	11	14	10	8	10	7	6	14	9	4
11	23	16	13	17	12	10	11	8	6	15	10	5
12	25	17	14	20	13	11	12	9	7	18	11	5
14	30	21	17	23	16	13	14	10	8	21	13	6
16	36	24	20	27	18	15	17	12	10	25	15	8
18	41	28	23	32	22	18	20	13	11	28	18	9
20	46	32	26	36	24	20	23	16	13	32	20	10
22	53	37	30	39	27	22	27	18	15	36	23	11
24	59	40	33	44	30	25	30	20	16	40	25	13
26	64	44	36	48	33	27	32	22	18	44	28	14
28	71	49	40	52	35	29	35	24	20	47	30	15
30	75	51	42	55	38	31	37	26	21	51	32	16
36	92	63	52	68	46	38	46	32	26	—	—	—
40	105	72	59	75	51	42	52	35	29	—	—	—
48	130	89	73	91	62	51	64	44	36	—	—	—

the drum booth, loop $3DEF43$. The pressure at 4 is also atmospheric. The head-loss equation can be $h_{3DE} = h_{34F}$. The application of the Hardy Cross method is illustrated in the next example.

Example 4-12

Ventilation requirements of the waste air works system of Figure 4-6 have been determined to be as follows: dump hopper = 20 m³/ min, bucket elevator = 6.0 m³/ min, ribbon blender = 8.0 m³/ min, and drum booth = 19.0 m³/ min. The duct lengths and fittings required are

TABLE 4-5 ENTRY, EXPANSION, AND CONTRACTION LOSSES IN DUCTS

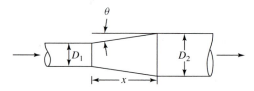

Taper angle, θ (deg.)	$\dfrac{x}{D_2 - D_1}$	Loss factor
$3\frac{1}{2}$	8.13	0.22
5	5.73	0.28
10	2.84	0.44
15	1.86	0.58
20	1.38	0.72
25	1.07	0.87
30	0.87	1.00
over 30		1.00

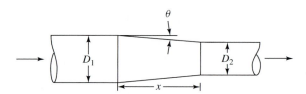

Taper angle, θ (deg.)	$\dfrac{x}{D_2 - D_1}$	Loss fraction of h_v difference	For abrupt contraction ($\theta > 60°$)	
			Ratio D_2/D_1	Factor K
5	5.73	0.05	0.1	0.48
10	2.84	0.06	0.2	0.46
15	1.86	0.08	0.3	0.42
20	1.38	0.10	0.4	0.37
25	1.07	0.11	0.5	0.32
30	0.87	0.13	0.6	0.26
45	0.50	0.20	0.7	0.20
60	0.29	0.30		

as follows:

Branch	Length (m)	Number of elbows (deg)	Number of branch entries (deg)
A	4.0	1, 90	0
B	1.3	0	1, 45
C	2.5	0	0
D	0.5	0	1, 30
E	1.5	0	0
F	3.7	1, 90	1, 30
G	4.0	1, 90	Expands into 5

Determine the head loss needed to size the fan H. Assume that $f = 0.03$ in all ducts and a throat radius for bends of $2D$. Assume that a minimum conveying velocity of 1000 m/min is sufficient. Although it cannot be neglected, neglect the head loss at the outfall main G and the losses in the bagfilter assembly 5.

Solution

$$g = 9.81 \; \frac{\text{m}}{\text{s}^2} = 9.81 \; \frac{\text{m}}{\text{s}^2} \left(\frac{60^2 \text{s}^2}{\text{min}^2} \right) = 35{,}316 \; \frac{\text{m}}{\text{min}^2}$$

$$k = f \frac{\ell}{2g D A^2} = 0.03 \left[\frac{\ell}{2(35{,}316)(D)(\pi D^2/4)^2} \right] = 6.88 \left(10^{-7} \right) \frac{\ell}{D^5}$$

Loop $1AB21$, first iteration (starting point $= 1$, ending point $=$ upper intersection of B):

(1) Duct	(2) Q (m³/min)	(3) A (m²)	(4) D (m)	(5) D_{cor} (in.)	(6) D_{cor} (m)	(7) A_{cor} (m²)
Ac	20	0.020	0.16	6	0.15	0.0182
Bcc	6.0	0.006	0.088	3	0.076	0.0046

(8) V_{cor} (m/min)	(9) ℓ_{eq} (m)	(10) ℓ_d (m)	(11) ℓ (m)	(12) k	(13) kQ^2 (m)	(14) h_e (m)	(15) h_t (m)
1099	1.83	4.0	5.83	0.053	21.16	8.55	29.71
1304	0.91	1.3	2.21	0.60	−21.6	−12.04	−33.64

$|-33.86 > 29.71|$.

Col. (1): With respect to starting point 1 or 2, A is clockwise and B is counterclockwise.

Col. (2): Ventilation rate, given.

Col. (3): Duct cross-sectional area at conveying velocity of 1000 m/min.

Col. (4): Duct diameter corresponding to area in col. (3), $D = 1.13\sqrt{A}$.

Col. (5): Duct diameter available according to Table 4-4, nearest lesser or equal to the D of col. (4) in inches obtained by 39.37 times [col. (4)]. The nearest lesser or equal diameter is chosen to maintain or improve the conveying velocity. Choosing the nearest larger diameter will decrease the conveying velocity, hence will decrease the collection efficiency of the system.

Col. (6): Duct diameter available in meters $= \dfrac{\text{col. (5)}}{12(3.281)}$.

Col. (7): Cross-sectional area corresponding to col. (6) $= 0.785[\text{col. (6)}]^2$.

Col. (8): $\dfrac{\text{Col. (2)}}{\text{Col. (7)}}$, actual ventilation velocity.

Col. (9): Equivalent length from Table 4-4 for the elbows and branch entries.

Col. (10): Length of duct, given.

Col. (11): col. (9) + col. (10)

Col. (12): $k = f\dfrac{\ell}{2gDA^2} = 6.89(10^{-7})\dfrac{\ell}{D^5}$.

Col. (13): From $h = kQ^2$. Note that from convention used in the derivation, kQ_{cc}^2 is negative while kQ_c^2 is positive. This is a loss due to skin friction of straight runs of pipe and equivalent lengths of bends and branch entries. Skin friction refers to the loss of head due to the rubbing of fluid with the interior surfaces of pipes.

Col. (14): Entrance loss, h_e. From Table 4-5, $K > 0.48$. Let $h_e = 0.5h_v$ (h_v = velocity head). This is a loss when air enters the duct from the hood.

Col. (15): Total loss, $h_t =$ col. (13) + col. (14). In this iteration, two head losses are obtained: 33.64 m and 29.71 m. To pull 6.0 m³/min through the B part of the loop (the counterclockwise part), the 33.64-m head loss must exist. To balance, this 33.86 m must also exist in the A part of the loop (the clockwise part); hence the larger head loss controls. From $Q \propto \sqrt{h_t}$, the new Q_A may be calculated as follows: $Q_A/20 = \sqrt{33.64}/\sqrt{29.71}$; $Q_B = 21.28$ m³/min. The second iteration will show that this new Q_B of 21.35 m³/min will balance the loop.

Loop $1AB21$, second iteration (starting point = 1, ending point = upper intersection of B):

(1) Duct	(2) Q m³/min	(3) A (m²)	(4) D (m)	(5) D_{cor} (in.)	(6) D_{cor} (m)	(7) A_{cor} (m²)
Ac	21.28	0.021	0.16	6	0.15	0.0182
Bcc	6	—	—	3	0.076	0.0046

(8) V_{cor} (m/min)	(9) ℓ_{eq} (m)	(10) ℓ_d (m)	(11) ℓ (m)	(12) k	(13) kQ^2 (m)	(14) h_e (m)	(15) h_t (m)
1169	1.83	4.0	5.83	0.053	24.0	9.67	33.67
1304	0.91	1.3	2.21	0.60	−21.6	−12.04	−33.64

Thus,

$$h_A = h_B = \frac{33.67 + 33.64}{2} = 33.66 \text{ m of air}$$

Loop $2BCD32$, first iteration (starting point = 2, ending point = upper intersection of D):

(16) Duct	(17) Q m³/min	(18) A (m²)	(19) D (m)	(20) D_{cor} (in.)	(21) D_{cor} (m)	(22) A_{cor} (m²)
Bc	6.0	—	—	3	0.076	0.0046
Cc	27.28	0.027	0.186	7	0.178	0.0249
Dcc	8.0	0.008	0.100	3	0.076	0.0046

(23) V_{cor} (m/min)	(24) ℓ_{eq} (m)	(25) ℓ_d (m)	(26) ℓ (m)	(27) k	(28) kQ^2 (m)	(29) h_e (m)	(30) h_t (m)
1304	0.91	1.3	2.21	0.60	21.6	12.04	33.64
1096	—	2.5	2.5	0.0096	5.73	—	7.14
							40.78
1739	0.61	0.5	1.11	0.30	−19.2	−21.41	−40.61

$$h_C = 7.14 \qquad h_D = \frac{40.78 + 40.61}{2} = 40.70 \text{ m of air}$$

Loop $3DEF43$, first iteration (starting point = 3, ending point = intersection of E and F):

(31) Duct	(32) Q (m³/min)	(33) A (m²)	(34) D (m)	(35) D_{cor} (in.)	(36) D_{cor} (m)	(37) A_{cor} (m²)
Dc	8.0	0.008	0.100	3	0.076	0.0046
Ec	35.28	0.035	0.21	8	0.203	0.032
Fcc	19.0	0.019	0.155	6	0.152	0.0182

(38) V_{cor} (m/min)	(39) ℓ_{eq} (m)	(40) ℓ_d (m)	(41) ℓ (m)	(42) k	(43) kQ^2 (m)	(44) h_e (m)	(45) h_t (m)
—	—	—	—	—	—	—	40.70
1103	—	2.0	2.0	0.0040	4.98	8.61	13.59
							54.29
1044	3.35	3.7	7.05	0.06	−21.66	−7.72	−29.38

$$|-29.38| < 54.29$$

Therefore, $Q_F/19 = \sqrt{54.29}/\sqrt{29.38}$, $Q_F = 25.83$ m³/m, $h_E = 13.59$ m of air, and $h_F = 54.29$ m of air.

$$h_A + h_C + h_E = 33.66 + 7.14 + 13.59 = 54.39 = h_B + h_C + h_E$$

$$h_D + h_E = 40.70 + 13.59 = 54.29$$

Therefore,

$$\text{head loss needed to design fan} = \frac{54.39 + 54.29}{2} = 54.34 \text{ m of air} \quad \textbf{Answer}$$

PUMPING STATIONS

The location where pumps are installed is a *pumping station*. There may be only one pump, or there may be several. Depending on the desired results, the pumps may be connected in parallel or in series. In *parallel connection*, the discharges of all the pumps are combined into one. Hence connecting the pumps in parallel increases the discharge from the pumping station. On the other hand, in *series connection*, the discharge of the first pump becomes the input of the second pump, and the discharge of the second pump becomes the input of the third pump, and so on. In this mode of operation, the head built up by the first pump is clearly added to the head built up by the second pump, and the head built up by the second pump is added to the head built by the third pump, and so on. Hence pumps connected in series increase the total head output from a pumping station by adding the heads of all pumps.

Figure 4-10 shows section and plan views of a sewage pumping station. Note that the pumps are connected in parallel and that the discharges are conveyed into a common manifold pipe. As indicated in the drawing, a *manifold pipe* has one or more branching pipes connected to a main, larger pipe.

The word *pump* is a general term used to designate the unit used to move a fluid from one point to another. The fluid may be contaminated such as air conveying fugitive dusts or water conveying sludge solids. There are two general classes of pumps: centrifugal and positive-displacement pumps. *Centrifugal pumps* move fluids by imparting to the fluid the tangential force of a rotating blade called an *impeller*. The motion of the fluid is a result of the *indirect action* of the impeller. *Displacement pumps*, on the other hand, literally push the fluid in order to move it. Hence the action is *direct* and moves the fluid *positively*, hence the name *positive-displacement pumps*. Figure 4-11 shows an example of a positive-displacement pump. Note that to move the wastewater, the screw pump literally pushes it.

Fluid machines that turn or tend to turn about an axis are called *turbomachines*. Thus centrifugal pumps are turbomachines, as are turbines, lawn sprinklers, ceiling fans, lawn mower blades, and turbine engines. The fan used to exhaust contaminated air in waste air works is a turbomachine. Of course, the fan is also a pump—a pump used in waste air works.

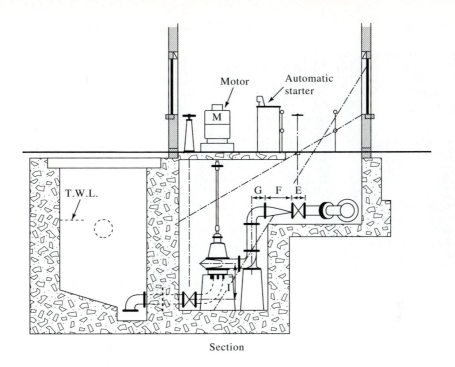

Motor

Automatic starter

M

T.W.L.

G F E

Section

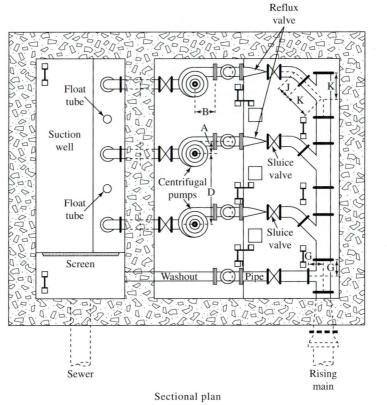

Reflux valve

Float tube

Suction well

Centrifugal pumps

Float tube

Screen

A

B

D

J K

K

Sluice valve

Sluice valve

G

G

Washout

Pipe

Sewer

Rising main

Sectional plan

Figure 4-10   Plan and section of pumping station.

Figure 4-11 Screw positive-displacement pump.

Pump Scaling Laws

When designing a pumping station, the engineer refers to a pump characteristic curve that defines the performance of a pump. Since there are several different sizes of pumps, theoretically, there should also be a number of these curves to correspond to each pump. In practice, however, this is not done. The characteristic performance of any other pump can be obtained from the curve of any one pump by the use of pump scaling laws, provided that the pumps are similar. The word *similar* will become clear later.

In pump operation, the following variables are encountered: the discharge Q, the pressure developed ΔP, the power requirement $\mathcal{P}$, and the efficiency η. The other variables are the diameter of impeller D, the rotational speed ω, the viscosity of the fluid μ, some roughness ϵ, the mass density of the fluid ρ, and some characteristic length ℓ of the chamber space. Some of these variables may not, of course, be important, depending on the type of dependent variable considered for the analysis.

D, ρ, ω, Q, μ, ϵ, and, ℓ are variables intrinsic to the pump system operation. They are the independent variables. ΔP, $\mathcal{P}$, and η are variables that are resultants to the effect of the intrinsic variables. They are the dependent variables. For ΔP the functional relationship is

$$\Delta P = \phi(\rho, \omega, D, Q, \mu, \epsilon, \ell) \tag{4-37}$$

$\mathcal{P}$ and η are not independent variables and are therefore not included. Since the friction factor f is a function of μ through the Reynolds number, the effect of μ on ΔP may be gaged on the effect of f on the ΔP. From the Moody diagram (Appendix 17), the value of f remains constant as the Reynolds number increases. Hence at high Reynolds numbers or high velocities as occurring in pump impellers, f does not affect ΔP; accordingly, μ does not affect ΔP either. Thus μ may be dropped. ℓ as a measure of the pump chamber space is already included in D. It may also be dropped. Finally, since the casing is too short, the effect of roughness ϵ is too small compared to the other causes of the ΔP. It may therefore also be dropped. Equation (4-37) now takes the form

$$\Delta P = \phi(\rho, \omega, D, Q) \tag{4-38}$$

Letting (x) represent "the dimensions of x," $(\Delta P) = F/L^2$, $(\rho) = Ft^2/L^4$, $(\omega) = 1/t$, $(D) = L$, and $(Q) = L^3/t$. By inspection of these dimensions, the number of reference dimensions is three. Since there are five variables, the number of Π terms is two (number of variables minus number of reference dimensions, $5 - 3 = 2$). The Π terms will be Π_1 and Π_2.

To eliminate the dimension F, divide ΔP by ρ. Hence

$$\frac{\Delta P}{\rho} \Rightarrow \left(\frac{\Delta P}{\rho}\right) = \frac{F/L^2}{Ft^2/L^4} = \frac{L^2}{t^2} \tag{4-39}$$

To eliminate t, divide by ω^2 as follows:

$$\frac{\Delta P}{\rho\omega^2} \Rightarrow \left(\frac{\Delta P}{\rho\omega^2}\right) = \frac{L^2}{t^2(1/t^2)^2} = L^2 \tag{4-40}$$

To eliminate dimensions completely, divide by D^2 as follows:

$$\frac{\Delta P}{\rho\omega^2 D^2} \Rightarrow \left(\frac{\Delta P}{\rho\omega^2 D^2}\right) = \frac{L^2}{L^2} \tag{4-41}$$

Thus,

$$\Pi_1 = \frac{\Delta P}{\rho\omega^2 D^2} \tag{4-42}$$

To get Π_2, operate on Q to obtain

$$\Pi_2 = \left(\frac{Q}{\omega D^3}\right) \tag{4-43}$$

The final functional relationship is

$$\frac{\Delta P}{\rho\omega^2 D^2} = \Psi\left(\frac{Q}{\omega D^3}\right) \tag{4-44}$$

But $\Delta P = \gamma H$, where H is a head loss or gain. Since $\gamma = \rho g$, substituting in equation (4-44) produces

$$\frac{Hg}{\omega^2 D^2} = \Psi\left(\frac{Q}{\omega D^3}\right) \tag{4-45}$$

$Hg/(\omega^2 D^2)$ is called the *head coefficient* C_H, while $Q/(\omega D^3)$ is called the *flow coefficient* C_Q. Since no one pump was chosen in the derivation of equation (4-45), the equation is general. For any number of pumps a, b, c, $\ldots$, n and using equation (4-45) we get the relationships

$$\left\{\frac{Hg}{\omega^2 D^2}\right\}_a = \left\{\frac{Hg}{\omega^2 D^2}\right\}_b = \left\{\frac{Hg}{\omega^2 D^2}\right\}_c = \cdots = \left\{\frac{Hg}{\omega^2 D^2}\right\}_n = \{C_H\}_n \tag{4-46}$$

$$\left\{\frac{Q}{\omega D^3}\right\}_a = \left\{\frac{Q}{\omega D^3}\right\}_b = \left\{\frac{Q}{\omega D^3}\right\}_c = \cdots = \left\{\frac{Q}{\omega D^3}\right\}_n = \{C_Q\}_n \tag{4-47}$$

Pumps that follow the relations above are called *similar* or *homologous pumps*. In particular, when the Π variable C_H, which involves force is equal in the series of pumps, the pumps are said to be *dynamically similar*. When the Π variable C_Q, which relates only to the motion of the fluid, is equal in the series of pumps, the pumps are said to be *kinematically similar*. Finally, when corresponding parts of the pumps are proportional, the pumps are said to be *geometrically similar*. The relationships of equations (4-46) and (4-47) are called *similarity* or *scaling laws*.

Considering the power $\mathcal{P}$ and the efficiency η as the dependent variables, similar dimensional analyses yield the following similarity relations:

$$\left\{ \frac{\mathcal{P}}{\rho \omega^3 D^5} \right\}_a = \left\{ \frac{\mathcal{P}}{\rho \omega^3 D^5} \right\}_b = \left\{ \frac{\mathcal{P}}{\rho \omega^3 D^5} \right\}_c = \cdots = \left\{ \frac{\mathcal{P}}{\rho \omega^3 D^5} \right\}_n = \{C_{\mathcal{P}}\}_n \quad (4\text{-}48)$$

$$\{\eta\}_a = \{\eta\}_b = \{\eta\}_c = \cdots = \{\eta\}_n \quad (4\text{-}49)$$

where $C_{\mathcal{P}}$ is called the *power coefficient*. Note that the efficiencies of similar pumps are equal. The similarity relations also apply to the same pump, where in this case the subscripts $a, b, c, \ldots, n$ represent different operating conditions of the pump.

Figure 4-12 shows typical pump characteristic curves indicating the head, discharge, power, rotational speed, and efficiency. The head is actually the head developed

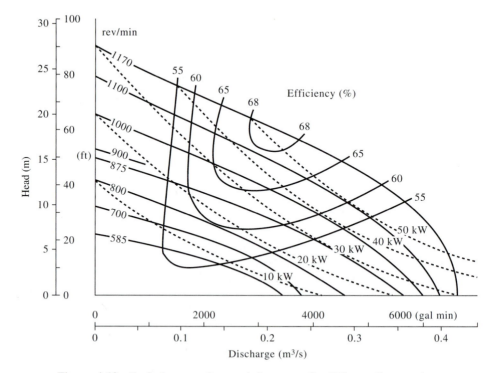

Figure 4-12 Typical pump characteristic curves for 375-mm-diameter impeller variable-speed pump.

by the pump TDH, which is equal to the energy head at the exit of the pump. The power is the brakepower. The efficiency is the ratio of the fluid power $\mathcal{P}_f$ to the brakepower $\mathcal{P}_s$.

Example 4-13

For the pump represented by Figure 4-12, determine (a) the discharge when the pump is operating at a head of 10 m and at a speed of 875 rpm, and (b) the efficiency and the power used.

Solution   (a) From the figure, $Q = 0.17$ m³/s   **Answer**
(b) $\eta = 63\%$ **Answer**; $\mathcal{P}_s = 26$ kW   **Answer**

Example 4-14

If the pump in Example 4-13 is operated at 1170 rpm, calculate the resulting H, Q, $\mathcal{P}_s$, and η.

Solution

$$\left\{\frac{Hg}{\omega^2 D^2}\right\}_b = \left\{\frac{Hg}{\omega^2 D^2}\right\}_a \Rightarrow \frac{H_b g}{\omega_b^2 D^2} = \frac{H_a g}{\omega_a^2 D^2} \Rightarrow H_b = 10\left(\frac{1170^2}{875^2}\right)$$

$$= 17.88 \text{ m}\quad \textbf{Answer}$$

$$\left\{\frac{Q}{\omega D^3}\right\}_b = \left\{\frac{Q}{\omega D^3}\right\}_a \Rightarrow \frac{Q_b}{\omega_b D^3} = \frac{Q_a}{\omega_a D^3} \Rightarrow Q_b = 0.17\left(\frac{1170}{875}\right)$$

$$= 0.23 \text{ m}^3/\text{s}\quad \textbf{Answer}$$

$$\left\{\frac{\mathcal{P}}{\rho\omega^3 D^5}\right\}_b = \left\{\frac{\mathcal{P}}{\rho\omega^3 D^5}\right\}_a \Rightarrow \frac{\mathcal{P}_{sb}}{\rho\omega_b^3 D^5} = \frac{\mathcal{P}_{sa}}{\rho\omega_a^3 D^5} \Rightarrow \mathcal{P}_{sb} = 26\left(\frac{1170^3}{875^3}\right)$$

$$= 62.16 \text{ kW}\quad \textbf{Answer}$$

$$\eta = 63\%\quad \textbf{Answer}$$

Example 4-15

If a homologous 30-cm pump is to be used for the problem in Example 4-14, calculate the resulting H, Q, $\mathcal{P}_s$ and η for the same rpm.

Solution   The diameter of the pump represented by Figure 4-12 is 375 mm.

$$\left\{\frac{Hg}{\omega^2 D^2}\right\}_b = \left\{\frac{Hg}{\omega^2 D^2}\right\}_a \Rightarrow \frac{H_b g}{\omega^2 D_b^2} = \frac{H_a g}{\omega^2 D_a^2} \Rightarrow H_b = 17.88\left(\frac{30^2}{37.5^2}\right)$$

$$= 11.447 \text{ m}\quad \textbf{Answer}$$

$$\left\{\frac{Q}{\omega D^3}\right\}_b = \left\{\frac{Q}{\omega D^3}\right\}_a \Rightarrow \frac{Q_b}{\omega D_b^3} = \frac{Q_a}{\omega D_b^3} \Rightarrow Q_b = 0.23\left(\frac{30^3}{37.5^3}\right)$$

$$= 0.12 \text{ m}^3/\text{s}\quad \textbf{Answer}$$

$$\left\{\frac{\mathcal{P}}{\rho\omega^3 D^5}\right\}_b = \left\{\frac{\mathcal{P}}{\rho\omega^3 D^5}\right\}_a \Rightarrow \frac{\mathcal{P}_{sb}}{\rho\omega^3 D_b^5} = \frac{\mathcal{P}_{sa}}{\rho\omega^3 D_a^5} \Rightarrow \mathcal{P}_{sb} = 62.16\left(\frac{30^5}{37.5^5}\right) =$$

20.37 kW   **Answer**

$$\eta = 63\% \quad \textbf{Answer}$$

Pump Specific Speed

Raising the flow coefficient $C_Q = Q/\omega D^3$ to the power $\frac{1}{2}$ and the head coefficient $C_H = gH/\omega^2 D^2$ to the power $\frac{3}{4}$, forming the ratio of the former to that of the latter, and simplifying, the expression for the specific speed N_s is obtained. The formula is

$$N_s = \frac{\omega\sqrt{Q}}{(gH)^{3/4}} \tag{4-50}$$

Since N_s is dimensionless, it can be used as a general characterization for an entire variety of pumps without reference to their sizes. Thus a certain range of the value of N_s would be a particular type of pump such as axial (no size considered), and another range would be another particular type of pump such as radial (no size considered). By characterizing all the pumps generally like this, the specific speed is of great importance in selecting the particular type of pump, whether radial or axial or any other type. For example, refer to Figure 4-13. The radial-vane pumps are in the range $N_s = 9.6$ to 19.2 (reading the metric units); the Francis vane pumps are in the range of 28.9 to 76.9; and so on. Therefore, if Q, ω, and H are known, N_s can be computed using equation (4-50) and thus, depending on the value obtained, the type of pump can be specified. In the United States, the equation is modified by using rpm as the unit for ω, gpm as the unit for Q, ft for the unit for H, and g is disregarded. Hence in U.S. practice, N_s is dimensional.

Pumping Station Heads

In the design of pumping stations, the engineer must be sure that the pumping system can deliver the fluid to the desired height. For this reason, energies are expressed in terms of heads. The various terminologies of heads are defined in Figure 4-14. The terms *suction* and *discharge* in the context of heads refer to the systems before and after the pumping station, respectively. *Static suction lift h_ℓ* is the vertical distance from the elevation of the inflow liquid level below the pump inlet to the elevation of the pump centerline (see the pump centerline in the figure). A lift is a negative head. *Static suction head h_s* is the vertical distance from the elevation of the inflow liquid level above the pump inlet to the elevation of the pump centerline. Static discharge head h_d is the vertical distance from the centerline elevation of the pump to the elevation of the discharge liquid level. *Total static head h_{st}* is the vertical distance from the elevation of the inflow liquid level to the elevation of the discharge liquid level. *Suction velocity head h_{vs}* is the entering velocity head at the suction side of the pump hydraulic system. In Figure 4-14, since the velocity in the wet well is practically zero, h_{vs} will also be practically zero. *Discharge velocity head h_{vd}* is the outgoing velocity head at the discharge side of the pump hydraulic system. In the figure it is also practically zero. *Total dynamic head* or *total developed head H* or TDH is the sum of the total static head, the friction head losses h_f, and the velocity head difference $h_{vd} - h_{vs}$. It is the total head developed by the pump or pump

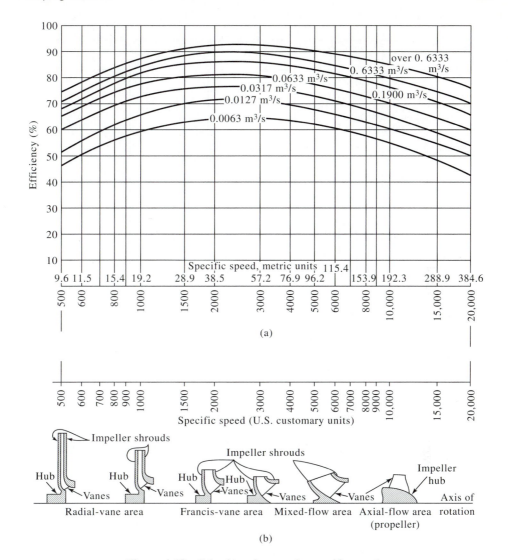

Figure 4-13   Selection of pumps by specific speeds.

assembly; it is the energy head given by the pump impeller to the pumped fluid. Friction head losses are the sum of the losses in straight runs of pipes h, transition losses, and fitting losses. Transition and fitting losses are called minor losses h_m. Friction losses are also composed of suction friction losses h_{fs} and discharge friction losses h_{fd}. In symbols, TDH is given by

$$TDH = h_{st} + h_f + (h_{vd} - h_{vs}) \tag{4-51}$$

$$h_f = h + h_m = h_{fs} + h_{fd} \tag{4-52}$$

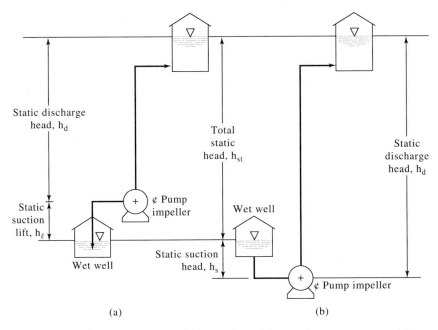

Figure 4-14   Head terms used in pumping: (a) pumping arrangement with suction lifts; (b) pumping arrangement with suction head.

The *inlet manometric head* h_i is the manometric level at the inlet to the pump; the outlet *manometric head* h_o is the manometric level at the outlet of the pump. (*Manometric level* is the height of liquid corresponding to the gage pressure.) In terms of these new variables and the velocity heads h_{vi} and h_{vo} for the pump inlet and outlet velocity heads, respectively, TDH may also be expressed as

$$\text{TDH} = h_o + h_{vo} - h_i - h_{vi} \tag{4-53}$$

For static suction lift conditions, h_i is negative and its theoretical limit is the negative of the difference between the prevailing atmospheric pressure and the vapor pressure of the liquid being pumped. Because of the suction action of the impeller and because the fluid is being lifted, the fluid column becomes "rubber-banded." Just like a rubber band, it becomes stretched as the pressure due to suction is progressively reduced; eventually, the liquid column ruptures. As the rupture occurs, the inlet suction pressure will actually have gone down to equal the vapor pressure, thus vaporizing the liquid and forming bubbles. This process is called *cavitation*. Cavitation can destroy hydraulic structures by the action of the outside atmospheric pressure, which has become greater than the pressure inside the structure. Also, as the bubbles that have been formed at a partial vacuum pressure at the inlet gradually progress along the impeller toward the outlet, the sudden increase in pressure causes an impact force. Continuous action of this force shortens the life of the impeller. The difference in pressure between the outside and the inside at incipient cavitation is the prevailing atmospheric pressure minus the vapor pressure of the liquid pumped.

The sum of the inlet manometric head and the inlet velocity head is called the *total inlet dynamic head* (TIDH). The sum of the outlet manometric head and the outlet velocity head is called the *total outlet dynamic head* (TODH). Of course, the TDH is also equal to the total outlet dynamic head minus the total inlet dynamic head.

$$TDH = TODH - TIDH \qquad (4\text{-}54)$$

Example 4-16

A designer wanted to recommend the use of an axial-flow pump to move wastewater to an elevation of 30 m above a sump. Friction losses and velocity at the sewage discharge level are estimated to be 20 m and 1.30 m/s, respectively. The operating drive is to be 1200 rpm. Suction friction losses are 1.03 m; the diameter of the suction and discharge lines are 250 and 225 mm, respectively. The vertical distance from the sump pool level to the pump centerline is 2 m. (a) Is the designer recommending the right pump? (b) If the temperature is 20°C, has cavitation occurred? (c) What are the inlet and outlet manometric heads? (d) What are the total inlet and outlet dynamic heads? From the values of the TIDH and TODH, calculate TDH.

Solution

(a)
$$N_s = \frac{\omega\sqrt{Q}}{(gH)^{3/4}}$$

$$TDH = h_{st} + h_f + (h_{vd} - h_{vs})$$

Take the sump pool level as point 1 and the sewage discharge level as point 2. $h_{vs} = 0$, since the pool velocity is zero.

$$h_{vd} = \frac{1.30^2}{2(9.81)} = 0.086 \text{ m}$$

$$TDH = 30 + 20 + 0.086 = 50.086 \text{ m}$$

$$\omega = \frac{1200(2\pi)}{60} = 125.67 \text{ rad/s}$$

$$A_d = \text{cross section of discharge pipe} = \frac{\pi D^2}{4} = \frac{\pi(0.225^2)}{4} = 0.040 \text{ m}^2$$

$$Q = (1.3)(0.040) = 0.052 \text{ m}^3/\text{s}$$

Thus,

$$N_s = \frac{125.67\sqrt{0.052}}{[(9.81)(50.086)]^{3/4}} = 0.27 \quad \text{falling outside the range of specific speeds in Figure 4-13}$$

However, the pump should not be an axial flow pump as recommended by the designer. **Answer**

(b)
$$\frac{V_1^2}{2g} + z_1 + \frac{P_1}{\gamma} + h_q - h_f + h_p = \frac{V_2^2}{2g} + z_2 + \frac{P_2}{\gamma}$$

Let the sump pool level be point 1 and the inlet to the pump be point 2.

$$A_2 = \frac{\pi (0.25^2)}{4} = 0.049 \text{ m}^2 \qquad V_2 = \frac{0.052}{0.049} = 1.059 \text{ m/s}$$

Thus,

$$0 + 0 + 0 + 0 - 1.03 + 0 = \frac{1.059^2}{2(9.81)} + 2 + \frac{P_2}{\gamma}$$

$$\frac{P_2}{\gamma} = -3.087 \text{ m}$$

At 20°C, $P_v = 2.34 \text{ kN/m}^2 = 0.239$ m of water. Assume a standard atmosphere of 1 atm = 10.34 m of water. Then

$$\text{theoretical limit of pump cavitation} = -(10.34 - 0.239) = -10.05 \text{ m} \ll -3.087$$

so cavitation has not been reached.   **Answer**

(c)   Inlet manometric head $= -3.087$ m   **Answer**

$$\text{TDH} = h_o + h_{vo} - h_i - h_{vi}$$

$$50.086 = h_0 + \frac{1.3^2}{2(9.81)} - (-3.087) - \frac{1.059^2}{2(9.81)}$$

$$h_0 = 46.97 \text{ m} = \text{outlet manometric head}   \textbf{Answer}$$

(d) $\text{TIDH} = h_i + h_{vi} = -3.087 + \dfrac{1.059^2}{2(9.81)} = -3.031 \text{ m}$ **Answer**

$$\text{TODH} = h_0 + h_{vo} = 46.97 + \frac{1.3^2}{2(9.81)} = 47.059 \text{ m}   \textbf{Answer}$$

$$\text{TDH} = \text{TODH} - \text{TIDH} = 47.059 - (-3.031) = 50.09 \text{ m}   \textbf{Answer}$$

Net Positive Suction Head

For a fluid to enter the pump, it must have sufficient energy to force itself toward the inlet. This means that a positive head (not negative head) must exist at the pump inlet. This head that must exist for pumping to be possible is termed the *net positive suction head* (NPSH). It is an absolute, not a gage, positive head.

Refer to Figure 4-14a. At the surface of the wet well, the net absolute pressure acting on the liquid is equal to the atmospheric pressure P_{atm} minus the vapor pressure of the liquid P_v. This net pressure is thus the atmospheric pressure corrected for vapor pressure, an "available energy" that forces the liquid to reach the inlet of the pump. Note that before the impeller can do its job, the fluid must reach it. Hence the need for a driving force at the inlet side. Since the surface of the well is below the pump, the available energy must be subtracted by h_ℓ before reaching the pump inlet. The other debits are the friction losses h_f. For Figure 4-14b, since the surface of the well is above the pump, h_s will be added, increasing the available energy. The losses will, again, be subtracted. In symbols,

$$\text{NPSH} = \frac{P_{atm} - P_v}{\gamma} - h_\ell \ (\text{or} + h_s) - h_f \tag{4-55}$$

It is instructive to derive equation (4-55) by applying the energy equation between the wet well pool surface and the inlet to the pump. Applying it to Figure 4-14a, the equation is derived as follows:

$$\frac{V_1^2}{2g} + z_1 + \frac{P_1}{\gamma} - h_f = \frac{V_2^2}{2g} + z_2 + \frac{P_2}{\gamma}$$

$$0 + 0 + \frac{P_{atm} - P_v}{\gamma} - h_f = \frac{V_2^2}{2g} + h_\ell + \frac{P_2'}{\gamma}$$

where V_1 is the velocity at the wet pool level, z_1 the elevation of the pool level with reference to a datum (the pool level, itself, in this case), P_1 the pressure at pool level, h_f the friction loss from pool level to the pump inlet, V_2 the velocity at the pump inlet, z_2 the elevation of the pump inlet with reference to a datum (the pool level), and P_2 the pressure at the inlet to the pump. P_2' is the "net" absolute pressure at the pump inlet (i.e., not a gage pressure but an absolute pressure with correction for vapor pressure). By also considering Figure 4-14b, the final equation after rearranging is

$$\frac{V_2^2}{2g} + \frac{P_2'}{\gamma} = \frac{P_{atm} - P_v}{\gamma} - h_\ell \ (\text{or} + h_s) - h_f \tag{4-56}$$

Therefore, NPSH is also

$$\text{NPSH} = \frac{V_2^2}{2g} + \frac{P_2'}{\gamma} = \frac{V_2^2}{2g} + \left(h_i + \frac{P_{atm} - P_v}{\gamma} \right) \tag{4-57}$$

In simple words, the NPSH is the amount of energy that the fluid possesses at the inlet of the pump. It is this energy that pushes the fluid into the pump impeller blades. Finally, NPSH and cavitation effects are related. If NPSH does not exist at the suction side, cavitation will obviously occur.

Example 4-17

In Example 4-16, what is the NPSH using (a) equation (4-57) and (b) equation (4-56)?

Solution

$$\text{(a)} \quad \text{NPSH} = \frac{V_2^2}{2g} + \frac{P_2'}{\gamma} = \frac{V_2^2}{2g} + \left(h_i + \frac{P_{atm} - P_v}{\gamma} \right)$$

$$= \frac{1.059^2}{2(9.81)} + [-3.087 + (10.34 - 0.239)] = 7.07 \text{ m} \quad \textbf{Answer}$$

(b) $$\text{NPSH} = \frac{P_{\text{atm}} - P_v}{\gamma} - h_\ell - h_f = \frac{P_{\text{atm}}}{\gamma} - \frac{P_v}{\gamma} - h_\ell - h_f$$

$$= 10.34 - 0.239 - 2 - 1.03 = 7.07 \text{ m} \quad \textbf{Answer}$$

Pumping Station Head Analysis

The *pumping station* and the *piping system* constitute the *pumping system.* In this system there are two types of characteristics: the *pump characteristics* and the *system characteristic.* The term *system characteristic* refers to the piping system characteristic. Specifically, *system characteristic* is the relation of Q and the associated head requirements. In the design of a pumping station, the pump characteristics and the system characteristic must be considered simultaneously.

Three types of heads are added in a system characteristic: total static head, friction head losses, and velocity head. These heads are the elements of the total developed head requirement of the system, TDHR. Note that TDH refers to the head produced by the pump and that the piping system "requires" this head in order for the fluid to flow (i.e., TDHR requires TDH).

It should be obvious that if the TDH of the pump assembly is less than the TDHR of the piping system, no fluid will flow. To ensure that the proper size of pumps is chosen for a given desired pumping rate, the TDH of the pumps must be equal to the TDHR of the piping system. This is easily done by plotting the pump head-discharge-characteristic curve and the system-characteristic curve on the same graph. The point of intersection of the two curves is the desired operating point. The principle of series or parallel connections of pumps must be used to arrive at the proper pump combination to suit the desired system characteristic requirement.

The piping system is composed of the suction piping, the pumping station piping, and the discharge piping system. For the purpose of system head calculations, it is convenient to disregard the head losses of the pumping station piping and the suction piping system. In this case the TDHR of the system characteristic will only include the total static head, the discharge piping losses, and the difference of velocity heads. The disregarded pumping station and suction piping losses are designated as station losses and applied as corrections to the pump head-discharge curve supplied by the manufacturer to obtain the *effective* pump head-discharge curve.

An illustration of this procedure is shown in Figure 4-15, which is a plot of the next example problem.

Example 4-18

Calculations for the system characteristic curve yield the following results:

Q (m³/s)	TDHR (m)	Q (m³/s)	TDHR (m)	Q (m³/s)	TDHR (m)
0.0	10.00	0.1	10.84	0.2	13.37
0.3	17.59	0.4	23.48	0.5	31.06

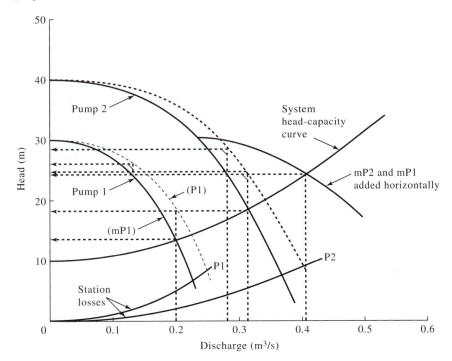

Figure 4-15 Use of pump characteristic head-discharge curve and system characteristic curve for sizing a pumping station.

The station losses are as follows:
Pump 1:

Q (m³/s)	h_f (m)	Q (m³/s)	h_f (m)	Q (m³/s)	h_f (m)
0.0	0.00	0.1	0.14	0.2	0.56

Pump 2:

Q (m³/s)	h_f (m)	Q (m³/s)	h_f (m)	Q (m³/s)	h_f (m)
0.0	0.00	0.1	0.06	0.2	0.26
0.3	0.58	0.4	1.03		

If two pumps with head-discharge characteristics plotted in Figure 4-15 (dashed lines) are to be used, determine (a) the pumping system discharges when each pump is operated separately and when both pumps are operated in parallel and (b) the head at which both pumps operated in series will deliver a discharge of 0.2 m³/s.

Solution   (a) The system head-discharge or head-capacity curve is plotted as shown in the figure. The pump head-discharge curves supplied by the manufacturer (dashed lines) are modified by the head losses as given above. The resulting effective head-discharge curves are drawn in solid lines designated as mP_1 and mP_2 for pumps 1 and 2, respectively. The intersection of the effective head-discharge curve and the system curve, when only pump 1 is operating, the pumping system discharge is 0.2 m^3/s.   **Answer**

When only pump 2 is operating, the system discharge is 0.31 m^3/s.   **Answer**

When both are operated in parallel, the effective characteristic curve for pump 1 is shifted horizontally to the right until the top end of the curve coincides with a portion of the effective curve of pump 2, as shown. This has the effect of adding the discharges for parallel operation. As indicated, when both are operated in parallel, the system discharge is 0.404 m^3/s.   **Answer**

(b) For the operation in series, the TDH for pump 1 for a system discharge of 0.2 m^2/s is 13 m; that of pump 2 is 32 m. Therefore, the system TDH is $32 + 13 = 45$ m. **Answer**

GLOSSARY

Aqueduct. *Aqua* means water and *duct* is a conduit; therefore, aqueduct is a conduit used to convey water.

Arterial main. The primary feeder.

Average daily flow. The yearly daily average flow of the appropriate probability of occurrence.

Branch sewer. Also called lateral sewer, the sewer that collects sewage from one or more buildings.

Building, house, or service pipe. The pipe that conveys water from the street main to the house or building.

Building or house sewer. The sewer that conveys sewage from the building or house to the branch or lateral sewer.

Canals. Open channels formed by a balanced cut-and-fill on the ground surface.

Capillary main. The distributor or street main.

Cavitation. A state of flow where the pressure in the liquid becomes equal to its vapor pressure.

Centrifugal pump. A pump that conveys fluid through the momentum created by a rotating impeller.

Combined sewer system. A system where stormwater and sanitary sewage are collected together.

Design period. A length of time for which the design is supposed to hold.

Discharge. In a pumping system, the arrangement of elements after the pumping station.

Discharge velocity head. The velocity head at the discharge of a pumping system.

Distribution reservoir. A reservoir constructed to equalize the supply and demand for treated water in a community and to provide a supply of water during emergency situations.

Domestic sewage. Sanitary sewage coming into homes.

Dry-weather flow. Sewage flow in a sanitary sewage collection system during periods of very low precipitation.

Fittings losses. Head losses in valves and fittings.

Flumes. Open channels formed of wood, masonry, and steels supported on or above the ground.

Force main. That part of a main where water is pumped or forced to a higher elevation.

Friction head loss. A head loss due to loss of internal energy.

Homologous pumps. Pumps that are similar. Similarity may be established dynamically, kinematically, or geometrically.

Impounding reservoir. A reservoir in the valley of a stream or river constructed for the purpose of holding streamflow; this flow may be released to satisfy demand during times of low stream flows.

Infiltration. That portion of precipitation that enters sewers through cracks and faulty joints.

Inflow. That portion of precipitation that enters sewers through holes in manholes and through roof leaders by illegal connection.

Inlet manometric head. The manometric level at the inlet to a pump.

Intercepting sewer. The sewer that accepts sewage from one or more trunk sewers.

Inverted siphon. A pressure conduit used to convey water across an obstruction.

Main. All pipes in the distribution system conveying water that ultimately reaches the consumer's premises.

Main sewer. The sewer that collects the sewage from one or more branch sewers.

Manhole. Manhole is an opening that allows a person to gain access to a structure.

Manifold pipe. A pipe with two or more pipes connected to it.

Maximum daily flow. The highest of the daily flows of the appropriate probability of occurrence.

Maximum hourly flow. The highest of the hourly flows of the appropriate probability of occurrence.

Net positive suction head (NSPH). The amount of energy possessed by a fluid at the inlet to a pump.

Outfall sewer. The sewer that conveys the sewage from a trunk or intercepting sewer into a sewage treatment plant; it is also the sewer that discharges the treated sewage into a receiving stream.

Outlet manometric head. The manometric level at the outlet of a pump.

Pneumatic system. A system used to convey air and other gases.

Positive-displacement pump. A pump that conveys fluid by moving it directly using a suitable mechanism such as a piston or a screw.

Primary feeder. Main that conveys water to the secondary feeder.

Probability distribution analysis. A method of statistical analysis where the probability of occurrence of events are arrayed in either descending or ascending order.

Pumping station. A location where one or more pumps are operated to convey fluids.

Pumping system. The pumping station and the piping system constitute the pumping system.

Return period or return interval. The average number of years that a particular event is equaled or exceeded.

Sanitary sewage. Sewage coming from the sanitary conveniences of homes, commercial establishments, industries, and the like.

Sanitary sewer system. The system for the collection of sanitary sewage.

Scaling laws. Mathematical equations that establish the similarity of homologous pumps.

Secondary feeder. Main that conveys water to the distribution mains.

Sewage. Liquid waste conveyed in a sewer.

Sewer. A conduit designed to convey liquid waste.

Sewerage. The collection, treatment, and disposal of sewage.

Sewerage works or sewage works. The equipment and physical structures required for the collection, treatment, and disposal of sewage.

Specific speed. A ratio obtained by manipulating the ratio of the flow coefficient to the head coefficient of a pump.

Standard fire stream. A fire flow of 250 gpm.

Standpipe. A vertical cylindrical tank whose height is considerably longer than its diameter.

Static discharge head. The vertical distance from the pump centerline to the elevation of the discharge liquid level.

Static suction head. The vertical distance from the elevation of the inflow liquid level above the pump centerline to the centerline of the pump.

Static suction lift. The vertical distance from the elevation of the inflow liquid level below the pump centerline to the centerline of the pump.

Storm sewage. Stormwater or water produced from rainfall.

Stormwater system. The system used for the collection of stormwater.

Street, service, or distributor main. Main laid along the street or some convenient place that actually distributes water to the consumers.

Suction. In a pumping system, the system of arrangement before the pumping station.

Suction velocity head. The velocity head at the suction side of a pumping system.

Total dynamic head or total developed head. The sum of the total static head, friction head losses, and the difference between the discharge velocity head and suction velocity head.

Total inlet dynamic head. The sum of the inlet velocity head and inlet manometric head of a pump.

Total outlet dynamic head. The sum of the outlet velocity head and outlet manometric head of a pump.

Total static head. The vertical distance between the elevation of the inflow liquid level and the discharge liquid level.

Transition losses. Head losses in expansions, contractions, and bends.

Trunk or main feeder. The main that conveys water to the primary feeder.

Trunk sewer. A large sewer that accepts sewage from one or more main sewers for discharge into an intercepting sewer or for discharge into a treatment plant through an outfall.

Turbomachine. Fluid machine that turns or tends to turn about an axis.

Ventilation volume. The rate of flow used to convey polluted air into a control device.

Wastewater. Spent water after homes, industries, commercial establishments, public places, and similar entities have used their water.

SYMBOLS

A_v	air contaminant capturing area of a virtual hood
C_H	head coefficient
C_P	power coefficient
C_Q	flow coefficient
C_V	velocity coefficient
d	partial depth of flow
D	diameter of pipe; depth at full depth of flow
D_c	diameter of a rising column of hot air a distance, y, from the surface of a hot bath source
D_f	diameter of a high canopy hood
D_s	diameter of a hot bath source
f	Darcy–Weisbach, Blasius, or Fanning friction factor
F	rate of flow in fire-flow formula
g	earth's gravitational acceleration
h	head loss in straight run of pipe, piezometric head
h_d	static discharge head
h_f	friction head loss
h_{fd}	discharge friction losses
h_{fs}	suction friction losses
h_i	inlet manometric head
h_ℓ	suction lift
h_o	outlet manometric head
h_s	suction head
h_{st}	total static head
h_{vd}	discharge velocity head
h_{vi}	inlet velocity head
h_{vo}	outlet velocity head
h_{vs}	suction velocity head
H	total dynamic or developed head
k	$(f\ell)/(2fDA^2)$, a head-loss coefficient in the Hardy Cross method
K	K_e = expansion loss coefficient, K_c = contraction loss coefficient, K_v = velocity head loss coefficient, K_b = bend loss coefficient, etc.
ℓ	length of pipe segment
mgd	million gallons per day

n	Manning coefficient of roughness; number of extraction pumps of equal sizes
N	Manning roughness coefficient at full flow in pressure flow
N_s	specific speed
NPSH	net positive suction head
p	wetted perimeter at partial flow
P	pressure
P_v	vapor pressure
q	rate of flow at partial flow
Q	rate of flow
Q_c, Q_{cc}	in the Hardy Cross method, clockwise and counterclockwise flow in a loop, respectively
$\dot{Q}_{in}$	rate of heat added to system
Q_v	rate of flow into a virtual hood
r	hydraulic radius at partial flow
R	hydraulic radius
Re	Reynolds number
s	energy slope at partial flow
S	energy slope at full flow
TDH	total dynamic or developed head
TIDH	total inlet dynamic head
TODH	total outlet dynamic head
v	velocity of flow in sewer at partial flow
v_s	self-cleaning velocity at partial flow
v_x	velocity at distance x measured parallel to an axis from hood entrance
Y	fraction of the hood entrance velocity
η	pump efficiency
μ	absolute or dynamic viscosity
$\bullet$	dot product operator

PROBLEMS

4-1. Treated wastewater at 20°C discharges from the secondary clarifier to an estuary below. The elevation at the discharge trough of the clarifier is 7 m. The water from the trough enters the circular discharge pipe at a rate of 2.5 m³/s. The length of the pipe is 50 m. Calculate the diameter of the pipe, assuming that it is made of reinforced concrete.

4-2. In Problem 4-1, calculate the entrance loss as the water enters the discharge pipe from the trough.

4-3. Water at a temperature of 22°C is to be conveyed from a reservoir with a water surface elevation of 70 m to another reservoir 300 m away with a water surface elevation of 20 m. Determine the diameter of a steel pipe if the flow is 2.5 m³/s. Assume a square-edged inlet and outlet and that there are two open gate valves in the system.

4-4. Assume that a turbine is installed between the two reservoirs in Problem 4-3. If the turbine incurred an energy loss equivalent to 0.2 of the velocity head, how much power did it develop?

4-5. Treated water from a water treatment plant is conveyed through a 4-in.-diameter steel main at an average velocity of 5.5 fps. A few feet downstream the 4-in. main is bypassed by a 1-in. steel pipe having an equivalent of 30 ft. The length of the main being bypassed is approximately 25 ft. Neglecting entrance and exit losses, determine the rate of flow of the water bypassed.

4-6. Solve Problem 4-5 by not neglecting the entrance and exit losses.

4-7. The estimated hourly requirement for the maximum day in a small city is tabulated below. Fire flows have been determined to be 65 L/s for a conflagration of 2 h. If pumping is to start at hour 0800 and to end at hour 1600, calculate the storage requirement. Determine the storage requirement by a method of storing the excess pumpage over the demand.

Hour ending	Demand (m³)	Hour ending	Demand (m³)
0100	160	1300	444
0200	129	1400	447
0300	146	1500	434
0400	138	1600	405
0500	159	1700	417
0600	196	1800	421
0700	274	1900	406
0800	390	2000	343
0900	480	2100	270
1000	500	2200	223
1100	473	2300	199
1200	451	2400	185

4-8. In Problem 4-7 plot the demand and pumping against time and determine the storage requirement.

4-9. A transmission main carries a flow of 180 mgd of treated water. It is made of reinforced concrete. The elevation at the distribution reservoir is 70 m and the elevation at the "foot of the hill" before the transmission main joins the primary main is 10 m at an available pressure of 415 kPa gage. Assume the combined friction losses to be two times the velocity head. Determine the diameter of the transmission main.

4-10. In Problem 4-9 a turbine is installed along the transmission line to charge an arrangement of electric batteries in an electric battery powerhouse. This powerhouse is used for electric power in a shop nearby. If the turbine has an energy loss of 0.2 of the velocity head, how much energy is stored as electric charge?

4-11. Hydrants are normally spaced to cover a radius of 60 m. Assuming that a rubber-lined hose of $2\frac{1}{2}$-in. diameter is used, calculate how high the water rise will be if the hose is aimed at an angle of 60° with the horizontal and (**a**) the hose is connected to a hydrant in a high-value district, and (**b**) the hose is connected to a hydrant in a residential area. Assume a *standard fire stream* of 250 gpm at a temperature of 20°C. Consider entrance and exit losses of the hose.

4-12. Draw a sketch of a fire hydrant.

4-13. If the city in Problem 4-7 is assumed to be built of buildings of ordinary construction as a general characterization with a total combined floor area of 2000 m^2 per story, compute the storage requirement. Assume that the average number of stories per building is two.

4-14. Solve Problem 4-13 for a noncombustible construction.

4-15. Solve Problem 4-13 for a fire-resistive construction.

4-16. For the given source and load in figure of the derivation of the Hardy Cross method, **(a)** determine the distribution of flows in the network and **(b)** the pressure at the load points B, H, and I if the pressure at the source is 60 psig using the starting and ending points below. Assume that all pipes are horizontal and that $f = 0.012$.

Loop	Starting point	Ending point
I	A	H
II	H	G
III	D	B

4-17. What is dry-weather flow?

4-18. The sewage inflow records of the sewage treatment plant in the town of Middletown were analyzed for the average flow of the driest months of each year (average of the flows during the successive months of June, July, and August of each year of record), yielding the following probability distribution.

Average dry-weather flow (L/capita-day)	Probability p that flow is equaled or exceeded
310	0.019
307	0.080
304	0.23
300	0.38
⋮	⋮
160	0.84
147	0.99

A representative section of the Middletown collection system was then selected for study for its dry-weather and wet-weather flow characteristics, producing the following results: average peak dry-weather flow = 205 L/s, average minimum dry-weather flow = 46 L/s, and average flow = 96 L/s. The combined infiltration inflow was found to be 70 L/s. If the midyear population of Middletown is 58,600, what are the present minimum, maximum, and average flows of the town that are equaled or exceeded 1% of the time?

4-19. An extensive study was done on the collection system of a town. The minimum and maximum flows were found to be 5037 m^3/day and 15,319 m^3/day, respectively. Assuming a self-cleaning velocity of 0.76 m/s, what size sewer should, at least, be discharging this flow at the present time?

4-20. At what slope should the sewer in Problem 4-19 be laid to ensure self-cleaning?

4-21. A spray booth has a 3-m by 2-m entrance. The 3 m is the vertical dimension and the 2 m is the horizontal dimension. A null point determined by experiment was located a distance of 2 m from the vertical centerline of the entrance, 1.5 m above the floor, and 2 m away from the face of the booth. If a capture velocity of 100 fpm is to be provided at the null point, what ventilation volume is required?

4-22. What is the face velocity at the entrance to the booth in Problem 4-21?

4-23. A plant is to be redesigned to increase the airflow from 10,000 cfm to 12,000 cfm. The existing fan runs at 1600 rpm, providing a pressure drop of 7.0 in. of water. Estimate the new fan speed.

4-24. What is the new pressure drop in Problem 4-23?

4-25. A slot hood is to be used to capture the emissions from a chrome-plating tank. The tank measures 2 m by 3 m on the surface. The slot is to be put on both of the 3-m sides only. From an actual experiment, the ventilation rate was determined to be 200 cfm/ft^2 of tank surface. Determine the slot hood dimensions.

4-26. A batch-galvanizing operation uses a batch kettle measuring 3 m in diameter. Because of interference of the crane used to load the kettle, the hood must be located 3.5 m above the molten zinc surface. The metal temperature is 885°F and the room temperature is approximately 80°F. Determine the dimensions of the hood.

4-27. Calculate the ventilation rate in Problem 4-26.

4-28. If the emissions from the galvanizing kettle of Problem 4-26 are to be captured by an enclosure, determine the total ventilation requirement. The enclosure is 8 m wide by 10 m long by 6 m high. The access opening is 3 m high by 10 m long and runs along the length of the enclosure. The total leakage area is approximately 0.093 m^2 and is located at the top of the enclosure. Assume an indraft velocity of 60 m/min. The average temperature of the air inside the hood is 66°C.

4-29. In the sketch for the waste air collection system discussed in this chapter, determine the head loss needed to size the fan H. Ventilation requirements have been determined to be as follows: dump hopper = 30 m^3/min, bucket elevator = 8.0 m^3/min, ribbon blender = 10.0 m^3/min, and drum booth = 25.0 m^3/min. The duct lengths and fittings required are as follows:

Branch	Length (m)	Number of elbows (deg)	Number of branch entries (deg)
A	4.0	1, 90	0
B	1.3	0	1, 45
C	2.0	0	0
D	1.7	0	1, 30
E	1.5	0	0
F	3.7	1, 90	1, 30
G	4.0	1, 90	Expands into 5

Assume that $f = 0.03$ in all ducts and a throat radius for bends of $2D$. Assume that a minimum conveying velocity of 1000 m/min is sufficient. Although it cannot be neglected, neglect the head loss at the outfall main G and the losses in the bagfilter assembly 5.

4-30. For the pump represented by the characteristic curve mentioned in this chapter, determine the discharge when the pump is operating at a head of 20 m at a speed of 1100 rpm.

4-31. Calculate the efficiency and power for the pump in Problem 4-30.

4-32. If the pump in Problem 4-30 is operated at 1170 rpm, calculate the resulting H, Q, $\mathcal{P}_s$, and η.

4-33. If a homologous 30-cm pump is to be used in Problem 4-30, calculate the resulting H, Q, $\mathcal{P}_s$, and η for the same rpm.

4-34. The outlet manometric head at the discharge of a pump is equal to the equivalent of 50 m of water. If the discharge velocity is 2.0 m/s, what is the total outlet dynamic head?

4-35. You are required to recommend the type of pump to be used to convey wastewater to an elevation of 8 m above a sump. Friction losses and the velocity at the sewage discharge level are estimated to be 3 m and 1.30 m/s, respectively. The operating drive is to be 1200 rpm. The suction friction loss is 1.03 m; the diameters of the suction and discharge lines are 250 and 225 m, respectively. The vertical distance from the sump pool level to the pump centerline is 2 m. What type of pump would you recommend?

4-36. In Problem 4-35, if the temperature is 20°C, has cavitation occurred?

4-37. Compute the inlet and outlet manometric heads in Problem 4-35.

4-38. In Problem 4-35, what are the inlet and outlet total dynamic heads? From the values of TIDH and TODH, calculate TDH.

4-39. In Problem 4-35, what is the NPSH using

(a) $\dfrac{V_2^2}{2g} + \dfrac{P_2'}{\gamma} = \dfrac{P_{\text{atm}} - P_v}{\gamma} - h_\ell \text{ (or } + h_s) - h_f$

(b) $\text{NPSH} = \dfrac{V_2^2}{2g} + \dfrac{P_2'}{\gamma} = \dfrac{V_2^2}{2g} + \left(h_i + \dfrac{P_{\text{atm}} - P_v}{\gamma} \right)$

4-40. In Example 4-18, calculate the percent errors in the answers if the characteristic curves of the pumps are not corrected for the station losses.

BIBLIOGRAPHY

COOPER, C. D., and F. C. ALLEY (1986). *Air Pollution Control: A Design Approach*. Waveland Press, Prospect Heights, IL.

DANIELSON, D. A. (1973). *Air Pollution Engineering Manual. Office of Air Quality Planning and Standards*, U.S. Environmental Protection Agency, Research Triangle Park, NC.

HAMMER, M. J. (1986). *Water and Wastewater Technology*. Wiley, New York.

McGHEE, T. J. (1991). *Water Supply and Sewerage*. McGraw-Hill, New York.

PEAVY, H. S., D. R. ROWE, and G. TCHOBANOGLOUS (1985). *Environmental Engineering*. McGraw-Hill, New York.

QASIM, S. R. (1985). *Wastewater Treatment Plants: Planning, Design, and Operation*. Holt, Rinehart and Winston, New York.

CHAPTER 5

Introduction to Environmental Quality Modeling

Environment connotes surroundings; and surroundings include the air, surface water, and soil. The environment of soil includes the vadose zone and the groundwater. *Environmental quality modeling* is defined as the *representation* of the various *characteristic quality processes* that occur in the air, surface water, and soil. In general, environmental quality modeling deals with the effect of waste discharges on the quality of the environment. These wastes may be hazardous or conventional in nature. Hazardous wastes include such wastes as pesticides and radioactive materials; conventional wastes are such wastes as the familiar biochemical oxygen demand.

The environmental qualities of air, surface water, and soil are the prototypes; the representations of these qualities are called *models*. In the form of mathematical equations, the models are called *mathematical models*. If actually representing smaller- or larger-scale versions of the prototypes, the models are called *physical models*. Mathematical models that are programmed in a computer by using various computer languages, such as FORTRAN, QuickBASIC, and Turbo Pascal, are called *computer models*. In this chapter we derive mathematical models of air quality, surface water quality, and subsurface water quality. Subsurface water refers to the underground waters, which, in turn, can be subdivided into the vadose or unsaturated zone and the groundwater zone.

GENERAL POLLUTANT MATERIAL BALANCE

From the Reynolds transport theorem, let ϕ, the property per unit mass, be designated as c', the mass of any pollutant per unit mass of the control mass. Then $c'(\rho) = c$, the mass of any pollutant per unit space volume, where ρ is the mass density of the control mass. (For discussions on the Reynolds transport theorem and the meaning of control mass, the reader should read these topics in Appendix 21 or in any book on fluid mechanics.) Substituting into the theorem and simplifying gives us

$$\frac{D}{Dt}c = \frac{\partial}{\partial t}c + \nabla \bullet c\mathbf{v} \tag{5-1}$$

where D/Dt is the material derivative and $\mathbf{v}$ is the velocity vector. The left-hand side of equation (5-1) is the Lagrangian method of describing the rate of change of c, while the right-hand side is the equivalent Eulerian method of describing the rate of change of the same property, c. (Again, see Appendix 21 for the meaning of the Lagrangian method and its counterpart, the Eulerian method. It should be noted here that by this theorem the boundaries of the control mass and the control volume coincide. The Lagrangian method addresses the control mass, and the Eulerian method addresses the control volume.)

SURFACE WATER QUALITY MODELING

Figure 5-1 shows a picture of a surface water body in the Everglades, Florida. Models of surface water similar to this are derived in this section. The basis of surface water quality modeling is the pollutant material balance equation, equation (5-1). But from the definition of the material derivative, $Dc/Dt = \partial c/\partial t + \nabla c \bullet \mathbf{v}$. Therefore, $c \nabla \bullet \mathbf{v} = 0$, and

$$\frac{Dc}{Dt} = \frac{\partial c}{\partial t} + \nabla c \bullet \mathbf{v} = \frac{\partial c}{\partial t} + u\frac{\partial c}{\partial x} + v\frac{\partial c}{\partial y} + w\frac{\partial c}{\partial z} \tag{5-2}$$

where x, y, and z, are the Cartesian coordinates and u, v, and w are the respective components of $\mathbf{v}$ in the x, y, and z directions.

The concentration c may be present in the dissolved form c_d and in the sorbed form c_{sp}. c_{sp} may be defined as the mass of solute per unit mass of solids or sediments suspended in water. The concentration of solids may, in turn, be designated as m_s, the mass of solids per unit volume of water. It has been observed that pollutants in water, particularly the hydrophobic organics, may be sorbed onto sediment particles, the c_{sp}.[1] This sorption phenomenon is due largely to the carbon content of the particles. Hence, corresponding to these two forms, in a unit of space volume, two material derivatives should be written: $D(\eta c_d)/Dt$ and $D(c_{sp}\eta m_s)/Dt$, where η, the porosity or volume fraction occupied by water, is used to convert concentrations to space values.

---

[1]USEPA (1988). *WASP4, A Hydrodynamic and Water Quality Model: Model Theory, User's Manual, and Programmer's Guide.* EPA/600/3-87/039. Environmental Research Laboratory, U.S. Environmental Protection Agency, Athens, Ga., p. 105.

Figure 5-1   Surface water body in the Everglades, Florida.

Equation (5-2) then becomes

$$\frac{D(\eta c_d)}{Dt} + \frac{D(c_{sp}\eta m_s)}{Dt} = \frac{\partial(\eta c_d)}{\partial t} + \frac{\partial(c_{sp}\eta m_s)}{\partial t} + \eta\left(u\frac{\partial c_d}{\partial x} + v\frac{\partial c_d}{\partial y} + w\frac{\partial c_d}{\partial z}\right) \quad (5\text{-}3)$$

The concentration c_d is expressed as the mass of solute per unit volume of water. c_d is larger than solute concentrations referred to space. The porosity η is used to convert c_d to its equivalent concentration referred to space. The same statement holds for m_s.

c_{sp} is expressed as the mass of the adsorbate pollutant per unit mass of the adsorbent particle. c_{sp} and c_d are related to each other by the *partition coefficient* K_{sp} as follows:

$$c_{sp} = K_{sp}c_d \quad (5\text{-}4)$$

K_{sp} is related to K_{sc} by

$$K_{sp} = f_{cp}K_{sc} \quad (5\text{-}5)$$

where f_{cp} is the mass fraction of carbon in the particles and K_{sc} is the partition coefficient of the solute between water and carbon. Equation (5-5) is interpreted to mean that sorption into the solid is due to the carbon fraction it contains.

The partition coefficient is a measure of the distribution of the pollutant between water and solid. It is determined by mixing the chemical in the system and determining the concentrations in the resulting phases after standing. The ratio of the chemical concentration in the solid phase to that in the liquid phase is the partition coefficient.

A unit volume of space is occupied by the water, pollutant solute, and sediment particles. c_d, which is a mass per unit volume of water, is equivalent to ηc_d when expressed in terms of mass per unit volume of space. Also, $c_{sp} = K_{sp}c_d$ is equivalent to $f_{cp}K_{sc}c_d m_s \eta$ in mass per unit volume of space. c is in mass per unit volume of space. In terms of c_d and $f_{cp}K_{sc}c_d m_s \eta (= c_{sp})$,

$$c = \eta c_d + f_{cp}K_{sc}c_d m_s \eta = c_d \eta (1 + f_{cp}K_{sc}m_s) \tag{5-6}$$

In terms of space volume, the fraction of the dissolved pollutant f_d is, therefore,

$$f_d = \frac{\eta c_d}{c} = \frac{\eta c_d}{c_d \eta (1 + f_{cp}K_{sc}m_s)} = \frac{1}{1 + f_{cp}K_{sc}m_s} \tag{5-7}$$

The corresponding fraction of pollutants sorbed onto the particles f_{sp} is

$$f_{sp} = \frac{c_{sp}m_s \eta}{c} = \frac{f_{cp}K_{sc}c_d m_s \eta}{c} = \frac{f_{cp}K_{sc}c_d m_s \eta}{c_d \eta (1 + f_{cp}K_{sc}m_s)} = \frac{f_{cp}K_{sc}m_s}{1 + f_{cp}K_{sc}m_s} \tag{5-8}$$

When samples of river water are taken and concentrations measured, the values obtained are not for c but for c_d. Hence c_d is the practical concentration to be used. In equation (5-3), c has been expressed in terms of c_d and c_{sp}. By the use of equations (5-4) and (5-5), c_{sp} may be expressed in terms of c_d as

$$c_{sp} = f_{cp}K_{sc}c_d \tag{5-9}$$

Substituting equation (5-9) into equation (5-3) and simplifying yields

$$\frac{D(\eta c_d)}{Dt}(1 + f_{cp}K_{sc}m_s) = \frac{\partial(\eta c_d)}{\partial t}(1 + f_{cp}K_{sc}m_s) + \eta \left(u\frac{\partial c_d}{\partial x} + v\frac{\partial c_d}{\partial y} + w\frac{\partial c_d}{\partial z} \right) \tag{5-10}$$

Equation (5-10) has been derived to predict the concentration of any dissolved pollutant in water. The corresponding equation for c_{sp} may be obtained by following similar mathematical steps. The result is

$$\frac{D(\eta c_{sp})}{Dt}\left(\frac{1}{f_{cp}K_{sc}} + m_s \right) = \frac{\partial(\eta c_{sp})}{\partial t}\left(\frac{1}{f_{cp}K_{sc}} + m_s \right)$$
$$+ \frac{\eta}{f_{cp}K_{sc}}\left(u\frac{\partial c_{sp}}{\partial x} + v\frac{\partial c_{sp}}{\partial y} + w\frac{\partial c_{sp}}{\partial z} \right) \tag{5-11}$$

Example 5-1

It is desired to model the concentration of suspended solids m_s in water. Write the model equation.

Solution

$$\frac{Dm_s}{Dt} = \frac{\partial m_s}{\partial t} + u\frac{\partial m_s}{\partial x} + v\frac{\partial m_s}{\partial y} + w\frac{\partial m_s}{\partial z} \quad \textbf{Answer}$$

Material Derivative, $D(\eta c_d)/Dt$

To complete equation (5-10), the material derivative should be evaluated. From fluid mechanics, the material derivative is the Lagrangian method of evaluating a rate of change (see Appendix 24). Hence the material derivative is a Lagrangian derivative. In the Lagrangian method, the control mass (water in the present case) is not allowed to pass through the control mass boundary; however, its properties, such as heat and momentum, can pass through. In this method of description, the control mass is followed everywhere it goes around in the surroundings, and the fate of any given property transpiring inside it is tracked. In the present case, this property is the concentration c_d. Since the control mass is not allowed to cross the boundary, the only way that this property can change is for c_d to diffuse or disperse across the boundary or for it to decay or grow in the control mass. Diffusion and dispersion here are collectively called *dispersion*.

Dispersion of pollutants or any substance can be from the outside to the inside of the control mass boundary, or vice versa. There are three possible cases that determine the direction of dispersion: the first case is when the substance is conservative, the second case is when the substance is nonconservative, and the third case is when the substance is growing inside the control mass. *Conservative* substances are those substances that do not decay, while *nonconservative* substances are those substances that do decay. An example of a conservative substance is salt, an example of a nonconservative substance is BOD, and an example of a growing substance is algae inside the control mass. Each of these possible cases will have have its own direction of dispersion and total derivative (the material derivative).

Ascertaining the direction of dispersion is very important, as it will govern the form of the differential equation that represents the process. The differential equation for processes where the direction of dispersion is from the inside to the outside of the control mass is different from those differential equations where the dispersion is from the outside to the inside of the control mass.

Consider a control mass as small as a point holding a conservative pollutant. As this mass moves about in its surroundings, the pollutant has two choices: remain in its place in the point control mass or move away. Hence, if dispersion is to occur, it would be away from the control mass toward the surroundings. Now, start with the same point control mass, again, but this time with a nonconservative pollutant. Since the pollutant is nonconservative, it will decay, and as it decays, a void will exist at its location. This void will clearly create a concentration gradient from the vicinity toward the point. Thus the direction of dispersion of nonconservative substances is from the outside of the control mass toward the inside.

In addition to the normal pollutants that are "floating" in the water column and diffusing toward the voids created by decay, pollutant sources may also exist distributed along the path of travel of the control mass. Although these sources will diffuse toward the voids, their motion toward the voids will be considered as a direct input G expressed in units of mass per unit time per unit volume of space. An example of these sources is the sediment oxygen demand, the organic substances deposited at streambeds, and leaves of trees falling into streams.

The third possible case, growth, is exemplified by the growing of algae inside the control mass. Since it is growing, algae will accumulate and therefore must disperse out of the control mass. Hence for growth processes, the direction of dispersion is from the inside toward the outside.

Inside the control mass, the decay or growth of c_d may be biological, chemical, or physical. An example of a biological reaction is the consumption of the pollutant by microorganisms; an example of a chemical reaction is the disappearance of the substance by hydrolysis; and an example of a physical reaction is settling. All these reactions may be considered separately, grouped according to similarity of reactions, or lumped into one. These reactions are called *kinetic reactions* or *kinetic transformations*. Kinetic reactions may decrease or increase c_d, as when the substance is one that is growing in the control mass.

The dispersion, kinetic transformation, G, and the like make up the member elements of the material derivative. Calling **j** the vector of dispersion, k the lumped decay coefficient, $\hat{n}$ the unit normal vector to the control surface A of control mass M, and μ the growth coefficient, the material derivative (total derivative) equations corresponding to the conservative, nonconservative, and growing substances are, respectively,

$$\int_M \frac{D(\eta c_d)}{Dt}\, d\Psi = \begin{cases} \oint_A (+\mathbf{j} \bullet \hat{n})\, dA & \text{conservative substance} \\[2ex] \oint_A (-\mathbf{j} \bullet \hat{n})\, dA + \int_M -k\eta c_d\, d\Psi & \\ \qquad\qquad + \int_M +G\, d\Psi & \text{nonconservative substances} \\[2ex] \oint_A (+\mathbf{j} \bullet \hat{n})\, dA + \int_M \mu\eta c_d\, d\Psi & \text{growing substances} \end{cases} \quad (5\text{-}12)$$

Ψ is the volume of the control mass, which by the Reynolds transport theorem, is also equal to the volume of the control volume since the respective boundaries of the two systems coincide.

When dispersion is going out of the control mass, **j** and $\hat{n}$ have the same signs, and when dispersion is going into the control mass, **j** and $\hat{n}$ have opposite signs. (The positive $\hat{n}$ is always normal to the control mass boundary, on the outside, and pointing in the direction away from the control-mass boundary surface. In the first and the last of equations (5-12), dispersion is going out of the control mass; hence the positive sign has been prefixed. In the second equation, dispersion is going toward the control mass; therefore, the negative sign has been prefixed. On the other hand, the kinetic coefficient is negative when c_d is disappearing or decaying and positive when c_d is increasing. To conform to these conventions, the negative sign has been used for k and the positive sign has been used for μ. Finally, G as a source is always positive. Using the Gauss–Green theorem (see Appendix 20) to the surface integrals and canceling $d\Psi$ in the limit, the total derivative differential equations are, respectively,

$$\frac{D(\eta c_d)}{Dt} = \begin{cases} (+\nabla \bullet j) & \text{conservative substances} \\ (-\nabla \bullet j) - k\eta c_d + G & \text{nonconservative substances} \\ (+\nabla \bullet j) + \mu\eta c_d & \text{growing substances} \end{cases} \quad (5\text{-}13)$$

j is equal to $E \bullet [-\nabla(\eta c_d)]$, where E is the dispersion tensor. For dispersion to occur, it must flow in the direction of decreasing concentration; thus the concentration gradient, $\nabla(\eta c_d)$, is inherently negative. A negative sign is therefore prefixed to make the entire gradient term positive. Also, η has been inserted in the argument (before ∇) since c_d is a concentration in the water while the gradient is with respect to the space. Substituting gives

$$\frac{D(\eta c_d)}{Dt} = \begin{cases} \nabla \bullet \{E \bullet [-\nabla(\eta c_d) & \text{conservative substances} \\ -\nabla \bullet \{E \bullet [-\nabla(\eta c_d)] - k\eta c_d + G & \text{nonconservative substances} \\ \nabla \bullet \{E \bullet [-\nabla(\eta c_d)] + \mu\eta c_d & \text{growing substances} \end{cases} \quad (5\text{-}14)$$

The second-order tensor E in the xyz coordinates is

$$E = E_{xx}\hat{i}\hat{i} + E_{xy}\hat{i}\hat{j} + E_{xz}\hat{i}\hat{k} + E_{yx}\hat{j}\hat{i} + E_{yy}\hat{j}\hat{j}$$
$$+ E_{yz}\hat{j}\hat{k} + E_{zx}\hat{k}\hat{i} + E_{zy}\hat{k}\hat{j} + E_{zz}\hat{k}\hat{k} \quad (5\text{-}15)$$

where $\hat{i}$, $\hat{j}$, and $\hat{k}$ are the unit vectors in the x, y, and z directions, respectively. Also, in the xyz coordinates ∇ is

$$\nabla = \hat{i}\frac{\partial}{\partial x} + \hat{j}\frac{\partial}{\partial y} + \hat{k}\frac{\partial}{\partial z} \quad (5\text{-}16)$$

Substituting these expressions for E and ∇ in the nonconservative substance equation of equation (5-14) and simplifying yields

$$\frac{D(\eta c_d)}{Dt} = \left[\frac{\partial}{\partial x}\left(E_{xx}\frac{\partial \eta c_d}{\partial x}\right) + \frac{\partial}{\partial x}\left(E_{xy}\frac{\partial \eta c_d}{\partial y}\right) + \frac{\partial}{\partial x}\left(E_{xz}\frac{\partial \eta c_d}{\partial z}\right)\right.$$
$$+ \frac{\partial}{\partial y}\left(E_{yx}\frac{\partial \eta c_d}{\partial x}\right) + \frac{\partial}{\partial y}\left(E_{yy}\frac{\partial \eta c_d}{\partial y}\right) + \frac{\partial}{\partial y}\left(E_{yz}\frac{\partial \eta c_d}{\partial z}\right) \quad (5\text{-}17)$$
$$\left. + \frac{\partial}{\partial z}\left(E_{zx}\frac{\partial \eta c_d}{\partial x}\right) + \frac{\partial}{\partial z}\left(E_{zy}\frac{\partial \eta c_d}{\partial y}\right) + \frac{\partial}{\partial z}\left(E_{zz}\frac{\partial \eta c_d}{\partial z}\right)\right] - k\eta c_d + G$$

Dispersion (or diffusion) is a process that passes normal to an area. Dispersion coefficients with mixed indices such as E_{xy} and E_{yz} signify dispersion parallel to an area, not normal to it; hence terms containing are zero. Therefore, equation (5-17) and the rest of equations (5-14) become

$$\frac{D(\eta c_d)}{Dt} = \begin{cases} -\left[\frac{\partial}{\partial x}\left(E_{xx}\frac{\partial \eta c_d}{\partial x}\right) + \frac{\partial}{\partial y}\left(E_{yy}\frac{\partial \eta c_d}{\partial y}\right) \right. \\ \qquad\qquad \left. + \frac{\partial}{\partial z}\left(E_{zz}\frac{\partial \eta c_d}{\partial z}\right)\right] \\[4pt] \left[\frac{\partial}{\partial x}\left(E_{xx}\frac{\partial \eta c_d}{\partial x}\right) + \frac{\partial}{\partial y}\left(E_{yy}\frac{\partial \eta c_d}{\partial y}\right) \right. \\ \qquad\qquad \left. + \frac{\partial}{\partial z}\left(E_{zz}\frac{\partial \eta c_d}{\partial z}\right)\right] - k\eta c_d + G \\[4pt] -\left[\frac{\partial}{\partial x}\left(E_{xx}\frac{\partial \eta c_d}{\partial x}\right) + \frac{\partial}{\partial y}\left(E_{yy}\frac{\partial \eta c_d}{\partial y}\right) \right. \\ \qquad\qquad \left. + \frac{\partial}{\partial z}\left(E_{zz}\frac{\partial \eta c_d}{\partial x_3}\right)\right] + \mu\eta c_d \end{cases}$$

conservative substances

nonconservative substances

growing substances

(5-18)

Equations (5-18) are the material derivatives of c_d. If the growing substances also decay, such as in death, the decay coefficient may also be incorporated.

Complete Material Balance Equations

Since the volume of the particles is extremely small, $\eta \approx 1$. Respectively substituting equations (5-18) into equation (5-10), considering that $\eta \approx 1$, and rearranging, we have

$$\frac{\partial c_d}{\partial t} + \frac{\partial}{\partial x}\left(E_{xx}\frac{\partial c_d}{\partial x}\right) + \frac{\partial}{\partial y}\left(E_{yy}\frac{\partial c_d}{\partial y}\right) + \frac{\partial}{\partial z}\left(E_{zz}\frac{\partial c_d}{\partial z}\right)$$

$$+ \frac{1}{1 + f_{cp}K_{sc}m_s}\left(u\frac{\partial c_d}{\partial x} + v\frac{\partial c_d}{\partial y} + w\frac{\partial c_d}{\partial z}\right) = 0 \qquad \text{conservative substances}$$

$$\frac{\partial c_d}{\partial t} - \frac{\partial}{\partial x}\left(E_{xx}\frac{\partial c_d}{\partial x}\right) - \frac{\partial}{\partial y}\left(E_{yy}\frac{\partial c_d}{\partial y}\right) - \frac{\partial}{\partial z}\left(E_{zz}\frac{\partial c_d}{\partial z}\right)$$

$$+ \frac{1}{1 + f_{cp}K_{sc}m_s}\left(u\frac{\partial c_d}{\partial x} + v\frac{\partial c_d}{\partial y} + w\frac{\partial c_d}{\partial z}\right)$$

$$+ kc_d - G = 0 \qquad\qquad\qquad\qquad \text{nonconservative substances}$$

$$\frac{\partial c_d}{\partial t} + \frac{\partial}{\partial x}\left(E_{xx}\frac{\partial c_d}{\partial x}\right) + \frac{\partial}{\partial y}\left(E_{yy}\frac{\partial c_d}{\partial y}\right) + \frac{\partial}{\partial z}\left(E_{zz}\frac{\partial c_d}{\partial z}\right)$$

$$+ \frac{1}{1 + f_{cp}K_{sc}m_s}\left(u\frac{\partial c_d}{\partial x} + v\frac{\partial c_d}{\partial y} + w\frac{\partial c_d}{\partial z}\right) - \mu c_d = 0 \qquad \text{growing substances}$$

(5-19)

Equations (5-19) are the complete pollutant material balance differential equation in the xyz coordinates.

Transport Terms and Determination of Dispersion Coefficients

The terms $\frac{\partial c_d}{\partial x}\left(E_{xx}\frac{\partial c_d}{\partial x}\right)$, $\frac{\partial}{\partial y}\left(E_{yy}\frac{\partial c_d}{\partial y}\right)$, $\frac{\partial}{\partial z}\left(E_{zz}\frac{\partial c_d}{\partial z}\right)$ and the term containing $u\frac{\partial c_d}{\partial x} + v\frac{\partial c_d}{\partial y} + w\frac{\partial c_d}{\partial z}$ of equation (5-19) are called *transport terms*. The terms containing the E's are the *dispersive transport* and the terms containing the velocity components are the *convective* or *advective transport*. For the dispersive transport terms, three types of dispersion coefficients are often encountered in practice. The first is the dispersion coefficient in streams and rivers, the second is the dispersion coefficient in tidal waters, and the third is the dispersion coefficient in lakes and reservoirs or open bodies of water. In streams and rivers, although there are strictly dispersions sidewise to the direction of flow, the predominant one is in the longitudinal dispersion in the direction of flow. Furthermore, as far as sidewise dispersion is concerned, after complete mixing downstream from a point of pollutant discharge, there is practically zero concentration gradient in this sidewise direction. Hence the three diffusive transport coefficients, E_{xx}, E_{yy}, and E_{zz}, reduce to only one. If the x coordinate is taken as the longitudinal direction, the only dispersion coefficient is E_{xx}.

Fisher and others[2] suggested that the following equation be used for estimating E_{xx}, in mi²/day, in streams and rivers.

$$E_{xx} = 3.4(10^{-5})\frac{U^2 B^2}{H U^*} \tag{5-20}$$

U is the mean velocity in fps, B the mean width in feet, H is the mean depth in feet, and U^* is the shear velocity in fps. From hydraulics, U^* is

$$U^* = \sqrt{gHs} \tag{5-21}$$

where $g = 32.17$ fps² and s is the energy slope.

Based on an evaluation of 18 streams and 40 time of travel studies with flows ranging from 35 to 33,000 cfs, river bed slopes ranging from 0.00015 to 0.0098, and Froude number F less than 0.5 ($F = U/\sqrt{gH}$), McQuivey and Keefer[3] proposed the following equation for predicting E_{xx}, in mi²/day:

$$E_{xx} = 1.8(10^{-4})\frac{Q}{s_0 B} \tag{5-22}$$

Q is the river flow in cfs, s_0 is the bed slope, and B is the mean river width in feet. *Time-of-travel studies* is the terminology given to a procedure of determining the average velocity of a stream. This is usually accomplished by dumping a tracer dye at an upstream

[2] H. B. Fisher, E. J. List, R. C. Y. Koh, J. Imberger, and N. H. Brooks (1979). *Mixing in Inland and Coastal Waters*. Academic Press, New York.

[3] R. S. McQuivey and T. N. Keefer (1974). "Simple Method for Predicting Dispersion in Streams." *Journal of the Environmental Engineering Division, Proceedings of the American Society of Civil Engineers*, 100(EE4), pp. 997–1011.

river point and taking note of the time taken for the dye to travel to a downstream river point of a known distance from the upstream point. From the data, the average velocity can thus be calculated. Note that equations (5-20) and (5-22) are dimensional equations.

Example 5-2

A river flow of 200 cfs has a mean depth 5 ft and a mean width of 40 ft. The bed slope is 0.0014. Using the two equations above, calculate E_{xx}.

Solution

$$E_{xx} = 3.4(10^{-5}) \frac{U^2 B^2}{HU^*}$$

$U^* = \sqrt{gHs}$; assume that s = bed slope of $s_0 = 0.0014$.

$$U = \frac{200}{5(40)} = 1 \text{ fps}$$

$$U^* = \sqrt{32.17(5)(0.0014)} = 0.475 \text{ fps}$$

Therefore,

$$E_{xx} = 3.4(10^{-5}) \frac{1^2 40^2}{5(0.475)} = 0.023 \text{ mi}^2/\text{day} \quad \textbf{Answer}$$

$$F = \frac{U}{\sqrt{gH}} = \frac{1}{\sqrt{32.17(5)}} = 0.079 < 0.5$$

$$E_{xx} = 1.8(10^{-4}) \frac{Q}{s_0 B} = 1.8(10^{-4}) \frac{200}{0.0014(40)} = 0.642 \text{ mi}^2/\text{day} \quad \textbf{Answer}$$

Note: These results show that values of E_{xx} depend on the source of the information.

A better method for predicting E_{xx} is as follows. Dump a conservative dye at some point in a stream and determine the resulting average dye concentrations at one or more downstream points. The average concentrations at these downstream points will have been reduced due to dispersion. Using equation (5-18), neglecting sidewise dispersion, the material balance equation with $\eta = 1$ is

$$\frac{dc_d}{dt} = -E_{xx} \frac{d^2 c_d}{dx^2} \tag{5-23}$$

In the Lagrangian method, the control mass is followed. Hence, for a differential strip distance of dx traversed, dt is equal to dx/u, where u is the longitudinal velocity. Substituting and rearranging, equation (5-23) becomes

$$E_{xx} \frac{d^2 c_d}{dx^2} + u \frac{dc_d}{dx} = 0 \tag{5-24}$$

Using the point where the dye is dumped as the origin results in the following boundary conditions: $c_d = c_{d0}$ when $x = 0$ and $c_d = 0$ when $x = \infty$. The solution is

$$c_d = c_{d0}e^{-(u/E_{xx})x} \tag{5-25}$$

Because of dispersion, the mass of dye will not arrive at the downstream points at the same time, but some will arrive sooner and some will arrive later. The time lag between the first arrival and the latest arrival is a proportionate measure of the dispersion coefficient E_{xx}. Any average dye concentration c_d at a downstream point may be obtained by taking the time-average concentration. The equation is

$$c_d = \frac{\sum c_{di} t_i}{t_2 - t_1} \tag{5-26}$$

where t_i is the time interval during which c_{di} is the average concentration, and t_1 and t_2 are the first and last times of arrival, respectively.

Put equation (5-25) in a form suitable for the field determination of E_{xx}. Rearranging this equation produces

$$E_{xx} = -\frac{ux}{\ln(c_d/c_{d0})} \tag{5-27}$$

Example 5-3

A dye study was conducted at the free-flowing portion of Grays Inn Creek in the town of Rock Hall in eastern Maryland on November 16, 1987. Approximately 6.7 g of Rhodamine WT in 100 mL of solution was dumped at the discharge point of the lagoon treatment plant, resulting in a concentration c_{d0} of $5.0(10^7)$ μg/L a few feet downstream. The resulting concentrations monitored over time at a distance of 1047 m downstream are shown in the following table. The stream time-of-travel velocity was 8373 m/day, and its flow was 900 m^3/day. Calculate E_{xx}.

Time, A.M.	Concentration (ppb)	Time, A.M.	Concentration (ppb)
09:20	0.0	09:30	0.0
09:40	0.03	09:45	0.17
09:50	0.70	09:55	3.0
10:00	9.0	10:05	19.0
10:10	30.0	10:15	39.0
10:20	43.5	10:25	50.0
10:30	52.0	10:35	52.0
10:40	50.0	10:45	48.0
10:50	45.0	10:55	42.0
11:00	39.0	11:15	31.0
11:30	24.0	11:45	19.0
12:00	16.0	12:30	12.0
13:00	6.4	14:00	4.7
15:00	1.8	16:00	0.70
17:00	0.0	18:00	0.0

Solution

$$c_{d2} = \frac{\sum c_{di} t_i}{t_2 - t_1} = \frac{1}{(17 - 9.5)(60)} [0.03(45 - 30) + 0.17(50 - 40) + 0.7(55 - 45)$$

$$+ 3.0(60 - 50) + 9.0(65 - 55) + 19(10 - 0.0) + 30(15 - 5) + 39(20 - 10)$$

$$+ 43.5(25 - 15) + 50(30 - 20) + 52(35 - 25) + 52(40 - 30) + 50(45 - 35)$$

$$+ 48(50 - 40) + 45(55 - 45) + 42(60 - 50) + 39(75 - 55) + 31(30 - 0)$$

$$+ 24(45 - 15) + 19(60 - 30) + 16(90 - 45) + 12(60) + 6.4(90) + 4.7(120)$$

$$+ 1.8(120) + 0.7(120)] = \frac{1}{450}(10,714.15) = 23.81 \text{ ppb} = 23.81 \ \mu g/L$$

$$E_{xx} = -\frac{ux}{\ln(c_d/c_{d0})} = -\frac{8373(1047)}{\ln[23.81/5.0(10^7)]} = 6.03(10^5) \ \text{m}^2/\text{day} \quad \textbf{Answer}$$

Diffusion is an intrinsic property of the pollutant diffusing, whereas dispersion is a property of the medium (water, in our case) through which the pollutant is dispersing—the pollutant is mechanically dispersed by turbulence. Since the pollutant is mechanically dispersed, this dispersion is called *mechanical dispersion*. In other words, dispersion is a function of the mechanical movement of the medium (i.e., on the degree of turbulence, not on the pollutant). Since diffusion and mechanical dispersion cannot be separated from each other, a single term called *hydrodynamic dispersion* is used to characterize these processes together. Since dispersion coefficient is not a function of the dispersing chemical, any conservative chemical may be used for its determination; the value thus obtained may henceforth be used as the coefficient for any substance or chemical dispersing in the particular stretch of water.

In tidal water waters, the flow behavior is such that water rushes toward inland during flood tides and empties toward the sea during ebb tides. Theoretically, one could obtain dispersion coefficients for each of the flooding and ebbing tide regimes. For practical purposes, however, this is not warranted, but rather, a tidally averaged value is determined. The corresponding velocity to be used is the *net advective velocity*, also called the *tidally averaged velocity*.

E_{xx} from a Natural Tracer

Develop a method for determining the tidally averaged E_{xx} by taking advantage of a natural tracer that exists in estuaries—salt. The value obtained will then be applicable to the particular stretch of estuary or tidal river.

Follow a parcel of water from the headwaters. At its origin the water is fresh; however, as it approaches the mouth of the estuary it becomes contaminated with salt: The direction of dispersion is therefore from the outside to the inside and the conservative equation of equation (5-18) must be modified. Neglecting sidewise dispersion, the conservative material balance equation after rearranging and transposing terms is thus

$$\frac{d}{dx}\left(E_{xx}\frac{dc_d}{dx}\right) - u\frac{dc_d}{dx} = 0 \tag{5-28}$$

Consider as an origin any point in the estuary where the concentration is some value c_{d0}. Imagine traversing in the direction upstream from this point until the concentration is zero. Hence, with the upstream direction negative, the boundary conditions are $c_d = c_{d0}$ at $x = 0$ and $c_d = 0$ at $x = -\infty$. The auxiliary equation of equation (5-28) is

$$E_{xx}m^2 - um = 0 \tag{5-29}$$

With this auxiliary equation along with the boundary conditions, the solution is

$$c_d = c_{d0}\exp\left(\frac{ux}{E_{xx}}\right) \Rightarrow E_{xx} = \frac{ux}{\ln(c_d/c_{d0})} \tag{5-30}$$

Example 5-4

Thirty miles upstream from the mouth of the Hudson Estuary, the following average chloride concentrations were obtained: 5000 mg/L, 5029 mg/L, and 4996 mg/L. At a point 48 miles upstream from the mouth, the chloride concentrations are 1800 mg/L and 1850 mg/L. The tidally averaged velocity is 0.025 fps. Calculate E_{xx}.

Solution

$$E_{xx} = \frac{ux}{\ln\left(\frac{c_d}{c_{d0}}\right)}$$

Use the 30-mile point as the origin:

$$c_{d30} = c_{d0} = \frac{5000 + 5029 + 4996}{3} = 5008.33 \text{ mg/L}$$

$$c_{d48} = c_d = \frac{1800 + 1850}{2} = 1825 \text{ mg/L at the 48-mile point}$$

Thus,

$$E_{xx} = \frac{0.025[-(48-30)(5280)]}{\ln(1825/5{,}008.33)} = 2353.59 \text{ ft}^2/\text{s} = 7.29 \text{ mi}^2/\text{day} \quad \textbf{Answer}$$

General Values of Dispersion Coefficients

For discharges into lakes and reservoirs or open bodies of water, since the hydrodynamic motion is conceivably in all directions, the most general equation [equation (5-19)] should be the one applied to the conservative tracer under consideration. The resulting equation should be solved numerically. For a conservative tracer dumped into the body of water, equation (5-19) becomes

$$\frac{\partial c_d}{\partial t} + \frac{\partial c_d}{\partial x}\left(E_{xx}\frac{\partial c_d}{\partial x}\right) + \frac{\partial}{\partial y}\left(E_{yy}\frac{\partial c_d}{\partial y}\right) + \frac{\partial}{\partial z}\left(E_{zz}\frac{\partial c_d}{\partial z}\right) + u\frac{\partial c_d}{\partial x} + v\frac{\partial c_d}{\partial y} + w\frac{\partial c_d}{\partial z} = 0 \tag{5-31}$$

To use equation (5-31) in solving the dispersion coefficients, the concentrations of the tracer c_d from several node points in the lake or open body of water must be sampled for. (*Node points* are the points in the water body that are utilized in a particular method of numerical analysis.) If u, v, and w are known and if the equation has been programmed into a computer, the dispersion coefficients may be determined by matching computer results of concentration using assumed coefficients with the field-sampled values. The assumed coefficient values that produce computer concentration results that match with the observed c_d field values are the correct dispersion coefficients. If the u, v, and w values are not known, the momentum equation must also be programmed into the computer and the velocities determined. The results are then coupled to equation (5-31) to determine the coefficients.

In lakes or reservoirs, sediments settle to the bottom and later decompose. The decomposition products then disperse from the bottom upward to the main body of water. Since the decomposition products are distributed all over the bottom, the concentration in any horizontal dimension in the main water body is constant or practically so; hence the concentration gradient is zero in this direction. The only direction of dispersion is therefore vertical. Once the vertical dispersion coefficient has been determined, the dispersal of the bottom-sediment decomposition products may then be modeled as only moving upward. Some values of tidal water longitudinal dispersion coefficients and lake vertical dispersion coefficients are shown in Table 5-1.

Kinetic Transformation Term, kc_d

The kinetic transformation that a pollutant or any substance may undergo can be one or more of the following: hydrolysis, photolysis, microbial degradation or growth, settling, and volatilization. Hydrolysis is the reaction of a chemical or substance with water; *photolysis* is the breakdown of a chemical due to the absorption of light energy; microbial degradation results when microorganisms utilize the chemical for food; *settling* is the action of gravity on the mass of particles; and *volatilization* is a process where molecules of the chemical escape from the liquid surface to enter into the gaseous phase. A transformation that decreases the concentration is a *degradation*. In general, excluding volatilization, k for the degradation of a given pollutant may then be written as

$$k = k_h + k_p + k_m + k_s \tag{5-32}$$

where k_h, k_p, k_m, and k_s correspond to the kinetic coefficients for hydrolysis, photolysis, microbial degradation, and settling, respectively. The reason for the exclusion of volatilization will be known in the latter part of this chapter.

Laboratory determinations of k values. Field determination of the k values is not possible because of the possible concurrent occurrences of other processes in the ambient environment. Hence individual k values should be determined in the laboratory where other variables can be controlled. For example, if k_h is to be determined, the acidity or basicity of the sample may be adjusted to a suitable pH. The disappearance of the test chemical can then be determined at several intervals of time and k_h, subsequently,

TABLE 5-1 TIDAL WATER LONGITUDINAL DISPERSION
COEFFICIENTS, E_{xx}, AND LAKE VERTICAL DISPERSION
COEFFICIENTS, E_{zz}

Water body	Type	E_{xx}(mi^2/day) or E_{zz}(cm^2/s)
Cape Fear River, NC	Tidal	7
Compton Creek, NJ	Tidal	1
Wappinger and Fishkill Creek, NY	Tidal	1
River Foyle, N. Ireland	Tidal	5
Upper Delaware River	Tidal	7
Lower Delaware River	Tidal	7
Upper Potomac River	Tidal	6
Lower Potomac River	Tidal	10
Hudson River, NY	Tidal	30
East River, NY	Tidal	15
Cooper, SC	Tidal	30
South River, NJ	Tidal	5
Houston Ship Channel, TX	Tidal	30
Southern San Francisco Bay	Tidal	6
Northern San Francisco Bay	Tidal	50
Rio Quayas, Ecuador	Tidal	25
Thames River, England, low flow	Tidal	3
Thames River, England, high flow	Tidal	11
Lake Vomb, Sweden	Lake	20
Lake Mein, Sweden	Lake	10
Lake Ontario, United States and Canada	Lake	10
Rochester Bay, Lake Ontario	Lake	2
Lake Erie, Central Basin	Lake	1
Cayuga Lake, NY	Lake	2
Atlantic Ocean, east of Barbados		
0 to 10-m depth	Ocean	7–45 cm^2/s
30 to 40-m depth	Ocean	0.25–0.85 cm^2/s

calculated. Imagine determining k_h in the field, where k_p, k_m, and k_s are happening at the same time.

Hydrolysis.   Hydrolysis reactions may be neutral, acid catalyzed, or base catalyzed. Recall that in previous equations the term containing k is a part of the material or Lagrangian derivative. In Chapter 2 this term was called R derivative. Hence the hydrolysis rate will be designated by the R Lagrangian derivative, R_h. In neutral hydrolysis, the hydrogen and hydroxyl ions balance out; however, in acid or base hydrolysis, there is excess of either ions. In acid or base hydrolysis, because of the excess, ions affect the rate of reaction; neutral reactions are not affected by these ions. The three rates for neutral R_{hn}, acid-catalyzed R_{ha}, and base-catalyzed reactions R_{hb} are presented in the respective equations

$$R_{hn} = -k_{hn}c_d \tag{5-33}$$

$$R_{ha} = -k_{ha}\{H^+\}c_d \tag{5-34}$$

$$R_{hb} = -k_{hb}\{OH^-\}c_d \tag{5-35}$$

where $\{\cdot\}$ is the symbol for "activity of," k_{hn} is the k_h for neutral hydrolysis, k_{ha} is the k_h for acid hydrolysis, and k_{hb} is the k_h for base-catalyzed hydrolysis. The rates for the acid-catalyzed and base-catalyzed reactions are assumed first order in each of the concentrations; hence the overall rates are second order. The minus signs are used since c_d will decrease.

Under neutral conditions, k_h is simply k_{hn}; at other conditions,

$$k_h = k_{ha}[H^+] + k_{hb}[OH^-] \tag{5-36}$$

The overall values of k_h may range from 10^{-1} to 10^{-7} per day.

k_h may be corrected for temperature using the Arrhenius law as follows:

$$\ln \frac{k_{h2}}{k_{h1}} = \frac{E(T_2 - T_1)}{R(T_1 T_2)} \Rightarrow k_{h2} = k_{h1}e^{[E(T_2-T_1)/RT_1 T_2]} = k_{h1}\theta^{T_2-T_1} \tag{5-37}$$

where the subscripts 1 and 2 refer to the first and second conditions, respectively. E and R are the activation energy and gas constant ($= 8.314$ J/mol-K), respectively. θ is the *temperature correction factor*.

Example 5-5

A laboratory experiment on a certain pollutant found the values of k_{ha} and k_{hb} to be $2.0(10^2)$ L/mol-day and 10 L/mol-day, respectively, at 25°C. Compute k_h at pH 6 and 25°C. What is R_h if c_d for the pollutant is 10^{-3} mol/L? If the activation energy is 26,000 J/mol, what is R_h at 30°C?

Solution

$$pOH = 14 - pH = 14 - 6 = 8$$

Thus,

$$[H^+] = 10^{-6} \text{ mol/L} \qquad [OH^-] = 10^{-8} \text{ mol/L}$$

$$k_h = k_{ha}[H^+] + k_{hb}[OH^-] = 2.0(10^2)(10^{-6}) + 10(10^{-8}) = 2.0(10^{-4}) \text{ per day} \quad \textbf{Answer}$$

$$R_h = -k_h c_d = 2.0(10^{-4})(10^{-3}) = -2.0(10^{-7}) \text{ mol/L-day} \quad \textbf{Answer}$$

$$\ln \frac{k_{h2}}{k_{h1}} = \frac{E(T_2 - T_1)}{R(T_1 T_2)}$$

$$T_2 = 30 + 273 = 303 \text{ K} \qquad T_1 = 25 + 273 = 298 \text{ K}$$

Therefore,

$$\ln \frac{k_{h2}}{2.0(10^{-4})} = \frac{26,000(303 - 298)}{8.314(303)(298)} = 0.173 \qquad k_{h2} = 2.38(10^{-4}) \text{ per day}$$

$$R_h = -2.38(10^{-4})(10^{-3}) = -2.0(10^{-7}) \text{ mol/L-day} \quad \textbf{Answer}$$

Photolysis.   When a photon of light energy $h\nu$ is absorbed by a molecule, the molecule becomes unstable and a number of reactions becomes possible. These possibilities are shown below.

$$
\begin{array}{l}
A_0 \xrightarrow{\;h\nu\ \text{absorption}\;} A^* \longrightarrow \text{intersystem crossing} \longrightarrow A'^* \\
\qquad\qquad\qquad\quad \longrightarrow A_0 + \text{heat (internal conversion)} \\
\qquad\qquad\qquad\quad \longrightarrow A_0 + h\nu \text{ (fluorescence)} \\[4pt]
\qquad\qquad\qquad\ + Q_0 \\
\qquad\qquad\qquad\quad \longrightarrow A_0 + \hat{Q} \text{ (quenching)} \\
\qquad\qquad\qquad\quad \longrightarrow \text{chemical reaction (photooxidation,} \\
\qquad\qquad\qquad\qquad\qquad \text{photoreduction, photoisomerization,} \\
\qquad\qquad\qquad\qquad\qquad \text{photosubstitution, photoaddition,} \\
\qquad\qquad\qquad\qquad\qquad \text{photofragmentation, and photohydrolysis)}
\end{array}
$$

As shown above, an original molecule in its ground state A_0 absorbs a photon of light of energy $h\nu$, converting the molecule into its excited state A^*. From A^*, five paths are possible. In the first path, the excess energy may be transferred to another molecule, forming the excited molecule A'^* through a process called *intersystem crossing*. A_o, in this case, only serves as a *sensitizer*. In the second path, the energy absorbed may simply be released, reverting the molecule back to the ground state in a process called *internal conversion*. In the third path, the excited molecule may re-release its photon of energy $h\nu$, reverting the molecule back to its original ground state. This process is called *fluorescence*. A second molecule Q_0 in its ground state may absorb the energy from the first molecule, transforming the second molecule into its heated state $\hat{Q}$. This process is called *quenching*. Or, in the last possible path, the excited molecule A^* may undergo several chemical reactions.

The excited molecule A'^* formed from the intersystem crossing undergoes further reactions in accordance with the following possible schemes:

$$
\begin{array}{l}
A^* \xrightarrow{\hspace{3cm}} A'^* \longrightarrow A_0' + \text{heat (internal conversion)} \\
\qquad\qquad\qquad\quad \longrightarrow A_0' + h\nu \text{ (phosphorescence)} \\[4pt]
\qquad\qquad\qquad\ + Q_0 \\
\qquad\qquad\qquad\quad \longrightarrow A_0 + \hat{Q} \\
\qquad\qquad\qquad\quad \longrightarrow \text{chemical reaction (photooxidation,} \\
\qquad\qquad\qquad\qquad\qquad \text{photoreduction, photoisomerization,} \\
\qquad\qquad\qquad\qquad\qquad \text{photosubstitution, photoaddition,} \\
\qquad\qquad\qquad\qquad\qquad \text{photofragmentation, and photohydrolysis)}
\end{array}
$$

The explanations for these possible paths are similar to the first set of possible schemes, except for phosphorescence. *Phosphorescence* is the term given for the second emission of $h\nu$. The first emission is called *fluorescence*. Heat energy released in phosphorescence is smaller than in fluorescence.

Let the R derivative of photolysis be designated by R_p and the amount of photon absorbed by Λ. R_p is a function of Λ and c_d; hence, assuming first order in each,

$$R_p = -k_{p\Lambda}\Lambda c_d \tag{5-38}$$

As it stands, Λ is not an easy parameter to determine. The easier way is to lump $k_{p\Lambda}\Lambda$ into k_p, which makes this a pseudo-first-order coefficient. k_p may be determined in the laboratory by measuring the decrease of c_d at intervals of time using a simulated sunlight and using equation (5-38). The value obtained may be used as the coefficient at the water surface in the ambient or environment k_{p0}. k_{p0} is the k_p at the surface which corresponds to the light intensity at the surface.

Light is attenuated as it penetrates below the surface according to the Lambert–Beer law. Hence, using the law, $dk_p/dz = -K_e k_p$, and k_p at any depth is derived as

$$k_p = k_{p0}\exp(-K_e z) \tag{5-39}$$

K_e is the light extinction coefficient. A factor may also have to be incorporated to consider the effect of cloud cover. An empirical relationship relating K_e to Secchi depth z_s was presented in Chapter 2. Table 5-2 shows some values of z_s. Example k_{p0} values for naphthalene and benzo[a]pyrene are 0.23 per day and 31 per day, respectively.

TABLE 5-2 VALUES OF SECCHI DEPTHS, z_S

Water body and location	z_s (m)
Delaware at Trenton, NJ	2.5
Delaware at Philadelphia, PA	0.8
Tahoe	90
Ontario	9.2
Sewage stabilization pond, Kadoka, SD	0.2

Example 5-6

If the average concentration of benzo[a]pyrene in Lake Ontario is 10^{-3} mmol/L and $k_{p0} = 31$ per day, what is the rate of decay due to photolysis R_p at a 4-m depth?

Solution

$$k_p = k_{p0}\exp(-K_e z)$$

$$K_e = \frac{1.8}{z_s} \qquad z_s = 9.2 \text{ m from Table 5-2}$$

Thus,

$$K_e = \frac{1.8}{9.2} = 0.2 \text{ m}^{-1}$$

$$k_p = (31) \exp[-0.2(4)] = 13.93 \text{ per day}$$

Therefore,

$$R_p = -13.93(10^{-6}) \text{ mol/L-day} \quad \textbf{Answer}$$

Microbial kinetics. The process of microbial degradation of pollutants, microbial growth, and microbial death is called *microbial kinetics*. Metabolism and cometabolism are responsible for the microbial degradation of a substance. *Metabolism* is the term used when the substance is used for food by microbes for growth and energy. These microbes may include bacteria, algae, and fungi. *Cometabolism* occurs when the substance is degraded, not as a food, but coincidentally used in the process of consuming the regular substances used for food. Microbial degradation follows directly from Monod's equation, as discussed elsewhere in this book. In water quality modeling, however, the equation is modified somewhat as shown in what follows. Utilizing Monod's concept, assuming that the pollutant in question is the limiting nutrient, the degradation is

$$R_m = -\frac{\mu_m c_d}{Y(K_{c_d} + c_d)}[X] \tag{5-40}$$

where μ_m is the maximum specific growth rate coefficient, $[X]$ is the concentration of microorganisms consuming the substrate, Y is the specific yield of organisms based on the limiting nutrient, and K_{c_d} is the half-saturation constant for the substrate. Although the food c_d could change with time, the concentration of organisms X may be considered constant or changing at a very, very slow rate compared to that of c_d. Also, if $c_d \ll K_{c_d}$, the denominator is practically constant and all the factors on the right-hand side of equation (5-40), except c_d, can be considered constant. All these factors may then be lumped into one single microbial decay coefficient k_m. Hence

$$R_m = -k_m c_d \tag{5-41}$$

It must be carefully noted that the basis of equation (5-41) is the assumption that $c_d \ll K_{c_d}$ and that the pollutant in question is the limiting nutrient. In some situations, however, the given pollutant may not be the one limiting, but other nutrients, such as nitrogen and phosphorus. In other words, $R_m = -\mu_m S/[Y(K_S + S)]$, where S is the concentration of the limiting nutrient and K_S is the half-saturation constant for the nutrient. In these cases it is prudent to assume that the rate of disappearance of the pollutant is proportional to the concentration of microorganisms present and the concentration of the pollutant (i.e., $R_m \propto c_d[X]$). At a given situation, the concentration of the organisms would be constant or changing at an exceedingly slow rate compared to that of c_d; thus $[X]$ in combination with the proportionality constant will be k_m. R_m would then still be given by equation (5-41). k_m may also be temperature corrected using the Arrhenius equation. Concomitant with the degradation of the pollutant is the resulting growth of the microorganisms and their eventual death. Microbial growth and death kinetics are discussed in Chapter 7.

Example 5-7

A wastewater treatment plant discharges treated sewage into a stream. If the concentration of BOD_5 1 mile downstream from the point of discharge is 3 mg/L and k_m is assumed to be 0.2 per day (in terms of BOD_5), what is the value of R_m at this point?

Solution

$$R_m = -k_m c_d = 0.2(3) = -0.6 \ \frac{mg/L}{day} \ \text{of } BOD_5 \quad \textbf{Answer}$$

Settling. This process applies only to solids suspended in water and is represented simply as a first-order process. Using R_s to designate rate of settling,

$$R_s = -k_s m_s \tag{5-42}$$

where m_s is the concentration of suspended solids. Again, as in other k values, k_s is very difficult to determine in the field because of competing processes that can occur all at the same time. Hence k_s should be determined in the laboratory by a settling test. Representative river samples should be taken. A method is indicated in the following example.

Example 5-8

A river sample is put in a settling column. At two points in time, samples are taken at the sampling port, producing the following results: at 0 day, $m_s = 20$ mg/L; at 2 days, $m_s = 15$ mg/L. Determine the settling coefficient k_s.

Solution

$$R_s = -k_s m_s$$

$$\frac{dm_s}{dt} = -k_s m_s \Rightarrow k_s = -\frac{1}{t} \ln \frac{m_s}{m_{s0}} = -\frac{1}{2} \ln \frac{15}{20} = 0.14 \text{ per day} \quad \textbf{Answer}$$

Volatilization. A pollutant dissolved in water with concentration c_d will volatilize into the atmosphere if the equilibrium concentration c_d^* in water corresponding to the concentration of the pollutant in the air c_a is less than c_d. By mass transfer theory, the rate of volatilization R_v is

$$R_v = -k_v(c_d - c_d^*) \tag{5-43}$$

where k_v is the *overall mass transfer coefficient*. c_d^* is related to c_a by the Henry's law constant K_H as

$$c_a = K_H c_d^* \tag{5-44}$$

Substituting in equation (5-43) gives us

$$R_v = -k_v \left(c_d - \frac{c_a}{K_H} \right) \tag{5-45}$$

From equation (5-45), R_v is not directly representable as a pseudo-first-order process, hence the exclusion in the general k coefficient expression as stated before.

Example 5-9

H$_2$S is liberated if an organic substance containing sulfur is degraded anaerobically. (a) If there is 10^{-7} mol/L of H$_2$S in water at equilibrium, what is its concentration c_a in air expressed in mole fraction and in atmospheres? (b) If the concentration in water of H$_2$S is 10^{-6} mol/L, with the concentration in air found in part (a), what is R_v under these concentration conditions? K_H for H$_2$S is $0.0483(10^4)$ atm/mol fraction at 20°C. Assume that temperature is 20°C.

Solution

(a)
$$c_a = K_H c_d^*$$

$$c_d^* = \frac{10^{-7}}{10^{-7} + 1000/18} = 1.8(10^{-9}) \text{ mol fraction}$$

Thus,

$$c_a = 0.0483(10^4)(1.8)(10^{-9}) = 9.0(10^{-7}) \text{ mol fraction} = 9.0(10^{-7})(1)$$
$$= 9.0(10^{-7}) \text{ atm} \quad \textbf{Answer}$$

(b)
$$R_v = -k_v \left(c_d - \frac{c_a}{K_H} \right) = -k_v \left[10^{-6} - \frac{9.0(10^{-7})}{0.0483(10^4)} \right]$$

$$= -9.98(10^{-7}) k_v \frac{\text{mol}}{\text{L}} \quad \textbf{Answer}$$

DO Sag

When an oxygen-consuming pollutant in large concentration is discharged into a stream, the DO will be initially consumed faster near the point of discharge. This will cause the oxygen concentration to decrease, which can, in severe cases, deplete the oxygen totally. Downstream from the discharge point, however, as the CBOD and NBOD are slowly used up and with the continuous reaeration from the atmosphere, the stream recovers and the DO goes back up again. The resulting profile of the dissolved oxygen concentration along the reach of the stream is called the *DO sag* or *DO profile*. Because of the presence of the oxygen-consuming pollutant, the DO can be below the saturation, O_s; hence the term *sag*. A plot of the DO sag is shown in Figure 5-2.

With η equals 1 and from the R derivative of DO (R_6; see the discussion of kinetics in Chapter 2), the Lagrangian derivative for dissolved oxygen is

$$
\frac{Dc_6}{Dt} = \frac{\partial}{\partial x}\left(E_{xx}\frac{\partial c_6}{\partial x}\right) + \frac{\partial}{y}\left(E_{yy}\frac{\partial c_6}{\partial y}\right) + \frac{\partial}{\partial z}\left(E_{zz}\frac{\partial c_6}{\partial z}\right) + R_6 + G_O
$$

$$
= \frac{\partial}{\partial x}\left(E_{xx}\frac{\partial c_6}{\partial x}\right) + \frac{\partial}{y}\left(E_{yy}\frac{\partial c_6}{\partial y}\right) + \frac{\partial}{\partial z}\left(E_{zz6}\frac{\partial c_6}{\partial z}\right) + k_2(O_s - c_6) \qquad (5\text{-}46)
$$

$$
+ P - R - k_c L_c - k_n L_n + G_O
$$

G_O is the G of DO, c_6 the concentration of dissolved oxygen, k_2 the reaeration coefficient, O_s the dissolved oxygen saturation, P the rate of photosynthesis, R the rate of respiration, k_c the CBOD deoxygenation coefficient, L_c the CBOD concentration, k_n the nitrogenous deoxygenation coefficient, and L_n the NBOD. An example of when G_O can occur is when the stream is aerated.

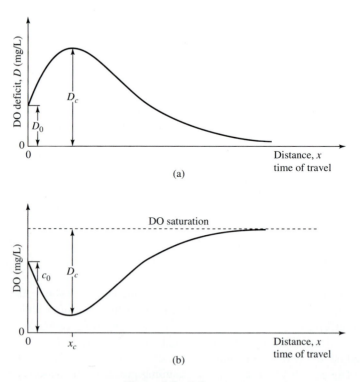

Figure 5-2   DO sag.

As originally derived by Streeter and Phelps,[4] the effects of dispersion, P, R, the NBOD, and G_O were neglected. Hence, equation (5-46) becomes simply

$$\frac{Dc_6}{Dt} = k_2(O_s - c_6) - k_c L_c \tag{5-47}$$

In Streeter and Phelps' derivation, the differential for L_c was assumed as DL_c/Dt (or $dL_c/dt) = -k'_c L_c$, which integrates to

$$L_c = L_{c0}e^{-k'_c t} \tag{5-48}$$

$k'_c = k_c + k_{s5}$, where k_{s5} is the rate of settling of CBOD. L_{c0} is L_c at $t = 0$. It is the oxygen equivalent of the waste. The oxygen consumed at time t is then equal to $L_{c0} - L_c = y = L_{c0} - L_{c0}\exp(-k'_c t) = L_{c0}[1 - \exp(-k'_c t)]$. Hence the original oxygen equivalent of the waste L_{c0} can be obtained by measuring the oxygen consumption y after any time t. Thus

$$L_{c0} = \frac{y}{1 - \exp(-k'_c t)} \tag{5-49}$$

From the definition of the deficit $D = O_s - c_6$, Dc_6/Dt is $-dD/dt$. Substituting in equation (5-47), along with equation (5-48), and using the boundary conditions $D = D_0$ at $t = 0$ and $D = 0$ at $t = \infty$, the original DO sag or Streeter–Phelps equation below is produced. This assumes that $k'_c = k_c$.

$$D = \frac{k_c L_{c0}}{k_2 - k_c}(e^{-k_c t} - e^{-k_2 t}) + D_0 e^{-k_2 t} \tag{5-50}$$

We should note that in equation (5-46) the Streeter–Phelps equation is derived as a Lagrangian equation. In other words, the equation is a model where the pollutant inside the control mass has been followed as it travels down the water body—the Lagrangian approach. By this approach the time t can be calculated by the distance the pollutant (control mass) has traveled divided by the velocity: $t = x/u$, where x is the distance traveled and u is the velocity.

In practice, the most useful application of equation (5-46) is its one-directional version. First, the Lagrangian derivative of L_c, L_n, P, and R must be simplified so that the results can be substituted into equation (5-46). DL_c/Dt is equal to $DL_c/(Dx/u) = u(dL_c/dx)$. In the x-coordinate direction, the Lagrangian derivative of L_c is, therefore,

$$u\frac{dL_c}{dx} = E_{xx}\frac{d^2L_c}{dx^2} - k'_c L_c + G_{Lc} \tag{5-51}$$

G_{Lc} is the G of L_c.

[4]H. W. Streeter and E. B. Phelps (1925). "A Study of the Pollution and Natural Purification of the Ohio River." *Public Health Bulletin 146*. Treasury Department, U.S. Public Health Service, Washington, D.C.

Using the boundary conditions $L_c = L_{c0}$ when $x = 0$ and $dL_c/dx = 0$ when $x = \pm\infty$, the solutions of equation (5-51) are

$$
L_c =
\begin{cases}
L_{c0}e^{q_{c-}x} + \dfrac{G_{Lc}}{k_c'}(1 - e^{q_{c-}x}) & x \geq 0 \\[3mm]
L_{c0}e^{q_{c+}x} + \dfrac{G_{Lc}}{k_c'}(1 - e^{q_{c+}x}) & x \leq 0
\end{cases}
\tag{5-52}
$$

where

$$
q_{c+} = \frac{u + \sqrt{u^2 + 4E_{xx}k_c'}}{2E_{xx}} \quad \text{and} \quad q_{c-} = \frac{u - \sqrt{u^2 + 4E_{xx}k_c'}}{2E_{xx}}
$$

Equations (5-52) are called the *downward* and *upward CBOD transport equations*, respectively.

The differential equation for L_n is similar to that of L_c. Manipulating this differential equation similar to that of L_c with the following boundary conditions: $L_n = L_{n0}$ when $x = 0$ and $dL_n/dx = 0$ when $x = \pm\infty$, the solutions are

$$
L_n =
\begin{cases}
L_{n0}e^{q_{n-}x} + \dfrac{G_{Ln}}{k_n}(1 - e^{q_{n-}x}) & x \geq 0 \\[3mm]
L_{n0}e^{q_{n+}x} + \dfrac{G_{Ln}}{k_n}(1 - e^{q_{n+}x}) & x \leq 0
\end{cases}
\tag{5-53}
$$

where

$$
q_{n+} = \frac{u + \sqrt{u^2 + 4E_{xx}k_n}}{2E_{xx}} \quad \text{and} \quad q_{n-} = \frac{u - \sqrt{u^2 + 4E_{xx}k_n}}{2E_{xx}}
$$

Equations (5-53) are called the *downward* and *upward NBOD transport equations*, respectively.

The production of oxygen by photosynthesis P and its utilization by respiration R is related to the concentration of chlorophyll a, c_4, in algae growth. Hence simplify the Lagrangian derivative of c_4 and use the result to calculate the corresponding P and R.

Since algae is growing inside the control mass, the direction of dispersion of chlorophyll a is from the inside toward the outside. As a result, the sign for the term containing E_{xx} is negative. This results in the equation for chlorophyll a corresponding to equation (5-46) as follows:

$$
u\frac{dc_4}{dx} = -\left(E_{xx}\frac{d^2c_4}{dx^2}\right) + \mu_p c_4 + [-k_d c_4 - k_{s4}c_4 = -(k_d + k_{s4})c_4 = -kk c_4] \tag{5-54}
$$

μ_p is the growth rate coefficient of algae, k_d the respiration rate coefficient, and k_{s4} is the settling rate coefficient of chlorophyll a, c_4. The boundary conditions are $x = 0$, $c_4 = c_{40}$ and $x = \pm\infty$, $dc_4/dx = 0$. With these boundary conditions, the solutions are

$$
c_4 =
\begin{cases}
c_{40}e^{q_\mu - x} & x \geq 0 \\
c_{40}e^{q_\mu + x} & x \leq 0
\end{cases}
\tag{5-55}
$$

where

$$q_{\mu-} = \frac{-u - \sqrt{u^2 + 4E_{xx}(\mu_p - kk)}}{2E_{xx}} \quad \text{and} \quad q_{\mu+} = \frac{-u + \sqrt{u^2 + 4E_{xx}(\mu_p - kk)}}{2E_{xx}}$$

Equations (5-55) are called the *downward* and *upward chlorophyll a transport equations*, respectively.

As derived in Chapter 2, P and R are, respectively,

$$P = \begin{cases} 1.24a_{pc}\mu_p c_{40}e^{q_{\mu-}x} & x \geq 0 \\ 1.24a_{pc}\mu_p c_{40}e^{q_{\mu+}x} & x \leq 0 \end{cases} \tag{5-56}$$

$$R = \begin{cases} 1.24a_{pc}k_d c_{40}e^{q_{\mu-}x} & x \geq 0 \\ 1.24a_{pc}k_d c_{40}e^{q_{\mu+}x} & x \leq 0 \end{cases} \tag{5-57}$$

Substituting the expressions for L_c, L_n, P, and R into the one-directional deficit form of equation (5-46) in the positive direction and rearranging gives,

$$E_{xx}\frac{d^2 D}{\partial x^2} - u\frac{dD}{dx} - k_2 D = 1.24a_{pc}\mu_p c_{40}e^{q_{\mu-}x} - 1.24a_{pc}k_d c_{40}e^{q_{\mu-}x}$$

$$- k_c\left[[L_{c0}e^{q_{c-}x} + \frac{G_{Lc}}{k_c}(1 - e^{q_{c-}x})] - k_n\left[L_{n0}e^{q_{n-}x} + \frac{G_{Ln}}{k_n}(1 - e^{q_{n-}x})\right] + G_O\right. \tag{5-58}$$

Using the boundary conditions $D = D_0$ at $x = 0$ and $dD/dx = 0$ at $x = x_\infty$, the solution for equation (5-58) is

$$D = D_0 e^{q_{0-}x} - \frac{1.24a_{pc}c_{40}(\mu_p - k_d)}{k_2 - E_{xx}q_{\mu-}^2 + uq_{\mu-}}(e^{q_{\mu-}x} - e^{q_{0-}x})$$

$$+ \frac{L_{c0}k_c}{k_2 - E_{xx}q_{c-}^2 + uq_{c-}}(e^{q_{c-}x} - e^{q_{0-}x}) + \frac{L_{n0}k_n}{k_2 - E_{xx}q_{n-}^2 + uq_{n-}}(e^{q_{n-}x} - e^{q_{0-}x})$$

$$+ G_{Lc}\left[\frac{1 - e^{q_{0-}x}}{k_2} - \frac{e^{q_{c-}x} - e^{q_{0-}x}}{(k_2 - E_{xx}q_{c-}^2 + uq_{c-})}\right] \tag{5-59}$$

$$+ G_{Ln}\left[\frac{1 - e^{q_{0-}x}}{k_2} - \frac{e^{q_{n-}x} - e^{q_{0-}x}}{(k_2 - E_{xx}q_{n-}^2 + uq_{n-})}\right]$$

$$- \frac{G_O}{k_2}(1 - e^{q_{0-}x}) \quad x \geq 0$$

$$q_{0-} = \frac{u - \sqrt{u^2 + 4E_{xx}k_2}}{2E_{xx}}$$

Equation (5-59) is the *downward deficit transport equation*.

The differential equation of the oxygen deficit in the negative direction is

$$
E_{xx}\frac{d^2 D}{\partial x^2} - u\frac{dD}{dx} - k_2 D = 1.24 a_{pc}\mu_p c_{40}e^{q_{\mu+}x} - 1.24 a_{pc}k_d c_{40}e^{q_{\mu+}x}
$$
$$
- k_c\left[L_{c0}e^{q_{c+}x} + \frac{G_{Lc}}{k_c}(1 - e^{q_{c+}x})\right] - k_n\left[L_{n0}e^{q_{n+}x} + \frac{G_{Ln}}{k_n}(1 - e^{q_{n+}x})\right] + G_O
$$

(5-60)

Doing similar manipulations and using the following boundary conditions $D = D_0$ at $x = 0$ and $dD/dx = 0$ at $x = -x_\infty$, we obtain

$$
D = D_0 e^{q_{0+}x} - \frac{1.24 a_{pc}c_{40}(\mu_p - k_d)}{k_2 - E_{xx}q_{\mu+}^2 + uq_{\mu+}}(e^{q_{\mu+}x} - e^{q_{0+}x})
$$
$$
+ \frac{L_{c0}k_c}{k_2 - E_{xx}q_{c+}^2 + uq_{c+}}(e^{q_{c+}x} - e^{q_{0+}x}) + \frac{L_{n0}k_n}{k_2 - E_{xx}q_{n+}^2 + uq_{n+}}(e^{q_{n+}x} - e^{q_{0+}x})
$$
$$
+ G_{Lc}\left(\frac{1 - e^{q_{0+}x}}{k_2} - \frac{e^{q_{c+}x} - e^{q_{0+}x}}{k_2 - E_{xx}q_{c+}^2 + uq_{c-}}\right)
$$
$$
+ G_{Ln}\left(\frac{1 - e^{q_{0+}x}}{k_2} - \frac{e^{q_{n+}x} - e^{q_{0+}x}}{k_2 - E_{xx}q_{n+}^2 + uq_{n+}}\right)
$$
$$
- \frac{G_O}{k_2}(1 - e^{q_{0+}x}) \qquad x \le 0
$$

(5-61)

$$
q_{0+} = \frac{u + \sqrt{u^2 + 4E_{xx}k_2}}{2E_{xx}}
$$

Equation (5-61) is the *upward deficit transport equation.*

Use of the deficit equation. As shown from the equations above, the deficit D at any point in the receiving water is a function of the downward as well as upward transports of pollutants. This means that to calculate the deficit at any point, both the downward and upward transport equations have to be used. Calculations become very long and tedious. As a part of this book, a diskette is therefore provided containing a computer program called *WAT25*. WAT25 is a program for all the transport equations derived above. The instructions to run this model are given in Appendix 22. WAT25 can be run in any IBM-compatible computer. The program has been made as user friendly as possible. To run it, just simply follow the instructions in the computer screen. In addition, the source code, WAT25.BAS, is included in the diskette. WAT25.BAS is written in QuickBasic v. 4.5.

WAT25 requires input values for k_c, k_n, μ_p, k_d, and k_2 be at 20°C. To find the values of these k's and μ_p at other temperatures, the Arrhenius relationships are used as follows:

$$k_T = k_{20}\theta^{T-20} \tag{5-62}$$

where k_T is k_c, k_n, μ_p, k_d or k_2 and θ is the Arrhenius temperature correction. For k_c, the correction is $\theta_c = 1.08$; for k_n, the correction is $\theta_n = 1.047$; for k_2, the correction is $\theta_{k2} = 1.028$; for μ_p and k_d, the corrections are 1.068 for both.

Concentrations at the point of discharge. Concentrations at the point of discharge or at confluence of two streams are computed by a simple material balance. Let Q_b be the background flow, c_{db} be the background concentration of any pollutant, Q_{trib} be the discharge flow of the treatment plant or the confluence stream, and c_{dtrib} be the pollutant concentration in Q_{trib}. The concentration c_{d0}, after mixing is

$$c_{d0} = \frac{c_{db}(Q_b) + c_{dtrib}(Q_{trib1})}{Q_b + Q_{trib}} \tag{5-63}$$

Example 5-10

A sodium chloride tracer resulting in a concentration of 5000 mg/L in the discharged effluent of a sewage treatment plant is used to determine the dispersion coefficient, E_{xx}. The concentration of sodium chloride in the background stream is 50 mg/L. If the plant flow is 200 m³/day and that of the stream is 2456 m³/day, what is the resulting salt concentration in the mixed stream?

Solution

$$c_{d0} = \frac{c_{db}(Q_b) + c_{dtrib}(Q_{trib1})}{Q_b + Q_{trib}} = c_{d0} = \frac{50(2456) + 5000(200)}{2456 + 200} = 423 \text{ mg/L} \quad \textbf{Answer}$$

Example 5-11

A community applied for a discharge permit from a state agency to discharge 15,000 m³/day of sewage into a free-flowing freshwater stream. The stream characteristics upstream of the point of discharge are as follows: flow rate = 15,000 m³/day, temperature = 22°C, BOD$_5$ = 3 mg/L, NH$_3$ = 2 mg/L as N, and DO = 8.0 mg/L. The state determined that the following information holds for the entire stretch of stream: $u = 0.2$ m/s (stream flow plus treated sewage), reaeration coefficient $k_2 = 10$ per day, $E_{xx} = 6.03(10^5)$ m²/day, and instream decay coefficients for CBOD and NH$_3$ are 0.34 per day and 0.4 per day, respectively. Also, the discharge effluent averages 25°C in temperature. Five miles downstream from the point of discharge the stream meets a very large river having of a discharge flow of 100 mgd. At 20°C, the bottle decay coefficient was determined to be 0.2 per day. What limitations should the state agency impose on the discharge permit if the state water quality standard for DO is 5.0 mg/L? Neglect P and R. There are no GL_c, GL_n, and G_O. Assume any other value required by the model.

Solution In practice, discharge limitations are set by assuming trial values and substituting them into model equations to see if the dissolved oxygen in the receiving stream does not fall below the state-adopted water quality standard. The procedure is so long and tedious that a computer model must be used. Prepare input to WAT25 and run the program.

Input. There are two types of inputs to WAT25: project constants and station parameters.

A. *Project constants*

Project name: Example 5-11
Modeler: Roscoe A. Sincero

Number of stations: 5 (this means that the stretch of stream from the point of
discharge to 5 miles downstream is divided into five stations)

k_c at 20°C (per day): 0.2
k_n at 20°C (per day): 0.4
k_d at 20°C (per day): 0
μ at 20°C (per day): 0
a_{pc} (dimensionless): 0
θ_c: 1.047
θ_n: 1.08
θ_{k2}: 1.028
θ_μ: 0
θ_{kd}: 0

L_b (background CBOD, mg/L): 3.0(1.5) = 4.5, 1.5 is a conversion factor from BOD$_5$ to CBOD
L_{nb} (background NBOD, mg/L): 9.14

$$NH_3 + 2O_2 \rightarrow NO_3^- + H^+ + H_2O$$

Therefore,

$$1 \text{ mg/L NH}_3\text{-N} = \frac{2(32)}{14} = 4.57 \text{ mg/L DO}$$

$$L_n = 2.0(4.57) = 9.14$$

O_b (background dissolved oxygen, mg/L): 8.0
c_{4b} (background chlorophyll *a*, mg/L): 0
temp$_b$ (background temperature, °C): 22
O_s (saturation dissolved oxygen, mg/L): 8.48

$$O_s \text{ at } 22°C = 8.72 \text{ mg/L} \qquad O_s \text{ at } 25°C = 8.24$$

$$\text{average } O_s = \frac{8.72(15,000) + 8.24(15,000)}{30,000} = 8.48$$

Q_b (background flow, m^3/day): 15,000

B. *Station parameters*

	Station				
	1	2	3	4	5
X (m, distance from station 1)	0	2,011	4,022	6,033	8,046.33
Note: 8046.33 m = 5-mile distance					
E_{xx} (m²/day)	603,000	603,000	603,000	603,000	603,000
u (m/day)	17,280	17,280	17,280	17,280	17,280
17,280 m/day = (0.2 m/s)(60 s/min)(60 min/h)(24 hr/day) = 17,280					
k_s (per day)	0.14	0.14	0.14	0.14	0.14
$k_c' = 0.34$ per day; $k_c = 0.20$ per day; $k_s = k_c' - k_c = 0.14$ per day					
k_2 (per day)	10	10	10	10	10
L_{trib} (mg/L, tributary L)	45	0	0	0	4.5
(20 is the CBOD from the plant)					
L_{ntrib} (mg/L, tributary L_n)	115	0	0	0	9.14
(40 is the NOBOD from the plant)					
O_{trib} (mg/L, tributary DO)	5.0	0	0	0	8.0
(5.0 is the DO from the plant)					
c_{4trib} (mg/L, tributary chlorophyll a)	0	0	0	0	0
$temp_{trib}$ (°C, tributary temperature)	25	0	0	0	22
Q_{evap} (m³/day, evaporation)	0	0	0	0	0
G_{Lc} (mg/L-day)	0	0	0	0	0
G_{Ln} (mg/L-day)	0	0	0	0	0
Q_{trib} (m³/day)	15,000	0	0	0	378,512.1
(15,000 is the plant discharge)					
378,512.1 m³/day = [100(10⁶) gal/day]/[(7.48 gal/ft³)(3.281² ft²/m²)					
G_O (mg/L-day)	0	0	0	0	0

Running WAT25 using the inputs above produces the following results.

Station	L (mg/L)	L_n (mg/L)	O (mg/L)
1	24.75	62.07	6.50
2	23.69	58.40	5.52
3	22.68	54.95	5.37
4	21.72	51.71	5.46
5	5.71	12.17	7.87

The lowest dissolved oxygen concentration is 5.37 mg/L, which is greater than the water quality standard of 5.0 mg/L, implying that secondary limits are acceptable. Therefore, referring to the station parameter input table above (see station 1), limitations of CBOD = 45 mg/L, NBOD = 115 mg/L, and DO = 5.0 will be sufficient to meet the water quality standard. In practice, these limitations are expressed in terms of BOD$_5$, TKN, and DO, repectively. Therefore, limitations

$$BOD_5 = \frac{45}{1.5} = 30 \text{ mg/L}$$

$$TKN \text{ (or ammonia)} = \frac{115}{4.57} = 25 \text{ mg/L} \quad \textbf{Answer}$$

$$\text{dissolved oxygen} = 5.0 \text{ mg/L}$$

Example 5-12

The community in Example 5-11, is investigating the possibility of discharging into a nearby tidal estuary which empties into an open sea. For practical purposes the water quality of the sea will not be affected by the discharge. The estuary has the following characteristics: depth = 8 m, net advective flow rate = 15,000 m^3/day, u = 0.02 m/s, reaeration coefficient k_2 = 2.0 per day, E_{xx} = 7.29 mi^2/day, BOD$_5$ = 3 mg/L, NH$_3$ = 2 mg/L as N, DO = 8.0 mg/L, temperature = 22°C, and instream decay coefficients for CBOD and NH$_3$ are 0.34 per day and 0.4 per day, respectively. Additional data: a_{pc} = 66.67, μ_p = 0.6 per day at the 4-m depth, c_4 = 10 μg/L at point of discharge, k_d = 0.05 per day, k_{s4} = 0.14 per day, G_{Lc} = 0.01 mg/L-day, G_{Ln} = 0.01 mg/L-day, and G_O = 0. What should be the limitations that the state impose? Which permit should the community adopt? Assume any reasonable values required by the model.

Solution   Prepare input to WAT25 and run the program.

A. *Project constants*

Project name: Example 5-12
Modeler: Arcadio A. Sincero Jr.

Number of stations: 5 (this means that the stretch of stream from the point of discharge to 5 miles downstream is divided into five stations)

k_c at 20°C (per day): 0.2
k_n at 20°C (per day): 0.4
k_d at 20°C (per day): 0.05
μ at 20°C (per day): 0.6
a_{pc} (dimensionless): 66.67
θ_c: 1.047
θ_n: 1.08
θ_{k2}: 1.028
θ_μ: 1.068
θ_{kd}: 1.068
L_b (background CBOD, mg/L): 3.0(1.5) = 4.5,   (1.5 is a conversion factor from BOD$_5$ to CBOD)
L_{nb} (background NBOD, mg/L): 9.14

$$NH_3 + 2O_2 \rightarrow NO_3^- + H^+ + H_2O$$

Therefore,

$$1 \text{ mg/L } NH_3\text{-N} = \frac{2(32)}{14} = 4.57 \text{ mg/L DO}$$

$$L_n = 2.0(4.57) = 9.14$$

O_b (background dissolved oxygen, mg/L): 8.0

c_{4b} (background chlorophyll a, mg/L): 0.01

temp_b (background temperature, °C): 22

O_s (saturation dissolved oxygen, mg/L): 8.48

$$O_s \text{ at } 22°C = 8.72 \text{ mg/L} \quad O_s \text{ at } 25°C = 8.24$$

$$\text{average } O_s \text{ after mixing} = \frac{8.72(15,000) + 8.24(15,000)}{30,000} = 8.48$$

Q_b (background flow, m^3/day): 15,000

B. *Station parameters*

	Station				
	1	2	3	4	5
X (m, distance from station 1)	0	2,011	4,022	6,033	8,046.33
Note: 8046.33 m = the 5-mile distance					
E_{xx} (m^2/day)		2589735.35	2589735.35	2589735.35	2589735.35	2589735.35
u (m/day)		1728	1728	1728	1728	1728
1728 m/day = (0.02 m/s)(60 s/min)(60 min/h)(24 h/day) = 1728					
k_s (per day)		0.14	0.14	0.14	0.14	0.14
$k_c' = 0.345$ per day; $k_c = 0.20$ per day; $k_s = k_c' - k_c = 0.14$ per day					
k_2 (per day)	2	2	2	2	2
L_{trib} (mg/L, tributary L)	13	0	0	0	4.5
$L_{n\text{trib}}$ (mg/L, tributary L_n)	30	0	0	0	9.14
O_{trib} (mg/L, tributary DO)	5.0	0	0	0	8.0
$c_{4\text{trib}}$ (mg/L, tributary chlorophyll a)	0	0	0	0	0.01
$\text{temp}_{\text{trib}}$ (°C, tributary temperature)	25	0	0	0	22
Q_{evap} (m^3/day, evaporation)	0	0	0	0	0
G_{Lc} (mg/L-day)	0.01	0.01	0.01	0.01	0.01
G_{Ln} (mg/L-day)	0.01	0.01	0.01	0.01	0.01
Q_{trib} (m^3/day)	15,000	0	0	0	1.0(10^{12})
1.0(10^{12}): assumed a large value as the equivalent Q_{trib} for the sea					
G_O (mg/L-day)	0	0	0	0	0

Running WAT25 using the foregoing inputs produces the following results.

Station	L (mg/L)	L_n (mg/L)	O (mg/L)
1	10.33	23.19	6.18
2	7.34	14.82	5.02
3	5.28	9.71	5.70
4	4.09	7.27	6.54
5	4.50	9.14	8.00

The dissolved oxygen concentration at station 2 is 5.02 mg/L, which is above the water quality standard of 5.0 mg/L. Therefore, referring to the station parameter input table above (see station 1), limitations of CBOD = 5.0 mg/L, NBOD = 15 mg/L, and DO = 5.0 will be sufficient to meet the water quality standard. In practice, these limitations are expressed in terms of BOD_5, TKN, and DO, repectively. Therefore, limitations

$$BOD_5 = \frac{13}{1.5} = 8.67 \quad \text{say, 8 mg/L}$$

$$\text{TKN (or ammonia)} = \frac{20}{4.57} = 4.38 \text{ say, 4 mg/L} \quad \textbf{Answer}$$

$$\text{dissolved oxygen} = 5.0 \text{ mg/L}$$

The community should adopt the permit for discharge to the free-flowing stream.   **Answer**

SUBSURFACE WATER QUALITY MODELING

Subsurface water consists of the water in the unsaturated or vadose zone and groundwater. The various transport and kinetic processes that occur in the surface water also occur to various degrees in subsurface water. For example, dispersion transport also occurs, although in a more complicated form due to the presence of soil grains. Pollutants also degrade both aerobically and anaerobically, as in surface waters. Of course, there are some processes occurring in surface waters that do not occur in subsurface waters, examples of which are settling and photolysis. Sunlight cannot penetrate the thickness of the soil.

Groundwater Quality Modeling

Subsurface water quality modeling is divided into two major divisions: groundwater quality modeling and vadose zone water quality modeling. These divisions differ as follows: groundwater is characterized by the absence of air, while the vadose zone is characterized by the presence of air and unsaturation by water.

For application to groundwater, m_s must be converted into a suitable form. m_s is the solids content per unit volume of water in groundwater; however, the solids content is analyzed as mass of solids per unit volume of space. Let m_d represent this solid concentration. m_s and m_d are then related by $\eta m_s = m_d$, and the Reynolds transport equation for c_d becomes

$$\frac{D(\eta c_d)}{Dt}\left(1 + \frac{f_{cp}K_{sc}m_d}{\eta}\right) = \frac{\partial(\eta c_d)}{\partial t}\left(1 + \frac{f_{cp}K_{sc}m_d}{\eta}\right) + \eta\left(u\frac{\partial c_d}{\partial x} + v\frac{\partial c_d}{\partial y} + w\frac{\partial c_d}{\partial z}\right)$$

(5-64)

The corresponding equations for conservative and nonconservative substances are then, respectively,

$$\frac{\partial(\eta c_d)}{\partial t} + \frac{\partial}{\partial x}\left(E_{xx}\frac{\partial\eta c_d}{\partial x}\right) + \frac{\partial}{\partial y}\left(E_{yy}\frac{\partial\eta c_d}{\partial y}\right) + \frac{\partial}{\partial z}\left(E_{zz}\frac{\partial\eta c_d}{\partial z}\right)$$

$$+ \frac{\eta}{1 + f_{cp}K_{sc}m_d/\eta}\left(u\frac{\partial c_d}{\partial x} + v\frac{\partial c_d}{\partial y} + w\frac{\partial c_d}{\partial z}\right) = 0 \qquad \text{conservative substances}$$

$$\frac{\partial(\eta c_d)}{\partial t} - \frac{\partial}{\partial x}\left(E_{xx}\frac{\partial\eta c_d}{\partial x}\right) - \frac{\partial}{\partial y}\left(E_{yy}\frac{\partial c_d}{\partial y}\right) - \frac{\partial}{\partial z}\left(E_{zz}\frac{\partial c_d}{\partial z}\right)$$

(5-65)

$$+ \frac{\eta}{1 + f_{cp}K_{sc}m_d/\eta}\left(u\frac{\partial c_d}{\partial x} + v\frac{\partial c_d}{\partial y} + w\frac{\partial c_d}{\partial z}\right)$$

$$+ k\eta c_d + (kc_{sp}m_d = kc_d f_{cp}K_{sc}m_d) - G = 0 \qquad \text{nonconservative substances}$$

An example of growing substances are the bacteria that grow on hazardous wastes; however, they would not have any transport equation since they will simply be attached to the soils. A differential equation was not therefore formulated for them. $kc_{sp}m_d$ represents the decay of the substance sorbed onto solids.

From fluid mechanics the general conservation equation is given by

$$\frac{\partial\rho}{\partial t} + \frac{\partial(\rho u)}{\partial x} + \frac{\partial(\rho v)}{\partial y} + \frac{\partial(\rho w)}{\partial z}$$

(5-66)

where ρ is the mass density of control mass and u, v, and w are the velocities in the x, y, and z directions, respectively. Applied to groundwater, this equation must be modified by multiplying ρ by η to convert ρ to a density referred to space. In addition, not all the water in the space will flow but only a fraction given by the specific yield s_y. Hence the conservation of mass for groundwater flow, assuming that ρ is constant, is

$$\rho\frac{\partial(\eta s_y)}{\partial t} + \eta\rho\left(\frac{\partial u}{\partial x} + \frac{\partial v}{\partial y} + \frac{\partial w}{\partial z}\right) = 0$$

(5-67)

From Darcy's law, $u = (K_{xx}/\eta)(\partial h/\partial x)$, $v = (K_{yy}/\eta)(\partial h/\partial y)$, and $w = K_{zz}/\eta)(\partial h/\partial z)$, where K_{xx}, K_{yy}, and K_{zz} are the hydraulic conductivities in the x, y, and z directions, respectively, and h is the piezometric head. Substituting in equation (5-67) and canceling ρ yields

$$\frac{\partial(\eta s_y)}{\partial t} + \eta\left(\frac{\partial\frac{K_{xx}}{\eta}\frac{\partial h}{\partial x}}{\partial x} + \frac{\partial\frac{K_{yy}}{\eta}\frac{\partial h}{\partial y}}{\partial y} + \frac{\partial\frac{K_{zz}}{\eta}\frac{\partial h}{\partial z}}{\partial z}\right) = 0$$

(5-68)

$\partial(\eta s_y)/\partial t$ is equal to $\left[\partial(\eta s_y)/\partial h\right]\left(\frac{\partial h}{\partial t}\right) \cdot \partial(\eta s_y)/\partial h$ is the volume fraction of water drained per unit drop in piezometric head, the storage coefficient s_c. Substituting in equation (5-68), we have

$$s_c \frac{\partial h}{\partial t} + K_{xx}\frac{\partial^2 h}{\partial x^2} + K_{yy}\frac{\partial^2 h}{\partial y^2} + K_{zz}\frac{\partial^2 h}{\partial z^2} = 0 \tag{5-69}$$

This equation can be solved for h; u, v, and w can then be computed using Darcy's law.

The total derivative in the positive x direction for the conservative and nonconservative substances are, respectively,

$$\frac{D(\eta c_d)}{Dt} = \begin{cases} u\dfrac{d(\eta c_d)}{dx} = -E_{xx}\dfrac{d^2\eta c_d}{dx^2} & \text{conservative substances} \\[2mm] u\dfrac{d(\eta c_d)}{dx} = +E_{xx}\dfrac{d^2\eta c_d}{dx^2} - k\eta c_d \\[2mm] \qquad\qquad - kc_d f_{cp} K_{sc} m_d & \text{nonconservative substances} \end{cases} \tag{5-70}$$

In the above, the distributed source G has been neglected. The solutions to equations (5-70) are, respectively,

$$c_d = \begin{cases} c_{d0}e^{(-u/E_{xx})x} & x \geq 0 & \text{conservative substances} \\ c_{d0}e^{q_- x} & x \geq 0 & \text{nonconservative substances} \end{cases} \tag{5-71}$$

$$q_- = \frac{u - \sqrt{u^2 + 4E_{xx}k(\eta)\left(1 + \frac{f_{cp}K_{sc}m_d}{\eta}\right)}}{2E_{xx}}$$

Equations for K_{sc}. Table 5-3 shows some equations that may be used to estimate K_{sc}. $K_{o,w}$ is the octanol–water coefficient. This is obtained by shaking an organic compound in a mixture of n-octanol and water and measuring the concentration of the

TABLE 5-3 EQUATIONS FOR ESTIMATING K_{SC}

Equation	Chemicals used in determination
$\log K_{sc} = 1.00 \log K_{o,w} - 0.21$	10 polyaromatic hydrocarbons
$K = 0.63 K_{o,w}$	Miscellaneous organics
$\log K_{sc} = 0.544 \log K_{o,w} + 1.377$	45 organics, mostly pesticides
$\log K_{sc} = 1.029 \log K_{o,w} - 0.18,\ r^2 = 0.91$	13 pesticides
$\log K_{sc} = 0.94 \log K_{o,w} + 0.22$	s-Trizines and dinitroanilines
$\log K_{sc} = 0.989 \log K_{o,w} - 0.346,\ r^2 = 0.991$	5 polyaromatic hydrocarbons
$\log K_{sc} = 0.937 \log K_{o,w} - 0.006$	Aromatics, polyaromatics, and triazines
$\ln K_{sc} = \ln K_{o,w} - 0.7301$	DDT, tetrachlorobiphenyl, lindane, 2,4,D, and dichloropropane
$\log K_{sc} = 0.72 \log K_{o,w} + 0.49,\ r^2 = 0.95,\ n = 13$	Methylated and chlorinated benzenes
$\log K_{sc} = 1.00 \log K_{o,w} - 0.317,\ r^2 = 0.98$	22 polynuclear aromatics

compound in the two phases that are formed. The ratio of the concentration in the octanol phase to that in the water phase is the octanol–water coefficient. Some values of $K_{o,w}$ are as follows: dichloroethane = 62, benzene = 135, trichloroethylene = 195, perchloroethylene = 760, naphthalene = 2350, and pyrene = 209,000.

Gaussian plume model. Figure 5-3 shows various plume configurations resulting from the discharge of a waste into groundwaters. The concentration at the center of the plume will be the highest; the concentration tapers off in the transverse directions. This transverse variation of concentration may be modeled by the Gaussian (normal) distribution (see the normal distribution at the bottom of the figure).

The normal distribution is defined as

$$Y = \frac{1}{\sigma\sqrt{2\pi}} e^{-\frac{1}{2}r^2/\sigma^2} \tag{5-72}$$

where Y is the probability coordinate of the normal curve, r the transverse distance from the center of the plume corresponding to a transverse concentration c, and σ the transverse distance from the center of the plume corresponding to the standard deviation of the distribution of the transverse concentrations.

At the center of the plume, r is equal to zero and Y equals $Y_c = 1/\sigma\sqrt{2\pi}$. As predicted by the transport model, the concentration c_d applies at the center of the plume. Therefore, the transverse concentration c is to c_d as Y is to Y_c. Or,

$$\frac{c}{c_d} = \frac{(1/\sigma\sqrt{2\pi}\,e^{-(1/2)r^2/\sigma^2}}{1/\sigma\sqrt{2\pi}} \Rightarrow c = c_d)e^{-(1/2)r^2/\sigma^2} \tag{5-73}$$

But $r^2 = y^2 + z^2$. Substituting in equation (5-73) yields

$$c = c_d e^{-(1/2)(y^2/\sigma^2 + z^2/\sigma^2)} = c_d e^{-(y/\sqrt{2}\sigma)^2} e^{-\left(z/\sqrt{2}\sigma\right)^2} \tag{5-74}$$

Using equations (5-71), the concentrations at any point x, y, z from the point of origin of the plume are given below for the conservative and nonconservative substances, respectively.

$$c = \begin{cases} c_{d0}e^{-(u/E_{xx})x}e^{-(y/\sqrt{2}\sigma_c)^2}e^{-(z/\sqrt{2}\sigma_c)^2} & x \geq 0 \quad \text{conservative substances} \\ c_{d0}e^{q-x}e^{-(y/\sqrt{2}\sigma_n)^2}e^{-(z/\sqrt{2}\sigma_n)^2} & x \geq 0 \quad \text{nonconservative substances} \end{cases} \tag{5-75}$$

Concentrations from relationship between E and σ. Express σ in terms of E. From equation (5-75), the following relation may be deduced:

$$e^{-(v/E_{yy})y} = e^{-y^2/2\sigma_c^2} \qquad \text{conservative substances} \tag{5-76}$$

Solving for $2\sigma_c^2$, yields

$$\sigma_c^2 = \frac{y E_{yy}}{2v} \Rightarrow \sigma_c^2 = \frac{z E_{zz}}{2w} \tag{5-77}$$

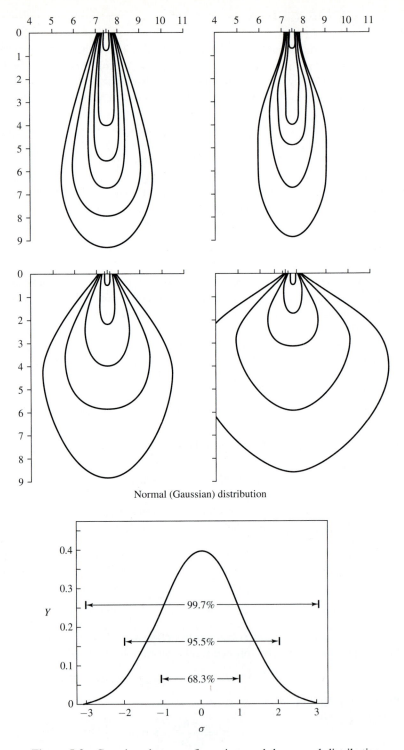

Normal (Gaussian) distribution

Figure 5-3 Gaussian plume configurations and the normal distribution.

Hence, from equation (5-77), the transverse concentrations may be normally distributed if y/v and z/w are constant. Also, for plume models, the transverse velocities are diffusion velocities; hence v and w are diffusion velocities.

Doing similar manipulations to the nonconservative substances gives, $\sigma_n^2 = -y/2q_y$ and $\sigma_n^2 = -z/2q_z$. Substituting the appropriate relations into equation (5-75) we have,

$$c = \begin{cases} c_{d0}e^{-(u/E_{xx}x)}e^{-(v/E_{yy})y}e^{-(w/E_{zz})z}] & x \geq 0 \quad \text{conservative substances} \\ c_{d0}e^{q_x-x}e^{q_y y}e^{q_z z} & x \geq 0 \quad \text{nonconservative substances} \end{cases}$$

(5-78)

$$q_y = \begin{cases} q_{y-} & y \geq 0 \\ q_{y+} & y \leq 0 \end{cases}$$

$$q_z = \begin{cases} q_{z-} & z \geq 0 \\ q_{z+} & z \leq 0 \end{cases}$$

$$q_{x-} = \frac{u - \sqrt{u^2 + 4E_{xx}k(\eta)(1 + f_{cp}K_{sc}m_d/\eta)}}{2E_{xx}}$$

$$q_{y-} = \frac{v - \sqrt{v^2 + 4E_{yy}k(\eta)(1 + f_{cp}K_{sc}m_d/\eta)}}{2E_{yy}}$$

$$q_{y+} = \frac{v + \sqrt{v^2 + 4E_{yy}k(\eta)(1 + f_{cp}K_{sc}m_d/\eta)}}{2E_{yy}}$$

$$q_{z-} = \frac{w - \sqrt{w^2 + 4E_{zz}k(\eta)(1 + f_{cp}K_{sc}m_d/\eta)}}{2E_{zz}}$$

$$q_{z+} = \frac{w + \sqrt{w^2 + 4E_{zz}k(\eta)(1 + f_{cp}K_{sc}m_d/\eta)}}{2E_{zz}}$$

Example 5-13

Benzene, whose solubility equals 0.07 part per 100 parts of water by weight, was spilled into an aquifer having a hydraulic conductivity $K_{xx} = 100$ m/day. What is the centerline concentration of the plume when it reaches the river bank 3300 m downstream? Field investigation showed that the elevation of the groundwater table is 30 m above the river. Use the following additional information: $f_{cp} = 0.1$, $\eta = 0.55$, formation specific gravity $= 2.65$, $E_{xx} = 3.76(10^{-4})$ cm^2/s, and $k = 0.07$ per day.

Solution

$$c = c_{d0}e^{q_x-x}e^{q_y y}e^{q_z z}$$

$$q_y = 0 \qquad q_z = 0$$

$$q_{x-} = \frac{u - \sqrt{u^2 + 4E_{xx}k(\eta)(1 + f_{cp}K_{sc}m_d/\eta)}}{2E_{xx}}$$

$$u = \frac{K_{xx}}{\eta}\frac{\partial h}{\partial x} = \frac{100}{0.55}\left(\frac{30}{3300}\right) = 1.65 \text{ m/day}$$

Use $\log K_{sc} = 0.72 \log K_{o,w} + 0.49$.

$$K_{o,w} = 135$$

Thus,

$$\log K_{sc} = 0.72 \log(135) + 0.49 = 2.02$$

$$K_{sc} = 104.7 \text{L/kg} = 0.1047 \text{ m}^3/\text{kg}$$

$$m_d = 2.65(10^3)(1.0 - 0.55) = 1.19(10^3) \text{ kg/m}^3$$

$$c_{d0} \approx \frac{0.07 \text{ g benzene}}{100 \text{ g H}_2\text{O}} = \frac{0.07(1000)}{0.1} = 700 \frac{\text{mg benzene}}{\text{L water}}$$

$$E_{xx} = 3.76(10^{-4})(10^{-2})^2(60)(60)(24) = 0.0032 \text{ m}^2/\text{day}$$

$$1 + \frac{f_{cp} K_{sc} m_d}{\eta} = 1 + \frac{0.1(0.1047)(1.19)(10^3)}{0.55} = 23.65$$

$$q_{x-} = \frac{1.65 - \sqrt{1.65^2 + 4(0.0032)(0.07)0.55(23.65)}}{2(0.0032)} = -0.55$$

$$c = 700e^{-0.55(3300)}(1)(1) = 0 \quad \textbf{Answer}$$

Vadose Zone Water Quality Modeling

In the vadose zone, three major phases exist: soil, water, and air. Because of the existence of three phases, three Lagrangian total derivatives will be written: $Dc_{sp}m_d/Dt$, Dc_d/Dt, and Dc_a/Dt. The subscript a refers to the concentration in air. Depending on whether or not the material is decaying, the differential equations converted to space concentrations are written below.

$$\frac{D(c_{sp}m_d)}{Dt} + \frac{D(\eta_d c_d)}{Dt} + \frac{D(\eta_a c_a)}{Dt} = -E_{xx}\frac{\partial^2(\eta_d c_d)}{\partial x^2} - E_{yy}\frac{\partial^2(\eta_d c_d)}{\partial y^2}$$

$$-E_{zz}\frac{\partial^2(\eta_d c_d)}{\partial z^2} - E_{axx}\frac{\partial^2(\eta_a c_a)}{\partial x^2} - E_{ayy}\frac{\partial^2(\eta_a c_a)}{\partial y^2}$$

$$-E_{azz}\frac{\partial^2(\eta_a c_a)}{\partial z^2} \qquad \text{conservative substances}$$

$$\tag{5-79}$$

$$\frac{D(c_{sp}m_d)}{Dt} + \frac{D(\eta_d c_d)}{Dt} + \frac{D(\eta_a c_a)}{Dt} = +E_{xx}\frac{\partial^2(\eta_d c_d)}{\partial x^2} + E_{yy}\frac{\partial^2(\eta_d c_d)}{\partial y^2}$$

$$+E_{zz}\frac{\partial^2(\eta_d c_d)}{\partial z^2} + E_{axx}\frac{\partial^2(\eta_a c_a)}{\partial x^2} + E_{ayy}\frac{\partial^2(\eta_a c_a)}{\partial y^2}$$

$$+E_{azz}\frac{\partial^2(\eta_a c_a)}{\partial z^2} - k(c_{sp}m_d) - k(\eta_d c_d)$$

$$-k(\eta_a c_a) \qquad \text{nonconservative substances}$$

η_d and η_a are the percent void volumes or porosities of water and air, respectively. $\eta_d + \eta_a = \eta$.

Equations (5-79) may be substituted into the Reynolds transport theorem, producing

$$-E_{xx}\frac{\partial^2(\eta_d c_d)}{\partial x^2} - E_{yy}\frac{\partial^2(\eta_d c_d)}{\partial y^2} - E_{zz}\frac{\partial^2(\eta_d c_d)}{\partial z^2} - E_{axx}\frac{\partial^2(\eta_a c_a)}{\partial x^2} - E_{ayy}\frac{\partial^2(\eta_a c_a)}{\partial y^2}$$

$$- E_{azz}\frac{\partial^2(\eta_a c_a)}{\partial z^2} = \frac{\partial(c_{sp}m_d)}{\partial t} + \frac{\partial(\eta_d c_d)}{\partial t} + \frac{\partial(\eta_a c_a)}{\partial t} + u\frac{\partial(\eta_d c_d)}{\partial x}$$

$$+ v\frac{\partial(\eta_d c_d)}{\partial y} + \frac{\partial(\eta_d c_d)}{\partial z} \qquad \text{conservative}$$

$$E_{xx}\frac{\partial^2(\eta_d c_d)}{\partial x^2} + E_{yy}\frac{\partial^2(\eta_d c_d)}{\partial y^2} + E_{zz}\frac{\partial^2(\eta_d c_d)}{\partial z^2} + E_{axx}\frac{\partial^2(\eta_a c_a)}{\partial x^2} + E_{ayy}\frac{\partial^2(\eta_a c_a)}{\partial y^2} \qquad (5\text{-}80)$$

$$+ E_{azz}\frac{\partial^2(\eta_a c_a)}{\partial z^2} - k(c_{sp}m_d) - k(\eta_d c_d) - k(\eta_a c_a)$$

$$= \frac{\partial(c_{sp}m_d)}{\partial t} + \frac{\partial(\eta_d c_d)}{\partial t} + \frac{\partial(\eta_a c_a)}{\partial t} + u\frac{\partial(\eta_d c_d)}{\partial x} + v\frac{\partial(\eta_d c_d)}{\partial y}$$

$$+ \frac{\partial(\eta_d c_d)}{\partial z} \qquad \text{nonconservative}$$

Vertical component equations.   In vadose zone modeling, the motion is vertical. Any sidewise direction may be accounted for by considering concentrations in this direction as normally distributed. In the z direction (positive z downward), equations (5-79) become, respectively,

$$\frac{D(c_{sp}m_d)}{Dt}\left[= w\frac{d(c_{sp}m_d)}{dz}\right] + \frac{D(\eta_d c_d)}{Dt}\left[= w\frac{d(\eta_d c_d)}{dz}\right] + \frac{D(\eta_a c_a)}{Dt}\left[= w\frac{d(\eta_a c_a)}{dz}\right]$$

$$= -E_{zz}\frac{d^2(\eta_d c_d)}{dz^2} - E_{azz}\frac{d^2(\eta_a c_a)}{dz^2} \qquad \text{conservative}$$

$$\frac{D(c_{sp}m_d)}{Dt}\left[= w\frac{d(c_{sp}m_d)}{dz}\right] + \frac{D(\eta_d c_d)}{Dt}\left[= w\frac{d(\eta_d c_d)}{dz}\right] + \frac{D(\eta_a c_a)}{Dt}\left[= w\frac{d(\eta_a c_a)}{dz}\right]$$

$$= E_{zz}\frac{\partial^2(\eta_d c_d)}{\partial z^2} + E_{azz}\frac{\partial^2(\eta_a c_a)}{\partial z^2} - k(c_{sp}m_d) - k(\eta_d c_d) - k(\eta_a c_a) \qquad \text{nonconservative}$$

$$(5\text{-}81)$$

As before, the most practical concentration to use is c_d. All concentrations above will therefore be expressed in terms of c_d. c_{sp} is equal to $f_{cp}K_{sc}c_d$. By Henry's law, $c_a = K_H c_d$, where K_H is the Henry's law constant. Since the dispersion coefficient is a function of the medium, E_{azz} is equal to E_{zz} also. Substituting all in equations (5-81) and solving yields,

$$c_d = \begin{cases} c_{d0}e^{-(fw/\xi_{zz})z} & z \geq 0 \quad \text{conservative substances} \\ c_{d0}e^{q_{vz-}z} & z \geq 0 \quad \text{nonconservative substances} \end{cases} \tag{5-82}$$

$$f = f_{cp}K_{sc}m_d + \eta_d + \eta_a K_H$$

$$\xi_{zz} = E_{zz}(7d + \eta_a K_H)$$

$$q_{vz-} = \frac{fw - \sqrt{(fw)^2 + 4\xi_{zz}fk}}{2\xi_{zz}}$$

Since w is a longitudinal velocity, it is an advective velocity.

Corresponding to equations (5-78), the general equations by extending equations (5-82) for any point x, y, and z assuming Gaussian plumes are

$$c = \begin{cases} c_{d0}e^{-(fw/\xi_{zz})z}e^{-(fu/\xi_{xx})x}e^{-(fv/\xi_{yy})y} & z \geq 0 \quad \text{conservative substances} \\ c_{d0}e^{q_{vz}z}e^{q_{vx}x}e^{q_{vy}y} & z \geq 0 \quad \text{nonconservative substances} \end{cases} \tag{5-83}$$

$$\xi_{xx} = E_{xx}(7d + \eta_a K_H)\delta \qquad \xi_{yy} = E_{yy}(7d + \eta_a K_H)$$

$$q_{vx} = \begin{cases} q_{vx-} & x \geq 0 \\ q_{vx+} & x \leq 0 \end{cases}$$

$$q_{vy} = \begin{cases} q_{vy-} & y \geq 0 \\ q_{vy+} & y \leq 0 \end{cases}$$

$$q_{vx-} = \frac{fu - \sqrt{(fu)^2 + 4\xi_{xx}fk}}{2\xi_{xx}} \qquad q_{vx+} = \frac{fu + \sqrt{(fu)^2 + 4\xi_{xx}fk}}{2\xi_{xx}}$$

$$q_{vy-} = \frac{fv - \sqrt{(fv)^2 + 4\xi_{yy}fk}}{2\xi_{yy}} \qquad q_{vy+} = \frac{fv + \sqrt{(fv)^2 + 4\xi_{yy}fk}}{2\xi_{yy}}$$

Since u and v are transverse velocities, they are diffusion velocities.

Example 5-14

A soil sample of 0.25 ft^3 is dried in an oven at 103°C to constant weight. The porosity of a similar sample from the same batch was originally determined to be 0.55. If 2691 g of moisture was liberated, calculate η_d and η_a.

Solution

$$\text{Volume of H}_2\text{O liberated} = \frac{2691}{454(62.4)} = 0.095 \text{ ft}^3$$

Thus,

$$\eta_d = \frac{0.095}{0.25} = 0.38 \quad \textbf{Answer;} \quad \eta_a = 0.55 - 0.38 = 0.17 \quad \textbf{Answer}$$

Buckingham–Darcy law.   The components u, v, and w in the foregoing equations must be determined. In groundwater flow, the interstitial velocity is determined by means of Darcy's law in conjunction with the porosity. In vadose zone flow, the counterpart is called the Buckingham–Darcy law. These two laws are identical except that the former deals with saturated conditions, whereas the latter deals with unsaturated conditions, which are under tension (i.e., the pressure is below atmospheric). Tension pressure is measured in the field by a *tensiometer* (Figure 5-4).

Tension pressure is also called *capillary* or *matric potential*. At any point in the soil formation Z distance above a reference datum, the total potential ϕ is given by the sum of the matric potential ψ and the gravitational potential Z. Hence the Buckingham–Darcy law is

$$\mathbf{v} = -\frac{\mathbf{K}}{\eta} \bullet \nabla\phi = -\frac{\mathbf{K}}{\eta} \bullet \nabla(\psi + \mathbf{Z}) \tag{5-84}$$

where $\mathbf{K}$ is a vector of unsaturated hydraulic conductivity. In the z direction, equation (5-84) becomes

$$w = -\frac{K_{zz}}{\eta}\frac{\partial}{\partial z}(\psi + Z) \tag{5-85}$$

Example 5-15

Tiny droplets of benzene are slowly leaking from an on-ground storage tank. Tensiometers as shown in Figure 5-4 show readings of -88 cm and -126 cm of water at depths of 30 cm and 100 cm, respectively, below the ground. The vertical unsaturated hydraulic conductivity K_{zz} is 5.4 m/day. The porosities are $\eta = 0.55$, $\eta_d = 0.38$, and $\eta_a = 0.17$. The other pertinent information are $f_{cp} = 0.1$, $K_{sc} = 104.7$ L/day, $m_d = 1.19(10^3)$ kg/m^3, and $E_{zz} = 0.0012$ m^2/day. The half-life of benzene is 10 days and its solubility in water is 0.07 part per 100 parts by weight. The vapor pressure of benzene is 1.0 mmHg. Calculate the benzene concentration 10 m below the surface.

Solution

$$c = c_{d0}e^{q_{vz-}z}e^{q_{vx}x}e^{q_{vy}y} \qquad z \geq 0 \quad \text{nonconservative substances}$$

$$q_{vx-} = q_{vy-} = 0$$

$$f = f_{cp}K_{sc}m_d + \eta_d + \eta_a K_{\mathrm{H}}$$

$$\xi_{zz} = E_{zz}(\eta_d + \eta_a K_{\mathrm{H}})$$

$$q_{vz-} = \frac{fw - \sqrt{(fw)^2 + 4\xi_{zz}fk}}{2\xi_{zz}}$$

$$w = -\frac{K_{zz}}{\eta}\frac{\partial}{\partial z}(\psi + Z)$$

$$K_{\mathrm{H}} = \frac{c_a}{c_d}$$

$$c_a = 1.0 \text{ mmHg in 760 mmHg of total air mixture} = \frac{1.0}{760} = 0.0013 \; \frac{\text{mol benzene}}{\text{mol air mixture}}$$

$$\mathrm{C_6H_6} = 12(6) + 1.008(6) = 78$$

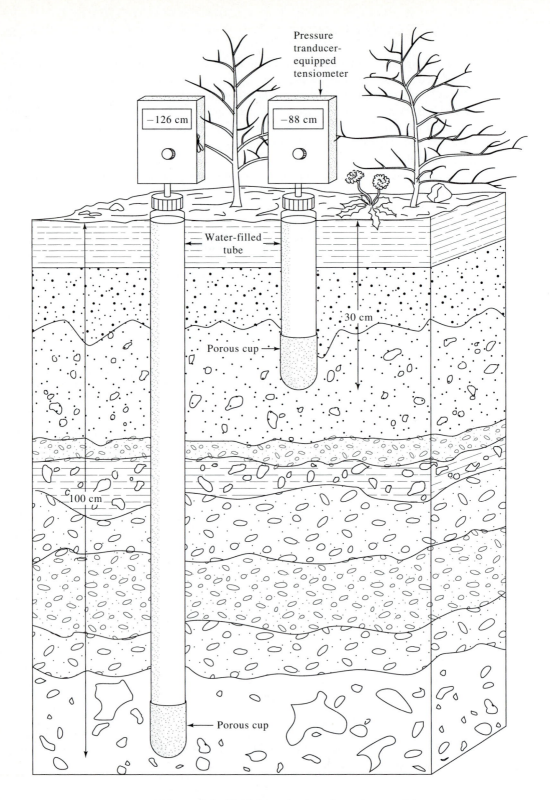

Figure 5-4 Field determination of matric potential.

Assuming a temperature of 20°C yields,

$$c_a = \frac{0.0013(78)(1000)}{22.4\,(273 + 20/273)} = 4.22 \quad \frac{\text{mg benzene}}{\text{L air}}$$

$$c_d = \frac{0.07 \text{ g benzene}}{100 \text{ g } H_2O} = \frac{0.07(1000)}{0.1} = 700 \quad \frac{\text{mg benzene}}{\text{L water}}$$

Then

$$K_H = \frac{4.22}{700} = 0.006 \quad \frac{\text{L water}}{\text{L air}} = 0.006 \quad \frac{m^3 \text{ water}}{m^3 \text{ air}}$$

$$f = 0.1(0.1047)(1.19)(10^3) + 0.38 + 0.17(0.006) = 12.84$$

$$\xi_{zz} = 0.0012[0.38 + 0.17(0.006)] = 0.00046 \text{ m}^2/\text{day}$$

$$k = \frac{\ln(c_d/c_{d0})}{t} = \frac{\ln 0.5}{\text{half-life}} = \frac{\ln 0.5}{10} = -0.07 \text{ per day} \Rightarrow 0.07 \text{ per day}$$

$$w = -\frac{5.4}{0.55}\left\{ \frac{[(-126)(10^{-2}) + (-100(10^{-2})] - [(-88)(10^{-2}) + (-30(10^{-2})]}{(100 - 30)(10^{-2})} \right\}$$

$$= 15.15 \text{ m/day}$$

$$q_{vz-} = \frac{12.84(15.15) - \sqrt{[12.84(15.15)]^2 + 4(0.00046)(12.84)(0.07)}}{2(0.00046)} = -0.0046 \text{ per day}$$

$$c_{d0} = 700 \text{ mg/L}$$

$$c_d = 700e^{-0.0046(10)}[1][1] = 700(0.99) = 669 \text{ mg/L} \quad \textbf{Answer}$$

AIR QUALITY MODELING

Equations for air quality models are similar to those that have already been derived, except that in air models no porosity is involved. The equations will simply be written at once.

$$c = \begin{cases} c_{d0}e^{-(u/E_{xx})x}e^{-(v/E_{yy})y}e^{-(w/E_{zz})z} & x \geq 0 \quad \text{conservative substances} \\ c_{d0}e^{q_{ax}-x}e^{q_{ay}y}e^{q_{az}z} & x \geq 0 \quad \text{nonconservative substances} \end{cases} \quad (5\text{-}86)$$

$$q_{ax-} = \frac{u - \sqrt{(u)^2 + 4E_{xx}k}}{2E_{xx}} \qquad q_{ay-} = \frac{v - \sqrt{(v)^2 + 4E_{yy}k}}{2E_{yy}} \qquad y \geq 0$$

$$q_{az-} = \frac{w - \sqrt{(w)^2 + 4E_{zz}k}}{2E_{zz}} \qquad z \geq 0$$

$$q_{ay+} = \frac{v + \sqrt{(v)^2 + 4E_{yy}k}}{2E_{yy}} \qquad y \leq 0$$

$$q_{az+} = \frac{w + \sqrt{(w)^2 + 4E_{zz}k}}{2E_{zz}} \qquad z \leq 0$$

In terms of the standard deviations σ_y and σ_z, equations (5-86) may be written as

$$c = \begin{cases} c_{d0} e^{-(u/E_{xx})x} e^{-(y/\sqrt{2}\sigma_c)^2} e^{-(z/\sqrt{2}\sigma_c)^2} & x \geq 0 \quad \text{conservative substances} \\ c_{d0} e^{q-x} e^{-(y/\sqrt{2}\sigma_n)^2} e^{-(z/\sqrt{2}\sigma_n)^2} & x \geq 0 \quad \text{nonconservative substances} \end{cases}$$

(5-87)

Lapse Rates and Dispersion

Dispersion of pollutants in the air is a function of lapse rate, which is the rate of change of temperature with height. The *dry adiabatic lapse rate* is the rate of change of temperature with height as a parcel of dry air rises or sinks without exchange of heat with the surroundings (adiabatic heat exchange). The *wetadiabatic lapse rate* is an adiabatic lapse rate with the atmosphere containing moisture. In this section we derive expressions for these lapse rates.

The ideal gas law applied to the air is $p = \rho RT/M_a$, where p is the pressure, ρ the density, R the universal gas constant, T the absolute temperature, and M_a the molecular weight of air (28.84). The rate of change of pressure with height is also given by $dp/dz = -\rho g$. Combining with equation (5-87) gives

$$\frac{dp}{p} = -\frac{gM_a}{RT} dz$$

(5-88)

The internal energy dU of the adiabatically rising dry air is given by the first law of thermodynamics: $dU = \delta Q \ (= 0$, since the process is adiabatic) $-p\,dV = -p\,dV$, where Q is the heat transfer and V is the volume of the parcel. dU is equal to $C_v\,dT$, where C_v is the heat capacity at constant volume. The ideal gas law for an air parcel of volume V is $pV = mRT/M_a$, where m is the mass of air. Hence $d(pV) = p\,dV + V\,dp = (mR/M_a)\,dT$. Substituting all into the first law, we obtain

$$C_v\,dT = V\,dp - \frac{mR}{M_a}\,dT = \frac{mRT}{M_a}\frac{dp}{p} - \frac{mR}{M_a}\,dT$$

Solving for dp/p gives

$$\frac{dp}{p} = \left(C_v\,dT + \frac{mR}{M_a}\,dT \right) \frac{M_a}{mRT}$$

(5-89)

Combining equation (5-89) with equation (5-88) produces the lapse rate dT/dz.

$$\frac{dT}{dz} = \Gamma = -\frac{g}{C_v/m + R/M_a} = -\frac{g}{c_v + R/M_a} = -\frac{g}{c_p}$$

(5-90)

c_v and c_p are the specific heats at constant volume and pressure, respectively. The average temperature in the troposphere of a standard atmosphere could range from $-63°C$ at 48 km above the surface of the earth to $27°C$ at the surface of the earth. From handbooks, c_p (for air) at this range is approximately 0.245 cal/g–°C = 245 cal/kg-°C. Use the following conversions: 1 cal = 4.184 J = 4.184 N-m. g = 9.81 m/s² = 9.81 N/kg. Therefore, the dry adiabatic lapse rate is

$$\frac{dT}{dz} = \Gamma = -\frac{9.81}{245(4.184)} = -0.0095700°C/m$$

If the atmosphere contains ω fraction of water vapor, the lapse rate becomes $-g/[(1-\omega)c_p + \omega c_{pw}]$, where c_{pw} is the specific heat of water vapor at constant pressure. From handbooks, at the range of -63 to $27°C$, c_{pw} is approximately 440 cal/kg-°C. The saturation absolute humidity at this temperature range is 0.00079 lb of water vapor per pound of dry air. Hence ω is $0.00079/(0.00079+1) = 0.00079$. The saturated adiabatic lapse rate is, therefore,

$$\frac{dT}{dz} = \Gamma = -\frac{9.81}{(1-0.00079)245 + 0.00079(440)(4.184)} = -0.009564°C/m$$

Thus the average lapse rate for the dry adiabatic is slightly larger than the average lapse rate for the saturated adiabatic process.

A parcel of heated polluted air when released from a smokestack will rise rapidly because of the large difference in density between the heated and the surrounding air. Because of this rapid rise, exchange of heat with the surroundings is negligibly small. A rapidly rising smoke is therefore practically under an adiabatic process. Whether or not the pollutant will be dispersed will depend on the actual ambient lapse rate of the surrounding air.

Figure 5-5 shows behaviors of plumes as functions of the ambient lapse rate. The adiabatic lapse rate of the rising plume is represented by the dashed line, and the actual ambient lapse rate is represented by the solid line. In the topmost figure, the ambient lapse rate is much faster than the adiabatic lapse rate. Hence the surrounding air is colder than the rising plume during transit and because the surrounding air has a greater density as a result, the plume will continue to rise until its temperature becomes equal to that of the surroundings. This type of ambient lapse rate is called *superadiabatic*, and the air is *unstable*. Because of this instability, pollutants will disperse very easily. As shown, the instability causes a *looping* condition.

When the ambient lapse rate is less than the adiabatic (second figure from the top), a condition called *subadiabatic*, the surrounding air becomes warmer during transit, and the plume released cannot rise. The lapse rate is weak and this condition is called *stable*, resulting in plumes that are coning.

During the night, the cool earth cools the overlying air. This results in an inverted lapse rate, which can persist even up to the morning hours until the sun breaks the inversion by heating the ground. This inverted lapse rate is called an *inversion*. Inversion inhibits vertical mixing, is stable, and causes the plume to spread horizontally. This spread is called *fanning*. When there is inversion below but a lapse rate above, the plume will disperse aloft but not below. This plume configuration is called *lofting*. When the reverse is true, that is, there is inversion above but lapse rate below, the plume will fumigate the ground. The configuration is called *fumigation*. When an inversion layer prevails below and above the plume, the plume is trapped. This configuration is called *trapping*.

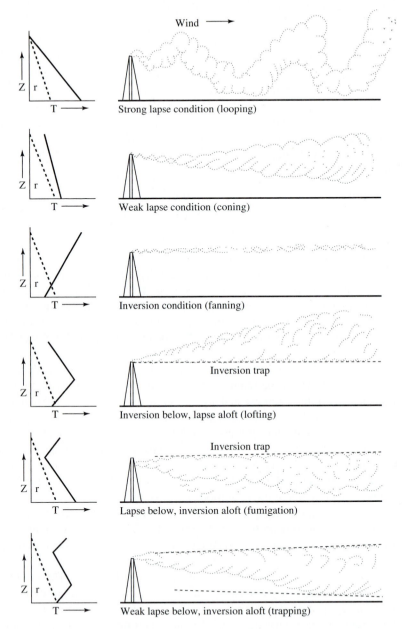

Figure 5-5 Types of plume behavior.

During calm days, the velocity of the air is zero. Under this condition, the plume will not move horizontally but straight up through buoyant forces. This type of configuration is called *neutral*, and the plume is a *neutral plume*.

Plume Model

Figure 5-6 shows a plume configuration with coordinate reference point indicated. Let W mass per unit time be the stack emission of pollutants. Normally, the drifting of the plume in the x direction is so much faster that the transport may be approximated solely by convection. In addition, because of the speed of propagation and the short time frames involved, the process can be considered conservative. By considering the transport as convective and conservative, the downstream average concentration $c_d = c_{d0}$ $\exp[(-u/E_{xx})x]$ may be approximated by $W/(uA_x)$, where A_x is the transverse cross-sectional area of the plume a distance x from the origin. At $x = 0$, A_x would be the cross-sectional area of the stack; at $x > 0$, A_x would expand laterally because of dispersion. Call this expanded cross section as the *equivalent stack section* at distance x from the origin. The extent to which A_x would expand can be computed from the variance of the distances (σ^2) from the plume centerline. From the property of the normal distribution, one standard deviation distance from the centerline of the plume represents 68.26% of the area of the equivalent stack section. Hence, at any distance x, A_x is therefore equal to $\pi\sigma^2/0.6826$. Because of uneven dispersion in the y and z directions, however, A_x will not be circular but more of an ellipse. (Note that the circle is a special type of ellipse with the major and the minor axes equal.) Hence, approximating as an ellipse, A_x is equal to $\pi\sigma_y\sigma_z/0.6826$, where σ_y and σ_z are the standard deviations corresponding to the variances of the distances in the y and z directions, respectively.

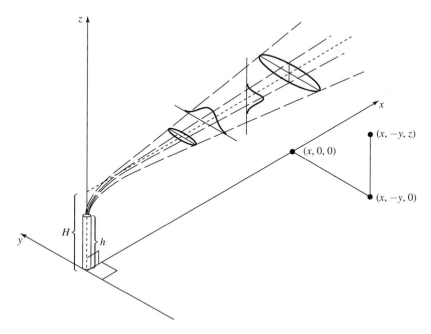

Figure 5-6 Plume dispersion coordinate system.

TABLE 5-4   VALUES OF a, c, d, AND f

Stability type	a	$x \leq 1$ km			$x \geq 1$ km		
		c	d	f	c	d	f
A	213	440.8	1.941	9.27	459.7	2.094	−9.6
B	156	106.6	1.149	3.3	108.2	1.098	2.0
C	104	61.0	0.911	0	61.0	0.911	0
D	68	33.2	0.725	−1.7	44.5	0.516	−13.0
E	50.5	22.8	0.678	−1.3	55.4	0.305	−34.0
F	34	14.35	0.740	−0.35	62.6	0.180	−48.6

At any distance x from the origin, the average concentration of pollutant at the equivalent stack section may now therefore be approximated by $0.6826W/u\pi\sigma_y\sigma_z$.

For turbulent flows in pipes, the ratio of the average velocity to the maximum velocity ranges from > 0.5 to 0.82. This gives an average of 0.66. Considering the plume as a turbulent pipe and approximating the ratio of the average concentration to the maximum concentration at the center of the plume as 0.66, the plume centerline concentration c_d is

$$\frac{0.6826W}{u\pi\sigma_y\sigma_z} = 0.66c_d \Rightarrow c_d = \frac{1.03W}{u\pi\sigma_y\sigma_z} \tag{5-91}$$

With c_d known, assuming a conservative process, and using the coordinates of Figure 5-6, equation (5-87) becomes

$$c = \frac{1.03W}{u\pi\sigma_y\sigma_z}e^{-(y/\sqrt{2}\sigma_y)^2}e^{-(z-H/\sqrt{2}\sigma_z)^2} \tag{5-92}$$

where H is the effective stack height.

The effective stack height is equal to the stack height h plus a correction Δh. The correction may be estimated by Holland's empirical equation,

$$\Delta h = \frac{v_s d}{u}\left[1.5 + 2.68(10^{-3})p\frac{\Delta T d}{T_s}\right] \tag{5-93}$$

where v_s is the stack gas velocity in m/s, p is the atmospheric pressure in mbar; ΔT is the stack gas temperature minus ambient air temperature in Kelvin, d is the inside stack diameter in meters, and T_s is the stack gas temperature in Kelvin.

Martin[5] curve-fitted empirical equations for σ_y and σ_z. They are

$$\sigma_y = ax^{0.894} \tag{5-94}$$

$$\sigma_z = cx^d + f \tag{5-95}$$

Martin's constants a, c, d, and f are shown in Table 5-4. The downwind distance x must be expressed in kilometers to obtain σ_y and σ_z in meters.

[5]D. O. Martin (1976). "The Change of Concentration Standard Deviation with Distance." *Journal of the Air Pollution Control Association*, 26(2).

Stability types A, B, C, D, E, and F are defined in Table 5-5. These stability types, which depend upon prevailing environmental conditions, were introduced by Pasquil,[6] hence are called *Pasquil stability types*.

TABLE 5-5   PASQUIL STABILITY TYPES[a]

Surface wind speed at 10 m above ground (m/s)	Day			Night	
	Incoming solar radiation:			Mostly overcast	Mostly clear
	Strong	Moderate	Slight		
<2	A	A–B	B	—	—
2	A–B	B	C	E	F
4	B	B–C	C	D	E
6	C	C–D	D	D	D
>6	C	D	D	D	D

[a]A, extremely unstable; B, moderately unstable; C, slightly unstable; D, neutral; E, slightly stable; F, moderately stable.

The surface wind speeds in Table 5-5 are measured 10 m above the ground. Wind speeds at other elevations are calculated using the power law written as

$$\frac{u_1}{u_2} = \left(\frac{z_1}{z_2}\right)^p \tag{5-96}$$

where u_1 and u_2 are the velocities at the lower elevation z_1 and higher elevation z_2, respectively. The wind exponent p is obtained as follows:

Stability type	p
A	0.15
B	0.15
C	0.20
D	0.25
E	0.40
F	0.60

Example 5-16

A power plant burns 10 tonnes/h of coal that contains 5.0% sulfur and discharges the combustion products through a stack 200 m high and 1.0-m inside diameter. A weather station anemometer located 10 m above the ground measures the wind speed at 3.0 m/s.

[6]F. Pasquil (1961). "The Estimation of the Dispersion of Windborne Material." *Meteorological Magazine* 90, p. 1063.

Other pertinent information is as follows: air temperature $= 15°C$, barometric pressure $= 1000$ mbar, stack gas velocity $= 10$ m/s, and stack gas temperature is $150°C$. The atmospheric conditions are moderately to slightly stable. (a) Determine the ground-level concentration of SO_2 at a distance 2 km downwind from the stack along the plume axis. (b) Assuming an effective stack height of 75 m but with the same wind velocity, determine the ground-level concentration at this distance. (c) What is the average concentration at the equivalent stack section at this point? (d) What is the concentration at the centerline of the plume at this point?

Solution

(a)
$$c = \frac{1.03W}{u\pi\sigma_y\sigma_z}e^{-(y/\sqrt{2}\sigma_y)^2}e^{-(z-H/\sqrt{2}\sigma_z)^2}$$

$$S + O_2 \longrightarrow SO_2$$

Then,

$$W = SO_2 \text{ produced} = 10(1000)(0.05)\frac{SO_2}{S} = 10(1000)(0.05)\left(\frac{32+32}{32}\right)$$

$$= 16{,}000 \text{ kg/h} = 4.44 \text{ kg/s}$$

$$\frac{u_1}{u_2} = \left(\frac{z_1}{z_2}\right)^p$$

The stability type is between E and F; hence, from the table, $p = 0.5$.

$$z_2 = H = h + \Delta h = 200 + \Delta h$$

$$\Delta h = \frac{v_s d}{u}\left[1.5 + 2.68(10^{-3})p\frac{\Delta T d}{T_s}\right]$$

$$v_s = 10 \text{ m/s} \qquad d = 1.0 \text{ m} \qquad \Delta T = 150 - 15 = 135°C = 135K$$

$$T_s = 150 + 273 = 423K \qquad p = 1000 \text{ mbar}$$

$$\Delta h = \frac{10(1)}{u}\left\{1.5 + 2.68(10^{-3})(1000)\left[\frac{135(1)}{423}\right]\right\} = \frac{23.55}{u} = \frac{23.55}{u_2}$$

$$\frac{3.0}{u_2} = \left(\frac{10}{200 + 23.55/u_2}\right)^{0.5} \qquad \text{let } X = \frac{3.0}{u_2} \Big/ \left(\frac{10}{200 + 23.55/u_2}\right)^{0.5}$$

u_2	X
10	1.35
x	1.0
13	1.03

$$\frac{x - 10}{13 - 10} = \frac{1.0 - 1.35}{1.03 - 1.35} \qquad x = 13.28 \text{ m/s} = u_2$$

$$H = h + \Delta h = 200 + \frac{23.55}{u_2} = 200 + \frac{23.55}{13.28} = 201.77 \text{ m}$$

$$\sigma_y = ax^{0.894}$$

$$\sigma_z = cx^d + f$$

From the table,

$$a = \frac{50.5 + 34}{2} = 42.25 \qquad c = \frac{55.4 + 62.6}{2} = 59$$

$$d = \frac{0.305 + 0.18}{2} = 0.24 \qquad f = \frac{-34.0 - 48.6}{2} = -41.3$$

$$\sigma_y = 42.25(2.0)^{0.894} = 78.51 \text{ m}$$

$$\sigma_z = 59.0(2.0)^{0.24} - 41.3 = 28.38 \text{ m}$$

Thus,

$$c = \frac{1.03(4.44)}{13.28\pi(78.51)(28.38)} e^{-(0/\sqrt{2}(78.51))^2} e^{-(0-201.77/\sqrt{2}(28.38))^2} = 5.13(10^{-15}) \text{ kg/m}^3$$

$$= 5.13(10^{-6}) \ \mu\text{g/m}^3. \quad \textbf{Answer}$$

(b) $$c = \frac{1.03(4.44)}{13.28\pi(78.51)(28.38)} e^{-(0/\sqrt{2}(78.51))^2} e^{-(0-75/\sqrt{2}(28.38))^2}$$

$$= 1.5(10^{-6}) \text{ kg/m}^3 = 1.50(10^3) \ \mu\text{g/m}^3 \quad \textbf{Answer}$$

(c) $$c = \frac{0.6826W}{u\pi\sigma_y\sigma_z} = \frac{0.6826(4.44)}{13.28\pi(78.51)(28.38)} = 3.26(10^{-5}) \text{ kg/m}^3$$

$$= 3.26(10^4) \ \mu\text{g/m}^3 \quad \textbf{Answer}$$

(d) $$c = \frac{1.03W}{u\pi\sigma_y\sigma_z} = \frac{1.03(4.44)}{13.28\pi(78.51)(28.38)} = 4.92(10^{-5}) \text{ kg/m}^3$$

$$= 4.92(10^4) \ \mu\text{g/m}^3 \quad \textbf{Answer}$$

GLOSSARY

Capillary potential. The matric potential (*see* Matric potential).

Computer models. Mathematical models programmed into a computer.

Conservative substance. A substance that does not decay.

Convective or advective transport. A transport of solute by the flow of the solvent.

Diffusion. A mass transfer due solely to the concentration gradient of the solute molecules.

Dispersive transport or dispersion. A mass transfer due to a gradient in concentration aided by the turbulence of the medium.

Dry adiabatic lapse rate. The rate of change of temperature with altitude as a parcel of dry air rises or sinks without exchange of heat with the surroundings.

Effective stack height. The height of a stack plus a correction.

Environmental quality modeling. The representation of the various characteristic quality processes that occur in the air, surface water, and the soil.

Fanning. Said of a plume that cannot disperse vertically but spreads horizontally.

Fluorescence. The re-release of an absorbed photon by the photon-absorbing excited molecule without involving another molecule.

Fumigation. The behavior of a plume when there is lapse rate below the point of exit of the plume but inversion above the point of exit of the plume.

Gaussian plume. Plume shaped like the normal distribution.

Hydrodynamic dispersion or, simply, dispersion. The dispersion of solute due to diffusion and mechanical dispersion.

Hydrolysis. The breakdown reaction of a substance with water.

Intersystem crossing. The transfer of excess energy from a photon-absorbing molecule to transform a second molecule to its excited state.

Inversion. A lapse rate where the temperature increases with altitude.

Lagrangian derivative. A total derivative of the control mass.

Lofting. Behavior of a plume where there is lapse rate above the point of exit of the plume but inversion below the point of exit of the plume.

Material derivative. A Lagrangian derivative.

Mathematical model. A model expressed in terms of mathematical equation.

Matric potential. The tension pressure in the vadose zone.

Mechanical dispersion. A dispersion of the solute due solely to the influence of the medium.

Model. The representation of a prototype.

Neutral plume. A plume formed in the absence of wind velocity.

Nonconservative substance. A substance that decays.

Oxygen deficit. The difference between the saturation dissolved oxygen (DO_{sat}) and the actual dissolved oxygen (DO).

Partition coefficient. A proportionality constant relating the concentration of solute in the solvent to the concentration of this solute in the solid.

Pheophytin. Dead chlorophyll a.

Phosphorescence. The simple release of energy without involving another molecule by the second excited molecule formed in the intersystem crossing.

Photolysis. The breakdown of a substance by absorption of light.

Physical model. A representation of the prototype by its smaller- or larger-scale version.

R Lagrangian derivative. That portion of the Lagrangian derivative representing the kinetic process.

Sensitizer. The molecule that absorbs a photon of energy but transfers it to second molecule to transform the molecule to an excited state.

Stable atmosphere. An atmosphere incapable of efficient dispersion of pollutants.

Subadiabatic lapse rate. The atmospheric condition where the actual ambient lapse rate is less than the adiabatic lapse rate.

Superadiabatic lapse rate. The atmospheric condition where the ambient lapse rate is greater than the adiabatic lapse rate of the rising parcel of air.

Tensiometer. A device used to measure the tension pressure in the vadose zone.

Tensor. A general term given to quantities such as scalar, vectors, and similar quantities of higher order.

Transport. The dispersion and convective terms of the material balance equation.

Trapping. The behavior of a plume when there are inversions both above and below the point of exit of the plume.

Unstable atmosphere. An atmosphere capable of efficient dispersion of pollutants.

Vadose zone. The zone from the surface of the earth to the groundwater surface.

Vector. A tensor of order 1.

Volatilization. The escape of molecules from a liquid or solid surface to enter the gaseous phase.

SYMBOLS

a_{cc}	CBOD/active chlorophyll a ratio
a_{nc}	nitrogen/active chlorophyll a ratio
a_{op}	oxygen/phytoplankton ratio
a_{pc}	phytoplankton/active chlorophyll a ratio
A_x	plume equivalent stack section
c	pollution concentration property of control mass per unit volume of space
c_d	concentration of pollutant dissolved in a unit volume of control mass
c_{sp}	concentration of pollutant sorbed per unit mass of sorbing solid or particulate
c_1	concentration of NH_3-N
c_2	concentration of NO_2-NO_3-N
c_3	concentration of orthophosphorus
c_4	concentration of phytoplankton chlorophyll a
c_5	concentration of CBOD
c_6	concentration dissolved oxygen
c_7	concentration of organic nitrogen
c_8	concentration of organic phosphorus
E	dispersion tensor
f_{cp}	fraction of carbon in particulate or solid
f_d	fraction of dissolved pollutant
f_{ON}	fraction of cellular nitrogen converted to organic nitrogen

f_{sp}	fraction of sorbed pollutant
G	distributed pollutant source (G_{Lc}, G_{Ln} the G of L_c and L_n, respectively)
h	piezometric head
H	effective stack height
j	dispersion vector
k	decay coefficient
k_c	CBOD decay coefficient
k_n	nitrogenous decay coefficient
K	hydraulic conductivity; K_{xx}, K_{yy}, and K_{zz}, the hydraulic conductivity in the x, y, and z directions
K_e	extinction coefficient
K_N	nitrogen half-saturation constant
K_P	phosphorus half-saturation constant
K_{sc}	partition coefficient of pollutant between water carbon
K_{sp}	partition coefficient of pollutant between water and solid
L_c	CBOD
L_n	nitrogenous oxygen demand, NBOD
m_d	mass of solid per unit volume of space
m_s	mass of solid per unit of control mass
$\hat{n}_i, \hat{n}_j, \hat{n}_k$	unit vectors; i, j, and k are indices
p	pressure
R	Lagrangian derivative kinetic component
s_c	storage coefficient
s_y	specific yield of aquifer
u, v, w	x, y, and z components of **v**
v	velocity vector
Ψ	space volume
z_s	Secchi depth
Z	gravitational potential
∇	nabla operator
Γ	lapse rate
η	porosity
μ	growth coefficient
ρ	mass density of control mass
σ	standard deviation of the transverse distances corresponding to the transverse concentrations c in a Gaussian plume; σ_y and σ_z, the σ in the y and z directions, respectively; σ_{cy} and σ_{cz}, the σ in the y and z directions, respectively, for conservative substances; σ_{ny} and σ_{nz}, the σ in the y and z directions, respectively, for nonconservative substances
ϕ	total potential
ψ	matric potential

PROBLEMS

5-1. At 30 miles upstream from the mouth of the Hudson Estuary, the following average chloride concentrations are obtained: 5000 mg/L, 5029 mg/L, and 4996 mg/L. At a point 48 miles upstream from the mouth, the chloride concentrations are 1800 mg/L and 1850 mg/L. The tidally averaged velocity is 0.025 fps. Calculate E_{xx} using the 48-mile point as the origin.

5-2. Calculate E_{xx} in Problem 5-1 by a graphical method.

5-3. In a dye study approximately 6.7 g of Rhodamine WT in 100 mL of solution was dumped at the discharge point of a lagoon treatment plant, resulting in a concentration c_{d0} of $5.0(10^7)$ μg/L a few feet downstream. The resulting concentrations monitored over time at a distance of 1047 m downstream are shown in the following table. The stream velocity was 8373 m/day, and its flow was 900 m^3/day. Determine the location of the centroid of the dye concentration expressed in time.

Time, A.M.	Concentration (ppb)	Time, A.M.	Concentration (ppb)
09:20	0.0	09:30	0.0
09:40	0.03	09:45	0.17
09:50	0.70	09:55	3.0
10:00	9.0	10:05	19.0
10:10	30.0	10:15	39.0
10:20	43.5	10:25	50.0
10:30	52.0	10:35	52.0
10:40	50.0	10:45	48.0
10:50	45.0	10:55	42.0
11:00	39.0	11:15	31.0
11:30	24.0	11:45	19.0
12:00	16.0	12:30	12.0
13:00	6.4	14:00	4.7
15:00	1.8	16:00	0.70
17:00	0.0	18:00	0.0

5-4. From the dye study of Problem 5-3, calculate the time of travel velocity of the stream.

5-5. In Problem 5-3, determine the ratio of (**a**) the maximum concentration to the minimum concentration, (**b**) the maximum concentration to the average concentration, and (**c**) the average concentration to the minimum concentration.

5-6. In Problem 5-3, calculate the ratios of the maximum, average, and minimum concentrations to the value of the centroidal concentration.

5-7. A laboratory experiment on a certain pollutant found the values of k_{ha} and k_{hb} to be $2.0(10^2)$ L/mol-day and 10 L/mol-day, respectively, at 25°C. Compute k_h at pH 8 and 25°C.

5-8. Given in Problem 5-7 that $c_d = 10^{-3}$ mol/L and the activation energy $= 26,000$ J/mol, calculate R_h at 30°C.

5-9. The average concentration of benzo[a]pyrene in Lake Ontario is 10^{-3} mmol/L. What is its rate of decay due to photolysis R_p at the 5-m depth if $k_{p0} = 31$ per day?

5-10. Assume that the concentration of the benzo[a]pyrene in Problem 5-9 is found at a given point of a free-flowing stream. Assuming that the stream velocity is 0.03 m/s, what would be the concentration 2.0 km downstream from the point? Assume first-order decay.

5-11. The BOD_5 of a waste is 225 mg/L and its CBOD is 325 mg/L. What is the decay coefficient to the base 10? What is it to the base e?

5-12. Assume that the BOD_5 of Problem 5-11 is found at the same point in the stream of Problem 5-9 that contains benzo[a]pyrene. Assuming first-order decay, what would be the BOD_5 concentration at the same 2.0 km downstream from the point?

5-13. A wastewater treatment plant discharges treated sewage into a stream. If the concentration of BOD_5 one mile downstream from the point of discharge is 5 mg/L and k_m is assumed to be 0.2 per day (in terms of BOD_5), what is the value of R_m at this point?

5-14. In Problem 5-13, the concentration of BOD_5 at the point is 5 mg/L. What was the concentration 1 mile upstream?

5-15. A sample is obtained at point A and another at point B along a stream. What is the term given to the proportionality constant for the decrease in CBOD concentration of the sample between the points? If sample A is analyzed for CBOD between two time periods, what is the term given to the proportionality constant?

5-16. Two students from the environmental engineering class at Morgan State University are studying Herring Run, a stream cutting through the property of the university. Their objective is to determine the distributed pollutant load (G_{Lc} and G_{Ln}) into the run. Write the differential equation that would be needed to meet the objective.

5-17. Integrate the differential equation in Problem 5-16.

5-18. Create a problem that will apply the formula derived in Problem 5-17.

5-19. The deoxygenation coefficient to the base 10 for a certain waste is 0.1 per day. What fraction is BOD_5 to CBOD of this waste?

5-20. Calculate the deoxygenation coefficient of Problem 5-19 to the base e.

5-21. The CBOD of a waste is 280 mg/L. Calculate the BOD_5 if the deoxygenation constant to the base e is 0.23.

5-22. Solve Problem 5-21 by first converting the deoxygenation constant to the base 10.

5-23. An estuarine sample is put in a settling column. At two points in time, samples are taken at the sampling port producing the following results: at 0 day, chlorophyll $a = 20$ μg/L, at 2 days, chlorophyll $a = 15$ μg/L. Determine the settling coefficient, k_s.

5-24. Calculate k_s in Problem 5-23 by graphical method.

5-25. A wastewater treatment plant discharges treated sewage into a stream. The characteristics of the effluent are: flow rate $= 15,000$ m^3/day, $BOD_5 = 10$ mg/L, $NH_3 = 5$ mg/L as N, DO $= 2.0$ mg/L, temperature $= 25°$C, and the bottle decay coefficient of the CBOD is 0.2 per day. The stream characteristics upstream of the point of discharge are as follows: flow rate $= 0.2$ m^3/s, $u = 0.2$ m/s (stream flow plus treated sewage), reaeration coefficient, $k_2 = 26$ per day, $E_{xx} = 6.03(105)$ mi^2/day, $BOD_5 = 3$ mg/L, $NH_3 = 2$ mg/L as N, DO $= 8.0$ mg/L, temperature $= 22°$C, and instream decay coefficient for CBOD, and $NH_3 = 0.3$ per day and 0.4 per day, respectively. Find the DO concentration 5 miles downstream from the point of discharge. Neglect P and R. There are no G_{Lc}, G_{Ln}, and G_O.

5-26. Using the stream characteristics of the stream in Problem 5-25, calculate the effluent limitations for a discharge of 20,000 m^3/day. Assume that the state water quality standard for DO is 5.0 mg/L.

5-27. Determine the DO 5 miles downstream if the treated sewage in Problem 5-26 is discharged into a tidal estuary with the following characteristics: depth $= 8$ m, net advective flow rate

$= 0.2$ m^3/s, u $= 0.002$ m/s (net advective velocity upstream of point of discharge), reaeration coefficient $k_2 = 3.0$ per day, $E_{xx} = 7.29$ mi^2/day, BOD$_5 = 3$ mg/L, NH$_3 = 2$ mg/L as N, DO $= 8.0$ mg/L, temperature $= 22°$C, and instream decay coefficient for CBOD, and NH$_3$ $= 0.3$ per day and 0.4 per day, respectively. Additional data: $a_{pc} = 66.67$, $\mu_p = 0.6$ per day at the 4-m depth, $c_4 = 100$ μg/L at point of discharge, $k_d = 0.05$ per day, $k_{s4} = 0.14$ per day, $G_{L_c} = 2$ mg/L-day, $G_{Ln} = 4$ mg/L-day, and $G_O = 0$.

5-28. What are the permit limitations if an effluent of 20,000 m^3/day is discharged to an estuary similar to the one in Problem 5-27? Assume that the state water quality standard for DO is 5.0 mg/L.

5-29. Search the literature to determine representative values of G_{Lc}, G_{Ln}, and G_O; from these values, resolve Problem 5-27.

5-30. A town plans to discharge 15,000 m^3/day of treated sewage into Lucky River. The characteristics of the stream are as follows: $k_c = 0.4$ per day and $k_2 = 2.3$ per day, both at 15°C; river temperature $= 22°$C; flow $= 6.0$ m^3/s; background BOD$_5 = 2.0$ mg/L; background DO $= 8.0$ mg/L; and velocity with discharge incorporated $= 0.40$ m/s. Assume that the river is very long. Neglecting the effect of NBOD, what BOD$_5$ can be discharged if the DO along the river should not fall below 5.0 mg/L? Assume that the DO at the effluent is set at 5.0 mg/L. If the ammonia content of the effluent is 25 mg/L and the background ammonia is 1.0 mg/L, what will be the lowest DO in the stream?

5-31. Benzene, of solubility equal to 0.07 part per 100 parts of water by weight, was spilled into an aquifer having a hydraulic conductivity $K_{xx} = 1000$ m/day. What is the centerline concentration of the plume when it reaches the river bank 3300 m downstream? Field investigation showed that the elevation of the groundwater table is 60 m above the river. Use the following additional information: $f_{cp} = 0.1$, $\eta = 0.55$, formation specific gravity $= 2.65$, $E_{xx} = 3.76(10^{-4})$ cm^2/s, and $k = 0.07$ per day.

5-32. Tiny droplets of benzene are leaking slowly from an on-ground storage tank. As shown in Figure 11-6, tensiometers show readings of -88 cm and -126 cm of water at depths of 30 cm and 100 cm, respectively, below the ground. The vertical unsaturated hydraulic conductivity K_{zz} is 10 m/day. The porosities are $\eta = 0.55$, $\eta_d = 0.38$, and $\eta_a = 17$. The other pertinent information are $f_{cp} = 0.1$, $K_{sc} = 104.7$, $m_d = 1.19(10^3)$ kg/m^3, and $E_{zz} = 0.0012$ m^2/day. The half-life of benzene is 10 days and its solubility in water is 0.07 part per 100 parts by weight. The vapor pressure of benzene is 1.0 mmHg. Calculate the benzene concentration 10 m below the surface.

5-33. Calculate the effective stack height using the following data: stack gas temperature $= 150°$C, stack gas velocity $= 9.2$ m/s, atmospheric pressure $= 29.92$ in. Hg, ambient air temperature $= 13°$C, wind velocity $= 4$ m/s, and physical stack height $= 200$ m with inside diameter of 1.0 m.

5-34. A power plant burns 5.45 tonnes/h of coal that contains 4.2% sulfur and discharges the combustion products through a stack 75 m high and a 1.0-m inside diameter. A weather station anemometer located 10 m above the ground measures the wind speed at 3.0 m/s. Other pertinent information is as follows: air temperature $= 15°$C, barometric pressure $= 1000$ mbar, stack gas velocity $= 10$ m/s, and stack gas temperature is 150°C. The atmospheric conditions are moderately to slightly stable. Determine the ground-level concentration of SO$_2$ at a distance of 2 km downwind from the stack along the plume axis.

5-35. In Problem 5-34, what is the average concentration at the equivalent stack section located at the 2-km point downwind from the stack?

5-36. In Problem 5-34, what is the concentration at the centerline of the plume at the 2-km point downwind from the stack?

5-37. A power plant burns coal containing 4.8% S at the rate of 6.5 tonnes/h. The combustion products are discharged through a stack having an effective height of 85 m. If atmospheric conditions are slightly unstable and the wind velocity is 8.0 m/s, calculate the maximum ground-level concentration of sulfur dioxide.

5-38. In Problem 5-37, at what distance from the stack does the maximum concentration occur?

5-39. A parcel of dry air is rising above a burning grass. Its temperature at 10 m is 60°C. Assuming adiabatic rising, what will be the temperature at 200 m?

5-40. Solve Problem 5-39 assuming a wet adiabatic lapse rate.

BIBLIOGRAPHY

DAVIS, M. L., and D. A. CORNWELL (1991). *Introduction to Environmental Engineering.* McGraw-Hill, New York.

FETTER, C. W. (1993). *Contaminant Hydrogeology.* Macmillan, New York.

FISHER, H. B., E. J. LIST, R. C. Y. KOH, J. IMBERGER, and N. H. BROOKS (1979). *Mixing in Inland and Coastal Waters.* Academic Press, New York.

MARTIN, D. O. (1976). "The Change of Concentration Standard Deviation with Distance." *Journal of the Air Pollution Control Association,* 26(2).

McQUIVEY, R. S., and T. N. KEEFER (1974). "Simple Method for Predicting Dispersion in Streams." *Journal of the Environmental Engineering Division, Proceedings of the American Society of Civil Engineers,* 100(EE4), pp. 997–1011.

PASQUIL, F. (1961). "The Estimation of the Dispersion of Windborne Material." *Meteorological Magazine,* 90, p. 1063.

PEAVY, H. S., D. R. ROWE, and G. TCHOBANOGLOUS (1985). *Environmental Engineering.* McGraw-Hill, New York.

STREETER, H. W. and E. B. PHELPS (1925). "A Study of the Pollution and Natural Purification of the Ohio River." *Public Health Bulletin 146.* Treasury Department, U.S. Public Health Service, Washington, D.C.

THOMANN, R. V., and J. A. MUELLER (1987). *Principles of Surface Water Quality Modeling and Control.* Harper & Row, New York.

USEPA (1988). *WASP4, A Hydrodynamic and Water Quality Model: Model Theory, User's Manual, and Programmers Guide.* EPA/600/3-87/039. Environmental Research Laboratory, U.S. Environmental Protection Agency, Athens, Ga.

CHAPTER 6

Conventional Water Treatment

Probably, about 90% of the population in the United States turn their faucets every morning, and probably, they never even wonder where the water came from or whether or not it is safe. The drinking water in the United States is relatively safe, but in developing countries, thousands of people die daily from waterborne diseases. Schistosomiasis and filariasis, major causes of blindness, are estimated to affect some 450 million people in more than 70 nations.[1] About 60% of the babies born in developing countries die of infantile gastritis. The localized Asiatic cholera epidemic in 1854 caused by London's Broad Street pump is well known. The epidemic continued until John Snow allegedly removed the handle of the pump. It is axiomatic that water for human consumption should be free of pathogens as well as deleterious chemicals. This means that unsafe water should be treated to meet drinking water standards (see Chapter 2).

Basically, the essence of a standard water treatment is removal of dissolved and undissolved materials. The basic conventional treatment of doing this is performed by two unit operations: settling and filtration and the unit process of disinfection. There are other methods of treating water such as reverse osmosis and carbon adsorption. These are advanced methods and are not discussed here. In this chapter we discuss only conventional treatment methods; advanced methods are treated in Chapter 9. Complementing the unit operations are unit processes the aims of which are to convert contaminants to solid forms that can easily be removed by settling and filtration. Examples of these unit processes are coagulation and softening; they are discussed in this chapter.

[1] From remarks of former USEPA Administrator Russel E. Train delivered before the Los Angeles World Affairs Forum, December 16, 1976.

Because of the application of the same physical principles, *unit operation* is an operation that is used in similar ways in different types of plants and factories. Settling and filtration are unit operations that are used in similar ways in water treatment plants, wastewater treatment plants, and other plants. Another example of a unit operation is the pumping of fluids as applied in water treatment, wastewater treatment, and chemical plants. A *unit process*, on the other hand, is a process that because of the application of the same chemical or biological principles is used in similar ways in different types of plants and factories. Coagulation and softening, mentioned above, are unit processes that are used in various types of plants, such as water and wastewater treatment.

Figures 6-1 and 6-2 are schematics of water treatment plants that treat groundwaters and surface waters, respectively. The aeration in Figure 6-1 is a physical process and is an example of a unit operation. Considering Figures 6-1 and 6-2 together, the following are the other unit operations: storage, presedimentation, mixing, flocculation, and settling. The unit processes are softening and disinfection.

The various unit operations and processes mentioned above are discussed in this chapter. The unit process called adsorption, which may be included in a process train, is discussed in a separate chapter. The end result of treatment is the production of water that meets safe drinking water standards and is therefore safe to drink. For discussions on the chemistry and biology of drinking water, see Chapter 2.

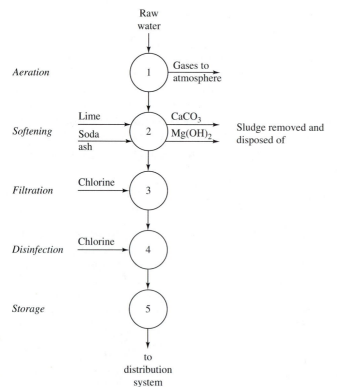

Figure 6-1 Typical plant flowsheet treating hard groundwater.

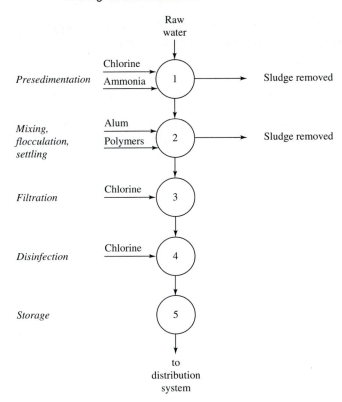

Presedimentation

Mixing, flocculation, settling

Filtration

Disinfection

Storage

Figure 6-2 Typical plant flowsheet treating turbid surface water.

RAW WATER

The raw water may be either hard or soft. If the raw water is hard, the treatment scheme of Figure 6-1 is applicable; if the raw water is soft, it is likely that the raw water is a surface water and the treatment scheme of Figure 6-2 is applicable. A general classification of raw water hardness is as follows:

Soft	< 50 mg/L as $CaCO_3$
Moderately hard	50–150 mg/L as $CaCO_3$
Hard	150–300 mg/L as $CaCO_3$
Very hard	> 300 mg/L as $CaCO_3$

SETTLING OR SEDIMENTATION

If the materials to be separated are already in solid form, the unit operation of settling is immediately applied to the materials; if the materials are dissolved or in a very minute form, means must be applied to transform the materials into solid forms that can be acted upon by gravity. In general, there are four types of settling: types 1 to 4. Type 1 settling involves the removal of discrete particles, type 2 settling involves the removal of flocculent particles, type 3 settling involves the removal of particles that settle in a contiguous zone, and type 4 settling involves settling where compression or compaction of the particle mass is occurring at the same time.

Type 1 Settling

When particles in suspension are dilute, they tend to act independently; hence their behaviors are therefore said to be *discrete* with respect to each other. Type 1 settling is also called *discrete settling*.

As a particle settles in a fluid, its body force f_g, bouyant force f_b, and drag force f_d act on it. Applying Newton's second law in the direction of settling,

$$f_g - f_b - f_d = ma \tag{6-1}$$

where m is the mass of the particle and a its acceleration. Calling ρ_p the mass density of the particle, ρ_w the mass density of water, V_p the volume of the particle, and g the acceleration due to gravity yields

$$f_g = \rho_p g V_p \tag{6-2}$$

$$f_b = \rho_w g V_p \tag{6-3}$$

Because particle will ultimately settle at its terminal settling velocity, the acceleration a is equal to zero. Substitutions in equation (6-1) produce

$$(\rho_p - \rho_w)g V_p - f_d = 0 \tag{6-4}$$

The drag stress is directly proportional to the dynamic pressure, $\rho_w v^2/2$, where v is the terminal settling velocity of the particle. Hence

$$f_d = C_D A_p \rho_w \frac{v^2}{2} \tag{6-5}$$

where C_D is the coefficient of proportionality called *drag coefficient* and A_p is the projected area of the particle that is causing the drag. Substituting equation (6-5) in equation (6-4) and solving for the velocity of spherical particles of diameter d (from $A_p = \pi d^2/4$, $d = $ diameter of spherical particle), we have

$$v = \sqrt{\frac{4}{3}g\frac{(\rho_p - \rho_w)d}{C_D \rho_w}} \tag{6-6}$$

The value of the coefficient of drag C_D varies with the flow regimes of laminar, transitional, and turbulent flows. The respective expressions are shown below.

$$C_D = \begin{cases} \dfrac{24}{Re} & \text{for laminar flow} \\[2mm] \dfrac{24}{Re} + \dfrac{3}{Re^{1/2}} + 0.34 & \text{for transitional flow} \\[2mm] 0.4 & \text{for turbulent flow} \end{cases} \tag{6-7}$$

where Re is the Reynolds number $= v\rho_w d/\mu$, and μ is the dynamic viscosity of water. Values of Re of less than 1 indicate laminar flow, while values greater than 10^4 indicate turbulent flow. Intermediate values indicate transitional flow.

Substituting the C_D for laminar flow ($C_D = 24/Re$) in equation (6-6) produces the Stokes equation,

$$v = \frac{g(\rho_p - \rho_w)d^2}{18\mu} \qquad (6\text{-}8)$$

To use the equations above for nonspherical particles, the diameter d must be the diameter of the equivalent spherical particle. The volume of the equivalent spherical particle, $V_s = 4/3\pi \, (d/2)^3$, must be equal to the volume of the nonspherical particle, $V_p = \beta d_p^3$, where β is a volume shape factor. Therefore,

$$\frac{4}{3}\pi \left(\frac{d}{2}\right)^3 = \beta d_p^3 \qquad (6\text{-}9)$$

$$d = 1.24\beta^{0.333} d_p \qquad (6\text{-}10)$$

The following values of sand volumetric shape factors β have been reported: angular $= 0.64$, sharp $= 0.77$, worn $= 0.86$, and spherical $= 0.52$.

Example 6-1

Determine the terminal settling velocity of a spherical particle having a diameter of 0.6 mm and specific gravity of 2.65. Assume that the settling is type 1 and the temperature of the water is 22°C.

Solution

$$v = \sqrt{\frac{4}{3}g \frac{(\rho_p - \rho_w)d}{C_D \rho_w}}$$

$$g = 9.81 \text{ m/s}^2$$

$$\rho_{w22} = 997 \text{ kg/m}^3$$

$$\rho_p = 2.65(1000) = 2650 \text{ kg/m}^3$$

$$d = 0.6(10^{-3}) \text{ m}$$

$$\mu_{22} = 9.2(10^{-4}) \text{ kg/m-s} = 9.2(10^{-4}) \text{ N-s/m}^2$$

Then

$$v = \sqrt{\frac{4}{3}(9.81)\frac{(2650 - 997)[0.6(10^{-3})]}{C_D(997)}} = \frac{0.114}{\sqrt{C_D}}$$

$$C_D = \begin{cases} \dfrac{24}{Re} & \text{(for Re} < 1) \\[2mm] \dfrac{24}{Re} + \dfrac{3}{Re^{1/2}} + 0.34 & \text{(for } 1 \le Re \le 10^4) \\[2mm] 0.4 & \text{(for Re} > 10^4) \end{cases}$$

$$Re = \frac{dv\rho}{\mu} = \frac{dv(997)}{9.2(10^{-4})} = 1{,}083{,}695.70vd = 650.22v$$

Solve by successive iteration.

C_D	v (m/s)	Re
1.0	0.114	74
1.0	0.114	74

$$v = 0.114 \text{ m/s} \quad \textbf{Answer}$$

Example 6-2

Determine the terminal settling velocity of a worn sand particle having a measured "diameter" of 0.6 mm and specific gravity of 2.65. Assume that the settling is type 1 and the temperature of the water is 22°C.

Solution

$$v = \sqrt{\frac{4}{3}g\frac{(\rho_p - \rho_w)d}{C_D\rho_w}}$$

$$g = 9.81 \text{ m/s}^2$$

$$\rho_{w22} = 997 \text{ kg/m}^3$$

$$\rho_p = 2.65(1000) = 2650 \text{ kg/m}^3$$

$$\mu_{22} = 9.2(10^{-4}) \text{ kg/m-s} = 9.2(10^{-4}) \text{ N-s/m}^2$$

$$d = 1.24\beta^{0.333}d_p; \qquad \text{for worn sands, } \beta = 0.86$$

Thus

$$d = 1.24(0.86^{0.333})(0.6)(10^{-3}) = 0.71(10^{-3}) \text{ m}$$

$$v = \sqrt{\frac{4}{3}(9.81)\frac{(2650 - 997)[0.71(10^{-3})]}{C_D(997)}} = \frac{0.123}{\sqrt{C_D}}$$

$$C_D = \begin{cases} \dfrac{24}{Re} & \text{(for Re < 1)} \\[2mm] \dfrac{24}{Re} + \dfrac{3}{Re^{1/2}} + 0.34 & \text{(for } 1 \le Re \le 10^4) \\[2mm] 0.4 & \text{(for Re > } 10^4) \end{cases}$$

$$Re = \frac{dv\rho}{\mu} = \frac{0.71(10^{-3})v(997)}{9.2(10^{-4})} = 769.42v$$

Solve by successive iteration.

C_D	v (m/s)	Re
1.0	0.123	94.6
0.90	0.129	99.64
0.88	0.115	88.78
0.93	0.119	91.20
0.91	0.118	90.6

$$v = 0.118 \text{ m/s} \quad \textbf{Answer}$$

In an actual treatment plant, there is, of course, no single particle but a multitude of particles settling in a column of water. For this reason, a new technique must be developed. A raw water that comes from a river is usually turbid. In some water treatment plants, a presedimentation basin is therefore constructed to remove some of the turbidities. This basin is called the *prototype*. To design this prototype properly, its performance should be simulated by a model. In environmental engineering, the model used is a column. Figure 6-3 shows a schematic of the settling column and the result of an analysis.

At time equals zero, let a particle of diameter d_0 be at the water surface of the column. After time t_0, let the particle be at the sampling port. Any particle that arrives at the sampling port at t_0 will be considered removed. In the prototype tank, this removal corresponds to the particle being deposited at the bottom of the tank. Call the time it takes a particle to travel from the surface to the port the *removal time* t_0. In the actual prototype sedimentation tank, this time corresponds to the time the particle is in residence inside the tank; hence it is also called the *retention* or *detention time*. The corresponding velocity of the particle is $v_0 = Z_0/t_0$, where Z_0 is the depth. Since particles with velocities equal to v_0 are removed, particles of velocities equal or greater than v_0 will all be removed. If x_0 is the fraction of all particles having velocities less than v_0, $1 - x_0$ is the fraction of all particles having velocities equal to or greater than v_0. Therefore, the fraction of particles that are removed with certainty is $1 - x_0$.

Assume that x_0 is put back into the column in Figure 6-3. During the removal time t_0, particles closer to the sampling port will be removed. Let the particles throughout the length of Z_p be the ones removed, and let their average velocity be v_{p0}. Since removal is proportional to velocity, the portion in x_0 that can be removed is $(v_{p0}/v_0)(x_0)$. Applying this expression, let dx be a differential of the remaining x_0 with average velocity of removable portion equals v_p. The partial removal in the fraction dx is therefore $(v_p/v_0) \, dx$ and the total removal R comprising all the particles with velocities equal to or greater than v_0 and all particles with velocities less than v_0 is then

$$R = 1 - x_0 + \int_0^{x_0} \frac{v_p}{v_0} \, dx \tag{6-11}$$

To evaluate the integral of equation (6-11) by numerical integration, we set

$$\int_0^{x_0} \frac{v_p}{v_0} \, dx = \frac{1}{v_0} \sum v_p \Delta x \tag{6-12}$$

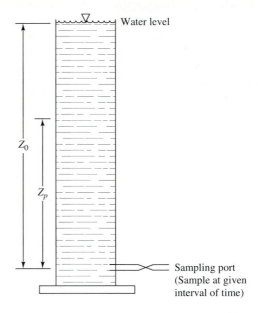

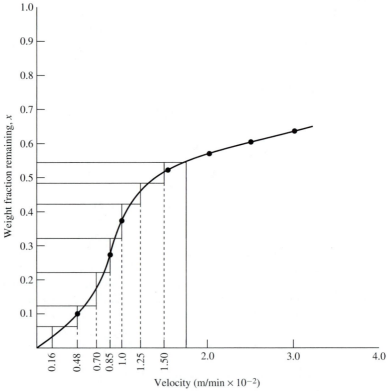

Figure 6-3 Settling column analysis of a type 1 suspension.

This equation requires the plot of v_p versus x. If the original concentration in the column is C_0 and, after a time of settling, the remaining concentration measured at the sampling port is C, the fraction of particles remaining in the water column adjacent to the port is

$$x = \frac{C}{C_0} \tag{6-13}$$

Corresponding to the fraction remaining x of equation (6-13), and referring to Figure 6-3, the average distance traversed by the particles from $Z_p = 0$ to Z_p is $Z_p/2$. Therefore, v_p is

$$v_p = \frac{Z_p}{2t_0} \tag{6-14}$$

The values x may now be plotted against the values v_p. From the plot, the numerical integration may be evaluated as explained in the next example.

Example 6-3

A certain municipality in Thailand plans to use the water from the Chao Praya River as a raw water for a contemplated water treatment plant. Since the river is very turbid, presedimentation is necessary. The result of a column test is as follows:

Time (min)	0	60	80	100	130	200	240	420
Conc. (mg/L)	299	190	179	169	157	110	79	28

What is the percentage removal of particles if the hydraulic loading rate is 25 m³/m²-day? The column is 4 m deep. (Hydraulic loading rate is the flow to the sedimentation tank divided by the surficial area of the tank.)

Solution

Time (min)	60	80	100	130	200	240	420
$x, \frac{C}{C_0} = \frac{C}{299}$	0.63	0.60	0.56	0.52	0.37	0.26	0.09
$v_p(10^2)$ (m/min)	3.3	2.5	2.0	1.55	1.0	0.83	0.48

The value of $v_p = 3.3(10^{-2})$ is obtained as follows:

$$v_p = \frac{Z_p}{2t_0} = \frac{4}{2(60)} = 0.033 = 3.3(10^{-2})$$

All the other v_p's are obtained similarly. The values of x versus v_p are plotted in Figure 6-3. 50 m³/m²-day $= 1.74(10^{-2})$ m/min $= v_0$. From the figure, $x_0 = 0.54$. Then

$$1 - x_0 = 0.46$$

$$\int_0^{x_0} \frac{v_p}{v_0} dx = \int_0^{0.54} \frac{v_p}{v_0} dx = \sum \frac{v_p}{v_0} \Delta x$$

From the figure,

$$\sum \frac{v_p}{v_0} \Delta x = \frac{1.5}{1.74}(0.06) + \frac{1.25}{1.74}(0.06) + \frac{1.0}{1.74}(0.1) + \frac{0.85}{1.74}(0.1) + \frac{0.7}{1.74}(0.1) + \frac{0.48}{1.74}(0.06)$$

$$+ \frac{0.16}{1.74}(0.06) = 0.26$$

Therefore,

$$R = 0.46 + 0.26 = 0.72 \Rightarrow 72\% \quad \textbf{Answer}$$

Type 2 Settling

Particles settling in a water column may have affinity toward each other and coalesce and form into flocs or aggregates. These larger flocs will now have more weight and settle faster, overtaking the smaller ones, thereby coalescing and growing still further into much more larger aggregates. The small particle that starts at the surface will end up as a large particle when it hits the bottom. The velocity of the floc will therefore not be terminal but changes as the size changes. The analysis used for discrete settling will accordingly not apply. Since the particles form into flocs, this type of settling is called *flocculent settling* or type 2 settling.

Since the velocity is terminal in the case of type 1 settling, only one sampling port was provided. In type 2 settling, however, to catch the changing behavior, multiple sampling ports are provided. Figure 6-4 shows the drawing of a column with multiple ports and the result of a settling analysis.

As before, let x_0 be the fraction remaining. Then r_0, the fraction removed, is $1 - x_0$. In general, $r = 1 - x$, where r is the fraction removed corresponding to the fraction remaining at any time t. The removal at the desired depth is, as before,

$$R = 1 - x_0 + \int_0^{x_0} \frac{v_p}{v_0} \, dx = r_0 + \int_0^{x_0} \frac{v_p}{v_0} \, dx \tag{6-15}$$

But $v_p/v_0 = (Z_p/2t_0)/(Z_0/t_0) = Z_p/(2Z_0)$ and $dx = -dr$. Also, from $1 - x = r$, when $x = 0$, $r = 1$, and when $x = x_0$, $r = r_0$. Therefore, equation (6-15) becomes

$$R = r_0 + \int_0^{x_0} \frac{v_p}{v_0} \, dx = r_0 - \int_1^{r_0} \frac{Z_p}{2Z_0} \, dr = r_0 + \int_{r_0}^{1} \frac{Z_p}{2Z_0} \, dr \tag{6-16}$$

The integral in equation (6-16) calls for the plot of Z_p/Z_0 against r. In performing the experiment using the column in Figure 6-4, samples are taken at the sampling ports at predetermined time intervals. Therefore, for each time interval, data are available for Z_p/Z_0 and r corresponding to the locations of the sampling ports. Hence a two-dimensional plot can be made using Z_p/Z_0 as the ordinate, r as the abscissa, and the time intervals as the parameter. From this plot, the integral of equation (6-16) may be integrated numerically for each of the time parameters.

Environmental engineers, however, prefer not to use the method of plotting discussed above. The preferred method is to plot the time not as a parameter but as the abscissa, and to use r as the parameter. Z_p/Z_0 is still the ordinate.

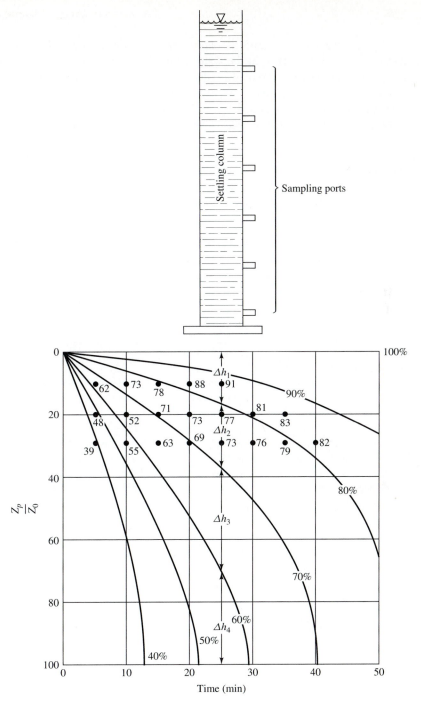

Figure 6-4 Settling of flocculent particles.

The time intervals used as parameters, mentioned above, are the times that the particles were in residence in the water column before reaching the respective sampling ports. These times are removal or detention times, which correspond to the detention times of the prototype tank. As recommended by Eckenfelder, the prototype detention time is normally obtained by scaling up the model removal time by 1.75; similarly, the prototype overflow rate is obtained by scaling down the model overflow rate by 0.65.[2]

Example 6-4

Assume that Anne Arundel County wants to expand its softening plant. A sample from their existing softening tank is prepared and a settling column test is performed. The initial solids concentration in the column is 250 mg/L. The results are as follows:

$\dfrac{Z_p}{Z_0}$	Sampling time (min)						
	5	10	15	20	25	30	35
0.1	95[a]	68	55	30	23		
0.2	129	121	73	67	58	48	43
0.3	154	113	92	78	69	60	53	45

[a] Values in the table are the results of the test for the suspended solids concentration at the given depths.

(a) If the settling column has a depth of 4 m, what is the model overflow rate corresponding to a model removal time of 25 min? (b) What are the corresponding prototype detention time and overflow rate, respectively? (c) What is the model removal efficiency at the 4-m depth? Is this equal to the prototype removal efficiency of a prototype sedimentation tank of 4-m depth?

Solution

(a) Model overflow rate $= \dfrac{4}{25} = 0.16$ m/min $= 0.16\dfrac{m^3}{m^2\text{-min}}$ **Answer**

(b) Prototype detention time $= 25(1.75) = 44$ min   **Answer**

Prototype overflow rate $= 0.16(0.65) = 0.10\dfrac{m^3}{m^2\text{-min}}$ **Answer**

(c)
$$R = r_0 + \int_{r_0}^{1} \frac{Z_p}{2Z_0}\, dr$$

[2] W. W. Eckenfelder (1980). *Principles of Water Quality Management.* CBI, Boston.

$\dfrac{Z_p}{Z_0}$	Sampling time (min)						
	5	10	15	20	25	30	35
0.1	62[a]	73	78	88	91		
0.2	48	52	71	73	77	81	83
0.3	39	55	63	69	73	76	79	82

[a] The percentage-removal values in the table are obtained as follows: For example, for 62,

$$62 = \left(1 - \frac{95}{250}\right)100$$

The table above has been plotted at the bottom of Figure 6-4, where time is the abscissa, Z_p/Z_0 the ordinate, and r the parameter. Isoremoval lines are then sketched as indicated by the solid curve lines.

From the figure, r_0 at t_0 of 25 min and Z_p/L_0 of 100% for the depth of 4 m is approximately 55%.

$$\int_{r_0}^{1} \frac{Z_p}{Z_0} dr = \left(\Delta h_1 + \Delta h_2 + \Delta h_3 + \frac{\Delta h_4}{2}\right)(60 - 55) + \left(\Delta h_1 + \Delta h_2 + \frac{\Delta h_3}{2}\right)(70 - 60)$$

$$+ \left(\Delta h_1 + \frac{\Delta h_2}{2}\right)(80 - 70) + \left(\frac{\Delta h_1}{2}\right)(100 - 80)$$

$$= (0.19 + 0.19 + 0.35 + 0.135)(60 - 55) + (0.19 + 0.19 + 0.175)(70 - 60)$$

$$+ (0.19 + 0.095)(80 - 70) + 0.095(100 - 80)$$

$$= 0.866(5) + 0.555(10) + 0.285(10) + 0.095(20) = 14.63$$

Then

$$R = 55 + \frac{14.63}{2} = 62.3\% = \text{model removal efficiency} \quad \textbf{Answer}$$

To obtain this removal efficiency in the prototype, the model removal time should be increased by 1.75.   **Answer**

Type 3 and Type 4 Settling

The principles involved in these types of settling apply to solids in a very concentrated suspension. We defer the discussion of this important topic until Chapter 7.

SETTLING OPERATIONS

In the production of potable water, the type of settling discussed above can be applied to remove suspended solids and dissolved solids that have been precipitated by chemical treatment. (*Potable* water is water absent substances in harmful concentrations. A potable water may not be *palatable*.) Settling units applied to water treatment are classified as rectangular basins, circular basins, and solids-contact clarifiers. The solids-contact clarifier is used to removed the solids that have been precipitated from a chemical reaction such as would occur in softening reactions to be discussed later.

Rectangular settling basins are basins that are rectangular in plans and cross sections. In plan, the length may vary from two to four times the width. The length may also vary from 10 to 20 times the depth. The depth of the basin may vary from 2.5 to 4 m. The influent is introduced at one end and allowed to flow through the length of the clarifier toward the other end. The solids that settle at the bottom are continuously scraped by a sludge scraper and removed. The clarified effluent flows out of the unit through a suitably designed effluent weir. Figure 6-5 is a photograph of a rectangular settling basin in Anne Arundel County, Maryland. A schematic of an effluent weir is shown in Figure 6-6.

Figure 6-5   Settling basin in Anne Arundel County, Maryland.

Circular settling basins are circular in plan. Unlike the rectangular basin, circular basins are easily upset by wind cross currents. Because of its rectangular shape, more energy is required to cause circulation in a rectangular basin; in contrast, the contents of the circular basin is conducive to circular streamlining. This condition may cause short circuiting of the flow. For this reason, circular basins are normally designed for diameters not to exceed 30 m in diameter.

The influent feed is introduced at the center feed pipe in a circular clarifier. From the center, the flow travels horizontally toward the outlet weir at the rim of the tank, dropping its suspended solids load along the way forming sludge at the bottom. As in

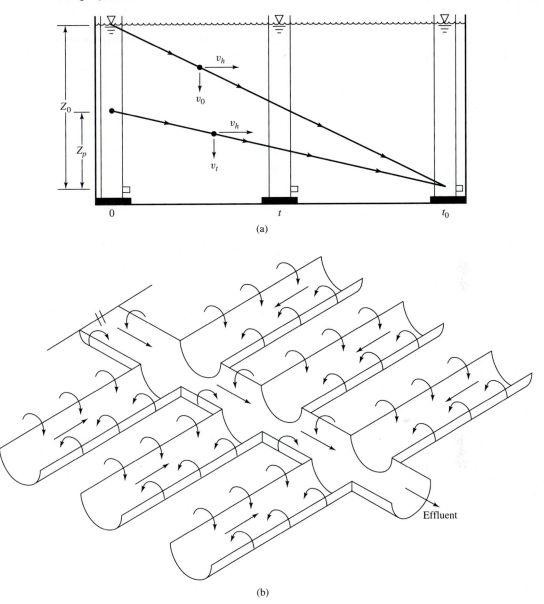

Figure 6-6 (a) Removal at the settling zone. (b) Weir arrangement at the outlet zone.

the case of the rectangular clarifier, the sludge is scraped continuously and subsequently withdrawn.

The solids-contact clarifier owes its name to the contacting of solids originally formed with the incoming influent during reaction. These contacting solids are sludge from the settling zone recycled to serve as nuclei in the reaction zone. There are two reaction zones: the *primary* and *secondary zones*. It is at these zones that the influent flow and the solids are in intimate contact. After contact, the mixture is introduced to the settling zone to clarify. The clarified water then moves out into the effluent weir.

There are generally four functional zones in a settling basin: the inlet zone, the settling zone, the sludge zone, and the outlet zone. The *inlet zone* provides a baffle aimed at properly introducing the inflow into the tank. For the rectangular basin, the baffle spreads the inflow uniformly across the influent vertical cross section; for the circular clarifier, a baffle at the tank center turns the inflow horizontally toward the rim of the clarifier; and for the solids-contact clarifier, draft tubes which also serve as baffles turn the inflow toward the rim of the clarifier. The length of the inlet zone is normally designed equal to the depth of the settling zone.

The *settling zone* is where the suspended solids load of the inflow is removed to be deposited into the *sludge zone* below. The *outlet zone* is where the effluent takes off into an effluent weir overflowing as a clarified liquid. The outlet zone is also normally designed equal to the depth of the settling zone. A settling zone and an outlet zone arrangement are shown in Figure 6-6.

Figure 6-6a shows the basic principles of removal of solids in the settling zone. The settling column discussed before is shown moving with the horizontal flow of the water from the inlet of the tank to the outlet. As the column moves, visualize the solids settling inside it; when the column reaches the end of the tank, these solids will have been deposited at the bottom already. The behavior of the solids outside the column will be similar to that inside. Thus the t_0 in the settling column is the same t_0 in the settling basin.

A particle possesses both downward terminal velocity v_0 or v_t, and a horizontal velocity v_h. For the particle to remain deposited at the sludge zone, v_h should be such as not to scour it. For light flocculent suspensions, v_h should not be greater than 9.0 m/h or 0.5 fpm; and for heavier, discrete-particle suspensions, it should not be more than 36 m/h or 2 fpm. If A is the vertical cross-sectional area, Q the flow, and W the width,

$$v_h = \frac{Q}{A}$$
$$= \frac{Q}{Z_0 W} \qquad (6\text{-}17)$$

For discrete particles, the detention time t_0 normally ranges from 1 to 4 h, while for flocculent suspensions, it normally ranges from 4 to 6 h. Calling V the volume of the tank and L the length, there are two ways that t_0 can be calculated: $t_0 = Z_0/v_0$ and $t_0 = V/Q = (W Z_0 L)/Q$. Therefore,

$$\frac{Z_0}{v_0} = \frac{W Z_0 L}{Q} \quad \text{and} \quad v_0 = \frac{Q}{LW} = \frac{Q}{A_s} = q_0 \qquad (6\text{-}18)$$

where A_s is the surface area of the tank and Q/A_s is called the *overflow rate*, q_0. *From this equation the overflow rate must be equal to the terminal settling velocity of the particles.*

In the design of circular basins, the horizontal velocity, called flow-through velocity v_h, is not normally considered. In these clarifiers, the particles starting from the center of the tank are progressively slowed down as they approached the rim. Hence there is no danger of the scouring velocity being exceeded. In the case of the rectangular clarifier, on the other hand, the length can be long, thus scouring velocities may occur in the settling zone.

In the outlet zone, weirs are provided for the effluent to take off. Even if v_h has been chosen properly but overflow weirs are not properly sized, flows can be turbulent at the weirs; this turbulence can entrain particles causing the design to fail. Overflow weirs should therefore be loaded with the proper amount of overflow. Weir overflow rates normally range from 6 m^3/h per meter of weir length for light flocs to 14 m^3/h per meter of weir length for heavier discrete-particle suspensions. When weirs constructed along the periphery of the tank are not sufficient to meet the weir loading requirement, inboard weirs may be constructed. One such example is shown in Figure 6-6b.

Example 6-5

The prototype detention time and overflow rate were calculated to be 44 min and 0.10 m/min, respectively. Design a rectangular settling basin using these values for a flow rate of 20,000 m^3/d.

Solution

$$Z_0 = v_0 t_0 = 0.1(44) = 4.4 \text{ m}$$

Use $v_h = 8$ m/h.

$$W Z_0 v_h = W(4.4)(8) = \frac{20{,}000}{24}; \qquad W = 23.67 \text{ m}$$

Use two tanks. Thus W per tank $= 23.67/2 = 11.83$, say 11 m. Then

$$WL = A_s = \text{overflow area} = 11L = \frac{20{,}000/[2(24)(60)]}{0.1} = 69.44; \qquad L = 6.3 \text{ m}$$

$$v_{h(\text{actual})} = \frac{20{,}000/[24(2)]}{11(4.4)} = 8.6 \text{ m/h} \qquad \text{O.K.}$$

Check for flow-through travel:

$$\text{flow-through travel} = v_h(t_0) = 8.6 \left(\frac{44}{60}\right) = 6.3 \qquad \text{O.K.}$$

Provide 4.4 m for each of inlet and outlet zones, 0.5 m freeboard, and 0.5 m sludge zone. 4.4 m is Z_0, the depth of settling zone. Therefore,

$$\text{length} = 6.3 + 2(4.4) = 15.1 \text{ m} \quad \textbf{Answer}$$

$$\text{width} = 11 \text{ m} \quad \textbf{Answer}$$

$$\text{depth} = 4.4 + 0.5 + 0.5 = 5.4 \text{ m} \quad \textbf{Answer}$$

Determine the overflow weir requirement. Use a weir loading of 6 m^3/h per meter of weir. Thus

$$\text{weir length} = \frac{20{,}000/[2(24)]}{6} = 69.44 \text{ m}$$

Since the width of one tank is only 11 m, an inboard weir arrangement such as that of Figure 6-6b should be provided.   **Answer**
Two sedimentation basins should be provided.   **Answer**

Example 6-6

Repeat Example $6 - 5$ for a circular settling basin.

Solution   First determine A_s.

$$A_s = \text{surface overflow area} = \frac{20{,}000/[24(60)]}{0.1} = 138.89 \text{ m}^2 \qquad \frac{\pi D^2}{4} = A_s = 138.89$$

$$D = \text{diameter of tank} = 13.30 \text{ m}$$

$$Z_0 = v_0 t_0 = 0.1(44) = 4.4 \text{ m}$$

Provide 4.4 m for each of inlet and outlet zones, 0.5 m freeboard, and 0.5 m sludge zone. Therefore,

$$D = \text{diameter of tank} = 13.3 + 2(4.4) = 22.1 \text{ m} \quad \textbf{Answer}$$

$$\text{depth} = 4.4 + 0.5 + 0.5 = 5.4 \text{ m} \quad \textbf{Answer}$$

Determine the overflow weir requirement. Use a weir loading of 6 m^3/h per meter of weir. Thus

$$\text{weir length} = \frac{20{,}000/24}{6} = 138.89 \text{ m}$$

Since the circumference of the tank is $\pi(22.1) = 69.43$ m only, an inboard weir arrangement such as that of Figure 6-6b should be provided.   **Answer**

FILTRATION

Normally, filtration follows settling. Figure 6-7 shows sketches of the various modes of filtration. In part (a), the influent water to be filtered is introduced at the bottom. The water, then, passes through the bed of sand in the upward direction and thus is filtered. In part (b), the influent is split into two flows; hence the operation is said to be in the *biflow mode*. In part (c), the influent is introduced at the top and is filtered through two beds of filter media in a downward flow; the filter is called a *dual-media, downflow filter*. In part (d), the filter is a triple-media filter.

If the influent introduced at the top is allowed to flow down the filter without the aid of a pump, the filter is said to be a gravity filter. In general, there are two types of gravity filters: slow-sand and rapid-sand filters. In the main, these filters are differentiated by their rates of filtration. *Slow-sand filters* normally operate at a rate of 1 to 10 mgad (million gallons per acre per day), while *rapid-sand filters* normally operate at a rate of 100 to 200 mgad. Shown in Figure 6-8 is a typical gravity filter.

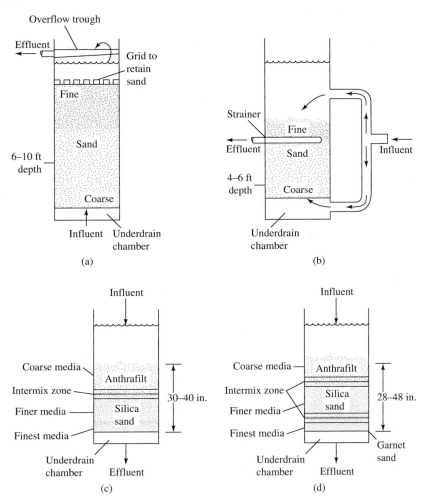

Figure 6-7 Various modes of filter operation: (a) upflow filter with surface grid; (b) biflow filter; (c) dual-media, downflow filter; (d) triple media (mixed media) filter.

Filter Medium Specification

The most important component of a filter is the medium. This medium must be of the appropriate size and, ideally, must be uniform. Small grain sizes tend to have higher head losses, while large grain sizes, although producing comparatively smaller head losses, are not as effective in filtering. The actual grain sizes are determined from what experience has found to be most effective. Since the actual medium is never uniform, grain sizes are specified in terms of effective size and uniformity coefficient. *Effective size* is defined as the size of sieve opening that passes the 10% finer of the medium sample. The effective

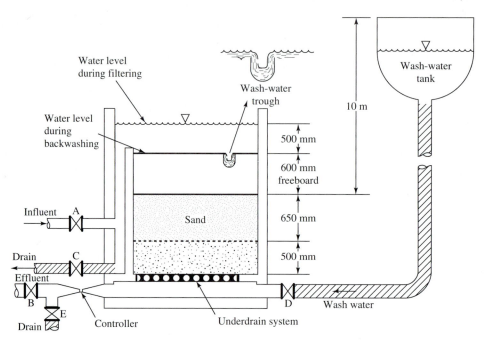

Figure 6-8 Typical gravity filter.

size is said to be the *10th percentile size* P_{10}. *Uniformity coefficient* is defined as the ratio of the size of the sieve opening that passes 60% finer of the medium sample to the size of the sieve opening that passes 10% finer of the medium sample. In other words, the uniformity coefficient is the ratio of the 60th percentile size (P_{60}) to the 10th percentile size (P_{10}). For slow-sand filters, the effective size ranges from 0.25 to 0.35 mm with uniformity coefficient ranging from 2 to 3. For rapid-sand filters, the effective size ranges from 0.45 mm and higher with uniformity coefficient ranging from 1.5 and lower.

     Preparation of filter medium.   Plot of a sieve analysis of a sample of run-of-bank sand is shown in Figure 6-9 by the segmented line labeled "stock sand. . . ." This sample may or may not meet the required effective size and uniformity coefficient specifications. To transform this sand into a usable sand, it must be given some treatment. The figure shows the cumulative percentages (represented by the "normal probability scale" on the ordinate) as a function of the increasing size of the sand (represented by the "size of separation" on the abscissa).

    Let p_1 be the percentage of the sample stock sand that is smaller than or equal to the desired P_{10} of the final filter sand, and p_2 be the percentage of the sample stock sand that is smaller than or equal to the desired P_{60} of the final filter sand. Since the percentage difference of P_{60} and P_{10} represents half of the final filter sand, $p_2 - p_1$ of the sample stock sand must represent half of the total sand in the final filter sand, also.

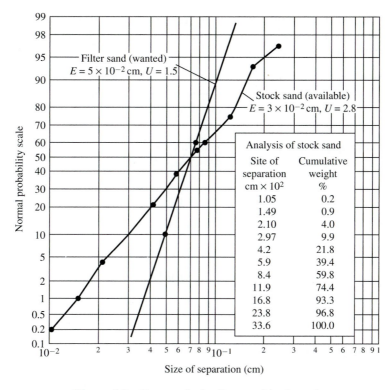

Figure 6-9 Sieve analysis of a run-of-bank sand.

Letting p_3 be the percentage of the stock sand that is transformed into the final usable sand,

$$p_3 = 2(p_2 - p_1) \tag{6-19}$$

Of this p_3, by definition, 10% must be the P_{10} of the final sand. Therefore, if p_4 is the percentage of the stock sand that is too fine to be usable,

$$p_4 = p_1 - 0.1p_3 = p_1 - 0.1(2)(p_2 - p_1) \tag{6-20}$$

Since the plot in the figure shows an increasing percentage as the size of separation increases, the sum of p_4 and p_3 must represent the percentage of the sample stock sand above which the sand is too coarse to be usable. Letting p_5 be this percentage, we have

$$p_5 = p_4 + p_3 \tag{6-21}$$

Now, to convert run-of-bank stock sand into a usable sand, an experimental curve such as Figure 6-9 is entered to determine the size of separation corresponding to p_4 and p_5. Having determined these sizes, the stock sand is washed in a sand washer that rejects the unwanted sand. The washer is essentially an upflow settling tank. By varying the upflow velocity of the water in the washer, the sand particles introduced into the tank are separated by virtue of the difference of their settling velocities. The lighter ones are carried into the effluent while the heavier ones settle at the bottom. The straight line in the figure represents the size distribution in the final filter sand when the p_4 and p_5 fractions have been removed.

Example 6-7

If the effective size and uniformity coefficient of a proposed filter are to be $5(10^{-2})$ cm and 1.5, respectively, perform a sieve analysis to transform the run-of-bank sand of Figure 6-9 into a usable sand.

Solution From Figure 6-9, for a size of separation of $5(10^{-2})$ cm, the percent p_1 is 30. Also, the P_{60} size is $5(10^{-2})(1.5) = 7.5(10^{-2})$ cm. From the figure, the percent p_2 corresponding to the P_{60} size of the final sand is 53. Then

$$p_3 = 2(p_2 - p_1) = 2(53 - 30) = 46\%$$

$$p_4 = p_1 - 0.1(p_3) = 30 - 0.1(46)$$

$$= 25.4\%; \quad \text{from the figure, the corresponding size of separation} = 4.5(10^{-2}) \text{ cm}$$

$$p_5 = 46 + 25.4 = 71.4\%; \quad \text{corresponding size of separation} = 1.1(10^{-1}) \text{ cm}$$

Therefore, the sand washer must be operated so that the p_4 sizes of $4.5(10^{-2})$ cm and smaller and the p_5 sizes of $1.1(10^{-1})$ cm and greater are rejected. **Answer**

Filter Head Loss

The motion of water through sand beds is just like the motion of water through a run of pipe. While the motion in the pipe is straightforward, however, the motion in the filter bed is tortuous. Nevertheless, the form of the friction head loss expression should remain the same. The motion of the water is either upward propelled by a pump or downward through the action of gravity. In either case, the flow has to be turbulent. Therefore, the head loss h is

$$h = f \frac{\ell V^2}{D \cdot 2g} \tag{6-22}$$

where f is the friction factor, ℓ the depth of the filter bed, V the insterstitial velocity, D the "diameter" of the cross section of flow, and g the gravity.

Since the filter is not actually a pipe, the diameter D must be expressed in some other form so that pertinent filter parameters may be used. One way to do this is to use the hydraulic radius, which for circular pipes is $D/4$. The equation then becomes

$$h = f \frac{\ell V^2}{4R \cdot 2g} \tag{6-23}$$

With D expressed in terms of R, pertinent filter parameters may now be used, and with the use of R, the equation has now become general. In other words, it can be used for any shape, since R is simply defined as the area of flow divided by the wetted perimeter. The expression *area of flow divided by the wetted perimeter* is the same as *volume of flow divided by the wetted area*, that is,

$$R = \frac{\text{volume of flow}}{\text{wetted area}} \tag{6-24}$$

Let N be the number of medium grains in the filter each of volume V_p. If η is the porosity, the volume of the filter is $NV_p/(1 - \eta)$. Therefore, the volume of flow is

$$\text{volume of flow} = \eta \frac{NV_p}{1 - \eta} \tag{6-25}$$

If A_p is the surface area of the particle, the wetted surface area in the filter is NA_p. Therefore, the hydraulic radius, R, is

$$R = \frac{\eta[NV_p/(1 - \eta)]}{NA_p}$$

$$= \frac{\eta}{1 - \eta} \frac{V_p}{A_p} \tag{6-26}$$

But $V_p = \beta d_p^3$ and $A_p = \alpha d_p^2$, where β is the volume shape factor and α is the area shape factor. Substituting yields

$$R = \frac{\eta}{1 - \eta} \frac{\beta}{\alpha} d_p \tag{6-27}$$

The velocity V is the interstitial velocity of the fluid through the pores of the bed. Compared to the superficial velocity V_s, it must be faster due to the effect of the porosity, η. (The cross-sectional area of flow is much constricted, hence the velocity is much faster.) In terms of η and V_s,

$$V = \frac{V_s}{\eta} \tag{6-28}$$

Substituting equations (6-28) and (6-27) in equation (6-23) and simplifying gives us

$$h = \frac{f\alpha}{8\beta} \frac{\ell(1 - \eta)V_s^2}{\eta^3 g d_p}$$

$$= f' \frac{\ell(1 - \eta)V_s^2}{\eta^3 g d_p} \tag{6-29}$$

The modified friction factor $f' = f\alpha/(8\beta)$ is given by

$$f' = 150\left(\frac{1 - \eta}{\text{Re}}\right) + 1.75 \tag{6-30}$$

The diameter in the definition of $\text{Re} = dV_s\rho_w/\mu$ must be the equivalent spherical diameter.

Equation (6-29) has been derived for a bed of uniform size. However, in some types of filtration plants such as those using rapid sand filters, the bed is backwashed every so often. This means that after backwashing and the grain particles allowed to settle, the grains deposit on the bed layer by layer of different "diameters." The bed is said to be stratified. To find the head loss across a stratified medium, equation (6-29) is applied layer by layer, each layer being converted to one single, average diameter. The head losses across each layer are then summed to produce the head loss across the bed as shown below.

$$h = \sum f_i' \frac{\ell_i(1-\eta)V_s^2}{\eta^3 g d_{pi}} \tag{6-31}$$

where the index i refers to the ith layer. If x_i is the fraction of the d_{pi} particles in the ith layer, ℓ_i equals $x_i\ell$. Assuming that the porosity η is the same throughout the bed, equation (6-31) becomes, after substitution,

$$h = \frac{\ell(1-\eta)V_s^2}{\eta^3 g} \sum \frac{f_i' x_i}{d_{pi}} \tag{6-32}$$

Example 6-8

A sharp filter sand has the sieve analysis shown below. The porosity of the unstratified bed is 0.39 and that of the stratified bed is 0.42. The lowest temperature anticipated of the water to be filtered is 4°C. Find the head loss if the sand is to be used in (a) a slow-sand filter 30 in. deep operated at 10 mgad and (b) a rapid-sand filter 30 in. deep operated at 125 mgad.

Sieve size	Average size (mm)	x_i
14–20	1.0	0.01
20–28	0.70	0.05
28–32	0.54	0.15
32–35	0.46	0.18
35–42	0.38	0.18
42–48	0.32	0.20
48–60	0.27	0.15
60–65	0.23	0.07
65–100	0.18	0.01

Solution

(a)
$$h = f' \frac{\ell(1-\eta)V_s^2}{\eta^3 g d_p}$$

$$d_{pave} = 1(0.01) + 0.7(0.05) + 0.54(0.15) + 0.46(0.18) + 0.38(0.18) + 0.32(0.20)$$

$$+ 0.27(0.15) + 0.23(0.07) + 0.18(0.01) = 0.40 \text{ mm}$$

$$10 \text{ mgad} = \frac{10(10^6)}{7.48(43,560)(3.281)(24)(60)(60)} = 10(1.08)(10^{-5}) = 1.08(10^{-4}) \text{ m/s}$$

$$f' = 150 \left(\frac{1 - \eta}{Re} \right) + 1.75$$

$$Re = \frac{dV_s \rho_w}{\mu}$$

$$d = 1.24\beta^{0.333} d_p; \qquad \beta = 0.77$$

Therefore,

$$d = 1.24(0.77^{0.333})(0.40)(10^{-3}) = 0.45(10^{-3}) \text{ m}$$

$$\mu = 15(10^{-4}) \text{ kg/m-s}$$

$$Re = \frac{0.45(10^{-3})(1.08)(10^{-4})(1000)}{15(10^{-4})} = 0.0324$$

$$f' = 150 \left(\frac{1 - 0.39}{0.0324} \right) + 1.75 = 2825.82$$

$$h = 2825.82 \left[\frac{30/[12(3.281)](1 - 0.39)[1.08(10^{-4})]^2}{0.39^3 (9.81)(0.4)(10^{-3})} \right] = 0.0658 \text{ m} \quad \textbf{Answer}$$

(b) $125 \text{ mgad} = 125(1.08)(10^{-5}) = 0.00135 \text{ m/s}$

$$h = \frac{\ell(1 - \eta)V_s^2}{\eta^3 g} \sum \frac{f_i' x_i}{d_{pi}}$$

$$Re_i = \frac{d_i (0.00135)(1000)}{15(10^{-4})} = \frac{1.24(0.77)^{0.333} d_{pi} (0.00135)(1000)}{15(10^{-4})} = 1022.98 d_{pi}$$

$$f_i' = 150 \left(\frac{1 - 0.42}{1022.98 d_{pi}} \right) + 1.75 = \frac{0.085}{d_{pi}} + 1.75$$

Sieve size	Average size (mm)	x_i	f_i'	$\dfrac{f_i' x_i}{d_{pi}}$
14–20	1.0	0.01	86.75	867.5
20–28	0.70	0.05	123.18	8,798.47
28–32	0.54	0.15	159.16	44,210.39
32–35	0.46	0.18	186.53	72,991.02
35–42	0.38	0.18	225.43	106,784.63
42–48	0.32	0.20	267.34	167,109.38
48–60	0.27	0.15	316.56	175,869.34
60–65	0.23	0.07	371.32	113,008.98
65–100	0.18	0.01	473.97	26,331.79
			$\sum$	715,971.50

$$h = \frac{30/[12(3.281)](1 - 0.42)(0.00135^2)}{0.42^3(9.81)}(715{,}971.50)$$

$$= 0.793 \text{ m} \quad \textbf{Answer}$$

Filter Head Loss Due to Deposited Materials

The head-loss expressions derived above pertain to head losses of clean filter beds. In actual operations, however, head loss is also a function of the amount of materials deposited in the pores of the filter. Letting q_i represent the deposited material per unit volume of bed in layer i, the corresponding head loss h_i is modeled as

$$h_i = a(q_i)^b \tag{6-33}$$

where a and b are constants. Taking logarithms, we have

$$\ln h_i = \ln a + b \ln q_i \tag{6-34}$$

This equation shows that plotting $\ln h_i$ against $\ln q_i$ will produce a straight line. An actual experiment performed by Tchobanoglous and Eliassen (1970) showed this to be so.[3]

In equation (6-33), when q_i equals zero, h_i equals zero—which is true. However, due to the bed, when q_i equals zero, a head loss must still exist. The total head loss of the filter bed, h, is therefore

$$h = h_0 + \sum a(q_i)^b \tag{6-35}$$

where h_0 is the clean bed head loss.

Now, develop a method to determine q_i. For simplicity, the index i will be dropped in subsequent treatment. Applying the Reynolds transport theorem (see Appendix 24) across a differential volume $d\Psi$ of the bed,

$$\frac{D \int_M q \, d\Psi}{Dt} = \frac{\partial \int_\Psi q \, d\Psi}{\partial t} + \oint_A C\mathbf{v} \bullet \hat{n} \, dA = 0 \tag{6-36}$$

where M is the control mass, Ψ the control volume, C the concentration of solids introduced into bed, $\mathbf{v}$ the velocity vector, $\hat{n}$ the unit normal vector, and A the bounding surface area of control mass and control volume.

Over a differential length $d\ell$ and cross-sectional area A of bed, we obtain

$$\oint_A C\mathbf{v} \bullet \hat{n}\hat{\eta} \, dA = -\dot{m} + \left(\dot{m} + \frac{\partial \dot{m}}{\partial \ell} \, d\ell \right)$$
$$= \frac{\partial \dot{m}}{\partial \ell} \, d\ell \tag{6-37}$$

where $\dot{m} = CQ$ and Q = flow, a constant through the bed. Therefore,

$$\oint_A C\mathbf{v} \bullet \hat{n}\hat{\eta} \, dA = Q\frac{\partial C}{\partial \ell} \, d\ell = V_s A \frac{\partial C}{\partial \ell} \, d\ell \tag{6-38}$$

---

[3]G. Tchobanoglous and E. D. Schroeder (1985). *Water Quality*. Addison-Wesley, Reading, Mass., p. 520.

Also, over the same differential length $d\ell$ and cross-sectional area A of the bed,

$$\frac{\partial \int_{\forall} q \, d\forall}{\partial t} = \frac{\partial (q \, d\forall)}{\partial t} = \frac{\partial q}{\partial t} \, dV = \frac{\partial q}{\partial t} A \, d\ell \tag{6-39}$$

Substituting equations (6-39) and (6-38) in equation (6-36) gives us

$$\frac{\partial q}{\partial t} A \, d\ell + V_s A \frac{\partial C}{\partial \ell} \, d\ell = 0 \tag{6-40}$$

Considering that the flow is one-dimensional, the partial notations will revert to total differential notation. Hence

$$\frac{dq_i}{dt} = -V_s \frac{dC_i}{d\ell_i} \tag{6-41}$$

The numerical counterpart of equation (6-41), using n as the index for time and m as the index for space, is

$$\frac{q_{i,n+1} - q_{i,n}}{\Delta t} = -V_s \frac{C_{i,n,m+1} - C_{i,n,m}}{\Delta \ell_i} \tag{6-42}$$

for the first time step. Solving for $q_{i,n+1}$, we have

$$q_{i,n+1} = q_{i,n} - \frac{\Delta t \, V_s}{\Delta \ell_i} (C_{i,n,m+1} - C_{i,n,m}) \tag{6-43}$$

Across a given layer i, the gradient, $(C_{i,n,m+1} - C_{i,n,m})/\Delta \ell_i$, practically does not vary with time of filtration. Letting the gradient be $\Delta C_{i,n}/\Delta \ell_i$, the equation becomes

$$q_{i,n+1} = q_{i,n} - \Delta t \, V_s \frac{\Delta C_{i,n}}{\Delta \ell_i} \tag{6-44}$$

The equation for the second time step is

$$q_{i,n+2} = q_{i,n+1} - \Delta t \, V_s \frac{\Delta C_{i,n}}{\Delta \ell_i}$$
$$= q_{i,n} - 2\Delta t \, V_s \frac{\Delta C_{i,n}}{\Delta \ell_i} \tag{6-45}$$

Using equation (6-45) as the model, the equation for the kth time step is

$$q_{i,n+k} = q_{i,n} - k\Delta t \, V_s \frac{\Delta C_{i,n}}{\Delta \ell_i} \tag{6-46}$$

The number of time steps k is given by

$$k = \frac{t}{\Delta t} \tag{6-47}$$

Therefore, the final equation becomes

$$q_{i,n+k} = q_{i,n} - t V_s \frac{\Delta C_{i,n}}{\Delta \ell_i} \tag{6-48}$$

The nature and sizes of particles introduced to the filter will vary depending on how they are produced. Hence the influent to a filter coming from a softening plant is different from the influent coming from a coagulated surface water. Also, the influent coming from a secondary-treated effluent is different from those of any of the drinking water treatment influent. *For these reasons, to use the foregoing equations for determining head losses, a pilot plant study should be conducted for a given type of influent.*

Example 6-9

To determine the values of *a* and *b* of equation (6-34), experiments were performed on uniform sands and anthracite media, yielding the following results:

	Diameter (mm)	Suspended solids removed (mg/cm)^3	Head loss (m)
Uniform sand	0.5	1.08	0.06
	0.5	7.2	4.0
	0.7	1.6	0.06
	0.7	11.0	4.0
	1.0	2.6	0.06
	1.0	17.0	4.0
Anthracite	1.0	6.2	0.06
(uniform)	1.0	43	4.0
	1.6	8.1	0.06
	1.6	58	4.0
	2.0	9.2	0.06
	2.0	68	4.0

Calculate the *a*'s and *b*'s corresponding to the respective diameters.

Solution   Uniform sand, diameter $= 0.5$ mm:

$$\ln h_i = \ln a + b \ln q_i$$

$$\ln 0.06 = \ln a + b \ln 1.08 \tag{1}$$

$$\ln 4.0 = \ln a + b \ln 7.2 \tag{2}$$

Solving yields $a = 0.50$, $b = 2.24$.   **Answer**

Uniform sand, diameter $= 0.7$ mm:

$$\ln 0.06 = \ln a + b \ln 1.60 \tag{1}$$

$$\ln 4.0 = \ln a + b \ln 11 \tag{2}$$

Solving yields $a = 0.0216$, $b = 2.18$.   **Answer**

Similar procedures are applied to the rest of the data. The following is the final tabulation of the answers:

	Diameter (mm)	a	b
Sand	0.5	0.5	2.24
	0.7	0.0216	2.18
	1.0	0.0133	2.24
Anthracite	1.0	0.00114	2.17
	1.6	0.00069	2.13
	2.0	0.00057	2.01

From the results of Example 6-9, the following equations with the values of b averaged are obtained.

$$h_i = \begin{cases} 0.50(q_i)^{2.22} \text{ for uniform sand, 0.5 mm in diameter} \\ 0.0216(q_i)^{2.22} \text{ for uniform sand, 0.7 mm in diameter} \\ 0.0133(q_i)^{2.22} \text{ for uniform sand, 1.0 mm in diameter} \\ 0.00114(q_i)^{2.10} \text{ for uniform anthracite, 1.0 mm in diameter} \\ 0.00069(q_i)^{2.10} \text{ for uniform anthracite, 1.6 mm in diameter} \\ 0.00057(q_i)^{2.10} \text{ for uniform anthracite, 2.0 mm in diameter} \end{cases} \quad (6\text{-}49)$$

The h_i's and q_i's in the equations above are in meters and mg/cm^3, respectively. Also, the head losses for diameters not included in the equations may be interpolated or extrapolated from the results obtained from the equations.

Example 6-10

The amount of suspended solids removed in a uniform anthracite medium 1.8 mm in diameter is 40 mg/cm^3. Determine the head loss due to suspended solids.

Solution

$$h_i = 0.00069(q_i)^{2.10} \quad \text{for uniform anthracite 1.6 mm in diameter}$$

$$= 0.00069(40)^{2.10} = 1.59 \text{ m}$$

$$h_i = 0.00057(q_i)^{2.10} \quad \text{for uniform anthracite 2.0 mm in diameter}$$

$$= 0.00057(40)^{2.10} = 1.32 \text{ m}$$

Interpolating, let x be the head loss corresponding to the 1.8-mm-diameter media.

$$\frac{1.59 - x}{1.59 - 1.32} = \frac{1.6 - 1.8}{1.6 - 2.0} \qquad x = 1.46 \text{ m} \quad \textbf{Answer}$$

Example 6-11

According to a recent modeling evaluation of the Pine Hill Run sewage discharge permit, a secondary-treated effluent of 20 mg/L can no longer be allowed into Pine Hill Run, an estuary tributary to the Chesapeake Bay. To meet the new, more stringent discharge requirement, the town decided to investigate filtering the effluent. A pilot study was conducted using a dual-media filter composed of anthracite as the upper 30-cm part and sand as the next lower 30-cm part of the filter. The results are shown below, where C_0 is the concentration of solids at the influent:

Q (L/m²-min)	Depth (cm)	$\frac{C}{C_0}$	Medium	Q (L/m²-min)	Depth (cm)	$\frac{C}{C_0}$	Medium
80	0	1	Anthracite	80	8	0.32	Anthracite
80	16	0.27	Anthracite	80	24	0.24	Anthracite
80	30	0.23	Anthracite	80	40	0.20	Sand
80	48	0.20	Sand	80	56	0.20	Sand
160	0	1	Anthracite	160	8	0.46	Anthracite
160	16	0.30	Anthracite	160	24	0.27	Anthracite
160	30	0.25	Anthracite	160	40	0.22	Sand
160	48	0.22	Sand	160	56	0.22	Sand
240	0	1	Anthracite	240	8	0.59	Anthracite
240	16	0.39	Anthracite	240	24	0.30	Anthracite
240	30	0.27	Anthracite	240	40	0.23	Sand
240	48	0.23	Sand	240	56	0.23	Sand

If the respective sizes of the anthracite and sand layers are 1.6 mm and 0.5 mm, what is the length of the filter run to a terminal head loss of 3 m at a filtration rate of 200 L/m²-min? Assume that the clean water head loss is 0.793 m. *Note*: Terminal head loss is the loss when the filter is about to be cleaned.

Solution   Since 200 L/m²-min is between 160 and 240 L/m⁴-min, only the data for these two flows will be analyzed. The length of the filter run for the 200-L/m³-min flow will be interpolated between the 160 and the 240.

	Filtration rate (L/cm²-min)					
	160			240		
Depth (cm)	(1) $\frac{C}{C_0}$	(2) C (mg/L)	(3) ΔC_i (mg/L)	(4) $\frac{C}{C_0}$	(5) C (mg/L)	(6) ΔC_i (mg/L)
0	1	20	—	1	20	—
8	0.46	9.2	−10.8	0.59	11.8	−8.2
16	0.30	6.0	−3.2	0.39	7.8	−4.0
24	0.27	5.4	−0.6	0.3	6.0	−1.8
30	0.25	5.0	−0.4	0.27	5.4	−0.6
40	0.22	4.4	−0.6	0.23	4.6	−0.8
48	0.22	4.4	0	0.23	4.6	0
56	0.22	4.4	0	0.23	4.6	0

Col. (1): given.

Col. (2): 20[col. (1)].

Col. (3): $C_{i,m+1} - C_{i,m}$; for example, $-10.8 = 9.2 - 20$.

$$q_{i,n+k} = q_{i,n} - tV_s \frac{\Delta C_{i,n}}{\Delta \ell_i}$$

$$160 \text{ L/m}^2\text{-min} = \frac{160(1000) \text{ cm}^3}{100^2 \text{ cm}^2\text{-min}} = 16 \text{ cm/min}$$

$$240 \text{ L/m}^2\text{-min} = \frac{240(1000) \text{ cm}^3}{100^2 \text{ cm}^2\text{-min}} = 24 \text{ cm/min}$$

$$1 \text{ mg/L} = \frac{1 \text{ mg}}{1000 \text{ cm}^3} = 10^{-3} \text{ mg/cm}^3$$

Thus

$$\Delta C_i \text{ mg/L} = \Delta C_i (10^{-3}) \text{ mg/cm}^3$$

$$q_{i,n+k} = q_{i,n} - tV_s \frac{\Delta C_{i,n}(10^{-3})}{\Delta \ell_i} = q_{i,n} - tV_s(10^{-3})\frac{\Delta C_{i,n}}{\Delta \ell_i} \quad \text{mg/cm}^3$$

$$h_i = \begin{cases} 0.00069(q_i)^{2.10} & \text{for uniform anthracite, 1.6 mm in diameter} \\ 0.50(q_i)^{2.22} & \text{for uniform sand, 0.5 mm in diameter} \end{cases}$$

For the 160 L/cm²-min:

| | | | Run length (min) | | | | | |
| | | | 600 | | 900 | | 1200 | |
Depth ΔC_i (cm) (mg/L)	$\Delta \ell_i$ (cm)	$\frac{\Delta C_i}{\Delta \ell_i}$	q_i (mg/cm³)	h_i (m)	q_i (mg/cm³)	h_i (m)	q_i (mg/cm³)	h_i (m)
0 —	—	—	—	—	—	—	—	—
8 −10.8	8	−1.35	12.96	0.15	19.44	0.35	25.92	0.64
16 −3.2	8	−0.4	3.84	0.01	5.76	0.027	7.68	0.05
24 −0.6	8	−0.075	0.72	0.000	1.08	0.000	1.44	0.001
30 −0.4	6	−0.07	0.64	0.000	0.96	0.000	1.28	0.001
40 −0.6	10	−0.06	0.576	0.15	0.864	0.36	1.15	0.68
48 0	—	—	—	—	—	—	—	—
48 0		Σ		0.31		0.737		1.37

The terminal head loss of 3 m is composed of the clean water head loss of 0.793 m and the head loss due to deposited solids of h_s:

$$h_s = 3 - 0.793 = 2.21 \text{ m}$$

900	0.737
1200	1.37
x	2.21

$$\frac{x - 1200}{1200 - 900} = \frac{2.21 - 1.37}{1.37 - 0.737}$$

$$x = 1598 \text{ min}$$

For the 240 L/cm^2-min:

Depth (cm)	ΔC_i (mg/L)	$\Delta \ell_i$ (cm)	$\dfrac{\Delta C_i}{\Delta \ell_i}$	Run length (min)					
				1500		2100		—	
				q_i (mg/cm^3)	h_i (m)	q_i (mg/cm^3)	h_i (m)	q_i (mg/cm^3)	h_i (m)
0	—	—	—	—	—	—	—	—	—
8	−8.2	8	−1.03	37.08	1.36	51.91	2.76	—	—
16	−4.0	8	−0.5	18	0.3	25.2	0.61	—	—
24	−1.8	8	−0.23	8.28	0.06	11.59	0.12	—	—
30	−0.6	6	−0.1	3.6	0.01	5.04	0.02	—	—
40	−0.8	10	−0.08	2.88	5.23	4.03	11.0	—	—
48	0	—	—	—	—	—	—	—	—
48	0		$\sum$		6.96		14.51		

$$
\begin{array}{cc}
x & 2.21 \\
1500 & 6.96 \\
2100 & 14.51
\end{array}
\qquad
\frac{x-1500}{1500-2100} = \frac{2.21-6.96}{6.96-14.51}
$$

$x = 1123$ min

$$
\begin{array}{cc}
160 & 1598 \\
200 & y \\
240 & 1123
\end{array}
\qquad
\frac{y-1598}{1123-1598} = \frac{200-160}{240-160}
$$

$y = 1360.5$ min   **Answer**

Filter Backwashing

In the early development of filters, units that had been clogged were renewed by scraping the topmost layers of sand. This necessitated designing the filters to clog purposely near the surface of the beds. The scraped sands were then cleaned by sand washers. In the nineteenth century, development of methods to clean sand in place rather than taking it out of the unit led to the development of the rapid-sand filter. This method of cleaning is called *backwashing*.

In backwashing, clean water is introduced at the filter underdrains at such a velocity as to expand the bed. The expansion frees the sand from clogging materials by causing the grains to rub against each other, dislodging any materials that have been clinging onto their surfaces. The dislodged materials are then discharged into washwater troughs.

To expand the bed, the excess of the weight of the grains over the buoyant forces exerted on the grains must be supported by the force due to the backwashing velocity. Per unit area of the filter, this backwashing force is simply equal to the pressure at the bottom of the filter, which, in turn, is equal to the specific weight of water γ_w times the head h_{fb} needed to suspend the grains. h_{fb} is also called the *backwashing head loss*. Also, per unit area of the filter, the excess of the weight over the buoyant force is

$\ell(1 - \eta_e)(\gamma_p - \gamma_w)$, where ℓ is the expanded depth of the bed, η_e is the expanded porosity of the bed and γ_p is the specific weight of the grain particles. Equating the backwashing force to the excess weight, we have

$$h_{fb} = \frac{\ell_e(1 - \eta_e)(\gamma_p - \gamma_w)}{\gamma_w} \tag{6-50}$$

Calling V_B the backwashing velocity, the commonly used expression for η_e is[4]

$$\eta_e = \left(\frac{V_B}{v}\right)^{0.22} \tag{6-51}$$

For stratified beds, equation (6-50) must be applied layer by layer. The backwashing head losses in each layer are then summed to produce the total head loss of the bed. Thus

$$h_{fb} = \sum h_{fb,i} = \sum \frac{\ell_{e,i}(1 - \eta_{e,i})(\gamma_p - \gamma_w)}{\gamma_w} \tag{6-52}$$

where the index i refers to the ith layer. If x_i is the fraction of the layer i, then $\ell_{e,i}$ equals $x_i \ell_e$. Hence equation (6-52) becomes

$$h_{fb} = \sum h_{fb,i} = \frac{\ell_e(\gamma_p - \gamma_w)}{\gamma_w} \sum x_i(1 - \eta_{e,i}) \tag{6-53}$$

The fractional expansion of the bed may be determined by considering that the total mass of expanded grains and the total mass of unexpanded grains are equal. Per unit surface area of the filter and assuming the porosity of the unexpanded bed to be constant throughout the depth, the material balance for each layer is

$$\ell_i(1 - \eta)\rho_p = \ell_{e,i}(1 - \eta_{e,i})\rho_p \tag{6-54}$$

where $\ell_{e,i}$ is the expansion per layer. Solving for $\ell_{e,i}$ gives

$$\ell_{e,i} = \frac{\ell_i(1 - \eta)}{1 - \eta_{e,i}} \tag{6-55}$$

Summing the expansion per layer, noting that $\ell_i = x_i \ell$, the fractional bed expansion is obtained as

$$\frac{\ell_e}{\ell} = (1 - \eta) \sum \frac{x_i}{1 - \eta_{e,i}} \tag{6-56}$$

Example 6-12

A sharp filter sand has the sieve analysis shown below. The average porosity of the stratified bed is 0.42. The lowest temperature anticipated of the water to be filtered is 4°C. Determine the percentage bed expansion and the backwashing head loss if the filter is to be backwashed at a rate of $0.4(10^{-2})$ m/s. $\rho_p = 2650$ kg/m^3. The bed depth is 0.7 m.

---

[4]Ibid., p. 528.

Sieve size	Average size (mm)	x_i
14–20	1.0	0.01
20–28	0.70	0.05
28–32	0.54	0.15
32–35	0.46	0.18
35–42	0.38	0.18
42–48	0.32	0.20
48–60	0.27	0.15
60–65	0.23	0.07
65–100	0.18	0.01

Solution

$$\frac{\ell_e}{\ell} = (1 - \eta) \sum \frac{x_i}{1 - \eta_{e,i}} = (1 - 0.42) \sum \frac{x_i}{1 - \eta_{e,i}} = (0.58) \sum \frac{x_i}{1 - \eta_{e,i}}$$

$$h_{fb} = \frac{\ell_e (\gamma_p - \gamma_w)}{\gamma_w} \sum x_i (1 - \eta_{e,i}) = \frac{\ell_e (9.81)(2650 - 1000)}{9.81(1000)} \sum x_i (1 - \eta_{e,i})$$

$$= 1.65 \ell_e \sum x_i (1 - \eta_{e,i})$$

$$\eta_e = \left(\frac{V_B}{v} \right)^{0.22} = \left[\frac{0.4(10^{-2})}{v} \right]^{0.22} = \frac{0.30}{v^{0.22}}$$

$$v = \sqrt{\frac{4}{3} g \frac{(\rho_p - \rho_w)d}{C_D \rho_w}} = \sqrt{\frac{4}{3} 9.81 \frac{(2,650 - 1000)d}{C_D(1000)}} = 4.65 \sqrt{\frac{d}{C_D}}$$

$$\beta = 0.77$$

$$d = 1.24 \beta^{0.333} d_p = 1.24(0.77)^{0.333} d_p = 1.14 d_p$$

Thus

$$v = 4.65 \sqrt{\frac{1.14 d_p}{C_D}} = 4.96 \sqrt{\frac{d_p}{C_D}}$$

$$C_D = \begin{cases} \dfrac{24}{\text{Re}} & \text{(for Re < 1)} \\[2mm] \dfrac{24}{\text{Re}} + \dfrac{3}{\text{Re}^{1/2}} + 0.34 & \text{(for } 1 \leq \text{Re} \leq 10^4) \\[2mm] 0.4 & \text{(for Re > } 10^4) \end{cases}$$

$$\text{Re} = \frac{dv\rho}{\mu} = \frac{1.14 d_p v(1000)}{15(10^{-4})} = 760,000 v d_p$$

Determine the settling velocities of the various fractions:

Sieve size	Average size, d_{pi} (mm)	x_i	C_{Di}	v_i (m/s)	Re_i
14–20	1	0.01	0.4	0.248	188
—	—	—	0.69	0.189	143
—	—	—	0.76	0.180	136
—	—	—	0.77	0.178	135
—	—	—	0.76	0.180	136
20–28	0.70	0.05	1.05	0.13	69
28–32	0.54	0.15	1.39	0.10	41
32–35	0.46	0.18	1.72	0.08	29
35–42	0.38	0.18	2.21	0.07	20
42–48	0.32	0.20	3.02	0.05	13
48–60	0.27	0.15	4.94	0.03	7
60–65	0.23	0.07	6.48	0.03	5
65–100	0.18	0.01	7.8	0.02	4

Sieve size	Average size, d_{pi} (mm)	x_i	v_i	$\eta_{e,i}$	$1 - \eta_{e,i}$	$\dfrac{x_i}{1 - \eta_{e,i}}$
14–20	1	0.01	0.18	0.44	0.56	0.020
20–28	0.70	0.05	0.13	0.47	0.53	0.09
28–32	0.54	0.15	0.10	0.50	0.50	0.30
32–35	0.46	0.18	0.08	0.52	0.48	0.38
35–42	0.38	0.18	0.07	0.54	0.46	0.39
42–48	0.32	0.20	0.05	0.58	0.42	0.48
48–60	0.27	0.15	0.03	0.65	0.35	0.43
60–65	0.23	0.07	0.03	0.65	0.35	0.20
65–100	0.18	0.01	0.02	0.71	0.29	0.03
					Σ	2.32

Thus

$$\text{percentage bed expansion} = \frac{\ell_e}{\ell}(100) = (0.58)\left(\sum \frac{x_i}{1 - \eta_{e,i}}\right)(100)$$

$$= 0.58(2.32)(100) = 134\% \quad \textbf{Answer}$$

Note: In practice, the bed is normally expanded to 150%.

Sieve size	Average size, d_{pi} (mm)	x_i	$1 - \eta_{e,i}$	$x_i(1 - \eta_{e,i})$
14–20	1	0.01	0.56	0.0056
20–28	0.70	0.05	0.53	0.0265
28–32	0.54	0.15	0.50	0.075
32–35	0.46	0.18	0.48	0.0864
35–42	0.38	0.18	0.46	0.0828
42–48	0.32	0.20	0.42	0.084
48–60	0.27	0.15	0.35	0.0525
60–65	0.23	0.07	0.35	0.0245
65–100	0.18	0.01	0.29	0.0029
			$\sum$	0.44

$$h_{fb} = 1.65 \ell_e \sum x_i(1 - \eta_{e,i}) = 1.65(1.34)(0.7)(0.44) = 0.68 \text{ m} \quad \textbf{Answer}$$

COAGULATION

In water treatment particles to be removed are colloids and in suspended forms. To remove them, colloids must be destabilized by coagulation. The degree of stability of a colloid particle is the result of the interaction of two forces: the potential force of repulsion and the *van der Waals force*. The van der Waals force is the natural attractive force that measures the tendency of like particles to join to form a larger mass. The joining of two colloidal particles is coagulation. Hence when the van der Waals force of attraction exceeds the potential force of repulsion, coagulation of the colloid particles results and the colloid is destabilized. There are four general ways of bringing about coagulation: (1) double-layer compression, (2) charge neutralization, (3) entrapment in a precipitate, and (4) intraparticle bridging.

Because the colloid is charged, opposite charges, called *counterions*, will surround it. The opposite charges will, in turn, be surrounded by charges opposite to them (the co-ions of the colloid charge), thus forming an electric double layer. When the concentration of counterions in the dispersion medium is smaller, the thickness of the electric double layer is larger. Since two approaching colloid particles cannot come closer to each other because of the thicker electric double layer, the colloid is stable. When the concentration is increased, more of the colloid charge will be neutralized, causing the double layer to shrink. The layer is then said to be *compressed*. As the layer is compressed sufficiently, a time will come when the van der Waals force exceeds the force of repulsion and coagulation will result.

The charge of a colloid can also be neutralized directly by the addition of ions of opposite charges that have the ability to adsorb directly to the colloid surface. For example, the positively charged dodecylammoniun, $C_{12}H_{25}NH_3^+$, tends to be hydrophobic and, as such, penetrates directly to the colloid surface and neutralizes it, if it happens to be negative. This is said to be a *direct charge neutralization*. Another direct charge neutralization method would be the use of a colloid of opposite charge.

A characteristic of some metal salt such as Al(III) and Fe(III) is that of forming a precipitate when added to water. For this precipitation to occur, a colloidal particle may provide the nucleation site, thus entrapping the colloid as the precipitate forms. Moreover, if several of these particles are entrapped and are close to each other, coagulation can result.

The last method of coagulation is intraparticle bridging. A bridging particle may attach one colloid particle to one of its active sites and another colloid particle to another site. An *active site* is a point in the molecule where particles may attach either by chemical bonding or by mere physical attachment. If the two sites are close to each other, coagulation of the colloids may occur; or the kinetic movement may loop the "bridge" assembly around, causing the attached colloids to "touch" each other. This can also bring coagulation.

Electrolytes and polyelectrolytes are used to coagulate colloids. *Electrolytes* are materials which when placed in solution cause the solution to be conductive to electricity because of charges they possess. They coagulate and precipitate colloids. *Polyelectrolytes* are polymers composed of more than one electrolytic site in the molecule, and *polymers* are molecules joined together to form larger molecules. The coagulating power of electrolytes is summed up in the *Schulze–Hardy rule*, which states: *The coagulation of a colloid is effected by that ion of an added electrolyte which has a charge opposite in sign to that of the colloidal particle; the effect of such an ion increases markedly with the number of charges carried.* Thus, comparing the effect of $AlCl_3$ and $Al_2(SO_4)_3$ in coagulating positive colloids, the latter is 30 times more effective than the former, since sulfate has two negative charges while the chloride has only one. In coagulating negative colloids, however, the two have about the same power of coagulation. Some of the reactions of common chemicals used in water treatment are shown below.

Aluminum sulfate (filter alum):

$$Al_2(SO_4)_3 \cdot 14.3H_2O + 3Ca(HCO_3)_2 \rightarrow 2Al(OH)_3\downarrow + 3CaSO_4 + 14.3H_2O + 6CO_2$$

$$Al_2(SO_4)_3 \cdot 14.3H_2O + 3Na_2CO_3 + 3H_2O \rightarrow 2Al(OH)_3\downarrow + 3Na_2SO_4 + 14.3H_2O + 3CO_2$$

Ferrous sulfate (copperas):

$$FeSO_4 \cdot 7H_2O + Ca(HCO_3)_2 \rightarrow Fe(HCO_3)_2 + CaSO_4 + 7H_2O$$

$$Fe(HCO_3)_2 + 2Ca(OH)_2 \rightarrow Fe(OH)_2\downarrow + 2CaCO_3\downarrow + 2H_2O$$

$$4Fe(OH)_2 + O_2 + 2H_2O \rightarrow 4Fe(OH)_3$$

Ferric salts:

$$2FeCl_3 + 3Ca(HCO_3)_2 \rightarrow 2Fe(OH)_3\downarrow + 3CaCl_2 + 6CO_2$$

$$2FeCl_3 + 3Ca(OH)_2 \rightarrow 2Fe(OH)_3\downarrow + 3CaCl_2$$

Lime:

$$Ca(OH)_2 + H_2CO_3 \rightarrow CaCO_3\downarrow + 2H_2O$$

$$Ca(OH)_2 + Ca(HCO_3)_2 \rightarrow 2CaCO_3 + 2H_2O$$

There are also other coagulants that have been used but, owing to high cost, their use is restricted to small installations. Examples of these are sodium aluminate, $NaAlO_2$; ammonia alum, $Al_2(SO_4)_3 \cdot (NH_4)_2 \cdot 24H_2O$; and potash alum, $Al_2(SO_4)_3 \cdot K_2SO_4 \cdot 24H_2O$. The reactions of sodium aluminate with aluminum sulfate and carbon dioxide are shown below.

$$6NaAlO_2 + Al_2(SO_4)_3 \cdot 14.3H_2O \rightarrow 8Al(OH)_3 + 3Na_2SO_4 + 2.3H_2O$$

$$2NaAlO_2 + CO_2 + 3H_2O \rightarrow 2Al(OH)_3\downarrow + Na_2CO_3$$

Coagulant Aids

Difficulties with sedimentation often occur because of flocs that are slow-settling and easily fragmented by the hydraulic shear in the sedimentation basin. For these reasons, coagulant aids are normally used. Acids and alkalis are used to adjust the pH to the optimum range. Typical acids used to lower the pH are sulfuric and phosphoric acids. Typical alkalis used to raise the pH are lime and soda ash. Polyelectrolytes are also used as coagulant aids. Polyelectrolytes exist in three forms: cationic (positive), anionic (negative), and nonionic (no charge) forms. The cationic form has been used successfully in some waters not only as a coagulant aid but also as the primary coagulant. In comparison to alum sludges, which are gelatinous and voluminous, sludges produced by using cationic polyelectrolytes are dense and easy to dewater for subsequent treatment and disposal. Anionic and nonionic polyelectrolytes are often used with primary metal coagulants to provide the particle bridging for effective coagulation. In general, the use of polyelectrolyte coagulant aids produces tougher and better settling flocs. Although, in theory, polyelectrolytes should be very effective as coagulants because of their many electrolytic or active sites, their main applications are still as coagulant aids rather than as the primary coagulants.

Activated silica and clays have also been used as coagulant aids. Activated silica is sodium silicate that has been treated with sulfuric acid, aluminum sulfate, carbon dioxide, or chlorine. When the activated silica is applied, a stable negative sol is produced. This sol unites with the positively charged primary-metal coagulant to produce tougher, denser, and faster-settling flocs.

Bentonite clays have been used as coagulant aids in conjunction with iron and alum primary coagulants in treating waters containing high color, low turbidity, and low mineral content. Low-turbidity waters are often hard to coagulate. Bentonite clay serves as a weighting agent that improves the settleability of the resulting flocs.

Jar Test

In practice, irrespective of what coagulant or coagulant aid is used, the optimum dose and pH are determined by a jar test. This consists of four to six beakers (such as 1000 mL in volume) filled with the raw water into which varying doses are administered. Each beaker is provided with a variable-speed stirrer capable of operating from 0 to 100 rpm. Upon introduction of the dose, the contents are rapidly mixed at a speed of about 60 to 80 rpm for a period of 1 min and then allowed to flocculate at a speed of 30 rpm for a period of 15 min. After the stirring is stopped, the nature and settling characteristics of the flocs are observed and recorded qualitatively as *poor, fair, good,* or *excellent.* A hazy sample denotes poor coagulation; a properly coagulated sample is manifested by well-formed flocs that settle rapidly with clear water in-between flocs. The lowest dose of chemicals and pH that produces the desired flocs and clarity represents the optimum. This optimum is then used as the dose in the actual operation of the plant.

Example 6-13

Using copperas as a coagulant, 30 g/m^3 was added to the raw water. (a) Calculate the minimum natural alkalinity required to react with the ferrous sulfate. (b) How much lime of 70% available CaO is required to convert the initial reaction product to $Fe(OH)_2$? (c) What concentration of dissolved O_2 is required to convert the $Fe(OH)_2$ to the ferric form?

Solution

(a) $$FeSO_4 \cdot 7H_2O + Ca(HCO_3)_2 \rightarrow Fe(HCO_3)_2 + CaSO_4 + 7H_2O \qquad (a)$$

$$\text{Eq. wt. of } FeSO_4 \cdot 7H_2O = \frac{FeSO_4 \cdot 7H_2O}{2} = \frac{55.85 + 32 + 16(4) + 7(18)}{2} = 138.93 \ \frac{g}{Eq}$$

$$\text{Eq. wt. of } Ca(HCO_3)_2 = \frac{Ca(HCO_3)_2}{2} = \frac{40 + 2(1.008 + 12 + 48)}{2} = 81.01 \ \frac{g}{Eq}$$

Thus

$$\text{equivalents of } Ca(HCO_3)_2 = \text{equivalents of } FeSO_4 \cdot 7H_2O$$

$$= \frac{30}{138.93} = 0.22 \ \frac{g\text{-}Eq}{m^3} = 0.22 \ \frac{mg\text{-}Eq}{L}$$

The interrelationship between $Ca(HCO_3)_2$ and $CaCO_3$ is shown in the following reaction:

$$CaCO_3 + CO_2 + H_2O \rightarrow Ca(HCO_3)_2$$

Therefore,

$$\text{eq. wt. of } CaCO_3 = \frac{CaCO_3}{2} = 50 \ \frac{mg}{mg\text{-}Eq}$$

Also,

$$\text{eq. wt. of } Ca(HCO_3)_2 = \frac{Ca(HCO_3)_2}{2} \qquad \text{as before}$$

and

$$\text{alkalinity required} = 0.22(50) = 11 \ \frac{mg}{L} \text{ as } CaCO_3 \quad \textbf{Answer}$$

(b) Initial reaction product is $Fe(HCO_3)_2$; hence

$$Fe(HCO_3)_2 + 2Ca(OH)_2 \rightarrow Fe(OH)_2\downarrow + 2CaCO_3\downarrow + 2H_2O \qquad (b)$$

From equation (a), equivalents of $Fe(HCO_3)_2$ for its equivalent weight of $Fe(HCO_3)_2/2 = 0.22$ mg-Eq/L.

$$2CaO + 2H_2O \rightarrow 2Ca(OH)_2$$

$$\text{eq. wt. of CaO} = \frac{2CaO}{2} = 40 + 16 = 56 \frac{mg}{mg\text{-Eq}}$$

Thus

$$\text{amount of 70\% lime needed} = \frac{0.22(56)}{0.7} = 17.6 \text{ mg/L} \quad \textbf{Answer}$$

$$\text{(c)} \quad 4Fe(OH)_2 + O_2 + 2H_2O \rightarrow 4Fe(OH)_3\downarrow \qquad \text{(c)}$$

From equation (b),

$$\text{amount of Fe(OH)}_2 = 0.22 \left[\frac{Fe(OH)_2}{2} \right] = 0.22 \left[\frac{55.85 + 2(17)}{2} \right] = 9.88 \text{ mg/L}$$

From equation (c), number of reference species $= 2(2) = 4$ mol of electrons. Thus

$$\text{eq. wt. of Fe(OH)}_2 = \frac{4Fe(OH)_2}{4} = 89.85 \text{ mg/mg-Eq}$$

$$\text{no. equiv. of Fe(OH)}_2 = \text{no. equiv. of O}_2 = \frac{9.88}{89.85} = 0.11 \text{ mg-Eq/L}$$

$$\text{eq. wt. of O}_2 = \frac{32}{4} = 8$$

Then

$$\text{concentration of O}_2 = 8(0.11) = 0.88 \text{ mg/L, to be maintained.} \quad \textbf{Answer}$$

Example 6-14

50 mg/L of alum is added to 50,000 m^3/day of raw water containing 60 mg/L of suspended solids. (a) Assuming that sufficient natural alkalinity is present, how many kilograms of sludge is produced per day? (b) Assuming that the specific gravity of the sludge is 1.04, how many cubic meters of sludge is produced per day? Assume that the removal efficiency of the settling basin is 65%.

Solution

$$\text{(a)} \quad Al_2(SO_4)_3 \cdot 14.3H_2O + 3Ca(HCO_3)_2 \rightarrow$$

$$2Al(OH)_3\downarrow + 3CaSO_4 + 14.3H_2O + 6CO_2$$

$$\text{Eq. wt. of alum} = \frac{Al_2(SO_4)_3 \cdot 14.3H_2O}{6} = \frac{2(26.98) + 3[32.1 + 4(16)] + 14.3(18)}{6}$$

$$= \frac{599.66}{6} = 99.94 \frac{mg}{mg\text{-Eq}}$$

$$\text{equiv. of alum} = \text{equiv. of Al(OH)}_3 = \frac{50}{99.94} = 0.50 \text{ mg-Eq/L}$$

$$\text{eq. wt. of Al(OH)}_3 = \frac{2Al(OH))_3}{6} = \frac{26.98 + 3(17)}{3} = 25.99 \text{ mg/mg-Eq}$$

$$\text{amount of alum precipitate} = 0.50(25.99) = 13.0 \text{ mg/L}$$

Thus

total amount of precipitate plus suspended solids from raw water

$$= 13.0 + 60 = 73 \text{ mg/L} = 73 \text{ g/m}^3$$

and

$$\text{kg of sludge produced} = 0.65(73)(50,000)(10^{-3}) = 2373 \text{ kg/day} \quad \textbf{Answer}$$

(b) Density, $\rho = 1.04(1000) = 1040 \text{ kg/m}^3$ and

$$\text{cubic meters of sludge} = \frac{2373}{1040} = 2.28 \text{ m}^3/\text{day} \quad \textbf{Answer}$$

RAPID-MIX AND FLOCCULATION

Coagulation is not sufficient to remove colloids. Coagulation must be followed by floc-culation. Also, coagulation will not be as efficient if the chemicals are not dispersed rapidly in the mixing tank. This process of rapidly mixing the coagulant in the vol-ume of the tank is called *rapid-mix*. Figure 6-10 shows a rapid-mix device (a), various impeller types for the rapid-mix (b), and rapid-mix and a flocculator unit together with flocculating paddles (c).

If the reaction is simply allowed to take place in one portion of the tank, then all four mechanisms of coagulation discussed previously will not be utilized. For exam-ple, charge neutralization will not be utilized in all portions of the tank, since by the time the coagulant arrives at the point in question, the reaction of charge neutralization will already have taken place. Hence rapid mixing is used to distribute the chemicals immediately.

The efficiency of a rapid-mix is measured in terms of the velocity gradient G. From fluid mechanics, $G = \partial u / \partial y$ in the x-direction and $\mathcal{P}_f = \Upsilon \omega$, where u is the velocity, y the coordinate, $\mathcal{P}_f$ the power given to fluid, Υ the fluid torque, and ω the fluid angular velocity. From its dimension, G is actually ω. Hence

$$\mathcal{P}_f = \Upsilon G = \Upsilon \frac{\Delta u}{\Delta y} \tag{6-57}$$

The torque Υ is equal to the shear stress τ times the area of shear A_s times the moment arm. Since the x-direction has been selected, the moment arm is Δy. From Newton's law of viscosity, $\tau = \mu(\Delta u / \Delta y)$. After substituting, equation (6-57) becomes

$$\mathcal{P}_f = \mu \frac{\Delta u}{\Delta y} A_s \Delta y \frac{\Delta u}{\Delta y} = \mu \mathcal{V} G^2 \tag{6-58}$$

where $\mathcal{V} = A_s(\Delta y)$, the volume of the fluid element.

Although equation (6-58) has been derived for the fluid element power in the x-dir-ection, it can be used as the element power in all directions and to the entire tank of

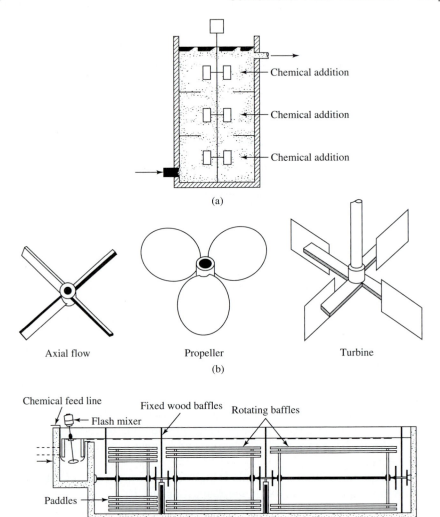

Figure 6-10 Rapid-mix and a flocculator unit: (a) rapid-mix; (b) basic impeller types; (c) rapid-mix, flocculator unit, and paddles.

volume content, V. Since $\mathcal{P}_f$ is fluid power, the input power to the mixer should be obtained by dividing $\mathcal{P}_f$ by the blade efficiency η and the motor efficiency $\mathcal{M}$. Solving for the velocity gradient,

$$G = \sqrt{\frac{\mathcal{P}_f}{\mu V}} \qquad (6\text{-}59)$$

The following G values with the corresponding rapid-mix detention time t_0 have been found to work in practice:

t_0 (s)	G (s^{-1})
0.5	4000
10–20	1500
20–30	950
30–40	850
40–130	750

For flocculators, equation (6-59) needs to be expressed in terms of flocculator parameters. Calling f_d the drag on the paddles having relative velocity with respect to the water of v_p and area perpendicular to the direction of motion of A_p, $\mathcal{P}_f$ may be expressed as

$$\mathcal{P}_f = f_d v_p = C_D A_p \rho_w \frac{v_p^2}{2} v_p = C_D A_p \rho_w \frac{v_p^3}{2} \tag{6-60}$$

Express v_p in terms of the absolute paddle velocity as $v_p = a v_{pa}$, where a is the fractional part of the absolute value of the paddle velocity v_{pa} that makes v_p. The value of a is about 0.75. In terms of the absolute paddle velocity, equation (6-60) becomes

$$\mathcal{P}_f = C_D A_p \rho_w \frac{a^3 v_{pa}^3}{2} \tag{6-61}$$

With this expression for $\mathcal{P}_f$, the equation of G becomes

$$G = \sqrt{\frac{C_D A_p \rho_w a^3 v_{pa}^3}{2 \Psi \mu}} \tag{6-62}$$

Higher values of G produce smaller flocs, while low values of G produce larger flocs. Although larger flocs are desirable, there is a time limit to which they are allowed to form. If they are allowed to form much longer than some ideal time, they will reach a critical size that due to shearing forces will simply break and crush to pieces. Since floc sizes obtained are a function of G and time, the parameter normally used as an index of flocculation is the product $G t_0$ along with G. G values are normally made larger at the head end of the flocculating tank and made smaller at the exit end to allow for larger flocs. The following $G t_0$ and G values have been found to work in practice:

Type of raw water	G (s^{-1})	$G t_0$ (dimensionless)
Low-turbidity, color removal	20–70	50,000–250,000
High-turbidity, solids removal	70–150	80,000–190,000

The literature recommends a value of 1.8 for C_D applicable to flat blades without, however, mentioning the corresponding value of the Reynolds number. For a Reynolds number of 10^5, the following equation for flat blades may be used:[5]

$$C_D = 0.008\frac{b}{D} + 1.3 \qquad (6\text{-}63)$$

where b is the length of the blade and D is the width. The Reynolds number is defined as $Re = av_{pa}D/v$, where v is the kinematic viscosity. The C_D predicted by equation (6-63) applies only to a single blade attached to radial arms. Work needs to be done to determine the value of C_D for multiple blades.

To prevent excessive velocity gradients between paddle tips, a minimum distance of 0.3 m should be provided between them. Also, a minimum clearance of 0.3 m should be provided between paddles and any structure inside the flocculator. As noted before, excessive velocity gradients can simply break the flocs to pieces. Paddle blade velocity should be less than 1.0 m/s.[6]

Example 6-15

A rapid-mix unit is to be designed for a 50,000-m³/day raw water. Using a detention time of 30 s, determine the size of the tank and the motor horsepower of the mixer, assuming a shaft and a motor efficiency of 75% and 90%, respectively. Assume a water temperature of 20°C.

Solution

$$\text{Volume, } V = \frac{50,000}{24(60)(60)}(30) = 17.36 \text{ m}^3 \quad \textbf{Answer}$$

Note: In an actual design, a second unit should be provided so that it can be put in service while the first unit is scheduled for repairs or has simply broken down.

$$\mathcal{P}_f = \mu V G^2 \qquad G = 800 \text{ s}^{-1} \qquad \mu = 10(10^{-4}) \text{ kg/m-s}$$

Thus

$$\mathcal{P}_i = \frac{\mathcal{P}_f}{\eta \mathcal{M}} = \frac{10(10^{-4})(17.36)(800^2)}{0.75(0.9)} = 16,459.85 \ \frac{\text{N-m}}{\text{s}} = 16,459.85 \text{ W}$$

As 1 hp = 745.7 W,

$$\mathcal{P}_i = \frac{16,459.85}{745.7} = 22.07 \text{ hp} \quad \textbf{Answer}$$

Example 6-16

A flocculator tank has three compartments, with each compartment having one paddle wheel. The flocculator is similar to the one shown in Figure 6-10c except for the number of paddle blades per paddle wheel. The ratio of the length of the paddle blades of the largest compartment to that of the length of the paddle blades of the middle compartment is 2.6:2, and

[5]B. R. Munson, D. F. Young, and T. H. Okiishi (1994). *Fundamentals of Fluid Mechanics*. Wiley, New York.

[6]H. S. Peavy, D. R. Rowe, and G. Tchobanoglous (1985). *Environmental Engineering*. McGraw-Hill, New York. Example problem pp. 147–151.

the ratio of the length of the paddle blades of the middle compartment to that of the length of the paddle blades of the smallest compartment is 2:1. The flocculator is to flocculate an alum-treated raw water of 50,000 m³/day at an average temperature of 20°C. Design for (a) the dimensions of the flocculator and flocculator compartments, (b) the power requirements assuming a $\mathcal{M}$ value of 90% and a η value of 75%, (c) the appropriate dimensions of the paddle slats, and (d) the rpm of the paddle wheels. Assume two flocculators in parallel and four paddle blades per paddle wheel.

Solution   (a) Assume an average G of 30 s^{-1} and a Gt_0 of 80,000. Then $t_0 = 44.44$ min.

$$\text{Volume of flocculator} = \left(\frac{50,000}{2}\right)(44.44)\left[\frac{1}{60(24)}\right] = 771.53 \text{ m}^3$$

To produce uniform velocity gradient, depth must be equal to width. Hence, assuming a depth of 5 m and letting L represent the length of tank, we obtain

$$5(5)L = 771.53 \qquad L = 30.86 \text{ m} \quad \textbf{Answer}$$

Thus

$$\text{length of compartment } 1 = \frac{1}{1+2+2.6}(30.86) = \frac{1}{5.6}(30.86) = 5.51 \text{ m} \quad \textbf{Answer}$$

width and depth of compartment 1 = width and depth of compartment 2

$$= \text{width and depth of compartment } 3 = 5 \text{ m, as assumed} \quad \textbf{Answer}$$

$$\text{length of compartment } 2 = \frac{2}{5.6}(30.86) = 11.02 \text{ m} \quad \textbf{Answer}$$

$$\text{length of compartment } 3 = 30.86 - 5.51 - 11.02 = 14.33 \text{ m} \quad \textbf{Answer}$$

$$\text{width of flocculator} = \text{depth of flocculator} = 5 \text{ m, as assumed} \quad \textbf{Answer}$$

(b) $\mathcal{P}_f = \mu \mathcal{V} G^2$. Assume the following distribution of G:

Compartment 1: $G = 40$ s^{-1}
Compartment 2: $G = 30$ s^{-1}
Compartment 3: $G = 20$ s^{-1}

$$\mu = 10(10^{-14}) \text{ kg/m-s} \qquad \mathcal{V}_1 = (5)(5)(5.51) = 137.75 \text{ m}^3$$

Thus

$$\mathcal{P}_i \text{ for compartment } 1 = \frac{10(10^{-4})(137.75)(40^2)}{0.75(0.9)} = 326.52 \text{ N-m/s} = 0.326 \text{ kW} \quad \textbf{Answer}$$

$$\mathcal{V}_2 = 25(11.02) = 275.50 \text{ m}^3$$

Thus

$$\mathcal{P}_i \text{ for compartment } 2 = \frac{10(10^{-4})(275.50)(30^2)}{0.75(0.90)} = 0.327 \text{ kW} \quad \textbf{Answer}$$

$$\mathcal{V}_3 = 25(14.33) = 358.25 \text{ m}^3$$

Thus

$$P_i \text{ for compartment 3} = \frac{10(10^{-4})(358.25)(20^2)}{0.75(0.90)} = 0.212 \text{ kW} \quad \textbf{Answer}$$

(c) $$P_f = C_D A_p \rho_w \frac{a^3 v_{pa}^3}{2} \qquad C_D = 0.008\frac{b}{D} + 1.3$$

Thus

$$P_f = \left(0.008\frac{b}{D} + 1.3\right)(nbD)(997)\frac{0.75^3 v_{pa}^3}{2} = \left(0.008\frac{b}{D} + 1.3\right)(nbD)(210.3)v_{pa}^3$$

where n is the number of paddle blades per paddle wheel.

$$\text{Re} = 10^5 = \frac{v_{pa}D}{\nu} \qquad v_{pa} = \frac{10^5(\nu)}{D}$$

$$P_f = \left(0.008\frac{b}{D} + 1.3\right)(nbD)(210.3)\left[\frac{10^5(\nu)}{D}\right]^3$$

$$= \left(0.008\frac{b}{D} + 1.3\right)(nbD)(210.3)\left[10^5(10^{-6})/D\right]^3 = \left(0.008\frac{b}{D} + 1.3\right)\left(\frac{nb}{D^2}\right)(0.21)$$

For compartment 1:

$$b = 5.51 - 2(0.3) = 4.91 \text{ m}$$

$$P_f = 326(0.75)(0.9) = 220.05 \text{ N-m/s}$$

Thus

$$220.05 = \left(0.008\frac{4.91}{D} + 1.3\right)\left[\frac{4(4.91)}{D^2}\right](0.21) = \left(0.008\frac{4.91}{D} + 1.3\right)\left(\frac{4.12}{D^2}\right)$$

$$= \frac{0.16}{D^3} + \frac{5.36}{D^2}$$

Let $0.16/D^3 + 5.36/D^2 = Y$.

D	Y
0.22	125.77
0.15	285.63

0.22	125.77	$\dfrac{y - 0.22}{0.15 - 0.22} = \dfrac{220.05 - 125.77}{285.63 - 125.77}$
y	220.05	
0.15	285.63	$y = 0.18$

Thus $D = 0.18$ m   **Answer**

For compartment 2:

$$b = 11.02 - 0.6 = 10.42 \text{ m}$$

$$\mathcal{P}_f = 327(0.75)(0.9) = 220.73 \text{ N-m/s}$$

$$220.73 = \left(0.008\frac{10.42}{D} + 1.3\right)\left[\frac{4(10.42)}{D^2}\right](0.21) = \left(0.008\frac{10.42}{D} + 1.3\right)\left(\frac{8.75}{D^2}\right)$$

$$= \frac{0.73}{D^3} + \frac{11.38}{D^2}$$

Let $0.73/D^3 + 11.38/D^2 = Y$.

D	Y
0.25	228.80
0.26	209.88

0.25	228.80	$\dfrac{y - 0.25}{0.26 - 0.25} =$	$\dfrac{220.73 - 228.80}{209.88 - 228.80}$
y	220.73		
0.26	209.88	$y = 0.25$	

Thus $D = 0.25$ m   **Answer**
    For compartment 3:

$$b = 14.33 - 0.6 = 13.73 \text{ m}$$

$$\mathcal{P}_f = 212(0.75)(0.9) = 143.1 \text{ N-m/s}$$

Thus

$$143.1 = \left(0.008\frac{13.73}{D} + 1.3\right)\left[\frac{4(13.73)}{D^2}\right](0.21) = \left(0.008\frac{13.73}{D} + 1.3\right)\left(\frac{11.53}{D^2}\right)$$

$$= \frac{1.27}{D^3} + \frac{14.99}{D^2}$$

Let $1.27/D^3 + 14.99/D^2 = Y$.

D	Y
0.30	213.59
0.36	142.88

0.30	213.59	$\dfrac{y - 0.30}{0.36 - 0.30} =$	$\dfrac{143.1 - 213.59}{142.88 - 213.59}$
y	143.1		
0.36	142.88	$y = 0.36$	

Thus

$$D = 0.36 \text{ m} \quad \textbf{Answer}$$

(d) For compartment 1:

$$v_{pa} = r\omega \qquad v_{pa} = \frac{10^5(v)}{D} = \frac{10^{-1}}{D} = \frac{10^{-1}}{0.18} = 0.56 \text{ m/s}$$

$$r = \frac{5 - 0.6}{2} = 2.2 \text{ m}$$

Thus

$$\omega = \frac{0.56}{2.2} = 0.25 \text{ rad/s} = 15.17 \text{ rads/min} = 2.43 \text{ rpm} \quad \textbf{Answer}$$

For compartment 2:

$$v_{pa} = \frac{10^{-1}}{0.25} = 0.40 \text{ m/s}$$

Thus

$$\omega = 1.74 \text{ rpm} \quad \textbf{Answer}$$

For compartment 3:

$$v_{pa} = \frac{10^{-1}}{0.36} = 0.28 \text{ m/s}$$

Thus

$$\omega = 1.22 \text{ rpm} \quad \textbf{Answer}$$

Example 6-17

Repeat Example 6-16 assuming a C_D value of 1.8. Assume a v_{pa} value of less than 1.0 m/s.

Solution

(a) The answers are the same as those in part (a) of Example 6-16.
(b) The answers are the same as those in part (b) of Example 6-16.

(c) $$\mathcal{P}_f = C_D A_p \rho_w \frac{a^3 v_{pa}^3}{2} \qquad C_D = 1.8$$

$$= (1.8)(4bD)(997)\frac{0.75^3 v_{pa}^3}{2} = 1514.19 b D v_{pa}^3$$

For compartment 1:

$$\mathcal{P}_f = 326(0.75)(0.9) = 220.05 \text{ N-m/s} \qquad b = 4.91 \text{ m}$$

Thus

$$220.05 = 1514.19 b D v_{pa}^3 = 7434.67 D v_{pa}^3$$

Assume that $v_{pa} = 0.9$ m/s; then

$$220.05 = 5419.87 D \qquad D = 0.04 \text{ m, too narrow!}$$

Assume that $v_{pa} = 0.5$ m/s; then

$$220.05 = 929.33 D \qquad D = 0.24 \text{ m} \quad \textbf{Answer}$$

For compartment 2:

$$P_f = 327(0.75)(0.9) = 220.73 \text{ N-m/s} \qquad b = 10.42 \text{ m}$$

$$220.73 = 1514.19 b D v_{pa}^3 = 15,777.86 D v_{pa}^3$$

Assume that $v_{pa} = 0.5$ m/s; then

$$220.73 = 1972.23 D \qquad D = 0.11 \text{ m} \quad \textbf{Answer}$$

For compartment 3:

$$P_f = 212(0.75)(0.9) = 143.1 \text{ N-m/s} \qquad b = 13.73 \text{ m}$$

$$143.1 = 1514.19 b D v_{pa}^3 = 20,789.83 D v_{pa}^3$$

Assume that $v_{pa} = 0.3$ m/s; then

$$143.1 = 561.33 D \qquad D = 0.25 \text{ m} \quad \textbf{Answer}$$

(d) For compartment 1:

$$v_{pa} = r\omega$$

$$r = \frac{5 - 0.6}{2} = 2.2 \text{ m}$$

Thus

$$\omega = \frac{0.50}{2.2} = 0.23 \text{ rad/s} = 13.64 \text{ rads/min} = 2.2 \text{ rpm} \quad \textbf{Answer}$$

For compartment 2:

$$\omega = 2.0 \text{ rpm} \quad \textbf{Answer}$$

For compartment 3:

$$\omega = \frac{0.30}{2.2} = 0.136 \text{ rad/s} = 8.18 \text{ rads/min} = 1.3 \text{ rpm} \quad \textbf{Answer}$$

WATER SOFTENING

Softening is the term given to the process of removing the hardness ions of water. This hardness is due to the presence of multivalent cations, mostly calcium and magnesium. Others that may be present but not in significant amounts are iron (Fe^{2+}), manganese (Mn^{2+}), strontium (Sr^{2+}), and aluminum (Al^{3+}).

In the process of cleansing using soap, lather is formed, causing the surface tension of water to decrease. This decrease in surface tension makes water molecules partially lose their mutual attraction toward each other, allowing them to wet "foreign" solids, instead thereby suspending the solids in water. As the water is rinsed out, the solids are removed from the soiled material.

In the presence of hardness ions, soap does not form lather immediately but reacts with the ions, thereby preventing the formation of lather and forming scum. Lather will form only when all the hardness ions are consumed. This means that hard waters are hard to lather. *Hard waters* may then be defined as those in which it is difficult to form lathers when reacted with soap.

Types of Hardness

There are two basic types of hardness: carbonate and noncarbonate hardness. When the hardness ions are associated with the CO_3^{2-} and HCO_3^- ions in water, the hardness is called *carbonate hardness*; otherwise, it is called *noncarbonate hardness*. In natural waters, carbonate hardness is usually associated with the natural HCO_3^- ions. An example of carbonate hardness is $Ca(HCO_3)_2$, and an example of noncarbonate hardness is $CaCl_2$.

In practice, when one addresses hardness removal, removal of the calcium and magnesium ions is usually meant. Our discussion of hardness removal will therefore conveniently be divided into calcium and magnesium hardness removal. Corresponding chemical reactions will be developed. If other specific hardness ions are also present and to be removed, such as iron, manganese, strontium, and aluminum, corresponding specific reactions would have to be developed for them.

Hardness removal is also called softening, since when hardness is removed, the water becomes soft. In general, there are three ways to soften water: chemical precipitation, ion exchange, and reverse osmosis. Only the chemical precipitation method is discussed in this chapter.

Softening of Carbonate Hardness of Calcium

Calcium carbonate hardness is present in the form of calcium bicarbonate. The calcium ion is removed by precipitating it in the form of calcium carbonate. From Chapter 2, the K_{sp} value of calcium carbonate is $5(10^{-9})$ at 25°C. This low value of K_{sp} means that calcium hardness can conveniently be removed. There is, however, always the solubility consideration for calcium carbonate. At 0°C, its solubility is 15 mg/L and, at 25°C its solubility is 14 mg/L. This solubility implies that no matter how much precipitant is applied to the water, there will always remain some 14 to 15 mg/L of $CaCO_3$.

There are three ways that carbonate hardness of calcium may be removed chemically: using lime (CaO), caustic soda (NaOH), and soda ash (Na_2CO_3). The corresponding reactions are depicted below. The downward-pointing arrows signify that $CaCO_3$ are precipitated.

$$Ca(HCO_3)_2 + CaO + H_2O \rightarrow 2CaCO_3\downarrow + 2H_2O \tag{6-64}$$

$$Ca(HCO_3)_2 + 2NaOH \rightarrow CaCO_3\downarrow + Na_2CO_3 + 2H_2O \tag{6-65}$$

$$Ca(HCO_3)_2 + Na_2CO_3 \rightarrow CaCO_3\downarrow + 2NaHCO_3 \tag{6-66}$$

The number of reference species in these reactions is 2 mol of positive or negative charges. Hence the equivalent weight of any of the participating species will be obtained by dividing the term of the affected species by 2.

Softening of Noncarbonate Hardness of Calcium

When calcium is paired with noncarbonate anion species such as Cl^-, SO_4^{2-}, and NO_3^-, the type of hardness is noncarbonate calcium hardness. This form of hardness can conveniently be removed by using soda ash (Na_2CO_3). The reaction with $CaCl_2$ is

$$CaCl_2 + Na_2CO_3 \rightarrow CaCO_3\downarrow + 2NaCl \qquad (6\text{-}67)$$

The number of reference species of all these reactions is 2 mol of positive or negative charges.

Softening of Carbonate Hardness of Magnesium

The carbonate hardness of magnesium is in the form of magnesium bicarbonate. From Chapter 2, the K_{sp} value of $Mg(OH)_2$ is a low value of $9(10^{-12})$. Hence the hardness is removed in the form of $Mg(OH)_2$. The solubility of $Mg(OH)_2$ at 0°C is 17 mg/L as $CaCO_3$; its solubility at 18°C is 15.5 mg/L as $CaCO_3$. As in the case of $CaCO_3$, these amounts will always remain in the treated effluent no matter how much precipitant is employed.

To remove the carbonate hardness of magnesium by chemical means, lime is used followed by caustic soda. The chemical reactions are

$$Mg(HCO_3)_2 + 2CaO + 2H_2O \rightarrow Mg(OH)_2\downarrow + 2CaCO_3\downarrow + 2H_2O \qquad (6\text{-}68)$$

$$Mg(HCO_3)_2 + 4NaOH \rightarrow Mg(OH)_2\downarrow + 2Na_2CO_3 + 2H_2O \qquad (6\text{-}69)$$

The number of reference species in all these reactions is 2 mol of positive or negative charges.

Softening of Noncarbonate Hardness of Magnesium

As in calcium, when the magnesium ion is paired with anions such as Cl^-, SO_4^{2-}, and NO_3^-, the type of magnesium hardness is called *noncarbonate*. The hardness can simply be removed by the use of caustic soda or a combination of lime and soda ash. The use of caustic soda will precipitate Mg^{2+} directly as $Mg(OH)_2$, the desired product. Lime will also precipitate Mg^{2+} as the hydroxide but will, in addition, also produce calcium noncarbonate hardness. This is the reason for the addition of soda ash to precipitate the resulting calcium ion which was added with the lime. The process of adding both lime and soda ash at the same time is called the *lime-soda ash process*. Representative chemical reactions are

$$MgSO_4 + 2NaOH \rightarrow Mg(OH)_2\downarrow + Na_2SO_4 \qquad (6\text{-}70)$$

$$MgSO_4 + CaO + H_2O \rightarrow Mg(OH)_2\downarrow + CaSO_4 \qquad (6\text{-}71)$$

$$MgCl_2 + CaO + H_2O \rightarrow Mg(OH)_2\downarrow + CaCl_2 \qquad (6\text{-}72)$$

The number of reference species in the reactions above is 2 mol of positive or negative charges. Hence the equivalent weight of all participating species should be divided by 2.

Order of Removal

Suppose that there are 4 gram-equivalents of Ca^{2+}, 1 gram-equivalent of Mg^{2+}, and 2.5 gram-equivalents of HCO_3^-. Suppose further that it has been decided to use lime. The question is: Which reaction takes precedence, equation (6-64) or equation (6-68)? Assuming that the former takes precedence, the amount of lime needed just to remove the bicarbonate portion of the hardness is $2.5(CaO/2) = 70$ g. Also, assuming that the latter takes precedence followed by the former, the corresponding amount of lime needed is $1.0(2CaO/2) + 1.5(CaO/2) = 98$ g. Since these two simple calculations produce two very different results, knowledge of the order of precedence of the reactions is very important. The order of precedence can be judged from the solubility of the precipitates: namely, $CaCO_3$ and $Mg(OH)_2$. Since $Mg(OH)_2$ is more insoluble than $CaCO_3$, the reaction producing this precipitate should take precedence; hence equation (6-68) predominates over equation (6-64). In the example above, the answer would be 98 g rather than 70 g.

Role of Natural CO_2 in Removal

In the removal of calcium and magnesium hardness, only three chemicals were used, either singly or in combination: lime, caustic soda, and soda ash. However, since the type of water treated comes from the ambient, CO_2 would be found dissolved. Therefore, in addition to the chemical reactions portrayed above, the precipitant chemicals must also react with carbon dioxide. This means that more chemicals than indicated in previous reactions would be needed as follows:

$$CO_2 + CaO + H_2O \rightarrow CaCO_3\downarrow + H_2O \tag{6-73}$$

$$CO_2 + 2NaOH \rightarrow Na_2CO_3 + H_2O \tag{6-74}$$

$$CO_2 + Na_2CO_3 + H_2O \rightarrow 2NaHCO_3 \tag{6-75}$$

Excess Treatment and Stabilization

The optimum pH for precipitation of $CaCO_3$ is 10.3. This can normally be accomplished by providing the stoichiometric amounts portrayed by the chemical reaction. The precipitation of $Mg(OH)_2$, however, requires raising the pH to 11; this, in turn, requires adding more lime if lime is the one used or more caustic if caustic is the one used. The excess chemical required for either lime or caustic is usually 1.25 mEq/L more than the stoichiometric amount.

The addition of excess lime causes the finished water to be scale forming. Also, because of inefficiency due to physical limitations of mixing and contact, the practical limit of softening can only bring down the concentration of hardness to 30 to 40 mg/L of $CaCO_3$ for the calcium hardness. Also, from its solubility, $Mg(OH)_2$ can only be removed down to 17 to 15.5 mg/L as $CaCO_3$.

As long as the concentrations of $CaCO_3$ and $Mg(OH)_2$ exceed their solubilities, continued precipitation can cause scale to form in distribution pipes. To prevent scale formation, they must be stabilized. Stabilization may be accomplished using one of several acids or using CO_2. When CO_2 is used, the process is called *recarbonation*. Illustrative chemical reactions are

$$CaCO_3 + CO_2 + H_2O \rightarrow Ca(HCO_3)_2 \qquad (6\text{-}76)$$

$$Mg(OH)_2 + 2CO_2 + 2H_2O \rightarrow Mg(HCO_3)_2 + 2H_2O \qquad (6\text{-}77)$$

$$2CaCO_3 + H_2SO_4 \rightarrow Ca(HCO_3)_2 + CaSO_4 \qquad (6\text{-}78)$$

If the pH has been raised to facilitate the removal of magnesium, the carbon dioxide and the acid used must first be expended to react with the hydroxide alkalinity before the foregoing reactions can take place.

A very soft water has a slimy feel. For example, rainwater, which is exceedingly soft, is slimy when used with soap. For this reason, hardness in water used for domestic purposes is not removed completely but to the level of 75 to 120 mg/L as $CaCO_3$.

Notes on Equivalent Weights

If all the softening reactions were reviewed, it would be observed that the equivalent weights of the hardness species are obtained by dividing the molecular weights by the respective total number of valences of the positive or negative ions, irrespective of the coefficients in the reaction. For example, in equation (6-70), the equivalent weight of $MgSO_4$ is obtained by dividing the molecular weight $MgSO_4$ by 2, the valence of Mg^{2+} or SO_4^{2-}. Also, in equation (6-68), the equivalent weight of $Mg(HCO_3)_2$ is obtained by dividing the molecular weight $Mg(HCO_3)_2$ by 2, the valence of Mg^{2+} or the total number of valences of HCO_3^-. These observations can be extended to all other species associated with the hardness in solution. *The equivalent weight of the hardness or associated species in solution is equal to the molecular weight divided by the total number of valences of the positive or negative ions of the species, irrespective of the coefficient.*

The findings above, however, cannot be generalized to the precipitant species or species other than the hardness and its associated species. For example, in equation (6-69), the valence of Na^+ and OH^- is 1. If the finding above is to be applied, the equivalent weight of NaOH is NaOH/1. This is not, however, correct. From the reaction the equivalent weight of NaOH is 4NaOH/2 or 2NaOH. Also, if the finding above is to be applied, the equivalent weight of CaO in equation (6-68) is CaO/2. However, the correct equivalent weight is 2CaO/2 or CaO. To conclude, *the equivalent weight of precipitant and species other than the hardness and its associated species cannot be generalized as molecular weight divided by the total number of valences of the species but must be deduced from the chemical reaction.* In fact, this statement holds for all other species except the hardness and associated species in the original solution. For emphasis, we make the following summary: *For the hardness ions and associated species, the equivalent weight is obtained by dividing the molecular weight by the total number of valences of the positive or negative charges of the species, irrespective of the coefficient; for all other species, the equivalent weight must be deduced from the balanced chemical reaction.*

Example 6-18

A raw water to be treated by the lime-soda process to the minimum hardness possible has the following characteristics: $CO_2 = 22.0$ mg/L, $Ca^{2+} = 80$ mg/L, $Mg^{2+} = 12.0$ mg/L, $Na^+ = 46.0$ mg/L, $HCO_3^- = 152.5$ mg/L, and $SO_4^{2-} = 216.0$ mg/L. (a) Check if the number of equivalents of positive and negative ions are balanced. (b) For a flow of 25,000 m^3/day, calculate the chemical requirements and the mass of solids and volume of sludge produced. Assume that the lime used is 90% pure and the soda ash used is 85% pure. Also, assume that the specific gravity of the sludge is 1.04.

Solution (a) Cations: $Ca^{2+} = 80/(40/2) = 4.0$ mEq/L; $Mg^{2+} = 12/(24/2) = 1.0$ mEq/L; $Na^+ = 46/23 = 2.0$ mEq/L. So total cations $= 4.0 + 1.0 + 2.0 = 7.0$ mEq/L.

$$\text{Anions: } HCO_3^- = \frac{152.5}{1.008 + 12 + 16(3)} = 2.5 \text{ mEq/L}$$

$$SO_4^{2-} = \frac{216.0}{[32 + 4(16)]/2} = 4.5 \text{ mEq/L}$$

Therefore, total anions $= 2.5 + 4.5 = 7.0$ mEq/L, so

the cations and anions are balanced. Answer

(b) *Chemical requirements*: By order of removal, molecules present are

$$Mg(HCO_3)_2 = 1.0 \Rightarrow HCO_3^- \text{ remaining} = 2.5 - 1 = 1.5$$

$$Ca(HCO_3)_2 = 1.5 \Rightarrow Ca^{2+} \text{ remaining} = 4.0 - 1.5 = 2.5$$

$$CaSO_4 = 2.5 \Rightarrow SO_4^{2-} \text{ remaining} = 4.5 - 2.5 = 2.0$$

$$Na_2SO_4 = 2.0$$

For CO_2:

$$CO_2 + CaO + H_2O \rightarrow CaCO_3\downarrow + H_2O$$

$$\text{pure lime required} = \frac{22}{CO_2/2}\left(\frac{CaO}{2}\right) = 28 \text{ mg/L}$$

For $Mg(HCO_3)_2$:

$$Mg(HCO_3)_2 + 2CaO + 2H_2O \rightarrow Mg(OH)_2\downarrow + 2CaCO_3\downarrow + 2H_2O$$

$$\text{pure lime required for removal} = (1 + 1.25)(40 + 16) = 126 \text{ mg/L}$$

For $Ca(HCO_3)_2$:

$$Ca(HCO_3)_2 + CaO + H_2O \rightarrow 2CaCO_3\downarrow + 2H_2O$$

$$\text{pure lime required for removal} = \frac{1.5(40 + 16)}{2} = 42 \text{ mg/L}$$

For $CaSO_4$:

$$CaSO_4 + Na_2CO_3 \rightarrow CaCO_3\downarrow + Na_2SO_4$$

$$\text{pure soda ash required for removal} = \frac{2.5[2(23) + 12 + 16(3)]}{2} = 132.5 \text{ mg/L}$$

$$\text{total lime requirement} = \frac{28 + 126 + 42}{0.9(1000)}(25,000) = 5444 \text{ kg/day} \quad \textbf{Answer}$$

$$\text{total soda ash requirement} = \frac{132.5}{0.85(1000)}(25,000) = 3897 \text{ kg/day} \quad \textbf{Answer}$$

Calculate sludge produced:

$$CO_2 + CaO + H_2O \rightarrow CaCO_3\downarrow + H_2O$$

$$\text{solids produced} = \frac{CaCO_3}{CaO}(28) = 50 \text{ mg/L}$$

$$Mg(HCO_3)_2 + 2CaO + 2H_2O \rightarrow Mg(OH)_2\downarrow + 2CaCO_3\downarrow + 2H_2O$$

$$\text{solids produced} = \left[\frac{Mg(OH)_2}{2CaO} + \frac{CaCO_3}{CaO}\right](126) = 290 \text{ mg/L}$$

$$Ca(HCO_3)_2 + CaO + H_2O \rightarrow 2CaCO_3\downarrow + 2H_2O$$

$$\text{solids produced} = \left[\frac{2CaCO_3}{CaO}\right](42) = 150 \text{ mg/L}$$

$$CaSO_4 + Na_2CO_3 \rightarrow CaCO_3\downarrow + Na_2SO_4$$

$$\text{solids produced} = \left[\frac{CaCO_3}{Na_2CO_3}\right](132.5) = 125 \text{ mg/L}$$

$$\text{total solids deposited} = [(50 + 290 + 150 + 125) - 35]\frac{25,000}{1000} = 14,500 \text{ kg/day}$$

35 mg/L is allowance for limit of technology.   **Answer**

$$\text{Assuming 2\% solids, volume of sludge produced} = \frac{14,500}{0.02(1.04)(1000)}$$

$$= 697 \text{ m}^3/\text{day.} \quad \textbf{Answer}$$

Example 6-19

In Example 6-18, (a) determine the amount of CO_2 needed for recarbonation; (b) determine the additional mass of solids produced after recarbonation; (c) determine the ionic composition of the finished water after the recarbonation; (d) show that the cations and anions are balanced.

Solution   (a) The amount of CO_2 needed for recarbonation is equal to the amount needed to transform $CaCO_3$ (= 35 mgL) to $Ca(HCO_3)_2$, the amount needed to neutralize the excess lime of 1.25 mEq/L and the amount needed to transform $Mg(OH)_2$ (= 16 mg/L) to $Mg(HCO_3)_2$.

$$CaCO_3 + CO_2 + H_2O \rightarrow Ca(HCO_3)_2$$

$$CO_2 \text{ needed} = \frac{35}{CaCO_3/2}\left(\frac{CO_2}{2}\right)(25,000) = 385,000 \text{ g/day} = 385 \text{ kg/day}$$

$$Mg(HCO_3)_2 + 2CaO + 2H_2O \rightarrow Mg(OH)_2\downarrow + 2CaCO_3\downarrow + 2H_2O$$

$$1.25 \text{ mEq/L CaO} = 1.25(56) = 70 \text{ mg/L}$$

$$CaO + CO_2 + H_2O \rightarrow CaCO_3\downarrow + H_2O$$

$$CO_2 \text{ needed } \frac{CO_2}{CaO}(70)\left(\frac{25,000}{1000}\right) = 1375 \text{ kg/day}$$

$$Mg(OH)_2 + 2CO_2 + 2H_2O \rightarrow Mg(HCO_3)_2 + 2H_2O$$

$$CO_2 \text{ needed } \frac{2CO_2}{Mg(OH)_2}(16)\left(\frac{25,000}{1000}\right) = 607 \text{ kg/day}$$

Thus

$$\text{total } CO_2 \text{ needed} = 385 + 1375 + 607 = 2367 \text{ kg/day} \quad \textbf{Answer}$$

(b) Additional mass of solids $= 0$.   **Answer**

(c) The finished water contains the cations Ca^{2+}, Mg^{2+}, and Na^+ and the anions HCO_3^-, and SO_4^{2-}, in addition to H^+ and OH^-, which are negligible in comparison.

Ca: $Ca(HCO_3)_2 \rightarrow Ca^{2+} + 2HCO_3^-$

$$[Ca(HCO_3)_2] = \frac{Ca(HCO_3)_2}{CaCO_3} = 56.7 \text{ mg/L} = \frac{56.7}{Ca(HCO_3)_2/2} = 0.7 \text{ mEq/L}$$

$$[Ca^{2+}] = 0.7 \text{ mEq/L} = [HCO_3]$$

Mg: $Mg(HCO_3)_2 + 2CaO + 2H_2O \rightarrow Mg(OH)_2\downarrow + 2CaCO_3\downarrow + 2H_2O$

$$Mg(OH)_2 + 2CO_2 + 2H_2O \rightarrow Mg(HCO_3)_2 + 2H_2O$$

$$Mg(HCO_3)_2 \rightarrow Mg^{2+} + 2HCO_3^-$$

$$[Mg(HCO_3)_2] = \frac{Mg(OH)_2}{2CaCO_3}(16)\left(\frac{Mg(HCO_3)_2}{Mg(OH)_2}\right) = \frac{146}{2(100)}(16) = 11.68 \text{ mg/L}$$

$$= \frac{11.68}{Mg(HCO_3)_2/2} = 0.16 \text{ mEq/L}$$

$$[Mg^{2+}] = 0.16 \text{ mEq/L} = [HCO_3]$$

Na: from added soda ash $= 2.5$ mEq/L

Cations: $Ca^{2+} = 0.7$

$Mg^{2+} = 0.16$

$Na^+ = 2.5 + 2.0 = 4.5$

Total $= 5.36$

Anions: $HCO_3 = 0.7 + 0.16 = 0.86$

$SO_4^{2-} = 4.5$

Total $= 5.36$ **Answer**

(d) From part (c), the ions are balanced.   **Answer**

The following are typical design criteria for softening systems:

Parameter	Mixer	Flocculator	Settling basin	Solids-contact basin
Detention time	5 min	30–60 min	2–4 h	2–4 h
Velocity gradient (s^{-1})	800	10–90	—	—
Flow-through velocity (fps)	—	0.15–0.5	0.15–0.5	—
Overflow rate (gpm/ft^2)	—	—	0.80–1.8	4.3

Split Treatment

Water with a high concentration of magnesium is often softened by a process called *split treatment*. Water softening may be done in either a single- or a two-stage treatment, or, theoretically, in more stages. In a single-stage treatment, all the chemicals are added in just one basin, whereas in a two-stage treatment, chemicals are added in two stages. In split treatment, the operation is in two stages. Part of the raw water is bypassed from the first stage (split). Excess lime to facilitate the precipitation of magnesium hydroxide to the limit of technology is added in the first stage, but instead of neutralizing this excess, it is used in the second stage to react with the calcium hardness of the bypassed flow that is introduced into the second stage.

Let Q be the rate of flow of water treated, Mgr be the concentration of magnesium in the raw water, and Mgf be the concentration of magnesium in the finished water. The amount of magnesium removed is then equal to $QMgr - QMgf$. If q_b is the bypassed flow, the amount of magnesium introduced to the second stage from the bypassed flow is $q_b Mgr$. The flow coming out of the first stage and flowing into the second stage is $Q - q_b$; the corresponding amount of magnesium introduced into the second stage is $(Q - q_b)Mg1$, where $Mg1$ is the concentration of magnesium in the effluent of the first stage. The total amount of magnesium introduced into the second stage is then the sum of those coming from the bypassed flow and those coming from the first stage and is equal to $q_b Mgr + (Q - q_b)Mgr1$. Since no removal of magnesium is effected in the second stage, this is equal to $QMgf$. Expressing this in terms of an equation, we have

$$QMgf = q_b Mgr + (Q - q_b)Mg1 \qquad (6\text{-}79)$$

Solving for the fraction of bypassed flow Q_x yields

$$Q_x = \frac{q_b}{Q} = \frac{Mgf - Mg1}{Mgr - Mg1} \qquad (6\text{-}80)$$

Example 6-20

For the raw water of Example 6-18, using the lime-soda process in the split-treatment mode to remove the total hardness to 120 mg/L containing a magnesium hardness of 30 mg/L as $CaCO_3$, calculate (a) the chemical requirements assuming recarbonation is done and (b) the ionic concentrations of the finished water. (c) Show that the ions are balanced.

Solution   As a basis of calculation, use 1.0 m^3 of raw water.

(a) From Example 6-18, molecules present are

$$Mg(HCO_3)_2 = 1.0 \text{ mEq/L}$$

$$Ca(HCO_3)_2 = 1.5 \text{ mEq/L}$$

$$CaSO_4 = 2.5$$

$$NaCl = 2.0$$

$$Mg(HCO_3)_2 + 2CaO + 2H_2O \rightarrow Mg(OH)_2\downarrow + 2CaCO_3\downarrow + 2H_2O$$

From the reaction,

$$1.5 \text{ mEq/L of } Mg(HCO_3)_2 = (1.0)\left(\frac{2CaCO_3}{2}\right) = 100 \text{ mg/L as } CaCO_3$$

The concentration of Mg in the effluent of the first stage is at the limit of technology corresponding to the solubility of 16 mg/L $Mg(OH)_2$ as $CaCO_3$. Therefore,

$$Q_x = \frac{q_b}{Q} = \frac{Mgf - Mg1}{Mgr - Mg1} = \frac{30 - 16}{100 - 16} = 0.17$$

First stage:

Lime used to remove $Mg(HCO_3)_2$ in the first stage $= \dfrac{(1 - 0.17)(1)(100 - 0)}{2(100)/2}\left(\dfrac{2CaO/2}{0.9}\right)$

$$= 52 \text{ g}$$

After this removal, 16 mg/L as $CaCO_3$ of $Mg(OH)_2$ is liberated. Hardness in plant effluent other than $Mg = 120 - 30 = 90$ mg/L as $CaCO_3$. Considering the limit of technology for $CaCO_3$ ($= 35$ mg/L), hardness in the effluent other than $CaCO_3$ and $Mg = 90 - 35 = 55$ mg/L as $CaCO_3$. Apportion the 55 mg/L as $CaCO_3$ between $Ca(HCO_3)_2$ and $CaSO_4$.

$$Ca(HCO_3)_2 + CaO + H_2O \rightarrow 2CaCO_3\downarrow + 2H_2O$$

$$1.5 \text{ mEq/L } Ca(HCO_3)_2 = 150 \text{ mg/L as } CaCO_3$$

$$CaSO_4 + Na_2CO_3 \rightarrow CaCO_3\downarrow + Na_2SO_4$$

$$2.5 \text{ mEq/L } CaSO_4 = 2.5(50) = 125 \text{ mg/L as } CaCO_3$$

$$Ca(HCO_3)_2 \text{ in finished water} = \frac{150}{150 + 125}(55) = 30 \text{ mg/L as } CaCO_3$$

$$CaSO_4 \text{ in finished water} = 55 - 30 = 25 \text{ mg/L as } CaCO_3$$

$$CO_2 + CaO + H_2O \rightarrow CaCO_3\downarrow + H_2O$$

total lime used to remove Mg hardness in first stage $=$ lime for $Mg(HCO_3)_2$

$+$ lime for $Ca(HCO_3)_2 +$ excess of 1.25 mEq/L $+$ lime for CO_2

$$= 52 + (1.0 - 0.17)(1)\left[(150 - 30)\frac{CaO}{2CaCO_3(0.9)} + (1.25)\frac{2CaO/2}{0.9} + \frac{22}{CO_2/2}(CaO/2)\left(\frac{1}{0.9}\right)\right]$$

$$52 + 121 = 173 \text{ g}$$

Note: Excess lime $= 65$ g; pure lime $= 58.1$ g.

Second stage: Chemicals needed in second stage = chemicals for bypassed $Mg(HCO_3)_2$ + chemicals for bypassed $Ca(HCO_3)_2$ + chemicals for bypassed $CaSO_4$ − excess lime of 65 g.

Bypassed $Mg(HCO_3)_2$:

$$Mg(HCO_3)_2 + 2CaO + 2H_2O \rightarrow Mg(OH)_2\downarrow + 2CaCO_3\downarrow + 2H_2O$$

$$\text{lime needed} = \frac{(0.17)(1)\,[100-(30-16)]}{2(100)/2}\left(\frac{2CaO/2}{0.9}\right) = 9.1 \text{ g}$$

Bypassed $Ca(HCO_3)_2$:

$$\text{lime needed} = \frac{(0.17)(1)(150-30)}{2(100)/2}\left(\frac{CaO/2}{0.9}\right) = 6.34 \text{ g}$$

$$CaSO_4 + Na_2CO_3 \rightarrow CaCO_3\downarrow + Na_2SO_4$$

$$\text{soda ash needed} = (1)(2.5)\left(\frac{Na_2CO_3/2}{0.85}\right) = 156 \text{ g}$$

Lime needed in second stage = 9.1 + 6.34 − 65 = −49.6 g; no additional lime required in second stage. Excess lime remaining = 49.6 g after second stage = 44.64 g pure.

$$\text{Total lime needed} = 173(25{,}000) = 4{,}325{,}000 \text{ g/day} = 4325 \text{ kg/day} \quad \textbf{Answer}$$

$$\text{Total soda as needed} = 156\left(\frac{25{,}000}{1000}\right) = 3900 \text{ kg/day} \quad \textbf{Answer}$$

CO_2 needed for recarbonation:

CO_2 needed = CO_2 needed to recarbonate $Mg(OH)_2$ from the limit of technology of

16 mg/L as $CaCO_3$ + CO_2 needed to recarbonate $CaCO_3$ from the limit of technology of

35 mg/L as $CaCO_3$ + CO_2 needed to recarbonate excess lime of 44.64 g pure.

$Mg(OH)_2$:

$$Mg(HCO_3)_2 + 2CaO + 2H_2O \rightarrow Mg(OH)_2\downarrow + 2CaCO_3\downarrow + 2H_2O$$

$$Mg(OH)_2\downarrow + CO_2 + H_2O \rightarrow Mg(HCO_3)_2$$

$$CO_2 \text{ needed} = \frac{16}{2CaCO_3/2}\left[\frac{Mg(OH)_2}{2}\right]\left[\frac{1}{Mg(OH)_2/2}\right]\left(\frac{CO_2}{2}\right)(1) = 3.52 \text{ g}$$

$CaCO_3$:

$$CO_2 + CaCO_3\downarrow + 2H_2O \rightarrow Ca(HCO_3)_2 + H_2O$$

$$CO_2 \text{ needed} = \frac{35}{50}\left(\frac{2CO_2}{2}\right)(1) = 15.4 \text{ g}$$

Excess lime:

$$CO_2 + CaO + H_2O \rightarrow CaCO_3\downarrow + H_2O$$

$$CO_2 + CaCO_3\downarrow + 2H_2O \rightarrow Ca(HCO_3)_2 + H_2O$$

$$\overline{2CO_2 + CaO + H_2O \rightarrow Ca(HCO_3)_2}$$

$$CO_2 \text{ needed} = 44.64\left(\frac{2CO_2}{CaO}\right)(1) = 70 \text{ g}$$

$$CO_2 \text{ required} = (3.52 + 15.4 + 70)\left(\frac{25,000}{1000}\right) = 2223 \text{ kg/day} \quad \textbf{Answer}$$

(b) The finished water contains the cations Ca^{2+}, Mg^{2+}, and Na^+ and the anions, HCO_3^- and SO_4^{2-}.

Ca^{2+}:

The ions come from limit of technology + recarbonation of excess lime

$$\text{ions come from limit of technology} = \frac{35}{50} = 0.7 \text{ mEq/L}$$

$$Ca^{2+} = 0.7 \text{ mEq/L} \qquad HCO_3^- = 0.7 \text{ mEq/L}$$

$$2CO_2 + CaO + H_2O \rightarrow Ca(HCO_3)_2$$

$$Ca(HCO_3)_2 = \frac{44.64}{CaO/2} = 1.59 \text{ mEq/L}$$

$$Ca^{2+} \text{ from recarbonation} = 1.59 \text{ mEq/L}, HCO_3^- = 1.59 \text{ mEq/L}$$

Mg^{2+}:

$$Mg(HCO_3)_2 + 2CaO + 2H_2O \rightarrow Mg(OH)_2\downarrow + 2CaCO_3\downarrow + 2H_2O$$

$$Mg(OH)_2\downarrow + CO_2 + H_2O \rightarrow Mg(HCO_3)_2$$

$$Mg(HCO_3)_2 \text{ from the limit of technology} = \frac{16}{100}\left[\frac{Mg(OH)_2/2}{Mg(OH)_2/2}\right] = 0.16 \text{ mEq/L}$$

$$Mg^{2+} = 0.16 \qquad HCO_3^- = 0.16 \text{ mEq/L}$$

Na^+:

Ions come from original sodium + soda ash

$$CaSO_4 + Na_2CO_3 \rightarrow CaCO_3\downarrow + Na_2SO_4$$

$$\text{pure soda ash needed} = (1)(2.5)(Na_2CO_3/2) = 132.5 \text{ g} = \frac{132.5}{106/2} = 2.5 \text{ mEq/L}$$

$$Na^+ = 2.5 \text{ mEq/L}$$

Positive ions:

$$Ca^{2+} = 0.7 + 1.59 = 2.29$$

$$Mg^{2+} = 0.16$$

$$Na^{2+} = 2 + 2.5 = 4.5$$

$$\text{Total} = 6.95 \text{ mEq/L} \quad \textbf{Answer}$$

Negatives:

$$HCO_3^- = 0.7 + 1.59 + 0.16 = 2.45$$

$$SO_4^{-2} = 4.5$$

$$\text{Total} = 6.95 \text{ mEq/L} \quad \textbf{Answer}$$

(c) From part (b) the ions are balanced.   **Answer**

Example 6-21

A sample of finished water containing 0.4 mEq/L of $CaCO_3$ is to be stabilized using H_2SO_4 or CO_2. What is the concentration of $CaCO_3$ in mg/L?

Solution   The respective stabilization reactions are:

$$2CaCO_3 + H_2SO_4 \rightarrow Ca(HCO_3)_2 + CaSO_4 \tag{a}$$

$$CaCO_3 + CO_2 + H_2O \rightarrow Ca(HCO_3)_2 \tag{b}$$

For stabilization using H_2SO_4, the equivalent weight of $CaCO_3$ is $2CaCO_3/2 = 100$. Therefore,

$$\text{concentration of } CaCO_3 = 0.4(100) = 40 \text{ mg/L}$$

For stabilization using CO_2, the equivalent weight of $CaCO_3$ is $CaCO_3/2 = 50$. Therefore,

$$\text{concentration of } CaCO_3 = 0.4(50) = 20 \text{ mg/L}$$

It is obvious that 40 mg/L $\neq$ 20 mg/L. *Therefore, concentrations given in equivalents are meaningless unless accompanied by the corresponding chemical reactions.*   **Answer**

Example 6-22

If the 0.4 mEq/L of $CaCO_3$ referred to in Example 6-21 is equivalent to 35 mg/L of $CaCO_3$, (a) determine the amount of H_2SO_4 needed to stabilize the water. (b) What is the amount if CO_2 is used?

Solution   As in Example 6-21, the respective stabilization reactions are:

$$2CaCO_3 + H_2SO_4 \rightarrow Ca(HCO_3)_2 + CaSO_4 \tag{a}$$

$$CaCO_3 + CO_2 + H_2O \rightarrow Ca(HCO_3)_2 \tag{b}$$

(a) For stabilization using H_2SO_4, the equivalent weight of $CaCO_3$ is $2CaCO_3/2 = 100$. Equivalent weight of $H_2SO_4 = H_2SO_4/2 = 49$. Therefore,

$$H_2SO_4 \text{ needed} = \frac{35}{100}(49) = 17.15 \text{ mg/L} \quad \textbf{Answer}$$

(b) For stabilization using CO_2, the equivalent weight of $CaCO_3$ is $CaCO_3/2 = 50$. Equivalent weight of $CO_2 = CO_2/2 = 22$. Therefore,

$$CO_2 \text{ needed} = \frac{35}{50}(22) = 15.4 \text{ mg/L} \quad \textbf{Answer}$$

DISINFECTION

Disinfection refers to the rendering of pathogenic organisms harmless. A companion term is sterilization. *Sterilization* refers to the killing of all organisms. Sterilization is not normally practiced in water treatment. The effluent after disinfection must meet the SWTR and the coliform rule.

Disinfectants

Historically, the disinfectant widely used is chlorine. Chlorine may be applied as a gas (Cl_2) or as the salts of hypochlorite [$Ca(OCl)_2$ and $NaOCl$]. The gas is supplied from liquid chlorine that is shipped in pressurized steel cylinders ranging in size from 100 lb to 1 ton. Calcium hypochlorite is available commercially in granular and powdered forms that contain 70% available chlorine. Sodium hypochlorite is handled in liquid form that contains from 5 to 15% available chlorine. (The term *available chlorine* will be defined later.) Other disinfectants that have been used are ozone, chlorine dioxide, and ultraviolet light (UV).

Ozone is produced on site by passing pure oxygen or dry, clean air in a high-strength electric field. The oxygen molecule first dissociates and the resulting nascent oxygen reacts with molecular O_2 to form ozone. The typical ozone dosage is 1.0 to 5.3 kg per 1000 m^3 of water with a power consumption of 10 to 20 kW/kg of ozone. The resulting ozone-treated water is environmentally safe but does not produce the residual disinfectant that is necessary in distribution systems to guard against after-treatment contamination. Ozone has been reported to be more effective than chlorine in inactivating resistant strains of bacteria and viruses.

As a strong oxidant, ClO_2 is similar to ozone. It does not form trihalomethanes, suspected carcinogens. Trihalomethanes are disinfection by-products of the use of chlorine. ClO_2 is particularly effective in destroying phenolic compounds. Like the use of chlorine, it produces measurable residual disinfectants. ClO_2 is a gas and its contact with light causes it to photooxidize, however. Hence it must be generated on-site. Although its principal application has been in wastewater disinfection, chlorine dioxide has been used in potable water treatment for oxidizing manganese and iron and for the removal of taste and odor. Its probable conversion to chlorate, a substance toxic to humans, makes its use for potable water treatment questionable.

Irradiation with ultraviolet light (UV) has also been used for disinfection. Although it provides no residual disinfectant, UV is effective in inactivating bacteria and viruses. UV is in the wavelength range 2000 to 3900 angstroms (Å) of the energy spectrum. The most effective band for disinfection, however, is from 2000 to 3000 Å. It can be generated with low-pressure mercury vapor lamps. A power input of 30 μW/cm^2 applied to thin sheets turbidity-free water would be sufficient. Being turbidity-free is to be stressed, since, if turbidity particles are present, bacteria, viruses and other pathogens can simply be hidden and be shielded against the UV irradiation.

Although very unlikely to be used for large-scale water treatment purposes, other disinfection products have been used. These include the halogens bromine and iodine, the metals copper and silver, and $KMnO_4$. Sonification, electric current, and heat have also been used. The use of electron beam as a method of disinfection is now under active research in the United States.[7] Gamma-ray irradiation is used in Germany and, along with the electron beam, is in active research in Japan.[8,9]

The chemical reactions of chlorine disinfectants are as follows:

$$Cl_2 + H_2O \rightarrow H^+ + Cl^- + HOCl \qquad (6\text{-}81)$$

$$Ca(OCl)_2 \rightarrow Ca^{2+} + 2OCl^- \qquad (6\text{-}82)$$

$$NaOCl \rightarrow Na^+ + Cl^- \qquad (6\text{-}83)$$

H^+, OCl^-, and $HOCl$ in reactions (6-81) to (6-83) relate to each other according to the following ionization reaction, which depends primarily on the pH of the solution:

$$HOCl \rightleftharpoons H^+ + OCl^- \qquad (6\text{-}84)$$

From this reaction, adding H^+ on the right pushes the reaction to the left. This means that at lower pH values, more $HOCl$ exists. It has been found that $HOCl$ is a stronger disinfectant than OCl^-. Hence to be more effective, the pH of the solution should be lowered. At pH 5.5 and below and at temperatures of 0 to 20°C, the concentration of $HOCl$ is practically 100%. At the same range of temperature, its percent composition is practically zero at pH 9 and above. The sum of the concentrations of chlorine, hypochlorous acid, and hypochlorite ion is collectively called *free chlorine*.

[7]E. H. Bryan (1990). "The National Science Foundation's Support of Research on Uses of Ionizing Radiation in Treatment of Water and Wastes." *Environmental Engineering, Proceedings of the 1990 Specialty Conference, ASCE*, Arlington, Va., pp. 47–54.

[8]W. Kawakami, S. Hashimoto, K. Nishimura, T. Miyata, and N. Suzuki (1978). "Electron-Beam Oxidation Treatment of a Commercial Dye by Use of a Dual-Tube Bubbling Column Reactor." *Environmental Science and Technology*, 12, pp. 189–194.

[9]S. Hashimoto, T. Miyata, and W. Kawakami (1980). "Radiation-Induced Decomposition of Phenol in Flow System." *Radiation and Physical Chemistry*, 16, pp. 59–65.

Reactions of chlorine with natural organics such as fulvic and humic acids produce undesirable disinfection by-products (DBPs) such as trihalomethanes (THM), the most predominant of which are chloroform and bromochloromethane. These THMs are suspected carcinogens, for which the Primary Drinking Water standard is 0.10 mg/L.

Minute quantities of phenolic compounds react with chlorine to create DBPs with severe taste and odor. To prevent the formation of DBPs, the organics in the raw water must be removed before disinfection, or if the DBPs have already been produced, they must be removed after formation. One way that this may be done is through the use of activated-carbon adsorption. The other way is to prevent the formation by avoiding the use of chlorine and substituting chloramines as the disinfectant. Chloramines do not react with phenols or phenolic compounds, hence avoiding the formation of DBPs of disagreeable taste and odor. Whether to substitute or not, would depend, however, on the actual operational situation.

Chloramines are substitution products of ammonia with chlorine. It can be formed by either direct reaction of chlorine and ammonia or by reaction of chlorine with amines, which are organic substances that contain nitrogen–carbon bonds. In the amine reaction, the ammonia is liberated from the amine compound to form the chloramines. In direct reaction with ammonia, chloramines can be produced by adding a quantity of ammonia followed by chlorine. The reactions are as follows:

$$NH_3 + HOCl \rightarrow NH_2Cl \text{ (monochloramine)} + H_2O \qquad (6\text{-}85)$$

$$NH_2Cl + HOCl \rightarrow NHCl_2 \text{ (dichloramine)} + H_2O \qquad (6\text{-}86)$$

$$NHCl_2 + HOCl \rightarrow NCl_3 \text{ (nitrogen trichloride)} + H_2O \qquad (6\text{-}87)$$

The amount of each species of chloramines produced in these reactions depends on the pH, temperature, and the concentration of HOCl used. The sum of the concentration of monochloramine and dichloramine is called *combined available chlorine*. These are the predominant species that exist during disinfection. Nitrogen trichloride is formed simply as a decomposition product as more HOCl is added. Hence reactions (6-85) and (6-86) are the ones effective in disinfection.

As chlorine is added to water, it reacts with constituents until residuals are produced. The residuals composed of the Cl_2, HOCl, and OCl^- species are called *free chlorine residuals*; the residuals composed of the chloramines are called *combined available chlorine residuals*. The sum of the residuals is simply called *total residual chlorine* (TRC).

Figure 6-11a shows the status of the residual as a function of the dosage of chlorine. From zero chlorine applied at the beginning of the abscissa to point *A*, the applied chlorine is immediately consumed by reducing such species as Fe^{2+}, Mn^{2+}, H_2S, and nitrites. As shown, no chlorine residual is produced. From *A* to *B*, chlorine reacts with organic compounds, ammonia, and amines to produce chloro-organic species and chloramines. Free chlorine is not formed. At this range of applied dosage, reactions (6-85) and (6-86) control the production of mono- and dichloramines. The distribution of these two forms depends on the pH and temperature. From *B* to the minimum, designated as the breakpoint, dichloramine is decomposed to nitrogen trichloride, reaction (6-87). In this range of chlorine dosage, the chloramines may also decompose to N_2 and N_2O. Possible

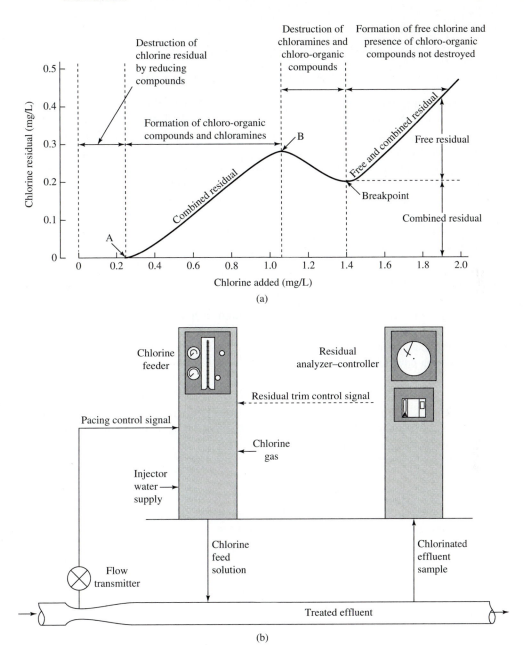

Figure 6-11 (a) Chlorine residual versus chlorine applied. (b) Automatic residual chlorine control of chlorine feeder.

reactions for these decompositions are

$$NH_2Cl + NHCl_2 + HOCl \rightarrow N_2O + 4HCl \tag{6-88}$$

$$2NH_2Cl + HOCl \rightarrow N_2 + H_2O + 3HCl \tag{6-89}$$

The point in the chlorine-dose-residual curve at which all destructible chloramines and chloro-organic compounds are decomposed and in which free chlorine residual begins to appear is called the *breakpoint*. Beyond the breakpoint, the residual chlorine is composed of free and combined residual.

Equations (6-88) and (6-89) show that hydrochloric acid is produced. This will require neutralizing this acid to the desired pH range. However, it is not possible to predict the amount of acid produced in an actual application, since the reaction transpiring is a combination of the two reactions. The practical way is to perform a test to determine the amount of base needed. By this method it has been found that 15.0 mg/L of alkalinity measured as $CaCO_3$ is required in normal situations.

Available Chlorine

The strength of a chlorine disinfectant is measured in terms of available chlorine. The *available chlorine* of a disinfectant is defined as the ratio of the mass of chlorine to the mass of the disinfectant that has the same unit of oxidizing power as chlorine. The unit of disinfecting power of chlorine may be found as follows in terms of 1 mol of electrons:

$$Cl_2 + 2e^- \rightleftharpoons 2Cl^- \tag{6-90}$$

From this equation, the unit of oxidizing power of Cl_2 is $Cl_2/2 = 35.5$. Consider another chlorine disinfectant such as NaOCl. To find its available chlorine, its unit of disinfecting power must also, first, be determined.

$$NaOCl + 2e^- + 2H^+ \rightleftharpoons Cl^- + Na^+ + H_2O \tag{6-91}$$

From this equation, the unit of disinfecting power of NaOCl is $NaOCl/2 = 37.24$. Therefore, the available chlorine of NaOCl is the ratio of the mass of chlorine to the mass of NaOCl that has the same unit of oxidizing power as chlorine, or

$$\text{available chlorine of NaOCl} = \frac{35.5}{37.24} = 0.95 \quad \text{or} \quad 95\%$$

In other words, NaOCl is 95% effective compared to chlorine.

Example 6-23

What is the available chlorine in dichloramine?

Solution   When dichloramine oxidizes a substance, its chlorine atom is reduced to chloride. And, as gleaned from its formula, the nitrogen must be converted to the NH_4^+ ion. Hence, the oxidation-reduction reaction using only half the reaction is

$$NHCl_2 + 4e^- + 3H^+ \rightleftharpoons 2Cl^- + NH_4^+$$

Thus

$$\text{unit of oxidizing power of dichloramine} = \frac{\text{NHCl}_2}{4} = 21.48$$

$$\text{available chlorine} = \frac{35.5}{21.48} = 1.65 \text{ or } 165\% \quad \textbf{Answer}$$

GLOSSARY

Aerosol. A colloid of solids and liquids dispersed in a gas.

Available chlorine. Of a given disinfectant, the mass of chlorine divided by the mass of the disinfectant having the same disinfecting power as the chlorine.

Breakpoint chlorination. The unit process of applying chlorine to the point where all the destructible chloramines and organic compounds are destroyed and where free chlorine residual begins to appear.

Carbonate hardness. Hardness associated with the carbonate and bicarbonate ions.

Coagulation. A unit process that destroys the mutual repulsive forces between particles.

Colloids. Particles so small that they are affected by the bombardment of atoms and molecules but not by the pull of gravity.

Combined available chlorine. The sum of the concentrations of monochloramine and dichloramine.

Copperas. Ferrous sulfate.

Counterions. The layer of opposite-charged ions surrounding the primary charge of a colloid.

Detention time, retention time, or residence time. The length of time that a particle of water is held up in a basin.

Discrete settling. Also called type 1 settling; a type of settling where the particles behave independently of each other.

Disinfection. The rendering of pathogenic organisms harmless.

Effective size. The sieve-size opening that passes the smaller 10% of a given sample.

Electrolyte. A material that when placed in solution will cause the solution to be conductive to electricity.

Flocculation. A unit operation that induces particles to coalesce and form bigger aggregate of particles.

Flocculent settling. Also called type 2 settling, the type of settling where the particles aggregate into flocs while settling.

Foam. A colloid of gas dispersed in a liquid.

Free chlorine. The sum of the concentrations of chlorine, hypochlorous acid, and the hypochlorite ion.

Hard water. Water that is difficult to lather when reacted with soap.

Hydrophilic colloid. Water-loving colloid.

Hydrophobic colloid. Water-hating colloid.

Inlet zone. That portion of a sedimentation basin where the inflow is allowed to position for proper entry into the settling zone.

Lime. Calcium oxide.

Lime-soda process. The softening of hardness using lime and sodium hydroxide.

Noncarbonate hardness. Hardness associated with ions other than the carbonate and bicarbonate ions.

Outlet zone. That portion of the sedimentation basin provided for proper takeoff of the effluent.

Overflow velocity or overflow rate. The flow to a tank divided by its surficial area.

Polyelectrolyte. Polymers with more than one electrolytic site in the molecule.

Polymer. A large molecule formed joining together smaller molecules of the same substance.

Potable water. Water that does not have substances in harmful concentrations.

Prototype. A scale-up from a physical model.

Rapid-mix. A process of rapidly distributing a coagulant throughout the volume of a tank.

Rapid-sand filter. A water filter operating at a rate of 100 to 200 mgad.

Recarbonation. The stabilization of a softened water by the use of carbon dioxide.

Settling zone. That portion in a sedimentation basin provided for the settling of particles.

Slow-sand filter. A water filter operating at a rate of 1 to 10 mgad.

Sludge zone. That portion of the sedimentation basin provided for the collection of settled sludge.

Split treatment. For waters high in magnesium concentration, a process where the raw water is split into two streams, the first stream is softened, and the second is bypassed and remixed with the softened first stream.

Sterilization. The killing of all organisms.

Turbidity. A measure of the extent to which suspended matter either absorbs or scatters light impinging on the suspension.

Uniformity coefficient. The ratio of the sieve-size opening that passes the smaller 60% of a given sample to the sieve-size opening that passes the smaller 10% of the sample.

Unit operation. An operation that, because of the application of the same physical principles, is used in similar ways in different types of plants and factories.

Unit process. A process that, because of the application of the same chemical or biological principles, is used in similar ways in different types of plants and factories.

van der Waals force. The natural mutual attractive force between two particles.

Water softening. A unit process of removing hardness ions from water.

SYMBOLS

a	fraction of absolute paddle velocity that makes v_p
A	flow-through cross-sectional area
A_p	projected area of particle; surface area of each grain of filter medium
A_s	overflow area; area of shear

b	length of paddle blade
C	concentration of particles remaining in the settling column at any time; concentration of materials in the flow to filter bed
C_D	coefficient of drag
C_0	concentration of particles in column at initial time
d	diameter of settling particle
d_p	diameter of a grain of filter medium
D	width of paddle blade
f	friction factor
f'	modified friction factor
f_b	bouyant force on a discrete particle
f_d	drag force on a discrete particle
f_g	body force on a discrete particle
g	acceleration due to gravity
G	$du/dy =$ velocity gradient
h	head loss across filter sand
h_{fb}	backwashing head loss
h_i	head loss across layer, i, of sand bed
kg/d	kilogram per day
ℓ	depth of bed
ℓ_e	expanded depth of bed
L	length of tank
m	inflow of material into bed
mEq	milligram equivalent
mEq/L	milligram equivalent per liter
$\mathcal{M}$	motor efficiency
N	number of grains in filter bed
p_1	percentage of a sample of stock sand that is equal to or less than the desired P_{10} in the final filter sand
p_2	percentage of a sample of stock sand that is equal to or less than the desired P_{60} in the final filter sand
p_3	percentage of stock sand that is transformed into final usable sand
p_4	percentage of stock sand that is too fine to be usable
p_5	percentage of stock sand above the size of which the sand is too coarse to be usable
P_{10}	10th percentile size of a sample of sand or grains
P_{60}	60th percentile size of a sample of sand or grains
$\mathcal{P}_f$	power given to fluid
q	material deposited per unit volume of sand bed
q_i	material deposited per unit volume of sand bed in layer i
Q	rate of flow
R	fraction of total particles removed; hydraulic radius

Re	Reynolds number
t	time
t_0	time of settling of particle from surface of settling column to sampling port; detention time of a settling tank
u	velocity in y-coordinate direction
v	terminal settling velocity of particle
v_h	flow-through velocity
v_p	velocities of particles comprised in x; it is less than v_0; paddle velocity relative to fluid
v_{pa}	absolute paddle velocity
v_0	Z_0/t_0
V	interstitial velocity through sand bed
$\forall$	volume of flocculation tank
V_B	backwashing velocity
V_p	volume of each grain of filter medium
V_s	superficial velocity
W	width of tank
x	fraction of particles not removed at any time
x_0	fraction of particles with velocities less than v_0
Z_p	vertical distance from any point on the settling column to sampling port; vertical distance from any point of the tank to its bottom
Z_0	vertical distance from surface of a settling column to sampling port; depth of settling basin
α	area shape factor
β	volume shape factor
γ_p	specific of particle
γ_w	specific weight of water
η	porosity; brake efficiency
η_e	expanded bed porosity
ρ_p, ρ_w	mass density of particle and water, respectively
Υ	fluid torque

PROBLEMS

6-1. Two spherical particles A and B have diameters of 0.6 mm and 0.9 mm, respectively. The particles have the same specific gravities of 2.65. Assume that the settling is type 1 and the temperature of water is 22°C. Determine the ratio of their terminal settling velocities.

6-2. Repeat Problem 6-1 for particles of sharp sand.

6-3. What is the impact of suspended solids on water treatment? How is it measured?

6-4. Determine the quantity of filterable residue on a sample if 200 mL of its filtrate is evaporated in an evaporating dish whose tare mass is 325.46 g. The total mass of the evaporating dish and residue after drying to constant weight is 325.57 g.

6-5. A sample of Tanjay River in Negros Oriental, the Philippines, is evaporated to dryness and sieve analysis performed on the residue. The result of the analysis is as follows:

Particle size (mm)	0.104	0.088	0.074	0.061	0.053	0.043
Weight fraction ≤ stated size	90	85	60	30	7	1

(a) For sedimentation analysis, convert this result to the percent remaining versus particle size and the percent remaining versus terminal settling velocity. **(b)** Compute the would-be percent removal in a sedimentation basin if the overflow rate is 25.0 m/day. Assume that the specific gravity of the particles is 2.0 and the temperature is 20°C.

6-6. What is type 1 suspension?

6-7. Determine the terminal settling velocity of a 100-μm particle in water at 22°C. The specific gravity of the particle is 2.5. Assume that $\beta = 0.9$. What is the velocity if the particle is settling in air at the same temperature?

6-8. Using any settling chart for flocculent particles, **(a)** determine the theoretical efficiency of a sedimentation basin with a depth of 3.5 m, a volume of 1500 m^3, and an inflow of 12,000^3/day. **(b)** What are the corresponding model detention time and model overflow rate, respectively?

6-9. For the particle in Problem 6-7, how long will it take for it to settle in a 2-m settling tank? Solve for both water and air.

6-10. A particle with a diameter of 1.1 mm and a specific gravity of 2.5 is released in water at a temperature of 20°C. How far will it travel in 3 s?

6-11. Particle A has a diameter of 0.4 mm and particle B has a diameter of 10 mm. If particle B is removed 100% in 9 s, what is the percent removal of particle A?

6-12. Name the types of settling basins employed in removing solids in water treatment.

6-13. Using a settling chart for flocculent particles, design a rectangular basin for an inflow of 30,000 m^3/day and a removal efficiency of 69.63%.

6-14. Design a rectangular settling basin to remove a type 2 suspension settling at a rate of 0.8 m/h. The flow rate is to be 20,000 m^3/day.

6-15. Repeat Problem 6-13 for a circular clarifier.

6-16. Repeat Problem 6-14 for a circular clarifier.

6-17. A consultant decided to recommend using an effective size of 0.55 mm and a uniformity coefficient of 1.65 for a proposed filter bed. Perform a sieve analysis to transform a run-of-bank sand you provide into a usable sand.

6-18. For the usable sand percentage distribution of Problem 6-17, what is the 50th percentile size?

6-19. For the sharp filter sand whose sieve analysis is shown in the following table, find the head loss if the sand is to be used in **(a)** a slow-sand filter 30 in. deep operated at 10 mgad and **(b)** a rapid-sand filter 30 in. deep operated at 125 mgad. The porosity of the unstratified bed is 0.39 and that of the stratified bed is 0.42. The lowest temperature anticipated of the water to be filtered is 20°C.

Sieve size	Average size (mm)	x_i
14–20	1.0	0.01
20–28	0.70	0.05
28–32	0.54	0.15
32–35	0.46	0.18
35–42	0.38	0.18
42–48	0.32	0.20
48–60	0.27	0.15
60–65	0.23	0.07
65–100	0.18	0.01

6-20. The amount of suspended solids removed in a uniform sand 0.8 mm in diameter is 2.0 mg/cm^3. Determine the total head loss due to suspended solids removed and the bed if the clean bed head loss is 0.793 m.

6-21. A 0.65-m-deep filter bed has a uniformly sized sand with a diameter of 0.45 mm, specific gravity of 2.65, and a volumetric shape factor of 0.87. If a hydrostatic head of 2.3 m is maintained over the bed, determine the flow rate at 20°C. Assume that the porosity is 0.4.

6-22. Determine the backwash rate at which the bed in Problem 6-21 will just begin to fluidize.

6-23. A sand filter is 0.65 m deep. It has a uniformly sized sand with a diameter of 0.45 mm, specific gravity of 2.65, and volumetric shape factor of 0.87. The bed porosity is 0.4. Determine the head required to maintain a flow rate of 10 m/h at 15°C.

6-24. A pilot study was conducted using a dual-media filter composed of anthracite as the upper 30-cm part and sand as the next lower 30-cm part of the filter. The results are shown below, where C_0 is the concentration of solids at the influent:

Q (L/m^2-min)	Depth (cm)	$\dfrac{C}{C_0}$	Medium	Q (L/m^2-min)	Depth (cm)	$\dfrac{C}{C_0}$	Medium
80	0	1	Anthracite	80	8	0.32	Anthracite
80	16	0.27	Anthracite	80	24	0.24	Anthracite
80	30	0.23	Sand	80	40	0.20	Sand
80	48	0.20	Sand	80	56	0.20	Sand
160	0	1	Anthracite	160	8	0.46	Anthracite
160	16	0.30	Anthracite	160	24	0.27	Anthracite
160	30	0.25	Sand	160	40	0.22	Sand
160	48	0.22	Sand	160	56	0.22	Sand
240	0	1	Anthracite	240	8	0.59	Anthracite
240	16	0.39	Anthracite	240	24	0.30	Anthracite
240	30	0.27	Sand	240	40	0.23	Sand
240	48	0.23	Sand	240	56	0.23	Sand

If the respective sizes of the anthracite and sand layers are 1.6 mm and 0.5 mm, what is the length of the filter run to a terminal head loss of 3 m at a filtration rate of 120 L/m^2-min? Assume that the clean water head loss is 0.793 m. *Note:* The terminal head loss is the loss when the filter is about to be cleaned.

6-25. The porosity of the unstratified bed composed of sharp filter sand is 0.39 and that of the stratified bed is 0.42. The lowest temperature anticipated of the water to be filtered is 20°C. Find the head loss if the sand is to be used in **(a)** a slow-sand filter 30 in. deep operated at 10 mgad and **(b)** a rapid-sand filter 30 in. deep operated at 125 mgad. The sieve analysis is as follows:

Sieve size	Average size (mm)	x_i
14–20	1.0	0.01
20–28	0.70	0.05
28–32	0.54	0.15
32–35	0.46	0.18
35–42	0.38	0.18
42–48	0.32	0.20
48–60	0.27	0.15
60–65	0.23	0.07
65–100	0.18	0.01

6-26. Pilot plant analysis on a mixed-media filter shows that a filtration rate of 15 m^3/m^2-h is acceptable. If a surface configuration of 5 m by 8 m is appropriate, how many filter units will be required to process 100,000 m^3/day of raw water?

6-27. To solve this problem do not use equivalent weights. Using copperas as a coagulant, 30 g/m^3 was added to the raw water. **(a)** Calculate the minimum natural alkalinity required to react with the ferrous sulfate. **(b)** How much lime of 70% available CaO is required to convert the initial reaction product to Fe(OH)$_2$? **(c)** What concentration of dissolved O$_2$ is required to convert the Fe(OH)$_2$ to the ferric form?

6-28. Determine the basin dimensions, power applied, and paddle configuration and rotational speed for a flocculator that processes 17,000 m^3/day of raw water. By jar test, the optimum Gt value was found to be 4.5(10^4).

6-29. A rapid mix tank blends 35 mg/L of alum with a raw water flow rate of 40,000 m^3/day. The detention time is to be 2 min. Determine the kilograms per day of alum added, the dimensions of the tank, and the power input necessary for a G value of 900 s^{-1}. The water temperature is 20°C.

6-30. The quantity 50 mg/L of alum is added to 50,000 m^3/day of raw water containing 60 mg/L of suspended solids, 80 mg/L of Ca^{2+}, 12 mg/L of Mg^{2+}, and 152.5 mg/L of HCO$_3^-$. **(a)** Is there sufficient alkalinity in the raw water? **(b)** If not, how much should be added? **(c)** What is the daily consumption of alum and the daily production of sludge?

6-31. For a water temperature of 4°C, design a rapid-mix unit to treat 50,000 m^3/day raw water. Use the following assumptions: detention time = 30 s, shaft and motor efficiency = 75% and 90%, respectively.

6-32. Describe how a jar test is performed and describe its importance in plant operation.

6-33. A 20,000-m^3/day flow at 20°C is flocculated in a flow-through flocculator whose paddles are arranged longitudinally. The optimum value of Gt determined by a jar test is 5.0(10^4). Determine **(a)** the basin dimensions, **(b)** the fluid power developed, and **(c)** assuming two cross-arms with one blade attached at the ends of the arms, determine the blade dimensions and the rotational speed.

6-34. Repeat Problem 6-33 using $C_D = 1.8$.

6-35. A raw water has the following constituents expressed in mEq/L: $Ca^{2+} = 4.6$, $Mg^{2+} = 1.0$, $Na^+ = 2.1$, $HCO_3^- = 2.4$, $SO_4^{2-} = 2.9$, $Cl^- = 2.4$, and $CO_2 = 0.6$. **(a)** Check if the ions are balanced. **(b)** What is the total hardness expressed as $CaCO_3$? **(c)** For a flow of 25,000 m^3/day, calculate the chemical requirements and the mass of solids and volume of sludge produced if the raw water is to be treated by the lime-soda process to a total hardness of 80 mg/L as $CaCO_3$, of which 50 mg/L as $CaCO_3$ is due to calcium. Assume that the lime used is 90% pure and the soda ash used is 85% pure. Also, assume that the specific gravity of the sludge is 1.04.

6-36. In Problem 6-35: **(a)** Determine the amount of CO_2 needed for recarbonation. **(b)** Determine the additional mass of solids produced after recarbonation.

6-37. In Problem 6-35: **(a)** Determine the ionic composition of the finished water after recarbonation. **(b)** Show that the ions are balanced.

6-38. As a rule of thumb, what hardness level indicates the need to soften water at the treatment plant?

6-39. A raw water containing the following constituent (in mEq/L) is to be treated by split treatment and the lime-soda process: $CO_2 = 0.5$, $Ca^{2+} = 3.8$, $Mg^{2+} = 3.0$, $Na^+ = 1.8$, $HCO_3^- = 2.5$, $SO_4^{2-} = 3.3$, and $Cl^- = 2.8$. **(a)** Calculate the chemical requirements assuming recarbonation. **(b)** Calculate the ionic concentrations of the finished water.

6-40. In Problem 6-39, show that the ions are balanced.

6-41. What is split treatment?

6-42. What is the percent available chlorine of the following: **(a)** $Ca(OCl)_2$, **(b)** $NaOCl$, **(c)** $HOCl$, and **(d)** NH_2Cl?

BIBLIOGRAPHY

BRYAN, E. H. (1990). "The National Science Foundation's Support of Research on Uses of Ionizing Radiation in Treatment of Water and Wastes." *Environmental Engineering, Proceedings of the 1990 Specialty Conference, ASCE*, Arlington, Va., pp. 47–54.

ECKENFELDER, W. W. (1980). *Principles of Water Quality Management*. CBI, Boston.

HASHIMOTO, S., T. MIYATA, and W. KAWAKAMI (1980). "Radiation-Induced Decomposition of Phenol in Flow System." *Radiation and Physical Chemistry*, 16, pp. 59–65.

KAWAKAMI, W., S. HASHIMOTO, K. NISHIMURA, T. MIYATA, and N. SUZUKI (1978). "Electron-Beam Oxidation Treatment of a Commercial Dye by Use of a Dual-Tube Bubbling Column Reactor." *Environmental Science and Technology*, 12, pp. 189–194.

METCALF & EDDY, INC. (1991). *Wastewater Engineering: Treatment, Disposal, Reuse*. McGraw-Hill, New York.

MUNSON, B. R., D. F. YOUNG, and T. H. OKIISHI (1994). *Fundamentals of Fluid Mechanics*. Wiley, New York.

PEAVY, H. S., D. R. ROWE, and G. TCHOBANOGLOUS (1985). *Environmental Engineering*. McGraw-Hill, New York.

TCHOBANOGLOUS, G., and E. D. SCHROEDER (1985). *Water Quality*. Addison-Wesley, Reading, Mass.

CHAPTER 7

Conventional Wastewater Treatment

In most developed countries, city residents used to put "night soil" in buckets along the streets where workers emptied the waste into "honey wagon" tanks. The waste was transported for disposal over agricultural lands. For rural communities in many underdeveloped countries, this method of disposal is still in use.

In the history of the developed nations of the world, the honey-wagon method of disposal was adequate since the volume of waste generated was relatively small. However, the development of the flush toilet, which produced large volumes of wastewater in the nineteenth century, overtaxed agricultural lands as a final receptor of the waste. This induced people to use natural channels to convey into the nearest watercourse the huge volumes of liquids generated by the flush toilet. The natural consequence of the use of natural channels was the construction of large sewers. *Combined sewers* came into being in the latter half of the nineteenth century.

Combined sewers and separate sanitary sewers simply discharged directly into streams, rivers, lakes, and estuaries. This practice resulted in the gross pollution of these receiving bodies of water. During the nineteenth century, for example, the River Thames was so grossly polluted that the House of Commons had to have rags wetted in lye stuffed into cracks in the windows of the Parliament buildings to reduce the stench. The practice of direct discharge without treatment is no longer allowed in the United States.

In this chapter we discuss conventional wastewater treatment, both primary and secondary. Discharge of effluent through diffusers is also discussed. For the overall effect of the effluent on surface waters, see Chapter 5.

EFFLUENT STANDARDS

The degree to which a wastewater is treated must meet the water quality standards of the receiving body of water into which the effluent is discharged. In the United States, under the *National Pollutant Discharge Elimination System* (NPDES), the Clean Water Act has mandated that the treatment must, at least, be to the level of secondary for the case of publicly owned treatment works (POTWs) of municipalities and to the level of best available treatment technology (BAT) for the case of industries. Before any wastewater treatment plant can operate, a discharge permit must be obtained from the permitting agency, which is normally the state where the plant is located. The numbers written in the permit, which are supposed to meet the water quality standards of the stream, become the effluent standards of operation of the plant, or, in short, the *effluent standards*. The operator of the plant must ensure that the effluent meet these standards.

COMAR (Code of Maryland Regulations) 26.08.01.01B(80), which conforms to the USEPA definition, defines secondary treatment as

(a) *Five-day biochemical oxygen demand*
  (i) 30 mg/L—average for a 30-day period
  (ii) 45 mg/L—average for a 7-day period
(b) *Total suspended solids*
  (i) 30 mg/L—average for a 30-day period
  (ii) 45 mg/L—average for a 7-day period
(c) *Bacterial control*: as required to meet water quality standards

COMAR 26.08.01.01B(10) also defines BAT, which conforms to the USEPA definition, as the best existing wastewater treatment technology economically achievable within an industrial category. BAT is also called BCT (best conventional pollutant control technology).

For municipal discharge permits, the secondary level of treatment is normally specified in terms of four parameters: BOD_5, total Kjeldahl nitrogen (TKN), suspended solids, and pH. As stated above, secondary standards for BOD_5 are 30 mg/L on a monthly average basis and 45 mg/L on an average weekly basis. The secondary standards for suspended solids are the same as those for BOD_5. For TKN, although not included in the definition of secondary standard, several authorities consider 20 to 25 mg/L as the secondary level of treatment. The federal standard for pH ranges from 6.0 to 9.0. In Maryland, the pH standard is stricter, 6.5 to 8.5.

The other effluent standard that must be met is coliform. The coliform standard is normally given in terms of total coliform and fecal coliform and set on the basis of stream classification. For example, the coliform standard for discharges to shellfish waters is different and more stringent from the standard for discharges to normal fresh free-flowing streams. In Maryland, the shellfish water standard is 14 fecal coliform per 100 mL as a median value, and the standard for fresh free-flowing stream is 200 fecal coliform per 100 mL as a logarithmic mean value. Along with the coliform standard is the total residual chlorine (TRC) standard. Chlorine is used to kill off the coliforms and pathogens, and must be limited, for it can damage the biota of the receiving stream if it is discharged in excessive amounts. The *Red Book* criterion and the Maryland standard for trout streams is 0.02 mg/L of TRC. (The Red Book is a book of water quality criteria published by the USEPA.)

In addition, municipal permits may include standards for toxic inorganic and organic substances. The toxic inorganic substances may include arsenic, barium, cadmium, chromium, copper, cyanide, lead, mercury, nickel, selenium, silver, and zinc; and, the toxic organic substances may include aldrin, benzidine, 1,1-DCE, DDT, dieldrin, endrin, lindane, PCP, PCE, 1,1,1-TCA, toxaphene, TCE, tributylin, and dioxin.

The permits for industrial discharges follow patterns similar to those for municipal permits, except that the two permits may also vary in several places. For example, if the industry is manufacturing a purely inorganic material, the permit may contain no limitation on BOD_5. In addition, the permit may also not contain limits on coliforms, nor may it contain limits on TKN.

A special requirement that is included in discharge permits for municipal treatment plants or POTWs is the *pretreatment program*. In its basic form, this program mandates that the POTW must require all industries that discharge to its collection system to pretreat to such a level that the discharged pollutants will not cause any upset in operation of the municipal treatment plant, will not simply pass through the plant without treatment, and will not contaminate the sludge.

Waters into which the discharge of secondary- or BAT-treated effluents will not cause violation of the water quality standards of the stream are called *effluent limited waters*; while those waters where secondary- or BAT-treated effluents will cause the violation of the standard are called *water quality limited waters*. For discharge into water quality limited waters, the effluent must be treated to a level more advanced than secondary. For BOD_5 removals, a treatment higher than secondary is called *advanced secondary treatment* (AST), normally considered as a level of treatment on the order of 10 to 15 mg/L of BOD_5 in the effluent. In general, treatment that produces effluent quality that is not attainable by ordinary secondary or BAT technology is called *advanced wastewater treatment*. In this chapter we discuss only secondary or BAT treatment levels.

PRIMARY TREATMENT

A schematic of typical primary treatment systems is shown in Figure 7-1a. In general, the removal of organic pollutants in wastewater may be divided into two stages. The first stage, which involves the removal of the wastewater solids, is called *primary treatment*; the second stage, which involves the removal of the colloids and dissolved organic matter, in wastewater, is called *secondary treatment*. In the primary treatment stage, the comminutor, screens, grit channel, and flowmeter constitute what are called the *preliminary primary treatment*, or simply, *preliminary treatment*, units. Preliminary treatment is composed of unit operations designed to remove large objects and grit materials. The top and middle portions of Figure 7-2 show drawings of bar screens and detritus tank.

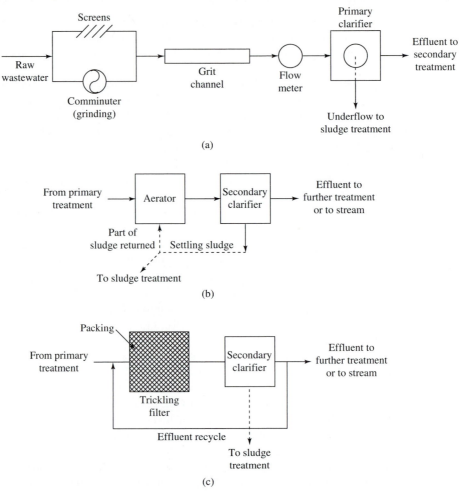

Figure 7-1 Primary and secondary treatment systems: (a) typical primary treatment system; (b) activated sludge secondary treatment system; (c) trickling filter secondary treatment system.

Screened solids are coated with organic matter, which makes them offensive-smelling, so they must be disposed of promptly. Although some treatment plants practice shredding the screenings for return to the main sewage flow for subsequent settlement in downstream units, this practice is not advisable. Since these solids have already been removed, they should not be returned to the stream flow only to be removed again later.

Bar screens are screens constructed of large bars. They can be cleaned either manually or automatically. The *detritus tank* is where grit is allowed to settle, along with some organic matter. Detritus is a mixture of grit and organic matter. The organic matter

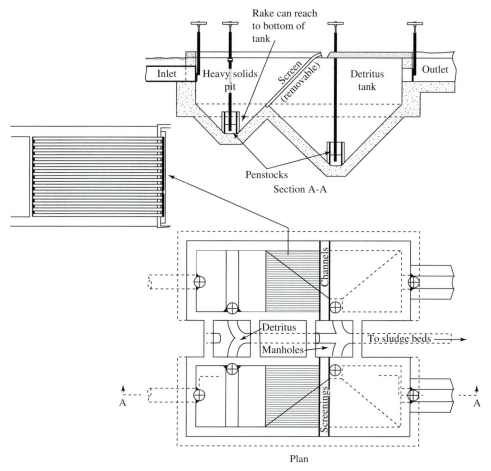

Figure 7-2   Top and middle portions: screens and detritus tanks; bottom portion: typical quantities of screenings.

is subsequently removed from the detritus by washing and the washings returned to the main stream. *Grit particles* are hard fragments of rock, sand, stone, bone chips, seeds, and coffee and tea grounds or similar particles. A long channel designed mainly to settle grit is called a *grit chamber*. Grit particles settle as a discrete entity. The mechanics of settling have been discussed in Chapter 6 and are not repeated here.

The design of grit tanks must be such that only the grit particle settles, leaving organic matter suspended and separated from the grit. To ensure this separation, the flow-through velocity in the channel must be carefully controlled at a specified value of approximately 0.3 m/s by grit-channel outlet flow-control devices such as proportional-flow weirs or critical-flow flumes. The cross-sectional shape of the grit channel will depend on the type of outlet flow-control device used. Figure 7-3b is a sectional view of a parabolic-sectioned grit channel controlled by a critical-flow flume. Figure 7-3a is a

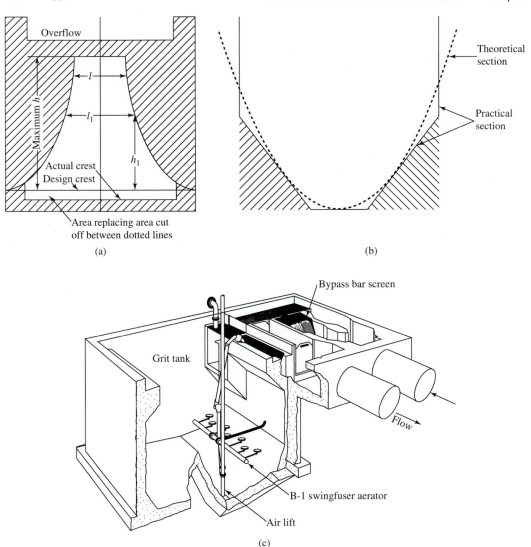

Figure 7-3 Velocity control sections of horizontal grit channels: (a) proportional flow weir; (b) section of a parabolic-sectioned grit channel; (c) aerated grit chamber.

proportional-flow weir that can be installed at the end of grit channels. This weir is simply a plate with the opening cut through it and the plate bolted or somehow attached to the downstream end of the channel.

Grit chambers may also be designed as aerated tanks, the airflow being adjusted so as to separate putrescible organic matters from grit. Figure 7-3c is an aerated grit chamber. A design criterion for airflow requirement ranges from 0.15 to 0.45 m^3/min per meter of tank length. A typical length/width ratio for the aerated grit chamber is 3:1.

Figure 7-4 shows devices used to shred screenings to pieces. The words *Griductor* and *Comminutor* are proprietary names. The basic parts of these devices are the screens and cutting teeth. Solids are cut by the teeth and are returned to the flow after passing through the holes or slots of the screens. To protect pumps, shredding devices should be put ahead of them. Grit removal should be done ahead of shredders to protect the cutting teeth, as well as ahead of pumps. In some situations, however, pumps are put ahead of grit-removal devices. Such is the case, for example, for grit chambers located at or above the ground with pumps required to elevate the sewage to them. In these situations, shredders are still put ahead of pumps but with no protection against grit, since the grit chambers are located after the pumps and thus after the shredders.

Grit Chamber Cross-Sections

As shown in Figure 7-3, the flow area of the proportional-flow weir is an orifice. From fluid mechanics, the flow Q through an orifice is given by

$$Q = K_0 \ell H^{3/2} \tag{7-1}$$

where K_0 is the orifice constant, ℓ the width of flow over the weir, and H the head over the weir crest. The parameter h is any height above the crest and ℓ is the corresponding width. There are several ways that the orifice can be cut through the plate; one way is to do it such that the flow Q will be linearly proportional to H. To fulfill this scheme, equation (7-1) will be revised by changing K_0 to K_0' and making $H^{3/2}$ a combination of H and h. The revised equation is

$$Q = K_0' \ell h^{1/2} H \tag{7-2}$$

Hence, to be linearly proportional, $K_0' \ell h^{1/2}$ must be a constant; or

$$K_0' \ell_1 h_1^{1/2} = K_0' \ell_2 h_2^{1/2} = \text{constant} \Rightarrow \ell h^{1/2} = \text{constant} \tag{7-3}$$

Equation (7-3) is the equation of the orifice opening of the proportional-flow weir in Figure 7-3. If the equation is followed strictly, however, the orifice opening will create two pointed corners that will probably clog. For this reason, for values of h less than 1 in., the side curves are terminated vertically to the weir crest. The area of flow lost by terminating at this point is of no practical significance; however, if terminated at an h value greater than 1 in., the area lost should be compensated for by lowering the actual crest below the design crest. This is indicated in the figure.

Since the height of the orifice crest from the bottom of the channel is small, H may be considered equal to the depth in the tank. Hence the general cross-sectional area of the tank may be represented by kwH, where k is a constant, w the width at a particular level, and H the depth in the tank. Now the flow through the tank is $Q = v_h(kwH)$, where v_h is the flow-through velocity to be made constant. This flow is also equal to the flow that passes through the control device at the end of the tank. From the equation of continuity,

$$v_h(kwH) = K_0'(\ell h^{1/2})H = \text{constant}(H) \tag{7-4}$$

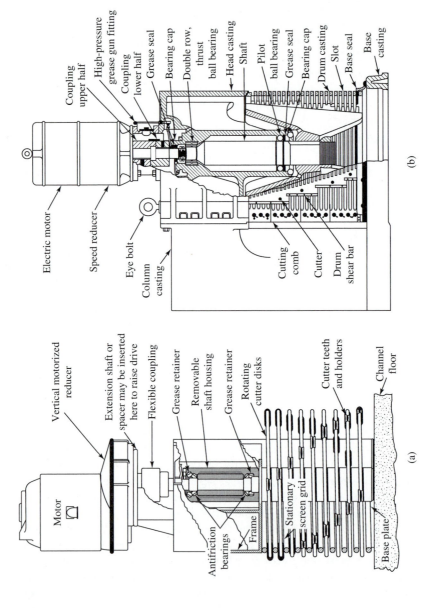

Figure 7-4 Screenings shredding devices: (a) Griductor (Infilco); (b) Comminutor (Chicago Pump Co.).

Electric motor
Speed reducer
Eye bolt
Column casting

Coupling upper half
High-pressure grease gun fitting
Coupling lower half
Grease seal
Bearing cap
Double row, thrust ball bearing
Head casting
Shaft
Pilot ball bearing
Grease seal
Bearing cap
Drum casting
Slot
Base seal
Base casting

Cutting comb
Cutter
Drum shear bar

(b)

Vertical motorized reducer
Extension shaft or spacer may be inserted here to raise drive
Flexible coupling
Grease retainer
Removable shaft housing
Grease retainer
Rotating cutter disks
Cutter teeth and holders
Channel floor

Motor

Antifriction bearings
Frame
Stationary screen grid
Base plate

(a)

Solving for w gives us

$$w = \frac{\text{constant}}{v_h k} = \text{constant} \tag{7-5}$$

Therefore, for grit chambers controlled by a proportional-flow weir, the width of the tank must be constant, which means that the cross section should be rectangular.

For grit chambers controlled by another critical-flow device, such as a Parshall flume (the proportional flow weir is also a critical-flow device), the flow through the device is also given by $Q = K_0 \ell H^{3/2}$. Hence the following equation may also be obtained:

$$v_h(kwH) = K_0 \ell H^{3/2} \tag{7-6}$$

Solving for H yields

$$H = \text{constant}(w^2) \tag{7-7}$$

which is the equation of a parabola. Thus for grit chambers controlled by Parshall flumes, the cross section of flow should be shaped like a parabola. For ease in construction, the parabola is not strictly followed but approximated. This is indicated in Figure 7-3b.

Example 7-1

Design the cross section of a grit removal unit consisting of four identical channels to remove grit for a peak flow of 80,000 m³/day, an average flow of 50,000 m³/day and a minimum flow of 20,000 m³/day. There should be a minimum of three channels operating at any time. Assume a flow-through velocity of 0.3 m/s and that the channels are to be controlled by Parshall flumes.

Solution   Four baseline cross-sectional areas must be considered, computed as follows:

$$A_{\text{peak, three channels}} = \frac{80,000}{3(0.3)(24)(60)(60)} = 1.03 \text{ m}^2$$

$$A_{\text{peak, four channels}} = \frac{80,000}{4(0.3)(24)(60)(60)} = 0.77 \text{ m}^2$$

$$A_{\text{ave}} = \frac{50,000}{4(0.3)(24)(60)(60)} = 0.48 \text{ m}^2$$

$$A_{\text{min}} = \frac{20,000}{4(0.3)(24)(60)(60)} = 0.19 \text{ m}^2$$

Since the channels are to be controlled by Parshall flumes, the cross sections are parabolic. Hence

$$H = \text{constant}(w^2), \text{ or using the depth } Z_0$$

$$Z_0 = \text{constant}(w^2) = Cw^2$$

Area of parabola $= \frac{2}{3} w Z_0$; assume top width at $A_{\text{peak, three chambers}} = 1.5$ m. Thus

$$1.03 = \frac{2}{3}(1.5)Z_0 \qquad Z_0 = 1.03 \text{ m}$$

$$C = \frac{1.03}{1.5^2} = 0.46$$

and equation of parabola: $Z_0 = 0.46w^2$. Determine the coordinates at the corresponding areas.

For $A_{\text{peak, four chambers}} = 0.77 \text{ m}^2$: $0.77 = \frac{2}{3}w(0.46w^2)$; $w = 1.36$ m and $Z_0 = 0.46(1.36^2) = 0.85$ m.

For $A_{\text{ave}} = 0.48 \text{ m}^2$: $0.48 = \frac{2}{3}w(0.46w^2)$; $w = 1.16$ m and $Z_0 = 0.46(1.16^2) = 0.62$ m.

For $A_{\text{min}} = 0.19 \text{ m}^2$: $0.19 = \frac{2}{3}w(0.46w^2)$; $w = 0.85$ m and $Z_0 = 0.46(0.85^2) = 0.33$ m.

Area $(\text{m})^2$	w (m)	Z_0 (m)	Q (m^3/day)	
1.03	1.5	1.03	80,000 in three channels	
0.77	1.36	0.85	80,000 in four channels	
0.48	1.16	0.62	50,000 in four channels	**Answer**
0.19	0.85	0.33	20,000 in four channels	
0	0	0		

Note: In practice, the coordinates above should be checked against the flow conditions of the chosen dimensions of the Parshall flumes. If the flumes are shown to be submerged, forcing them not to be at critical flows, other coordinates of the parabolic cross sections must be tried until the flumes will show critical flow conditions or be unsubmerged.

Example 7-2

Repeat Example 7-1 for grit channels controlled by proportional flow weirs.

Solution   For grit channels controlled by proportional weirs, the cross sections should be rectangular. Hence

$$w = \frac{\text{constant}}{v_h k} = \text{constant} \qquad \text{assume that } w = 1.5 \text{ m}$$

Hence the depths, Z_0, and other parameters for various flow conditions are as follows (for a constant flow-through velocity of 0.30 m/s):

Area $(\text{m})^2$	w (m)	Z_0 (m)	Q (m^3/day)	
1.03	1.5	0.69	80,000 in three channels	
0.77	1.5	0.51	80,000 in four channels	
0.48	1.5	0.32	50,000 in four channels	**Answer**
0.19	1.5	0.13	20,000 in four channels	
0	1.5	0		

Primary Sedimentation

Referring to Figure 7-1, the primary clarifier or primary sedimentation tank is the last unit in primary treatment. The effluent from primary sedimentation then becomes the influent for secondary treatment. Primary sedimentation in present-day practice may involve *simple* or *plain sedimentation* or addition of chemicals. The same quality of effluent produced by primary sedimentation may also be obtained by using fine screens. Using screens this way qualifies the treatment to be a primary treatment.

The nature of the solids removed in primary sedimentation is flocculent. Hence the theory involved is that of type 2 or flocculent settling. Since this theory has been discussed in Chapter 6, it is not repeated here. The settling units in the wastewater and water systems are also similar (i.e., either a flow-through or a circular tank is used); hence, again, the discussion is not repeated. A major difference between sedimentation in water and wastewater treatment, however, is the amount of scum produced. Whereas in water treatment there is practically no scum, in wastewater treatment, such a large amount of scum is produced that an elaborate scum-skimming device is used at the outlet end of the clarifier. Some design criteria for primary sedimentation tanks are shown in Table 7-1. Figure 7-5a shows a diagram of an outlet weir and scum removal arrangement.

TABLE 7-1   DESIGN CRITERIA FOR PRIMARY SEDIMENTATION TANKS

	Value	
Parameter	Range	Typical
Detention time (h)	1.5–2.5	2.0
Overflow rate (m/day)		
Average flow	30–50	
Peak flow	80–120	90
Weir loading (m^3/m^2-day)	120–450	200
Dimensions (m)		
Rectangular		
Depth	2–6	3.5
Length	15–100	30
Width	3–30	10
Sludge scraper speed (m/min)	0.5–1.5	1.0
Circular		
Depth	3–5	4.5
Diameter	3.0–60	30
Bottom slope (mm/m)	60–160	80
Sludge scraper speed (rpm)	0.02–0.05	0.03

Figure 7-5b shows typical BOD_5 and suspended solids removal efficiency curves for sedimentation basins as functions of overflow rate. A smaller overflow rate produces a longer detention time. Although longer detention times tend to effect more solids removal in water treatment, in primary sedimentation a longer detention time causes septic conditions. Because of the formation of gases, septicity makes solids rise, resulting in inefficiency of the basin. Hence, in practice, there is a practical range of values of 1.5 to 2.5 h for primary settling detention times wastewater treatment.

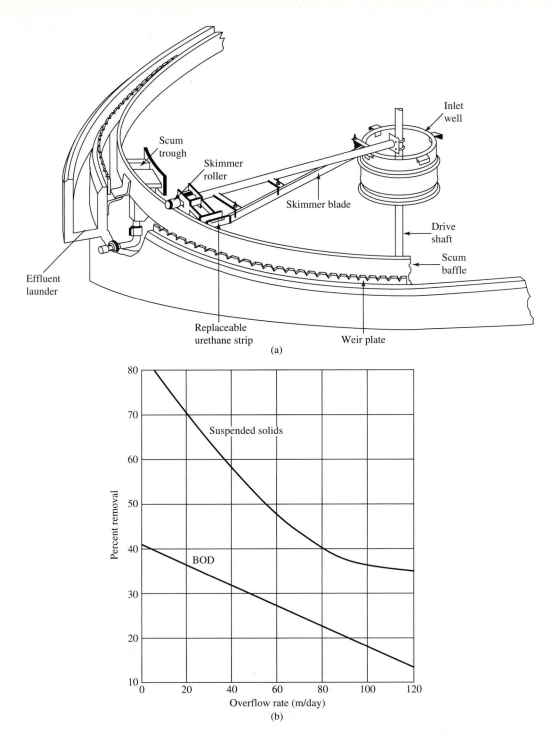

Figure 7-5 (a) Outlet weir and scum skimming arrangements (Infilco Degremont, Inc.). (b) Suspended solids and BOD_5 removal as a function of overflow rate.

Example 7-3

To meet effluent limits, it has been determined that the design of a primary sedimentation basin must remove 65% of the influent suspended solids. The average influent flow is 6000 m^3/day, the peak daily flow is 13,000 m^3/day, and the minimum daily flow is two-thirds of the average flow. Using Figure 7-5 and other pertinent information, design a circular basin to meet the effluent requirement at any cost.

Solution   From Figure 7-5, the overflow rate corresponding to 65% efficiency = 28 m/day. Therefore,

$$\text{overflow area at average flow} = \frac{6000}{28} = 214.29 \text{ m}^2 \qquad \text{diameter } D = 16.52 \text{ m}$$

$$\text{overflow area at peak flow} = \frac{13,000}{28} = 464.29 \qquad \text{diameter } D = 24.31 \text{ m}$$

$$\text{overflow area at minimum flow} = \frac{(2/3)6000}{28} = 142.86 \qquad \text{diameter } D = 13.49 \text{ m}$$

To meet effluent limits at any cost, the basin must be designed on the basis of the peak flow. Then, using a detention time of 1.5 h (refer to Table 7-1),

$$\text{settling zone of tank} = \frac{13,000}{24}(1.5) = 812.50 \text{ m}^3$$

$$\text{height of settling zone} = \frac{812.50}{464.29} = 1.75 \text{ m} = \frac{28}{24}(1.5)$$

and

$$\text{diameter of tank} = \text{settling zone diameter} + \text{inlet zone} (= \text{settling zone depth})$$

$$+ \text{ outlet zone} (= \text{settling zone depth})$$

$$= 24.31 + 2(1.75) = 27.81 \text{ m}, \quad \text{say 28 m} \quad \textbf{Answer}$$

$$\text{depth of tank} = \text{settling zone depth} + \text{free board} (= 0.5 \text{ m}) + \text{sludge zone} (= 0.5 \text{ m})$$

$$= 1.75 + 2(0.5) = 2.75 \text{ m}, \quad \text{say 3 m} \quad \textbf{Answer}$$

Check for detention times for average and minimum flows:

$$\text{detention time for average flow} = \frac{812.50}{6000/24} = 3.25 \text{ h}$$

$$\text{detention time for minimum flow} = \frac{812.50}{(2/3)(6000/24)} = 4.87 \text{ h}$$

From Table 7-1, all the detention times above will exceed the practical limit of 2.5 h if the respective flow durations exceed this time. Hence provision must be made in design to recycle primary effluent to reduce the detention time, if needed, to practical limits.
    Check the weir overflow rates:

$$\text{weir overflow rate at peak flow} = \frac{13,000}{\pi(28)} = 147.8 \text{ m}^3/\text{m-day} \ll 250$$

However, since the overriding consideration is meeting the effluent limit at any cost, this will be acceptable.   **Answer**

SECONDARY TREATMENT

Primary treatment of wastewaters merely removes the solids originally suspended in the waste. Before the environmental revolution, which started in 1970, most municipalities in the United States were merely treating waste to the primary stage. However, starting in the 1970s, dischargers were required to meet water quality standards. Meeting water quality standards necessitated performing mathematical water quality modeling of receiving streams. Using the model of Streeter and Phelps developed in 1927, the results often showed that more stringent limits than primary were required. In fact, in many cases, more stringent limits than secondary were even often stipulated in the permits as a result of the use of these models. Moreover, the Clean Water Act now requires the minimum effluent standards of secondary to be set in discharge permits.

The secondary treatment of waste involves removing the "leftovers" from primary treatment. These leftovers are composed of colloidal and dissolved organic matters. Since their forms are colloidal and dissolved, they can no longer be removed by simple sedimentation. They must be transformed into solids that can then easily be settled. This transformation involves feeding them to microorganisms, mostly bacteria. As the bacteria feed on the colloidal and dissolved organic matters, they grow and multiply, thus converting those which were once colloidal and dissolved into solids that are capable of settling.

Kinetics of Growth and Food Utilization

The kinetics of the microbial process may conveniently be divided into kinetics of growth and kinetics of food (or substrate) utilization. Let X be the mixed population of microorganisms utilizing the organic waste. The rate of increase of the [X] is normally modeled as a first-order process as follows:

$$\frac{d[X]}{dt} = \mu[X] \tag{7-8}$$

where μ is the *specific growth rate* of the mixed population in units of per unit time and t is the time.

When an organism is surrounded by an abundance of food, the growth rate represented by equation (7-9) is at its maximum. This growth rate is said to be *logarithmic*. When the food is in short supply or depleted, the organisms will cannibalize each other. The growth rate at these conditions is said to be *endogenous*.

Monod[1] discovered that in pure cultures μ is a function of or is limited by the concentration [S] of a limiting substrate or nutrient and formulated the following equation:

$$\mu = \mu_{max} \frac{[S]}{K_s + [S]} \tag{7-9}$$

where μ_{max} is the maximum μ. In equation (7-9), if μ is made equal to one-half of μ_{max} and K_s solved, K_s will be found equal to [S]. Therefore, K_s is equal to the concentration

---

[1]J. Monod (1949). "The Growth of Bacterial Cultures." *Annual Review Microbiology III.*

of the substrate that will make the specific growth rate equal to one-half the maximum growth rate and hence is called the *half-velocity* constant. Although the equation applies only to pure cultures, if average population values of μ_{max} are used, it can be applied to the kinetics of mixed population as well.

In the dynamics of any population, some members are born, some members die, and some members simply grow in mass. In addition, in the absence of food, the population may cannibalize each other. The dynamics or kinetics of death and cannibalization may simply be represented mathematically as a decay of the population $k_d[X]$, where k_d is the rate of decay. Incorporating Monod's concept and the kinetics of death into equation (7-9), the model for the rate of increase of [X] now becomes

$$\frac{d[X]}{dt} = \mu_{max} \frac{[S]}{K_s + [S]}[X] - k_d[X] \qquad (7\text{-}10)$$

Substrate kinetics may be established by noting that as organisms grow, substrates are consumed. Therefore, the rate of decrease of the concentration of the substrate is proportional to the rate of increase of the concentration of the organisms. The first term on the right-hand side of equation (7-10) is the rate of increase of microorganisms that corresponds to the rate of decrease of the substrate. Hence the rate of decrease of the substrate, $-dS/dt$, is

$$-\frac{d[S]}{dt} = U\mu_{max} \frac{[S]}{K_s + [S]}[X] = \frac{1}{Y}\mu_{max} \frac{[S]}{K_s + [S]}[X] \qquad (7\text{-}11)$$

U is a proportionality constant called the specific substrate utilization rate; it is the reciprocal of Y, the specific yield of organisms. *Specific yield* simply means the mass of organisms produced per unit mass of substrate consumed, and *specific substrate utilization rate* simply means the mass of substrate consumed per unit mass of organisms produced.

For more discussions on kinetics and the chemistry of biological oxidation, see Chapter 2.

Culture Systems in Practical Applications

In the practical application of the foregoing kinetics, microorganisms are either suspended or attached. The activated sludge system and ponds and lagoons are examples of a *suspended-culture system*, while trickling filters, biotowers, and rotating biological contactors are examples of an *attached-culture system*. Figure 7-6a shows an activated sludge unit. The three units in Figure 7-6b are an aerobic pond, trickling filter, and aerated contact plates. The pond is a suspended-culture system, while the trickling filter and contact plates are attached-culture systems. The bottom portion shows a rotating biological contactor (RBC), which is also an attached-culture system. The term *suspended-culture system* means that the microorganisms are intermixed with, and hence convected by, the flow field. The term *attached-culture system*, on the other hand, means that the microorganisms are attached to a stratum or solid object, such as rocks and stones and disks.

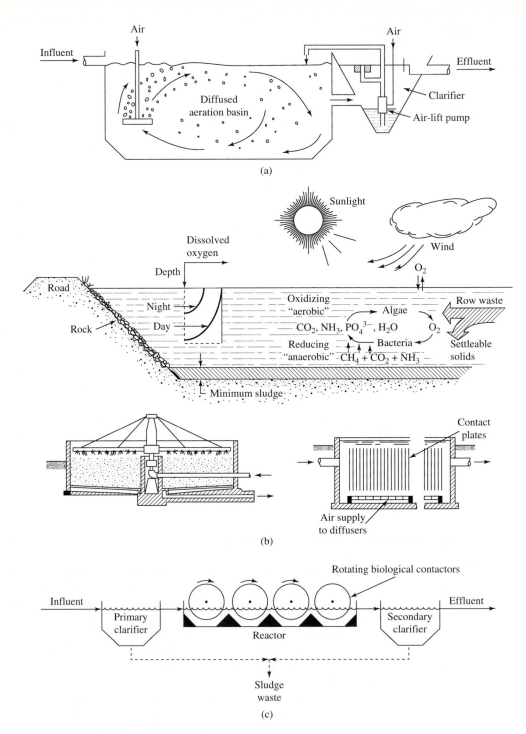

Figure 7-6 (a) Activated sludge unit. (b) Pond, trickling filter, and aerated contact plates. (c) Rotating biological contactor plant system.

The Activated Sludge System

Figure 7-7 shows schematics of the activated sludge system. Primary effluent is introduced into the reactor, where air is supplied for consumption by the aerobic microorganisms. In the reactor, the solubles and colloids are transformed into microbial masses. The effluent of the reactor then goes to the secondary clarifier, where the microbial masses are settled and separated from the clarified water. The clarified water is then discharged as effluent to a receiving stream. The settled microbial masses form the sludge. Some of the settled sludges are wasted and some are returned to the reactor. When once again exposed to the air in the reactor, the organisms become reinvigorated or activated—hence the term *activated sludge*. The wasting of the sludge is necessary so as not to cause a buildup of mass in the reactor.

In general, there are two extreme types of activated sludge processes: completely mixed and plug-flow. The schematics of these processes are shown in Figure 7-7. The two schematic diagrams are the same except for the concentrations of the parameters in the respective reactors. In the *completely mixed system*, there is only one concentration value in the reactor for a given parameter. For example, [X] at the influent end is equal to the [X] at the effluent end. The reason for this is that the process is completely mixed. On the other hand, in the *plug flow reactor*, there is no forward and backward intermixing of the flow field in the longitudinal direction, but every parcel of flow marches uniformly forward, in sequence, from the influent end toward the effluent end. The concentration in a plug-flow reactor is therefore not uniform.

Completely mixed process.   Develop the mathematics of the activated sludge process by performing a material balance on the completely mixed process, first. Let the symbols depicted in Figure 7-7 hold true where only Q_0 and V have not been defined before. Q_0 represents the rate of flow to the plant and V the volume of the reactor. The meanings of the various subscripts should be obvious.

The material balance is to be performed inside the domain surrounded by the dashed lines. This domain is the control volume and the control mass. (Note that from the Reynolds transport theorem, the boundaries of the control mass and the control volume are indentical; see Appendix 21.) Since the efficiency of the reactor is tied to the performance of the clarifier, the material balance is normally performed considering the reactor and clarifier together. Now, start with the material balance on X.

From fluid mechanics (or see Appendix 21) the Reynolds transport theorem is

$$\frac{D \int_M \phi \rho \, d\Psi}{Dt} = \frac{\partial \int_\Psi \phi \rho \, d\Psi}{\partial t} + \oint_A \phi \rho \mathbf{v} \bullet dA \qquad (7\text{-}12)$$

where ϕ is the property per unit mass, ρ the mass per unit volume of space of the control mass, M the control mass representation, Ψ the control volume, and $\mathbf{v}$ the velocity vector. In the present case, $\phi \rho = [X]$. Also, since the reaction occurs only in the reactor, $\Psi = V$. Hence

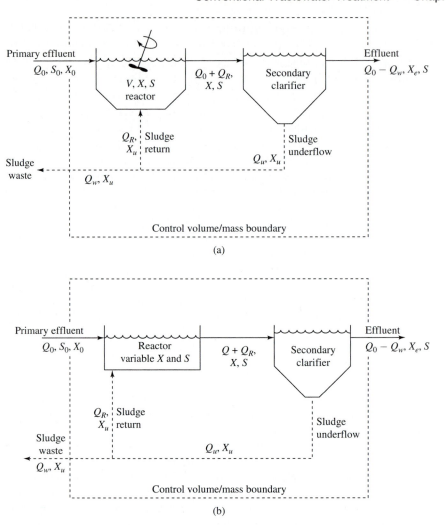

Figure 7-7   Schematics of the two general types of activated sludge systems: (a) completely mixed; (b) plug flow.

$$\frac{d \int_M \phi \rho d \forall}{dt} = \frac{d[X]}{dt} V \tag{7-13}$$

Equation (7-13) expresses the Lagrangian rate of increase of [X] in the reactor volume; it is a total derivative. The total derivative of [X] was found previously; combining it with equation (7-13) produces

$$\frac{d \int_M \phi \rho d \forall}{dt} = \frac{d[X]}{dt} V = \mu_{max} \frac{[S]}{K_s + [S]} [X] V - k_d [X] V \tag{7-14}$$

$\partial \int_{CV} \phi\rho d\text{V}/\partial t$ is equal to $(\partial[X]/\partial t)V$; it is a *local* rate of change of $[X]$—a partial derivative. By definition, at steady state, the partial derivative (not total derivative) of any parameter with respect to time is zero. Hence, assuming steady state, the local rate of change of $[X]$ is zero.

The convective rate of change of $[X]$ is

$$\oint_A \phi\rho\mathbf{v} \bullet dA = (-Q_0[X_0] + (Q_0 - Q_w)[X_e] + Q_w[X_u]) \tag{7-15}$$

Now, substituting everything into equation (7-12), we have

$$\mu_{\max}\frac{[S]}{K_s + [S]}[X]V - k_d[X]V = \frac{\partial[X]}{\partial t}V(=0) + \{(-Q_0[X_0] + (Q_0 - Q_w)[X_e] + Q_w[X_u]\} \tag{7-16}$$

After some algebraic manipulation, considering that $V(\partial[X]/dt) = 0$,

$$\mu_{\max}\frac{[S]}{K_s + [S]} = \frac{Q_w[X_u] + (Q_0 - Q_w)[X_e] - Q_0[X_0]}{V[X]} + k_d \tag{7-17}$$

The same mathematical operations may be applied to the substrate. These will not be repeated but the result will be written at once, which is

$$-V\frac{d[S]}{dt} = \frac{\mu_{\max}[S]}{YK_s + [S]}[X]V$$

$$= -\left(\frac{\partial[S]}{\partial t}V \ (= 0 \text{ at steady state}) + \{-Q_0[S_0] + (Q_0 - Q_w)[S] + Q_w[S]\}\right) \tag{7-18}$$

Equation (7-18) may be manipulated to produce

$$\mu_{\max}\frac{[S]}{K_s + [S]} = \frac{Q_0 Y}{V[X]}([S_0] - [S]) \tag{7-19}$$

Equating the right-hand sides of equations (7-17) and (7-19) and rearranging gives us

$$\frac{Q_w[X_u] + (Q_0 - Q_w)[X_e] - Q_0[X_0]}{V[X]} = \frac{Q_0 Y}{V[X]}([S_0] - [S]) - k_d \tag{7-20}$$

The ratio $V[X]/\{Q_w[X_u] + (Q_0 - Q_w)[X_e] - Q_0[X_0]\}$ is the biomass (or mixed-liquor volatile suspended solids, MLVSS—the numerator) in the reactor divided by the net rate of biomass wasting (the denominator). This ratio θ_c represents the average time that the biomass is in residence in the reactor. It is called by various names, such as *mean cell residence time* (MCRT), *sludge retention time* (SRT), and *sludge age*. The ratio $V/Q_0 = \theta$ is called the *nominal hydraulic retention time* (NHRT). The word *nominal* is used here, since θ is not the actual detention time of the tank. The actual detention time is $V/(Q_0 + Q_R)$, where Q_R is the recirculated flow. Using θ_c and θ in equation (7-20) and solving for $[X]$ gives

$$[X] = \frac{\theta_c Y([S_0] - [S])}{\theta(1 + k_d\theta_c)} \tag{7-21}$$

Equation (7-21) may also be solved for [S], producing

$$[S] = [S_0] - \frac{[X]\theta(1 + k_d\theta_c)}{\theta_c Y} \tag{7-22}$$

Equation (7-21) may also be transformed as follows:

$$\frac{[S_0] - [S]}{\theta[X]} = \frac{Q_0([S_0] - [S])}{V[X]} = \frac{1 + k_d\theta_c}{\theta_c Y} = U_{\text{net}} \tag{7-23}$$

U_{net} is the *net specific substrate utilization rate*. A term related to U is the *food/microorganism ratio*, F/M, defined as $Q_0[S_0]/V[X]$.

Another important parameter in the design and operation of an activated sludge plant is the *recirculation ratio* $R = Q_R/Q_0$. R may be obtained by performing a material balance around the secondary clarifier. Since the clarifier is not aerated, $d[X]/dt = 0$. Also, at steady state, $\partial[X]/\partial t = 0$. Adopting these facts and performing the material balance produces

$$(Q_0 + Q_R)[X] = (Q_0 - Q_w)[X_e] + (Q_R + Q_w)[X_u] \tag{7-24}$$

Solving for the ratio gives

$$\frac{Q_R}{Q_0} = R = \frac{Q_0([X] - [X_e]) - Q_w([X_u] - [X_e])}{Q_0([X_u] - [X])} \tag{7-25}$$

Plug flow process.   As indicated in Figure 7-7b, [X] and [S] are variable along the length of a plug flow reactor. Analyzing equation (7-21), however, in conjunction with Figure 7-7 shows that [S$_0$] and [S] are concentrations crossing the control boundary. To repeat, [S] is the concentration of S that has already crossed the control surface and must not be confused with the concentration of S in the bioreactor, which may be designated as [S̄], representing an average concentration. Hence since they are outside the control surface, [S$_0$] and [S] are insensitive to the form of the [X] inside the control volume. What they do is simply give the value of [X] inside the control volume and nothing more. Hence any form of [X] may be specified. Although [X] can have several values throughout the length of the reactor, a single value can surely represent these several values. This single value is nothing but the average. Hence designate [X] as the average and use the symbol [X̄]. Therefore, for the plug flow process, equation (7-21) becomes

$$[\overline{X}] = \frac{\theta_c Y([S_0] - [S])}{\theta(1 + k_d\theta_c)} \tag{7-26}$$

Accordingly, for plug-flow processes,

$$[S] = [S_0] - \frac{[\overline{X}]\theta(1 + k_d\theta_c)}{\theta_c Y} \tag{7-27}$$

$$\frac{Q_R}{Q_0} = R = \frac{Q_0([\overline{X}] - [X_e]) - Q_w([X_u] - [X_e])}{Q_0([X_u] - [\overline{X}])} \tag{7-28}$$

Generalization of formulations. From the foregoing discussions of the completely mixed and plug flow processes, the forms of the equations are similar. For this reason, equations for plug flow or completely mixed processes and for any variation of these processes (as discussed below) can be used generally to apply to the general activated sludge process. Also, unless otherwise specified, effluent [S] is always the soluble fraction. For example, effluent BOD_5 is always soluble BOD_5.

Variations of the activated sludge process. The completely mixed and plug flow processes are the extremes of the activated sludge process. In between these two extremes, however, are several variations: step aeration, tapered aeration, contact stabilization, pure-oxygen activated sludge, oxidation ditch, high rate, and extended aeration. In *step aeration*, the influent is introduced at intermediate points to provide more uniform BOD loading along the reactor tank. In *tapered aeration*, the amount of air is introduced proportionally to the amount of BOD still needed to be stabilized. This means adding more air at the influent end of the tank and progressively decreasing toward the effluent end. In the *contact stabilization* or *biosorption process*, the biomass adsorbs organics in a contact basin for a very short time of 15 to 60 min, the sludge settled out in a secondary clarifier, the settled sludge reaerated for approximately 90 min, and returned to the contact basin, where the biomass again adsorbs organics. The *oxidation ditch* variation is a closed-loop open channel shaped primarily like an ellipse. The wastewater is propelled around the loop by surface brush aerators called *Kessener* aerators. The *high-rate* variation is characterized by a short detention time and a high F/M ratio designed to operate at the logarithmic growth rate. The *extended-aeration* variation is the opposite of the high rate designed to operate at long detention times, low F/M ratios, and under the endogenous phase of growth. Figure 7-8 is a photograph of an activated sludge reactor at Back River sewage treatment plant in Baltimore, Maryland. Table 7-2 shows some design criteria for the several variations of the activated sludge process.

TABLE 7-2 SOME DESIGN PARAMETERS FOR VARIATIONS OF THE ACTIVATED SLUDGE PROCESS

Type of process	θ_c (days)	F/M (kg BOD_5 to kg MLSS)	θ (h)	MLSS (mg/L)	BOD^5 efficiency	Air supplied (m³/kg BOD_5)
Tapered aeration	4–15	0.3	5	1500–3000	90	50
Conventional	4–15	0.3	5	1500–3000	90	50
Step aeration	4–15	0.3	4	2000–3500	90	50
Completely mixed	4–15	0.3	4	3000–6000	90	50
Biosorption	—	0.5	0.3	1000–3000	85	—
High rate	4–15	0.5	1	4000–10,000	80	30
Pure oxygen	8–20	0.5	2	6000–8000	90	—
Extended aeration	20–30	0.1	20	3000–6000	85	100

Figure 7-8 Activated sludge reactor at Back River sewage treatment plant, Baltimore, Maryland.

Example 7-4

An activated sludge process is set to operate at an MCRT of 10 days and at a NHRT of 4 h. The volume of the bioreactor is 1600 m^3, and the underflow concentration X_u is 10,000 mg/L. If the MCRT and NHRT are to be maintained, at what rate should the sludge be wasted to maintain the [X] at 4000 mg/L of MLVSS? Neglect the influent and effluent biomass concentrations.

Solution

$$\theta_c = \frac{V[X]}{Q_w[X_u] + (Q_0 - Q_w)[X_e] - Q_0[X_0]}$$

$$\theta_c = \frac{V[X]}{Q_w[X_u]} \qquad 10 = \frac{1600(4000)}{Q_w(10,000)} \qquad Q_w = 64 \text{ m}^3/\text{day} \quad \textbf{Answer}$$

Example 7-5

Pilot-plant kinetic studies have established the following values: $Y = 0.5$ kg MLVSS/kg BOD$_5$, $k_d = 0.05$ per day. Q_0 is equal to 10,000 m^3/day, $[X_u]$ is equal to 10,000 mg/L, and Q_w is equal to 64 m^3/day. The plant is to be operated at a mean cell residence time of 10 days and a NHRT of 4 h. (a) If the influent to the bioreactor is 150 mg/L of BOD$_5$, what MLVSS should be maintained in the bioreactor to meet an effluent limit of 5 mg/L of BOD$_5$? (b) At what recirculation ratio is the plant to be operated?

Solution

(a)
$$[\overline{X}] = \frac{\theta_c Y([S_0] - [S])}{\theta(1 + k_d \theta_c)} = \frac{10(0.05)(0.15 - 0.005)}{(4/24)[(1 + 0.05(10)]}$$

$$= 0.29 \text{ kg/m}^3 = 2900 \text{ mg/L} \quad \textbf{Answer}$$

(b)
$$\frac{Q_R}{Q_0} = R = \frac{Q_0([\overline{X}] - [X_e]) - Q_w([X_u] - [X_e])}{Q_0([X_u] - [\overline{X}])}$$

Neglecting $[X_e]$ as small in comparison to $[\overline{X}]$ and $[X_u]$ gives

$$R = \frac{10{,}000(2900) - 64(10{,}000)}{10{,}000(10{,}000 - 2900)} = 0.40 \quad \textbf{Answer}$$

Resolve this problem assuming that $[X_e]$ is not negligible.

Example 7-6

(a) In Example 7-5, if the recirculation ratio is set at 10, what will be the resulting MLVSS? (b) What will be the resulting BOD$_5$ at the effluent? (c) What is the least amount of $[\overline{X}]$ that will make the effluent BOD$_5 = 0$?

Solution

(a)
$$\frac{Q_R}{Q_0} = R = \frac{Q_0([\overline{X}] - [X_e]) - Q_w([X_u] - [X_e])}{Q_0([X_u] - [\overline{X}])}$$

$$10 = \frac{10{,}000[\overline{X}] - 64(10{,}000)}{10{,}000(10{,}000 - [\overline{X}])} \qquad [\overline{X}] = 99{,}964 \text{ mg/L}$$

This number is theoretically possible if the dissolution of oxygen is unlimited. **Answer**

(b) $[S] = [S_0] - \dfrac{[\overline{X}]\theta(1 + k_d\theta_c)}{\theta_c Y} = 150 - \dfrac{99{,}964(4/24)[(1 + 0.05(10)]}{10(0.5)}$

$$= 150 - 4998 = -4848 \approx 0 \quad \textbf{Answer}$$

(c) $\qquad 0 = 150 - \dfrac{[\overline{X}](4/24)(1.5)}{5} \qquad [\overline{X}] = 3000 \text{ mg/L} \quad \textbf{Answer}$

Example 7-7

The influent of 10,000 m^3/day to a secondary reactor has a BOD$_5$ of 150 mg/L. It is desired to have an effluent BOD$_5$ of 5 mg/L, an MLVSS of 3000 mg/L, and an underflow concentration of 10,000 mg/L. Using the kinetic constants given in Example 7-5, design for the volume of the reactor. What are the volume and mass flows of sludge wasted per day?

Solution   Assume that $\theta_c = 10$ days. Therefore,

$$[\overline{X}] = \frac{\theta_c Y([S_0] - [S])}{\theta(1 + k_d \theta_c)} = \frac{Q_0 \theta_c Y([S_0] - [S])}{V(1 + k_d \theta_c)}$$

$$3 = \frac{10{,}000(10)(0.5)(0.15 - 0.005)}{V[(1 + 0.05(10)]} \qquad V = 1611 \text{ m}^3 \quad \textbf{Answer}$$

$$\theta_c = \frac{V[X]}{Q_w[X_u]}, \qquad \text{neglecting influent and effluent biomass concentrations}$$

$$10 = \frac{1611(3)}{Q_w[X_u]}; \; Q_w[X_u] = 483.3 \text{ kg/day} \quad \textbf{Answer}$$

$$Q_w = \frac{483.3}{10} = 48.33 \text{ m}^3/\text{day} \quad \textbf{Answer}$$

Example 7-8

For the reactor in Example 7-9, what recirculation ratio is required to produce an effluent BOD$_5$ of zero?

Solution $[\overline{X}]$ to produce a zero $[S] = 3{,}000$ mg/L, so

$$R = \frac{Q_0([\overline{X}] - [X_e]) - Q_w([X_u] - [X_e])}{Q_0([X_u] - [\overline{X}])} \qquad \text{neglect } [X_e]$$

$$= \frac{10{,}000(3000) - 64(10{,}000)}{10{,}000(10{,}000 - 3000)} = 0.42 \quad \textbf{Answer}$$

Determination of kinetic constants.   The following equation has been derived:

$$[S] = [S_0] - \frac{[\overline{X}]\theta(1 + k_d \theta_c)}{\theta_c Y} \tag{7-29}$$

Although this equation was derived for the plug flow process, it nevertheless represents any activated sludge process. It may be rearranged to produce

$$\frac{[S_0] - [S]}{[\overline{X}]\theta} = \frac{1}{\theta_c Y} + \frac{k_d}{Y} \tag{7-30}$$

Equation (7-30) is an equation of a straight line. $1/Y$ is the slope and k_d/Y is the y-intercept. Hence two pairs of values of $([S_0] - [S])/[\overline{X}]\theta$ and $1/\theta_c$ are all that are needed to obtain the parameters Y and k_d of this equation.

The following equation below has also been derived:

$$\mu_{max}\frac{[S]}{K_s + [S]} = \frac{Q_0 Y}{V[X]}([S_0] - [S]) \tag{7-31}$$

The parameters K_s and μ_{max} may be obtained from this equation, since this equation may be rearranged to give

$$\frac{\theta[X]}{[S_0] - [S]} = \frac{Y K_s}{\mu_{max}[S]} + \frac{Y}{\mu_{max}} \tag{7-32}$$

which is also an equation of a straight line. K_s/μ_{max} is the slope and Y/μ_{max} is the y-intercept. Also, from this equation only two pairs of values of $\theta[X]/([S_0] - [S])$ and $Y/[S]$ are needed to obtain the values of K_s and μ_{max}.

Ponds and lagoons. Ponds and lagoons are earthen basins in which wastewaters are introduced and retained for a period of time to effect the necessary degree of treatment. They are generally called *waste stabilization basins.* Although the terms are used interchangeably in practice, *ponds* are basins in which the supply of oxygen is provided through photosynthesis, whereas *lagoons* are basins in which the supply of oxygen is provided by artificial means. Ponds may be classified as aerobic and facultative. *Aerobic ponds* are shallow of depths usually less than 1 m, thereby assuring that dissolved oxygen is always present throughout the depths. In contrast, *facultative ponds* are deeper, normally measuring from 1 to 2.5 m. In this design three depth zones exist: the top fully aerobic zone, the bottom fully anaerobic zone, and the intermediate zone, which can be aerobic or anaerobic, depending on the degree of activity of biological decomposition and photosynthesis. Accordingly, in facultative ponds, facultative bacteria exist.

Lagoons may be classified as aerobic, facultative, and anaerobic lagoons. *Aerobic lagoons* are those in which artificial aeration is used to provide oxygen at all depths at all times. *Facultative lagoons*, also provided with artificial aeration, are similar to facultative ponds in that dissolved oxygen cannot be found at the deeper portion of the basin, where anaerobic conditions exist. However, as in facultative ponds, aerobic zones exit at the upper portions. Owing to the high loading of organic matter received, *anaerobic lagoons* are simply basins that are devoid of oxygen at all depths. Naturally, there is no need to provide artificial aeration. The principles of aeration are discussed later in this chapter. A symbiotic relationship exists between algae and bacteria in a pond. By virtue of photosynthesis, algae produce oxygen which bacteria then utilize in the degradation of waste. As bacteria decompose the waste, nutrients are liberated which algae can use—a symbiotic relationship.

Theoretically, all formulas derived for the activated sludge process should apply to ponds and lagoons—were it not that the MLVSS in the basin is composed of, in addition to the normal BOD-degrading bacteria, a large concentration of algae. The MLVSS in the formulas derived are only for microorganisms directly responsible for waste stabilization. The algae in the basin are not directly responsible; hence the MLVSS from algae cannot be substituted into the formulas. In theory, to apply the formula, the MLVSS due to the bacteria must be separated from the MLVSS due to algae. This, of course, is a formidable undertaking.

Because of the drawback noted above and because the operation of the waste stabilization basin is largely controlled by the weather, designs are normally based on loading factors. Depending on the type of stabilization basin, one can find design criteria and factors tabulated in the literature. These are based on actual field experiences. Table 7-3 shows some design parameters for facultative ponds and lagoons. Another drawback is the high concentration of suspended solids in the effluent of stabilization basins. The effluent suspended solids of 35 to 100 mg/L will surely not meet the secondary standard requirement of the Clean Water Act. Some means must be applied to remove the algae in the effluent. Algae are responsible for the high suspended solids. When discharged into a receiving stream, these algae decompose, resulting in a high dissolved oxygen demand.

TABLE 7-3 SOME DESIGN PARAMETERS FOR FACULTATIVE PONDS AND LAGOONS

Parameter	Facultative pond	Facultative lagoon
Pond size (ha)	0.5–5	0.5–5
Operation	Series of parallel	Series or parallel
Detention time (days)	30	30
Depth (m)	1.5	1.5
pH	6.5–9.0	6.5–9.0
BOD_5 specific loading rate (kg/ha-day)	15–100	50–200
BOD_5 destruction	90%	90%
Algal concentration (mg/L)	15–90	20
Effluent suspended solids (mg/L)	33–100	30–100

Example 7-9

Design an oxidation pond to treat 5000 m^3/day of municipal wastewater. The average influent BOD_5 is 150 mg/L and it is desired to have an average effluent BOD_5 of 5.0 mg/L.

Solution   From Table 7-3 assume a detention time of 30 days; then

$$\text{volume of pond} = 30(5000) = 150,000 \ m^3$$

Assuming a depth of 1 m, surface area = 150,000 m^2 = 15 ha.

$$1 \ mg/L = 10^3 \ kg/m^3 \qquad 150 \ mg/L \ BOD_5 = 0.15 \ kg/m^3 \ BOD_5$$

$$BOD_5 \ \text{load} = 0.15(5000) = 750 \ kg/day$$

So BOD_5 loading rate = 750/15 = 50 kg/ha-day. From Table 7-3, the BOD_5 loading rate = 15 to 90 kg/ha-day. Since the range is so wide, there is no guarantee that this design will work. One alternative would be to perform a pilot study in the given area. The results may then be used as the loading factor for your design.

$$\text{Pond area} = 15 \ \text{ha} \quad \textbf{Answer}$$

$$\text{Pond depth} = 1 \ \text{m} \quad \textbf{Answer}$$

Trickling Filters, Biotowers, and RBCs

These three types of systems have similar principles of treatment, and for simplicity, let us call them attached-growth processes. They rely on microorganisms attached and living on some medium to extract the nutrient contained in a wastewater flow passing by them.

Since the [X] in the formulation for the activated sludge process is admixed in the liquid, while the [X] in the attached-growth process is not but is attached on a growth medium, a new derivation is needed. Again, this derivation will apply the Reynolds transport theorem (see Appendix 21).

Figure 7-9 shows an element of filter (or attached-growth) medium, microbial, and liquid films. w is the width of microbial and liquid films considered, h the thickness of the flow film passing by the microbial film, and dz the differential length of the films in the direction of flow. These dimensions define the volume and boundary of the control mass.

Consider the rate of decrease of the nutrients inside the control mass. There are two ways that this change can occur: by diffusion of the nutrients out of the control mass and by decay. Since this change occurs in the control mass, it is the Lagrangian derivative: $D[S]/Dt$. The consumption of the substrate, however, does not take place in the liquid film but in the microbial film. Hence this decay may be considered zero.

On the other hand, the nutrients in the liquid film will diffuse toward the microbial film as long as they are consumed by the microorganisms. This rate of diffusion dM_s/dt is proportional to the microbial activity in the film and to the concentration of substrate [S] in the liquid film, where M_s is the mass of substrate diffusing. For a diffusion area of $w\,dz$, dM_s/dt is given by

$$-\frac{dM_s}{dt} = K[X_f][S]w\,dz \tag{7-33}$$

where K is a proportionality constant.

Once the microbial film is established, $[X_f]$ remains constant, and equation (7-33) can be revised to

$$-\frac{dM_s}{dt} = K'[S]w\,dz \tag{7-34}$$

The Lagrangian derivative becomes

$$wh\,dz\left(-\frac{D[S]}{Dt}\right) = -\frac{dM_s}{dt} = K'[S]w\,dz \tag{7-35}$$

Now, consider the Eulerian derivative. The local derivative is $\partial([S]wh\,dz)/\partial t$, and the convective derivative is $-Q_a[S] + Q_a([S] + (\partial[S]/\partial z)dz)$. The Eulerian derivative is equal to the local derivative plus the convective derivative. Hence by the Reynolds transport theorem (Lagrangian derivative = Eulerian derivative),

$$-K'[S]w\,dz = -\frac{\partial}{\partial t}([S]wh\,dz) + \left\{-Q_a[S] + Q_a\left([S] + \frac{\partial[S]}{\partial z}, dz\right)\right\} \tag{7-36}$$

At steady state, the partial derivative with respect to time is equal to zero. Also, the flow process may be considered one-dimensional in the direction of z, where z is the flow direction not necessarily straight but may be curvilinear. Thus all partial derivatives become total derivatives. Simplifying equation (7-36) and integrating, the result is

$$\ln\frac{[S]}{[S_a]} = -\frac{K'wz}{Q_a} = -\frac{K'A_s}{Q_a} = -\frac{K'}{Q_a/A_s} \tag{7-37}$$

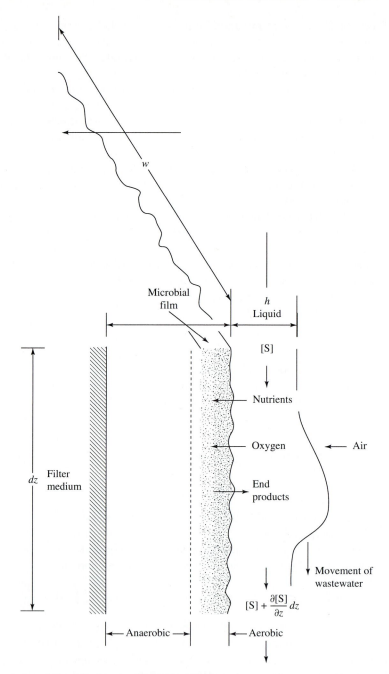

Figure 7-9 Schematic representation of exchanges taking place along surfaces of biological slimes in trickling filters and rotating biological contactors.

$A_s = wz$ is the total surface of the microbial film. The ratio Q_a/A_s is called the *actual specific hydraulic loading rate* (ASHLR). ASHLR is different from the *nominal specific hydraulic loading rate* (NSHLR), which is the plant hydraulic loading Q_0 divided by the surficial area. When the hydraulic loading or specific hydraulic loading rate is mentioned, the nominal is usually what is meant. The organic loading corresponding to ASHLR is the *actual specific organic loading rate* (ASOLR), while that for NSHLR is the *nominal specific organic loading rate* (NSOLR). $[S_a]$ is the influent substrate concentration resulting from the mixing of the plant influent flow Q_0 and the recycle flow Q_R. If a material balance is performed on $[S]$ between Q_0 and Q_R, $[S_a]$ will be found to be equal to $[S_a] = ([S_0] + R[S])/(1 + R)$. $Q_0 + Q_R$ is equal to Q_a.

As opposed to suspended-growth processes, recirculation in attached-growth processes does not improved efficiency. In addition, the settled sludge cannot be recirculated, since the solids can clog the filter. Recirculation of the effluent is, however, used to provide a more uniform hydraulic loading and to effect a more thorough removal of the BOD.

The rate of diffusion into the microbial film is a function of the film activity. Since microbial activity is a function of temperature, it follows that K' is also a function of temperature. In addition, K' must also be a function of the diffusional characteristics of the particular waste molecules. The temperature dependency of K' is given by the Arrhehius equation,

$$K'_T = K'_{20}(1.035)^{T-20} \qquad (7\text{-}38)$$

The value of K' can vary over a wide range, depending on the type of waste. For this reason, for a given specific design, it is advisable to perform a pilot-plant study.

Table 7-4 shows some design parameters for trickling filters. Of course, biotowers are trickling filters that are tall, and would have similar characteristics as the conventional

TABLE 7-4 TYPICAL DESIGN CRITERIA FOR TRICKLING FILTERS

Item	Low-rate	Intermediate	High-rate	Super-rate
NSHLR[a] (m^3/m^2-day)	1–4	4–10	10–40	40–200
NSOLR[a], (kg BOD_5 per m^3/day)	0.01–0.1	0.2–0.5	0.3–1.0	1–6
Depth (m)	1–3	1–3	1.0–2.0	5–10
Recirculation ratio	0	0–1	1–3	1–4
Filter media	Rock, slag, etc.	Rock, slag, etc.	Rock, slag, synthetics	Synthetics
Power (kW/10^3 m^3)	2–5	2–5	5–10	10–20
Filter flies	Many	Intermediate	Few, larvae are washed away	Few or none
Sloughing	Intermittent	Intermittent	Continuous	Continuous
Dosing intervals	Not more than 5 min	10–50 s	<15 s	Continuous
Effluent	Usually fully nitrified	Partially nitrified	Nitrified at low loadings	Nitrified at low loadings

[a]Based on surficial areas.

trickling filters. As indicated in the table, trickling filters are classified as low-rate, intermediate, high rate, and super-rate (roughing) filters. Table 7-5 shows some properties of filter media that have been used. Finally, Table 7-6 shows performance characteristics of an RBC from a chart of Autotrol Corporation. It must be emphasized that the final design of RBCs should be based on results of pilot plant studies for a particular waste rather than using generalized values such as those in Table 7-6.

RBC is a unique adaptation of the attached-growth process. The solid objects upon which the microbial films are attached are flat disks that range in diameter from 2 to 4 m and up to 1 cm in thickness. The disks are mounted at spacings of about 30 to 40 mm on a common shaft that rotates at approximately 1 to 2 rpm. One shaftful of media is called a *module*.

For computation of the specific hydraulic loading rate, the surficial area and the actual surface area are equal and given by the microbial surface area. This area corresponds to the sum of the areas of the two circular surfaces on the two sides of the disk multiplied by the total number of disks mounted on the shaft. The holes cut off by the

TABLE 7-5 PHYSICAL PROPERTIES OF TRICKLING FILTER MEDIA

Medium	Nominal size (mm)	Mass/unit bulk volume (kg/m^3)	Specific surface area (m^2/m^3)	Void space (%)
River rock				
Small	30–50	1200–1500	55–70	40–50
Large	100–130	800–1000	40–50	50–60
Blast furnace slag				
Small	40–80	800–1000	50–80	45–55
Large	80–130	800–1000	50–65	55–65
Plastic				
Conventional	600×600×1200	30–100	80–100	95–97
High specific surface	600×600×1200	30–100	100–200	95–97
Redwood	1200×1200×500	150–175	40–50	75–80

TABLE 7-6 PERFORMANCE CHARACTERISTICS OF A ROTATING BIOLOGICAL CONTACTOR FOR TEMPERATURES GREATER THAN 55°C

$[S_0]$ (mg/L)	$[S]$ (mg/L)	$[S]/[S_0]$	NSHLR (L/day-m^2)	$[S_0]$ (mg/L)	$[S]$ (mg/L)	$[S]/[S_0]$	NSHLR (L/day-m^2)
150	20	0.13	72	150	15	0.10	61
150	5	0.03	32	100	20	0.20	108
100	15	0.15	94	100	5	0.05	48
60	20	0.33	179	60	15	0.25	156
60	5	0.08	83	40	20	0.50	268
40	15	0.38	240	40	5	0.13	126
30	20	0.67	360	30	15	0.50	324

shaft through the mounted disks and the area on the rim edges may be neglected. A module 3.7 m in diameter and 7.6 m long can contain some 10,000 m^2 of microbial film surface. For the trickling filter, the microbial surface area may be computed with the aid of Table 7-5, which gives the specific surface area for a particular medium.

Example 7-10

(a) A trickling filter bed composed of small river rocks is 3 m deep and 6 m in diameter. If the plant inflow Q_0 is 10,000 m^3/day and the recirculation ratio is 2, what are the NSHLR and the ASHLR? (b) Using the same data as in part (a), what are the NSHLR and the ASHLR if an RBC is used? There are 40 modules used with disks measuring 3.7 m in diameter, 5 mm in thickness, and spaced on the shaft at 30 mm apart. The length of the module is 7.6 m.

Solution (a) From Table 7-5, the specific surface area $= 55$ to 70 m^2/m^3, say 60.

$$A_s = (60) \left[\frac{\pi(6^2)}{4} \right] (3) = 5089.38 \text{ m}^2 \qquad \text{surficial area} = \frac{\pi(6^2)}{4} = 28.27$$

$$\text{NSHLR} = \frac{10,000}{28.27} = 353.73 \frac{\text{m}^3}{\text{m}^2\text{-day}} \quad \textbf{Answer}$$

$$\text{ASHLR} = \frac{10,000 + 2(10,000)}{5089.38} = 5.89 \frac{\text{m}^3}{\text{m}^2\text{-day}} \quad \textbf{Answer}$$

(b)   Number of disks per module $= \dfrac{7600 - 2\left(\frac{5}{2}\right)}{30} + 1$

$$= 253.17 + 1, \text{ say } 254$$

$$A_s = 40(254) \left(2 \left[\frac{\pi(3.7^2)}{4} \right] \right) = 218,482.69 \text{ m}^2$$

$$\text{NSHLR} = \frac{10,000}{218,482.69} = 0.046 \frac{\text{m}^3}{\text{m}^2\text{-day}} \quad \textbf{Answer}$$

$$\text{ASHLR} = \frac{10,000 + 2(10,000)}{218,482.69} = 0.137 \frac{\text{m}^3}{\text{m}^2\text{-day}} \quad \textbf{Answer}$$

Example 7-11

A biotower composed of conventional plastic medium is used to treat the effluent of a primary clarifier with a BOD$_5$ of 150 mg/L. The flow from the clarifier is 20,000 m^3/day. From a pilot-plant study, K' was determined to be 0.117 m/day. Two towers are to be used, each with a square surface and separated by a common wall and a depth of 7.0 m. A recirculation ratio of 2 is to be used. Calculate the dimensions of the units that will produce an effluent of 10 mg/L of soluble BOD$_5$.

Solution

$$\ln \frac{[S]}{[S_a]} = -\frac{K'}{Q_a/A_s}$$

$$Q_a = 20{,}000 + 2(20{,}000) = 60{,}000 \text{ m}^3/\text{day}$$

$$[S_a] = \frac{[S_0] + R[S]}{1 + R} = \frac{150 + 2(10)}{3} = 56.67 \text{ mg/L}$$

$$\ln \frac{10}{56.67} = -\frac{0.117}{60{,}000/A_s} \qquad A_s = 889{,}569.17 \text{ m}^2$$

From Table 7-5, specific surface area = 80 to 100 m²/m³, say 90. Then

$$\text{total bed volume} = \frac{889{,}569.17}{90} = 9884.10 \text{ m}^3$$

$$\text{volume per tower} = \frac{9884.10}{2} = 4942.05 \text{ m}^3$$

$$\text{surface area} = \frac{4{,}942.05}{7} = 706.0 \text{ m}^2$$

$$\text{length} = \text{width} = 26.57 \text{ m} \quad \textbf{Answer}$$

Example 7-12

From the data for the RBC in Table 7-6, determine K'.

Solution

$$\ln \frac{[S]}{[S_a]} = -\frac{K'}{Q_a/A_s} \qquad K' = -\frac{Q_a}{A_s} \ln \frac{[S]}{[S_a]}$$

Assume that NSHLR = ASHLR = Q_a/A_s.

$$\text{Average} \ln \frac{[S]}{[S_0]} = \frac{\ln 0.133 + \ln 0.033 + \cdots + \ln 0.125 + \ln 0.50}{14} = \frac{-24.6}{14} = -1.72$$

$$\text{Average ASHLR} = \frac{72 + 32 + \cdots + 126 + 324}{14} = \frac{2151}{14} = 153.64$$

$$K' = -153.64(-1.72) = 264.04 \ \frac{\text{L}}{\text{m}^2\text{-day}} = 0.264 \ \frac{\text{m}}{\text{day}} \quad \textbf{Answer}$$

Example 7-13

Determine the surface area of an RBC required to treat the wastewater in Example 7-11. Use the K' value obtained in Example 7-12. If the surface area A_s per module is 10,000 m², determine the number of modules to be used.

Solution

$$\ln \frac{[S]}{[S_a]} = -\frac{K'}{Q_a/A_s}$$

$$\ln\frac{10}{56.67} = -\frac{0.264}{60,000/A_s} \qquad A_s = 394,240.88 \text{ m}^2 \quad \textbf{Answer}$$

$$\text{Number of modules} = \frac{394,240.88}{10,000} = 39.42, \quad \text{say 40 modules} \quad \textbf{Answer}$$

PRINCIPLES OF AERATION

Oxygen is a necessary nutrient. In attached-growth processes, air is easily absorbed into the wastewater that flows by the microbial film. In trickling processes, for example, due to the heat of reaction, air in the bed becomes lighter than the surrounding ambient air and thus is convected upward through the filter. This upward convection is met with the countercurrent downward flow of the wastewater, which due to the turbulence caused in flowing through the "rocks," induces aeration. In the RBC, a similar process occurs; as the shaft rotates a falling thin film of water glides along the faces of the disks, which due to the turbulence of the flow and a sufficiently thin film of water, induces aeration.

In suspended-growth processes, however, air must literally be forced into the liquid. In wastewater treatment practice, three types of aeration units are used: the bubble diffuser type, the turbine aerator type, and the surface aerator type. These types of units are shown in Figure 7-10.

Small-bubble diffusion units such as porous media, plates, or tubes are made of silicon dioxide (SiO_2) or aluminum oxide (Al_2O_3) grains fused together by a ceramic binder to form a porous mass. Compressed air is blown through the small orifice openings in the units, creating fine bubbles that, owing to the high surface area created in the bubbles, provide oxygen transfer from the gaseous phase to the liquid phase. The bubbles may also be created by large orifices, creating larger bubbles. Large-bubble diffusers, however, are not as efficient as small-bubble diffusers. The turbine-type aerator has an air sparger below the turbine agitator. As the air is sparged, bubbles rise toward the turbine impeller, the blades shearing the larger bubbles into smaller ones for more effective oxygen transfer. The surface aerator has a propeller that draws water from just beneath the surface of the water. The water is then sprayed into the air, aerating it.

The basic mathematical treatment of oxygen mass transfer from air to water is based on the concept of a driving force. As in any other situation, mass will always tend to fill a void space. If water is void of oxygen, oxygen will "flow" into it; the greater the void, the faster will be the rate. This void is the driving force. The greatest void is when the dissolved oxygen is zero, and the smallest void is when the water is almost saturated with oxygen. In between these two extremes of voidness is the difference between the saturation value of the dissolved oxygen $[C_s]$ and the actual concentration of dissolved oxygen $[C]$. This difference, $[C_s] - [C]$, is a void; it is the driving force for oxygen mass transfer to water at any time. The rate at which the concentration of oxygen will

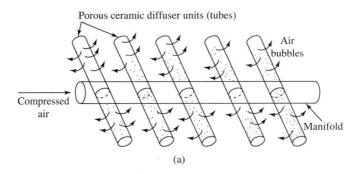

(a)

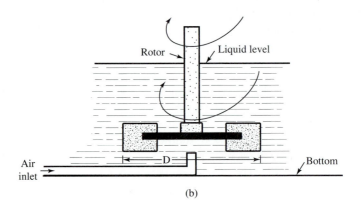

(b)

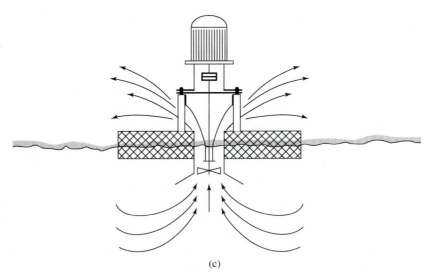

(c)

Figure 7-10   Types of aerators used in wastewater treatment: (a) bubble diffusers; (b) turbine aerator; (c) surface aerator.

increase is $(d[C]/dt)$. Similar to reaeration in streams encountered in Chapter 5, this total differential increase is also equal to $k_2([C_s] - [C])$. In equation form,

$$\frac{d[C]}{dt} = k_2([C_s] - [C]) \tag{7-39}$$

where k_2 is the reaeration coefficient. k_2 is the reaeration coefficient used in modeling streams. In wastewater treatment, however, k_2 is changed to $K_L a$. Equation (7-39) may also be written for point values. These are indicated by putting a caret above the symbols. Using $K_L a$ for k_2, the equation becomes

$$\frac{d[\hat{C}]}{dt} = K_L a([C_s] - [\hat{C}]) \tag{7-40}$$

For the volume of the reactor, the concentration to be used should be the average. Hence the equation becomes

$$\frac{d[\overline{C}]}{dt} = K_L a([C_s] - [\overline{C}]) \tag{7-41}$$

where the overbars designate average values.

Equipment Specification

The rating of aeration equipment is reported at standard conditions defined as 20°C, 1 atm pressure, and 0 mg/L of dissolved oxygen concentration in tap water (or distilled water). Under these conditions, equation (7-41) becomes

$$\frac{d[\overline{C}]}{dt} = (K_L a)_{\ell,20}[C_{s,20,sp}] \tag{7-42}$$

$(K_L a)_{\ell,20}$ = the $K_L a$ at standard conditions and $C_{s,20,sp}$ = the saturation DO at 20°C and standard pressure. Equation (7-42) is the standard oxygen rate (SOR). Equipment are specified in terms of SOR. Since testing is not normally done at standard conditions, $(K_L a)_{\ell,20}$ must be obtained from the $K_L a$ at the condition of testing by the Arrhenius temperature relation,

$$K_L a = (K_L a)_{\ell,20}(\theta^{T-20}) \tag{7-43}$$

where θ is the temperature correction factor and T is the temperature in degrees Celsius at testing conditions. This equation assumes that the effect of pressure on $K_L a$ is negligible.

The ability of an aeration equipment to transfer oxygen at field conditions is, using equation (7-41),

$$\frac{d[\overline{C}]}{dt} = (K_L a)_w([C_{s,w}] - [\overline{C}]) \tag{7-44}$$

where $(K_L a)_w$ and $[C_{s,w}]$ are the $K_L a$ and $[C_s]$ of the wastewater at field conditions, respectively. Equation (7-44) represents the actual oxygenation rate (AOR). $(K_L a)_w$ may also be expressed in terms of its value at 20°C, $(K_L a)_{w,\ell,20}$, by the temperature relation

$$(K_L a)_w = (K_L a)_{w,\ell,20}(\theta^{T-20}) \tag{7-45}$$

From equations (7-42), (7-44), (7-45), we find that

$$SOR = \frac{AOR}{\alpha[(\beta[C_s] - [\overline{C}])/C_{s,20,sp}]\theta^{T-20}} \qquad (7\text{-}46)$$

where $\alpha = (K_La)_{w,\ell,20}/(K_La)_{\ell,20}$.

The method of determining AOR is discussed later. Once AOR is obtained, SOR can be calculated using equation (7-46); thus the size of the equipment can be specified.

Determination of Aeration Parameters

The environmental engineer specifies aeration equipment based on standard laboratory tests performed by equipment manufacturers. In other words, although the engineer can easily determine the AOR, AOR must still be converted to SOR to match with the standard manufacturer's value. As shown in equation (7-46), to perform this conversion, the α and β parameters must be determined.

   Determination of β.   The determination of β is very simple. Take a liter jar and fill it halfway with the sample. The jar is then vigorously shaken to saturate the sample with air or oxygen and the dissolved oxygen concentration measured. Table 7-7 shows the saturated concentrations of dissolved oxygen in clean water at 1 atm at several temperatures. From this table, at the temperature corresponding to the temperature of the experiment, the saturation DO for clean water can be obtained. From this, along with the saturated DO of the sample determined in the experiment $C_{s,w}$, β may be calculated as

$$\beta = \frac{[C_{s,w}]}{[C_s]} \qquad (7\text{-}47)$$

TABLE 7-7   SATURATION DO IN CLEAN WATER AT 1 ATM PRESSURE

Temperature (°C)	Saturation dissolved oxygen (mg/L)	Temperature (°C)	Saturation dissolved oxygen (mg/L)
0	14.60	18	9.45
5	12.75	19	9.26
6	12.43	20	9.07
7	12.12	21	8.90
8	11.83	22	8.72
9	11.55	23	8.56
10	11.27	24	8.40
11	11.01	25	8.24
12	10.76	26	8.09
13	10.52	27	7.95
14	10.29	28	7.81
15	10.07	29	7.67
16	9.85	30	7.54
17	9.65		

If the test is being conducted at elevation other than zero or 1 atm of barometric pressure, the $[C_s]$ obtained from Table 7-7 must be corrected for the pressure of the test. From fluid mechanics, the variation of atmospheric pressure P_b with altitude z in the trophosphere is

$$P_b = P_{b0} \left(1 - \frac{Bz}{T_0} \right)^{g/RB} \tag{7-48}$$

where P_{b0} is the barometric pressure at $z = 0$; B (the temperature lapse rate) = 0.00650 K/m for standard atmosphere; T_0 is the temperature in K at $z = 0$; $g = 9.81$ m/s^2; and R (the gas constant for air) = 286.9 N-m/kg-K. This equation assumes that the temperature varies linearly with altitude. Therefore, if the elevation where the test is conducted is known, the corresponding barometric pressure can be found. Since $[C_s]$ varies directly as the pressure and if the condition of test is at barometric pressure, the $[C_s]$ at test conditions will be

$$[C_s] = \frac{P_b}{P_s} [C_{s,sp}] \tag{7-49}$$

where $[C_{s,sp}]$ is the $[C_s]$ from Table 7-7 at the standard pressure of $P_s (= 760$ mmHg).

Pressures corresponding to $[C_s]$.   In sizing aerators, what pressure to use when determining $[C_s]$ corresponds to the type of aeration device used. For surface aerators, the corresponding pressure should be taken at the surface of the tank. For the case of bubble-diffusion and turbine-type aerators, however, since the point of release of the air is submerged, the pressure must correspond to the average depth of submergence. If the submergence depth is Z_d, the corresponding average pressure P is

$$P = P_b + \frac{Z_d}{2} \gamma \tag{7-50}$$

where γ is the specific weight of the mixed liquor.

Determination of α.   To determine α, equations (7-41) and (7-44) are used, with equation (7-44) modified for aerobic respiration. From these equations a method can be devised to determine the $K_L a$ values. The parameter α may then be calculated. First, determine $(K_L a)_{\ell,20}$. Integrating equation (7-41), we have

$$\ln \frac{[C_s] - [\overline{C}]}{[C_s]} = -K_L a t \tag{7-51}$$

This equation is an equation of a straight line between the logarithmic function and t. To determine $K_L a$, assume that there are n pairs of experimental values of $[\overline{C}]$ and t. The equation may then be summed for these pairs of values to obtain

$$\sum_{m=1}^{m=n} \left(\ln \frac{[C_s] - [\overline{C}]}{[C_s]} \right)_m = -K_L a \sum_{m=1}^{m=n} t_m \tag{7-52}$$

From this equation, $K_L a$ may be evaluated, which can then be corrected to obtain $(K_L a)_{\ell,20}$ using the temperature relation.

Since the organisms in wastewater respire, oxygen utilization must be incorporated. Calling the respiration rate $\bar{r}$, equation (7-44) is modified to

$$\frac{d[\overline{C}]}{dt} = (K_La)_w([C_{s,w}] - [\overline{C}]) - \bar{r} \qquad (7\text{-}53)$$

Equation (7-53) may be rewritten as

$$\frac{d[\overline{C}]}{dt} = (K_La)_w([C_{s,w}] - [\overline{C}]) - \bar{r} = (K'_La)_w([C_{s,w}] - [\overline{C}]) \qquad (7\text{-}54)$$

where $(K'_La)_w$ is an apparent overall mass transfer coefficient. It encompasses both $(K_La)_w$, the true overall mass transfer coefficient, and $\bar{r}$.

The second part of equation (7-54) is similar in form to equation (7-41). It can therefore be manipulated to obtain an equation similar to equation (7-52) only using to $(K'_La)_w$,

$$\sum_{m=1}^{m=n} \left(\ln \frac{[C_{s,w}] - [\overline{C}]}{[C_{s,w}]} \right)_m = -(K'_La)_w \sum_{m=1}^{m=n} t_m \qquad (7\text{-}55)$$

$(K'_La)_a$ may now be solved.

For n values of $[\overline{C}]$, equation (7-54) may be written as

$$K_{La,w} \sum_{m=1}^{m=n} ([C_{s,w}] - [\overline{C}])_m - \sum_{m=1}^{m=n} \bar{r}_m = (K'_La)_w \sum_{m=1}^{m=n} ([C_{s,w}] - [\overline{C}]))_m \qquad (7\text{-}56)$$

from which $(K_La)_w$ may be solved once $\bar{r}$ is determined in a separate test and once $(K'_La)_w$ has been calculated using equation (7-55). $(K_La)_w$ may now be corrected to obtain $(K_La)_{w,\ell,20}$ using the temperature relation. Once $(K_La)_{\ell,20}$ and $(K_La)_{w,\ell,20}$ are known, α can be computed.

The actual laboratory experimentation involves deaerating the sample. This is done by purging nitrogen gas into the sample to drive off the dissolved oxygen or by consuming the dissolved oxygen using sodium sulfite (Na_2SO_3) with cobalt chloride ($CoCl_2$) added as a catalyst. The sulfite converts to sulfate when reacted with the dissolved oxygen. From the stoichiometry of the reaction, about 7.9 ppm of the sulfite is needed per ppm of the dissolved oxygen. A 10 to 20% excess is normally used. Cobalt chloride has been used in concentration of 1.5 ppm. As soon as the sample is completely deoxygenated, reaeration is allowed to take place using the type of aeration system to be employed in the prototype, such as bubble-diffusion, turbine, or surface-aeration type. The increase in dissolved oxygen concentration with respect to time is monitored. The data obtained are then used to calculate the aeration parameters.

Calculation of oxygen requirement, the AOR.   Imagine a control mass laden with dissolved oxygen and pollutants. The concentration of dissolved oxygen in this mass will change according to the rate of aeration, $(K_La)_w([C_{s,w}] - [\overline{C}]) = \text{AOR}$, and the rate of respiration by the organisms, $\bar{r}$, in consuming the pollutants. For this process, the following equation has been derived:

$$\frac{d[\overline{C}]}{dt} = (K_L a)_w ([C_{s,w}] - [\overline{C}]) - \overline{r} \tag{7-57}$$

As a Lagrangian derivative, equation (7-58) is

$$\frac{D[\overline{C}]}{Dt} = (K_L a)_w ([C_{s,w}] - [\overline{C}]) - \overline{r} = \text{AOR} - \overline{r} \tag{7-58}$$

Now, derive the Eulerian derivative [counterpart to equation (7-58)]. The local derivative is $\partial[\overline{C}]/\partial t$ and the convective derivative is $-Q_0[\overline{C}_0]] + (Q_0 - Q_w)[\overline{C}_e]$, where the Q's had been defined before; $[\overline{C}_0]$ is the inflow dissolved oxygen concentration to the reactor; and $[\overline{C}_e]$ is the outflow dissolved oxygen concentration from the secondary clarifier. By the Reynolds transport theorem,

$$(V)\frac{D[\overline{C}]}{Dt} = V(\text{AOR} - \overline{r}) = (V)\frac{\partial[\overline{C}]}{\partial t} - Q_0[\overline{C}_0] + (Q_0 - Q_w)[\overline{C}_e] \tag{7-59}$$

At steady state, the partial derivative is zero; also, $[\overline{C}_0]$ and $[\overline{C}_e]$ are equal to zero. Hence

$$\text{AOR} = \overline{r} \tag{7-60}$$

The respiration rate, $\overline{r}$, is due to the consumption of BOD. BOD is composed of CBOD and NBOD. Let $\overline{r}_{s0}$ be the respiration due to CBOD and $\overline{r}_{n0}$ be the respiration due to NBOD. $\overline{r}$ is then

$$\overline{r} = \overline{r}_{s0} + \overline{r}_{n0} \tag{7-61}$$

Now, apply the Reynolds transport theorem to the fate of the CBOD. The Lagrangian rate of decrease of CBOD in the control mass, $-D[S]/Dt$, is due to its consumption to produce energy by respiration, $\overline{r}_s$, and to its consumption in synthesis for growth or to replace dead cells, syn_s:

$$-\frac{D[S]}{Dt} = \overline{r}_s + \text{syn}_s \tag{7-62}$$

For the Eulerian derivative, the local derivative is $\partial[S]/\partial t$ and the convective derivative is $-Q_0[S_0] + Q_0[S]$. At steady state, the partial derivative is equal to zero. Hence from the Reynolds transport theorem, the Lagrangian derivative is equal to the Eulerian derivative:

$$(V)\frac{D[S]}{Dt} = (-\overline{r}_s - \text{syn}_s)V = -Q_0[S_0] + Q_0[S] \tag{7-63}$$

$$V(\overline{r}_s) = Q_0([S_0] - [S]) - V(\text{syn}_s) \tag{7-64}$$

$V(\text{syn}_s)$ is the amount of CBOD in the waste sludge, $Q_w[X_u]$. Let f_s be the factor for converting the CBOD in the waste sludge to the oxygen equivalent of $V(\text{syn}_s)$. Also, let the concentrations above be in terms of CBOD_5. If f is the reciprocal factor for converting CBOD_5 to CBOD_u (ultimate CBOD), equation (7-64) may be written in terms of its equivalent oxygen equation as

$$V(\bar{r}_{s0}) = \frac{Q_0([S_0] - [S])}{f} - f_s Q_w[X_u])$$ (7-65)

$\bar{r}_{s0}$ is the oxygen equivalent of $\bar{r}_s$. f is normally taken as 0.68; however, to do an accurate job for a given specific waste, it should be determined experimentally.

By analogy with equation (7-65), the respiration rate due NBOD, $\bar{r}_{n0}$ is

$$V(\bar{r}_{n0}) = f_N Q_0([N_0] - [N]) - f_n Q_w[X_u]$$ (7-66)

where f_N is the factor for converting nitrogen concentrations to oxygen equivalent, $[N_0]$ and $[N]$ are the nitrogen concentrations in the influent and effluent, respectively, and f_n is the factor for converting the nitrogen in the wasted sludge to the oxygen equivalent.

Having determined the expressions for $\bar{r}_{s0}$ and $\bar{r}_{n0}$, the equation for AOR is

$$V(AOR) = V(\bar{r}) = \frac{Q_0([S_0] - [S])}{f} - f_s Q_w[X_u] + f_N Q_0([N_0] - [N]) - f_n Q_w[X_u]$$

(7-67)

Determination of f_s, f_N, and f_n.   The formula for microorganisms has been given as $C_5H_7NO_2$.[2] To find the oxygen equivalent of the mass synthesized, f_s, react this "molecule" with oxygen as follows:

$$C_5H_7NO_2 + 5O_2 \rightarrow 5CO_2 + 2H_2O + NH_3$$ (7-68)

From this equation, the oxygen equivalent of the mass synthesized is 1.42 mg of O_2 per mg of $C_5H_7NO_2$. Therefore, $f_s = 1.42$.

The reduction in the concentration of nitrogen is also brought about by reaction with oxygen for energy and for the requirement for synthesis. NBOD is actually in the form of NH_3. Reaction with O_2 yields

$$NH_3 + 2O_2 \rightarrow HNO_3 + H_2O$$ (7-69)

From this reaction, the oxygen equivalent per mg of NH_3-N is 4.57 mg. Hence f_N is equal to 4.57.

The nitrogen for synthesis goes with the sludge wasted, which can be expressed in terms of the total Kjeldahl nitrogen (TKN). Bacteria (VSS) contain approximately 14% nitrogen. (Protein contains approximately 16% nitrogen.) Hence the equivalent of the nitrogen in the sludge wasted is $4.57(0.14)Q_w[X_u] = 0.64Q_w[X_u]$ and the value of f_n is 0.64.

The total actual oxygen requirement (AOR) is therefore

$$V(AOR) = \left(Q_0\frac{[S_0] - [S]}{f} - 1.42Q_w[X_u]\right) + (4.57Q_0([N_0] - [N]) - 0.64Q_w[X_u])$$

(7-70)

The AOR of equation (7-70) is used to find SOR in order to size the aerator needed.

[2]M. G. Mandt and B. A. Bell (1982). *Oxidation Ditches in Wastewater Treatment* Ann Arbor Science, Ann Arbor, Mich., p. 48.

Example 7-14

A sample of wastewater is tested for β. The temperature is 25°C and the plant elevation is 1000 ft. The dissolved oxygen concentration after shaking the jar vigorously is 7.5 mg/L. Calculate β.

Solution

$$\beta = \frac{[C_{s,w}]}{[C_s]}$$

$[C_s]$ at 25°C and 1 atm = 8.24 mg/L = $C_{s,sp}$. $P = P_0(1 - Bz/T_0)^{g/RB}$; $P_0 = 1$ atm; $B = 0.0065$ K/m; $z = 1000/3.281 = 304.79$ m. $R = 286.9$ N-m/kg-K; $T_0 = 25 + 273 = 298$ K. Thus

$$P = 1\left[1 - \frac{0.0065(304.79)}{298}\right]^{9.81/[286.9(0.0065)]} = 0.97 \text{ atm} = 737.2 \text{ mmHg}$$

$$[C_s] = \frac{P}{P_s}[C_{s,sp}] = \frac{737.2}{760}(8.24) = 7.99 \text{ mg/L}$$

$$\beta = \frac{7.5}{7.99} = 0.94 \quad \textbf{Answer}$$

Example 7-15

A settling column 4 m in height is used to determine the α of the same wastewater tested in Example 7-14. The wastewater is to be aerated using a fine-bubble diffuser in the prototype aeration tank. The laboratory diffuser releases air at the bottom of the tank. The result of the unsteady-state aeration test is shown below. Assume that $\beta = 0.94$, $\bar{r} = 1.0$ mg/L-h, and the plant elevation is 1000 ft. For practical purposes, assume that the mass density of water = 1000 kg/m³. Calculate α.

Tap water at 5.5°C		Wastewater at 25°C	
Time (min)	$[\bar{C}]$ (mg/L)	Time (min)	$[\bar{C}]$ (mg/L)
3	0.5	3	0.9
6	1.7	6	1.8
9	3.1	9	2.4
12	4.4	12	3.3
15	5.5	15	4.0
18	6.1	18	4.7
21	7.1	21	5.3

Solution

$$\sum_{m=1}^{m=n} \ln\left(\frac{[C_s] - [\bar{C}]}{C_s}\right)_m = -K_L a \sum_{m=1}^{m=n} t_m$$

$$P = P_b + \frac{Z_d}{2}\gamma$$

From Example 7-14, $P_b = 0.97$ atm $= 0.97(101,330) = 98,290.1$ N/m^2. Then

$$P = 98,290.1 + \frac{4}{2}(1000)(9.81) = 117,913.01 \text{ N/m}^2$$

$[C_s]$ at 5.5°C and 1 atm ($= 101,330$ N/m^2) $= 12.59$ mg/L

$[C_s]$ at 5.5°C and 117,913.01 N/m$^2 = 12.59\left(\dfrac{117,913.01}{101,330}\right) = 14.65$ mg/L

Tap water at 5.5°C:

Time (min)	$[\overline{C}]$ (mg/L)	$\ln\dfrac{[C_s] - [\overline{C}]}{[C_s]}$
3	0.5	−0.0347
6	1.7	−0.1233
9	3.1	−0.2378
12	4.4	−0.3572
15	5.5	−0.4707
18	6.1	−0.5385
21	7.1	−0.6629
84		−2.4251 ← $\sum$

$$K_La = \frac{2.4251}{84} = 0.0289 \text{ per minute}$$

$$(K_La)_{\ell,20} = \frac{0.0289}{1.024^{5.5-20}} = 0.041 \text{ per minute} = 2.46 \text{ per hour}$$

$[C_s]$ at 25°C and 1 atm($= 101,330$ N/m^2) $= 8.24$ mg/L

$[C_s]$ at 25°C and 117,913.01 N/m$^2 = 8.24\left(\dfrac{117,913.01}{101,330}\right) = 9.59$ mg/L

$$[C_{s,w}] = 0.94(9.59) = 9.01 \text{ mg/L}$$

Wastewater at 25°C:

Time (min)	$[\overline{C}]$(mg/L)	$\ln\dfrac{[C_{s,w}] - [\overline{C}]}{[C_{s,w}]}$	$[C_{s,w}] - [\overline{C}]$
3	0.9	−0.1052	8.11
6	1.8	−0.2229	7.21
9	2.4	−0.3098	6.61
12	3.3	−0.4561	5.71
15	4.0	−0.5869	5.01
18	4.7	−0.7374	4.31
21	5.3	−0.8873	3.71
84		−3.3056	40.67 ← $\sum$

$$(K'_L a)_w = \frac{3.3056}{84} = 0.0394 \text{ per minute} = 2.36 \text{ per hour}$$

$$(K_L a)_w \sum_{m=1}^{m=n} \big([C_{s,w}] - [\overline{C}]\big)_m - \sum_{m=1}^{m=n} \bar{r}_m = (K'_L a)_w \sum_{m=1}^{m=n} \big([C_{s,w}] - [\overline{C}]\big)_m$$

$$(K_L a)_w (40.67) - 7(1.0) = 2.36(40.67)$$

$$(K_L a)_w = 2.53 \text{ per hour}$$

$$(K_L a)_{w,\ell,20} = \frac{2.53}{1.024^{25-20}} = 2.25 \text{ per hour} \qquad \alpha = \frac{2.25}{2.46} = 0.91 \quad \textbf{Answer}$$

Example 7-16

The influent of 10,000 m³/day to a secondary reactor has a BOD₅ of 150 mg/L. It is desired to have an effluent BOD₅ of 5 mg/L, an MLVSS of 3000 mg/L, and an underflow concentration of 10,000 mg/L. The effluent suspended solids concentration is 7 mg/L at 71% volatile suspended solids content. The volume of the reactor is 1611 m³ and $\theta_c = 10$ days. Calculate the SOR. Assume the aerator to be of the fine-bubble diffuser type with an $\alpha = 0.55$; depth of submergence equals 8 ft. Assume that the β of liquor is 0.90. The influent TKN is 25 mg/L and the desired effluent NH₃-N concentration is 5.0 mg/L. $f = 0.7$. The average temperature of the reactor is 25°C and is operated at an average of 1.0 mg/L of dissolved oxygen.

Solution

$$\text{AOR}(V) = \left(Q_0 \frac{[S_0] - [S]}{f} - 1.42 Q_w [X_u] \right) + (4.57 Q_0([N_0] - [N]) - 0.64 Q_w [X_u])$$

$$\theta_c = 10 = \frac{V[X]}{Q_w[X_u] + (Q_0 - Q_w)[X_e] - Q_0[X_0]}$$

$$= \frac{1611(3)}{Q_w[X_u] + (10{,}000 - Q_w)(0.71)(0.007) - Q_0(0)}$$

neglecting Q_w compared to 10,000.

$$Q_w[X_u] = 433.33 \text{ kg/day}$$

$$\text{AOR}(V) = \left[10{,}000 \left(\frac{0.15 - 0.005}{0.7} \right) - 1.42(433.3) \right]$$

$$+ [4.57(10{,}000)(0.025 - 0.005) - 0.64(433.3)]$$

$$= 2092.8 \text{ kg/day} \qquad \text{AOR} = 1.30 \frac{\text{kg}}{\text{m}^3\text{-day}}$$

$$\text{SOR} = \frac{\text{AOR}}{\alpha\{(\beta[C_s] - [\overline{C}])/[C_{s,20sp}]\}\theta^{T-20}}$$

At 25°C, $[C_{s,sp}] = 8.24$ mg/L; $[C_{s,20sp}] = 9.07$ mg/L. $P = P_b + (Z_d/2\gamma)$; assume that the plant is located at sea level; then

$$P = 101{,}330 + \frac{8/3.281}{2}[997(9.81)] = 113{,}253.89 \text{ N/m}^2$$

$$C_s = \frac{P}{P_s}C_{s,sp} = \frac{113{,}253.89}{101{,}330}(8.24) = 9.21 \text{ mg/L}$$

$$\text{SOR} = \frac{1.30}{0.55\{[0.9(9.21) - 1.0]/9.07\}(1.024^{25-20})} = 2.6 \text{ kg/day-m}^3 \quad \textbf{Answer}$$

EFFLUENT DISPOSAL

As mentioned at the beginning of this chapter, effluents before discharge must meet effluent standards. One of these standards is that for coliforms. Effluents must therefore be disinfected. The disinfectant commonly used is chlorine, although others, such as UV, ozone, and bromine chloride, have been used. Disinfection was discussed in Chapter 6 and will not be repeated here. As a result of disinfection, residual chlorine can have a detrimental effect on stream biota. Limits on total residual chlorine (TRC) are therefore included in the discharge permit. To meet these limits, effluents are dechlorinated using SO_2 and sulfites.

The treated wastewater is disposed of through an effluent disposal conduit called an *outfall*. In shallow, free-flowing streams, this is simply a pipe that protrudes on the side of the streambank. For discharges into deeper receiving waters, the effluent may be dispersed through an elaborate diffuser system to ensure complete mixing with the stream. In estuarine tidal and coastal systems, in addition to the diffuser-system calculations, a hydraulic profile calculation must be made to ensure that the tide will not flood downstream units of the treatment plant. In fact, the hydraulic profile of the entire treatment plant should be calculated to ascertain the elevations of the hydraulic gradient through the plant corresponding to the desired design flow. Changing river stages also affect the hydraulic profile of the treatment plant.

Diffuser Calculations

Figure 7-11 shows schematics of diffuser systems and an outfall. At the bottom figure is the outfall with diffuser manifold attached at the terminus. If a cross section is cut through the manifold, the sections shown Figure 7-11a will result. These figures are either jets or plumes. If the configurations have been formed by pure jet action from the jet at the exit branch pipe or port, the configurations are *jets*. However, if the configurations have been formed by the difference in density between the ambient water and the wastewater that exits from the branch pipe or port, the configurations are *plumes*. In practical situations, the effect of most of the jet action is diminished a short distance after the jet emerges from the exit opening. The reason for this is the tremendous resistance that the surrounding ambient water exerts on the jet. Without the density difference that forms the plume, the configuration will simply disappear just as soon as it is created. Hence, in applications, the configuration is assumed to be a plume. Our discussion will therefore consider only the plume.

There are two types of plumes: the solitary plume and the plane plume. A *solitary plume* is the configuration that forms from a single discharge pipe or port, while a *plane plume* is the configuration that forms from a discharge from a slot exit. Figure 7-11b shows solitary plumes arranged in a line, a distance L apart along the diffuser manifold. If the distance L is made shorter, the plumes generated will merge together, forming a *compound plume* as if the source were coming from a slot. Therefore, treat an arrangement of solitary plumes in a straight line forming a compound plume as a plane plume. The top schematics of the figure can be viewed as the cross section of either a solitary plume or a plane plume. Discharge pipes and ports usually vary in diameter from 2 to 10 in. and are spaced 5 to 50 ft apart.

The right-hand side of Figure 7-11a is a plume formed in receiving waters having a density gradient $d\rho_a/dz$, where ρ_a is the density of the ambient water. The plume rises until it reaches a maximum height z_{max}, where the ambient density is equal to the density of the mixture of wastewater and ambient water. From this height the plume spreads out laterally. The most important parameter in the discussion of plumes is the average dilution $\overline{S}$ that the wastewater receives. At a height z above the port exit, $\overline{S}$ is given empirically for solitary plumes and for plane plumes, respectively, as follows:[3]

$$\overline{S} = \frac{0.15 \left(g \frac{\rho_a - \rho}{\rho} Q \right)^{1/3} z^{5/3}}{Q} \tag{7-71}$$

$$\overline{S} = \frac{0.38 \left(g \frac{\rho_a - \rho}{\rho} q \right)^{1/3} z}{q} \tag{7-72}$$

where g is the acceleration due to gravity, ρ the density of wastewater, Q the rate of flow coming out of the port, and q the discharge per unit length of plane plume.

For the plumes discharged into waters having a density gradient, z_{max} is given empirically by[4]

$$z_{max} = 3.6 \left(g \frac{\rho_a - \rho}{\rho} q \right)^{1/3} \left(\frac{-g}{\rho_{a,av}} \frac{d\rho_a}{dz} \right)^{-1/2} \tag{7-73}$$

where $\rho_{a,av}$ is the average ambient fluid density.

The formulas derived for plumes above are for wastewaters discharged straight up from the discharge pipe or port. In practice, however, the discharge direction is also

[3]J. A. Roberson, J. J. Cassidy, and M. H. Chaudry (1988). *Hydraulic Engineering*. Houghton Mifflin, Boston, p. 418.

[4]R. B. Wallace and S. J. Wright (1984). "Spreading Layer of Two-Dimensional Buoyant Jet." *Journal of the Hydraulic Division, ASCE*, 110(6).

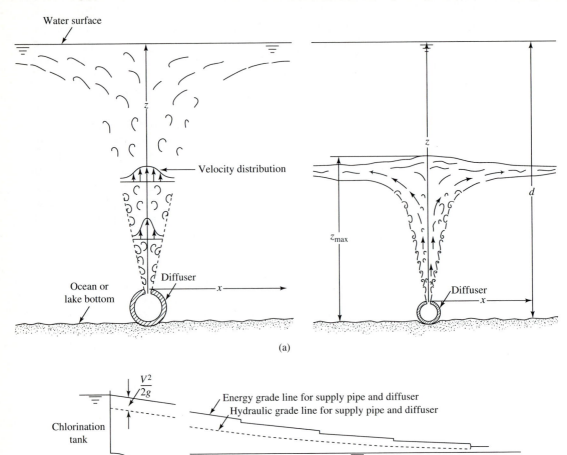

Figure 7-11 Schematics of diffuser systems and outfall pipe. (a) Left-hand side, cross-sectional view of diffuser-pipe manifold and rising jet or plume in uniform-density receiving water; right-hand side, cross-sectional view of diffuser-pipe manifold and rising jet or plume in a receiving water with density gradient. (b) Outfall or supply pipe with diffuser system connected at terminus.

made at an angle. When discharge is made at an angle more dilution is received by the wastewater, since more time will be spent in diluting the wastewater. Hence the formulas derived above incorporate some factor of safety.

Example 7-17

(a) 10,000 m³/day of treated effluent is to be discharged through a single diffuser. What depth of water is required to effect a dilution of 200? (b) If the treated effluent is to be discharged through a manifold having 10 diffuser pipes and of a length of 12 m, what is the depth required to effect the same degree of dilution? Assume $(\rho_a - \rho)/\rho = 0.025$.

Solution

(a)
$$\bar{S} = \frac{0.15\left(g\frac{\rho_a-\rho}{\rho}Q\right)^{1/3} z^{5/3}}{Q}$$

$$200 = \frac{0.15\,[9.81(0.025)(10,000/24(60)(60))]^{1/3}\, z^{5/3}}{10,000/[24(60)(60)]}$$

$$z = 41.92 \text{ m} \quad \textbf{Answer}$$

(b)
$$\bar{S} = \frac{0.38\left(g\frac{\rho_a-\rho}{\rho}q\right)^{1/3} z}{q}$$

$$q = \frac{10,000}{24(60)(60)(12)} = 0.0096 \text{ m}^3/\text{m-s}$$

$$200 = \frac{0.38\,[9.81(0.025)(0.0096)]^{0.33333}\, z}{0.0096} \qquad z = 37.92 \text{ m} \quad \textbf{Answer}$$

Example 7-18

Determine the height z_{max} that the plume will rise for a discharge of 0.02 m³/s-m. The density at the bottom of the ocean floor is 1025.00 kg/m³ and at a point 60 m from the ocean floor, the density is 1024.28 kg/m³. Assume that $(\rho_a - \rho)/\rho = 0.025$. Also determine $\bar{S}$ at this height.

Solution

$$z_{max} = 3.6\left(g\frac{\rho_a - \rho}{\rho}q\right)^{1/3}\left(\frac{-g}{\rho_{a,av}}\frac{d\rho_a}{dz}\right)^{-1/2}$$

$$\frac{1}{\rho_{a,av}}\frac{d\rho_a}{dz} = \frac{1}{(1024.28+1025.00)/2}\left(\frac{1024.28-1025.00}{60}\right) = -0.0000117 \text{ kg/m}^3\text{-m}$$

$$z_{max} = 3.6\,[9.81(0.025)(0.02)]^{0.3333}\,[-9.81(-0.0000117)]^{-0.5} = 57.08 \text{ m} \quad \textbf{Answer}$$

$$\bar{S} = \frac{0.38\left[g(\frac{\rho_a-\rho}{\rho})q\right]^{1/3} z}{q}$$

$$= \frac{0.38\,[9.81(0.025)(0.02)]^{0.3333}\,(57.08)}{0.02} = 184.27 \quad \textbf{Answer}$$

Hydraulic Profile Calculations

Since it is the receiving water elevation that profoundly affects the hydraulic profile, calculation is best started from the terminus of the discharge port. The profile calculation then works its way up through the diffuser manifolds, the supply pipe, the effluent chlorinator discharge channel, the chlorinator, the discharge conduit from the secondary clarifier to the chlorinator, the secondary effluent discharge channel, and so on, up the flowsheet of the treatment plant. Calculations involve the application of the energy equation and the calculation of head losses along the path of wastewater flow through straight runs of pipes, bends, fittings, transitions, and the like. Open-channel-flow calculations for discharge troughs and weirs may also be used. The principles of calculations of head losses and the application of the energy equation are discussed in any book on fluid mechanics and will not be repeated here. The hydraulic profile that will be addressed in this section will involve the outfall pipe and the diffuser system. The remaining hydraulic profile calculations above and beyond the outfall are similar in principles in that the basic task is simply the calculations of head losses and the application of the energy equation to determine the various heads. Once all the heads are calculated, the profile can be drawn.

Referring to Figure 7-11b, call q_n the flow through any diffuser port n. Thus, starting from the lower end of the diffuser manifold, the first port has a q_n of q_1; the second port, a q_n of q_2; and so on. Let h_{p1} be the head loss at port 1 as the wastewater exits into the ambient fluid. In terms of h_{p1} and the orifice equation $q_1 = Ka_1\sqrt{2gh_{p1}}$, which, in general, for any port q_n is

$$q_n = Ka_n\sqrt{2gh_{pn}} \tag{7-74}$$

where a_n is the cross-sectional area of port n with head loss h_{pn} and K is given empirically by[5]

$$K = 0.675\sqrt{1 - \frac{V_n^2}{2gh_{pn}}} \tag{7-75}$$

V_n is the velocity in the manifold pipe at port n. V_n is

$$V_n = \frac{V_{n,n-1} + V_{n,n+1}}{2} \tag{7-76}$$

where $V_{n,n-1}$ is the velocity before port n and $V_{n,n+1}$ is the velocity after it. Equations (7-74) and (7-75) may be combined to yield

$$q_n = 0.675a_n\sqrt{2gh_{pn} - V_n^2} \tag{7-77}$$

Let H be the available head loss from the surface of the last unit in the treatment plant (e.g., chlorinator) to the level of the receiving water below. Also, let h_1 be the combined head loss in the supply pipe and the manifold pipe up to port 1. Applying the

[5]K. Subramanya and S. C. Awasthy (1970). "Discussion of the Paper by Vigander." *Journal of the Hydraulics Division, ASCE*, 102(HY5).

energy equation from the surface of the last unit to the surface of the receiving water, $h_{p1} = H - h_1 - h_d$, where h_d is the head loss due to the depth of water from the port to the surface of the receiving water. For any one port n, the general head loss, h_{pn}, is

$$h_{pn} = H - h_n - h_d \tag{7-78}$$

where h_n is the head loss up to port n.

Example 7-19

Secondary-treated wastewater at 15°C is to be discharged into an estuary. Design a manifold diffuser system to discharge 0.11 m³/s. Assume that the head available between the last unit in the treatment plant and the first diffuser port is 9.62 m and $h_d = 2$ m. Also, assume that the length of the supply pipe is 122 m. Construct the hydraulic profile.

Solution   To ensure a self-cleaning action in the diffuser manifold, assume a velocity of 0.76 m/s in the spacing L before the first port (see Figure 7-11). Assuming four diffuser ports spaced 3 m apart, $q = 0.11/4 = 0.0275$ m³/s. Therefore, the manifold pipe cross-sectional area at port 1 = 0.0275/0.76 = 0.036 m² and $D = 0.214$ m = 0.70 ft. Use this as the diameter of the supply pipe and the whole manifold pipe and assume a PVC pipe. Since the pipe is made of PVC, assume it to be smooth.
Compute the head loss at the outfall supply pipe:

$$\text{Re} = \frac{DV\rho}{\mu} \qquad \rho = 998 \text{ kg/m}^3 \qquad \mu = 11(10^{-4}) \text{ kg/m-s}$$

$$V = 0.11/0.036 = 3.06 \text{ m/s}$$

$$\text{Re} = \frac{0.214(3.06)(998)}{11(10^{-4})} = 5.93(10^5)$$

From the Moody chart (Appendix 17), $f = 0.013$; thus

$$h = f\frac{\ell}{D} \cdot \frac{V^2}{2g} = 0.013\left(\frac{122}{0.214}\right)\left[\frac{3.06^2}{2(9.81)}\right]$$

$$= 3.54 \text{ m} \qquad \text{loss in supply pipe length of 122 m}$$

Compute the entrance loss, h_e, to supply pipe from last unit of treatment plant: $h_e = K_e(V^2/2g)$. Assuming that the entrance is well rounded, $K_e = 0.04$, from any book on fluid mechanics. Then

$$h_e = 0.04\left[\frac{3.06^2}{2(9.81)}\right] = 0.019 \text{ m}$$

$$\text{total loss in supply pipe} = 3.54 + 0.019 = 3.56 \text{ m}$$

Compute head loss in manifold pipe:

$$\text{length of diffuser manifold pipe} = (4 - 1)(3) + 0.5 = 9.5 \text{ m}$$

For h_{p1}:

$$\text{Re} = \frac{DV\rho}{\mu} \qquad \rho = 998 \text{ kg/m}^3 \qquad \mu = 11(10^4) \text{ kg/m-s}$$

$$V_1 = \frac{0.11/0.036 + 0.0275/0.036}{2} = 1.91 \text{ m/s}$$

$$\text{Re} = \frac{0.214(1.91)(998)}{11(10^{-4})} = 3.71(10^5)$$

From the Moody chart (Appendix 17), $f = 0.0138$.

$$h = f\frac{\ell}{D} \cdot \frac{V^2}{2g} \qquad \ell = 9 \text{ m}$$

$$= 0.0138\left(\frac{9}{0.214}\right)\left[\frac{1.91^2}{2(9.81)}\right] = 0.11 \text{ m}$$

Thus $h_1 = 3.56 + 0.11 = 3.67$ m and $h_{p1} = 9.62 - 3.67 - 2 = 3.95$ m. Repeat for h_{p2}.

Compute the head loss in the manifold pipe:

For h_2:

$$V_2 = \frac{0.11/0.036 + 2(0.0275)/0.036}{2} = 2.29 \text{ m/s}$$

$$\text{Re} = \frac{0.214(2.29)(998)}{11(10^{-4})} = 4.45(10^5)$$

From the Moody chart (Appendix 17), $f = 0.0134$.

$$h = f\frac{\ell}{D} \cdot \frac{V^2}{2g} \qquad \ell = 6 \text{ m}$$

$$= 0.0134\left(\frac{6}{0.214}\right)\left[\frac{2.29^2}{2(9.81)}\right] = 0.10 \text{ m}$$

Thus $h_2 = 3.56 + 0.10 = 3.66$ m and $h_{p2} = 9.62 - 3.66 - 2 = 3.96$ m. Assume that $h_{p3} = 3.97$ m and $h_{p4} = 3.98$ m.

$$q_1 = 0.675a_1\sqrt{2gh_{p1} - V_1^2}$$

$$V_1 = \frac{0.0275/0.036 + 0}{2} = 0.38 \text{ m/s}$$

$$0.0275 = 0.675a_1\sqrt{2(9.81)(3.95) - 0.38^2} \qquad a_1 = 0.0046 \text{ m}^2$$

$$d_p \text{ (= port diameter)} = 0.077 \text{ m} = 7.7 \text{ cm}$$

$$V_2 = \frac{2(0.0275)/0.036 + (0.0275)/0.036}{2} = 1.15 \text{ m/s}$$

$$0.0275 = 0.675a_1\sqrt{2(9.81)(3.96) - 1.15^2} \qquad a_2 = 0.0047 \text{ m}^2$$

$$d_p \text{ (= port diameter)} = 0.077 \text{ m} = 7.7 \text{ cm}$$

$$V_3 = \frac{3(0.0275)/0.036 + 2(0.0275)/0.036}{2} = 1.91 \text{ m/s}$$

$$0.0275 = 0.675a_1\sqrt{2(9.81)(3.97) - 1.91^2} \qquad a_3 = 0.0047 \text{ m}^2$$

$$d_p \text{ (= port diameter)} = 0.077 \text{ m} = 7.7 \text{ cm}$$

$$V_4 = \frac{4(0.0275)/0.036 + 3(0.0275)/0.036}{2} = 2.67 \text{ m/s}$$

$$0.0275 = 0.675a_1\sqrt{2(9.81)(3.98) - 2.67^2} \qquad a_4 = 0.0048 \text{ m}^2$$

$$d_p \text{ (= port diameter)} = 0.078 \text{ m} = 7.8 \text{ cm}$$

Let all d_p's be = 7.77 cm. **Answer**
Diameter of manifold pipe = 0.214 m; if not available, calculation must be redone to conform to pipe available. **Answer**
Diffuser spacing = 3.0 m. **Answer**
The hydraulic gradient is plotted in Figure 7-11.

GLOSSARY

Activated sludge. The reinvigorated microorganisms sludge.

Activated sludge process. A process of treating waste where activated sludge is recirculated to the aeration tank for treatment of the waste.

Actual oxygenation rate (AOR). The rate of transfer of oxygen as is actually occurring in an operating unit; it is equal to the respiration rate.

Actual specific hydraulic loading rate (ASHLR). The ratio of the actual flow to the actual surface area of microbial film.

Actual specific organic loading rate (ASOLR). The ratio of the organic loading to the actual surface area of microbial film.

Alpha. The ratio of the $K_L a$ of the waste at 20°C to the $K_L a$ of tap water at 20°C.

Best available technology economically achievable (BAT). Also called *best conventional pollutional control technology* (BCT); equivalent to secondary treatment.

Beta. The ratio of the dissolved oxygen saturation of a given wastewater to the dissolved oxygen saturation of tap water which is at the same temperature and pressure as that of the wastewater.

Biochemical oxygen demand. A measure of the strength of an organic substance indicated by the amount of equivalent oxygen the substance consumed in its biological decomposition utilizing oxygen as a reactant.

Carbonaceous biochemical oxygen demand (CBOD). The biological oxygen demand corresponding to the carbon content of a substance; $CBOD_5$, CBOD after 5 days; $CBOD_u$, ultimate CBOD.

Chemical oxygen demand. A measure of the total ability of a substance to consume oxygen, determination normally done using chemical oxidizers such as potassium dichromate.

Compound plume. Plume formed from the merging of two or more plumes.

Contact stabilization or biosorption process. A variation of the activated process where an aerated sludge is allowed to contact the influent for a short period of 15 to 60 min.

Detritus. A mixture of grit and organic matter.

Dissolved or filtrable solids. The fraction of total solids that passes through a filter.

Effluent limited waters. Waters where discharge of effluent meeting the secondary or BAT standards does not violate water quality standards.

Effluent standard. The values of parameters that a plant must meet at its discharge effluent.

Extended-aeration process. A variation of the activated sludge process characterized by long detention times and low F/M ratio and operated at the endogenous phase of growth.

Filter yield. In cake filtration, the amount of cake formed per unit area of filter cloth per unit time.

Five-day biochemical oxygen demand (BOD$_5$). A biochemical oxygen demand exerted in 5 days.

Fixed solids. The solids that remain after the sample is decomposed at $600°C$.

Free ammonia. Hydrolysis product of organic nitrogen.

Grit. Rough and hard particles of detritus.

Half-velocity constant. The concentration of the limiting nutrient that will make the specific growth rate equal to half the maximum specific growth rate.

High-rate activated sludge process. A variation characterized by short detention time and a high F/M ratio and operated at the logarithmic growth phase.

Jet. The configuration formed as fluid exits from an opening into another or the same fluid through the effect of its momentum.

Kinetics. The rate of change of the concentration of a substance with respect to time.

Kjeldahl nitrogen. The sum of the concentrations of ammonia and organic nitrogen.

Lagoon. A basin used to hold waste.

Mean cell residence time (MCRT), **sludge retention time** (SRT), **or sludge age.** The average time that microorganisms are in residence in a reactor.

Nitrogenous biochemical oxygen demand (NBOD). The biochemical oxygen demand corresponding to the nitrogen content of a substance.

Nominal specific hydraulic loading rate (NSHLR). The plant influent flow divided by the surficial microbial area of the treatment unit.

Nominal specific organic loading rate (NSOLR). The ratio of the organic loading to the surficial microbial area of the treatment unit.

Organic nitrogen. Nitrogen content of an organic substance.

Plain plume. The plume formed from a slot exit.

Plain sedimentation. A simple, unaided sedimentation of particles.

Plume. The configuration formed due to the difference in densities when a fluid exits into another or the same fluid.

Pond. A basin where stabilization of the waste is effected through the use of oxygen supplied by photosynthesis.

Preliminary treatment. The treatment of sewage without the use of the primary settling tank.

Primary treatment. The treatment of sewage where the major method of removing the pollutant is by settling.

Secondary effluent standard. An effluent standard that requires meeting a BOD₅ of 30 mg/L monthly and 45 mg/L weekly and meeting suspended solids of 30 mg/L monthly and 45 mg/L weekly.

Secondary treatment. Treatment that meets secondary standards.

Settleable solids. The volume of solids that settles after a predetermined period of time.

Solitary plume. The plume formed from a single discharge port.

Specific growth rate. The proportionality constant in the kinetic growth expression for organisms.

Specific substrate utilization rate. The reciprocal of the specific yield.

Specific yield. The amount of organisms produced per unit amount of substrate consumed.

Stabilization basin. Ponds and lagoons used to stabilize waste.

Standard oxygenation rate (SOR). The rate of oxygenation of tap (or distilled) water by an equipment at 20°C, 1 atm pressure, and zero concentration of oxygen.

Step aeration. A variation of the activated sludge where the influent is introduced at various points along the aeration tank to provide a more uniform BOD loading to the reactor.

Suspended or nonfiltrable solids. The fraction of total solids that does not pass through a filter.

Tapered aeration. A variation of the activated sludge process where the air is introduced along the length of the tank in amounts proportional to the amount of BOD remaining to be stabilized.

Total nitrogen. The sum of the concentrations of organic, free ammonia, nitrite, and nitrate nitrogens.

Total organic carbon. The total carbon content of a substance.

Total solids. The material left after water has been evaporated from the sample.

Ultimate biochemical oxygen demand (BOD$_u$). A biochemical oxygen demand exerted after an infinite time.

Underflow concentration. The concentration in the underflow of a sedimentation basin or thickener.

Volatile solids. The solids that disappear after the sample is decomposed at 600°C.

Water quality limited waters. Waters of which the discharge of secondary or BAT-treated effluents will cause the violation of water quality standards.

Water quality standards. Stream values of parameters that must not be violated by any discharge.

SYMBOLS

a_n	cross-sectional area of a manifold port
A_s	total surface area of attached-growth microbial process
B	temperature lapse rate
C	concentration of dissolved oxygen

$\hat{C}$	point concentration of dissolved oxygen
$\overline{C}$	average dissolved oxygen concentration
C_s	saturation dissolved oxygen concentration
$C_{s,sp}$	C_s at standard pressure of P_s (1 atm)
ΔE	energy loss through a manifold port
f	fraction of BOD_5 in CBOD
F/M	food to microorganisms ratio
g	acceleration due to gravity
K	orifice coefficient of a manifold port
$K_L a$	oxygen overall mass transfer coefficient
$(K_L a)_{\ell,20}$	$K_L a$ at standard conditions of 20°C
$(K_L a)_w$	$K_L a$ of wastewater
$(K_L a)_{w,\ell,20}$	$(K_L a)_w$ at standard conditions of 20°C
K_s	half-velocity constant
M_s	mass of substrate diffusing to the microbial film from an element of flowing wastewater in a trickling filter
N	concentration of NH_3-N of TKN
N_0	influent N
P_s	standard pressure
P_0	atmospheric pressure at elevation, z, equals zero
q	discharge per unit length of plane plume
q_n	discharge in a manifold port
Q	rate of flow; discharge in a diffuser port
Q_a	the sum of Q_o and Q_R
Q_e	effluent discharge
Q_R	recirculation flow
Q_w	volumetric rate of sludge wasting
Q_0	plant inflow
$\overline{r}$	wastewater respiration rate
R	gas constant
$\overline{S}$	average dilution
[S]	concentration of substrate
[S_0]	inflow concentration of substrate
t	time
U	specific substrate utilization rate; specific crop utilization rate for a particular fertilizer constituent
v_h	flow-through velocity
V	volume of reactor
V_n	velocity in the manifold cross section directly opposite the manifold port n
w	width of grit chamber; overall characteristic width of an attached-growth microbial film
[X]	concentration of microorganisms
[$\overline{X}$]	average concentration of microorganisms

$[X_e]$	effluent concentration of microorganisms
$[X_f]$	concentration of microbial film in a trickling filter
$[X_u]$	sedimentation tank underflow concentration of microorganisms
$[X_0]$	influent concentration of microorganisms
Y	specific yield of microorganisms
z	overall characteristic depth of an attached-growth microbial film; elevation in atmosphere; distance above a diffuser port
Z_d	submergence depth
Z_0	depth of flow
α	$\dfrac{(K_L a)_{w,\ell,20}}{(K_L a)_{\ell,20}}$
β	ratio of the saturated dissolved oxygen concentration of wastewater at wastewater conditions of temperature and pressure to that of tap water at the same wastewater conditions of temperature and pressure
η	porosity
θ	nominal hydraulic retention time; temperature correction factor
θ_c	mean cell residence time
μ	specific growth rate; dynamic viscosity
$\mu_{\max}$	maximum μ
ρ	mass density; density of discharged effluent
ρ_a	ambient water density
$\rho_{a,av}$	average ρ_a
ϕ	property per unit mass

PROBLEMS

7-1. A 10-mL sample is pipetted directly into a 300-mL incubation bottle. The initial DO of the diluted sample is 9.0 mg/L and its final DO is 2.0 mg/L. The dilution water is incubated in a 200-mL bottle, and the initial and final DOs are, respectively, 9.0 and 8.0 mg/L. If the sample and the dilution water are incubated at 20°C for 5 days, what is the BOD_5 of the sample at this temperature?

7-2. A suspended solids analysis is run on a sample. The tare mass of the crucible and filter is 55.3520 g. A sample of 260 mL is then filtered and the residue dried to constant mass at 103°C. The constant mass of the crucible, filter, and the residue is 55.3890 g. The residue is then decomposed at 600°C to drive off the volatile matter. Assume that the filter does not decompose at this temperature. After decomposition, the constant weight of the crucible and the residue was again determined and was found to be equal to 30.3415 g. What is the percent volatile matter in the sample?

7-3. What is an effluent-limited stream? A water-quality limited stream?

7-4. What is unit operation? Unit process?

7-5. If there should be a minimum of one channel operating at any time, design the cross section of a grit-removal unit consisting of four identical channels to remove grit for a peak flow of 80,000 m³/day, an average flow of 50,000 m³/day and a minimum flow of 20,000 m³/day. Assume a flow-through velocity of 0.3 m/s and that the channels are to be controlled by Parshall flumes.

7-6. Solve Problem 7-5 for grit channels controlled by proportional flow weirs.

7-7. A grit chamber is designed for a flow-through velocity of 0.30 m/s, a depth of 0.9 m, and a length of 10.5 m. If inorganic particles have a specific gravity of 2.65, estimate the diameter of the largest particle that can be removed 100% at 20°C.

7-8. Calculate the total airflow for an aerated grit chamber processing 25,000 m³/day of municipal wastewater. Also calculate the appropriate dimensions.

7-9. What are the common devices used to measure flows in wastewater treatment plants?

7-10. The treatment plant flows are as follows: average influent flow = 6000 m³/day, the peak daily flow = 13,000 m³/day, and the minimum daily flow is two-thirds of the average flow. To meet effluent limits, it has been determined that the design of a primary sedimentation basin must remove 65% of the influent suspended solids. Using any chart having the relationship of percent removal and overflow rate, design a sedimentation basin that should meet detention time requirement, as well as effluent limits.

7-11. Determine the following for a primary clarifier with a 30-m diameter: surface overflow rate and the approximate removal rate for BOD_5 and suspended solids at average flow and peak flow. The other information are as follows: wastewater flow = 15,000 m³/day, and peak flow is 1.80 times average. The wastewater contains 190 BOD_5 and 212 mg/L of suspended solids at average flow and 230 mg/L BOD_5 and 370 mg/L suspended solids at peak flow.

7-12. In Problem 7-11, calculate the kilograms per day of sludge for the average and peak flows.

7-13. The design criteria for overflow rates for sedimentation basins has been set as 40 m³/m²-day for average conditions and 110 m³/m²-day for peak flow conditions. The average flow to the plant is 25,000 m³/day and the peak flow is 40,000 m³/day. Determine the dimensions of the primary clarifier if it is (**a**) a circular basin and (**b**) a rectangular basin.

7-14. In Problem 7-13, determine the dimensions of the primary clarifier if it is a square cross-flow tank.

7-15. An activated sludge process is operated at an MCRT of 10 days and an NHRT of 4 h. The volume of the reactor is 1600 m³ and the rate of sludge wasting is 64 m³/day. If the MCRT and the NHRT are maintained, what is the sludge underflow concentration corresponding to an MLVSS of 4000 mg/L?

7-16. What are the two extremes of the activated sludge process?

7-17. The following kinetic parameters have been determined by pilot-plant studies: $Y = 0.57$ and $k_d = 0.06$ per day. Using these parameters, a tapered aeration plant is to be designed to treat 13,000 m³/day of primary settled sewage. The effluent from the primary settling tank has a BOD_5 of 150 mg/L and a suspended solids of 130 mg/L. The effluent is to contain 5.0 mg/L of BOD_5 and 30 mg/L of suspended solids. The plant is to operate at a MLVSS of 2500 mg/L at a MCRT of 10 days. The ratio of BOD_5 to suspended solids in the effluent is 0.65. If the reactor is 5.5 m wide and 5.5 m deep, determine its length.

7-18. In Problem 7-17, determine the total BOD_5 in the effluent.

7-19. The total BOD_5 in the effluent is composed of the BOD of the residual substrate and the BOD of the microbial cells. Analysis of the effluent shows that the total BOD_5 is 30 mg/L and the BOD_5 corresponding to the microbial cells is 13.8 mg/L. Q_0 is equal to 10,000 m^3/day; $[X_u]$ is equal to 10,000 mg/L; and Q_w is equal to 64 m^3/day. The plant is to be operated at an MCRT and NHRT of 10 days and 4 h, respectively. **(a)** If the influent to the bioreactor is 150 mg/L of BOD_5, what MLVSS should be maintained in the bioreactor to meet a BOD_5 value of 30 mg/L? **(b)** At what recirculation ratio is the plant to be operated? Use the kinetic parameters of Problem 7-17.

7-20. A completely mixed activated sludge process is to be designed to treat $15,000^3$ per day of industrial waste containing 1250 mg/L of BOD_5. The permitting agency of the state requires that the effluent be treated to a level of 30 mg/L. A treatability analysis found that a mean cell residence time of 25 days is sufficient and that the unit could operate at a MLVSS of 6000 mg/L. $Y = 0.07$ g/g and the value of $k_d = 0.04$ per day. The underflow concentration is 10,000 mg/L. Calculate the volume of the reactor and the mass and volume of solids wasted per day.

7-21. In Problem 7-20, what is the recirculation ratio?

7-22. **(a)** In Problem 7-20, if the recirculation ratio is set at 1.0, what will be the resulting MLVSS? **(b)** What will be the resulting BOD_5 at the effluent?

7-23. In Problem 7-20, what is the recirculation ratio that will make the effluent $BOD_5 = 0$?

7-24. Describe three aerating devices used in wastewater treatment. Can the MLVSS be increased without limit? Why?

7-25. The influent of 10,000 m^3/day to a secondary reactor has a BOD_5 of 150 mg/L. It is desired to have an effluent total BOD_5 of 30 mg/L with a BOD_5 for the microbial cells of 13.8 mg/L, an MLVSS of 3000 mg/L, and an underflow concentration of 10,000 mg/L. Using the kinetic parameters used in Problem 7-17, calculate the volume of the reactor. What are the volume and mass flows of sludge wasted per day?

7-26. A sewage flow of 6000 m^3/day is to be treated using a facultative pond having a depth of 2 m and a surface area of 25 ha. The waste contains a soluble BOD_5 of 160 mg/L and a decay rate of 0.35 per day. What is the soluble BOD_5 in the effluent?

7-27. For the reactor in Problem 7-25, what recirculation ratio is required to produce an effluent BOD_5 of zero?

7-28. The kinetic parameters K_s, μ_{max}, Y, and k_d are determined using a bench-scale activated-sludge reactor without recycle. The results are shown below. Determine the parameters.

Run	$[S_0]$ (mg/L BOD_5)	$[S]$ (mg/L BOD_5)	θ (days)	$[X]$ (mg VSS/L)
1	300	7	3.1	127
2	300	12	2.1	126
3	300	19	1.7	135
4	300	31	1.2	130

7-29. It is required to design an oxidation pond to treat 5000 m^3/day of municipal sewage. Develop design criteria. Using these criteria, size the pond.

7-30. A trickling filter 6 m in diameter and 3 m high is to be packed with filter medium high. If the plant inflow Q_0 is 10,000 m³/day and the recirculation ratio is 1, which is more efficient, a trickling filter with plastic packing or one with rock packing?

7-31. How much error is introduced by neglecting microbial film area at the edges of the disks of an RBC? There are 40 modules used with disks measuring 3.7 m in diameter, 5 mm in thickness, and spaced on the shaft at 30 mm apart. The length of the module is 7.6 m.

7-32. Examine the effect of recirculation ratios varying from 0 to 5 on a trickling filter process. Assume that $K' = 0.25$ m/day. When the recirculation ratio is zero, $Q_a/A_s = 7.0$ m/day.

7-33. Calculate the dimensions of biotower units that will produce an effluent of 10 mg/L of soluble BOD₅. The tower is to be composed of high-specific surface plastic medium and is to treat the effluent of a primary clarifier with a BOD₅ of 150 mg/L. The flow from the clarifier is 20,000 m³/day. From a pilot-plant study, K' was determined to be 0.117 m/day. Two towers are to be used, each with a square surface and separated by a common wall and a depth of 7.0 m. A recirculation ratio of 2 is to be used.

7-34. Design a biofilter to treat an average domestic flow rate of 9000 m³/day from a small community in which is located a small vegetable cannery. The seasonal flow from the cannery is 6000 m³/day which occurs from May to October. The average domestic BOD₅ is 215 mg/L and the combined domestic and cannery BOD₅ is 560 mg/L. The average low temperature from May to October is 22°C and the average low temperature during the lowest-temperature month of January is 0°C. The BOD₅ removal constant at 25°C is 0.09 m/day with a temperature correction factor, θ, of 1.08. The specific filter area of the packing material is 87.2 m²/m³. The effluent BOD₅ requirement is 30 mg/L. The height of the tower is restricted to 11 m.

7-35. Determine the surface area of an RBC required to treat the wastewater of Problem 7-34. Use the value $K' = 0.264$ m/day. If the surface area A_s per module is 10,000 m³, determine the number of modules to be used.

7-36. The β value of a sample of wastewater tested at 25°C and at an elevation of 5000 ft is 0.94. What would be its dissolved oxygen concentration at mean sea level at a temperature of 20°C?

7-37. A laboratory diffuser in an unsteady-state reaeration test releases air at the bottom of a tank having a 4-m depth of water column. The results are shown below. Assume that $\beta = 0.94$, $\bar{r} = 1.0$ mg/L-h, and the plant elevation is 1000 ft. For practical purposes, assume that the mass density of water is 1000 kg/m³. Calculate α.

Tap water at 5.5°C		Wastewater at 25°C	
Time (min)	$[\overline{C}]$ (mg/L)	Time (mg/L)	$[\overline{C}]$ (mg/L)
3	0.5	3	0.9
9	3.1	9	2.4
15	5.5	15	4.0
21	7.1	21	5.3

7-38. Calculate the SOR for a fine-bubble diffuser aerator with $\alpha = 0.55$. The depth of submergence is to be 5 ft. The influent flow is 10,000 m³/day and has a BOD₅ of 150 mg/L. It is desired to have an effluent BOD₅ of 5 mg/L, an MLVSS of 3000 mg/L, and an underflow concentration of 10,000 mg/L. The volume of the reactor is 1611 m^3 and $\theta_c = 10$ days. Assume a β value of liquor of 0.90. The influent TKN is 25 mg/L and the desired effluent NH$_3$-N concentration is 5.0 mg/L. $f = 0.7$. The average temperature of the reactor is 25°C and is operated at an average of 1.0 mg/L of dissolved oxygen.

7-39. (**a**) 10,000 m^3/day of treated effluent is to be discharged through two separate diffuser pipes. What depth of water is required to effect a dilution of 200? (**b**) If the treated effluent is to be discharged through a manifold having 20 diffusers and a length of 12 m, what is the depth required to effect the same degree of dilution? Assume that $(\rho_a - \rho)/\rho = 0.025$. Draw conclusions.

7-40. Determine the height z_{max} that the plume will rise to for a discharge of 0.02 m^3/s-m. The density at the bottom of the ocean floor is 1024.28 kg/m^3, and at a point 60 m from the ocean floor, the density is 1025 kg/m^3. Assume that $(\rho_a - \rho)/\rho = 0.025$. Also determine $\overline{S}$ at this height.

7-41. Secondary-treated wastewater at 15°C is to be discharged into an estuary. Design a manifold diffuser system using two ports to discharge 0.11 m^3/s. Assume that the head available between the last unit in the treatment plant and the first diffuser port is 7.62 m. Also, assume that the length of the supply pipe is 122 m. Construct the hydraulic profile.

BIBLIOGRAPHY

APHA, AWWA, and WEF (1992). *Standard Methods for the Examination of Water and Wastewater.* American Public Health Association, Washington, D.C.

MANDT, M. G., and B. A. BELL (1982). *Oxidation Ditches in Wastewater Treatment.* Ann Arbor Science, Ann Arbor, Mich.

METCALF & EDDY, INC. (1991). *Wastewater Engineering: Treatment, Disposal, Reuse.* McGraw-Hill, New York.

MONOD, J. (1949). "The Growth of Bacterial Cultures." *Annual Review of Microbiology III.*

ROBERSON, J. A., J. J. CASSIDY, and M. H. CHAUDRY (1988). *Hydraulic Engineering.* Houghton Mifflin, Boston.

SAWYER, C. N. and P. L. McCARTY (1978). *Chemistry for Environmental Engineers.* McGraw-Hill, New York.

SUBRAMANYA, K., and S. C. AWASTHY (1970). "Discussion of the Paper by Vigander." *Journal of the Hydraulics Division, ASCE,* 102(HY5).

WALLACE, R. B., and S. J. WRIGHT (1984). "Spreading Layer of Two-Dimensional Buoyant Jet." *Journal of Hydraulic Division, ASCE,* 110(6).

CHAPTER 8

Sludge Treatment and Disposal

A by-product in the treatment of wastewater is the huge volume of sludge that must be disposed of. The problem of disposing sludge is complex because (1) it is largely composed of the substances that make wastewater objectionable, (2) it is composed of organic matter that can decompose making it objectionable, and (3) the sludge solid itself is only a small portion of the total sludge, resulting in a large sludge volume with only little solids to dispose of. Sludge contains only from 0.25 to 16% solids. To dispose sludge economically, its volume must therefore be reduced. In this chapter we discuss sludge treatment methods aimed at reducing its volume. For discussions on the chemistry of aerobic and anaerobic decomposition, see Chapter 2.

OVERVIEW OF SLUDGE TREATMENT

Strictly, sludge treatment starts when water is settled in the sedimentation basin. There are two methods of clarification: the first is the normal settling of solids as discussed in Chapters 6 and 7 and the second is the process called flotation. After these clarification steps, the sludge is withdrawn for further treatment.

Figure 8-1a shows a generalized flow sheet for the various sludge treatment methods. At the bottom, flotation is indicated. *Flotation* is the reverse of settling. Instead of settling down, solids settle up, so to speak. In this process, pressurized air is introduced into the flotation tank, forming air bubbles. These bubbles make the solids adhere to them, causing the solids to rise to the surface along with the air bubbles to form the *floats*. The

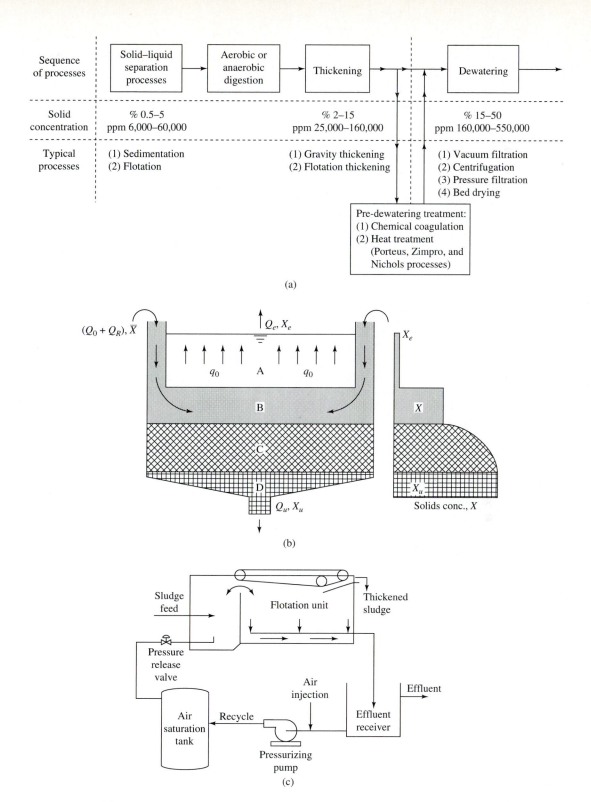

Figure 8-1 Schematics of (a) flowsheet for various methods of sludge treatment; (b) gravity thickener; (c) flotation plant.

floats are then skimmed off for further treatment. Sludges produced by clarification using sedimentation and flotation are approximately 0.5 to 5% solids.

Next to sedimentation and flotation are the processes of thickening. The aim of thickening is to expel additional water to increase the percent solids content of the sludge, hence thickening it. The production of so much percent solids in normal sedimentation or flotation for clarifying water is, in fact, due to the thickening at the bottom of the sedimentation basin or surface of the flotation tank. In general, there are two methods of thickening: gravity and flotation thickening. *Flotation thickening* is the same flotation process discussed above, only this time, the aim is solely thickening, without regard to clarification. *Gravity thickening* relies on gravity to pull down the solids. As the sludge settles, the percent solids content is increased by expulsion of water from the sludge matrix due to compression and consolidation. Water expulsion is also aided by vertical studs mounted on a revolving bottom scraper. The studs cause agitation in the sludge, helping to release the water the sludge traps due to its flocculent structure.

Figure 8-1b may be considered as a schematic section of a gravity thickener or secondary clarifier. Gravity thickeners operate in much the same way as the secondary clarifier used to settle sludge from the biological reactor. The major difference between the two is that the aim of the secondary clarifier is to clarify the effluent liquor from the reactor, while the aim of the gravity thickener is to thicken the sludge. Thickening that can occur at the bottom of the clarifier is only of secondary importance, although it generally controls the design. Thickeners are normally deeper than secondary clarifiers, providing more volume to accommodate sludge. As indicated in the drawing, the solids concentration increases from top to bottom of the thickener. Thickening can increase the percent solids 2 to 15%. After thickening, or in some cases, before thickening, the sludge may be subjected to either aerobic or anaerobic digestion. In aerobic digestion, a large portion of the VSS is converted to CO_2, NH_3, and H_2, while in anaerobic digestion, a large portion of this VSS is converted to CO_2, CH_4, and small quantities of H_2S and H_2. Hence, in digestion, the sludge is largely reduced in volume, this in addition to effecting stabilization, making the sludge less offensive.

Figure 8-2a shows the schematic section of a conventional anaerobic digester. The section is divided into several zones: the gas dome, where the product gas is collected for withdrawal; the scum layer, where the undigested portions float to the top, forming the scum; the supernatant, which is the water from the sludge introduced and the water from the reaction; the active layer, where the anaerobic reaction takes place; and the stabilized solids zone, where the digested sludge is stored temporarily.

A high-rate anaerobic digester is depicted in Figure 8-2b. This digester is characterized by a level of sludge loading higher than that of a conventional digestor, with the incorporation of mixing. Since mixing is incorporated, the zones of the conventional digester are not established. The gas dome, however, is still present. Anaerobic digesters may be designed to operate in two stages. The first stage would be the high-rate stage; the second would be similar to that in the conventional digester in that separation of the supernatant and digested sludge takes place here.

Thickening may be done before or after digestion. Which one precedes the other would depend on specific circumstances and should be investigated. However, since

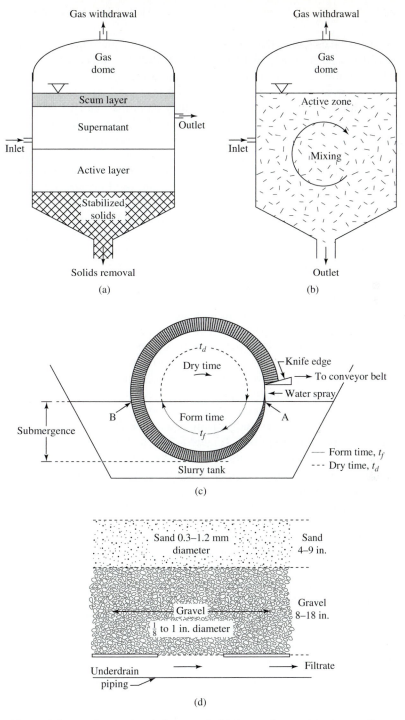

Figure 8-2 (a) Conventional digester; (b) high-rate digester (c); vacuum filter; (d) sludge-drying bed.

digestion takes longer than the detention time of a thickener, the volume of the digester would be much, much smaller when put after the thickener than when put before it. Although this arrangement is more economical, there are installations in which digesters are put ahead of thickeners[1] (see Figure 8-1).

Thickened and digested sludges may be further reduced in volume by dewatering. Various dewatering operations are used, including vacuum filtration, centrifugation, pressure filtration, belt filters, and bed drying. In *vacuum filtration*, a drum wrapped in filter cloth rotates slowly while the lower portion is submerged in a sludge tank (Figure 8-2). A vacuum applied to the underside of the drum sucks the sludge into the filter cloth, separating the filtrate and thus dewatering the sludge. In *centrifugation*, centrifugal action is used to force the solids toward the sides of a rotating bowl, thus separating them from the supernate, which stays at the center of the bowl (Figure 8-3). A rotating mechanism scrapes the solids from the side of the bowl for discharge through an outlet port, while the supernate at the center discharges through another port. In *pressure filtration*, which operates in a cycle, the sludge is pumped through the unit, forcing its way into *filter plates* wrapped in *filter cloths* (Figure 8-3). The plates are held in place by *filter frames*. As the sludge is forced through the plates, the filtrate passes through the filter cloth, leaving the solids on the cloth to accumulate in the recess of the frames. As determined by the cycle, the press is opened to remove the accumulated and dewatered sludge. In *belt filter operation*, sludges are put on belts, where they are allowed to dewater by gravity, application of pressure, or by capillary action. In the *application of pressure*, the sludge is sandwiched between two belts (Figure 8-3), which apply pressure. In *capillary action*, the sludge on a screen belt is placed over a second belt capable of delivering a capillary action on the sludge (Figure 8-3). The second belt is actually made of a spongy material that sucks water from the sludge, thereby dewatering it. *Bed drying* is an operation where the sludge is put on sand beds (Figure 8-2). The filtrate is then allowed to drain through the bed and be collected in underdrain pipings. To avoid the effects of the elements, it is best to enclose the beds in some housing. Both enclosed and open sludge-drying beds are used in practice.

Frequently, dewatering of sludges is difficult, especially when gelatinous. For example, it is extremely difficult to dewater gelatinous sludges using vacuum filtration. Pre-dewatering treatment is therefore recommended. Figure 8-1 indicates two methods of pre-dewatering: chemical coagulation and heat treatment using the Porteus, Zimpro, and Nichols processes. The coagulants normally used are $FeCl_3$, lime, and polyelectrolytes. Coagulation treatment may be preceded by elutriation, which removes much of the alkalinity, thus reducing the amount of chemicals needed. The *Porteus process* involves wet oxidation of sludge by dissolved oxygen under a steam pressure of 180 to 210 psig and a temperature of 290 to 390°F. With a detention time in the reactor of about 30 min, an 80 to 90% oxidation of organic matter is accomplished in this process. The *Zimpro process* is similar to the Porteus process but air (rather than steam) is injected at a pressure of 150 to 300 psig at a maximum operating temperature of 300 to 600°F.

---

[1] W. Viessman, Jr., and M. J. Hammer (1993). *Water Supply and Pollution Control.* Harper & Row, New York, p. 647.

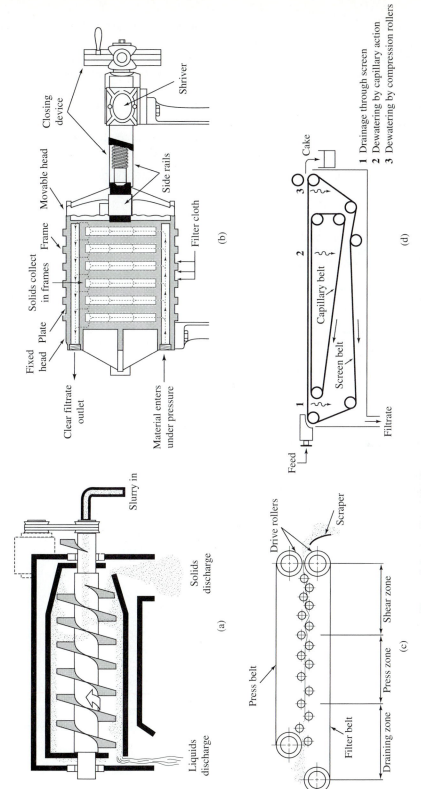

Figure 8-3 (a) Centrifuge. (b) Plate-and-frame press. (c) Belt pressure filter. (d) Capillary dewatering belt.

This process can also accomplish 80 to 90% oxidation of organic matter. The *Nichols process* is a simple thermomechanical system whereby no air or steam is added but the sludge is simply heated to temperatures of 395°F for a period of 30 min. This is enough to destroy the gelatinous nature of the sludge, making it amenable to dewatering. Of course, oxidation of the sludge is negligible, if it occurs.

Tables 8-1 to 8-8 relate to the various parameters of sludge.

TABLE 8-1 SPECIFIC GRAVITIES OF SLUDGE SOLIDS AND SLUDGES FROM VARIOUS PROCESSES

Treatment processes	Sludge solids	Sludge
Activated sludge (waste sludge)	1.3	1.00
Aerated lagoon (waste sludge)	1.3	1.01
Chemical addition to primary clarifiers for phosphorus removal		
Low lime (350–500 mg/L)	2.0	1.04
High lime (800–1600 mg/L)	2.0	1.05
Extended aeration (waste sludge)	1.3	1.02
Primary sedimentation	1.4	1.02
Trickling filter (waste sludge)	1.5	1.03
Algae removal	1.2	1.01
Suspended-growth denitrification	1.2	1.01
Roughing filters	1.3	1.02

MASS–VOLUME RELATIONSHIPS

As shown in the preceding tables, the mass and volume of sludge depend largely on the water content. Letting P_s be the fraction of solids, $1 - P_s$ will be the fraction of water. The relationship of the two fractions to the mass of the sludge $M_{s\ell}$ is

$$\frac{M_{s\ell}}{S_{s\ell}(1000)} = \frac{P_s M_{s\ell}}{S_s(1000)} + \frac{(1 - P_s)M_{s\ell}}{1000} \qquad (8\text{-}1)$$

Solving for S_s gives

$$S_s = \frac{P_s S_{s\ell}}{1 - S_{s\ell}(1 - P_s)} \qquad (8\text{-}2)$$

where $S_{s\ell}$ is the specific gravity of the sludge and S_s is the specific gravity of the sludge solids. If M_s is the mass of solids, the volume of sludge $V_{s\ell}$ is

$$V_{s\ell} = \frac{M_s}{1000 S_{s\ell} P_s} = \frac{M_{s\ell}}{1000 S_{s\ell}} \qquad (8\text{-}3)$$

During the sludge digestion, volatile solids are those acted upon and hence the ones that disappear. The fixed solids are inert to the reaction, and for a given amount of original sludge, they remain after digestion.

TABLE 8-2 EXPECTED SLUDGE SOLIDS PERCENT CONCENTRATIONS FROM VARIOUS PROCESSES

Process	Range	Typical
Secondary settling tank		
Waste activated sludge with primary settling	0.5–1.5	0.7
Waste activated sludge without primary settling	0.8–2.5	1.3
Pure-oxygen activated sludge with primary settling	1.3–3	2
Pure-oxygen activated sludge without primary settling	1.4–4	3
Trickling filter humus sludge	1–3	2
Primary settling tank		
Primary sludge	4–10	6
Primary sludge to a cyclone	0.5–3	2
Primary and waste activated sludge	3–10	4
Primary and trickling filter humus	4–10	5
Primary sludge with iron addition for phosphorus removal	5–15	8
Primary sludge with low lime addition for phosphorus removal	2–8	5
Primary sludge with high lime addition for phosphorus removal	4–16	10
Scum	3–12	6
Gravity thickener		
Primary sludge	6–12	8
Primary and waste activated sludge	3–12	5
Primary and trickling filter humus	4–10	5
Flotation thickener		
Waste activated sludge	3–6	4
Anaerobic digester		
Primary sludge	5–12	6
Primary and waste activated sludge	3–8	4
Primary and trickling filter humus	3–8	4
Aerobic digester		
Primary sludge	3–7	4
Waste activated sludge	1–3	1
Waste activated and primary sludge	20–4	3

TABLE 8-3 COMPARISON OF THICKENED AND UNTHICKENED SLUDGES AND SPECIFIC SOLIDS LOADING RATE TO GRAVITY THICKENERS

Type of sludge	Solids concentration (%)		Specific solids loading rate $(kg/m^2\text{-day})$
	Unthickened	Thickened	
Separate			
Activated sludge	1–3	1–4	20–45
Pure-oxygen sludge	1–4	2–5	30–50
Primary sludge	4–12	7–12	120–150
Trickling filter sludge	1–3	4–10	45–50
Combined			
Primary and trickling filter sludge	4–12	4–12	70–100
Primary and air-activated sludge	3–10	3–12	50–90

TABLE 8-4 SPECIFIC SOLIDS LOADING RATE FOR
DISSOLVED-AIR FLOTATION UNITS

Type of sludge	Loading (kg/m^2-day)
Air activated sludge (mixed-liquor)	30–80
Air activated sludge (settled)	60–100
Pure-oxygen activated sludge (settled)	50–170
50% primary + 50% activated sludge (settled)	100–200
Primary	to 270

TABLE 8-5 DESIGN CRITERIA FOR AEROBIC DIGESTERS

Parameter	Value
Hydraulic detention time (days at 20°C)	
Waste activated sludge	10–15
Activated sludge without primary settling	10–15
Primary plus activated or trickling filter sludge	15–20
Specific solids loading rate (kg VS/m^3-day)	1.5–4.5
Oxygen requirement (kg/kg BOD_5 destroyed)	1.5–20
Energy requirements for mixing	
Mechanical aerators ($kW/10^3$ m^3)	20–40
Air mixing ($m^3/10^3 m^3$-min)	20–40
Dissolved oxygen level in liquid (mg/L)	1–2

TABLE 8-6 EXPECTED PERFORMANCE OF VACUUM FILTERS

Type of sludge	Yield (kg/m^2-h)	Cake solids (%)
Fresh solids		
Primary	20–50	30
Primary and trickling filter	25–35	25
Primary and air activated	20–25	20
Primary and oxygen activated	20–30	25
Air activated	15–20	20
Pure-oxygen activated	15–20	20
Digested sludge (with or without elutriation)		
Primary	25–35	25
Primary and trickling filter	20–25	20
Primary and air activated	20–25	15
Primary and oxygen activated	25–30	20

TABLE 8-7 EXPECTED PERFORMANCE OF
SOLID-BOWL CENTRIFUGES

Type of sludge	Cake solids (%)
Untreated	
Primary	30
Primary and trickling filter	20
Primary and air activated	15
Trickling filter	15
Air activated	10
Pure-oxygen activated	15
Digested	
Primary	30
Primary and trickling filter	20
Primary and air activated	15

TABLE 8-8 TYPICAL AREA REQUIREMENTS FOR OPEN SLUDGE-DRYING
BEDS

Type of sludge	Area $(m^2/10^3)$ persons	Specific sludge loading rate (kg solids/m^2-yr)
Primary digested	100–140	130–210
Primary and humus digested	100–170	110–150
Primary and activated digested	150–280	50–100
Primary and chemically precipitated	180–240	110–150

Example 8-1

Determine the volume of sludge before and after digestion and the percent reduction for 600 kg of primary sludge solids with the characteristics shown in the following table. Assuming that the digested sludge can be further dewatered to 30% solids, how much more percent reduction in volume will be realized?

	Primary	Digested
Solids (%)	5	10
VS (%)	55	90% of original VS destroyed
Sp. gravity	1.01	1.04

Solution

$$V_{s\ell} = \frac{M_s}{1000 S_{s\ell} P_s} = \frac{600}{1000(1.01)(0.05)} = 11.88 \text{ m}^3 \quad \textbf{Answer}$$

$$\text{Fixed solids} = 600(1.0 - 0.55) = 270 \text{ kg}$$

$$\text{Total solids in digested sludge} = 270 + 600(0.55)(0.1) = 303 \text{ kg}$$

$$\text{Volume of digested sludge} = \frac{303}{1000(1.04)(0.1)} = 2.91 \text{ m}^3 \quad \textbf{Answer}$$

$$\% \text{ reduction} = \frac{11.88 - 2.91}{11.88}(100) = 75.51\% \quad \textbf{Answer}$$

$$S_s = \frac{P_s S_{s\ell}}{1 - S_{s\ell}(1 - P_s)} = \frac{0.1(1.04)}{1 - 1.04(1 - 0.1)} = 1.625$$

$$\frac{M_{s\ell}}{S_{s\ell}(1000)} = \frac{P_s M_{s\ell}}{S_s(1000)} + \frac{(1 - P_s)M_{s\ell}}{1000} \qquad \text{apply this to the dewatered digested sludge}$$

$$\frac{1}{S_{s\ell}} = \frac{0.3}{1.625} + (1 - 0.3); \; S_{s\ell} \text{ of the dewatered sludge} = 1.13$$

Assuming the contents of the solids filtrate to be negligible, the volume of dewatered sludge is

$$\frac{303}{1000(1.13)(0.3)} = 0.88 \text{ m}^3$$

Thus

$$\% \text{ reduction} \; \frac{11.88 - 0.88}{11.88}(100) = 92.6\%$$

$$\text{additional } \% \text{ reduction} = 92.6 - 75.51 = 17.09 \quad \textbf{Answer}$$

SECONDARY CLARIFICATION AND GRAVITY THICKENING

As mentioned earlier, thickening starts in the settling tank (and in the flotation tank). For this reason, secondary clarification was not discussed in Chapter 7 but was postponed until this chapter. Sludges resulting from trickling filters and other attached-growth processes are very different from those obtained from the activated sludge reactor. Trickling filter sludges behave more like primary sludges; the design of the secondary clarifier for trickling filter solids is therefore similar to primary sedimentation design. This design has already been discussed; in this chapter we discuss only design of secondary clarifiers for the activated sludge process.

The sole purpose of secondary clarification is to remove the biological flocs. However, coincident with the process is the inevitable thickening that occurs at the bottom of the tank. Hence the secondary clarifier performs two important functions: clarification and thickening. Figure 8-1b may be considered a schematic section of either a secondary clarifier or a thickener. As shown, there are four zones: A, B, C, and D. Zone A is the *clarification zone*. In zone B the concentration of the solids is constant, as indicated by X, the biosolids coming out of the bioreactor $[\overline{X}]$. Since the concentration is constant throughout, settling in this zone is called *zone settling*. In zone C the concentration varies from low to high in going from the upper to the lower part of the zone. This is where thickening occurs. In zone D, the concentration is constant, again; the concentration in this zone is called the *underflow concentration* and is the concentration that is withdrawn from the bottom of the tank.

In sizing a secondary clarifier, the clarifier and the thickener functions are addressed. In the gravity thickener, although there is also some clarification, the thickening function is the major design parameter. In designing the clarifier portion of the clarifier, the subsidence velocity of zone B must be determined. This is modeled by allowing a sample to settle in a cylinder and following the movement of the interface between zones A and B over time. This subsidence velocity is equated to the overflow velocity to size the clarifier area A_c.

The design of the thickener area considers zone C. As indicated, the concentration of solids in this zone is variable. Hence the behavior of the solids is modeled by making several dilutions of the sludge to conform to the several concentrations in the zone. The ultimate aim of the experiment is to be able to determine the *solids loading* into the thickener zone. The experiment is similar to that of the clarifier design except that several concentrations are modeled in this instance, and it is the relation between subsidence velocity and concentration that is sought. When the velocity is multiplied by the concentration, the result is the solids loading called the *solids flux*, the parameter that sizes the thickener portion.

Performing a material balance at any elevation section of the thickener area, two solid fluxes must be accounted for: the flux due to the gravity settling of the solids and the flux due to the conveyance effect of the withdrawal of the sludge in the underflow of the tank. Calling the total flux G_t, the gravity flux $V_c[X_c]$, and the conveyance flux $V_u[X_c]$, the material balance equation is

$$G_t = V_c[X_c] + V_u[X_c] \tag{8-4}$$

where V_c is the subsidence velocity at the section of the thickening zone, $[X_c]$ the solids concentration at the section, and V_u is the underflow velocity computed at the section as Q_u/A_t, where Q_u is the underflow rate of flow and A_t is the thickener area at the elevation section considered. Some value of this flux G_t is the one that will be used to design the thickener area and therefore must be determined.

Because the solids concentration in the thickening zone is variable, G_t in the zone is also variable. Of these several values of G_t, there is only one value that would make the solids loading at the corresponding elevation section equal to the rate of withdrawal of sludge in the underflow. This particular G_t value is called the *limiting flux* $G_{t\ell}$, since it is the one flux that corresponds to the underflow withdrawal rate. Since $G_{t\ell}$ corresponds to the rate of sludge withdrawal, the thickener area determined by it will be the thickener area for design. At the limiting flux condition, equation (8-4) is revised to

$$G_{t\ell} A_t = Q_u[X_u] = V_c[X_{c\ell}]A_t + V_u[X_{c\ell}]A_t \tag{8-5}$$

where ℓ is added as an index to emphasize the limiting condition. Dividing all throughout by A_t gives

$$G_{t\ell} = V_u[X_u] = V_c[X_{c\ell}] + V_u[X_{c\ell}] \tag{8-6}$$

In equation (8-6), several values of $V_u[X_u]$ will result, corresponding to the various combinations of the values of V_c, $X_{c\ell}$, and V_u. The smallest of these values is the one chosen for design. Solving for V_u, we obtain

$$V_u = \frac{V_c[X_{c\ell}]}{[X_u] - [X_{c\ell}]} \tag{8-7}$$

Having determined $G_{t\ell}$, the thickener area A_t can be determined as follows:

$$A_t = \frac{(Q_0 + Q_R)[\overline{X}]}{G_{t\ell}} \tag{8-8}$$

where Q_0 is the effluent flow from the primary clarifier to the reactor and Q_R is the recirculation flow. $Q_0 + R_R$ is also the total flow to the clarifier or thickener. A_t is then compared with the clarifier area A_c and the larger of the two chosen for the design. For gravity thickeners, A_t is used automatically.

Example 8-2

The activated sludge bioreactor facility of a certain plant is to be expanded. The results of a settling cylinder test of the existing bioreactor suspension are as follows:

MLSS (mg/L)	1410	2210	3000	3500	4500	5210	6510	8210
V_c (m/h)	2.93	1.81	1.20	0.79	0.46	0.26	0.12	0.084

$Q_0 + Q_R$ is 10,000 m³/day and the influent MLSS is 3500 mg/L. Determine the size of the clarifier that will thicken the sludge to 10,000 mg/L of underflow concentration.

Solution

$$V_u = \frac{V_c[X_{c\ell}]}{[X_u] - [X_{c\ell}]}$$

MLSS (mg/L)	1410	2210	3000	3500	4500	5210	6510	8210
V_c (m/h)	2.93	1.81	1.20	0.79	0.46	0.26	0.12	0.084
V_u (m/h)	0.48	0.51	0.51	0.43	0.38	0.28	0.22	0.39
$V_u[X_u]$ (m/h)·(mg/L)	4809	5100	5100	4300	3800	2800	2200	3900

The smallest $V_u[X_u]$ is 2200 (m/h)·(mg/L); Thus $G_{t\ell} = 2200$ (m/h)·(mg/L) (*Note:* To be more accurate, $V_u[X_u]$ may be plotted against MLSS and the minimum $V_u[X_u]$ obtained from the curve. The other way to have more accurate results is to make the MLSS values much closer to each other.)

$$A_t = \frac{(Q_0 + Q_R)[\overline{X}]}{G_{t\ell}} = \frac{(10,000/24)(3500)}{2200} = 662.88 \ \text{m}^2$$

For an MLSS of 3500, 0.79 m/h. Assuming that the solids content in the effluent is negligible, the solids content in the underflow is

$$10,000(3.5) = 35,000 \ \text{kg/day} = 1458.33 \ \text{kg/h}$$

Thus

$$Q_u = \frac{1458.33}{10} = 145.83 \ \text{m}^3/\text{h}$$

$$Q_e = \frac{10,000}{24} - 145.83 = 270.84 \ \text{m}^3/\text{h}$$

$$A_c = \frac{270.84}{0.79} = 342.83 \ \text{m}^2 < 662.88 \ \text{m}^2$$

Thus the thickening function controls and the area of the thickener is 662.88 m². **Answer**

Flotation

As mentioned before, flotation may be used in lieu of the normal clarification by solids-downward-flow sedimentation basins. It can also be used for thickening. The mathematical treatments for both flotation clarification and flotation thickening are the same. Referring to Figure 8-1c, the dissolved air concentration $\overline{C}_{aw,t}$ of the wastewater in the air saturation tank is

$$\overline{C}_{aw,t} = f C_{asw,t,sp} \frac{P}{P_s} = f\beta C_{ast,sp} \frac{P}{P_s} \tag{8-9}$$

where f is the fraction of saturation achieved (0.5 to 0.8), $C_{asw,t,sp}$ is the dissolved air saturation value of wastewater in the air saturation tank (at temperature equal to the wastewater temperature in the tank and at standard barometric pressure P_s), and P is the pressure in the tank. $C_{ast,sp}$ is the dissolved air saturation value of tap water or distilled water (at the temperature of the wastewater in the saturation tank and pressure P_s). $C_{ast,sp}$ can be obtained from the corresponding value for oxygen ($C_{st,sp}$) by multiplying $C_{st,sp}$ by $(28.84)/[0.21(32)]$, where 28.84 is the molecular weight of air, 0.21 is the mole fraction of oxygen in air, and 32 is the molecular weight of oxygen. Hence

$$C_{ast,sp} = 4.29 C_{st,sp} \tag{8-10}$$

The total amount of air introduced into the flotation tank comes from the air saturation tank and from the influent feed (sludge feed in Figure 8-1). The air from the influent feed is $Q_0 C_{asw,o,sp} P_a/P_s$, where P_a is the barometric pressure of operation of the flotation unit and $C_{asw,o,sp}$ is the dissolved air saturation value of the wastewater in Q_0 (at the temperature of Q_0 and standard barometric pressure P_s). $Q_0 C_{asw,o,sp} P_a/P_s$ is also equal to $Q_0 \beta C_{aso,sp} P_a/P_s$, where $C_{aso,sp}$ is the dissolved air saturation value of tap water at the temperature of Q_0 and P_s. Hence the total air A_{in} is

$$A_{in} = RQ_0 f \beta C_{ast,sp} \frac{P}{P_s} + Q_0 \beta C_{aso,sp} \frac{P_a}{P_s} \qquad (8\text{-}11)$$

where R is the recycle ratio. Substituting equation (8-10) into equation (8-11),

$$A_{in} = \frac{4.29 Q_0 \beta}{P_s} (Rf C_{st,sp} P + C_{so,sp} P_a) \qquad (8\text{-}12)$$

where $C_{st,sp}$ is the tap water oxygen saturation value at standard condition pressure of P_s and temperature of the air saturation tank, and $C_{so,sp}$ is the tap water oxygen saturation value at standard condition pressure of P_s and temperature of the influent feed. P_s is equal to 760 mmHg, 1 atm of pressure, 101,330 N/m^2, and so on, depending on the system of units used.

As the pressurized flow from the air saturation tank is released into the flotation unit, the pressure reduces to atmospheric, P_a. It would be accurate to assume that the condition at this point in the flotation unit is saturation at the prevailing temperature and pressure. Hence, after pressure release, the remaining dissolved air A_0 is

$$A_0 = \frac{4.29 Q_0 \beta}{P_s} (RC_{so,sp} P_a + C_{so,sp} P_a) \qquad (8\text{-}13)$$

and the air utilized for flotation is $A_{in} - A_0 = A_{used}$; or

$$A_{used} = \frac{4.29 Q_0 \beta}{P_s} R(f C_{st,sp} P - C_{so,sp} P_a) \qquad (8\text{-}14)$$

The solids in the influent is $Q_0 C_0$, where C_0 is the suspended solids influent concentration. The air used/solids ratio A/S is then

$$A/S = \frac{A_{used}}{Q_0 C_0} = \frac{4.29 \beta R \big(f C_{st,sp} P - C_{so,sp} P_a \big)}{P_s C_0} \qquad (8\text{-}15)$$

For operations without recycle, $Q_0 R$ is simply Q_0. From this, equation (8-15) becomes

$$A/S = \frac{A_{used}}{Q_0 C_0} = \frac{4.29 \beta \big(f C_{st,sp} P - C_{so,sp} P_a \big)}{P_s C_0} \qquad (8\text{-}16)$$

Laboratory determination of design parameters. To design a flotation unit, the overflow area, the pressure in the air saturation tank, and the recirculation ratio must be determined. The overflow velocity needed to estimate the overflow area may be determined in the laboratory. Figure 8-4a shows a laboratory flotation device. A sample of the sludge is put in the pressure tank and pressurized. The tank is then shaken to ensure that the sludge is saturated with air under the corresponding pressure. The pressure and temperature readings in the tank are noted. A valve leading to the flotation cylinder is then opened to allow the pressurized sludge to flow into the cylinder. The rate of rise of the sludge interface in the flotation cylinder is followed with respect to time; this gives the rise velocity. This is equated to the overflow velocity to compute the overflow area.

The value of the air saturation tank pressure and the recirculation ratio may be designed depending on the A/S ratio to be employed in the operation of the plant. This

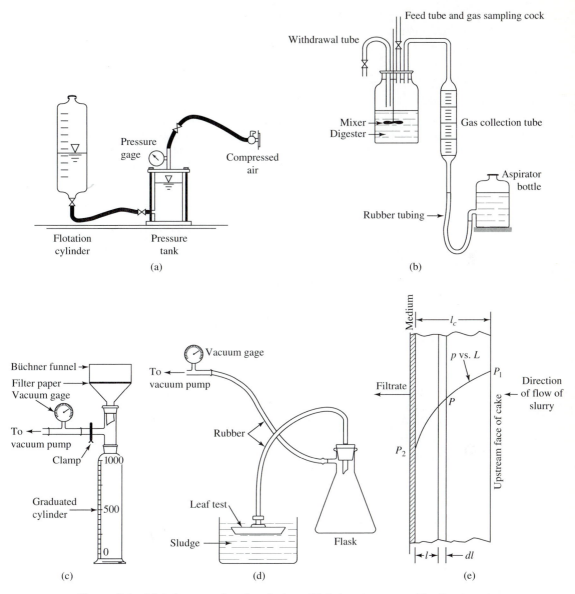

Figure 8-4 (a) Laboratory flotation device. (b) Laboratory anaerobic digester setup. (c) Büchner funnel filtration assembly. (d) Leaf filter assembly. (e) Mechanics of cake filtration.

A/S ratio, in turn, is determined by the laboratory flotation experiment just described. A sample of subnatant is taken from the bottom of the flotation cylinder and the clarity or turbidity determined. Hence, by performing several runs for different values of A/S, corresponding values of clarity will be obtained, thereby producing a relationship between

A/S and clarity. The A/S corresponding to the desired clarity is then chosen for the design calculations.

In performing the laboratory float experiment, no recirculation is used. Hence the formula to be used to compute the A/S ratio is equation (8-16). Ensuring during the experiment that the pressure tank is fully saturated, f can be considered unity.

Example 8-3

A laboratory experiment is performed to obtain the air to solids ratio A/S to be used in the design of a flotation unit. The pressure gage reads 40 psig and the temperature of the sludge and the subnatant in the flotation cylinder is 20°C. The prevailing barometric pressure is 14.6 psi. The total solids in the sludge is 10,000 mg/L and β was originally determined to be 0.95. Determine the A/S ratio.

Solution

$$A/S = \frac{4.29\beta \left(f C_{st,sp} P - C_{so,sp} P_a\right)}{P_s C_0}$$

$$C_{st,sp} \text{ at } 20°C = 9.08 \text{ mg/L}$$

$$C_{so,sp} \text{ at } 20°C = 9.08 \text{ mg/L}$$

$$A/S = \frac{4.29(0.95)\,[(1)(9.08)(40 + 14.6) - 9.08(14.6)]}{14.7(10,000)} = 0.010 \quad \textbf{Answer}$$

Example 8-4

It is desired to thicken an activated sludge liquor from 3000 mg/L to 4% using a flotation thickener. A laboratory study indicated that an A/S ratio of 0.010 is optimal for this design. The subnatant flow rate was determined to be 8 L/m²-min. The barometric pressure is assumed to be the standard of 14.7 psi and the design temperature is to be 20°C. Assume that $f = 0.5$ and $\beta = 0.95$. The sludge flow rate is 400 m³/day. Design the thickener with and without recycle.

Solution   (a) *Without recycle*:

$$A/S = \frac{4.29\beta\left(f C_{st,sp} P - C_{so,sp} P_a\right)}{P_{st} C_0}$$

$$0.010 = \frac{4.29(0.95)[(0.5)9.08P - 9.08(14.7)]}{14.7(3,000)} \qquad P = 29.4 \text{ psia} \quad \textbf{Answer}$$

Assuming that the solids content in the subnatant is negligible, the solids constant in the float is

$$3.0(400) = 1200 \text{ kg/day}$$

$$4\% \text{ solids} = 40,000 \text{ mg/L} = 40 \text{ kg/m}^3$$

$$\text{float rate of flow, } Q_u = \frac{1200}{40} = 30 \text{ m}^3/\text{day}$$

$$\text{subnatant flow, } Q_e = 400 - 30 = 370 \text{ m}^3/\text{day}$$

$$\text{subnatant flow area, } A_s = \frac{370}{8(10^{-3})(60)(24)} = 32.12 \text{ m}^2 \quad \textbf{Answer}$$

(b) *With recycle*: Use the same operating pressure of 38.53 psia.

$$A/S = \frac{4.29\beta Rf\left(\overline{C}_{st,sp}P - \overline{C}_{so,sp}P_a\right)}{P_{st}C_0}$$

$$0.010 = \frac{4.29(0.95)(R)(0.5)\,[9.08(38.53) - 9.08(14.7)]}{14.7(3000)} \qquad R = 1 \quad \textbf{Answer}$$

$$\text{Subnatant flow area, } A_s = \frac{2(400) - 30}{8(10^{-3})(60)(24)} = 66.84 \text{ m}^2 \quad \textbf{Answer}$$

Note: This example shows that recycling is of no use. It increases the overflow area and it needs additional piping for the recycling.

AEROBIC DIGESTION

The principles involved in aerobic digestion are very similar to that in the activated sludge process; the only difference is that the microorganisms in the digestion process are being starved and, in fact, cannibalizing each other, while in the activated sludge process, they are being fed. Hence since no food is being "served," $\mu_{\max}\{[S]/K_s + [S]\}[X]V$ is equal to zero and the conservation equation is simply (see Chapter 7).

$$-k_d[\overline{X}]V = -\frac{\partial[\overline{X}]}{\partial t}V + \{-Q_0[X_0] + (Q_0 - Q_w)[X_e] + Q_w[X_u]\} \qquad (8\text{-}17)$$

At steady state, equation (8-17) may be solved for k_d, obtaining

$$\frac{1}{k_d} = \frac{[\overline{X}]V}{Q_0[X_0] - (Q_0 - Q_w)[X_e] - Q_w[X_u]} \qquad (8\text{-}18)$$

$(Q_0 - Q_w)[X_e]$ is equal to $Q_e[X_e]$, where Q_e is the effluent discharge flow and $Q_w[X_u]$ is equal to zero (no sludge wasting in the recycle flow, since there is no recycle). Thus equation (8-18) becomes

$$\frac{1}{k_d} = \frac{[\overline{X}]V}{Q_0[X_0] - Q_e[X_e])} = \theta_d \qquad (8\text{-}19)$$

By analogy with θ_c, which is called the mean cell retention time (MCRT), call θ_d the mean cell destruction time (MCDT), which is the time it takes to digest the cell in the digester. For a given sample of wastewater, MCDT may be obtained by performing an aerobic digestion experiment.

The oxygen requirement may be determined by performing an oxygen uptake rate along with the digestion experiment. This will involve putting the sample in a bottle, aerating it, and following the decrease in the concentration of the dissolved oxygen in the bottle with time.

Example 8-5

Aerobic digestion and oxygen uptake experiments are performed producing the results below. Estimate the MCDT for a 90% destruction of the VSS and the average oxygen requirement.

Time of aeration (days)	VSS remaining (mg/L)	$\bar{r}$ (mg/L-hr)
0	5550	—
1	5190	34.0
2	4940	27.0
5	4410	18
8	4159	15
10	3861	11.8
15	3490	7.8
20	3250	5.2
25	3190	4.0

Solution

20	3250	5.2
25	3190	4.0
?	x	0

$$\frac{x - 3190}{3190 - 3250} = \frac{0 - 4.0}{4.0 - 5.2}$$

$x = 2990$ mg/L $\rightarrow$ nonbiodegradable VSS

Hence revise the table to show the biodegradable VSS by subtracting 2990 from the respective VSS.

Time of aeration (days)	Biodegradable VSS remaining (mg/L)	$\bar{r}$ (mg/L-hr)
0	2560	—
1	2200	34.0
2	1950	27.0
5	1420	18
8	1169	15
10	871	11.8
15	500	7.8
20	260	5.2
25	200	4.0

For 90% destruction, biodegradable VSS remaining = 0.1(2560) = 256 mg/L. Hence

20	260
x	256
25	200

$$\frac{x - 20}{25 - 20} = \frac{256 - 260}{200 - 260}$$

$x = 20.33$ days = MCDT   **Answer**

Time of aeration (days)	$\bar{r}$ (mg/L-hr)	Δt	$\bar{r}_{ave}$	$\Delta t \bar{r}_{ave}$
0	—	—	—	—
1	34.0	—	—	—
2	27.0	1	30.5	30.5
5	18	3	22.5	67.5
8	15	3	16.5	49.5
10	11.8	2	13.4	26.8
15	7.8	5	9.8	49.0
20.33	5.2	5.33	6.5	34.65
	$\sum$	19.33		257.95

Thus

$$\text{oxygen requirement} = \frac{257.95}{19.33} = 13.34 \text{ mg/L-h} \quad \textbf{Answer}$$

Example 8-6

Using the values obtained in Example 8-5, calculate the volume of the digester and the oxygen requirement to digest 6000 lb/day of volatile sludge solids. Assume that 3500 mg/L of average MLVSS is to be maintained in the digester.

Solution

$$\frac{[\overline{X}]V}{Q_0([X_0] - Q_e[X_e]} = \theta_d$$

$$6000 \text{ lb/day} = 6000(454) = 2,724,000 \text{ g/day} = 2724 \text{ kg/day}$$

From the data of Example 8-5, fraction biodegradable VSS $= 1 - 2990/5550 = 0.46$. For 90% destruction,

$$Q_0([X_0] - Q_e[X_e] = 2724(0.46)(0.9) = 1127.74 \text{ kg/day}$$

Thus

$$\frac{3.5V}{1127.74} = 20.33 \qquad V = 6550.56 \text{ m}^3 \quad \textbf{Answer}$$

Determine average laboratory VSS:

Time of aeration (days)	VSS (mg/L)	Δt	VSS_{ave}	$\Delta t(VSS_{ave})$
0	—	—	—	—
1	5,190	—	—	—
2	4,940	1	5,065	5,065
5	4,410	3	4,675	14,025
8	4,159	3	4,284.5	12,853.5
10	3,861	2	4,010	8,020
15	3,490	5	3,675.5	18,377.5
20.33	3,250	5.33	3,370	17,962.1
	$\sum$	19.33		76,303.1

Thus

$$\text{average VSS} = \frac{76,303.1}{19.33} = 3947.39 \text{ mg/L}$$

$$\text{specific oxygen utilization rate} = \frac{13.34}{3947.39}(3500) = 11.83 \text{ mg/L-h}$$

$$\text{oxygen requirement} = 11.83(6550.56)(1000) = 1.86(10^9) \text{ mg/day}$$

$$= 1859.53 \text{ kg/day} \quad \textbf{Answer}$$

ANAEROBIC DIGESTION

In the design of the anaerobic digester, equation (8-18) may be modified as follows: Referring to Figure 8-2 for the conventional anaerobic digester, Q_w may be considered as the rate of flow of the digested sludge from the active layer zone toward the stabilized solids zone. Designating Q_w as Q_{st} and $[X_u]$ and $[X_{st}]$, the MCDT for the anaerobic digester may then be written as

$$\theta_d = \frac{[\overline{X}]V}{Q_0[X_0] - Q_e[X_e] - Q_{st}[X_{st}]} \tag{8-20}$$

The MCDT θ_d may be determined, experimentally, by performing an anaerobic digestion experiment using a setup similar to that shown in Figure 8-4b. As indicated by the gas collection tube, the stopping of the gas production signals the end of the digestion period. Data similar to those collected for the aerobic digester may be gathered and the size of the digester determined (see examples of aerobic digester calculations).

In the design of the conventional anaerobic digester (also called the standard-rate digester), provision can be made to store digested sludge temporarily at the bottom of the tank. This is not the same as the aerobic digester, or the high-rate anaerobic digester, where no storage can be made since the tank contents are mixed. The total volume of the active layer and the stabilized solids zone of the standard-rate digester V_t may be estimated as

$$V_t = \frac{Q_0 + Q_{st}}{2}\theta_d + Q_{st}t_2 \tag{8-21}$$

$$= V + Q_{st}t_2$$

V can also be calculated from equation (8-20) and t_2 is the digested sludge storage detention time.

In a two-stage high-rate digester, the first stage is where the reaction takes place and the volume may be calculated using equation (8-20) with Q_{st} equals zero. The second stage is used for dewatering and storage and the volume may be calculated using the first of equations (8-21). Of course, in this instance, Q_{st} is not equal to zero. A digester is shown in Figure 8-5.

Figure 8-5 Digesters at Back River sewage treatment plant, Maryland.

Example 8-7

A total of 6000 lb/day of VSS at a flow rate of 80 m³/day is to be anaerobically digested. The results of a laboratory digestion study show an MCDT of 10 days. The biodegradable fraction of the VSS is 46%, of which it is desired to destroy 90%. The VSS is 70% of the total solids. The digested sludge is 5% solids at a specific gravity of 1.02. Using a two-stage high-rate digester system, compute the total volume required, excluding gas, supernatant, and scum volume storage. Assume a mean MLVSS of 3500 mg/L, a dewatering time of 3 days, and a sludge storage detention time of 90 days.

Solution Calculate the first-stage volume:

$$\theta_d = \frac{[\overline{X}]V}{Q_0[X_0] - Q_e[X_e] - Q_{st}[X_{st}]}$$

$$6000 \text{ lb/day} = 2724 \text{ kg/day}$$

$$Q_0[X_0] - Q_e[X_e] - Q_{st}[X_{st}] = 2724(0.46)(0.9) = 1127.74 \text{ kg/day}$$

$$10 = \frac{3.5V}{1127.74} \qquad V = 3222.11 \text{ m}^3$$

Calculate the second-stage volume:

$$V_t = \frac{Q_0 + Q_{st}}{2}\theta_d + Q_{st}t_2$$

$$\text{digested sludge solids} = 2724(1 - 0.46) + 2724(0.46)(0.1) + \left(\frac{2724}{0.7} - 2724\right)$$

$$= 2763.69 \text{ kg/day}$$

$$Q_{st} = \frac{2763.69}{1.02(1000)(0.05)} = 54.19 \text{ m}^3/\text{day}$$

$$V_t = \frac{80 + 54.19}{2}(3) + 54.19(90) = 5078.39 \text{ m}^3$$

$$\text{total volume} = 3222.11 + 5078.39 = 8300.50 \text{ m}^3 \quad \textbf{Answer}$$

Example 8-8

Solve Example 8-7 using a standard-rate anaerobic digester. Assume that MCDT $= 30$ days to produce the same quality of digested sludge.

Solution

$$\theta_d = \frac{[\overline{X}]V}{Q_0[X_0] - Q_e[X_e] - Q_{st}[X_{st}]}$$

$$30 = \frac{3.5V}{1127.74} \qquad V = 9666.34 \text{ m}^3$$

$$V_t = V + Q_{st}t_2$$

$$= 9666.34 + 54.19(90) = 14,543.44 \text{ m}^3$$

or

$$V_t = \frac{Q_0 + Q_{st}}{2}\theta_d + Q_{st}t_2 = \frac{80 + 54.19}{2}(30) + 54.19(90) = 6889.95 \text{ m}^3$$

The larger volume controls; therefore, $V_t = 14,543.44 \text{ m}^3$.   **Answer**

CAKE FILTRATION

In dewatering using vacuum filtration or the plate-and-frame press, a cake is formed on the surface of the filter cloth. Hence these processes of dewatering may be called *cake filtration*. In cake filtration, the flow of the filtrate in its most elementary form may be considered as through a tightly packed bank of small crooked tubes across the cake. The rightmost bottom drawing of Figure 8-5 is a schematic section of a filter cake and filter cloth. Because of pressure, the solids are tightly packed and the flow is laminar. From fluid mechanics, the Hagen–Poiseuille equation may be written as

$$\Delta P = \frac{32\mu\ell V}{(2R)^2} \tag{8-22}$$

where μ is the dynamic viscosity, ℓ the thickness of the cake, V the superficial velocity of filtration, and R the hydraulic radius. But $R = [\eta/(1 - \eta)](V_p/A_p)$, where η is the porosity of cake, V_p the volume of cake particles, and A_p the surface area of particles. Hence substituting gives us

$$\Delta P = \frac{8(1-\eta)^2 A_p^2}{\eta^2 V_p^2}\mu\ell V \tag{8-23}$$

Applying to a differential thickness of cake yields

$$dP = \frac{8(1-\eta)^2 A_p^2}{\eta^2 V_p^2}\mu V d\ell \tag{8-24}$$

If ρ_p is the density of solids, the mass dm in the differential thickness of cake $d\ell$ is $dm = A\, d\ell(1-\eta)\rho_p$, where A is the superficial area of filtration. Solving this for $d\ell$ and substituting, equation (8-24) becomes

$$dP = \frac{8(1-\eta)A_p^2 \mu V}{\eta^2 V_p^2 \rho_p A}dm = \alpha\frac{\mu V}{A}dm \tag{8-25}$$

where α is called the *specific cake resistance*. Equation (8-25) may be integrated from the filter cloth ($P = P_2$) to the surface of the cake ($P = P_1$). Using an average value of α and integrating gives

$$-\Delta P = \bar{\alpha}\frac{\mu V}{A}m_c \tag{8-26}$$

where $\bar{\alpha}$ is the average α over the cake thickness and m_c is the total mass of cake collected on the cloth. Including the resistance of the filter cloth R_m, the total pressure drop may be written as

$$-\Delta P = \mu V\left(\frac{\bar{\alpha}m_c}{A} + R_m\right) \tag{8-27}$$

Determination of $\bar{\alpha}$

Calling V the volume of the filtrate collected at any time t, V is $(dV/dt)/A$; also, expressing m_c as cV (where c is the mass of cake collected per unit volume of filtrate), substituting in equation (8-27), and integrating yields

$$\frac{t}{V} = \frac{\mu c\bar{\alpha}}{2(-\Delta P)A^2}V + \frac{\mu R_m}{(-\Delta P)A} \tag{8-28}$$

which is an equation of a straight line between t/V and V. The value of $\mu c\bar{\alpha}/2(-\Delta P)A^2$ can be determined, hence, also, $\bar{\alpha}$.

The laboratory experiment involves using a Büchner funnel using the setup shown in Figure 8-5. For a given pressure difference $(-\Delta P)$, the amount of filtrate collected is recorded with time. The data collected gives the relationship between t/V and V as called for by equation (8-28). The cake collected is also weighed to determine c; μ is determined from the temperature of the filtrate.

Example 8-9

A Büchner funnel experiment to determine the specific cake resistance of a certain sludge is performed. The results are as follows:

Volume of filtrate (mL)	Time (sec)	t/V
25	48	1.92
50	150	3.0
75	308	4.12
100	520	5.2

$-\Delta P = 20$ in. Hg, filter area $= 550$ cm^2, $\mu = 15(10^{-4})$ kg/m-s, and $c = 0.25$ g/cm^3. Determine $\bar{\alpha}$.

Solution

$$\frac{t}{V} = \frac{\mu c \bar{\alpha}}{2(-\Delta P)A^2} V + \frac{R_m}{(-\Delta P)A}$$

$$= mV + b$$

$$m = \frac{(1.92 + 3.0)/2 - (4.12 + 5.2)/2}{(25 + 50)/2 - (75 + 100)/2} = 0.044 \frac{\text{s}}{(\text{mL})^2} = 0.044 \frac{\text{s}}{[(10^{-3})(10^{-3})]^2}$$

$$= 4.4(10^{10}) \frac{\text{s}}{\text{m}^6}$$

$$-\Delta P = \frac{20}{29.92}(101,330) = 67,733.96 \frac{\text{N}}{\text{m}^2}$$

$$\text{filter area} = 550(10^{-2})^2 = 0.055 \text{ m}^2$$

$$c = 0.25 \left[\frac{10^{-3}}{(10^{-2})^3} \right] = 250 \frac{\text{kg}}{\text{m}^3}$$

$$\frac{\mu c \bar{\alpha}}{2(-\Delta P)A^2} = 4.4(10^{10}) \qquad \bar{\alpha} = 4.8(10^{13}) \frac{1}{\text{kg-m}} \qquad \textbf{Answer}$$

Design Cake Filtration Equation

In an actual filtration installation, be it a vacuum filter or a plate-and-frame press, the resistance of the filter medium is practically negligible; it may therefore be neglected. Considering this fact, equation (8-28) may be written as

$$\frac{t}{V} = \frac{\mu c \bar{\alpha}}{2(-\Delta P)A^2} V \tag{8-29}$$

Since it is important to be able to calculate the amount of dewatered sludge that is finally to be disposed of, it is convenient to express equation (8-29) in a form that will give this amount directly without considering the filtrate. In practice, this is usually done by utilizing the specific loading rate, also called the *filter yield* L_f, defined as

$$L_f = \frac{c\mathcal{V}}{At} \tag{8-30}$$

From this definition, L_f is the amount of cake formed per unit area of filter cloth per unit of time. In vacuum filtration (see Figure 8-2), the only time that the cake is formed is when the drum is submerged in the tank. In pressure filtration, the only time that the cake is formed is when the sludge is pumped into the plates. Hence, t in the equation above is called the form time t_f. Also, both the vacuum filter and the plate-and-frame press operate on a cycle. Calling the cycle time t_c, t_f may be expressed as a fraction f of t_c. Hence $t = t_f = f t_c$ may be substituted in equation (8-30), and if this equation is substituted in equation (8-29) and the result simplified and rearranged, we have

$$L_f = \sqrt{\frac{2(-\Delta P)c}{\mu \bar{\alpha} f t_c}} \tag{8-31}$$

For incompressible cakes, equation (8-31) is the design cake filtration equation. In vacuum filtration, f is equal to the fraction of submergence of the drum. Also, for the pressure filter, f is the fraction that the sludge is pumped over the total cycle time. (In the operation of a pressure filter, the sludge is first pumped into the filter plates, the filter assembly is opened to remove the cake collected, the filter cloth is put back into the filter plates, the assembly is closed and tightened, and the sludge is pumped again, completing the cycle.)

For compressible cakes such as those of sewage sludges, $\bar{\alpha}$ is not constant. Equation (8-31) must therefore be modified for the expression of the specific cake resistance. The usual form used is

$$\bar{\alpha} = \bar{\alpha}_0 (-\Delta P)^s \tag{8-32}$$

where s is a measure of cake compressibility. If s is zero, the cake is incompressible and $\bar{\alpha}$ equals $\bar{\alpha}_0$. Substituting equation (8-32) in equation (8-31), we have

$$L_f = \sqrt{\frac{2(-\Delta P)^{1-s}c}{\mu \bar{\alpha}_0 f t_c}} \tag{8-33}$$

Determination of Cake Filtration Parameters

To use equation (8-33) in design, L_f, $-\Delta P$, f, and t_c are specified. μ is specified from the temperature of filtration. The only parameters to be determined, then, are c, $\bar{\alpha}_0$, and s. c can easily be determined by collecting the volume of the filtrate and weighing the amount of cake retained on the filter. Taking the logarithms of both sides of equation (8-32) gives

$$\ln \bar{\alpha} = \ln \bar{\alpha}_0 + s \ln(-\Delta P) \tag{8-34}$$

Hence, from this equation, $\bar{\alpha}_0$ and s may be determined from two pairs of values of $\bar{\alpha}$ and $-\Delta P$.

The experimental procedure may be performed using the leaf filter assembly of Figure 8-4d, although the Büchner experiment may also be used. The leaf filter is immersed in the sludge and a vacuum is applied to suck the filtrate during the duration of the form time t_f. For a given $-\Delta P$, the volume of filtrate over the time t_f is collected. This will give one value of $\bar{\alpha}$. To satisfy the requirement of equation (8-34), at least two runs are made at two different pressure drops. From these pairs, the parameters $\bar{\alpha}_0$ and s may be calculated.

Example 8-10

A leaf-filter experiment is run to determine $\bar{\alpha}_0$ and s for a $CaCO_3$ slurry in water producing the results below. $\mu = 15(10^{-4})$ kg/m-s. The filter area is 440 cm^2, the mass of solid per unit volume of filtrate is 23.5 g/L, and the temperature is 25°C. Calculate $\bar{\alpha}_0$ and s.

Test number:	I	II
$-\Delta P$ (psi):	6.7	16.2
V (L)	Time (s)	
0.5	17.3	6.8
1.0	41.3	19.0
1.5	72.0	34.6
2.0	108.3	53.4
2.5	152.1	76.0
3.0	210.7	102.0

Solution For $-\Delta P = 6.7$ psi:

Volume of filtrate (L)	Time (s)	t/V
0.5	17.3	34.6
1.0	41.3	41.3
1.5	72.0	48
2.0	108.3	54.15
2.5	152.1	60.84
3.0	201.7	67.33

$$\frac{t}{V} = \frac{\mu c \bar{\alpha}}{2(-\Delta P)A^2} V + \frac{R_m}{(-\Delta P)A}$$

$$= mV + b$$

$$m = \frac{(34.6 + 41.3 + 48)/3 - (54.15 + 60.84 + 67.33)/3}{(0.5 + 1.0 + 1.5)/3 - (2.0 + 2.5 + 3.0)/3} = 12.97 \frac{s}{L^2} = 1.30(10^7) \frac{s}{m^6}$$

$$-\Delta P = \frac{6.7}{29.92}(101{,}330) = 22{,}690.88 \frac{N}{m^2}$$

$$\text{filter area} = 440(10^{-2})^2 = 0.044 \text{ m}^2$$

$$c = 23.5 \left(\frac{10^{-3}}{10^{-3}} \right) = 23.5 \frac{kg}{m^3} \qquad \mu = 8.9(10^{-4}) \text{ kg/m-s}$$

$$\frac{\mu c \bar{\alpha}}{2(-\Delta P)A^2} = 1.30(10^7) \qquad \bar{\alpha} = 5.46(10^{10}) \frac{1}{kg\text{-}m}$$

For $-\Delta P = 16.2$ psi:

Volume of filtrate (L)	Time (s)	t/V
0.5	6.8	13.6
1.0	19.0	19.0
1.5	34.6	23.07
2.0	53.4	26.7
2.5	76.0	30.4
3.0	102.0	34

$$m = \frac{(13.6 + 19.0 + 23.07)/3 - (26.7 + 30.4 + 34)/3}{(0.5 + 1.0 + 1.5)/3 - (2.0 + 2.5 + 3.0)/3} = 7.87 \frac{s}{L^2} = 7.87(10^6) \frac{s}{m^6}$$

$$-\Delta P = \frac{16.2}{29.92}(101,330) = 54,864.51 \frac{N}{m^2}$$

$$\text{filter area} = 440(10^2)^2 = 0.044 \text{ m}^2$$

$$c = 23.5 \left(\frac{10^{-3}}{10^{-3}} \right) = 23.5 \frac{kg}{m^3} \qquad \mu = 8.9(10^{-4}) \text{ kg/m-s}$$

$$\frac{\mu c \bar{\alpha}}{2(-\Delta P)A^2} = 7.87(10^6) \qquad \bar{\alpha} = 7.99(10^{10}) \frac{1}{kg\text{-}m}$$

$$\ln \bar{\alpha} = \ln \bar{\alpha}_0 + s \ln(-\Delta P)$$

$$\ln 5.46(10^{10}) = \ln \bar{\alpha}_0 + s \ln 22,690.88 \qquad \text{(a)}$$

$$\ln 7.99(10^{10}) = \ln \bar{\alpha}_0 + s \ln 54,864.51 \qquad \text{(b)}$$

Solving, $\bar{\alpha}_0 = 7.29(10^8)$ in MKS units   **Answer**

$$s = 0.43 \quad \textbf{Answer}$$

Example 8-11

A $CaCO_3$ sludge with cake filtration parameters determined in Example 8-10 is to be dewatered in a vacuum filter under a vacuum of 630 mmHg. The mass of filtered solids per unit volume of filtrate is to be 60 kg/m^3. The filtration temperature is determined to be 25°C and the cycle time, half of which is form time, is 5 min. Calculate the filter yield.

Solution

$$L_f = \sqrt{\frac{2(-\Delta P)^{1-s}c}{\mu \bar{\alpha}_0 f t_c}}$$

$$-\Delta P = \frac{630}{760}(101,330) = 83,997.24 \text{ N/m}^2 \qquad c = 60 \text{ kg/m}^3 \qquad s = 0.43$$

$$\mu = 8.9(10^4) \text{ kg/m-s} = 0.0534 \text{ kg/m-min}; \qquad \bar{\alpha}_0 = 7.29(10^8)1/\text{kg-m}$$

$$f = 0.5 \qquad t_c = 5 \text{ min}$$

$$L_f = \sqrt{\frac{2(83,997.24)^{1-0.43}(60)}{0.0534(7.29)(10^8)(0.5)(5)}} = 0.028 \text{ kg/m}^2\text{-min} \quad \textbf{Answer}$$

COMPOSTING

Another method of sludge treatment is composting. *Composting* is an accelerated process of oxidizing with oxygen a solid organic material. There are two general methods of composting: the windrow and the static-pile method. The *windrow* method is done by piling the waste in a row or windrows and turning the piles once or twice per week for a more effective aeration and for releasing excess heat. Composting is an exothermic process producing temperatures of 140 to 160°F or higher. The *static-pile* method involves piling the waste in individual locations and sucking air through the pile by means of a suction blower to aerate it. Since the sludge is too wet, means must be employed to increase the porosity of the mass for effective aeration in both composting methods. This is normally done by mixing the sludge with wood chips.

The ratio of dewatered sludge cake to wood chips on a volumetric basis is between 1 and 2 sludge cake to 2 to 3 wood chips; the mixing should reduce the moisture content to about 60%. This produces the desired porosity. Composting takes approximately 6 weeks with additional 2 or more weeks for curing and complete stabilization. After stabilization, the wood chips are separated for reuse and the compost bagged and sold as a low-grade fertilizer, soil conditioner, or used for land reclamation. For a discussion on the chemistry of composting, see Chapter 2.

SLUDGE DISPOSAL

In the treatment of wastewaters, there are two streams that must be disposed of properly: the treated effluent and the sludge. The disposal of treated effluent to deep and shallow waters has already been discussed in this book. Now turn to the disposal of sludge. In general, the following may be regarded as methods of sludge disposal: incineration, lagooning, dumping, landfilling, and land spreading for use as fertilizer or soil conditioner. *Incineration* is simply the combustion of sludge converting the material into gases and residues. Incineration involves a heat balance, and the calculation considers the heat needed to evaporate the water, heat loss due to radiation and combustion residues, the sensible heat of the combustion gases, and the heating value of the sludge solids. The principle of heat balance calculation is discussed in Chapter 11. Material balance calculations for incineration or combustion are discussed in Chapter 10.

Strictly, incineration is not an ultimate disposal. It is only a disposal in the sense that some of the materials are disposed of into the air as gases. However, the process leaves some residue that must still be disposed of. In addition, because of the carbon dioxide and sulfur dioxide constituents in the discharged gases, incineration should be avoided if other methods are available. CO_2 is an agent of the greenhouse effect and SO_2 is a precursor of acid rain.

The *lagoon* is an earthen basin. If untreated and digested sludges are deposited here, the basin becomes a place of disposal or treatment, as the case may be, in a method called *lagooning*. The processes of thickening and dewatering discussed previously occur in the lagoon. If the sludge is untreated, digestion also occurs in the basin. The sludge may be stored indefinitely in the lagoon, which in this case becomes an ultimate point of disposal. However, another place must be found for the additional sludges that remain to be disposed of, or the sludge may be emptied from the lagoon for draining and drying from time to time. In this instance the lagoon becomes a part of the treatment flow sheet, and the sludge removed must be disposed of somewhere else. Lagoons should be lined to prevent contaminating the groundwater.

Dumping of sludge by barging into the sea was considered an acceptable method of disposal in the past. At present, however, it is being discouraged because of its effect on the environment surrounding the discharge point. For example, marine scientists have discovered that the practice of York City of dumping sludge into the sea created a deposit of "black mayonnaise" at the seafloor, with obvious consequences to bottom dwellers. However, dumping of sludge not into the sea but in other places may be acceptable, depending on the particular situation. For example, if an abandoned mine quarry is available, it may be a good place simply to dump the sludge. Of course, the site must be far from residential areas and highways to avoid the problem of odors, and hydrogeologic investigation must be performed to ascertain that groundwaters will not be polluted. If these conditions are not met, dumping should not be allowed.

Where municipal solid wastes are landfilled, sludges may be deposited in the same manner as the solid waste in the sanitary landfill. In a *sanitary landfill*, wastes are deposited in a designated area, compacted with a roller or a tractor, and covered with soil at the end of a day's operation. If done according to the textbook, the operation should be clean. However, in practice, a landfill looks just like a landfill with trash scattered all around. This is due mainly to mismanagement. (Landfilling of solid wastes is discussed further in Chapter 11.)

Land spreading is distributing or scattering the sludge on the ground. The idea is for the sludge ultimately to be incorporated in the soil to act as fertilizer or soil conditioner. Soil conditioning means the making of a more porous soil, one that retains more moisture and enables plants to grow better. Dewatered and composted sludges may be distributed in thin layers over the ground by a truck spreader and plowed under after they have dried. The truck spreader is also a common method of distributing liquid sludge over the ground, although modified for pumping liquids. The liquid sludge may also be distributed over the ground by spraying, just as in spray irrigation. Finally, the liquid sludge may also be distributed by the *ridge-and-furrow method*. In this method, the sludge flows in the furrows in between ridges of crops. In the foregoing methods, the rate of

TABLE 8-9 NITROGEN AND PHOSPHORUS REQUIREMENTS FOR SOME TYPES OF CROPS

Type of crop	Nitrogen (kg/ha-yr)	Phosphorus (kg/ha-yr)
Forage		
Alfalfa	504	39
Bromegrass	190	30
Coastal bermuda	540	70
Reed canary	360	40
Ryegrass	240	70
Sweet clover	180	20
Tall fescue	130	25
Field		
Barley	70	20
Corn	170	20
Cotton	70	15
Milomaize	95	20
Soybeans	100	15

application that the soil can assimilate must be determined properly; otherwise, ponding problems may occur. Examples of application rates are 2.0 to 6.0 cm/yr.

Fertilizing Value of Sludge

The fertilizer value of a material is determined by its nitrogen, phosphorus, and potassium content. Typical values of these nutrients as percentages of total solids in a typical digested sludge are nitrogen, 3.0; phosphorus as P_2O_5, 2.5; and potassium as K_2O, 0.5. These analyses are referred to as the NPK values of the fertilizer. Thus the NPK values of a typical digested sludge are 3.0, 2.5, and 0.5. The NPK values of Milorganite, a heat-dried sludge of an activated-sludge plant in Milwaukee, Wisconsin are 6.0, 2.0, and 0.0. A typical analysis of a Hechinger[3] lawn fertilizer is 28, 3, 3. The phosphorus content of sludge is comparable to that of a regular fertilizer; however, for the case of nitrogen, a comparison is nowhere to be found. The comparison between the potash contents is not as bad as the comparison between the nitrogen contents. This "imbalance" in the ratio of nitrogen to phosphorus in digested sludges is a major drawback in their use.

The nitrogen and phosphorus requirements for some types of forage and field crops are shown in Table 8-9. Fertilizer applications must not exceed a great percentage over the values given in this table, especially for nitrogen. Excessive applications of nitrogen fertilizer will result in the conversion by some processes of the excess nitrogen to nitrate. The nitrate may, then, contaminate groundwaters. For sludge fertilizers, the organic nitrogen portion must first degrade to ammonia to be usable to the plants. Hence not all of the nitrogen in the sludge will be consumed at a single setting, but only a fraction at a time. Some example of degradation rates are 20 to 35% in the first year, 5 to 15% in the second year, and 1 to 5% in the succeeding years.

[3]Hechinger, a lumber distributor in the United States, also sells fertilizers.

Let N be the specific yearly application rate of sludge, D the fraction of a particular fertilizer constituent in the sludge, U the specific yearly requirement of the crop for a particular fertilizer constituent, and $r_1, r_2, \ldots, r_n$, be the fractions of the particular fertilizer constituent that are being used by the crop in the first, second, $\ldots$, and nth years of application. U for the first, second, third, $\ldots$, and nth years are, respectively, $U_1, U_2, U_3, \ldots$, and U_n. N for the first, second, third, $\ldots$, and nth year are, also, respectively, $N_1, N_2, N_3, \ldots$, and N_n. The corresponding equations are

$$U_1 = N_1 r_1 D \tag{8-35}$$

$$U_2 = N_1 D(1 - r_1)r_2 + N_2 r_1 D \tag{8-36}$$

$$U_3 = N_1 D(1 - r_1)(1 - r_2)r_3 + N_2 D(1 - r_1)r_2 + N_3 r_1 D \tag{8-37}$$

$$U_4 = N_1 D(1-r_1)(1-r_2)(1-r_3)r_4 + N_2 D(1-r_1)(1-r_2)r_3 + N_3 D(1-r_1)r_2 + N_4 r_1 D \tag{8-38}$$

Knowing that $N_1 = N_2 = N_3 = N_4$, equation (8-38) may be solved just for N_4, producing

$$N_4 = \frac{U_4}{D\left[(1 - r_1)(1 - r_2)(1 - r_3)r_4 + (1 - r_1)(1 - r_2)r_3 + (1 - r_1)r_2 + r_1\right]} \tag{8-39}$$

For n years,

$$N_n = \frac{U_n}{D\left[(1 - r_1)(1 - r_2)\cdots(1 - r_{n-1})r_n + (1 - r_1)\cdots(1 - r_{n-2})r_{n-1} + \cdots + (1 - r_1)r_2 + r_1\right]} \tag{8-40}$$

where $N_n = N$ and $U_n = U$.

Example 8-12

A farmer is growing alfalfa for forage in a 5-ha farm. Considering only the nitrogen and phosphorus requirements, determine (a) the yearly application rate if Milorganite is used and (b) the yearly application rate if Hechinger's fertilizer is used. (c) If Milorganite costs $6.99 for a bag of 18.4 kg and Hechinger's fertilizer costs $10.99 for a bag of 15.52 lb, compare the cost of fertilizing the farm using separate fertilizers. Assume that for Milorganite 35% of the applied nitrogen is consumed in the first year, 15% in the second year, 5% in the third and fourth years, and negligible percentages in years thereafter. The nitrogen in Hechinger's fertilizer may be assumed to be utilizable by the crop immediately. (*Note:* The cost figures cited above are actual prices as of this writing, January 13, 1992.)

Solution   (a) From Table 8-9, alfalfa needs 504 kg/ha-yr of nitrogen and 39 kg/ha-yr of phosphorus.

$$N_n = \frac{U_n}{D\left[(1 - r_1)(1 - r_2)\cdots(1 - r_{n-1})r_n + (1 - r_1)\cdots(1 - r_{n-2})r_{n-1} + \cdots + (1 - r_1)r_2 + r_1\right]}$$

Nitrogen requirement:

$$n = 4 \qquad r_1 = 0.35 \qquad r_2 = 0.15 \qquad r_3 = r_4 = 0.5 \qquad D = 0.06$$

$$N_4 = \frac{504}{0.06\left[(1-0.35)(1-0.15)(1-0.05)(0.05) + (1-0.35)(1-0.15)(0.05) + (1-0.35)(0.15) + 0.35\right]}$$

$$= 16{,}755.10 \text{ kg/ha-yr} \rightarrow 16{,}755.10(5) = 83{,}775.48 \text{ kg/yr}$$

Phosphorus requirement:

$$2\% \text{ as } P_2O_5 = \frac{2P}{P_2O_5}(2) = 0.44(2) = 0.87\% \text{ as } P \rightarrow D$$

Assuming that all P is used: $r_1 = 1.0$, $n = 1.0$. Thus

$$N_1 = \frac{39}{0.0087(1.0)} = 4482.76 \text{ kg/ha-yr} \rightarrow 22{,}413.79 \text{ kg/yr} < 83{,}775.48 \text{ kg/yr}$$

$$\text{milorganite to be used} = 83{,}775.48 \text{ kg/yr} \quad \textbf{Answer}$$

(b) For the Hechinger fertilizer, the nitrogen content is utilized immediately by the plant. Hence

$$N_1 \text{ for nitrogen} = \frac{504}{0.28(1.0)} = 1800 \text{ kg/ha-yr} \rightarrow 9000 \text{ kg/yr}$$

$$N_1 \text{ for phosphorus} = \frac{39}{0.44(0.03)(1.0)} = 2954.55 \text{ kg/ha-yr} \rightarrow 14{,}772.73 \text{ kg/yr} > 9000$$

$$\text{fertilizer requirement} = 14{,}772.73 \text{ kg/yr} \quad \textbf{Answer}$$

(c) For Milorganite, cost $= \dfrac{6.99}{18.4}(83{,}775.48) = \$31{,}825.58 \text{ /yr}$

For Hechinger's fertilizer, cost $= \dfrac{10.99}{15.52}(14{,}772.73)(2.2) = \$23{,}013.86 \text{ /yr}$

Thus

$$\text{milorganite is } \frac{31{,}825.58 - 23{,}013.86}{23{,}013.86}(100)$$

$$= 38.29\% \text{ more expensive than the regular fertilizer.} \quad \textbf{Answer}$$

Note: This example illustrates that for sludge fertilizers to be competitive with regular fertilizers, they have to be amended by adding more nitrogen constituent, perhaps using ammonium nitrate.

Objections to the Use of Sludge

There are two major components that make the use of sludge and sludge products objectionable: the heavy metal and toxic elements content and pathogens. Heavy metals are characterized by their high specific gravities. For example, Cd, Pb, and Hg are heavy metals; they have specific gravities of $8.65^{20/20}$, $11.337^{20/20}$, and $13.546^{20/20}$, respectively. (The numerators of the exponents refer to the temperature in degrees Celsius at which the density of the material is obtained; the denominators of the exponents refer to the temperature in degrees Celsius at which the density of the reference water is obtained.) Heavy metals at certain concentrations are toxic and can bioaccumulate. The aforementioned heavy metals are toxic and can bioaccumulate, for example, and be assimilated into the food chain. The metalloid arsenic has these properties. Typical concentrations for some of the other heavy metals and toxic elements found in sludges are shown in Table 8-10. There may be no other way to eliminate these metals from the sludge but by the success of the pretreatment program.

TABLE 8-10 TYPICAL CONCENTRATIONS OF HEAVY METALS AND TOXIC ELEMENTS IN SLUDGES

Metal or element	Specific gravity[a]	Concentration (mg/L)		
		Range	Mean	Median
Silver	$10.5^{20/20}$	?–950	230	95
Arsenic	$5.727^{14/14}$	10–20	10	10
Boron	2.32	200–1500	420	340
Barium	3.5	?–3000	1500	1250
Beryllium	1.816	?	?	?
Cadmium	$8.65^{20/20}$	?–1000	90	20
Cobalt	$8.9^{20/20}$	?–850	360	110
Chromium	7.1	20–30,000	1700	700
Copper	8.92	50–17,000	1300	700
Mercury	$13.546^{20/20}$	0.5–90	10	5
Manganese	$7.2^{20/20}$	100–9000	2000	450
Nickel	$8.90^{20/20}$	?–3000	400	100
Lead	$11.337^{20/20}$	80–30,000	2000	500
Strontium	2.6	?–2000	450	200
Selenium	$4.26^{25/25}$	10–200	30	20
Vanadium	5.96	?–2000	500	500
Zinc	7.14	50–30,000	4000	2000

[a]Referred to room temperature when no exponent is affixed (15 to 20°C).

Wastewaters generally contain four major types of pathogens: bacteria, protozoa, viruses, and helminths. The concentrations of these pathogenic organisms in wastewater depend on the health condition of the community. In times of epidemics, the concentrations will be high. The community would be very fortunate if it can have no inhabitant sick at a given time. It will therefore be expected that pathogens will be present in the wastewater at all times. Since these pathogens come from a relatively large volume of wastewater, when ending up in the sludge, they, in general, become concentrated and be very infectious. Some disease-producing protozoa and helminths in wastewaters and sludges are shown in Table 8-11. Table 8-12 shows survival times of various pathogens in soil and on plant surfaces.

Helminth ova are definitely a big problem. Bacteria and viruses can also survive for months; protozoan cysts die off in a matter of days. For as long as these pathogens are not inactivated, promotion in the use of sludge for crops intended for human consumption is not a logical decision and would be a hypocrisy.

Aerobic digestion can reduce viral and bacterial pathogens by 90%. Helminth ova are reduced to varying degrees. Anaerobic digestion can also reduce bacterial and

TABLE 8-11 SOME DISEASE-PRODUCING PROTOZOA AND HELMINTHS CONTAINED IN WASTEWATERS AND SLUDGES

Organism	Disease/symptoms
Protozoa	
Cryposporidium	Gastroenteritis
Entamoeba histolytica	Acute enteritis
Giardia lamblia	Giardiasis (including diarrhea, abdominal cramps, weight loss)
Balantidium coli	Diarrhea and dysentery
Toxoplasma gondii	Toxoplasmosis
Helminths	
Ascaris lumbricoides	Digestive and nutritional disturbances, abdominal pain, vomiting, restlessness
Ascaris suum	May produce symptoms such as coughing, chest pain, and fever
Trichuris trichiura	Abdominal pain, diarrhea, anemia, weight loss
Toxocara canis	Fever, abdominal discomfort, muscle aches, neurological symptoms
Taenia saginata	Nervousness, insomnia, anorexia, abdominal pain, digestive disturbances
Taenia solium	Nervousness, insomnia, anorexia, abdominal pain, digestive disturbances
Necatur americanus	Hookworm disease
Hymenolepis nana	Taeniasis

TABLE 8-12 SURVIVAL TIMES OF PATHOGENS IN SOIL AND ON PLANT SURFACES

Pathogen	Soil		Plants	
	Absolute maximum	Common maximum	Absolute maximum	Common maximum
Bacteria	1 year	2 months	6 months	1 month
Viruses	6 months	3 months	2 months	1 month
Protozoan cysts	10 days	2 days	5 days	2 days
Helminth ova	7 years	2 years	5 months	1 month

viral pathogens by 90%. Helminth ova are not substantially reduced even at mesophilic digestion but may be substantially reduced by thermophilic digestion. Stabilization of the sludge by lime can reduce viral and bacterial pathogens by well over 90%. Some helminth ova may be destroyed by this treatment, but some species are not affected substantially. Air drying may also reduce bacterial and viral pathogens by 90%, but, again, helminth ova remain substantially unaffected. Composting can also reduce bacterial and viral pathogens by at least 90%. Helminth ova populations are diminished but not necessarily eliminated. Heat drying and use of the electron beam and gamma-ray irradiation can reduce bacterial and viral pathogens and helminth ova to below detectable levels. Heat treatments such as the Porteus and Zimpro processes, if carried out properly, effectively destroy pathogenic viruses, bacteria, and helminth ova.

GLOSSARY

Activated sludge. The reinvigorated microorganisms sludge.

Aerobic sludge digestion. A process of reducing volume of sludge by allowing the microorganisms in the sludge to undergo endogenous respiration in the presence of air.

Air-to-solids ratio (A/S). In flotation thickening the ratio of the amount of flotation air used to the inflow total solids.

Anaerobic sludge digestion. A process of reducing volume of sludge by allowing the microorganisms in the sludge to undergo endogenous respiration in the absence of air.

Biodegradable VSS. The biodegradable portion of the volatile suspended solids (VSS).

Cake filtration. The dewatering of sludge where water is removed through passing across a cake formed during the process.

Composting. An accelerated process oxidizing a solid organic material using atmospheric oxygen.

Filter yield. In cake filtration, the amount of cake formed per unit area of filter cloth per unit time.

Flotation thickening. A method of removing solids where the solids float to the top of the unit.

Heavy metals. Metals of high specific gravity.

Lagoon. A basin used to hold waste.

Limiting flux. The solids flux that determines the area of a thickener.

Mean cell destruction time (MCDT). The average time it takes to destroy the cells in a digester.

Nichols process. A thermomechanical process of treating where sludge is subjected to heating at high temperature.

Porteus process. A wet oxidation of sludge using steam to dissolve oxygen at high pressure and temperature.

Solids flux. At a given point in a settling tank, the settling velocity of the solids multiplied by the concentration is called the solids flux.

Static-pile composting. A method of composting where the materials are piled in separate locations and air is sucked through the piles.

Underflow concentration. The concentration in the underflow of a sedimentation basin or thickener.

Vacuum filtration. A form of filtration where water is removed by the application of a vacuum.

Volatile solids. The solids that disappear after the sample is decomposed at $600°C$.

Windrow composting. A composting method where the materials are piled in rows or windrows.

Zimpro process. A wet oxidation of sludge similar to the Porteus process but using air at high pressure to dissolve oxygen at high temperature.

Zone settling. That portion in a settling basin where solids settle under one contiguous, constant concentration.

SYMBOLS

A	area of filter cloth
A_c	area required for clarification
A_p	surface area of a particle
A_t	area required for thickening
c	mass of filter cake collected per unit volume of filtrate
C	concentration of dissolved oxygen
$\overline{C}_{a,w}$	average air concentration of waste
$C_{aso,sp}$	saturation dissolved air concentration of tap water at standard pressure and the temperature influent flow
$C_{ast,sp}$	saturation dissolved air concentration of tap water at standard pressure and temperature of the saturation tank
$C_{asw,t,sp}$	saturation dissolved air concentration of the waste at standard pressure and temperature of the saturation tank
C_s	saturation dissolved oxygen concentration
$C_{so,sp}$	saturation dissolved oxygen concentration of tap water at standard pressure and temperature of the influent
$C_{st,sp}$	saturation dissolved oxygen concentration of tap water at standard pressure and temperature of the air saturation tank
D	fraction of a particular fertilizer constituent in sludge
f	fraction of air saturation achieved; fraction that the form time, t_f, is compared to the cycle time, t_c
G_t	total solids flux in a thickening process
$G_{t\ell}$	limiting total solids flux
L_f	specific filter loading rate or filter yield
m_c	mass of deposited filter cake
$M_{s\ell}$	mass of sludge
P_a	atmospheric pressure
P_s	fraction of solids in sludge
P_s	standard pressure
Q_e	effluent discharge
Q_R	recirculation flow

Q_{st}	rate of flow of sludge into storage at bottom of digestion tank
Q_w	volumetric rate of sludge wasting
Q_0	plant inflow
r	fraction of a particular fertilizer constituent in sludge used by the crop in the first year, r_1; in the second year, r_2; in the third year, r_3; etc.
R_m	resistance of filter medium
s	exponent of cake compressibility
S_s	specific gravity of sludge solids
$S_{s\ell}$	specific gravity of sludge
t_c	cycle time of cake filtration
t_f	cake form time
U	specific substrate utilization rate; specific crop utilization rate for a particular fertilizer constituent
V	volume of reactor; superficial velocity in cake filtration
$\mathcal{V}$	volume of filtrate
V_p	volume of a particle
V_u	velocity at a section of the thickener due to sludge withdrawal in the underflow
$[\overline{X}]$	average concentration of microorganisms
$[X_c]$	solids concentration at a section of a thickener
$[X_e]$	effluent concentration of microorganisms
$[X_{st}]$	microbial concentration in digested sludge
$[X_u]$	sedimentation tank underflow concentration of microorganisms
$[X_0]$	influent concentration of microorganisms
Y	specific yield of microorganisms
θ_d	mean cell destruction time

PROBLEMS

8-1. A sample of sludge containing 1.0% solids is put in a 1-L graduated cylinder up to the 1-L mark. It is then allowed to settle until the percent solids increases to twice the original. The supernatant is decanted in the process. Calculate the percentage reduction in volume.

8-2. In Problem 8-1, if the specific gravity of the solid particles is 2.1, what is the porosity of the mixture?

8-3. Calculate the volume of water in Problem 8-1.

8-4. A sludge contains 99% moisture and is run onto a sludge drying bed to a depth of 14 in. Calculate the approximate percentage moisture when the moisture has evaporated down to 7 in.

8-5. In Problem 8-4, what will be the depth of sludge on the bed when the percentage moisture is 92%?

8-6. The results of a settling cylinder test of a bioreactor suspension are as follows:

MLSS (mg/L)	1410	2210	3000	3500	4500	5210	6510	8210
V_c (m/h)	2.93	1.81	1.20	0.79	0.46	0.26	0.12	0.084

$Q_0 + Q_R$ is 10,000 m³/day and the underflow concentration is 10,000 mg/L. Determine the MLVSS needed to operate the bioreactor efficiently.

8-7. In Problem 8-6, why is the minimum of the products of MLVSS and V_c chosen as the design solid flux?

8-8. What is the maximum solid flux in Problem 8-6?

8-9. What are the basic dimensions of solid fluxes?

8-10. A laboratory experiment found the A/S ratio to be equal to 0.010. The temperature of the sludge and the subnatant in the flotation cylinder is 20°C. The prevailing barometric pressure is 14.6 psi. The total solids in the sludge is 10,000 mg/L, and β was previously determined to be equal to 0.95. What is the gage pressure under which the flotation cylinder is operated?

8-11. Solve Problem 8-10 if the plant is located at an elevation of 2000 ft above mean sea level.

8-12. Aerobic digestion and oxygen uptake experiments are performed, producing the results shown below, expressed in terms of total suspended solids (TSS). For a VSS of 70%, estimate the MCDT for a 90% destruction of the VSS.

Time of aeration (days)	VSS remaining (mg/L)	$\bar{r}$ (mg/L-hr)
0	5550	—
1	5190	34.0
2	4940	27.0
5	4410	18
8	4159	15
10	3861	11.8
15	3490	7.8
20	3250	5.2
25	3190	4.0

8-13. Using the values obtained in Problem 8-12, calculate the volume of the digester required to digest 6000 lb/day of sludge total solids containing 70% volatile matter. Assume 3500 mg/L of average MLVSS is to be maintained in the digester operated at a recirculation ratio of 1.

8-14. In Problem 8-12, estimate the average oxygen requirement.

8-15. In Problem 8-2, calculate the oxygen requirement.

8-16. A digested sludge is 1.5% solids at a specific gravity of 1.01. A total of 6000 lb/day of VSS at a flow rate of 80 m³/d is to be anaerobically digested. The results of a laboratory digestion study show an MCDT of 10 days. The biodegradable fraction of the VSS is 46%, of which it is desired to destroy 90%. The VSS is 70% of the total solids. Using a two-stage high-rate digester system, compute the total volume required, excluding gas, supernatant, and scum volume storage. Assume a mean MLVSS of 3,500 mg/L, a dewatering time of 3 days, and a sludge storage detention time of 90 days.

8-17. Solve Problem 8-16 using a standard-rate anaerobic digester. Assume that MCDT = 30 days to produce the same quality of digested sludge.

8-18. Is it possible to solve Problem 8-16 for an MLVSS of 6000 mg/L? Why?

8-19. Is it possible to solve Problem 8-17 for an MLVSS of 6000 mg/L? Why?

8-20. Calculate the volume of a semispherical gas holder to make available 1500 m^3 of gas at 24°C and 5 cm of water above atmospheric. The holder is to be constructed as a dome over an anaerobic digester.

8-21. The sludge entering a digestion chamber contains 76% volatile matter; the resulting digested sludge contains 55% volatile matter. Calculate the percent reduction in volatile matter during digestion.

8-22. Given the following information: population = 10,000; per capita daily dry solids in sludge = 0.077 kg; percent dry solids in raw sludge = 3; sludge specific gravity = 1.02; sludge added daily = 2% by volume. Calculate the required digester volume assuming that the tank is kept full.

8-23. The results of a Büchner funnel experiment to determine the specific cake resistance of a certain sludge are as follows:

Volume of filtrate (mL)	Time (s)	t/V
25	48	1.92
100	520	5.2

$-\Delta P = 20$ in. Hg, filter area = 550 cm^2, $\mu = 15(10^{-4})$ kg/m-s, and $c = 0.25$ g/cm^3. Determine $\bar{\alpha}$.

8-24. What is the resistance of the filter medium, R_m, in Problem 8-23?

8-25. In Problem 8-23, what is the volume of filtrate expected in 308 s?

8-26. In cake filtration, how does viscosity affect the volume of filtrate?

8-27. For the results of a leaf-filter experiment on a CaCO$_3$ slurry, tabulated below, calculate $\bar{\alpha}_0$ and s. $\mu = 15(10^{-4})$ kg/m-s. The filter area is 440 cm^2, the mass of solid per unit volume of filtrate is 23.5 g/L, and the temperature is 25°C.

Test number:	I	II
$-\Delta P$ (psi):	6.7	16.2
V (L)	Time (s)	
1.0	41.3	19.0
1.5	72.0	34.6
2.5	152.1	76.0
3.0	210.7	102.0

8-28. Calculate the filter yield for a CaCO$_3$ sludge with cake filtration parameters determined in Problem 8-27. The cake is to be dewatered in a vacuum filter under a vacuum of

630 mmHg. The mass of filtered solids per unit volume of filtrate is to be 60 kg/m^3. The filtration temperature is determined to be 25°C and the cycle time, one-fourth of which is form time, is five min.

8-29. What is the resistance of the filter medium, R_m, in Problem 8-27?

8-30. In Problem 8-27, what is the volume of filtrate expected in 308 s for test I?

8-31. Filtered sludge containing 20% moisture and weighing 15 kg was analyzed obtaining the following ultimate analysis on a dry basis: C = 52.2%, O = 38.0%, H = 2.5%, and the rest nitrogen. How many kilograms of oxygen will be required for combustion?

8-32. In Problem 8-31, how many kilograms of air is required?

8-33. On the basis of satisfying the nitrogen requirement, determine the rate of application of dry sludge to ryegrass. Assume that the soil has an initial nitrogen content of zero and that the sludge has 3% nitrogen by weight. The decay rate of nitrogen is as follows: 30% for the first year, 15% the second year, and 5% for the third and subsequent years.

8-34. Solve Problem 8-33 on the basis of satisfying the phosphorus requirement.

8-35. From the results of Problems 8-33 and 8-34, recommend the rate of application of dry sludge.

8-36. Considering only the nitrogen and phosphorus requirement, determine the yearly application rate if Milorganite is used. If Milorganite costs $6.99 for a bag of 18.4 kg and Hechinger's fertilizer costs $10.99 for a bag of 15.52 lb, compare the cost of fertilizing the farm using separate fertilizers. Assume that for Milorganite 35% of the applied nitrogen is consumed in the first year, 15% in the second year, 5% in the third and fourth years, and negligible percentages in years thereafter. The nitrogen in Hechinger's fertilizer may be assumed to be utilizable by the crop immediately. The farmer is growing barley for forage in a 5-ha farm.

8-37. In Problem 8-36, determine the yearly application rate if a Hechinger fertilizer is used.

8-38. Assuming the limiting application of cadmium to the soil is to be 9 kg/ha, what is the application rate for a sludge containing 40 ppm of cadmium?

8-39. The size of the farm in Problem 8-36 is 100 ha. Design the number of tank vehicles and the sizes of the vehicles and the pumps to deliver the sludge. Assume that the application is to be completed in one week.

8-40. If carried out properly, what processes can destroy helminth ova?

8-41. What can you say about the future of the application of electron-beam irradiation as a method of disinfecting sludge?

BIBLIOGRAPHY

BASTIAN, R. K. (1977). "Municipal Sludge Management," in R. C. Loehr (ed.), *Land as a Waste Management Alternative*. Ann Arbor Science, Ann Arbor, Mich., pp. 673–689.

LILEY, P. E., and W. R. GAMBILL (1973). "Physical and Chemical Data," in R. H. Perry and C. H. Chilton (eds.), *Chemical Engineers' Handbook*. McGraw-Hill, New York, pp. 3-1 to 3-44.

METCALF & EDDY, INC. (1991). *Wastewater Engineering: Treatment, Disposal, Reuse*. McGraw-Hill, New York.

PEAVY, H. S., D. R. ROWE, and G. TCHOBANOGLOUS (1985). *Environmental Engineering*. McGraw-Hill, New York.

RAMALHO, R. S. (1977). *Introduction to Wastewater Treatment Processes.* Academic Press, New York.

USEPA (1985). *Health Effects of Land Application of Municipal Sludge.* EPA/600/1-85/015. USEPA Health Effects Laboratory, Research Triangle Park, N.C.

USEPA (1989). *Control of Pathogens in Municipal Wastewater Sludge.* EPA/625/10-89/006. Center for Environmental Research Information, Cincinnati, Ohio.

VESILIND, P. A. (1974). *Treatment and Disposal of Wastewater Sludges.* Ann Arbor Science, Ann Arbor, Mich.

VIESSMAN, W., JR., and M. J. HAMMER (1993). *Water Supply and Pollution Control.* Harper & Row, New York.

CHAPTER 9

Advanced Wastewater and Water Treatment and Land Treatment Systems

As population increases, so does the quantity of wastewater produced. However, the capacity of the stream to assimilate the waste is finite. For this reason, whereas before a given stream may be effluent limited, now it has become water quality limited, thus requiring more stringent treatment. Advanced wastewater treatment systems and the related land treatment systems are discussed in this chapter. Additionally, some of the advanced treatment methods discussed may be applied to the treatment of water for drinking purposes. These include carbon adsorption, ion exchange, and the membrane processes, such as reverse osmosis, ultrafiltration, and electrodialysis. Whenever these units are discussed in this chapter, wastewater would also mean water, and vice versa. Biological nutrient processes for advanced methods of wastewater treatment are discussed. For discussions on the chemistry of biological processes, read Chapter 2.

Sand filtration is also an advanced method of wastewater treatment. Since this was discussed in Chapter 6, it is not repeated here. Suffice it to say that sand filtration is used to produce 15 mg/L of BOD$_5$ and 10 mg/L of TKN in a process called *advanced secondary treatment* (AST). Figure 9-1 shows drawings of units used in advanced wastewater treatment and water treatment systems.

CARBON ADSORPTION

Solids are formed because of the attraction of the component atoms within the solid toward each other. In the interior of a solid, attractive forces are balanced among the various atoms making up the lattice. At the surface, however, the atoms are subjected to unbalanced forces—those toward the interior are attracted, but those at the surface are not. Because of this unbalanced nature, any particle that lands on the surface may be attracted by the solid. This is the phenomenon of *adsorption*. Adsorption is the process of concentrating solute at the surface of a solid by virtue of this attraction.

Adsorption may be *physical* or *chemical*. Physical adsorption is also called *van der Waals adsorption*, and chemical adsorption is also called *chemisorption*. Adsorption is a surface-active phenomenon, which means that larger surface areas exposed to the solutes result in higher adsorption. The solute is called the *adsorbate*; the solid that adsorbs it is called the *adsorbent*. The adsorbate is said to be sorbed onto the adsorbent when it is adsorbed, and it is said to be desorbed when it passes into solution.

Adsorption capacity is enhanced by activating the surfaces. The process of activation is accomplished by subjecting a prepared char of carbon material such as wood or coal to an oxidizing steam at high temperatures of about 1700°F, resulting in the water gas reaction, $C + H_2O \rightarrow H_2 + CO$. The gases released develop a very porous structure in the char. This high porosity increases the area for adsorption. One gram of char can produce about 1000 m^2 of adsorption area. After activation, the char is then further processed into two types of finished product: a powdered form called PAC (*powdered activated carbon*) and a granular form called GAC (*granular activated carbon*). PAC is normally less than 200 mesh, and GAC is normally greater than 0.1 mm in diameter.

Adsorption Capacity

The adsorption capacity of activated carbon may be determined by the use of an adsorption isotherm. The *adsorption isotherm* is an equation relating the amount of solute adsorbed onto the solid and the equilibrium concentration of the solute in solution at a given temperature. The following are isotherms that have been developed: Freundlich; Langmuir; and Brunauer, Emmet, and Teller (BET isotherm). The most commonly used isotherms for the application of activated carbon in wastewater treatment are the Freundlich and Langmuir isotherms, written, respectively, as

$$\frac{X}{M} = k[C]^{1/n} \tag{9-1}$$

$$\frac{X}{M} = \frac{ab[C]}{1 + b[C]} \tag{9-2}$$

where X is the mass of adsorbate adsorbed onto the mass of adsorbent M; $[C]$ is the concentration of adsorbate in solution in equilibrium with the adsorbate adsorbed; k, a, and b are constants.

The rate r_s at which solute is adsorbed onto an adsorbent is proportional to its concentration and the amount of solute-adsorbing capacity remaining in the adsorbent $(X/M)_{ult} - X/M$, where "ult" stands for ultimate. Also, the rate r_d at which the sorbed materials are desorbed is proportional to the amount of material already sorbed, X/M. At equilibrium, r_s is equal to r_d. Calling k_s and k_d the sorption and desorption proportionality

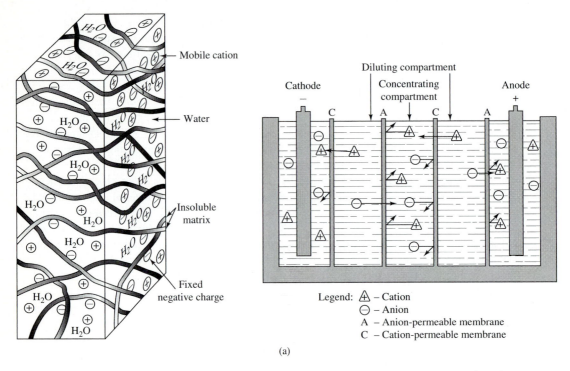

(a)

Figure 9-1   Advanced wastewater treatment systems: (a) electrodialysis unit.

constants, respectively, we have

$$k_s[C]\left[\left(\frac{X}{M}\right)_{\text{ult}} - \frac{X}{M}\right] = k_d\left(\frac{X}{M}\right) \tag{9-3}$$

Solving for X/M produces equation (9-2), where $a = (X/M)_{\text{ult}}$ and $b = k_s/k_d$.

The straight-line forms of the Freundlich and Langmuir isotherms are, respectively,

$$\ln\frac{X}{M} = \ln k + \frac{1}{n}\ln[C] \tag{9-4}$$

$$\frac{[C]}{X/M} = \frac{1}{ab} + \frac{1}{a}[C] \tag{9-5}$$

Since the equations are for straight lines, only two pairs of values of the respective parameters are required to solve for the constants. In equation (9-4), the required pairs of values are for the parameters $\ln(X/M)$ and $\ln[C]$; in equations (9-5), the required pairs are for the parameters $[C]/(X/M)$ and $[C]$.

To use equation (9-4) or (9-5), constants are empirically determined by running an experiment. This is done by adding increasing amounts of the adsorbent to a sample of adsorbate solution in a container. For each amount of adsorbent, the equilibrium concentration $[C]$ is determined. The pairs of experimental values can then be manipulated

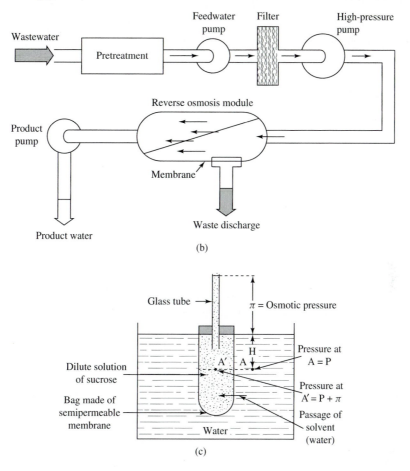

Figure 9-1 (continued)   Advanced wastewater treatment systems: (b) reverse osmosis system; (c) osmosis process.

to obtain the desired parameter values from which the constants are determined. Once the constants are determined, the resulting model is used to determine M, the amount of adsorbent (activated carbon) that is needed. From the equation, the adsorption capacity of activated carbon, $(X/M)_{ult}$, may also be determined. The absorption capacity of activated carbon is the X/M that produces the lowest residual concentration of the adsorbate possible. This is illustrated in Example 9-1.

Example 9-1

A wastewater containing 25 mg/L of phenol is to be treated using PAC to produce an effluent concentration of 0.10 mg/L. The PAC is simply added to the stream and the mixture subsequently settled in the following sedimentation tank. The constants of the Langmuir equation are determined by running a jar test producing the results shown below. The volume of waste subjected to each test is 1 L. If the flow rate Q_0 is 0.11 m³/s, calculate the quantity of PAC needed for the operation. What is the adsorption capacity of the PAC?

Calculate the quantity of PAC needed to treat the influent phenol to the ultimate residual concentration.

Test	PAC added (g)	Equilibrium concentration of phenol (mg/L)
1	0.25	6.0
2	0.32	1.0
3	0.5	0.25
4	1.0	0.09
5	1.5	0.06
6	2.0	0.06
7	2.6	0.06

Solution

$$\frac{[C]}{X/M} = \frac{1}{ab} + \frac{1}{a}[C]$$

Neglect tests 5 to 7, since the 0.06 values represent the ultimate residual concentrations and would not conform to the Langmuir equation.

Test	PAC added (g)	Equilibrium concentration of phenol (mg/L)	$\dfrac{[C]}{X/M}$ (mg/L)
1	0.25	6.0	78.95[a]
2	0.32	1.0	13.33
3	0.5	0.25	5.05
4	1.0	0.09	3.61

[a]$78.95 = \dfrac{6}{[(25-6)(1)]/[0.25(1000)]}$; other values in the column are computed similarly.

$$\frac{6.0 + 1.0}{2} = 3.5 \qquad \frac{78.95 + 13.33}{2} = 46.14$$

$$\frac{0.25 + 0.09}{2} = 0.17 \qquad \frac{5.05 + 3.61}{2} = 4.33$$

$$\frac{1}{a} = \frac{46.14 - 4.33}{3.5 - 0.17} = 12.56$$

$$\frac{1}{ab} = 46.14 - (12.56)(3.5) = 2.20$$

Equation of isotherm: $\dfrac{[C]}{X/M} = 2.20 + 12.56[C]$

When $[C] = 0.10$ mg/L,

$$\frac{0.10}{X/M} = 2.20 + 12.56(0.10) \qquad X/M = 0.0289 \frac{\text{kg phenol}}{\text{kg C}}$$

Total phenol to be removed $= 0.11(0.025 - 0.0001) = 0.00274$ kg/s

$$\text{PAC required} = \frac{0.00274}{0.0289}(60)(60)(24) = 8181.60 \text{ kg/day} \quad \textbf{Answer}$$

The lowest concentration of phenol is 0.06 mg/L. Thus when $[C] = 0.06$ mg/L,

$$\frac{0.06}{(X/M)_{\text{ult}}} = 2.20 + 12.56(0.06)$$

$$(X/M)_{\text{ult}} = 0.020 \frac{\text{kg phenol}}{\text{kg C}} = \text{adsorption capacity} \quad \textbf{Answer}$$

Total phenol to be removed to the ultimate residual concentration $= 0.11(0.025 - 0.00006)$

$$= 0.00274 \text{ kg/s}$$

$$\text{PAC required} = \frac{0.00274}{0.020}(60)(60)(24) = 11{,}837 \text{ kg/d} \quad \textbf{Answer}$$

Bed Adsorption

In bed adsorption, the water to be treated is passed through a bed of activated carbon. The method of introduction of the influent may be made similar to sand filtration. In addition, the bed may be moving countercurrent or co-current to the flow of influent.

Figure 9-2a shows a schematic of an adsorption bed of depth L. Although the influent feed is shown introduced at the top, the following analysis applies to other modes of introduction, as well, including moving beds. Hence the figure should be viewed as a relative motion of the feed and the bed. The curve lines represent the configuration of the variation of the concentration of the adsorbate at various times as the adsorbate passes through the column. Thus the curve labeled t_1 is the configuration at time t_1. The lower end of the curve is indicated by a zero concentration (or any concentration that represents the limit of removal) and the upper end is indicated by the influent concentration $[C_0]$. Hence the volume of bed above curve t_1 is an exhausted bed. Below the curve, the bed is clean (i.e., no adsorption is taking place, since all adsorbables had already been removed by the portion of the bed at curve t_1 and above). The zone of bed encompassed by the curve represents a bed partially exhausted.

The curve t_1 then advances to form the curve t_2 at time t_2. The curve shows a breakthrough of concentration of some fraction of $[C_0]$. This breakthrough concentration appears in the effluent. Finally, t_2 advances to t_3 at time t_3. At this time the bed is almost exhausted. The profile concentration represented by curve t_1 is called an *active zone*. This zone keeps on advancing until the entire bed becomes totally exhausted.

Figure 9-2b shows the movement of the active zone represented by the length δ as it advances through the bed at various times. At the beginning of breakthrough, at which the lower end of δ barely touches the bottom of the column, the total volume

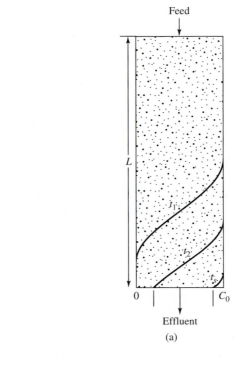

(a)

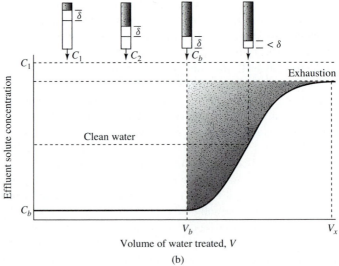

Volume of water treated, V

(b)

Figure 9-2   (a) Active zone of an adsorption or ion-exchange column. (b) Break-through curve.

of treated water is represented by V_b. The shaded portion in the curve represents the total breakthrough mass of adsorbate before exhaustion. The total volume of wastewater treated at exhaustion is designated by V_x.

Perform a mass balance on the active zone during breakthrough. The total mass of pollutant that escaped removal from the beginning to the completion of breakthrough is

$$\sum (V_{n+1} - V_n) \frac{[C_{n+1}] + [C_n]}{2} \qquad (9\text{-}6)$$

where the indices n and $n + 1$ refer to the volume or concentration that broke through δ at a time step of Δt from t_n to t_{n+1}. Also, at the completion of breakthrough, the total mass of pollutant introduced into the influent of δ is $(V_x - V_b)[C_0]$. Hence the mass of pollutants retained in the active zone M_r is

$$M_r = (V_x - V_b)[C_0] - \sum (V_{n+1} - V_n) \frac{[C_{n+1}] + [C_n]}{2} \qquad (9\text{-}7)$$

M_r is also given by $A_s \delta \rho_p (X/M)_{\text{ult}}$, where A_s is the superficial area of the bed, ρ_p the packed density of the carbon bed, and the others are as defined before. Combining with equation (9-7) and solving for δ gives

$$\delta = \frac{(V_x - V_b)[C_0] - \sum (V_{n+1} - V_n)\{([C_{n+1}] + [C_n])/2\}}{A_s \rho_p (X/M)_{\text{ult}}} \qquad (9\text{-}8)$$

Example 9-2

A breakthrough experiment is conducted for phenol, producing the results shown below. Determine the length δ of the active zone. The diameter of the column used is 1 in. and the packed density of the bed is 45 lb/ft^3. $[C_0]$ is equal to 25 mg/L. Use the $(X/M)_{\text{ult}}$ value from Example 9-1.

C (mg/L)	V (L)
0.06	1.0
1.0	1.24
6.0	1.31
10	1.43
15	1.48
18	1.58
20	1.72
23	1.83
25	2.00

Solution

$$\delta = \frac{(V_x - V_b)[C_0] - \sum (V_{n+1} - V_n)\{([C_{n+1}] + [C_n])/2\}}{A_s \rho_p (X/M)_{\text{ult}}}$$

C (mg/L)	V (L)	$\dfrac{V_{n+1} - V_n}{1000}$ (m³)	$\dfrac{[C_{n+1}] + [C_n]}{2(1000)}$ (kg/m³)	$\dfrac{V_{n+1} - V_n}{1000}\dfrac{[C_{n+1}] + [C_n]}{2(1000)}$ (kg)
0.06	1.0	—	—	—
1.0	1.24	0.00024	0.00053	0.00000013
6.0	1.31	0.00007	0.0035	0.00000025
10	1.43	0.00012	0.008	0.00000096
15	1.48	0.00005	0.0125	0.00000063
18	1.58	0.0001	0.0165	0.0000017
20	1.72	0.00014	0.019	0.0000027
23	1.83	0.00011	0.0215	0.0000024
25	2.00	0.00017	0.024	0.0000041
				$\sum$ 0.0000128

$$A_s = \frac{\pi[1/12(3.281)]^2}{4} = 0.00051 \text{ m}^2$$

$$\rho_p = \frac{45(0.454)}{(1/3.281)^3} = 721.58 \frac{\text{kg}}{\text{m}^3}$$

$$\delta = \frac{(0.002 - 0.001)(0.025) - 0.0000128}{0.00051(721.58)(0.020)} = 0.0017 \text{ m} = 1.7 \text{ mm} \quad \textbf{Answer}$$

As soon as the bed is exhausted, as determined by the breakthrough of concentrations, the carbon may be replaced. The replaced carbon may be reactivated again for reuse. Up to 30 or more reactivations may be made without appreciable loss of adsorptive power of the reactivated carbon.

Example 9-3

A wastewater containing 25 mg/L of phenol and having the characteristic breakthrough of Example 9-2 is to be treated by adsorption onto an activated carbon bed. The $(X/M)_{\text{ult}}$ value of the bed for the desired effluent of 0.06 mg/L is 0.02 kg solute per kilogram of carbon. If the flow rate is 0.11 m³/s, design the absorption column. Assume that the influent is introduced at the top of the bed. The packed density of the carbon bed is 721.58 kg/m³.

Solution The design will include the determination of the amount of activated carbon needed, the dimensions of the column, and the interval of activated carbon replacement.

Amount of phenol to be removed $= 0.11(0.025 - 0.00006)(60)(60)(24) = 237$ kg/day

$$\text{Amount of activated carbon needed} = \frac{237}{0.029} = 11,850 \frac{\text{kg}}{\text{day}} \quad \textbf{Answer}$$

From Example 9-2, adsorbate retained in δ length of test column at exhaustion $= (0.002 - 0.001)(0.025) - 0.0000128 = 0.0000122$ kg. For A_s of the test column $= 0.0005$ m² and $\delta = 0.0017$ m, adsorbate retained per unit volume in $\delta = 14.35$ kg.

Assume that the diameter of the actual column $= 4$ m.

$$A_S = \frac{\pi(4^2)}{4} = 12.57 \text{ m}^2$$

Adsorbate retained in δ length of two actual columns in parallel

$$= 2(12.57)(0.0017)(14.35) = 0.62 \text{ kg}$$

Assuming carbon replacement is to be done during every week of operation, the amount of adsorbate retained in a column length of length $L - \delta$ is

$$237(7) - 0.62 = 1658 \text{ kg}$$

$$\text{carbon required for 1658 kg of adsorbate} = \frac{1658}{0.02} = 82,919 \text{ kg} = 41,460 \text{ kg/column}$$

Thus

$$L = \frac{41,460/721.58}{12.57} + 0.0017 = 4.57,$$

say 4.8 m, including freeboard and other allowances   **Answer**

Diameter $= 4$ m   **Answer**

Interval of carbon replacement $=$ once every week   **Answer**

Example 9-4

Design the column of Example 9-3 if the feed is introduced at the bottom and the carbon is removed continuously at the bottom and added continuously at the top. Due to the countercurrent operation, assume that the bed expands by 40%.

Solution   The design will include, in addition to those in Example 9-3, the determination of the carbon removal and addition rates at the bottom and top of the column, respectively. These rates are determined from the length of the column and the interval of replacement of Example 9-3.

$$\text{Removal rate} = \text{addition rate} = \frac{4.57}{7(24)}(12.57)(721.58) = 247 \text{ kg/h}   \textbf{Answer}$$

The superficial velocity, from Example 9-3 $= 0.11/12.57 = 0.0088$ m/s, velocity relative to the stationary bed. In the countercurrent operation, this relative velocity must be maintained if the breakthrough curve is to be applicable. Hence in the present design, considering the 40% expansion,

$$V_s - \left[-\frac{4.57(1.4)}{7(24)(60)(60)} \right] = 0.0088 \qquad V_s = 0.0088 \text{ m/s}$$

$$A_s = \frac{0.11}{0.0088} = 12.5 \text{ m}^2 \Rightarrow \text{diameter} = 4 \text{ m}$$

$L = 1.4(4.57) = 6.38$ m, say 4.6 m, including freeboard and other allowances   **Answer**
$D = 4.0$ m   **Answer**

ION EXCHANGE

Ion exchange is the displacement of one ion by another. The displaced ion is originally a part of an insoluble material, and the displacing ion is originally in solution. At the completion of the process, the two ions are in reversed places: the displaced ion moves into solution and the displacing ion becomes a part of the insoluble material.

There are two types of ion-exchange materials: the cation-exchange material and the anion-exchange material. The *cation-exchange material* exchanges cations, while the *anion-exchange material* exchanges anions. The insoluble part of the exchange material is called the *host*. If R^{-n} represents the host part and C^{+m} the exchangeable cation, the cation-exchange material may be represented by $(R^n)_r(C^{+m})_c$, where r is the number of active sites in the insoluble material and c is the number of charged exchangeable particles attached to the host material. (Note that rn must equal cm.) On the other hand, if R^{+o} represents the host part of the anion-exchange material and A^{-p} its exchangeable anion, the exchange material may be represented by $(R^{+o})_r(A^{-p})_a$, where the subscripts are defined similar to those for the cation-exchange material. Letting C_s^{+q} be the displacing cation from solution, the cation-exchange reaction is

$$(R^{-n})_r(C^{+m})_c + c_s C_s^{+q} \rightleftharpoons (R^{-n})_r(C_s^{+q})_{c_s} + cC^{+m} \qquad (9\text{-}9)$$

where c_s is the moles per liter of C_s in solution. Also, letting A_s^{-t} be the displacing anion from solution, the anion-exchange reaction may be represented by

$$(R^{+o})_r(A^{-p})_a + a_s A_s^{-t} \rightleftharpoons (R^{+o})_r(A_s^{-t})_{a_s} + aA^{-p} \qquad (9\text{-}10)$$

where a_s is also the moles of A_s in solution per liter. As shown by the equations above, ion-exchange reactions are governed by equilibrium. For this reason, effluents from ion-exchange processes never yield pure water.

Table 9-1 shows the *displacement series* for ion-exchange materials. When an ion species high in the table is in solution, it can displace ion species in the insoluble material below it in the table and thus be removed from solution.

Originally, natural and synthetic aluminosilicates called *zeolites* were the only ones used as exchange materials. They have now largely been replaced by synthetic resins. *Synthetic resins* are insoluble polymers to which are added, by certain chemical reactions, acidic and basic groups called *functional groups*. These groups are capable of performing reversible exchange reactions with ions in solution. The total number of these groups determine the *exchange capacity* of the exchange material, while the type of functional group determines ion selectivity. When the exchange capacity of the exchange material is exhausted, the exchanger may be regenerated by the reverse reactions above. The principles of regeneration are discussed next.

Sodium and Hydrogen Cycle

As shown in Table 9-1, sodium, lithium, and hydrogen are the logical choices for the exchangeable ions. In practice, however, sodium and hydrogen are the ions of choice. The cation-exchange resin using sodium may be represented by $(R^{-n})_r(\text{Na}^+)_{rn}$. Its

TABLE 9-1
DISPLACEMENT
SERIES FOR AN
ION-EXCHANGE
MATERIAL

Cations	Anions
La^{3+}	SO_4^{2-}
Y^{3+}	CrO_4^{2-}
Ba^{2+}	NO_3^{-}
Sr^{2+}	AsO_4^{3-}
Ca^{2+}	PO_4^{3-}
Mg^{2+}	MoO_4^{2-}
Cs^{+}	I^{-}
Rb^{+}	Cl^{-}
K^{+}	F^{-}
Na^{+}	OH^{-}
Li^{+}	
H^{+}	

exchange reaction with Ca^{2+} and similar ions is

$$(R^{-n})_r(Na^+)_{rn} + \frac{rn}{2}Ca^{2+} \rightleftharpoons (R^{-n})_r(Ca^{2+})_{rn/2} + rnNa^+ \qquad (9\text{-}11)$$

As shown, Ca^{2+} has become embedded in the resin, thus, removed from solution, and Na^+ has become solubilized. Similar reactions may be formulated for the rest of the ions in Table 9-1.

As soon as the resin is exhausted, it may be regenerated. As shown in equation (9-11), by the *law of mass action*, the reaction may be driven to the left by increasing the concentration of the sodium ion on the right. In practice, this is what is actually done. The resin is regenerated by using a concentration of NaCl of about 5 to 10%, thus driving the reaction to the left. Operations where regeneration is done using NaCl are said to run on the *sodium cycle*. Regeneration may also be made using acids, such as H_2SO_4. Where regeneration is through the use of acids, the cycle is called the *hydrogen cycle* (from the proton or hydrogen ion content of acids).

Table 9-2 shows approximate exchange capacities and regeneration requirements for ion exchangers. As shown, the values have great ranges. Hence, in practice, one must have to perform an actual experiment or obtain data from the manufacturer for a particular ion exchanger to determine the exchange capacity and regeneration requirement. The capacity of an ion exchanger varies with the nature and concentration of ions in solution. This is much the same as the characteristics of activated carbon. Hence the experiment procedure is practically the same as that of the activated carbon discussed previously.

TABLE 9-2 APPROXIMATE EXCHANGE CAPACITIES AND
REGENERATION REQUIREMENTS OF ION EXCHANGERS

Exchanger and cycle	Exchange capacity (gEq/ft^3)	Regenerant	Regenerant requirement (gEq/ft^3)
Cation exchangers			
Natural zeolite, Na	5–10	NaCl	4–7
Synthetic zeolite, Na	10–20	NaCl	2–3
Resin, Na	10–50	NaCl	2–4
Resin, H	10–50	H_2SO_4	2–4
Anion exchanger			
Resin, OH	20–30	NaOH	5–8

Production of Pure Water

Theoretically, it would seem possible to produce pure water by combining the cation exchanger operating on the hydrogen cycle and the anion exchanger operating on the OH cycle. This is shown in the sequence of reactions below using Ca^{2+} as the displacing cation and Cl^- as the displacing anion:

$$(R^{-n})_r(H^+)_{rn} + \frac{rn}{2}Ca^{2+} \rightleftharpoons (R^{-n})_r(Ca^{2+})_{rn/2} + rnH^+ \tag{9-12}$$

$$(R^{+o})_r(OH^-)_{ro} + roCl^- \rightleftharpoons (R^{+o})_r(Cl^-)_{ro} + roOH^- \tag{9-13}$$

Equation (9-12) would be the reaction in a first unit producing effluent H^+ ions and all anions not removed by the cation exchanger. All cations represented by Ca^{2+} are, of course, removed by the exchanger. The effluent from the first unit is now introduced and used as the influent to the second unit, whose reaction is represented by equation (9-13), producing OH^-. All the anions are removed by this second unit; the H^+ is, of course, not removed. The H^+ and OH^- then combine to produce H_2O.

On the surface, one may conclude that, indeed, pure water has been produced. However, because of the restraint imposed by equilibrium, pure water is never produced. In addition, the H^+ produced in equation (9-12) and the OH^- produced in equation (9-13) do not altogether disappear to produce H_2O but that some H^+ or OH^- are left unreacted. The explanation is in equation (9-14), which shows $(rn - ro)$ moles of H^+ or OH^- remaining unreacted.

$$rnH^+ + roOH^- \rightleftharpoons roH_2O + (rn - ro)H^+ \quad \text{or}$$
$$rnH^+ + roOH^- \rightleftharpoons rnH_2O + (ro - rn)OH^+ \tag{9-14}$$

Design of Bed Exchangers

The principle of design of the bed exchanger is exactly the same as that of the activated carbon bed. In fact, the top and bottom drawings of Figure 9-2 can represent an ion-exchange bed and breakthrough curve, respectively, as they can with the activated carbon bed. All parameters are similar: active zone, exchange capacity, δ, breakthrough curve, and the like. The treatment will therefore no longer be discussed here.

Example 9-5

Using a bed exchanger, 20,000 gallons of water is to be treated for hardness removal between regenerations having intervals of 8 h. The raw water contains 400 mg/L of hardness as $CaCO_3$. The exchanger is a resin of exchange capacity of 40 gEq/ft^3 based on 50 as the equivalent weight of $CaCO_3$. By a breakthrough experiment, δ is determined to be about 1 ft. Determine the dimensions of the exchanger. Assume that the equilibrium concentration of hardness in the effluent is negligible.

Solution

$$\text{Equivalents of hardness} = \frac{400}{50(1000)} = 0.008 \text{ gEq/L} = 8 \text{ gEq/m}^3$$

$$\text{Total exchange required} = \frac{20,000}{7.48(3.281)^3}(8) = 605.61 \text{ gEq}$$

Let the diameter of the bed be equal to 2 ft. Then

$$\frac{\pi(2^2)}{4}(1)\left(\frac{1}{2}\right)(40) + \frac{\pi(2^2)}{4}(L-1)(40) = 605.61$$

$L = 5.32$ ft; including freeboard and other allowances. Let $L = 7$ ft.   **Answer**

Diameter $= 2$ ft   **Answer**

MEMBRANE PROCESSES

In general, there are two types of membrane processes used to treat water: *electrodialysis membrane* and *pressure membrane* processes. The electrodialysis membrane process is discussed first.

Electrodialysis Membrane Process

Figure 9-1a (left), shows a cut section of an electrodialysis membrane. Electrodialysis membranes are sheetlike barriers made out of high-capacity, highly cross-linked ion-exchange resins that allow passage of ions but not of water. There are two types: *cation membranes*, which transmit only cations, and *anion membranes*, which allow passage of only anions. The cut section in the figure is a cation membrane composed of an *insoluble matrix* and *water* in the pore spaces. *Negative charges* are fixed into the insoluble matrix, and *mobile cations* reside in the pore spaces occupied by water. It is the residence of these mobile cations that gives the membrane the property of allowing cations to pass through it. These cations will go out of the structure if they are replaced by other cations that enter the structure. If the entering cations came from water external to the membrane, the cations are removed from the water. In anion membranes, the mechanics just described are reversed. The mobile ions in the pores are the *anions*; the ions fixed to the insoluble matrix are the *cations*. The entering and replacing ions are anions from the water external to the membrane.

The process of removal of the ions is portrayed in Figure 9-1a (right). Inside the tank, cation and anion membranes are installed alternate to each other. Two electrodes are put on each side of the tank. By impressing electricity on these electrodes, the positive anode attracts negative ions in solution, while the negative cathode attracts positive ions in the solution. This impression of electricity is the reason why the respective ions replace their like ions in the membranes. As shown in the figure, two compartments become "cleaned" of ions and one compartment (in the middle) becomes "dirty" with ions. The two compartments are diluting compartments and the middle compartment is a concentrating compartment. The water in the diluting compartment is then withdrawn as the product water. The concentrated solution in the concentrating compartment is discharged to waste.

Power Requirement of Electrodialysis Units

From Figure 9-1a, the membranes are arranged C A C A from left to right. In compartments C A, the water is deionized, while in compartment A C, the water is not deionized. The number of deionizing compartments is two. Also, note that the membranes are always arranged in pairs; that is, cation membrane C is always paired with anion membrane A. Hence the number of membranes in a unit is always even. If the number of membranes is increased from four to six, the number of deionizing compartments will increase from two to three; if increased from six to eight, the number of deionizing membranes will increase from three to four; and so on. Hence if m is the number of membranes in a unit, the number of deionizing compartments is equal to $m/2$, and the number of deionizing-concentrating compartment pairs is also equal to $m/2$.

Since 1 equivalent of a substance is equal to 1 equivalent of electricity, in electrodialysis calculations, concentrations are conveniently expressed in terms of equivalents per unit volume. Let the flow to the electrodialysis unit be Q_0. The flow per deionizing-concentrating compartment or cell is then equal to $Q_0/(m/2)$. If the influent ion concentration (positive or negative) is $[C_0]$ equivalents per unit volume, the total rate of inflow of ions is $[C_0]Q_0/(m/2)$ equivalents per unit time. One equivalent is also equal to 1 faraday. Since a faraday or equivalent is equal to 96,494 coulombs, assuming a coulomb efficiency of η, the amount of electricity needed to remove the ions in one cell is equal to $96{,}494[C_0]Q_0\eta/(m/2)$ coulombs per unit time. If time is expressed in seconds, coulomb per second is ampere. Therefore, $96{,}494[C_0]Q_0\eta/(m/2)$ amperes of current must be impressed on the membranes of the cell to effect the removal of the ions. The same amount of current also passes through all the cells of the unit.

In electrodialysis calculations, a term called *current density* (CD) is often used. Current density is the current in milliamperes that flows through a square centimeter of membrane perpendicular to the current direction. A ratio called *current density to normality* (CD/N) is also normally used, where N is the normality. A high value of this ratio means that there is insufficient charge to carry the current. When this occurs, a localized deficiency of ions on the membrane surfaces may occur. This occurrence is called *polarization*. In commercial electrodialysis units CD/N of up to 1000 are utilized.

The electric current I that is impressed at the electrodes is not necessarily the same current that passes through the cells or deionizing compartments. The actual current that passes through successfully is a function of the current efficiency, which varies with the nature of the electrolyte, its concentration in solution, and the membrane system. Call $\mathcal{M}$ the current efficiency. The amperes passing through the solution is equal to the amperes required to remove the ions, that is,

$$I\mathcal{M} = \frac{96{,}494[C_0]Q_0\eta}{m/2} \tag{9-15}$$

The emf E across the electrodes is given by Ohm's law as shown below, where R is the resistance across the unit.

$$E = IR \tag{9-16}$$

If I is in amperes and R is in ohms, E is in volts. From basic electricity and using equations (9-15) and (9-16), the power P is

$$P = EI = I^2R = 3.72(10^{10})\left(\frac{[C_0]Q_0\eta}{m\mathcal{M}}\right)^2 R \tag{9-17}$$

If I is in amperes, E is in volts, and R is in ohms, P in equation (9-17) is in watts. Of course, the combined units of N and Q_0 must be in corresponding consistent units.

Example 9-6

A brackish water of 100,000 gallons per day containing 4000 mg/L of ions expressed as NaCl is to be deionized using an electrodialysis unit. There are 400 membranes in the unit, each measuring 18 by 20 in. Resistance across the unit is 6 Ω and the current efficiency is 90%. CD/N to avoid polarization is 700. Estimate the impressed current and voltage, the Coulomb efficiency, and the power requirement.

Solution

$$[C_0] = [\text{NaCl}] = \frac{4.0}{\text{NaCl}} = \frac{4}{23 + 35.45} = 0.068\frac{\text{gEq}}{\text{L}} = 68\frac{\text{gEq}}{\text{m}^3} \qquad N = 0.068$$

$$CD = 700(0.068) = 47.6 \text{ mA/cm}^2$$

$$I = \frac{47.6(18)(20)(2.54^2)}{1000(0.9)} = 122.84 \text{ A} \quad \textbf{Answer}$$

$$E = IR = 122.84(6) = 737.04 \text{ V} \quad \textbf{Answer}$$

$$I\mathcal{M} = \frac{96{,}494[C_0]Q_0\eta}{m/2}$$

$$122.84(0.9) = \frac{96{,}494(68)\{100{,}000/[7.48(3.281)^3(24)(60)(60)]\}\eta}{400/2}$$

$$\eta = 0.77 \quad \textbf{Answer}$$

$$P = EI = I^2R = 122.84^2(6) = 90{,}534 \text{ W} \quad \textbf{Answer}$$

Pressure Membrane Processes

According to Jacangelo,[1] there are three allied pressure-membrane processes: ultrafiltration (UF), nanofiltration (NF), and reverse osmosis (RO). He states that UF removes particles ranging in size from 0.001 to 10 μm and RO can remove particles ranging in sizes from 0.0001 to 0.001 μm. As far as size removals are concerned, NF stays in between UF and RO, being able to remove particles in the approximate size range of 0.001 μm. UF is normally operated in the range 15 to 75 psig; NF, in the range 75 to 250 psig; and RO, in the range 200 to 1200 psig. Microfiltration (MF) may be added to this list. MF retains larger particles than UF and operates at a lesser pressure (10 psig). While the nature of membrane retention of particles in UF is molecular screening, that in MF is solid screening. UF screens salts, while MF screens dissolved solids. On the other hand, comparing RO and UF, RO presents a diffusive transport barrier and possible molecular screening. Diffusive transport refers to the diffusion of solute across the membrane. RO creates a barrier to this diffusion.

The basics of a normal osmosis process are shown in Figure 9-1c. A bag of semipermeable membrane is shown placed inside a bigger container full of pure water. Inside the membrane bag is a solution of solute, sucrose. Because sucrose has osmotic pressure, it "sucks" water from outside the bag, causing the water to pass through the membrane. Introduction of the water into the membrane bag, in turn, causes the solution level to rise as indicated by the height π in the figure. The height π is the measure of the osmotic pressure. It follows that if sufficient pressure is applied to the tip of the tube in excess of that of the osmotic pressure the height π will be suppressed and the flow of water through the membrane will be reversed (i.e., it would be from inside the bag toward the outside into the bigger container). Sucrose in a concentration of 1000 mg/L has an osmotic pressure of 1.05 psia. Hence the reverse pressure to be applied must be, theoretically, in excess of 1.05 psia for a sucrose concentration of 1000 mg/L. For NaCl, its osmotic pressure in a concentration of 35,000 mg/L is 398 psia. Hence, to reverse the flow in a NaCl concentration of 35,000 mg/L, a reverse pressure in excess of 398 psia should be applied. The operation just described (i.e., applying sufficient pressure to the tip of the tube to reverse the flow of water) is the fundamental description of the *basic reverse osmosis process*. UF, NF, MF, and RO are all reverse osmosis processes. The middle drawing of Figure 9-1 is a schematic of an RO plant. Figure 9-3 is a photograph of a bank of modules in the Sanibel–Captiva reverse osmosis plant in Florida. This plant treats water for drinking purposes.

Over the course of development of the membrane technology, RO modular designs as shown in Figure 9-4 evolved. They are tubular, plate-and-frame, spiral wound, and hollow fine-fiber modules. In the *tubular* design, the membrane is lined inside the tube, which is made of ordinary tubular material. Water is allowed to pass through the inside of the tube under excess pressure, causing the water to permeate the membrane and to collect at the outside of the tube as the *product* or *permeate*. The portion of the influent that did not permeate becomes concentrated. This is called the *concentrate* or

[1]J. C. Jacangelo (1989). "Membranes in Water Filtration." *Civil Engineering*, 59(5), pp. 68–71.

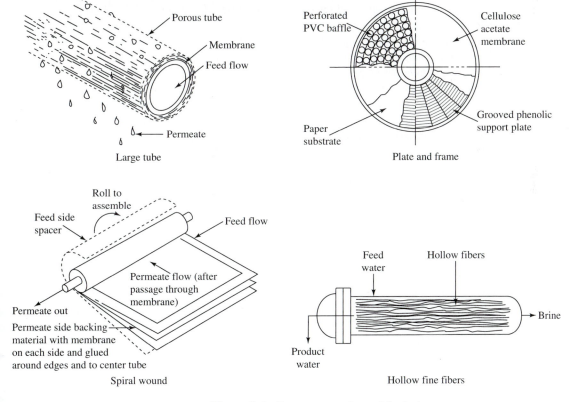

Figure 9-3 Bank of modules at the Sanibel–Captiva reverse osmosis plant, Florida.

Porous tube

Membrane

Feed flow

Permeate

Large tube

Perforated PVC baffle

Cellulose acetate membrane

Paper substrate

Grooved phenolic support plate

Plate and frame

Roll to assemble

Feed side spacer

Feed flow

Permeate flow (after passage through membrane)

Permeate out

Permeate side backing material with membrane on each side and glued around edges and to center tube

Spiral wound

Feed water

Hollow fibers

Brine

Product water

Hollow fine fibers

Figure 9-4 Reverse osmosis module designs.

reject. The *plate-and-frame* design is similar to the plate-and-frame press. In the case of RO, the semipermeable membrane replaces the filter cloth. The *spiral-wound* design consists of two flat sheets of membranes separated by porous spacers. The two sheets are sealed on the three sides; the fourth side is attached to a central collector pipe; and the entire sealed sheets are rolled around the central pipe. As the sheets are rolled around the pipe, a second spacer, called an *influent spacer*, is provided between the sealed sheets. In the final configuration, the spiral-wound sealed membrane looks like a cylinder. Water is introduced into the influent spacer, thereby allowing it to permeate through the membrane into the spacer between the sealed membrane. The permeate, now inside the sealed membrane, flows toward the central pipe and exits through the fourth unsealed side into the pipe. The permeate is collected as the *product water*. The concentrate or reject continues to flow along the influent spacer and be discharged as the effluent reject or effluent concentrate. This concentrate, which may contain hazardous molecules, poses a problem for disposal.

In the *hollow fine-fiber* design, the hollow fibers are a bundle of thousands of parallel, self-supporting, hairlike fibers enclosed in a fiberglass or epoxy-coated steel vessel. Water is introduced into the hollow bores of the fibers under pressure. The permeate water exits through one or more module ports. The concentrate also exits in a one or more separate module ports, depending on the design. All these module designs may be combined into banks of modules and may be connected in parallel or in series. In addition, since the efficiency of rejection is not 100%, solutes break through the membrane to contaminate the permeate.

As mentioned above, the reason the product or the permeate contains solute (that ought to be removed) is that the solute has broken through the membrane surface along with the water. It may be said that as long as the solute stays away from the membrane surface, only water will pass through into the product side and the permeate will be solute-free. However, it is not possible to exclude the solute from contacting the membrane surface; hence it is always liable to breakthrough. The efficiency at which solute is rejected is therefore a function of the interaction of the solute and the membrane surface. As far as solute rejection and breakthrough are concerned, a review of literature revealed the following conclusions:[2]

1. Percentage removal is a function of functional groups present in the membrane.
2. Percentage removal is a function of the nature of the membrane surface. For example, solute and membrane may have the tendency to bond by hydrogen bonding. If such is the case, the solute would easily permeate to the product side.
3. In a homologous series of compounds, percentage removal increases with the molecular weight of the solute.
4. Percentage removal is a function of the size of the solute molecule.

---

[2]A. P. Sincero (1989). *Reverse Osmosis Removal of Organic Compounds: A Preliminary Review of Literature*. Contract DAALO-86-D-0001. U.S. Army Biomedical Research and Development Laboratory, Fort Detrick, Frederick, Md. pp. 6–40.

5. Percentage removal increases as the percent dissociation of the solute molecule increases. Since the degree of dissociation of a molecule is a function of pH, percentage removal is therefore also a function of pH.

This review also found that the percentage removal of a solute is affected by the presence of other solutes. For example, methyl formate experienced a drastic change in percentage removal when mixed with ethyl formate, methyl propionate, and ethyl propionate. When alone, it was removed by only 14%, but when mixed with the others, the removal increased to 66%. Therefore, design of RO processes should be done by obtaining criteria utilizing laboratory or pilot-plant testing on the given influent.

Solute–water separation theory. The sole purpose of using the membrane is to separate the solute from the water molecules. While MF, UF, and NF may be viewed as normal filtration processes, only done on high-pressure modes compared to regular filtration, the RO process is thought to proceed in a somewhat different way. Several theories have been advanced as to how the separation in RO is effected. Of these theories, the one suggested by Sourirajan with schematics shown in Figure 9-5a is the most plausible.

Sourirajan's theory is called the *preferential sorption–capillary flow theory*. This theory asserts that there is a competition between the solute and the water molecules for the surface of the membrane. Since the membrane is an organic substance, several hydrogen bonding sites exist on its surface which preferentially bond water molecules to them. (The hydrogen end of water molecules bonds by hydrogen bonding to other molecules.) As shown on the left of Figure 9-5a, H_2O molecules are shown layering over the membrane surface, excluding the solute ions of Na^+ and Cl^-. Hence this exclusion brings about an initial separation.

On the right-hand side, a pore through the membrane is postulated, accommodating two diameters of water molecules. This pore size designated as $2t$, where t is the diameter of the water molecule, is called the *critical pore diameter*. From this conception the final separation of the water molecules and the solutes is materialized by applying pressure, thus, pushing H_2O through the pores.

As the process progresses, solutes build and line up on the membrane surface, creating a concentration boundary layer. This layer concentration is much larger than in the bulk solution and, also much larger, of course, than the concentration in the permeate side. This concentration difference creates a pressure for diffusive transport. The membrane, however, creates a barrier to this diffusion, thus retaining the solute and not allowing it to pass through. Eventually, the solute will diffuse out and leak to the permeate side.

Types of membranes. Probably, the first membrane put to practical use was the cellulose acetate (CA) membrane. The technique of preparation was developed by Sourirajan and Loeb and consisted of casting step, evaporation step, gelation step, and shrinkage step. The *casting step* involves casting a solution of cellulose acetate in acetone containing an additive into flat or tubular surfaces. The additive (such as magnesium

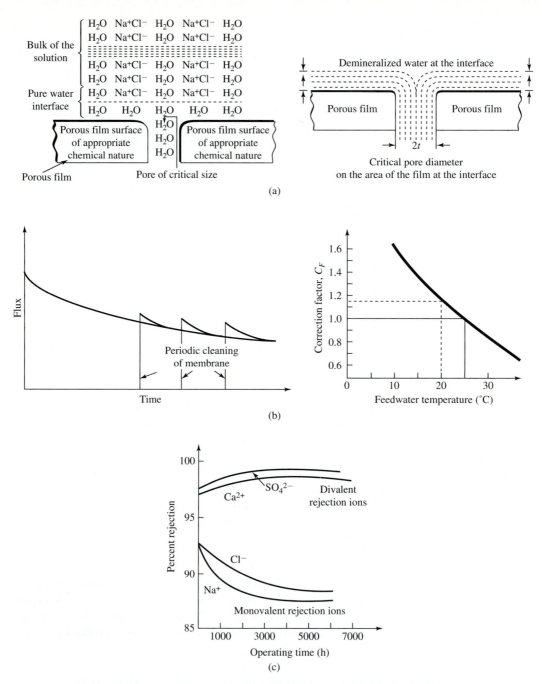

Figure 9-5 (a) Schematic representation of preferential sorption–capillary flow theory (left); critical pore diameter for separation (right). (b) Flux decline with time (left); correction factor for surface area of cellulose acetate membranes (right). (c) Solute rejection as a function of operating time. (a) Reprinted with permission from Sourizajan, S. and Agrawal, J.P. (1969) *Industrial Eng. Chem*, 61 by the American Chemical Society.

perchlorate) must be soluble in water, so that it will easily leach out in the gelation step, creating a porous film. After casting, the solvent acetone is *evaporated*. The material is then subjected to the *gelation step*, where it is immersed in cold water. The film material sets to a gel and the additive leaches out. Finally, the film is subjected to the *shrinkage step* which determines the size of the pores, depending on the temperature used in shrinking. High temperatures create smaller pores.

After this first development of the CA membrane, different types of membranes followed: CAB, CTA, PBIL, and PA membranes. CAB is membrane of *cellulose acetate butyrate*; CTA is *cellulose triacetate*. The PBIL membrane is a *polybenzimidazolone* polymer (Figure 9-6). Polyethylene amine reacted with tolylene diisocyanate produces the *NS-100 membrane* (NS stands for "nonpolysaccharide"); when reacted with iso-phthaloyl chloride, the membrane is called the *PA-100 membrane*. Epiamine (a polyether amine) reacted with isophthaloyl chloride produces the *PA-300 membrane*. The PAs in the foregoing prefixes stand for "polyamide." Hence the membranes referred to are *polyamide membranes. meta*-Phenylenediamine reacted with trimesoyl chloride produces the *FilmTec FT-30* membrane. The formula for the NS-100 and 300 membranes and the FT-30 membranes also contains the amide group; hence they are also polyamide membranes. 2-Hydroxymethyl furan when dehydrated using H_2SO_4 produces the *NS-200*. Another membrane formed from 2-hydroxymethyl furan is *Toray PEC-1000*. NS-200 is not a polyamide membrane.

Possible reactions for the synthesis of these membranes are shown in Figure 9-6. For example, in the reaction of polyethylene amine with tolylene diisocyanate forming NS-100, the H of the *n* repeating units of polyethylene amine moves to the N of tolylene diisocyanate, destroying the double bonding between N and C. The C of the carboxyl group of tolylene diisocyanate then bonds with the N of the amine. The sketch shown in the figure is just one unit of the structure of the product. The final structure would be a "mesh" of cross-linked assembly, thus creating molecular "pores." The ethylene repeating units form the backbone of the membrane, and the benzene rings form the cross-linking mechanism that tie together the ethylene backbones. The reader should further scrutinize the remainder of the synthesis reactions.

The ethylene units and the benzene rings are nonpolar regions, while the peptide bonds and the amines are polar regions of the NS-100 membrane. In the NS-100, nonpolar regions exceed polar regions; hence this membrane is said to be apolar. On the hand, CA is more polar, since the OH^- region exceeds the acetyl regions. The OH^- of CA are the polar regions, while the acetyl groups are the nonpolar regions. The polarity or apolarity of any membrane is very important in characterizing its property to reject solutes.

Membrane Performance Characterization

The performance of a given membrane may be characterized according to its product flux and purity of product. Flux, which is a rate of flow per unit area of membrane, is a function of membrane thickness, chemical composition of feed, membrane porosity, time of operation, pressure across membrane, and feedwater temperature. Product purity, in turn, is a function of the rejection ability of the particular membrane.

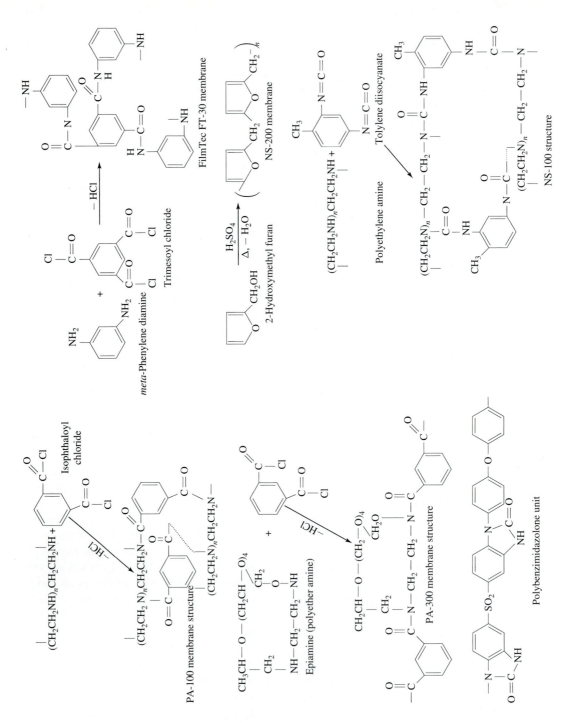

Figure 9-6 Reactions for synthesis of the various reverse osmosis membranes.

Flux decline. The left-hand drawing in Figure 9-5b shows the decline in flux with time of operation for a given membrane and membrane pressure differential. The lower solid curve is the actual decline without the effect of cleaning. The sawtoothed configuration is the effect of periodic cleaning. Although the relationship is shown as a curve line, experience has shown that if the data are plotted in a log-log paper, the relationship will be a straight line. Hence, empirically, the following equation may be obtained:

$$\ln F = m \ln t + K \tag{9-18}$$

where F is the flux, t the time, m the slope of the line, and K is a constant. Equation (9-18) may be used to estimate the ultimate flux of a given membrane at the end of its life (which could be 1 to 2 years).

Example 9-7

A long-term experiment for a CA membrane module operated at 400 psi using a feed of 2000 mg/L of NaCl at 25°C produces the results shown below. What is the expected flux at the end of 1 year of operation? What is the expected flux at the end of 2 years?

Time (h)	1	10,000	25,000
Flux (gal/ft^2-day)	16	12	11

Solution

$$\ln F = m \ln t + K$$

$$\frac{1 + 10,000}{2} = 5000.5 \qquad \frac{16 + 12}{2} = 14$$

$$\frac{10,000 + 25,000}{2} = 17,500 \qquad \frac{12 + 11}{2} = 11.5$$

$$m = \frac{\ln 14 - \ln 11.5}{\ln 5000.5 - \ln 17,500} = -0.157$$

$$\ln 14 = -0.157 \ln 5000.5 + K \qquad K = 3.98$$

equation is $\ln F = -0.157 \ln t + 3.98$

$$\ln F_{1year} = -0.157 \ln[365(24)] + 3.98 \qquad F_{1year} = 12.87 \text{ gal/ft}^2\text{-day} \quad \textbf{Answer}$$

$$\ln F_{2year} = -0.157 \ln[2(365)(24)] + 3.98 \qquad F_{2year} = 11.54 \text{ gal/ft}^2\text{-day} \quad \textbf{Answer}$$

Flux through membrane. The flow of permeate through a membrane may be considered as a "microscopic" form of cake filtration, where the solute that polarizes at the feed side of the membrane may be considered as the cake. From Chapter 8, the volume of filtrate V that passes through the cake in time t is given by

$$\frac{t}{V} = \frac{\mu c \bar{\alpha}}{2(-\Delta P)A^2} + \frac{\mu R_m}{(-\Delta P)A} \tag{9-19}$$

where μ is the absolute viscosity of filtrate, c the mass of cake per unit volume of filtrate collected, $\bar{\alpha}$ the specific cake resistance, $-\Delta P$, the pressure drop across the cake and filter, A, the filter area, and R_m, the filter resistance. In RO, c corresponds to the solute collected per unit volume of permeate at the layer adjacent to the membrane (called the concentration boundary layer) and R_m to the resistance of the membrane. All the other parameters have meanings similar to those provided in Chapter 8.

The volume flux F is $V/t\,A$. Using this and solving equation (9-19) for F, we have

$$\frac{V}{tA} = F = \frac{2}{\mu c \bar{\alpha} + 2A\mu R_m}(-\Delta P)A \qquad (9\text{-}20)$$

Initially neglecting the resistance of the solute in the concentration boundary layer gives

$$\frac{V}{tA} = F = \frac{1}{\mu R_m}(-\Delta P) \qquad (9\text{-}21)$$

As in cake compressibility, incorporate the compressibility of the membrane as the applied pressure is increased. Considering the resistance of the solute, designate the combined effect of compressibility, membrane resistance R_m, and solute resistance as $\bar{\alpha}_m$. Analogous to cake filtration, call the specific membrane resistance $\bar{\alpha}_m$. Hence

$$\frac{V}{tA} = F = \frac{1}{\mu\bar{\alpha}_m}(-\Delta P) \qquad (9\text{-}22)$$

where $\bar{\alpha}_m$ is given by

$$\bar{\alpha}_m = \bar{\alpha}_{mo}(-\Delta P)^s \qquad (9\text{-}23)$$

where s is a measure of membrane compressibility. When s is equal to zero, $\bar{\alpha}_m$ is equal to $\bar{\alpha}_{mo}$, the constant of proportionality of the equation. The straight-line form of equation (9-23) is

$$\ln \bar{\alpha}_m = \ln \bar{\alpha}_{mo} + s \ln(-\Delta P) \qquad (9\text{-}24)$$

Calling the pressure in the feed side P_f, the net pressure P_{fn} acting on the membrane in the feed side is

$$P_{fn} = P_f - \pi_f \qquad (9\text{-}25)$$

where π_f is the osmotic pressure in the feed side. Also, calling P_p the pressure in the permeate side, the net pressure P_{pn} acting on the membrane in the permeate side is

$$P_{pn} = P_p - \pi_p \qquad (9\text{-}26)$$

where π_p is the osmotic pressure in the permeate side. Hence

$$-\Delta P = P_{fn} - P_{pn} = (P_f - \pi_f) - (P_p - \pi_p) = (P_f - P_p) - (\pi_f - \pi_p) \qquad (9\text{-}27)$$

and the flux F is

$$F = \frac{1}{\mu\bar{\alpha}_m}[(P_f - P_p) - (\pi_f - \pi_p)]$$

$$= \frac{1}{\mu\bar{\alpha}_{mo}(-\Delta P)^s}[(P_f - P_p) - (\pi_f - \pi_p)] = \frac{1}{\mu\bar{\alpha}_{mo}}[(P_f - P_p) - (\pi_f - \pi_p)]^{1-s}$$

$$F = \frac{1}{\mu\bar{\alpha}_{mo}}(-\Delta P)^{1-s}$$

(9-28)

Table 9-3 shows osmotic pressures of various solutes.

TABLE 9-3 TYPICAL OSMOTIC PRESSURES AT 25°C

Compound	Concentration (mg/L)	Osmotic pressure (psi)
NaCl	35,000	400
NaCl	1,000	11
NaHCO$_3$	1,000	13
Na$_2$SO$_4$	1,000	6
MgSO$_4$	1,000	4
MgCl$_2$	1,000	10
CaCl$_2$	1,000	8
Sucrose	1,000	1
Dextrose	1,000	2

Referring to Table 9-3, some generalizations may be made. For example, comparing the osmotic pressures of 1000 mg/L of NaCl and 1000 mg/L of Na$_2$SO$_4$, NaCl has about 1.9 that of the osmotic pressure of Na$_2$SO$_4$. In solution for the same masses, NaCl yields about 1.6 times more particles than Na$_2$SO$_4$. From this it may be concluded that osmotic pressure is a function of the number of particles in solution. Comparing the 1000 mg/L concentrations of Na$_2$SO$_4$ and MgSO$_4$, the osmotic pressure of Na$_2$SO$_4$ is about 1.7 times that of MgSO$_4$. In solution Na$_2$SO$_4$ yields about 1.3 more particles than MgSO$_4$. The same conclusions will be arrived at if other comparisons are made; hence osmotic pressure depends on the number of particles in solution. From this finding, osmotic pressure is, therefore, additive.

Example 9-8

The feedwater to an RO unit contains 3000 mg/L of NaCl, 300 mg/L of CaCl$_2$, and 400 mg/L of MgSO$_4$. The membrane used is cellulose acetate and the results of a certain study are shown below. Determine the specific membrane resistance as a function of pressure drop. What will be the flux rate if the pressure applied is increased to 700 psig? Assume that for the given concentrations the osmotic pressures are NaCl = 34.2 psi, CaCl$_2$ = 2.49 psi, and MgSO$_4$ = 1.44 psi. Also assume that the temperature during the experiment is 25°C.

Applied pressure (psig)	Flux (gal/ft^2-day)
250	3.0
600	4.6

Solution

$$F = \frac{1}{\mu\bar{\alpha}_m}[(P_f - P_p) - (\pi_f - \pi_p)]$$

$$\mu = 9.0(10^{-4}) \text{ kg/m-s} = 77.76 \text{ kg/m-day}$$

$$\pi_f = 34.2 + 2.49 + 1.44 = 38.13 \text{ psi} = \frac{101,330}{14.7}(38.13) = 262,837.62 \text{ N/m}^2$$

$$P_p = 14.7 \text{ psi} = 101,330 \text{ N/m}^2 \qquad \pi_p = 0, \quad \text{assumed}$$

$$250 \text{ psig} = \frac{101,330}{14.7}(250 + 14.7) = 1,824,629.3 \text{ N/m}^2$$

$$600 \text{ psig} = \frac{101,330}{14.7}(600 + 14.7) = 4,237,248.4 \text{ N/m}^2$$

$$3.0 \text{ gal/ft}^2\text{-day} = \frac{3}{7.48(3.281)} = 0.122 \text{ m}^3/\text{m}^2\text{-day}$$

$$4.6 \text{ gal/ft}^2\text{-day} = \frac{4.6}{7.48(3.281)} = 0.19 \text{ m}^3/\text{m}^2\text{-day}$$

$$0.122 = \frac{1}{77.76\bar{\alpha}_m}[(1,824,629.3 - 101,330) - (262,837.62 - 0)] = \frac{1,460,461.9}{77.76\bar{\alpha}_m}$$

$$= \frac{18,781.66}{\bar{\alpha}_m} \qquad \bar{\alpha}_m = 153,948 \text{ m}^{-1}$$

$$0.17 = \frac{1}{77.76\bar{\alpha}_m}[(4,237,248.8 - 101,330) - (262,837.62 - 0)] = \frac{3,873,081}{77.76\bar{\alpha}_m} = \frac{49,808.14}{\bar{\alpha}_m}$$

$$\bar{\alpha}_m = 292,989.1 \text{ m}^{-1}$$

$$\ln \bar{\alpha}_m = \ln \bar{\alpha}_{mo} + s \ln(-\Delta P)$$

$$\ln 153,948 = \ln \bar{\alpha}_{mo} + s \ln 1,460,461.9 \qquad\qquad\qquad \text{(a)}$$

$$\ln 292,989.1 = \ln \bar{\alpha}_{mo} + s \ln 3,873,081 \qquad\qquad\qquad \text{(b)}$$

Solving yields $s = 0.66; \bar{\alpha}_{mo} = 13.142$

$$\bar{\alpha}_m = 13.142(-\Delta P)^{0.66} \quad \textbf{Answer}$$

$$700 \text{ psig} = \frac{101,330}{14.7}(700 + 14.7) = 4,926,568.1 \text{ N/m}^2$$

$$F = \frac{1}{\mu\bar{\alpha}_{mo}}(-\Delta P)^{1-s} = \frac{1}{77.76(13.142)}[(4,926,568.1 - 101,330) - (262,837.62 - 0)]^{1-0.66}$$

$$= 0.18 \text{ m}^3/\text{m}^2\text{-day} = 4.42 \text{ gal/ft}^2\text{-day} \quad \textbf{Answer}$$

Effect of temperature on permeation rate. As shown in equation (9-28), the flux is a function of the dynamic viscosity μ. Since μ is a function of temperature, the flux or permeation rate is therefore also a function of temperature. As temperature increases, the viscosity of water decreases. Hence, from equation (9-28), the flux is expected to increase with increase in temperature. Correspondingly, it is also expected that the flux would decrease as the temperature decreases. The middle graph of Figure 9-5 shows the correction factor C_f for membrane surface area (for CA membranes) as a function of temperature relative to 25°C. As shown, lower temperatures have larger correction factors. This is due to the increase of μ as the temperature decreases. The opposite is true for the higher temperatures. These correction factors are applied to the membrane surface area to produce the same flux relative to 25°C.

Percent solute rejection or removal. The other parameter important in the design and operation of RO units is the percent rejection or removal of solutes. Let Q_0 be the feed inflow, $[C_0]$ be the feed concentration of solutes, Q_p be the permeate outflow, C_p be the permeate concentration of solutes, Q_c be the concentrate outflow, and C_c be the concentrate concentration of solutes. By mass balance of solutes, the percent rejection R is

$$R = \frac{\sum Q_0 C_{0i} - \sum Q_p C_{pi}}{\sum Q_0 C_{0i}}(100) = \frac{\sum Q_c C_{ci}}{\sum Q_0 C_{0i}}(100) \qquad (9\text{-}29)$$

where the index i refers to the solute species i.

Figure 9-5c shows the effect of operating time on percent rejection. As shown, this particular membrane rejects divalent ions better than it does the monovalent ions. In general, percent rejection increases with value of charge of ion.

Example 9-9

A laboratory RO unit 60 in. in length and 12 in. in diameter has an active surface area of 1100 ft^2. It is used to treat a feedwater with the following composition: NaCl = 3000 mg/L, CaCl$_2$ = 300 mg/L, and MgSO$_4$ = 400 mg/L. The product flow is 15 gal/ft^2-day and contains 90 mg/L NaCl, 6 mg/L CaCl$_2$, and 8 mg/L MgSO$_4$. The feedwater inflow is 27,500 gallons per day. (a) What is the percent rejection of NaCl? (b) What is the overall percent rejection of ions?

Solution

(a) $\quad R = \dfrac{\sum Q_0 C_{0i} - \sum Q_p C_{pi}}{\sum Q_0 C_{0i}}(100) = \dfrac{27{,}500(3000) - 15(1100)(90)}{27{,}500(3000)}(100)$

$\qquad = 98.2\%$ for NaCl **Answer**

(b) $\quad R = \dfrac{27{,}500(3000 + 300 + 400) - 15(1100)(90 + 6 + 8)}{27{,}500(3000 + 300 + 400)}(100)$

$\qquad = 98.3\%$ for all ions **Answer**

NUTRIENT REMOVAL

One of the reasons for removing nutrients is the prevention of eutrophication of receiving streams. Of concern are the nutrients phosphorus and nitrogen. The absence of any one of these nutrients in the water column will prevent eutrophication. Thus to control the process of eutrophication, only one of them may be removed from the effluent of treatment plants. The question is which one. In most cases, since treatment to remove phosphorus is much cheaper than treatment to remove nitrogen, removal of phosphorus is implemented. To control the eutrophication problem of the Chesapeake Bay in Maryland, the total phosphorus limit for all discharges exceeding a certain flow to the upper reaches of the bay is 2.0 mg/L. More recently, however, pressure has been building up to include the control of nitrogen. In situations where the natural concentration ratio of phosphorus to nitrogen in the environment is much, much higher than normally occurs in algal cells, phosphorus removal may not be effective. Hence nitrogen removal may be warranted. However, there is always the possibility that nitrogen fixers may simply manufacture the nitrogen nutrient from the nitrogen contained in the air and dissolved in water, nullifying the effort of removing nitrogen. Nonetheless, the technology for the removal of nitrogen exists and the removal of nitrogen is being practiced. In this discussion of nutrient removal, we discuss nitrogen removal first, followed by phosphorus removal.

Nitrogen Removal

There are two general methods of removing nitrogen: a physical method utilizing stripping columns to remove ammonia and a biological method utilizing the processes of nitrification and denitrification—biological nitrogen removal (BNR). *Nitrification* is the process of oxidizing nitrogen to nitrate, and *denitrification* is the process of converting the nitrate to nitrogen gas, thus removing nitrogen.

    Removal of ammonia by stripping.   The reaction of NH_3 with water may be represented by

$$NH_3 + H_2O \rightleftharpoons NH_4{}^+ + OH^- \tag{9-30}$$

From this reaction, increasing the pH will drive the reaction to the left, increasing the concentration of NH_3. This makes ammonia more easily removed by stripping. In practice, the pH is increased to about 10 to 11 by using lime. Stripping is done by introducing the wastewater at the top of the column and allowing it to flow down countercurrent to the flow of air introduced at the bottom. The stripping medium inside the column may be composed of packings or fillings such as Raschig rings and Berl saddles, or sieve trays and bubble caps. The liquid flows in thin sheets around the medium, thereby allowing more intimate contact between liquid and the stripping air.

    Let G be the moles of ammonia-free air, L the moles of ammonia-free water, Y_1 the moles of ammonia per mole of ammonia-free air at the bottom, Y_2 the moles of ammonia per mole of ammonia-free air at the top, X_1 the moles of ammonia per mole ammonia-free water at the bottom, and X_2 the moles of ammonia per mole of ammonia-free water at the top. At any section between the bottom and the top of the column, the X's and Y's are simply X and Y. By mass balance, the ammonia stripped from the wastewater is equal to the ammonia absorbed by the air. Hence

$$G(Y_2 - Y_1) = L(X_2 - X_1) \tag{9-31}$$

From the equilibrium between ammonia in water and ammonia in air, the following equations for the relationships between X and Y at various temperatures and 1 atm of total pressure may be derived (read literature).[3]

$$\text{At } 10°\text{C:} \quad Y = 0.469X \tag{9-32}$$

$$\text{At } 20°\text{C:} \quad Y = 0.781X \tag{9-33}$$

$$\text{At } 30°\text{C:} \quad Y = 1.25X \tag{9-34}$$

$$\text{At } 40°\text{C:} \quad Y = 2.059X \tag{9-35}$$

$$\text{At } 50°\text{C:} \quad Y = 2.692X \tag{9-36}$$

The coefficients of X in these equations plot a straight line with the temperatures T (in degrees Celsius). Letting these coefficients be C produces the regression equation,

$$C = 0.0585T - 0.338 \tag{9-37}$$

Therefore, in general, the relationship between X and Y is

$$Y = (0.0585T - 0.338)X \tag{9-38}$$

In equation (9-31), L, X_1, X_2, and Y_1 are known values. Assuming that the ammonia in the air leaving the top of the tower is in equilibrium with the ammonia content of the entering wastewater and using equation (9-38), Y_2 may be written as

$$Y_2 = (0.0585T - 0.338)X_2 \tag{9-39}$$

Hence substituting equation (9-39) in equation (9-31) and solving for G/L, the moles of air per mole of wastewater needed for stripping is

$$\frac{G}{L} = \frac{X_2 - X_1}{(0.0585T - 0.338)X_2 - Y_1} \tag{9-40}$$

G in equation (9-40) is the theoretical amount of air needed to strip ammonia in the wastewater from a concentration of X_2 to a concentration of X_1. In practice, the theoretical G is often multiplied by a factor of 1.5 to 2.0 to account for nonideality or nonequilibrium conditions. Once G has been established, the cross-sectional area of the tower may be computed using the equation of continuity. The superficial velocity through the tower should be less than the velocity that will cause flooding or boiling up of the incoming wastewater. The method for estimating the tower height will be deferred until Chapter 12.

[3] Metcalf & Eddy, Inc. (1991). *Wastewater Engineering: Treatment, Disposal, Reuse.* McGraw-Hill, New York.

Example 9-10

A treated wastewater containing 25 mg/L of NH_3-N is to be completely stripped. The temperature of operation is 15°C and the flow is 0.3 mgd. Determine the amount of air needed assuming a multiplying factor of 1.5 for the theoretical G. Calculate the cross-sectional area of the tower. Assume a superficial velocity of 1 fps.

Solution

$$X_2 = 25 \text{ mg/L} = 0.00179 \text{ gmol/L} = \frac{0.00179}{(1000 - 0.025)/18} = 3.22(10^{-5}) \frac{\text{gmol } NH_3\text{-N}}{\text{gmol ammonia-free water}}$$

$$X_1 = 0 \qquad Y_1 = 0$$

Assuming equilibrium at the top of the tower, we have

$$\frac{G}{L} = \frac{X_2 - X_1}{(0.0585T - 0.338)X_2 - Y_1} = \frac{3.22(10^5) - 0}{[0.0585(15) - 0.338](3.22)(10^{-5}) - 0} = 1.85$$

$$\text{Actual } G/L = 1.85(1.5) = 2.78$$

$$0.3 \text{ mgd} = \frac{0.3(10^6)}{7.48(3.281)^3} = 1135.54 \text{ m}^3/\text{day} = 1.135(10^9) \text{ g/day} = 6.30(10^7) \text{ gmol/day}$$

$$\text{air needed} = 2.78(6.30)(10^7) = 1.75(10^8) \text{ gmol/day}$$

$$= 1.75(10^8)(22.4)\left(\frac{15 + 273}{273}\right)\left(\frac{1}{1000}\right) = 4.14(10^6) \text{ m}^3/\text{day at 15°C} \quad \textbf{Answer}$$

$$1 \text{ fps} = \frac{1}{3.281}(60)(60)(24) = 26{,}333.44 \text{ m/day}$$

$$\text{cross-sectional of tower} = \frac{4.14(10^6)}{26{,}333.44} = 157.2 \text{ m}^2$$

$$\text{Using three towers, cross-sectional area} = \frac{157.2}{3} = 52.41 \text{ m}^2. \quad \textbf{Answer}$$

Biological nitrogen removal.   In BNR two steps are required: oxidation of nitrogen to nitrate and subsequent reduction of the nitrate to gaseous nitrogen, N_2, called *denitrification*. In the oxidation step, nitrogen is first oxidized to the nitrite form by *Nitrosomonas*, followed by the oxidation of the nitrite to nitrate by *Nitrobacter*. This process is called *nitrification*. In the denitrification step, the process occurs under anaerobic conditions utilizing the normal heterotrophic organisms. Next we derive the biochemical equations that represent these steps.

Nitrification: Nitrosomonas Stage.   From Chapter 2, the generalized donor reaction, acceptor reaction, and synthesis reaction mediated by *Nitrosomonas* are, respectively, shown by the following half-cell reactions.

$$\tfrac{1}{6}NH_4^+ + \tfrac{1}{3}H_2O \rightarrow \tfrac{1}{6}NO_2^- + \tfrac{4}{3}H^+ + e^- \qquad \text{donor reaction} \qquad (9\text{-}41)$$

$$\tfrac{1}{4}O_2 + H^+ + e^- \rightarrow \tfrac{1}{2}H_2O \qquad \text{acceptor reaction} \qquad (9\text{-}42)$$

$$\tfrac{1}{5}CO_2 + \tfrac{1}{20}HCO_3^- + \tfrac{1}{20}NH_4^+ + H^+ + e^- \rightarrow$$
$$\tfrac{1}{20}C_5H_7O_2N + \tfrac{9}{20}H_2O \qquad \text{synthesis reaction} \qquad (9\text{-}43)$$

In the discussion that follows, equivalents will be used. It must be remembered that the number of equivalents of a substance varies depending on the reaction to which the substance is referred.

If the equations above are simply added without modification, there will remain electrons free to roam in solution; this is not possible. In actual reactions, the foregoing equations must be modified to make the electrons given up by equation (9-41) balanced by the electrons accepted by equations (9-42) and (9-43). Starting with 1 gram-equivalent of NH_4-N referred to equation (9-43), assume m equivalent is incorporated into the cell of *Nitrosomonas*, $C_5H_7NO_2$. Hence the NH_4-N equivalent remaining for the donor reaction of equation (9-41) is $1 - m$, equals $(1 - m)[(1/20)N/1](1/N) = \tfrac{1}{20}(1 - m)$ moles. [The $\tfrac{1}{20}$ came from equation (9-43), used to compute the equivalent weight of NH_4-N.] Hence only $\tfrac{1}{20}(1 - m)$ moles of NH_4-N is available for the donor reaction to donate electrons. By equation (9-41), the modified donor reaction is then

$$\frac{1/6}{1/6}\left(\frac{1}{20}\right)(1 - m)NH_4^+ + \frac{1/3}{1/6}\left(\frac{1}{20}\right)(1 - m)H_2O \rightarrow$$
$$\frac{1/6}{1/6}\left(\frac{1}{20}\right)(1 - m)NO_2^- + \frac{4/3}{1/6}\left(\frac{1}{20}\right)(1 - m)H^+ + \frac{1}{1/6}\left(\frac{1}{20}\right)(1 - m)e^- \qquad (9\text{-}44)$$

From equation (9-43), the m equivalent of NH_4-N is $m[(1/20)N/1](1/N) = m/20$ moles. Hence the synthesis reaction becomes

$$\frac{1/5}{1/20}\left(\frac{m}{20}\right)CO_2 + \frac{1/20}{1/20}\left(\frac{m}{20}\right)HCO_3^- + \frac{1/20}{1/20}\left(\frac{m}{20}\right)NH_4^+ + \frac{1}{1/20}\left(\frac{m}{20}\right)H^+$$
$$+ \frac{1}{1/20}\left(\frac{m}{20}\right)e^- \rightarrow \frac{1/20}{1/20}\left(\frac{m}{20}\right)C_5H_7O_2N + \frac{9/20}{1/20}\left(\frac{m}{20}\right)H_2O \qquad (9\text{-}45)$$

From equations (9-44) and (9-45), the electron-mole left for the acceptor reaction is

$$\frac{1}{1/6}\left(\frac{1}{20}\right)(1 - m) - \frac{1}{1/20}\left(\frac{m}{20}\right) = \frac{3 - 13m}{10}$$

Hence the acceptor reaction, equation (9-42), modifies to

$$\frac{1}{4}\left(\frac{3 - 13m}{10}\right)O_2 + \frac{3 - 13m}{10}H^+ + \frac{3 - 13m}{10}e^- \rightarrow \frac{1}{2}\left(\frac{3 - 13m}{10}\right)H_2O \qquad (9\text{-}46)$$

Adding equations (9-44), (9-45), and (9-46) produces the overall reaction for the *Nitrosomonas* reaction shown below. Note that after addition, the electrons e^- are gone. The overall reaction should indicate no electrons in the equation, since electrons cannot just roam around in the solution but must be taken up by some atom.

$$\frac{1}{20}NH_4^+ + \frac{1}{40}(3 - 13m)O_2 + \frac{m}{20}HCO_3^- + \frac{m}{5}CO_2 \rightarrow$$

$$\frac{m}{20}C_5H_7NO_2 + \frac{1}{20}(1 - m)NO_2^- + \frac{1 - 2m}{20}H_2O + \frac{1 - m}{10}H^+ \tag{9-47}$$

Nitrification: Nitrobacter Stage.   Now derive the *Nitrobacter* reaction. Again, from Chapter 2, the generalized donor reaction, acceptor reaction, and synthesis reaction mediated by *Nitrobacter* are, respectively, shown by the half-cell reactions below. For the case of *Nitrobacter*, there are two competing synthesis reactions: the one that involves the nitrate ion and the one that involves the ammonium ion. When these two ions are present, the organism prefers to use the ammonium to the nitrate ion. Only when the ammonium is reduced to low levels will the organism use the nitrate ion. Since the process involved is nitrification, there is an abundance of the ammonium ion in solution. Hence, as shown below, the synthesis reaction involves the NH_4^+.

$$\tfrac{1}{2}NO_2^- + \tfrac{1}{2}H_2O \rightarrow \tfrac{1}{2}NO_3^- + H^+ + e^- \qquad \text{donor reaction} \tag{9-48}$$

$$\tfrac{1}{4}O_2 + H^+ + e^- \rightarrow \tfrac{1}{2}H_2O \qquad \text{acceptor reaction} \tag{9-49}$$

$$\tfrac{1}{5}CO_2 + \tfrac{1}{20}HCO_3^- + \tfrac{1}{20}NH_4^+ + H^+ + e^- \rightarrow$$

$$\tfrac{1}{20}C_5H_7O_2N + \tfrac{9}{20}H_2O \qquad \text{synthesis reaction} \tag{9-50}$$

In the BNR process, the *Nitrobacter* reaction follows the *Nitrosomonas* reaction. From equation (9-47), $\frac{1}{20}(1-m)$ moles of NO_2^--N has been produced. These nitrite ions serve as the elector donor for *Nitrobacter*. Hence the donor reaction of equation (9-48) becomes

$$\frac{1/2}{1/2}\left[\frac{1}{20}(1 - m)\right]NO_2^- + \frac{1/2}{1/2}\left[\frac{1}{20}(1 - m)\right]H_2O \rightarrow$$

$$\frac{1/2}{1/2}\left[\frac{1}{20}(1 - m)\right]NO_3^- + \frac{1}{1/2}\left[\frac{1}{20}(1 - m)\right]H^+ + \frac{1}{1/2}\left[\frac{1}{20}(1 - m)\right]e^- \tag{9-51}$$

Let n, based on equation (9-50), be the equivalents of *Nitrobacter* cells produced. This quantity, n equivalents, is equal to $n(C_5H_7O_2N/20)/C_5H_7O_2N = n/20$ moles. Modifying, the synthesis reaction becomes

$$\frac{1/5}{1/20}\left(\frac{n}{20}\right)CO_2 + \frac{1/20}{1/20}\left(\frac{n}{20}\right)HCO_3^- + \frac{1/20}{1/20}\left(\frac{n}{20}\right)NH_4^+ + \frac{1}{1/20}\left(\frac{n}{20}\right)H^+$$

$$+ \frac{1}{1/20}\left(\frac{n}{20}\right)e^- \rightarrow \frac{1/20}{1/20}\left(\frac{n}{20}\right)C_5H_7O_2N + \frac{9/20}{1/20}\left(\frac{n}{20}\right)H_2O \tag{9-52}$$

Subtracting the electrons used in equation (9-52), $(1/\frac{1}{20})(n/20)$, from the electrons donated in equation (9-51), $(1/\frac{1}{2})\left[\frac{1}{20}(1 - m)\right]$, produces the amount of electrons available for the acceptor (energy) reaction, $(1 - m - 10n)/10$. Modifying the acceptor reaction gives us

$$\frac{1}{4}\left(\frac{1-m-10n}{10}\right)O_2 + \left(\frac{1-m-10n}{10}\right)H^+ + \left(\frac{1-m-10n}{10}\right)e^- \rightarrow$$
$$\frac{1}{2}\left(\frac{1-m-10n}{10}\right)H_2O \tag{9-53}$$

Adding equations (9-51), (9-52), and (9-53) produces the overall reaction for *Nitrobacter*,

$$\frac{1-m}{20}NO_2^- + \frac{n}{20}NH_4^+ + \frac{n}{5}CO_2 + \frac{n}{20}HCO_3^- + \frac{1-m-10n}{40}O_2 \rightarrow$$
$$\frac{n}{20}C_5H_7NO_2 + \frac{1-m}{20}NO_3^- + \frac{n}{20}H_2O \tag{9-54}$$

Overall nitrification.   The *Nitrosomonas* reaction, equation (9-47), and the *Nitrobacter* reaction, equation (9-54), may now be added to produce the overall nitrification reaction.

$$\frac{1+n}{20}NH_4^+ + \frac{n+m}{20}HCO_3^- + \frac{n+m}{5}CO_2 + \frac{2-7m-5n}{20}O_2 \rightarrow$$
$$\frac{n+m}{20}C_5H_7NO_2 + \frac{1-m}{10}H^+ + \frac{1-m}{20}NO_3^- + \frac{1+n-2m}{20}H_2O \tag{9-55}$$

The literature reports cell yields for *Nitrosomonas* of 0.04 to 0.29 mg/mg NH_4-N. To convert these values to number of equivalents, the equivalent weight to be used for the cells should be $C_5H_7O_2N/20$ and that for nitrogen should be $N/20$. This will enable m to be solved, as will be shown in Example 9-11. The literature also reports cell yields of *Nitrobacter* of 0.02 to 0.084 mg/mg NO_2-N.[4] From the *Nitrobacter* reaction, moles of cells per mole of NO_2-N is $n/(1-m)$. Also, for nitrification to be the dominant reaction, the BOD_5/TKN ratios should be less than 3. At these ratios, the nitrifier population is about 10% and higher. For BOD_5/TKN ratios of greater than 5.0, the process may be considered a combined carbonaceous–nitrification reaction. At these ratios, the nitrifier population is less than 4%. In addition, to ensure complete nitrification, the dissolved oxygen concentration should be at least about 2.0 mg/L.

Example 9-11

A wastewater containing 60 mg/L of TKN is to be treated to a level of 20 mg/L of TKN. If the flow is 3.0 mgd, determine the amount of alkalinity needed per day to buffer the reactor to a pH of 7.0. Assume cell yields of 0.15 mg VSS/mg NH_3-N for *Nitrosomonas* and 0.02 mg VSS/mg NO_2-N for *Nitrobacter*.

---

[4]M. G. Mandt and B. A. Bell (1982). *Oxidation Ditches in Wastewater Treatment*. Ann Arbor Science, Ann Arbor, Mich., p. 48.

Solution

$$\frac{1+n}{20}NH_4^+ + \frac{n+m}{20}HCO_3^- + \frac{n+m}{5}CO_2 + \frac{2-7m-5n}{20}O_2 \rightarrow$$

$$\frac{n+m}{20}C_5H_7NO_2 + \frac{1-m}{10}H^+ + \frac{1-m}{20}NO_3^- + \frac{1+n-2m}{20}H_2O$$

$$0.15 \text{ mgVSS/mg NH}_3\text{-N} = \frac{0.15/(C_5H_7NO_2/20)}{1/(N/20)} = \frac{0.15/(113/20)}{1/(14/20)} = 0.01858\frac{\text{Eq VSS}}{\text{Eq NH}_3\text{-N}}$$

$$= 0.01858\frac{\text{Eq NH}_3\text{-N converted to cells}}{\text{Eq original NH}_3\text{-N}} = m$$

$$0.02 \text{ mg VSS/mg NO}_2\text{-N} = \frac{0.02/C_2H_5NO_2}{1/N} = 0.00248\frac{\text{mol VSS}}{\text{mol NO}_2\text{-N}} = \frac{n}{1-m}$$

$$n = 0.00243$$

$$\text{coefficient of } H^+ = \frac{1-m}{10} = \frac{1-0.01858}{10} = 0.0981$$

Hence from the reaction, $1 + n/20 = 0.050$ mol of NH$_4$-N (or TKN) produces 0.0981 mol of H$^+$, or 1 mol of TKN produces 1.96 mole of H$^+$.

$$\text{Mol of TKN destroyed} = \frac{3.0(10^6)}{7.48(3.281)^3}[1000](0.060 - 0.020)\frac{1}{14} = 32{,}443.89 \text{ mol/day}$$

$$H^+ \text{ produced} = 32{,}443.89(1.96) = 63{,}590.024 \text{ mol/day}$$

$$[H^+] = \frac{63{,}590.024}{3.0(10^6)/7.48(3.281)^3} = 5.60 \text{ mol/m}^3 = 5.60(10^3) \text{ mol/L} \qquad \text{pH} = 2.25$$

$$H^+ + CaCO_3 \rightarrow Ca^{2+} + HCO_3^-$$

moles H$^+$ to be destroyed to raise pH to 7.0

$$= 5.60(10^{-3}) - 10^{-7} = 5.60(10^{-3})/L = \text{moles alkalinity needed} = 560 \text{ mg/L as CaCO}_3$$

$$= 560\left[\frac{3.0(10^6)}{7.48(3.281)^3}\right]\left(\frac{1000}{10^6}\right) = 6359 \text{ kg/day} \quad \textbf{Answer}$$

Denitrification: Heterotrophic Side Reaction.   Aside from the normal anoxic reaction of denitrification, there are two side reactions that can occur: continued oxidation using the leftover dissolved atmospheric oxygen from the nitrification reaction (heterotrophic stage) and nitrite reduction. Immediately after nitrification, a large concentration of dissolved oxygen still exists in the reactor. Since the heterotrophic bacteria are fast growers compared to *Nitrosomonas* and *Nitrobactera*, they overwhelm the reaction, and the overall chemical process is the normal heterotrophic carbonaceous reaction in this last stage of oxidation. (The heterotrophic bacteria are present continually during nitrification.) By control of the process, it is possible to balance the growth rates of the nitrifiers and the heterotrophs such that they can be growing together. However, as soon as the oxygen supply is cut off, the nitrifiers cannot compete against the heterotrophs in the ever-decreasing concentration of dissolved oxygen. Hence the activities of the nitrifiers "fade away" and the heterotrophs predominate in the last stage of oxidation after aeration is cut off.

The other side reaction is the reduction of nitrite to the nitrogen gas. Although the process is aimed at oxidizing nitrogen to the nitrate stage, some nitrite can still be found. In the absence of oxygen, the nitrite ion is reduced to the nitrogen gas. This will be discussed further after we discuss the regular denitrification reaction.

Now, derive the chemical reaction for the heterotrophic stage. Let r be the equivalents of O_2, based on the oxygen acceptor reaction, used after the end of the nitrification step. Using sewage ($C_{10}H_{19}NO_3$) as the electron donor for the denitrification reaction, the ammonium ion is produced (Chapter 2). As mentioned before, given NO_3^- and NH_4^+ in solution, organisms prefer to use NH_4^+, first, before using NO_3^-. Therefore, from the two competing synthesis reactions, the one using the ammonium is favored. Letting q be the equivalents of cells, based on the synthesis reaction, produced during the last stage of the aerobic reaction, the synthesis reaction may be modified as follows:

$$\frac{q}{5}CO_2 + \frac{q}{20}HCO_3^- + \frac{q}{20}NH_4^+ + qH^+ + qe^- \rightarrow \frac{q}{20}C_5H_7NO_2 + \frac{9q}{20}H_2O \tag{9-56}$$

From the r equivalents of O_2, the acceptor reaction is

$$\frac{r}{4}O_2 + rH^+ + re^- \rightarrow \frac{r}{2}H_2O \tag{9-57}$$

From these equations the total electron moles needed from the donor is $r + q$. Hence the donor reaction using sewage ($C_{10}H_{19}NO_3$) as the electron donor is modified as follows:

$$\frac{r+q}{50}C_{10}H_{19}NO_3 + \frac{9(r+q)}{25}H_2O \rightarrow$$

$$\frac{9(r+q)}{50}CO_2 + \frac{r+q}{50}NH_4^+ + \frac{r+q}{50}HCO_3^- + (r+q)H^+ + (r+q)e^- \tag{9-58}$$

Adding equations (9-56), (9-57), and (9-58), the overall aerobic reaction is

$$\frac{r+q}{50}C_{10}H_{19}NO_3 + \frac{r}{4}O_2 \rightarrow \frac{q}{20}C_5H_7NO_2$$

$$+ \frac{9r-q}{50}CO_2 + \frac{2r-3q}{100}NH_4^+ + \frac{2r-3q}{100}HCO_3^- + \frac{14r+9q}{100}H_2O \tag{9-59}$$

From equation (9-59), if Y_c is the cell yield in moles per mole of sewage, $(q/20)/[r + q/50]$, then

$$q = \frac{2rY_c}{5 - 2Y_c} \tag{9-60}$$

Normal Anoxic Denitrification.   Let s, based on the synthesis reaction, be the equivalents of cells produced for the regular anoxic denitrification reaction. Again, since there is an abundance of NH_4^+, the synthesis reaction will involve this ion; thus

$$\frac{s}{5}CO_2 + \frac{s}{20}HCO_3^- + \frac{s}{20}NH_4^+ + sH^+ + se^- \rightarrow \frac{s}{20}C_5H_7NO_2 + \frac{9s}{20}H_2O \tag{9-61}$$

Also, let p, based on the NO_3^- acceptor reaction, be the equivalents of NO_3^- utilized in the anoxic reaction. The revised acceptor reaction is

$$\frac{p}{5}NO_3^- + \frac{6p}{5}H^+ + pe^- \rightarrow \frac{p}{10}N_2 + \frac{3(p)}{5}H_2O \tag{9-62}$$

From equations (9-61) and (9-62), the total e^- needed from the donor reaction is $s + p$. Hence, the donor reaction using sewage is

$$\frac{p+s}{50}C_{10}H_{19}NO_3 + \frac{9(p+s)}{25}H_2O \rightarrow$$
$$\frac{9(p+s)}{50}CO_2 + \frac{p+s}{50}NH_4^+ + \frac{p+s}{50}HCO_3^- + (p+s)H^+ + (p+s)e^- \tag{9-63}$$

Adding equations (9-61), (9-62), and (9-63), the overall reaction for denitrification is produced,

$$\frac{p}{5}NO_3^- + \frac{p}{5}H^+ + \frac{p+s}{50}C_{10}H_{19}NO_3 \rightarrow$$
$$\frac{s}{20}C_5H_7NO_2 + \frac{p}{10}N_2 + \frac{2p-3s}{100}NH_4^+ + \frac{2p-3s}{100}HCO_3^- \tag{9-64}$$
$$+ \frac{9p-s}{50}CO_2 + \frac{24p+9s}{100}H_2O$$

From equation (9-64), if Y_n is the cell yield in moles per mole of nitrate, $(s/20)/(p/5)$, then

$$s = 4pY_n \tag{9-65}$$

Denitrification: NO_2^- Reduction Side Reaction.   Now, derive the overall reaction for the reduction of nitrite. Based on NO_2^- as the electron acceptor, let t be the equivalents of NO_2^- utiliized. The acceptor reaction is

$$\frac{t}{3}NO_2^- + \frac{4t}{3}H^+ + te^- \rightarrow \frac{t}{6}N_2 + \frac{2t}{3}H_2O \tag{9-66}$$

Because of the abundance of the nitrate ion, the synthesis reaction will be based on nitrate as the nitrogen source. Let u, based on the synthesis reaction, be the equivalents of cells formed. The synthesis reaction then becomes

$$\frac{5u}{28}CO_2 + \frac{u}{20}NO_3^- + \frac{29u}{20}H^+ + ue^- \rightarrow \frac{u}{20}C_5H_7NO_2 + \frac{11u}{20}H_2O \tag{9-67}$$

From equations (9-66) and (9-67), the total moles of e^- needed from the electron donor is $t + u$. Hence, the donor reaction using sewage is

$$\frac{t+u}{50}C_{10}H_{19}NO_3 + \frac{9(t+u)}{25}H_2O \rightarrow$$

$$\frac{9(t+u)}{50}CO_2 + \frac{t+u}{50}NH_4^+ + \frac{t+u}{50}HCO_3^- + (t+u)H^+ + (t+u)e^- \tag{9-68}$$

Adding equations (9-66), (9-67), and (9-68) produces the overall reaction for the nitrite reduction,

$$\frac{t}{3}NO_2^- + \frac{u}{28}NO_3^- + \frac{28t+3u}{84}H^+ + \frac{t+u}{50}C_{10}H_{19}NO_3 \rightarrow \frac{u}{20}C_5H_7NO_2$$

$$+ \frac{t}{6}N_2 + \frac{t+u}{50}NH_4^- + \frac{126t+u}{700}CO_2 + \frac{t+u}{50}HCO_3^- + \frac{644t+69u}{2100}H_2O \tag{9-69}$$

If Y_{dr} is the cell yield in moles per mole of sewage, $(u/20)/[(t+u)/50]$, then

$$u = \frac{2tY_{dr}}{5 - 2Y_{dr}} \tag{9-70}$$

Table 9-4 shows some values of Y_c, Y_n, and Y_{dr} that may be used to compute first approximations to q, s, and u.

Total Carbon Source Requirement. From the derivations above, pertinent reactions using domestic sewage as the source for the heterotrophic side reaction, normal anoxic denitrification reaction, and the nitrite-reduction side reaction are rewritten, respectively, below.

$$\frac{r+q}{50}C_{10}H_{19}NO_3 + \frac{r}{4}O_2 \rightarrow \frac{q}{20}C_5H_7NO_2 + \frac{9r-q}{50}CO_2 + \frac{2r-3q}{100}NH_4^+$$

$$+ \frac{2r-3q}{100}HCO_3^- + \frac{14r+9q}{100}H_2O \tag{9-71}$$

TABLE 9-4   VALUES OF Y_C, Y_N, AND Y_{DR}

Carbon source	Y_c	Y_n	Y_{dr}
Domestic waste (mg VSS/mg BOD$_5$)	0.50–0.7		
Domestic waste (mg VSS/mg COD)	0.7		
Soft-drink waste (mg VSS/mg COD)	0.4		
Skim milk (mg VSS/mg BOD$_5$)	0.5		
Pulp and paper (mg VSS/mg BOD$_5$)	0.5		
Shrimp processing (mg VSS/mg BOD$_5$)	0.5		
Methanol (mg VSS/mg NO$_3$-N)		0.7–1.5	
Domestic sludge (mg VSS/mg BOD$_5$)			0.04–0.1
Fatty acid (mg VSS/mg BOD$_5$)			0.04–0.07
Carbohydrate (mg VSS/ mg BOD$_5$)			0.02–0.04
Protein (mg VSS/mg BOD$_5$			0.05–0.09

$$\frac{p}{5}NO_3^- + \frac{p}{5}H^+ + \frac{p+s}{50}C_{10}H_{19}NO_3 \rightarrow$$

$$\frac{s}{20}C_5H_7NO_2 + \frac{p}{10}N_2 + \frac{2p-3s}{100}NH_4^+ \tag{9-72}$$

$$+ \frac{2p-3s}{100}HCO_3^- + \frac{9p-s}{50}CO_2 + \frac{24p+9s}{100}H_2O$$

$$\frac{t}{3}NO_2^- + \frac{u}{28}NO_3^- + \frac{28t+3u}{84}H^+ + \frac{t+u}{50}C_{10}H_{19}NO_3 \rightarrow$$

$$\frac{u}{20}C_5H_7NO_2 + \frac{t}{6}N_2 + \frac{t+u}{50}NH_4^+ \tag{9-73}$$

$$+ \frac{126t+u}{700}CO_2 + \frac{t+u}{50}HCO_3^- + \frac{644t+69u}{2100}H_2O$$

From these equations the amount of carbon needed in terms of moles of total sewage SEW is

$$SEW = \frac{4(r+q)}{50r}[DO] + \frac{5(p+s)}{50p}[NO_3\text{-}N] + \frac{3(t+u)}{50t}[NO_2\text{-}N] \tag{9-74}$$

where the dissolved oxygen DO and NO_3-N and NO_2-N are all expressed in moles. They are the amounts actually consumed during the entire denitrification process, which involves the side reactions and the normal anoxic reaction. SEW may be expressed in terms of the equivalent oxygen demand.

Consider the following overall oxidation reaction for sewage that represents the biochemical reaction in a BOD bottle:

$$2C_{10}H_{19}NO_3 + \tfrac{1}{4}O_2 \rightarrow 18CO_2 + 2NH_4^+ + 2HCO_3^- + 14H_2O \tag{9-75}$$

From this equation, moles of O_2 per mole of sewage $= \left(\tfrac{1}{4}\right)/2$. Or, using f, the conversion factor from ultimate BOD to BOD_5 ($CBOD_u$ to $CBOD_5$), moles of $CBOD_5$ per mole of sewage $= f\left(\tfrac{1}{4}\right)/2 = f/8$. In terms of BOD_5, moles BOD_5 of sewage required is

$$BOD_5 = \frac{f(SEW)}{8} = \frac{f}{8}\left(\frac{4(r+q)}{50r}[DO] + \frac{5(p+s)}{50p}[NO_3\text{-}N] + \frac{3(t+u)}{50t}[NO_2\text{-}N]\right)$$

$$\tag{9-76}$$

Net Effluent Nitrogen and Use of Alternative Carbon Source. Although the entire denitrification process indicates that nitrates have been denitrified, some ammonia has also been produced. These are shown in equations (9-71), (9-72), and (9-73), from which

$$NH_4\text{-}N = \frac{4(2r-3q)}{100r}[DO] + \frac{5(2p-3s)}{100p}[NO_3\text{-}N] + \frac{3(t+u)}{50t}[NO_2\text{-}N] \tag{9-77}$$

This nitrogen will appear in the effluent of the denitrification unit.

For the ammonium to be removed it must be nitrified and denitrified using an alternative carbon source that does not contain nitrogen. An example is methanol. Methanol was the carbon source used early in the study of denitrification. To compute the carbon requirement, reactions parallel to the ones above should be derived for the alternative carbon source. Equation (9-76) may, however, be used as a very crude approximation.

Example 9-12

After cutting off the aeration to a nitrification plant, the dissolved oxygen concentration is 2.0 mg/L. How much sewage is still needed for the carbonaceous reaction in this last stage of aerobic reaction before denitrification sets in? How much NH_4-N is produced?

Solution

$$\frac{r+q}{50}C_{10}H_{19}NO_3 + \frac{r}{4}O_2 \rightarrow \frac{q}{20}C_5H_7NO_2 + \frac{9r-q}{50}CO_2 + \frac{2r-3q}{100}NH_4^+$$

$$+ \frac{2r-3q}{100}HCO_3^- + \frac{14r+9q}{100}H_2O$$

From Table 9-4, assume that $Y_c = 0.60$ mg VSS/mg BOD_5; assuming that $f = 0.68$, moles of BOD_5 per mole of sewage $= 0.68/8 = 0.085$, or moles of sewage per mole of BOD_5 = 11.76. Also note that BOD_5 is oxygen and a mole of BOD_5 is 32 mass units of oxygen.

$$C_5H_7NO_2 = 5(12) + 7(1.008) + 14 + 2(16) = 113$$

$$C_{10}H_{19}NO_3 = 10(12) + 19(1.008) + 14 + 3(16) = 201.2$$

$$Y_c = \frac{0.60/113}{(1/32)(11.76)} = 0.014\frac{\text{mol } C_5H_7NO_2}{\text{mole } C_{10}H_{19}NO_3}$$

$$q = \frac{2rY_c}{5 - 2Y_c}$$

From energy reaction, the equivalent weight of $O_2 = O_2/4 = 8$. Thus

$$r = \frac{2}{8} = 0.25 \text{ mEq/L} \Rightarrow 0.0625 \text{ mmol/L of } O_2$$

$$q = \frac{2(0.25)(0.014)}{5 - 2(0.014)} = 0.0014 \text{ mEq/L}$$

$$\text{mole of sewage needed} = \frac{(r+q)/50}{r/4} = \frac{(0.25+0.0014)/50}{0.25/4} = 0.080 \text{ mmol/L}$$

$$= 16.20 \text{ mg/L} \quad \textbf{Answer}$$

$$NH_4\text{-N produced} = \frac{(2r-3q)/100}{(r+q)/50}(0.080) = \frac{[2(0.25) - 3(0.0014)]/100}{(0.25+0.0014)/50}(0.080)$$

$$= 0.079 \text{ mmol/L} = 1.1 \text{ mg/L} \quad \textbf{Answer}$$

Example 9-13

A domestic wastewater with a flow of 20,000 m^3/day is to be denitrified. The effluent from the nitrification tank contains 30 mg/L of NO_3-N, 2.0 mg/L of dissolved oxygen, and 0.5 mg/L of NO_2-N. Assume that an effluent total nitrogen not to exceed 3.0 mg/L is required by the permitting agency. (a) Calculate the quantity of domestic sewage to be provided to satisfy the carbon requirement. (b) What is the corresponding BOD_5 to be provided?

Solution

(a) $$SEW = \frac{4(r+q)}{50r} DO + \frac{5(p+s)}{50p} NO_3\text{-N} + \frac{3(t+u)}{50t} NO_2\text{-N}$$

$$NH_4\text{-N} = \frac{4(2r-3q)}{100r} DO + \frac{5(2p-3s)}{100p} NO_3\text{-N} + \frac{3(t+u)}{50t} NO_2\text{-N}$$

$$t = \frac{0.5}{NO_2\text{-N}/3} = 0.11 \text{ mEq/L}$$

From Table 9-4, use $Y_{dr} = 0.5$ mg VSS/mg BOD_5; assume that $f = 0.68$. Then

$$Y_{dr} = \frac{0.5/113}{(1/32)(11.76)} = 0.012 \frac{\text{mol}}{\text{mol sewage}}$$

$$u = \frac{2tY_{dr}}{5 - 2Y_{dr}} = \frac{2(0.11)(0.012)}{5 - 2(0.012)} = 0.0005 \text{ mEq/L}$$

$$r = \frac{2.0}{32/4} = 0.25 \text{ mEq/L}$$

From Table 9-4, use $Y_c = 0.6$ mg VSS/mg BOD_5. Then

$$Y_c = \frac{0.6/113}{11.76/32} = 0.014 \frac{\text{mol}}{\text{mol sewage}}$$

$$q = \frac{2rY_c}{5 - 2Y_c} = \frac{2(0.25)(0.014)}{5 - 2(0.014)} = 0.0014 \text{ mEq/L}$$

Referring to Table 9-4, use

$$Y_n = 0.80 \frac{\text{mg} VSS}{\text{mgNO}_3\text{-N}} = \frac{0.80/113}{1/14} = 0.10 \frac{\text{mol VSS}}{\text{mol N}}$$

Then

$$s = 4pY_n = 4(0.10)p = 0.40p$$

$$p \text{ mEq/L of NO}_3\text{-N} = p(N/5)(1/N) = p/5 \text{ mmol of NO}_3\text{-N}$$

$$NO_3\text{-N in effluent} = \frac{30}{14} - \frac{p}{5} = (2.14 - 0.2p) \text{ mmol/L}$$

$$\text{effluent NH}_4\text{-N} = \frac{4[2(0.25) - 3(0.0014)]}{100(0.25)} \left(\frac{2}{32}\right) + \frac{5[2p - 3(0.40p)]}{100p} \left(\frac{p}{5}\right)$$

$$+ \frac{3(0.11 + 0.0005)}{50(0.11)} \left(\frac{0.5}{14}\right) = 0.0048 + 0.008p + 0.002 = 0.0068 + 0.008p$$

$$\frac{3}{14} = (2.14 - 0.2p) + (0.0068 + 0.008p) = 2.15 - 0.192p$$

$p = 10.1 \Rightarrow 2.02$ mmol/L of NO_3-N. Note that the equivalent mmoles of p cannot exceed $30/14 = 2.14$ mmol/L. Thus

$$s = 0.40(10.1) = 4.04$$

$$\text{SEW} = \frac{4(0.25 + 0.0014)}{50(0.25)}\left(\frac{2}{32}\right) + \frac{5(10.1 + 4.04)}{50(10.1)}\left(\frac{10.1}{5}\right) + \frac{3(0.11 + 0.0005)}{50(0.11)}\left(\frac{0.5}{14}\right)$$

$$= 0.005 + 0.28 + 0.002 = 0.287 \text{ mmol/L} = 58 \text{ mg/L}$$

$$= 58(10^{-3})(20{,}000) = 1155 \text{ kg/day} \quad \textbf{Answer}$$

(b) Molecular weight of BOD_5 = molecular weight of CBOD = (O_2) = 32.

$$BOD_5 = \frac{f(\text{SEW})}{8} = \frac{0.68(0.287)}{8}(32) = 0.78 \text{ mg/L}$$

$$= 0.078(10^{-3})(20{,}000) = 16 \text{ kg/day} \quad \textbf{Answer}$$

Nitrification kinetics. The kinetics of nitrification is a function of several factors the most important of which include pH, temperature, and the concentrations of ammonia and dissolved oxygen. Experience has shown that the optimum pH of nitrification lies between 7.2 and 8.8. Outside this range, the rate becomes limited. As shown in the nitrification reaction, acidity is produced. However, carbonaceous reactions produce alkalinity that serves to neutralize this acidity produced. Design calculations must therefore be made such that alkalinity production of the carbonaceous reaction balances the acidity production of the nitrification reaction. Temperature affects the half-velocity constant K_n of the Monod equation. It also affects maximum growth rates. Growth rates for *Nitrosomonas* and *Nitrobacter* are affected differently by change in temperature. At elevated temperatures *Nitrosomonas* growth rates are accelerated while those of *Nitrobacter* are depressed. *Nitrosomonas* is, however, so slow growing compared to *Nitrobacter* that the kinetics is controlled by *Nitrosomonas*. Expressing the effects of these predominant factors on the specific growth rate μ_n, the following Monod equation is obtained:

$$\mu_n = \mu_{nmax}\frac{[N]}{K_n + [N]}\frac{[O_2]}{K_{O2} + [O_2]}C_{pH} \tag{9-78}$$

where μ_{nmax} is the maximum μ_n, $[N]$ is the concentration of NH_4-N (*note:* since TKN hydrolyzes to produce NH_4^+, $[N]$ may also represent TKN), $[O_2]$ is the concentration of dissolved oxygen, and K_{O2} is the half-velocity constant for oxygen. K_n, μ_{nmax}, and C_{pH} have been determined experimentally and are given, respectively, by[5]

$$K_n = 10^{0.051T - 1.158} \tag{9-79}$$

$$\mu_{nmax} = 0.47e^{0.098(T-15)} \tag{9-80}$$

$$C_{pH} = 1 - 0.833(7.2 - pH) \tag{9-81}$$

---

[5] Ibid., p. 51.

The values of K_{O2} range from 0.25 to 2.46 mg/L. T is in degrees Celsius, K_n is in mg/L, and μ_{nmax} is per day. C_{pH} is a fractional correction factor.

For nitrification to be successful, the mean cell retention time (MCRT) of the nitrifiers must control the design over that of the carbonaceous organisms. Naturally, since the nitrifiers are slow growing, their MCRTs are longer; hence if they are not made to control the design, they will simply be washed out of the reactor without ever being given the chance to multiply.

In the operation of the nitrification unit, the dissolved oxygen concentration, pH, and temperature are maintained at practically constant levels. Hence equation (9-78) may be written as

$$\mu_n = \mu'_{nmax} \frac{[N]}{K_n + [N]} \tag{9-82}$$

where

$$\mu'_{nmax} = \mu_{nmax} \frac{[O_2]}{K_{O2} + [O_2]} C_{pH} = \text{a constant}$$

The equation for the kinetics of the activated sludge reactor under steady state derived in Chapter 7 is

$$\mu_{max} \frac{[S]}{K_s + [S]} = \frac{Q_w[X_u] + (Q_0 - Q_w)[X_e] - Q_0[X_0]}{V[X]} + k_d \tag{9-83}$$

where [S] is the concentration of substrate, μ_{max} the maximum specific growth rate cofficient, K_s the half-saturation of the substrate, Q_w the sludge wasting flow rate, $[X_u]$ the underflow concentration of sludge, Q_0 the flow to the reactor, $[X_e]$ the effluent concentration of microorganisms, X_0 the inflow concentration of microorganisms, V the volume of reactor, [X] the concentration of microorganisms in the reactor (MLVSS), and k_d the death rate coefficient.

But

$$\mu_{max} \frac{[S]}{K_s + [S]} = \mu \quad \text{and} \quad \frac{Q_w[X_u] + (Q_0 - Q_w)[X_e] - Q_0[X_0]}{V[X]} = \frac{1}{\theta_c}$$

where θ_c is the mean cell retention time (MCRT). Hence the equation becomes

$$\mu = \frac{1}{\theta_c} + k_d \tag{9-84}$$

Using equation (9-84) as the model, the corresponding equation for the nitrifiers is

$$\mu_n = \frac{1}{\theta_n} + k_{dn} \tag{9-85}$$

In equation (9-85), k_{dn} is the coefficient of decay, μ_n the specific growth rate, and θ_n is the MCRT. A typical range of values for k_{dn} is 0.04 to 0.08 per day. Since θ_n is the larger, in design, θ_c must be made equal to θ_n. Also, a safety factor SF is applied to the minimum θ_n $(= \theta_n^m)$ to obtain the design θ_n $(= \theta_n^d)$. In equation form this is written

$$SF = \frac{\theta_n^d}{\theta_n^m} = \frac{\theta_c^d}{\theta_c^m} \tag{9-86}$$

In equation (9-85), the theoretical limit of μ_n may be approached when θ_n approaches infinity. At this limiting condition, μ_n is equal to k_{dn}. Combining this condition with equation (9-82) and solving for [N], the limit concentration $[N_{lim}]$ is

$$[N_{lim}] = \frac{k_{dn} K_n}{\mu'_{nmax} - k_{dn}} \tag{9-87}$$

Once the MCRT has been determined, the volume of the reactor may be determined. From Chapter 7, the formula is reproduced below, where the cell yield Y has been changed to Y_c. Note that as mentioned before, θ_c is made equal to θ_n, or, more specifically, θ_c is made equal to θ_n^d.

$$\frac{Q_0([S_0] - [S])}{V[X]} = \frac{1 + k_d \theta_c}{\theta_c Y_c} \tag{9-88}$$

In nitrification, heterotrophs and nitrifiers are coexisting. The process is still the activated sludge process, where the major objective is the satisfaction of the carbonaceous demand and nitrification considered incidental. Hence, to size the volume of the reactor using equation (9-88), $[S_0]$ and $[S]$ must be for carbonaceous concentrations. The proportion of the respective population varies with the ratio of BOD_5 to TKN as shown in Table 9-5.

Example 9-14

Design an activated sludge process to nitrify an influent with the following characteristics to 2.0 mg/L ammonia: $Q_0 = 1$ mgd, $BOD_5 = 200$ mg/L, TKN = 60 mg/L, and minimum operating temperature = 10°C. The effluent is to have a BOD_5 of 5.0 mg/L. The DO is to be maintained at 2.0 mg/L. Apply an SF of 2.0 and assume that $K_{O2} = 1.3$ mg/L. Calculate the MCRT needed to nitrify to the limiting concentration of ammonia.

Solution

$$\mu_n = \frac{1}{\theta_n} + k_{dn}$$

$$K_n = 10^{0.051T - 1.158} = 10^{0.051(10) - 1.158} = 0.225 \text{ mg/L}$$

$$\mu'_{nmax} = \mu_{nmax} \frac{[O_2]}{K_{O2} + [O_2]} C_{pH}$$

$$\mu_{nmax} = 0.47 e^{0.098(T - 15)} = 0.47 e^{0.098(10 - 15)} = 0.29 \text{ per day}$$

TABLE 9-5 RELATIONSHIP BETWEEN THE FRACTION OF NITRIFYING ORGANISMS AND THE BOD_5/TKN RATIO

BOD_5/TKN	Nitrifier fraction	BOD_5/TKN	Nitrifier fraction
0.5	0.40	5	0.05
1	0.20	6	0.04
2	0.15	7	0.04
3	0.08	8	0.03
4	0.06	9	0.03

$$\mu'_{n\text{max}} = 0.29 \left(\frac{2.0}{1.3 + 2.0} \right) [1 - 0.833(7.2 - 7.2)] = 0.176 \text{ per day}$$

Assume that $k_{dn} = 0.06$ per day.

$$\mu_n = \mu'_{n\text{max}} \frac{[N]}{K_n + [N]} = 0.176 \left(\frac{2}{0.225 + 2.0} \right) = 0.16 \text{ per day}$$

$$0.16 = \frac{1}{\theta_n} + 0.06; \quad \theta_n = 10.0 \text{ days}$$

$$\theta_n^d = \theta_c^d = 2(10) = 20 \text{ days}$$

$$\frac{\text{BOD}_5}{\text{TKN}} = \frac{200}{60} = 3.33$$

From Table 9-5, nitrifier population:

$$\frac{x - 0.083}{0.064 - 0.083} = \frac{3.33 - 3}{4 - 3} \qquad x = 0.077$$

Assume that MLVSS = 2500 mg/L. Then

$$\text{heterotrophs} = (1 - 0.077)(2500) = 2307 \text{ mg/L}$$

Assume that $k_d = 0.05$ per day at 10°C and $Y_c = 0.60$ mg VSS/ mg BOD$_5$.

$$\frac{Q_0([S_0] - [S])}{V[X]} = \frac{1 + k_d \theta_c}{\theta_c Y_c}$$

$$\frac{[1(10^6)/7.48(3.281)^3](0.20 - 0.005)}{V(2.307)} = \frac{1 + 0.05(20)}{20(0.60)} \qquad V = 1920 \text{ m}^3 \quad \textbf{Answer}$$

$$[N_{\text{lim}}] = \frac{k_{dn} K_n}{\mu'_{n\text{max}} - k_{dn}}$$

$$= \frac{0.06(0.225)}{0.176 - 0.06} = 0.12 \text{ mg/L}$$

$$\mu_n = \mu'_{n\text{max}} \frac{[N]}{K_n + [N]} = 0.176 \left[\frac{0.12}{0.225 + 0.12} \right] = 0.061217 \text{ per day}$$

$$0.061217 = \frac{1}{\theta_n} + 0.06 \qquad \theta_n = 822 \text{ days} \quad \textbf{Answer}$$

Denitrification kinetics. In denitrification, the parameters that can limit the kinetics are the carbon source and the nitrate ions. The Monod equation may be written as

$$\mu_{dn} = \mu_{dn\text{max}} \frac{[N]}{K_{dn} + [N]} \frac{[S]}{K_s + [S]} \qquad (9\text{-}89)$$

where μ_{dn} is the specific growth rate, $\mu_{dn\max}$ the maximum μ_{dn}, [N] the concentration of the nitrate ion, K_{dn} the half-velocity constant for denitrification, [S] the concentration of the carbon source, and K_s the half-velocity constant for the carbon source. Values for $\mu_{dn\max}$ range from 0.3 to 0.9 per day, those for K_{dn} range from 0.06 to 0.14 mg/L, and those for K_s range from 25 to 100 mg/L as BOD_5.[6,7]

By analogy with nitrification kinetics, the expressions for μ_{dn} and $[N_{\lim}]$ are written, respectively, below.

$$\mu_{dn} = \frac{1}{\theta_{dn}} + k_{ddn} \tag{9-90}$$

$$[N_{\lim}] = \frac{k_{ddn} K_{dn}}{\mu'_{dn\max} - k_{ddn}} \tag{9-91}$$

where $\mu'_{dn\max} = \mu_{dn\max}\{[S]/(K_s + [S])\}$, $\theta_{dn} = $ MCRT for the denitrifiers, $k_{dd} = k_d$ for denitrifiers.

Now size the reactor volume. Since the overriding consideration in denitrification is the reduction of the nitrogen concentration, the [S]'s must be changed to the nitrogen concentration [N]. Hence the equation is

$$\frac{Q_0([N_0] - [N])}{V[X_{dn}]} = \frac{1 + k_{ddn}\theta_{dn}}{\theta_{dn} Y_n} \tag{9-92}$$

Values for k_{ddn} range from 0.04 to 0.08 per day.

There are two possible calculations for the volume of the denitrification reactor: calculation where denitrification is allowed to occur in the tank where nitrification is occurring, and calculation where denitrification is done in a separate tank or in a separate portion of the same tank, as in the oxidation ditch. Where denitrification and nitrification occur in the same tank, the process is a batch process, an example of which is the SBR (sequencing batch reactor). In this instance, the larger of the two volume requirements (denitrification or nitrification) will control the design. Where denitrification and nitrification are done in separate tanks or in a separate portion of the same tank, each process will require its own reactor volume.

Example 9-15

A domestic wastewater with a flow of 20,000 m^3/day is nitrified. The effluent from the nitrification tank contains 30 mg/L of NO_3-N, 2.0 mg/L of dissolved oxygen, and 0.5 mg/L of NO_2-N. (a) Assuming that an effluent total nitrogen not to exceed 3.0 mg/L and an effluent BOD_5 not to exceed 10 mg/L are required by the permitting agency, calculate the reactor volume required for denitrification. (b) How long does it take to attain the limiting nitrate ion concentration?

---

[6] Metcalf & Eddy, *Wastewater Engineering.*

[7] Mandt and Bell, *Oxidation Ditches in Wastewater Treatment*, p. 64.

Solution

(a)
$$\mu_{dn} = \frac{1}{\theta_{dn}} + k_{ddn}$$

Required NO_3-N concentration $= 3.0 - 0.5 = 2.5$ mg/L

$$\mu_{dn} = \mu_{dn\text{max}} \frac{[N]}{K_{dn} + [N]} \frac{[S]}{K_s + [S]}$$

Assume that $\mu_{dn\text{max}} = 0.6$ per day, $K_s = 25$ mg/L of BOD_5, $k_{ddn} = 0.07$ per day and $K_{dn} = 0.1$ mg/L. Then

$$\mu_{dn} = 0.6 \left(\frac{2.5}{0.1 + 2.5} \right) \left(\frac{10}{25 + 10} \right) = 0.165 \text{ per day}$$

$$0.165 = \frac{1}{\theta_{dn}} + 0.07 \qquad \theta_{dn} = 10.52, \qquad \text{say 21 days using SF} = 2$$

$$\frac{Q_0([N_0] - [N])}{V[X_{dn}]} = \frac{1 + k_{ddn}\theta_{dn}}{\theta_{dn} Y_n}$$

Assume that $[X_{dn}] = 2500$ mg/L and $Y_n = 0.8$ mg VSS/mg NO_3-N. Then

$$\frac{20,000(0.03 - 0.0025)}{V(2.5)} = \frac{1 + 0.07(21)}{21(0.8)}; V = 1496.4 \text{ m}^3 \quad \textbf{Answer}$$

(b)
$$[N_{\text{lim}}] = \frac{k_{ddn} K_{dn}}{\mu'_{dn\text{max}} - k_{ddn}}$$

$$\mu'_{dn\text{max}} = \mu_{dn\text{max}} \frac{[S]}{K_s + [S]} = 0.6 \left(\frac{10}{25 + 10} \right) = 0.17 \text{ per day}$$

$$[N_{\text{lim}}] = \frac{0.07(0.1)}{0.17 - 0.07} = 0.07 \text{ mg/L}$$

$$\mu_{dn} = 0.6 \left(\frac{0.07}{0.1 + 0.07} \right) \left(\frac{10}{25 + 10} \right) = 0.070588 \text{ per day}$$

Thus

$$0.070588 = \frac{1}{\theta_{dn}} + 0.07 \qquad \theta_{dn} = 1700 \text{ days} \quad \textbf{Answer}$$

Phosphorus Removal

As mentioned before, to control eutrophication of receiving streams, the amounts of nutrients to be discharged are controlled. In most cases, only phosphorus is targeted to be removed; in some very unusual circumstances, both nutrients, phosphorus and nitrogen, are required to be removed. The latter case is usually dictated by politics, although in some situations, some authorities claim that the requirement can be defended scientifically.[8] Federal and state authorities have required total phosphorus limits ranging from 0.1 to 2.0 mg/L.

There are two general ways to remove phosphorus: by biological uptake and by chemical precipitation. Assume that the percent phosphorus content of biological sludge is known. If the rate of sludge wasting is given, the rate of phosphorus removal can be calculated. From knowledge of the flow rate and the influent phosphorus concentration, the effluent phosphorus concentration can then be estimated. The formula of bacterial cells in the wasted sludge, including phosphorus, has been written as $C_5H_7NO_2P_{0.074}$.[9]

The other method of biological removal of phosphorus is by *luxury uptake*. In this process, after the sludge is removed by sedimentation, it is subjected to an anaerobic condition. This condition causes the organism to release its phosphorus content, thereby producing a sludge that is highly deficient in phosphorus. As the sludge is returned to the reactor, the organisms, because of having experienced a deficiency, tend to assimilate an extraordinarily large amount of phosphorus. Hence the term *luxury uptake*. Authorities claim that an effluent concentration of 1.0 mg/L can be achieved without further treatment, such as chemical addition.[10] However, it would very dangerous simply to rely on this information. Due to unforeseen variables that can creep into the process, it would be best that a pilot study be conducted for a particular waste. Numerous times, values claimed by the literature and by so-called authorities are nowhere near the values obtained in practice.

Chemical precipitation is normally accomplished by the use of lime and salts of iron and aluminum. Iron is usually in the form of ferric chloride or sulfate. Depending on the pH values, phosphorus in wastewater is present in the various forms of the dissociation products of phosphoric acid and in the forms of *polyphosphate* or *condensed phosphate* and *organic phosphorus*. In the condition of pH prevailing in wastewaters, the predominant form of the acid dissociation product (collectively called *orthophosphate*) is HPO_4^{2-}. Since the phosphorus can be present in various forms, it makes no sense to write the chemical reactions involved when determining the amount of chemicals needed for precipitation. Suffice it to say that the precipitates formed are metal phosphates, except for the use of lime where the precipitate formed is hydroxyapatite, $Ca_5(OH)(PO_4)_3$.

The iron and aluminum precipitants produce acids, while lime produces a basic condition. These are indicated by the reactions below. Alum and iron reactions are buffered between pH 5.5 and 7.0. The use of lime is normally carried out at pH above 11.

$$Fe^{3+} + HPO_4^{2-} \rightarrow FePO_4 + H^+ \tag{9-93}$$

$$5Ca(OH)_2 + 3HPO_4^{2-} \rightarrow Ca_5(OH)(PO_4)_3 + 6OH^- + 3H_2O \tag{9-94}$$

$$Al^{3+} + HPO_4^{2-} \rightarrow AlPO_4 + H^+ \tag{9-95}$$

---

[8]C. D'Elia of the University of Maryland Chesapeake Bay Laboratory, personal communication.

[9]R. S. Ramalho (1977). *Introduction to Wastewater Treatment Processes*. Academic Press, New York, p. 171.

[10]T. J. McGhee (1991). *Water Supply and Sewerage*. McGraw-Hill, New York, p. 531.

Points of addition of the precipitants can be ahead of the primary tank, in the reactor, or after secondary treatment. As shown in the previous equations, however, for the chemical reactions to occur, the phosphorus must be in the form of orthophosphorus. Hence since phosphorus can also be in the condensed and organic forms, additions other than after secondary treatment will effect a lesser degree of removal. After secondary treatment, the condensed and organic forms will have been hydrolyzed to the ortho form, hence precipitation reaction can be carried out effectively. In addition, lime cannot be added in the bioreactor, since the pH can rise to undesirable levels to inhibit biological activity. Also, in addition to the precipitation reactions, removal of phosphorus may be effected by adsorption of the condensed and organic species to precipitate flocs. Representative doses of the aluminum and iron salts are shown in Table 9-6. Actual doses should be determined by on-site testing.

TABLE 9-6   REPRESENTATIVE DOSES OF
ALUMINUM AND IRON SALTS

Percent P	Al:P mole ratio	Fe:P mole ratio
95	2.4:1	1.70:1
90	—	1.30:1
85	1.7:1	—
80	—	0.90:1
75	1.5:1	—
70	—	0.65:1

Example 9-16

A domestic wastewater with a flow of 20,000 m^3/day contains 15 mg/L of total phosphorus. (a) If 6000 lb/day of sludge solids containing 75% VSS is wasted, calculate the concentration of total phosphorus in the effluent. Assume that the sludge contains 2.0% solids and that the water density is 997 kg/m^3. (b) If the process train takes advantage of the luxury uptake (1.0 mg/L total P in effluent), how much total P is wasted in the sludge?

Solution

(a)   Total input of total phosphorus $= (0.015)(20,000) = 300$ kg/day

$$C_5H_7NO_2P_{0.074} = 5(12) + 7(1.008) + 14 + 2(16) + 0.074(31) = 115.35$$

$$\text{Total P in wasted sludge} = 6000(0.75)\left[\frac{0.074(31)}{115.35}\right] = 89.49 \text{ kg/day}$$

$$\text{H}_2\text{O in sludge wasted} = \frac{(6000/0.02) - 6000}{997} = 294.88 \text{ m}^3/\text{day}$$

$$\text{Effluent total P} = \frac{300 - 89.49}{20,000 - 294.88} = 0.0106 \text{ kg/m}^3 = 10.6 \text{ mg/L} \quad \textbf{Answer}$$

(b)   Total P in effluent $= 0.001(20,000 - 294.88) = 19.71$ kg/day

Total P in wasted sludge $= 300 - 19.71 = 280.29$ kg/day   **Answer**

Example 9-17

How much aluminum salt is needed to reduce the total P in Example 9-16 to a level of 2.0 mg/L in the effluent?

Solution

Total P to be removed $= 300 - (20,000 - 294.88)(0.002) = 260.59$ kg/day $= 8.41$ mol/day

$$\% \text{ removal} = \frac{260.59}{300}(100) = 86.96$$

From Table 9-6, aluminum salt needed:

$$\frac{x - 1.7}{2.3 - 1.7} = \frac{86.96 - 85}{95 - 85} \qquad x = 1.82 \text{ mol/mol total P}$$

Thus total moles required $= 1.82(8.41) = 15.29$ mol/day $= 15.29(27) = 412.72$ kg/d of Al   **Answer**

LAND TREATMENT SYSTEMS

In general, the following may be considered as land treatment systems: irrigation or slow-rate treatment system, rapid infiltration, overland flow, wetlands, and septic tank-leaching field treatment system. The use of aquaculture and leaching cesspools may also be added to this list. Figures 9-7 and 9-8 show the schematics of these land treatment systems.

Irrigation, Rapid Infiltration, and Overland Flow Systems

The *irrigation* or *slow-rate land treatment system* involves the use of wastewater to irrigate three classes of water-tolerant plants: forage and field crops, landscape vegetation, and serviculture (the growing of woodlands). Examples of *forage crops* are reed canary grass, brome grass, tall fescue, perennial rye grass, and coastal bermuda grass. Examples of *field crops* are barley, sorghum, corn, and milo. Application to *landscape vegetation* includes highway median and border strips, airport strips, golf courses, parks and recreational areas, and wildlife areas. Plants utilize nutrients from the waste for their growing needs. Thus, depending on the type of plant, nitrogen and phosphorus are removed.

Uptake of various plants was considered in Chapter 8 in connection with the discussion on sludge used as fertilizer. Wastewaters are applied either by sprinklers or sprays or by surface-applicator arrangements such as gated pipes distributing sewage to ridge-and-furrow systems. Figure 9-9 is a photograph of a spray irrigation system using reclaimed wastewater in a highway.

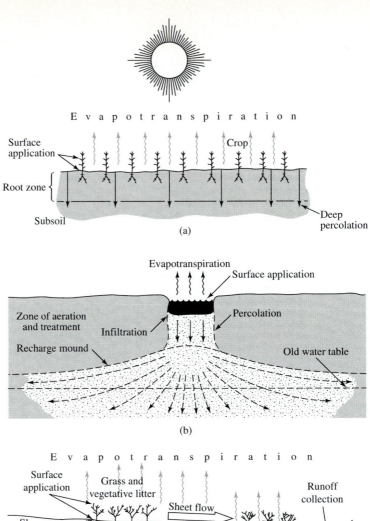

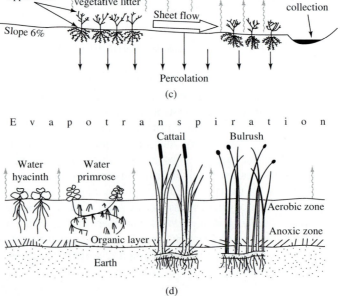

Figure 9-7 Four land treatment systems: (a) irrigation; (b) rapid infiltration; (c) overland flow; (d) wetland.

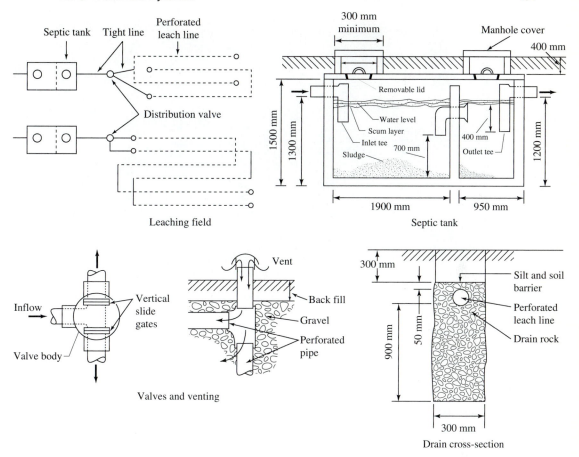

Figure 9-8   Septic tank–leaching field system.

Rapid-infiltration systems involve applying wastewater at rates much faster than the irrigation system. The method does not rely on the ability of plants to remove nutrients but is designed to hasten the rate of infiltration into the ground. Systems objective can include groundwater recharge and recovery of the treated wastewater after natural treatment by the process. In this system, nitrogen is removed by denitrification.

The *overland flow* method essentially applies biological treatment to the waste as it is applied over the upper reaches of sloped, vegetated terraces and allowed to flow across to runoff collection ditches. Water-tolerant grasses such as bermuda, tall fescue, and red canary are planted to minimize channeling and erosion. Bacteria growing in thatches that accumulate on the soil surface carry out oxidation, nitrification, and denitrification.

The mechanisms of treatment of the three land treatment systems discussed above are infiltration and percolation, evaporation, and biological and physical treatment. For infiltration and percolation, the role of sodium merits consideration. High sodium concentrations in clay-bearing soils disperse soil particles, decreasing soil permeability. A

Figure 9-9   Spray irrigation system using reclaimed wastewater in a highway.

permeability characterization of soil may be made by using the sodium absorption ratio (SAR) developed by the U.S. Department of Agriculture Salinity Laboratory:

$$SAR = \frac{[Na]}{\sqrt{([Ca] + [Mg])/2}} \tag{9-96}$$

where the concentrations are all expressed in mEq/L. A SAR value above 9 may adversely affect the permeability of fine-textured soils.

Wetland Treatment System

A *wetland treatment system* (Figure 9-7d) consists of natural or artificial wetlands to which wastewater is applied for the purpose of treatment. To avoid nuisance conditions, the system is normally shallow, typically less than 0.6 m in depth. Water hyacinth, water primrose, cattail, and bulrushes are grown. Wetlands are effective for treatment of wastewaters for a number of reasons. Bacteria attached to the submerged roots and stems of aquatic plants remove soluble and colloidal BOD from the waste. The adsorption and filtration ability of the roots and stems, the ion exchange and absorption ability of the wetland sediment, and the submerged condition of plants that reduce the perturbing effects of climate contribute to the effectiveness of wetland treatment systems.

A problem in the use of wetlands, however, is the potential proliferation of mosquitoes. The wastewater marsh containing polluted water under stagnant conditions, combined with submerged, floating, and emergent vegetation, creates an ideal habitat for mosquito breeding. To solve this problem, system design should allow natural predators of mosquito larvae to thrive, such as mosquito fish, dragonfly and damselfly nymphs; and a variety of water beetles. In addition, the system should not be overloaded.

Septic Tank–Leaching Field System

This system is an on-site method of disposing wastewater, where dwellings, commercial establishments, and the like are located in unsewered areas. As shown in Figure 9-8, the system is composed of septic tank and drain or leaching fields (top two figures). The design and installation of the tank is indicated in the third figure from the top. The tank is buried underground. Settleable solids from the wastewater settle and form solids at the bottom of the tank. Lighter solids such as grease float to the top to form scums. Partially treated wastewater flows through the space between the scum and the sludge at the bottom into the effluent outlet.

The purpose of the leaching field is to disperse the effluent from the septic tank and to allow it to percolate through the ground in such a manner as not to clog the soil. The field is composed of perforated drain pipes buried in trenches as indicated at the bottom. The trench is filled with sand and gravel so as to permit unobstructed passage of water into the soil. For efficient operation of the leaching field, two parallel fields are desirable. While one field is used, the other could be rested. The middle drawing shows the valve that is used to alternate wastewater flow between two drain-field systems. Also indicated on the right-hand side are the vent and inspection port.

Current design practice makes use of the results of percolation (perc or perk) tests. In the perc test, holes varying in diameter from 150 to 300 mm and in depths varying from 0.6 to 1.0 m or to the bottom of the proposed trench are dug in the ground and soaked for a period of 24 h. After this period, the percolation rate is determined by measuring the time it takes the water surface in the hole to lower to a predetermined level, or the depth that the water surface will fall over a specified period of time. From the results of the test, the allowable hydraulic loading rate is then determined by using some table prepared for the purpose (may be obtained from a county department of public works).

However, the method described above is not accurate. For one thing, soaking for 24 h before percolation, which would be true in actual operation, is far from a state of equilibrium conditions. Also, wastewaters tend to build up solids and mats on the surface of the trench, thus blocking infiltration; in the test, water is used. A better method would be to determine the equilibrium rate and to use this value for design. A study conducted on loam and sandy loam soils yielded an equilibrium value of 20 mm/day after about 253 days.[11] This study also indicated that for a given wastewater, equilibrium values are the same regardless of the initial permeabilities of the soil. Others[12] have also found the following "equilibrium" values: 0.3 in./h for water after 120 days and 0.3 ft/day for sewage after 70 days. In mm/day, these values are, respectively, 183 and 91. This shows a large discrepancy between the use of water and sewage.

[11]O. B. Kaplan (1991). *Septic Systems Handbook*. Lewis Publishers, Chelsea, Mich., p. 67.

[12]L. W. Canter and R. C. Knox (1985). *System Tank Effects on Ground Water Quality*. Lewis Publishers, Chelsea, Mich., p. 34.

The maximum limit of percolation is that of the hydraulic conductivity of the soil formation. From Darcy's law,

$$V = K \frac{\partial h}{\partial \ell} \tag{9-97}$$

Initially, the length of percolating water ℓ, called the *advancing front*, is zero, and the velocity of percolation V is infinite. After a time, ℓ increases and V decreases. After an infinitely long time, h and ℓ will become equal, making the derivative equal to 1; hence the velocity of percolation or percolation rate becomes equal to the hydraulic conductivity. Normally, a period of greater than 250 days may be considered as "the infinitely long time."

Design and Operating Parameters

The following criteria, which are found to work in practice, are critical to the successful design of land treatment systems. These criteria, among which are hydraulic and organic loading rates, should not be exceeded. Table 9-7 shows design and operating parameters of land treatment systems.

To determine the hydraulic loading rate that a system can accept, a water balance must be performed. The sum of the wastewater applied and the design precipitation must equal the sum of evaporation and percolation. The values of precipitation and evaporation to use must be obtained from a probability distribution analysis of the highest precipitaion and lowest evaporation. A wet year in a return interval of 10 years may be used. Probability distribution analysis was discussed in Chapter 3 and is not repeated here.

The organic loading consideration includes BOD as well as nutrient balance. The key principle in nutrient balance is that sufficient amounts of nutrients must be provided for the needs of the plant. As far as BOD loading is concerned, the key point is that critical loading rates must not be exceeded, to ensure effective operation of the system. In an actual design, an extensive search of the literature and/or a pilot study must be made to ascertain the appropriate loading rates to be used. An example problem on nutrient balance was given in Chapter 8. The same technique may be used.

Example 9-18

An investigation into the capability of a certain piece of land for land treatment yielded the results shown in the table on page 456. If a flow of 1.0 mgd with a BOD_5 of 40 mg/L is to be applied, what are the land requirements for (a) irrigation, (b) rapid infiltration, (c) overland flow, and (d) wetland methods of treatment?

Solution   (a) *Irrigation:* Assume that irrigation is possible only from May to October.

Total evapotranspiration and percolation

$$= (13.2 + 16.5 + 17.8 + 16.5 + 11.2 + 9.9) + (25.4)(6) = 237.5 \text{ cm} = 0.2375 \text{ m}$$

Total precipitation $= 1.0 + 0.5 + 0.3 + 0.5 + 1.5 = 3.8 \text{ cm} = 0.0038 \text{ m}$

TABLE 9-7 DESIGN AND OPERATING PARAMETERS OF LAND TREATMENT SYSTEMS

Parameter	Irrigation	Rapid infiltration	Overland flow	Wetlands	Septic tank-leach field
Application method	Sprinkler or surface	Surface	Sprinkler or surface	Surface	Subsurface
Minimum pretreatment	Sedimentation	Sedimentation	Grit removal	Sedimentation	Sedimentation
Disposition of wastewater	Evapotranspiration	Percolation	Surface runoff and evapotranspiration	Surface runoff and evapotranspiration	Percolation
Vegetation	Required	Optional	Required	Required	
Hydraulic loading (m^3/m^2-week)	0.02–0.1	0.1–2.5	0.1–0.4	1–30 m/yr	
Organic loading g (BOD_5/m^2-week)	1–20	45–400	15–60	50–80	
Effluent BOD_5 (g/m^3)	1–6	5–15	10–15	5–25	
Effluent SS (g/m^3)	1–6	1–5	10–25	5–30	

Month	Evapotranspiration	Percolation (cm)	Precipitation (cm)
January	1.8	25.4	5.8
February	3.8	25.4	5.8
March	7.9	25.4	5.3
April	9.9	25.4	4.1
May	13.2	25.4	1.0
June	16.5	25.4	0.5
July	17.8	25.4	0.3
August	16.5	25.4	Trace
September	11.2	25.4	0.5
October	9.9	25.4	1.5
November	3.8	25.4	2.6
December	2.0	25.4	5.6

Wastewater that can be applied $= 0.2375 - 0.0038 = 0.2337$ m in 6 months. This must be equal to the total wastewater in 1 year.

$$\text{Total wastewater flow} = \frac{1(10^6)}{7.48(3.281)^3}(365) = 1{,}381{,}569 \text{ m}^3/\text{yr, to be applied in 6 months}$$

Thus,

$$\text{area requirement} = \frac{1{,}381{,}569}{0.2337} = 5{,}911{,}720.10 \text{ m}^2 = 591 \text{ ha, based on hydraulics}$$

Total BOD_5 loading $= (0.040)(1{,}381{,}569) = 55{,}262.76$ kg/yr, to be applied in 6 months

From Table 9-7 use a BOD_5 loading rate of 10 g/m^2-wk $= (0.010)(4)(6) = 0.24$ kg/m^2 in 6 months. Then

$$\text{area requirement} = \frac{55{,}262.76}{0.24} = 230{,}261.5 \text{ m}^2 = 23.26 \text{ ha, based on organic loading}$$

Hence the recommendation would be based on hydraulics; area requirement = 591 ha. **Answer**

(b) *Rapid infiltration:* Sewage will be assumed applied year-round. Adding the yearly evapotranspiration, percolation, and precipitation, the following are obtained:

$$\text{Yearly evapotranspiration} = 114.3 \text{ cm} = 0.1143 \text{ m}$$

$$\text{Yearly percolation} = 304.8 \text{ cm} = 0.3048 \text{ m}$$

$$\text{Yearly precipitation} = 33.0 \text{ cm} = 0.033 \text{ m}$$

Thus the total sewage that can be applied $= (0.1143 + 0.3048) - 0.033 = 0.3861$ m/yr, too low for rapid infiltration. Compare with Table 9-7.

$$\text{Area requirement} = \frac{1{,}381{,}569}{0.3861} = 3{,}578{,}267.30 \text{ m}^2 = 357 \text{ ha, based on hydraulics}$$

From Table 9-7 use a BOD_5 loading rate of 300 g/m^2-week $= (0.30)(4)(12) = 14.4$ kg/m^2-yr.

$$\text{Area requirement} = \frac{55{,}262.76}{14.4} = 3837.69 \text{ m}^2 = 0.38 \text{ ha, based on organic loading}$$

Hence the recommendation would be based on hydraulics; area requirement = 357 ha. **Answer**

(c) *Overland flow:* Since overflow treatment depends on microbial activity, assume that operation is limited to the months of May to October as in the case of irrigation system. From hydraulic balance, the amount of wastewater that can be applied by normal irrigation is 0.2337 m in 6 months. This figure will be modified in overland flow design. From Table 9-7 use a hydraulic loading rate of 0.20 m^3/m^2-week = 0.20(4)(6) = 4.8 m in 6 months. Then

$$\text{area requirement} = \frac{1{,}381{,}569}{4.8} = 287{,}826.88 \text{ m}^2 = 28.78 \text{ ha, based on hydraulics}$$

From Table 9-7 use a BOD$_5$ loading rate of 40 g/m^2-week = (0.040)(4)(6) = 0.96 kg/m in 6 months. Then

$$\text{area requirement} = \frac{55{,}262.76}{0.96} = 57{,}563.38 \text{ m}^2 = 5.76 \text{ ha, based on organic loading}$$

Hence the recommendation would be based on hydraulics; area requirement = 28.78 ha. **Answer**

(d) *Wetland:* Sewage will be assumed applied year-round. If the land is to be converted into a wetland, assume that the data on percolation will no longer apply. Hence the total sewage that can be applied to satisfy hydraulic balance of evapotranspiration and precipitation only is 0.1143 − 0.033 = 0.0813 m/yr. From Table 9-7, use hydraulic loading rate of 20 m/yr. Then

$$\text{area requirement} = \frac{1{,}381{,}569}{20 + 0.0813} = 68{,}798.78 \text{ m}^2 = 6.88 \text{ ha, based on hydraulics}$$

From Table 9-7 use a BOD$_5$ loading rate of 70 g/m^2-week = (0.070)(4)(12) = 3.36 kg/m^2-yr. Then

$$\text{area requirement} = \frac{55{,}262.76}{3.36} = 16{,}447.25 \text{ m}^2 = 1.65 \text{ ha, based on organic loading}$$

Hence the recommendation would be based on hydraulics; area requirement = 6.88 ha. **Answer**

Example 9-19

Design an on-site septic tank system for a residential home in an unsewered area. Assume an occupancy of 8 persons with a sewage production of 380 L-capita/day. Also, assume an equilibrium percolation rate of 0.02 mm/day.

Solution

$$\text{Sewage production} = 0.38(8) = 3.04 \text{ m}^3/\text{day}$$

Assuming a detention time of 4 days, volume of tank = 12.16, say 12 m^3 **Answer**

$$\text{Trench area required} = \frac{3.04}{0.02} = 152 \text{ m}^2$$

Assuming a width of 1.0 m, trench length = 152/1(1) = 152 m. Use a dual system of trenches, as shown in Figure 9-7. Also, construct trenches on 2-m centers. **Answer**

GLOSSARY

Activated carbon. Carbon whose adsorbing surfaces are increased by a process called activation.

Active zone. That region of an adsorption bed that is actually absorbing at a given instant of time.

Adsorbate. The solute that is adsorbed onto the surface of a solid.

Adsorbent. The solid that adsorbs the adsorbate.

Adsorption. The concentration of solute on the surface of a solid due to the unequal attractive forces of lattice atoms at the surface compared to that at the interior of the solid.

Adsorption capacity. The ratio of the mass of the adsorbate to the mass of the adsorbent that produces the lowest possible concentration of the adsorbate in the equilibrium solution.

Adsorption isotherm. The equation relating the equilibrium concentration of the adsorbate to the ratio of the adsorbate to the adsorbent at a given temperature.

Advanced secondary treatment (AST). An advanced treatment producing 15 mg/L BOD_5 and 10 mg/L TKN.

Advanced tertiary treatment. Treatment above the secondary level.

Anion exchanger. Ion exchanger that exchanges anions between the exchanger bed and the water.

Anion membrane. Electrolytic membrane that allows only the passage of anions.

Cation exchanger. Ion exchanger that exchanges cations between the exchanger bed and the water.

Cation membrane. Electrolytic membrane that allows only the passage of cations.

Cell yield. Mass of organisms produced per unit mass of substrate utilized.

Chemical adsorption. Also called chemisorption, the type of adsorption that is due to a chemical reaction between the solute and the solid surface.

Concentrate. The high-concentration effluent that did not pass through the membrane in a reverse osmosis process.

Current density. The current in milliamperes that flows through a square centimeter of electrolytic membrane perpendicular to the current direction.

Denitrification. The process of converting nitrate to nitrogen gas.

Electrolytic membrane. Sheetlike barrier, made out of high-capacity, highly cross-linked ion-exchange resins that allow the passage of ions but not water under the influence of electricity.

Granular activated carbon (GAC). Activated carbon in granular form.

Host. The insoluble portion in an ion-exchange material.

Hydrogen cycle. The ion-exchange process that uses the hydrogen ion as the exchangeable cation.

Ion exchange. The displacement of one ion by another.

Ion exchanger. A unit that treats water by the exchange of ions.

Irrigation or slow-rate land treatment system. A method of waste treatment where wastewater is used to irrigate water-tolerant plants for forage and field crops, landscape vegetation, or serviculture.

Luxury uptake. The extraordinary large assimilation of nutrient after subjecting the organism to a deficiency of the nutrient.

Microfiltration. A reverse osmosis process operated at a pressure of less than 10 psig.

Nanofiltration (NF). A reverse osmosis process operated at a pressure of 75 to 250 psig.

Nitrification. The process of oxidizing nitrogen to nitrate.

Osmotic pressure. The pressure exerted by a solvent in passing through a semipermeable membrane toward the solute.

Overland flow. A method of waste treatment where wastewater is allowed to run off over the surface of the ground.

Permeate. The product that passes through the membrane in a reverse osmosis process.

Physical adsorption. Also called van der Waals adsorption, a type of adsorption that is due to the van der Waals force.

Polarization. A localized deficiency of ions on an electrolytic membrane surface due to a low current density/normality ratio.

Powdered activated carbon (PAC). Activated carbon in powdered form.

Rapid infiltration. A method of waste treatment where wastewater is applied to land at rates faster than irrigation.

Reverse osmosis. A process of treating water by applying pressure greater than the osmotic pressure of the solute on one side of a membrane.

Sodium cycle. The ion-exchange process that uses sodium as the exchangeable cation.

Synthetic resin. Polymers to which are added certain functional groups that make the polymer an ion exchange material.

Ultrafiltration (UF). A reverse osmosis process operated at a pressure of 15 to 75 psig.

Wetland treatment system. Natural or artificial wetlands used for treatment of wastewater.

Zeolite. Natural or synthetic aluminosilicates used as ion-exchange material.

SYMBOLS

c	solute collected per unit volume of permeate
$[C]$	equilibrium concentration of adsorbate in solution
$[C_n]$	concentration of wastewater treated at time n during breakthrough
$[C_{n+1}]$	concentration of wastewater treated at time, $n + 1$, during breakthrough
C_{pH}	pH correction factor for μ_{nmax}
f	factor to convert CBOD to BOD_5
F	flux in reverse osmosis process
G	moles of ammonia-free air
k_{dn}	decay constant for nitrifiers
K_{dn}	half-saturation constant for nitrogen in denitrification
K_n	half-saturation constant for nitrogen in nitrification
K_{O2}	half-saturation constant for oxygen

K_s	half-saturation constant of carbonaceous nutrient
L	moles of ammonia-free water
m	equivalent of NH_4-N incorporated to *Nitrosomonas* per equivalent of original NH_4-N
M	mass of adsorbent
n	equivalents of *Nitrobacter* synthesized per equivalent of original NH_4-N
p	equivalents of NO_3-N used in the anoxic reaction
P_f	pressure in feed side of RO
P_{fn}	net pressure in feed side of RO
P_p	pressure in permeate side of RO
P_{pn}	net pressure in permeate side of RO
q	equivalents of cells synthesized during last stage of aerobic reaction
r	equivalents of oxygen remaining after nitrification
R	percent solute rejection
R_m	membrane resistance
s	equivalent cells produced during regular anoxic reaction
s	coefficient of membrane compressibility
SEW	total moles of sewage requirements
t	equivalents of NO_2-N reduced during denitrification
u	equivalents of cells synthesized during NO_4-N reduction in denitrification
V_b	volume of wastewater treated at breakthrough
V_n	volume of wastewater treated at time n during breakthrough
V_{n+1}	volume of wastewater treated at time $n + 1$ during breakthrough
V_x	volume of wastewater treated at exhaustion of bed
X	mass of adsorbate; moles of ammonia per mole of ammonia-free water, L
Y	moles of ammonia per mole of ammonia-free air, G
Y_c	aerobic cell yield in moles per mole of sewage
Y_{dr}	nitrite reduction cell yield in moles per mole of sewage
Y_n	cell yield in denitrification in moles per mole of nitrate-N
$\overline{\alpha}_m$	specific membrane resistance
δ	length of active zone
θ_n	MCRT for nitrifiers
μ	permeate absolute viscosity
μ_{dn}	specific growth rate for denitrification
$\mu_{dn\text{max}}$	maximum μ_{dn}
μ_n	specific growth rate for nitrification
$\mu_{n\text{max}}$	maximum μ_n
π_f	osmotic pressure in feed side of RO
π_p	osmotic pressure in permeate side of RO

PROBLEMS

9-1. Solve this problem by using the Freundlich isotherm. A wastewater containing 25 mg/L of phenol is to be treated using PAC to produce an effluent concentration of 0.10 mg/L. The PAC is simply added to the stream and the mixture subsequently settled in the following sedimentation tank. The results of the test are shown below. The volume of waste subjected to each test is one liter. If the flow rate Q_0 is 0.11 m³/s, calculate the quantity of PAC needed for the operation. What is the adsorption capacity of the PAC?

Test	PAC added (g)	Equilibrium concentration of phenol (mg/L)
1	0.25	6.0
2	0.32	1.0
3	0.5	0.25
4	1.0	0.09
5	1.5	0.06
6	2.0	0.06
7	2.6	0.06

9-2. Solve this problem using the $(X/M)_{ult}$ valued obtained in Problem 9-20. A breakthrough experiment is conducted for phenol, producing the results shown below. Determine the length δ of the active zone. The diameter of the column used is 1 in., and the packed density of the bed is 45 lb/ft³. $[C_0]$ is equal to 25 mg/L.

C (mg/L)	V (L)
0.06	1.0
1.0	1.24
6.0	1.31
10	1.43
15	1.48
18	1.58
20	1.72
23	1.83
25	2.00

9-3. In Problem 9-1, calculate the quantity of PAC needed to treat the influent phenol to the ultimate residual concentration.

9-4. A waste containing 50 mg/L of phenol was subjected to an equilibrium test to determine the constants in the Langmuir isotherm. One-liter samples were dosed with powdered activated carbon in four jars, and when equilibrium was reached the equilibrium concentrations of phenol were analyzed. From the results shown below, determine the constants.

Jar	Carbon added (g)	Equilibrium concentration (mg/L)
1	0.52	6.2
2	0.62	1.1
3	1.2	0.30
4	2.3	0.79

9-5. In Problem 9-4, calculate the dosage required to produce an effluent concentration of 0.15 mg/L.

9-6. By performing a breakthrough experiment, δ was found to be equal to 10 cm. The X/M value of the bed, corresponding to 1.0 mg/L phenol, was found to be 0.061 kg solute per kilogram of carbon. The influent phenol concentration is 35 mg/L and is to be reduced to 1.0 mg/L. If the influent flow is 0.11 m^3/s and regeneration is to be done every week, design the carbon bed. Assume that the packed density of the bed is 721.58 kg/m^3.

9-7. Verify if the carbon adsorption data shown below follow the Langmuir and Freundlich isotherms.

C (mg/L)	X/M (g/g)
10	0.21
20	0.26
30	0.35

9-8. What kinds of pollutants are removed by activated carbon columns? How are they regenerated?

9-9. A wastewater containing 25 mg/L of phenol and having the characteristic breakthrough of Problem 9-2 is to be treated by adsorption onto an activated carbon bed. The X/M value of the bed for the desired effluent of 0.1 mg/L is 0.029 kg of solute per kilogram of carbon. If the flow rate is 0.11 m^3/s, design the absorption column. Assume that the influent is introduced at the middle of the column. The packed density of the carbon bed is 721.58 kg/m^3.

9-10. What percent of the total surface area of activated carbon is the internal surface?

9-11. A water softener tank is 2.0 m high and 0.5 m in diameter. The tank is filled with ion-exchange resin with an exchange capacity of 57 kg/m^3 as $CaCO_3$, allowing 0.5 ft of free-board. The occupants use 2500 L/day. If the influent water has a hardness of 280 mg/L as $CaCO_3$ and is to be softened to 80 mg/L, how much should be bypassed? What is the regeneration cycle? Assume that δ is 0.5 ft and that the effluent from the softener before mixing with the bypass flow is virtually free of hardness.

9-12. A sample of water has the following analysis: carbon dioxide = 0.41 mEg/L, total alkalinity = 2.0 mEg/L, total hardness = 4.3 mEg/L, and Mg hardness = 1.0 mEg/L. Calculate the amount of lime and soda ash required to soften the water.

9-13. A brackish water of 100,000 gal/day containing 4000 mg/L of ions expressed as NaCl is to be deionized using an electrodialysis unit. A membrane size of 18 by 20 in. is available. The

CD/N value to be used to avoid polarization is 700. Assuming that the coulomb and current efficiencies are, respectively, 69% and 90%, design the unit. Assume that the resistance through the unit is 6 Ω.

9-14. An electrodialysis unit consists of 300 membranes measuring 20 by 24 in. The ion concentration measured in terms of NaCl in 50,000 gal of brackish water to be demineralized is 3500 mg/L. The resistance through the unit is 6 Ω and the current efficiency is 90%. The CD/N value to be used to avoid polarization is 700. Estimate the removal efficiency and the power consumption.

9-15. A city of 400,000 population plans to use a brackish raw water supply containing 20,000 mg/L as sodium chloride at a rate of 150 gal/capita-day. If the proposed process will result in 30% recovery, what is the annual volume of waste brine? What is the weight of equivalent sodium chloride in this annual volume of waste brine?

9-16. A long-term experiment for a CA membrane module operated at 400 psi using a feed of 2000 mg/L of NaCl at 25°C produces the results below. Assuming that the membrane does not deteriorate, how long will it take to reduce the flux to zero?

Time (h)	1	10,000	25,000
Flux (gal/ft^2-day)	16	12	11

9-17. The feedwater to an RO unit contains 3000 mg/L of NaCl, 300 mg/L of CaCl$_2$, and 400 mg/L of MgSO$_4$. What is the flux rate if the pressure applied is 1400 psig? Assume that for the given concentrations the osmotic pressures are NaCl = 34.2 psi, CaCl$_2$ = 2.49 psi, and MgSO$_4$ = 1.44 psi. Also, assume that the temperature during the experiment is 25°C.

9-18. For the particular location of the city in Problem 9-15, the annual evaporation rate of the waste brine is 10 in. If solar evaporation is to be used to dispose of this liquid waste, what is the surface area of the evaporation pond required for this purpose?

9-19. Calculate the ion concentrations in the concentrate of a laboratory RO unit 60 in. in length and 12 in. in diameter. The RO has an active surface area of 1100 ft^2 and is used to treat a feedwater with the following composition: NaCl = 3000 mg/L, CaCl$_2$ = 300 mg/L, and MgSO$_4$ = 400 mg/L. The product flow is 15 gal/ft^2-day and contains 90 mg/L of NaCl, 6 mg/L of CaCl$_2$, and 8 mg/L of MgSO$_4$. The feedwater inflow is 27,500 g/day.

9-20. Determine the amount of air needed to strip completely a wastewater containing 25 mg/L of NH$_3$-N. The temperature of operation is 22°C and the flow is 0.3 mgd. Assume a multiplying factor of 1.5 for the theoretical G. Also, calculate the cross-sectional area of the tower. Assume a superficial velocity of 1 fps.

9-21. An RO system is designed to treat 120,000 gal/day of brackish water. Records in a nearby plant indicates that the lowest temperature is 65°F. The membrane to be used is cellulose acetate with an expected average flux of 15 gal/day-ft^2 over an estimated life of 2 years operating at 600 psig. If the average flux was determined at a base temperature of 80°F, determine the required membrane area for the operating temperature of 65°F.

9-22. A wastewater containing 60 mg/L of TKN is to be treated to a level of 20 mg/L. Each day, 66,170 mol of sewage are consumed. The influent contains 100 mg/L of alkalinity as CaCO$_3$. If the flow is 3.0 mgd, determine the amount of alkalinity needed to buffer the reactor to a pH of 7.0. Assume cell yields of 0.15 mg VSS/mg NH$_4$-N for *Nitrosomonas* and 0.02 mg VSS/mg NO$_2$-N for *Nitrobacter*.

9-23. The effluent requirement of an oxidation ditch system is 15 mg/L of BOD_5 and 15 mg/L of suspended solids. If the effluent suspended solids is 65% volatile, determine the soluble effluent BOD_5 concentration used for design.

9-24. If, after nitrification in Problem 9-22 and after the aeration has been cut off, the dissolved oxygen concentration remaining is 2.0 mg/L, what will be the resulting alkalinity after completion of this final stage of the aerobic reaction?

9-25. A total of 1 mgd of wastewater having a BOD_5 of 200 mg/L is to be treated. The state permitting agency requires that the effluent be treated to 10 mg/L. Kinetic studies yielded the following values: $Y = 0.65$ mg VSS/mg BOD_5 and $k_d = 0.55$ per day. $\theta_c = 6$ days. If the MLSS containing 70% volatile is 3000 mg/L, calculate the volume of the reactor.

9-26. Calculate the quantity of methanol and the corresponding BOD_5 to be provided to denitrify a domestic wastewater with a flow of 20,000 m^3/day. The effluent from the nitrification tank contains 30 mg/L of NO_3-N, 2.0 mg/L of dissolved oxygen, and 0.5 mg/L of NO_2-N. Assume that an effluent total nitrogen not to exceed 3.0 mg/L is required by the permitting agency.

9-27. A domestic sewage has the following characteristics: $BOD_5 = 200$ mg/L, suspended solids = 200 mg/L, TKN = 25 mg/L, alkalinity = 370 mg/L, and minimum temperature = 10°C. Design a reactor to treat 1.5 mgd of this waste. Assume the dissolved oxygen to be 2.0 mg/L, that the discharge permit requires the BOD_5 to be not more than 5.0 mg/L, and that the TKN is all used for synthesis.

9-28. An activated sludge process to nitrify an influent with the following characteristics to 2.0 mg/L ammonia is to be designed: $Q_0 = 1$ mgd, $BOD_5 = 200$ mg/L, TKN = 60 mg/L, and minimum operating temperature = 20°C. The effluent is to have a BOD_5 of 5.0 mg/L. The DO is to be maintained at 2.0 mg/L. Apply an SF of 2.0 and assume $K_{O2} = 1.3$ mg/L. Calculate the MCRT needed to nitrify to the limiting concentration of ammonia.

9-29. The effluent from a reactor nitrifying 20,000 m^3/day of domestic wastewater contains 30 mg/L of NO_3-N, 2.0 mg/L of dissolved oxygen, and 0.5 mg/L of NO_2-N. (**a**) Assuming an effluent total nitrogen not to exceed 2.0 mg/L and an effluent BOD_5 not to exceed 4 mg/L are required by the permitting agency, calculate the reactor volume required for denitrification. (**b**) How long does it take to attain the limiting nitrate ion concentration?

9-30. If in Problem 9-29 the reactor is an oxidation ditch, design the anoxic section to reduce the NO_3-N concentration to 5 mg/L. Assume that all influent TKN is converted to NO_3-N.

9-31. The total phosphorus content of a domestic wastewater with a flow of 20,000 m^3/day is 7 mg/L. (**a**) If 6000 lb/day of sludge solids containing 75% VSS is wasted, calculate the concentration of total phosphorus in the effluent. Assume that the sludge contains 2.0% solids and water density is 997 kg/m^3. (**b**) If the process train takes advantage of the luxury uptake (1.0 mg/L of total P in effluent), how much total P is wasted in the sludge?

9-32. A wastewater of 12,000 m^3/day contains 10 mg/L of ammonia nitrogen and no organic nitrogen. Estimate the methanol requirement and cell production for complete assimilation of the ammonia.

9-33. A nitrified effluent contains 15 mg/L of nitrate nitrogen, 1.6 mg/L of nitrite nitrogen, and 2.0 mg/L of dissolved oxygen. Compute the methanol requirement for denitrification. How will the effluent BOD_5 affect the methanol requirement?

9-34. How much iron salt as a coagulant is needed to reduce the total P of Problem 9-31 to 2.0 mg/L?

9-35. The effluent limit for total phosphorus has been set at 1.0 mg/L. If the soluble phosphorus in the effluent is 10 mg/L, calculate the alum requirement that will be required to meet the effluent limit. What is the volume of sludge to be disposed of per day if the percent solids of the alum sludge is 6%, the specific gravity is 1.05, and plant flow rate is 50,000 m^3/day?

9-36. A treated wastewater contains 25 mg/L of ammonia. This ammonia is to be stripped using 5550 m^3 of air per cubic meter of wastewater. If the temperature of operation is 17°C, evaluate the efficiency of the operation in terms of the air/liquid ratio.

9-37. Nitrate ions are to be removed using an anionic resin ion exchanger. The exchange capacity of the resin is 55 kg expressed as $CaCO_3$ per cubic meter. Determine the design volume of resin to be used to treat 1500 m^3/day of wastewater containing 25 mg/L of nitrate nitrogen.

9-38. A 20,000-gal/day sonozone plant is treating domestic waste containing 9.19 mol/L of BOD_5. How many kWh of electricity is required by the ozone generator to destroy the BOD completely? Assume 20 kWh/kg of ozone.

9-39. As a result of the following study, what treatment will you recommend? The investigation is about the capability of a certain piece of land for land treatment. The flow is 1.0 mgd with a BOD_5 of 40 mg/L.

Month	Evapotranspiration	Percolation (cm)	Precipitation (cm)
January	1.8	25.4	5.8
February	3.8	25.4	5.8
March	7.9	25.4	5.3
April	9.9	25.4	4.1
May	13.2	25.4	1.0
June	16.5	25.4	0.5
July	17.8	25.4	0.3
August	16.5	25.4	Trace
September	11.2	25.4	0.5
October	9.9	25.4	1.5
November	3.8	25.4	2.6
December	2.0	25.4	5.6

9-40. Assuming a four-person occupancy, design an on-site septic tank system for a wealthy residential home in an unsewered area. The sewage production is 380 L-capita/day. Assume an equilibrium percolation rate of 0.02 mm/day.

9-41. An irrigation system is used to accept 35 L/s of effluent for an application rate of 6 cm/week. For a year-round operation, calculate the field area. If the system is designed for only a 36-week/yr operation, calculate the field area.

9-42. A sprinkler system is used for irrigation using wastewater. The sprinkler is arranged in a rectangular grid of 15 m by 20 m. If each nozzle discharges 2 L/s, what is the application rate in cm/h? How many hours must the system be operated in a single area of application to satisfy the requirement of 6 cm/week?

BIBLIOGRAPHY

CANTER, L. W., and R. C. KNOX (1985). *System Tank Effects on Ground Water Quality*. Lewis Publishers, Chelsea, Mich.

JACANGELO, J. C. (1989). "Membranes in Water Filtration." *Civil Engineering,* 59(5), pp. 68–71.

KAPLAN, O. B. (1991). *Septic Systems Handbook*. Lewis Publishers, Chelsea, Mich.

MANDT, M. G., and B. A. BELL (1982). *Oxidation Ditches in Wastewater Treatment*. Ann Arbor Science, Ann Arbor, Mich.

McGHEE, T. J. (1991). *Water Supply and Sewerage*. McGraw-Hill, New York.

METCALF & EDDY, INC. (1991). *Wastewater Engineering: Treatment, Disposal, Reuse*. McGraw-Hill, New York.

RAMALHO, R. S. (1977). *Introduction to Wastewater Treatment Processes*. Academic Press, New York.

SINCERO, A. P. (1989). *Reverse Osmosis Removal of Organic Compounds: A Preliminary Review of Literature*. Contract DAALO#-86-D-0001. U.S. Army Biomedical Research and Development Laboratory, Fort Detrick, Frederick, Md.

TCHOBANOGLOUS, G., and E. D. SCHROEDER (1985). *Water Quality*. Addison-Wesley, Reading, Mass.

CHAPTER 10

Pollution from Combustion and Atmospheric Pollution

The term *combustion* refers to any chemical reaction with oxygen that generates energy. The energy may manifest in the form of heat or light or as energy stored in the bonds of ATP. There are two types of combustion: rapid and slow or biological combustion. *Rapid* combustion releases the energy in the form of intense heat and light; *slow* or *biological* combustion releases the energy in the form stored in ATP. In the process of metabolism, organisms mediate biological combustion. The role of rapid combustion in producing pollutants is discussed in this chapter. In addition, automotive emission control, atmospheric pollution resulting from photochemical smog, acid rain, the destruction of the ozone layer, and the greenhouse effect are also discussed.

RAPID COMBUSTION

Rapid combustions that generate pollutants are of interest to the environmental engineer. These pollutants include CO, CO_2, SO_x, and NO_x.

Products of combustion of any fuel reflect its constituents and the degree of completion of the combustion. The incomplete combustion of a hydrocarbon fuel produces carbon monoxide and water. C_xH_y may represent any hydrocarbon and the following equation may represent its incomplete combustion:

$$C_xH_y + \frac{2x + y}{4}O_2 \rightarrow xCO + \frac{y}{2}H_2O \tag{10-1}$$

On the other hand, the complete combustion of hydrocarbons produces CO_2, the desirable aspect of rapid combustion, although CO_2 is also an agent of global warming. The following equation shows the reaction:

$$C_xH_y + \frac{4x+y}{4}O_2 \rightarrow xCO_2 + \frac{y}{2}H_2O \qquad (10\text{-}2)$$

Equation (10-1) not only represents production of the pollutant CO but also a loss of energy that the molecule of CO still contains. From a practical standpoint, discounting the disastrous effect of CO_2 on global warming, equation (10-2) is the desirable reaction. This reaction liberates all the heats of reaction of the hydrocarbon.

The study of combustion frequently uses the terms *percent excess air, stoichiometric ratio*, and *equivalence ratio. Percent excess air* is the difference between the actual mass of air used and the theoretical divided by the theoretical times 100. The *stoichiometric ratio* is the ratio of the theoretical fuel/air mass ratio to the actual fuel/air mass ratio. The *equivalence ratio* is the reciprocal of the stoichiometric ratio. Air is 21% O_2 and 79% N_2.

Example 10-1

A solid-waste-to-energy recovery plant fires solid waste represented by $C_{50}H_{100}O_{40}N$. The plant burns 100 tonnes/day of solid waste, producing 20 tonnes/day of residue containing 5% C. (a) If the solid waste burns under conditions of insufficient air, how much CO is produced per day? (b) If the solid waste burns at 20% excess, how much CO_2 is produced per day? (c) What are the stoichiometric and equivalence ratios? Assume that fuel nitrogen converts to NO.

Solution   (a) Since the ratio of H to O in H_2O is 2:1, the H in H_2O of the solid waste is $2(40) = 80$, from the formula of $C_{50}H_{100}O_{40}N$. Therefore, the H of the hydrocarbon portion of the solid waste is $100 - 80 = 20$. Hence, $C_xH_y = C_{50}H_{20}$.

$$C_{50}H_{20} + 30O_2 \rightarrow 50CO + 10H_2O$$

The oxidation state of O has been changed from 0 to -2, involving a total of $30(2)(2) = 120$ mol of electrons. Hence the number of reference species $= 120$ mol of electrons.

$$\text{Weight fraction of C in solid waste} = \frac{50C}{C_{50}H_{100}O_{40}N}$$

$$= \frac{50(12)}{50(12) + 100(1.008) + 40(16) + 14}$$

$$= \frac{600}{1354.8} = 0.44$$

Carbon burned to gases $= 100(0.44) - 20(0.05) = 43$ tonnes/day (i.e., the carbon burned to gases in the C_{50} of $C_{50}H_{20}$).

$$\text{Eq. wt. of } C_{50} \text{ in } C_{50}H_{20} = \frac{C_{50}}{120} = \frac{50(12)}{120} = 5.0\frac{\text{tonnes}}{\text{tonne-Eq}}$$

Thus,

$$\text{tonnes of CO/day} = \frac{43}{5.0}\left[\frac{50(CO)}{120}\right] = \frac{43}{5.0}\left[\frac{50(12+16)}{120}\right] = 100.3 \quad \textbf{Answer}$$

(b)
$$C_{50}H_{20} + 55O_2 \rightarrow 50CO_2 + 10H_2O$$

$$\text{References species} = 55(2)(2) = 220 \text{ mol of electrons}$$

$$\text{Eq. wt. of } C_{50} \text{ in } C_{50}H_{20} = \frac{C_{50}}{220} = \frac{50(12)}{220} = 2.73\frac{\text{tonnes}}{\text{tonne-Eq}}$$

$$\text{Tonnes of CO}_2\text{/day} = \frac{43}{2.73}\left[\frac{50(CO_2)}{220}\right] = \frac{43}{2.73}\left[\frac{50(12+32)}{220}\right] = 157.51 \quad \textbf{Answer}$$

(c)
$$C_{50}H_{100}O_{40}N + \frac{111}{2}(=55.5)O_2 \rightarrow 50CO_2 + 50H_2O + NO$$

$$\text{Stoichiometric fuel/air ratio} = \frac{C_{50}H_{100}O_{40}N}{55.5O_2 + (79/21)(55.5)N_2} = \frac{1354.8}{55.5(32) + (79/21)(55.5)(28)}$$

$$= \frac{1354.8}{7622} = 0.18$$

$$\text{Actual fuel/air ratio} = \frac{1354.8}{7622(1.2)} = 0.15$$

$$\text{Stoichiometric ratio} = \frac{0.18}{0.15} = 1.2 \quad \textbf{Answer}$$

$$\text{Equivalence ratio} = \frac{1}{\text{stoichiometric ratio}} = \frac{1}{1.2} = 0.83 \quad \textbf{Answer}$$

C_xH_yS or, more generally, R–S may represent fuels containing sulfur, where R stands for any organic group that may contain not only C and H but also other elements as well. Considering fuel as simply C_xH_yS, the following equation may represent the complete combustion of fuels:

$$C_xH_yS + \frac{4x+y+4}{2}O_2 \rightarrow xCO_2 + \frac{y}{2}H_2O + SO_2 \tag{10-3}$$

The other sulfur combustion product produced along with SO_2 is a small amount of SO_3, sulfur trioxide. In addition, SO_2, when released to the atmosphere, converts to SO_3 as follows:

$$SO_2 + OH\cdot \rightarrow HOSO_2\cdot \tag{10-4}$$

$$HOSO_2\cdot + O_2 \rightarrow SO_3 + HO_2\cdot \tag{10-5}$$

A group of atoms that retains its identity in the course of a reaction is called a *radical*. When an electron bond is broken during the formation of the radical, an electron becomes unpaired. When the radical formed contains one or more of these unpaired electrons or when an atomic species is formed containing one or more of these unpaired electrons, the radical or the atomic species is called a *free radical*. When the radical or the atomic species contains only one unpaired electron, it is a *monoradical*; when it contains two, it is a *biradical*; and so on.

The $OH\cdot$, $HOSO_2\cdot$, and $HO_2\cdot$ are species containing unpaired electrons; they are therefore free radicals. The unpaired electron is designated by a center dot ($\cdot$). In this designation no attempt is made to identify the atom that contains the unpaired electron, nor is it attempted to identify the number of unpaired electrons. The dot simply signifies that the species is a free radical (whether mono, bi, etc., is immaterial). The presence of these unpaired electrons makes the free radicals very reactive.

The trioxide in the presence of H_2O produces sulfuric acid, an important agent in the production of acid rain. The following equation shows the acid production:

$$SO_3 + H_2O \rightarrow H_2SO_4 \tag{10-6}$$

SO_2 also reacts with H_2O to produce sulfurous acid,

$$SO_2 + H_2O \rightarrow H_2SO_3 \tag{10-7}$$

Reduction with hydrogen may desulfurize fuels as indicated in the following equation:

$$R\text{--}S + H_2 \rightarrow H_2S + R$$

Example 10-2

A power plant burns 6.0 tonnes of coal per hour. The coal contains 4.0% sulfur. Assuming that all the sulfur converts to SO_3 and that enough moisture is present in the atmosphere to react with the trioxide, calculate the amount of acid produced per hour.

Solution

$$S + O_2 \rightarrow SO_2$$

Reference species = 4 mol of electrons (deduced from reduction of O_2)

$$\text{Eq. S} = \frac{6(0.04)}{S/4} = 0.03 \text{ tonne-Eq/h} = \text{eq. } SO_2$$

$$\text{Mass } SO_2 \text{ produced per hour} = 0.03 \left(\frac{SO_2}{4} \right) = 0.03 \left(\frac{32 + 32}{4} \right) = 0.48 \text{ tonne}$$

$$SO_2 + OH\cdot \rightarrow HOSO_2\cdot \tag{a}$$

$$HOSO_2\cdot + O_2 \rightarrow SO_3 + HO_2\cdot \tag{b}$$

$$SO_3 + H_2O \rightarrow H_2SO_4 \tag{c}$$

From equations (a) and (b), SO_2 and SO_3 interact through 1 mol of O_2. Hence reference species are the four mol of electrons involved in the reduction of O_2.

$$\text{Eq. wt. of } SO_3 = \frac{SO_3}{4}$$

$$\text{Mass } SO_3 \text{ produced per hour} = 0.03 \left(\frac{SO_3}{4} \right) = 0.03 \left(\frac{32 + 48}{4} \right) = 0.6 \text{ tonne}$$

Since equation (c) is not an oxidation-reduction reaction, equivalent weights will be based on the number of moles of positive or negative charges. For SO_3, S has an oxidation state of $+6$. So, on this basis, the number of reference species equals 6 mol of positive charges. Also on the same molecule, O has a total oxidation state of -6. Therefore, on this basis, the number of reference species equals 6 mol of negative charges. Considering H_2O, O has a total oxidation state of -2; hence the number of reference species equals 2 mol of negative charges. For the case of hydrogen in H_2O, the number of reference species equals 2 mol of positive charges. By convention, adopt the smaller and the number of reference species $= 2$ mol of positive or negative charges. From equation (c),

$$\text{amount of acid mist} = \frac{0.6}{SO_3/2} \left(\frac{H_2SO_4}{2} \right)$$

$$= \frac{0.6}{32 + 48}[2(1) + 32 + 4(16)] = 0.735 \text{ tonne/h} \quad \textbf{Answer}$$

Using the balanced chemical reactions, this may also be solved simply as follows:

$$\text{amount of acid} = \frac{H_2SO_4}{SO_2} \left(\frac{SO_2}{S} \right) [0.04(6)] = 0.735 \text{ tonne/h} \quad \textbf{Answer}$$

The oxides of sulfur are collectively called SO_x. There are six gaseous oxides of S: sulfur monoxide (SO), sulfur dioxide (SO_2), sulfur trioxide (SO_3), sulfur tetroxide (SO_4), sulfur sesquioxide (S_2O_3), and sulfur heptoxide (S_2O_7). SO_x is mainly SO_2 and SO_3.

Rapid combustion also produces oxides of nitrogen, collectively called NO_x. There are also six known gaseous oxides of nitrogen: nitric oxide (NO), nitrogen dioxide (NO_2), nitrous oxide (N_2O), nitrogen sesquioxide (N_2O_3), nitrogen tetroxide (N_2O_4), and nitrogen pentoxide (N_2O_5). The oxides of primary importance in environmental engineering are NO and NO_2.

There are two types of NO_x: thermal and fuel NO_x. Air is 79 mol % N_2. In combustion, N_2 oxidizes to NO_x. On the other hand, the fuel itself may contain nitrogen. The nitrogen in the fuel when burnt also converts to NO_x. The NO_x formed from the combustion of atmospheric nitrogen is called *thermal* NO_x; the NO_x formed from the combustion of fuel nitrogen is called *fuel* NO_x.

The formation of free radicals is important in combustion. In 1946, Zeldovich proposed the free-radical chain formation of thermal NO in air at high combustion temperatures. The representation of the reaction chain follows:

$$O_2 + M \rightleftharpoons 2O\cdot + M \qquad (10\text{-}8)$$

$$O\cdot + N_2 \rightleftharpoons NO\cdot + N\cdot \qquad (10\text{-}9)$$

$$N\cdot + O_2 \rightleftharpoons NO\cdot + O\cdot \qquad (10\text{-}10)$$

In equation (10-8), oxygen, under the influence of a third body, M, dissociates into the free radical, $O\cdot$. This radical is then available to react with molecular N_2 in accordance with equation (10-9), producing thermal $NO\cdot$ and another free radical, $N\cdot$. This newly formed $N\cdot$ reacts further with molecular O_2, producing another thermal $NO\cdot$ as shown in equation (10-10).

Notice that the bullet designations for free radicals are not always used. Thus NO is NO·. Other free radicals that can use the bullet designation are $NO_2·$ and $ClO_2·$. These bullets may be removed with the understanding that the species are free radicals. Thus NO· is NO; $NO_2·$ is NO_2; $ClO_2·$ is ClO_2; N· is N.

Another possible species formed during combustion is the NO_2. Its concentration relative to NO is, however, small. The equilibrium constants of these two species will easily confirm this fact, as will be shown shortly.

Adding reactions (10-9) and (10-10) produces the following overall reaction:

$$N_2 + O_2 \rightleftharpoons 2NO \tag{10-11}$$

The equilibrium expression for this reaction is

$$K_{NO} = \frac{y_{NO}}{(y_{N_2})^{1/2}(y_{O_2})^{1/2}} = 4.69 \exp\left(\frac{-21,600}{RT}\right) \tag{10-12}$$

where R is the universal gas constant, equal to 1.987 cal/mol-K and T is the temperature in kelvin. The y's are the mole fractions of the species.

As shown in equation (10-12), the equilibrium constant is a function of temperature. The equation also shows a very high energy of activation (21,600 cal/mol). Energy of activation is the amount of energy that must be put into the reaction before the reaction can start. It should be noted that the exponential function, e (or exp, the base of natural logarithms), is large when its exponent is negatively small. In equation (10-12) the exponent will be negatively small when T is large. Since the activation energy is large, the value of T must be very large to make the exponent negatively small. At large temperatures, the equilibrium constant, then, is large. The expression for the equilibrium constant places y_{NO} at the numerator, indicating that it would be high for large equilibrium constants. Therefore, it is concluded that high combustion temperatures produce large concentrations of thermal NO. (Flame-zone temperatures range from 1600 to 1900°C.)[1]

Now, examine the case of thermal NO_2. NO may react with molecular O_2, producing NO_2 as shown in the equation

$$NO + \frac{1}{2}O_2 \rightleftharpoons NO_2 \tag{10-13}$$

The expression for the equilibrium constant is

$$K_{NO_2} = \frac{y_{NO_2}}{y_{NO}(y_{O_2})^{1/2}} = 2.5 \times 10^{-4} \exp(13,720/RT) \tag{10-14}$$

All symbols are similar to those defined before.

In equation (10-14) the sign of the exponent of e (or exp) is positive. This means that at high temperatures, K_{NO_2} is small. Since the equilibrium expression locates y_{NO_2} in the numerator, at high temperatures the concentration of NO_2 is also small. The coefficient, 2.5×10^{-4}, is already very small. This, coupled with a high temperature inside the exponential function, will produce a negligibly small value of the equilibrium constant,

---

[1] C. D. Cooper and F. C. Alley (1986). *Air Pollution Control: A Design Approach*. Waveland Press, Prospect Heights, Ill.

K_{NO_2}. Conclude, then, that at combustion temperatures, thermal NO_2 is negligibly small and that the major NO_x formed during combustion is NO.

Nitrogen bound in the fuel also converts to NO_x. Fuel-bound nitrogen can account for over 50% of the total NO_x produced in combustion. Not all of the fuel-bound nitrogen, however, converts to NO_x. Depending on the fuel equivalence ratio, percent conversion could range from 5% for equivalence ratios greater than 2.5 to approximately 60% for an equivalence ratio of 0.2. The other possible end product of fuel nitrogen combustion is molecular N_2.

Basic Elements Affecting Rapid Combustion

There are four basic elements that must be set to the optimum for an efficient combustion to occur: oxygen concentration, turbulence, time, and temperature. There should be sufficient oxygen to combust carbon to its highest oxidation state of CO_2. More oxygen, however, will burn nitrogen and sulfur to their oxides, which is undesirable in the first place. To determine up to what degree air should be added would therefore need a careful balance among production of CO_2, SO_x, and NO_x. Complete mixing of oxygen and fuel to ensure efficient combustion requires turbulence. Turbulence causes rapid contact between oxygen and fuel, thus ensuring efficient combustion. Combustion should also provide enough residence time to complete the reaction, although as in the case of oxygen, more time would also produce large quantities of NO_x and SO_x.

The fourth element, temperature, is rather unique. Before it can have a pronounced effect, the energy of activation must first reach some high threshold value. Take the case of CO that results from the incomplete combustion of C. The equilibrium expression for CO at 1 atm pressure is [2]

$$CO_2 \rightleftharpoons CO + \tfrac{1}{2}O_2 \tag{10-15}$$

The equilibrium expression is

$$K_{CO_2} = \frac{y_{CO}(y_{O_2})^{1/2}}{y_{CO_2}} = 3 \times 10^4 e^{-67,000/RT} \tag{10-16}$$

Again, R is the universal gas constant (equals 1.987 cal/mol-K), and T is the temperature in kelvin. Because of the high activation energy, equation (10-16) states that a certain high temperature must be reached before obtaining CO. Deducing from this equation, high temperatures favor the production of CO.

Kinetics of NO Formation and Ideal Combustion Temperature

The equilibrium expressions for NO and NO_2 above do not really accurately predict the combustion concentrations of NO. The mixture of fuel and air resides at the combustion zone for a mere split second (on the order of milliseconds in the combustion cylinder of

[2] J. H. Seinfeld (1986). *Atmospheric Chemistry and Physics of Air Pollution.* Wiley, New York.

internal combustion engines and about 0.5 s in normal furnaces), giving the species no time to attain equilibrium. For this reason, equilibrium expressions are not sufficient to predict the combustion concentrations of thermal NO_x at the flame zone.

Kinetic theory has been used to explain the discrepancy when using equilibrium to predict thermal NO_x concentrations. According to Ammann and Timmins,[3] the forward reaction velocity coefficient for reaction (10-9), which is responsible for the production of NO, is

$$k_{f9} = 1.16 \times 10^{-10} \exp(-75,500/RT) \; \frac{cm^3}{molecule\text{-}s} \tag{10-17}$$

and the corresponding backward reaction velocity coefficient is

$$k_{b9} = 2.57 \times 10^{-11} \; \frac{cm^3}{molecule\text{-}s} \tag{10-18}$$

The corresponding forward and backward reaction velocity coefficients for reaction (10-10) are, respectively,

$$k_{f10} = 2.21 \times 10^{-14}(T) \exp(-7080/RT) \; \frac{cm^3}{molecule\text{-}s} \tag{10-19}$$

$$k_{b10} = 5.3 \times 10^{-15}(T) \exp(-39,100/RT) \; \frac{cm^3}{molecule\text{-}s} \tag{10-20}$$

Again, as in the equilibrium constant, before a reaction can take place, the temperature must reach a threshold value for those expressions containing the activity coefficient. Once reached, reactions (10-9) and (10-10) can proceed to produce NO. However, as noted before, the residence time in the combustion zone is only on the order of seconds. In the immediate surroundings of the zone of combustion are cooler temperatures. This means that the very short residence time subjects the reaction species immediately to decreasing temperatures as soon as they leave the combustion zone (i.e., to a *quenching process*).

Referring to reaction (10-9) and the reaction velocity coefficients given above, as the quenching process proceeds, the forward reaction progressively decreases until it stops when the temperature reaches the threshold corresponding to the activation energy. The backward reaction, however, since it is independent of temperature, continues to operate. This means that the NO reserve concentration formed during the combustion zone reaction decreases. Reaction (10-9) produces NO only when the temperature is high (at the combustion zone) but uses NO at a point in time during the quenching process.

Meanwhile, as the NO concentration decreases, some of the free-radical N reacts with O_2 according to reaction (10-10). Since the forward activation energy of this reaction is very low (7080 cal/mol), its forward reaction to produce NO continues even down to a relatively low temperature value. On the other hand, considering the backward reaction, the activation energy is very large. This means that the backward reaction speed is high

[3]P. R. Ammann and R. S. Timmins (1966). "Chemical Reactions during Thermal Quenching of Oxygen–Nitrogen Mixtures from Very High Temperatures." *AlChE Journal*, 12, pp. 956–963.

only at elevated temperatures, and practically zero at low temperatures when the forward reaction still continues to be effective. Hence the backward reaction does not impede the production of NO during the quenching process. To summarize, at low temperatures, the forward reaction of reaction (10-10) continues to produce NO while its backward reaction practically stops. Hence, in the quenching process, as the species depart from the combustion zone, reaction (10-10) continuously produces NO while reaction (10-9) depletes it. Overall, at the combustion exit, a substantial amount of the NO originally produced at the combustion zone remains.

The qualitative description above of the rate of thermal NO formation using Zeldovich's equations may be written mathematically as

$$\frac{d\{NO\}}{dt} = R_{f9} + R_{f10} - R_{b9} - R_{b10} \tag{10-21}$$

where R_{f9}, R_{b9}, R_{f10}, and R_{b10} are the forward and backward reactions of equations (10-9) and (10-10), respectively. Subscripts f and b refer to "forward" and "backward," respectively, and t is time. In words, equation (10-21) states: The rate of formation of NO is equal to the forward reaction rates of equations (10-9) and (10-10) minus their respective backward reaction rates.

The rate of formation of N is

$$\frac{d\{N\}}{dt} = R_{f9} - R_{b9} + R_{b10} - R_{f10} \tag{10-22}$$

The forward reaction velocity of equation (10-9), which contains a high activation energy and is a function of temperature, restricts R_{f9} to satisfying both the activation energy requirement and the attainment of the proper threshold temperature. On the other hand, the backward reaction velocity coefficient is neither a function of temperature nor does it contain an activation energy. This means that R_{b9} operates at every instant. Furthermore, equation (10-10) consumes N as soon as equation (10-9) produces it.

Overall, as soon as the temperature attains the proper value, R_{b9} and R_{f10} immediately consume whatever N R_{f9} produces. N attains a pseudo-steady-state condition. Hence, equating equation (10-22) to zero and substituting the result in equation (10-21) produces

$$\frac{d\{NO\}}{dt} = 2(R_{f10} - R_{b10}) \tag{10-23}$$

The concentration of O is a function of equations (10-8) to (10-10). Although O_2 broke up because of the influence of the high temperature in the combustion flame, its tendency to recombine is stronger than its tendency to react with molecular N_2. Atmospheric N_2 has a high bond energy (225,000 cal/mol) that O would prefer to recombine than to break the N–N bond to form NO. Hence although O is a function of the three equations, assume that only the equilibrium of equation (10-8) controls its concentration. The forward and backward reaction velocity coefficients for this equation are, respectively,[4] $k_{f8} = 1.876(10^{-6})T^{-1/2} \exp(-18,000/RT)$ and $k_{b8} = 2.6(10^{-33})$, with units of

---

[4] Seinfield, *Atmospheric Chemistry and Physics of Air Pollution.*

cm^3/molecule-s. Proceeding with the assumption, we have

$$\frac{d\{O\}}{dt} = 2R_{f8} - 2R_{b8} = 0 \tag{10-24}$$

Substituting pertinent expressions for the R's and simplifying yields

$$\{O\} = \sqrt{\frac{k_{f96}}{k_{b96}}} \sqrt{\{O_2\}}$$

$$= \sqrt{K} \sqrt{\{O_2\}} \tag{10-25}$$

where

$$K = \frac{1.876(10^{-6})T^{-1/2}\exp(-118,000/RT)}{2.6(10^{-33})}$$

$$= 7.2(10^{26})T^{-1/2}\exp(-118,000/RT) \tag{10-26}$$

Also, from equating $d\{N\}/dt$ to zero,

$$R_{f9} - R_{b9} = -R_{b10} + R_{f10} \tag{10-27}$$

Substituting the respective reaction rates in equation (10-27) and simplifying for N, we have

$$\{N\} = \frac{\sqrt{K}\sqrt{\{O_2\}}(k_{f9}\{N_2\} + k_{b10}\{NO\})}{k_{f10}\{O_2\} + k_{b9}\{NO\}} \tag{10-28}$$

Combining equations (10-23), (10-9), and (10-28), we obtain

$$\frac{d\{NO\}}{dt} = \frac{2k_{f98}k_{f97}\sqrt{K}(\sqrt{\{O_2\}})^3\{N_2\} - 2k_{b98}k_{b97}\sqrt{K}\sqrt{\{O_2\}}\{NO\}^2}{k_{f98}\{O_2\} + k_{b97}\{NO\}} \tag{10-29}$$

The first term in the numerator on the right-hand side of equation (10-29) represents a rate of NO formation, while the second term represents a rate of NO destruction. If the rate of formation balances against the rate of destruction, there will be no increase in NO concentration above that produced from the fuel nitrogen. Since this represents an ideal situation, it is instructive to find the corresponding temperature. The temperature at which this occurs can be found by equating the numerator to zero. Hence

$$k_{f98}k_{f97}\sqrt{K}(\sqrt{\{O_2\}})^3\{N_2\} - k_{b98}k_{b97}\sqrt{K}\sqrt{\{O_2\}}\{NO\}^2 = 0 \tag{10-30}$$

Substituting the expressions for the coefficients and solving for T produces the equation below. Call this temperature the *ideal temperature of combustion*.

$$T_{\text{ideal}} = \frac{(1/R)(43,480)}{2.93 - \ln(\{NO\}^2/\{O_2\}\{N_2\})} \tag{10-31}$$

NO_x Equilibrium Away from Combustion Zone

No matter what happened in the combustion zone, the laws of equilibrium eventually control the characteristics of the combustion gases away from the flames. Now summarize the equilibrium expressions for the phenomenon of NO_x formation that takes place outside the flame of combustion. The reaction for NO formation is

$$N_2 + O_2 \rightleftharpoons NO_2 \tag{10-32}$$

The equilibrium expression for this reaction was given before. The reaction for the formation of NO_2 from the NO and O_2 is

$$NO + \tfrac{1}{2}O_2 \rightleftharpoons NO_2 \tag{10-33}$$

Also, the equilibrium expression for this reaction was given before. Adding these two equations produces

$$N_2 + 2O_2 \rightleftharpoons 2NO_2 \tag{10-34}$$

Adding the two equations is equivalent to multiplying their respective equilibrium constants. This produces the overall constant

$$K_{NO_2(overall)} = \frac{y_{NO_2}}{(y_{N_2})^{1/2} y_{O_2}} = 11.73(10^{-4}) \exp(-7880/RT) \tag{10-35}$$

Equation (10-35) represents the equilibrium for NO_2. For completeness, reproduce the expression for the equilibrium constant for the formation of NO,

$$K_{NO} = \frac{y_{NO}}{(y_{N_2})^{1/2}(y_{O_2})^{1/2}} = 4.69 \exp(-21,600/RT) \tag{10-36}$$

For the combustion gases away from the combustion zone, equations (10-35) and (10-36) apply.

Example 10-3

Analysis of flue gas at a temperature of 200°C and 1 atm pressure found 75% N_2, 4% O_2, 9% CO_2 and the rest H_2O. Calculate the equilibrium concentrations of CO, NO, and NO_2.

Solution

$$K_{CO_2} = \frac{y_{CO}(y_{O_2})^{1/2}}{y_{CO_2}} = 3 \times 10^4 e^{-67,000/RT}$$

$$= 3(10^4) \exp\left[-\frac{67,000}{1.987(200 + 273)}\right] = 3(10^4)(1.10)(10^{-31}) = 3.3(10^{-27})$$

$$y_{CO} = \frac{3.3(10^{-27})(0.09)}{\sqrt{0.04}} = 1.48(10^{-27}) \quad \text{or} \quad 1.48(10^{-25})\% \quad \textbf{Answer}$$

$$K_{NO} = \frac{y_{NO}}{(y_{N_2})^{1/2}(y_{O_2})^{1/2}} = 4.69 \exp(-21,600/RT)$$

$$= 4.69 \exp\left[-\frac{21,600}{1.987(200+273)}\right] = 4.69[1.04(10^{-10})] = 4.90(10^{-10})$$

$$y_{NO} = 4.90(10^{-10})(\sqrt{0.75})(\sqrt{0.04}) = 8.48(10^{-11}) \quad \text{or} \quad 8.48(10^{-9})\% \quad \textbf{Answer}$$

$$K_{NO_2(\text{overall})} = \frac{y_{NO_2}}{(y_{N_2})^{1/2}y_{O_2}} = 11.73(10^{-4})\exp(-7880/RT)$$

$$= 11.73(10^{-4})\exp\left[-\frac{7,880}{1.987(200+273)}\right]$$

$$= 11.73(10^{-4})(0.000228) = 0.00267(10^{-4})$$

$$y_{NO_2} = 0.00267(10^{-4})\sqrt{0.75}(0.04) = 9.25(10^{-9}) \quad \text{or} \quad 9.25(10^{-7})\% \quad \textbf{Answer}$$

or,

$$K_{NO_2} = \frac{y_{NO_2}}{y_{NO}(y_{O_2})^{1/2}} = 2.5 \times 10^{-4}\exp(13,720/RT)$$

$$= 2.5(10^{-4})\exp\left[\frac{13,720}{1.987(200+273)}\right] = 2.5(10^{-4})(2,187,037.90) = 546.76$$

$$y_{NO_2} = 546.76(8.48)(10^{-11})\sqrt{0.04} = 9.27(10^{-9}) = 9.27(10^{-7})\% \quad \textbf{Answer}$$

Example 10-4

A flue gas in equilibrium at a temperature of 1900°C and 1 atm pressure contains 75% N_2, 4% O_2, 9% CO_2 and the rest H_2O. Calculate the equilibrium concentrations of CO, NO, and NO_2.

Solution

$$K_{CO_2} = \frac{y_{CO}(y_{O_2})^{1/2}}{y_{CO_2}} = 3 \times 10^4 e^{-67,000/RT}$$

$$= 3(10^4)\exp\left[-\frac{67,000}{1.987(1900+273)}\right] = 3(10^4)(1.82)(10^{-7}) = 5.46(10^{-3})$$

$$y_{CO} = \frac{5.46(10^{-3})(0.09)}{\sqrt{0.04}} = 0.00246 \quad \text{or} \quad 0.246\% \quad \textbf{Answer}$$

$$K_{NO} = \frac{y_{NO}}{(y_{N_2})^{1/2}(y_{O_2})^{1/2}} = 4.69 \exp(-21,600/RT)$$

$$= 4.69 \exp\left[-\frac{21,600}{1.987(1900+273)}\right] = 4.69(0.0067) = 0.031$$

$$y_{NO} = 0.031(\sqrt{0.75})(\sqrt{0.04}) = 0.0054 \quad \text{or} \quad 0.54\% \quad \textbf{Answer}$$

$$K_{NO_2 \text{(overall)}} = \frac{y_{NO_2}}{(y_{N_2})^{1/2} y_{O_2}} = 11.73(10^{-4}) \exp(-7880/RT)$$

$$= 11.73(10^{-4}) \exp\left[-\frac{7,880}{1.987(1900 + 273)}\right]$$

$$= 11.73(10^{-4})(0.16) = 0.000189$$

$$y_{NO_2} = 0.000189\sqrt{0.75}(0.04) = 3.27(10^{-5}) = 3.27(10^{-3})\% \quad \textbf{Answer}$$

The results of Examples 10-3 and 10-4 exemplify the importance of flue gas temperature. Whereas when the stack temperature was only 200°C, the NO concentration was $8.48(10^{-9})\%$, when the temperature was 1900°C, the NO concentration increased to 0.54%. Similar conclusions may be drawn for the other species.

Generalized Fate of Fuel Components in Rapid Combustion

$C_xH_yO_zN_aS_b$ may represent the generalized form of a fuel formula including nitrogen and sulfur. The conditions of combustion determine the fate of the constituent elements. Hydrogen, of course, converts to H_2O. The H_2O resulting from the mere combination of H and O contained in the fuel is called *combined water*.

The nitrogen atom in fuel is, for the most part, bonded to hydrogen. In incomplete combustion, it is reasonable to assume that it would remain bonded; the compound that satisfies this is ammonia. In complete combustion, nitrogen converts to NO and N_2, as mentioned before. Incomplete combustion of S will be assumed to produce H_2S. The complete combustion produces SO_2. Combustion also forms NO_2 and SO_3 but since they are negligibly small in amount, they will be neglected in the formulation of the chemical reaction below. The respective chemical reactions follow.

$$C_xH_yO_zN_aS_b + \frac{2x + y - 2b - 3a - 2z}{4}O_2 \rightarrow xCO$$

$$+ \frac{y - 2b - 3a}{2}H_2O + aNH_3 + bH_2S \qquad \text{incomplete} \qquad (10\text{-}37)$$

$$C_xH_yO_zN_aS_b + \frac{4x + y + 2c + 4b - 2z}{4}O_2 \rightarrow xCO_2$$

$$+ \frac{y}{2}H_2O + cNO + \frac{a - c}{2}N_2 + bSO_2 \qquad \text{complete} \qquad (10\text{-}38)$$

The fuel formula shown above can be determined by performing elemental analysis. (Elemental analysis is called *ultimate analysis*.) Hence, in equation (10-37), once x, y, z, a, and b are known, all the terms of the equation can be determined. In equation (10-38), however, simple knowledge of the fuel formula cannot determine all the terms of the equation. The coefficient c of NO on the right-hand side is unknown. Knowledge of the percent conversion to NO of the fuel nitrogen is necessary before c can be determined. Figure 10-1 shows the percent conversion of fuel nitrogen to NO_x. In general, this figure

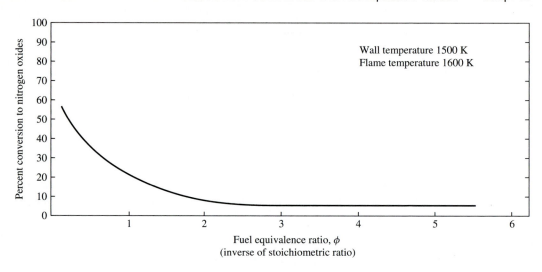

Figure 10-1   Conversion of fuel nitrogen to NO_x (for certain pulverized coals).

states that as more fuel is used relative to air, the percent conversion of fuel nitrogen to NO_x decreases.

Example 10-5

Ultimate analysis of a solid waste used to fire a solid-waste-to-energy recovery plant shows the following results: $x = 50$, $y = 100$, $z = 40$, and $a = 1$. The plant burns 100 tonnes/day of solid waste, producing 20 tonnes/day of residue containing 5% C. Calculate the combustion products for incomplete combustion. The humidity of the air is 0.016 mole of H_2O per mol of dry air.

Solution   From the results of the ultimate analysis, the fuel formula is $C_{50}H_{100}O_{40}N$.

$$C_xH_yO_zN_aS_b + \frac{2x + y - 2b - 3a - 2z}{4}O_2 \rightarrow xCO + \frac{y - 2b - 3a}{2}H_2O + aNH_3 + bH_2S$$

$$\frac{2x + y - 2b - 3a - 2z}{4} = \frac{2(50) + 100 - 2(0) - 3(1) - 2(40)}{4} = 29.25$$

$$\frac{y - 2b - 3a}{2} = \frac{100 - 2(0) - 3(1)}{2} = 48.5$$

$$C_{50}H_{100}O_{40}N + 29.25O_2 \rightarrow 50CO + 48.5H_2O + NH_3$$

$$\text{Reference species} = 29.25(4) = 117 \text{ tonne-mol of electrons}$$

Equivalent weights:

$$C_{50}H_{100}O_{40}N = \frac{C_{50}H_{100}O_{40}N}{117} = \frac{12(50) + 1.008(100) + 16(40) + 14}{117} = \frac{1354.8}{117}$$

$$= 11.58\frac{\text{tonnes}}{\text{tonne-Eq}}$$

$$O_2 = \frac{29.25O_2}{117} = \frac{29.25(32)}{117} = 8.0 \frac{\text{tonnes}}{\text{tonne-Eq}}$$

$$CO = \frac{50CO}{117} = \frac{50(12 + 16)}{117} = 11.97 \frac{\text{tonnes}}{\text{tonne-Eq}}$$

$$H_2O = \frac{48.5H_2O}{117} = \frac{48.5[1.008(2) + 16]}{117} = 7.45 \frac{\text{tonnes}}{\text{tonne-Eq}}$$

$$NH_3 = \frac{NH_3}{117} = \frac{17}{117} = 0.145 \frac{\text{tonne}}{\text{tonne-Eq}}$$

$$\text{weight fraction of C in fuel} = \frac{50C}{C_{50}H_{100}O_{40}N} = \frac{50(12)}{1354.8} = 0.44$$

$$\text{C gasified} = 100(0.44) - 20(0.05) = 43 \text{ tonnes/day}$$

$$\text{eq. wt. of } C_{50} = \frac{12(50)}{117} = 5.13 \frac{\text{tonnes}}{\text{tonne-Eq}}$$

$$\text{eq. } C_{50} \text{ gasified} = \frac{43}{5.13} = 8.38 \text{ tonne-Eq/day}$$

$$\text{tonnes/day CO} = 8.38(11.97) = 100.31 \Rightarrow \frac{100.31}{CO} = \frac{100.31}{12 + 16}$$

$$= 3.58 \text{ tonne-mol/day}$$

$$\text{tonnes/day H}_2\text{O} = 8.38(7.45) = 62.43 \Rightarrow \frac{62.43}{18} = 3.47 \text{ tonne-mol/day}$$

$$\text{tonnes/day NH}_3 = 8.38(0.145) = 1.22 \Rightarrow \frac{1.22}{17} = 0.072 \text{ tonne-mol/day}$$

$$\text{moles of air used in combustion} = \frac{8.38(8)}{O_2} \left(\frac{100}{21}\right) = 9.98 \text{ tonne-mol/day}$$

$$\text{moisture from air} = 0.016(9.98) = 0.16 \text{ tonne-mol/day}$$

$$N_2 \text{ from air} = 0.79(9.98) = 7.88 \text{ tonne-mol/day}$$

Constituent	Tonne-mol/day	Percent	
CO	3.58	23.61	
H_2O	3.63	23.94	
NH_3	0.07	0.46	**Answer**
N_2	7.88	51.98	
Σ	15.16	100.0	

Example 10-6

Ultimate analysis of a solid waste used to fire a solid-waste-to-energy recovery plant shows the following results: $x = 50$, $y = 100$, $z = 40$, and $a = 1$. The plant burns 100 tonnes/day of solid waste producing 20 tonnes/day of residue containing 5% C. Calculate the combustion products at the combustion zone for a complete combustion using 20% excess air and neglecting thermal NO. Assume that Figure 10-1 applies and an incoming air humidity of 0.016 mol of H_2O per mol of dry air.

Solution   From the results of the ultimate analysis, the fuel formula is $C_{50}H_{100}O_{40}N$.

$$C_xH_yO_zN_aS_b + \frac{4x + y + 2c + 4b - 2z}{4}O_2 \rightarrow xCO_2 + \frac{y}{2}H_2O + cNO + \frac{a-c}{2}N_2 + bSO_2$$

To find the value of c, we use Figure 10-1.

$$\text{Mol air used per mole of fuel} = \frac{100}{21}\left(\frac{4x + y + 2c + 4b - 2z}{4}\right)$$

$$= \frac{100}{21}\left[\frac{4(50) + 100 + 2c + 4(0) - 2(40)}{4}\right]$$

$$= 1.19(220 + 2c)$$

$$\text{Molecular weight of air} = \frac{21(32) + 79(28)}{100} = 28.84$$

$$\text{Stoichiometric fuel/air ratio} = \frac{C_{50}H_{100}O_{40}N}{1.19(220 + 2c)(28.84)} = \frac{39.48}{220 + 2c}$$

$$\text{Actual fuel/air ratio} = \frac{C_{50}H_{100}O_{40}N}{1.19(220 + 2c)(28.84)(1.2)} = \frac{39.48}{(220 + 2c)(1.2)} = \frac{32.9}{(220 + 2c)}$$

$$\text{Stoichiometric ratio} = \frac{39.48/(220 + 2c)}{32.9/(220 + 2c)} = 1.2$$

$$\text{Equivalence ratio} = \frac{1}{\text{stoichiometric ratio}} = \frac{1}{1.2} = 0.83$$

From Figure 10-1, for equivalence ratio = 0.83, percent fuel nitrogen conversion to NO_x = 25. Thus, $c/a = 0.25$ and $c = 0.25a = 0.25(1) = 0.25$ and

$$\frac{4x + y + 2c + 4b - 2z}{4} = \frac{4(50) + 100 + 2(0.25) + 4(0) - 2(40)}{4} = 55.125$$

$$C_{50}H_{100}O_{40}N + 55.125O_2 \rightarrow 50CO_2 + 50H_2O + 0.25NO + 0.375N_2$$

$$\text{Reference species} = 55.125(4) = 220.5 \text{ tonne-mol of electrons}$$

Equivalent weights:

$$C_{50}H_{100}O_{40}N = \frac{C_{50}H_{100}O_{40}N}{220.5} = \frac{12(50) + 1.008(100) + 16(40) + 14}{220.5} = \frac{1354.8}{220.5}$$

$$= 6.14\frac{\text{tonnes}}{\text{tonne-Eq}}$$

$$O_2 = \frac{55.125 O_2}{220.5} = \frac{55.125(32)}{220.5} = 8.0 \frac{\text{tonnes}}{\text{tonne-Eq}}$$

$$CO_2 = \frac{50 CO_2}{220.5} = \frac{50(12 + 32)}{220.5} = 9.98 \frac{\text{tonnes}}{\text{tonne-Eq}}$$

$$H_2O = \frac{50 H_2O}{220.5} = \frac{50[1.008(2) + 16]}{220.5} = 4.09 \frac{\text{tonnes}}{\text{tonne-Eq}}$$

$$NO = \frac{0.25 NO}{220.5} = \frac{0.25(14 + 16)}{220.5} = 0.034 \frac{\text{tonne}}{\text{tonne-Eq}}$$

$$N_2 = \frac{0.375 N_2}{220.5} = \frac{0.375(28)}{220.5} = 0.048 \frac{\text{tonne}}{\text{tonne-Eq}}$$

$$\text{weight fraction of C in fuel} = \frac{50 C}{C_{50}H_{100}O_{40}N} = \frac{50(12)}{1354.8} = 0.44$$

$$C \text{ gasified} = 100(0.44) - 20(0.05) = 43 \text{tonnes/day}$$

$$\text{eq. wt. of } C_{50} = \frac{12(50)}{220.5} = 2.72 \frac{\text{tonnes}}{\text{tonne-Eq}}$$

$$\text{eq. of } C_{50} \text{ gasified} = \frac{43}{2.72} = 15.80 \text{ tonne-Eq/day}$$

$$\text{tonnes/day } CO_2 = 15.80(9.98) = 157.68 \Rightarrow \frac{157.68}{CO_2} = \frac{157.68}{12 + 32}$$

$$= 3.58 \text{ tonne-mol/day}$$

$$\text{tonnes/day } H_2O = 15.80(4.09) = 64.62 \Rightarrow \frac{64.62}{18} = 3.59 \text{ tonne-mol/day}$$

$$\text{tonnes/day } NO = 15.80(0.034) = 0.54 \Rightarrow \frac{0.54}{30} = 0.018 \text{ tonne-mol/day}$$

$$\text{tonnes/day } N_2 = 15.80(0.048) = 0.76 \Rightarrow \frac{0.76}{28} = 0.027 \text{ tonne-mol/day}$$

$$\text{theoretical moles of air used in combustion} = \frac{15.80(8.0)}{O_2} \left(\frac{100}{21} \right)$$

$$= 18.81 \text{ tonne-mol/day}$$

$$\text{moisture from air} = 0.016(18.81)(1.2) = 0.36 \text{ tonne-mol/day}$$

$$N_2 \text{ from air} = 0.79(18.81)(1.2) = 17.83 \text{ tonne-mol/day}$$

$$O_2 \text{ not combusted} = 18.81 \left(\frac{21}{100} \right)(1.2) - \frac{15.80(8)}{32} = 0.79 \text{ tonne-mol/day}$$

To summarize:

$$CO_2 = 3.58 \text{ tonne-mol/day}$$

$$H_2O = 3.59 + 0.36 = 3.95 \text{ tonne-mol/day}$$

$$NO = 0.018 \text{ tonne-mol/day}$$

$$N_2 = 0.027 + 17.83 = 17.86 \text{ tonne-mol/day}$$

$$O_2 = 0.79 \text{ tonne-mol/day}$$

Constituent	Tonne-mol/day	Percent	
CO_2	3.58	13.66	
H_2O	3.95	15.08	
NO	0.018	0.07	**Answer**
N_2	17.86	68.18	
O_2	0.79	3.01	
$\sum$	26.198	100.0	

Example 10-7

Ultimate analysis of a solid waste used to fire a solid-waste-to-energy Ultimate analysis of a solid waste used to fire a solid-waste-to-energy recovery plant shows the following results: $x = 50$, $y = 100$, $z = 40$, and $a = 1$. The plant burns 100 tonnes/day of solid waste producing 20 tonnes/day of residue containing 5% C. Calculate the ideal temperature of combustion using the same combustion conditions as in Example 10-6.

Solution

$$T_{\text{ideal}} = \frac{1}{R} \frac{43,480}{2.93 - \ln(\{NO\}^2/\{O_2\}\{N_2\})}$$

From Example 10-6:

$$\{N_2\} = \frac{17.86}{26.2}\left(\frac{10^6}{10^6}\right) = \frac{17.86}{26.2}(10^6) \text{ ppm} = 0.68(10^6) \text{ ppm}$$

$$\{O_2\} = \frac{0.79}{26.2}(10^6) = 0.03(10^6) \text{ ppm}$$

$$\{NO\} = \frac{0.018}{26.2}(10^6) = 687.02 \text{ ppm}$$

$$T_{\text{ideal}} = \frac{1}{1.987}\left\{\frac{43,480}{2.93 - \ln(687.02)^2/[0.03(10^6)(0.68)(10^6)]}\right\}$$

$$= 0.503(3196.10) = 1607.64\text{K} \quad \textbf{Answer}$$

Note: This neglects thermal NO. A two-stage combustion, which affords a lower first-stage temperature, may minimize NO formation, hence result in lower NO production.

AUTOMOTIVE EMISSION CONTROL

Basically, CO, NO_x, and hydrocarbons are the emissions that need to be controlled in automobiles or other internal combustion engines. As atmospheric pollutants, these emissions are precursors of photochemical smog. CO also combines with hemoglobin, precluding the formation of oxyhemoglobin, thus suffocating a person. Although NO_x has been linked to increased incidence of bronchitis, the major reason for its control is the role it plays in disturbing the NO_2 photolytic cycle (to be discussed later).

To begin the study of automotive emission control, it is best to discuss how the engine performs. Figure 10-2 shows the sequence of operation of a four-cycle engine (a) and a longitudinal section of an engine (b). Locate the piston in the engine longitudinal section and see that there are four cylinders in this engine. Note that in the rightmost cylinder, the piston is at the top of its travel (top dead center, TDC) and that in the second cylinder from the left, the piston is at the bottom of its travel (bottom dead center, BDC). Figure 10-2a illustrates the performance of each of these four cylinders.

At the start of the *intake stroke*, the inlet valve opens, allowing a mixture of air and fuel from an *intake manifold* to enter the cylinder. As soon as the intake stroke completes, the *compression stroke* starts, the piston compressing the air–fuel mixture against the closed intake and exhaust valves. At the top of the piston travel, the spark plug fires, igniting the compressed mixture. This creates the downward *power stroke*. At the end of the power stroke, the exhaust valve opens and the exhaust stroke begins expelling the combustion gases into the *exhaust manifold*. (See the schematics of intake, compression, power, and exhaust.)

The expanding gases in the power stroke transmit the force created to the face of the piston, to the connecting rod, and then to the crankshaft. Locate the connecting rod and crankshaft at the bottom of the figure; the crankshaft is indicated by the curved arrows at the top, with the connecting rod wobbling about it. The crankshaft then relays the force to the *flywheel*. The flywheel serves to stabilize the rotation. The flywheel, in turn, transmits the force to the *transmission assembly* (see Figure 10-3b). Transmission is the place where speed and power are interconverted through gearings or torque conversion. For example, the first gear has high power but low speed, while the fourth gear has low power but high speed.

From the transmission, the force goes through a *drive line* composed of universal joints, differentials, and driving and driven shafts. From here the force continues to be submitted to the *driving axle*, which, ultimately, turns the *wheels*.

To transmit force to the axle, *driving* and *driven shafts* are used. Because of gearings the driving shaft turns, but because the engine body bounces up and down, the force does not transmit directly to the axle. Rather, the force goes from the driving shaft to the driven shaft through a *universal joint*. This joint stabilizes or harmonizes the transmission of force as it accommodates the bouncing of the engine. From the driven shaft, the force goes to the differential before reaching the axle. The *differential* permits one drive wheel to turn at different speeds from the other, which is important when the car is negotiating a curve.

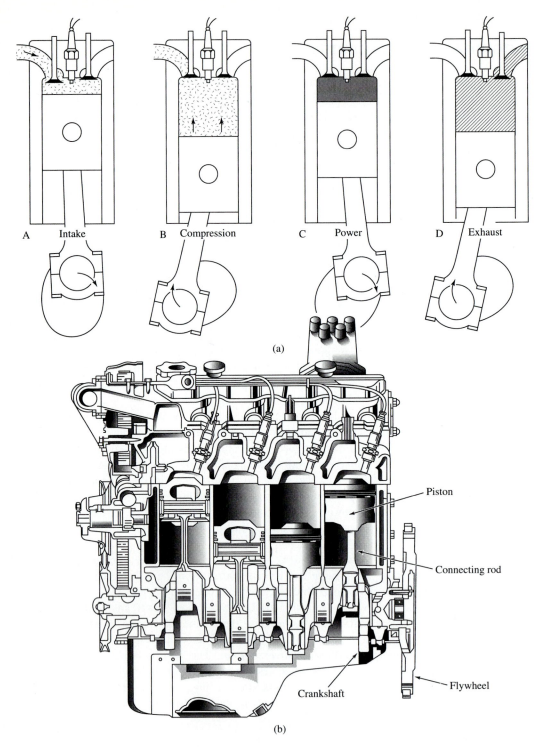

A   Intake   B   Compression   C   Power   D   Exhaust

(a)

Piston

Connecting rod

Flywheel

Crankshaft

(b)

Figure 10-2 (a) Sequence of operation of four-cycle engine. (b) Section of engine. (Reprinted by permission of Chrysler Corporation.)

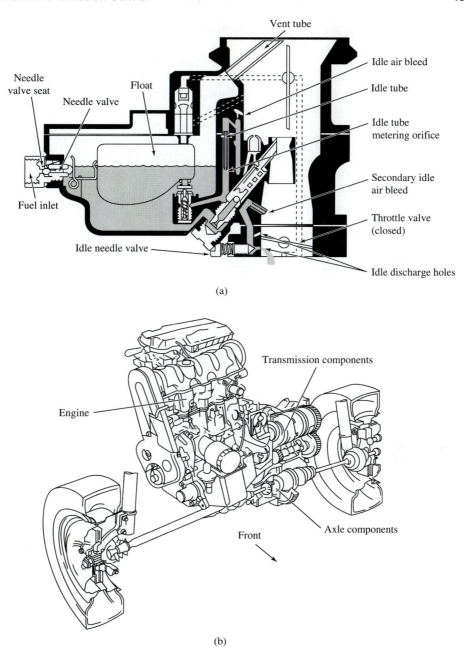

Figure 10-3   (a) Section of carburetor. (b) View of engine with transmission and axle components. (Reprinted by permission of Chrysler Corporation.)

Control of Hydrocarbon Emissions

Potential emissions of hydrocarbons are at the crankcase (Figure 10-4), at the carburetor float chamber (Figure 10-3), at the gasoline tank, at the tailpipe (Figure 10-4), the engine timing, and the general state of tune-up of the engine.[5,6] As the piston moves up and down the cylinder, leakage occurs between the *piston rings* and the walls of the cylinder. This leakage, called *blowby*, escapes into the crankcase of the crankshaft. To prevent escape to the atmosphere, the crankcase vents into the intake manifold through a *PCV* (positive crankcase ventilation) *valve*. This valve acts like a check valve that allows flow in one direction only. It adjusts the rate of removal of blowby gases to match the changing air intake requirements of the engine. Blowby gases may also contain NO_x and CO but mostly hydrocarbons.

Refer to Figure 10-3a. The right portion of this drawing is the carburetor barrel. Locate the throttle valve; this is shown closed. Stepping on the gas pedal makes this valve open, and because the cylinder piston is an effective positive-displacement pump, air is sucked through the carburetor barrel. At the middle of the barrel is a venturi element; this element speeds up the air velocity, decreasing the pressure at the point. The low pressure causes the gasoline to be sucked up from the float chamber, mixing with the air at the throat. This intense mixing causes the gasoline to be atomized with the air. This process is called *carburetion*; hence the entire unit assembly is called a *carburetor*.

For the engine to run smoothly, control of the ratio of air to fuel must be precise; otherwise, the engine will stall. To get the ratio, the carburetor must be calibrated, which means that the flow of gasoline from the float chamber into the venturi throat must be precise, hence installation of the float in the chamber. This float functions in exactly the same way as the float in the ordinary tank used to flush the toilet bowl. This element maintains the fuel head in the chamber, ensuring precise delivery of gasoline. However, to maintain the head, the chamber must be vented. Figure 10-4a shows the venting of the chamber to a carbon canister.

Since evaporation is very dependent on temperature, emission from the float chamber is especially severe right after the engine turns off. Emissions during this period are called *hot soak* emissions. As the engine turns off, the activated carbon adsorbs the vapors. On restarting, a flow of filtered air through the canister purges these vapors. The vapor mixture then goes through one or more tubes feeding into the carburetor and/or carburetor air filter.

One way to control carburetor hydrocarbon emissions is to replace the carburetor with a fuel injection system. Figure 10-4b shows one version of this system. A pump conveys the fuel from the gasoline tank, through the fuel filter, and into the fuel rail. The figure shows the fuel rail distributing gasoline into six fuel injector valves. These valves inject the gasoline directly into the intake manifold. Excess gasoline then flows back into

---

[5] W. K. Toboldt and L. Johnson (1983). *Automotive Encyclopedia*. Goodheart–Willcox, South Holland, Ill., p. 784.

[6] P. A. Vesilind, J. J. Peirce, and R. F. Weiner (1988). *Environmental Engineering*. Butterworth, Boston, p. 446.

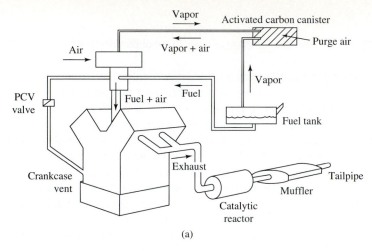

(a)

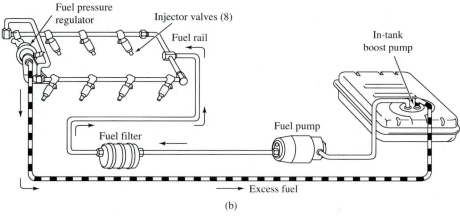

(b)

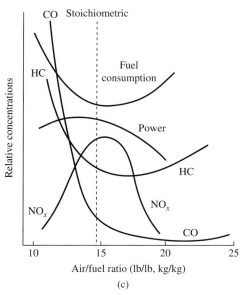

(c)

Figure 10-4 (a) Control points of pollutants in automobile. (b) Schematic of fuel injection system. (c) Effect of air/fuel ratio on emissions, power, and fuel economy.

the gasoline tank through a fuel pressure regulator valve, which reduces the fuel pressure from the injector to a pressure slightly above that prevailing in the gasoline tank. In the case of the carburetor, the mixture of air and gasoline is carbureted through the venturi before going to the intake manifold. In the fuel injection system, on the other hand, the mixture is carbureted right at the intake manifold by an injector atomizing system. The design is such that right after contact between the atomized gasoline and air, the mixture becomes carbureted. Hence the design effectively eliminates the evaporative emission at the float chamber. There is no float chamber in the fuel injection system.

Other potential hydrocarbon emissions come from the *fuel tank*. As shown in Figure 10-4, a carbon canister adsorbs the vapors through a vent line installed from the tank to the canister. The air then purges these vapors from the canister and the mixture ultimately introduced into the engine for combustion.

The third emission source of hydrocarbons is the *tailpipe*. Combustion happens so fast that there is no time to burn the fuel completely. Unburned gases therefore end up as tailpipe emissions. Two possible methods of control are the use of thermal reactors and catalytic reactors.

The *thermal reactor* is essentially an afterburner. Upon exit from the exhaust manifold, a pump conveys the gases to the unit, where excess air is introduced to burn hydrocarbons and CO completely to carbon dioxide and water (around $1100°C$). The other method is the use of a *catalytic reactor*, specifically, the three-way *catalytic converter* (operating at around $800°C$). The three-way catalyst is composed of platinum and rhodium. The unit oxidizes hydrocarbons and CO simultaneously to carbon dioxide and water while reducing NO_x to N_2 at the same time. The equation that follows represents the unbalanced reaction:

$$HC + CO + NO_x \xrightarrow{\text{catalyst}} H_2O + CO_2 + N_2 \tag{10-39}$$

Improper ignition timing is also an important source of hydrocarbon emission. When the engine is running but the car is not, the engine is *idling*. At idling, the throttle valve (see Figure 10-3) opens only partially, thus admitting less air. The air–fuel mixture becomes *richer* (i.e., more fuel than air). With rich mixtures, not all the fuel will burn, and more emissions will result at the exhaust pipe. To produce fewer emissions, opening the throttle valve wider makes the mixture *leaner* (more air than fuel); however, the presence of more mixture than required also makes the engine idle faster. Retarding the spark plug firing will correct this idling. Since retarded ignition means late firing, it also means less power; hence the engine will not idle as fast.

Refer to Figure 10-2. Right at the top of a cylinder (*cylinder head*) is the spark plug. This plug connects to the spark plug wire, which, in turn, connects to the distributor. In the figure the distributor projects at the top of the engine (bottom drawing). This distributor controls the sequence or timing of firing. Hence, during tune-up, the mechanic rotates the distributor to adjust the sequence to a retarded firing at idle, if needed. Tightening an adjustment screw will then ensure that the adjustment will not change. Also, for idling conditions, the mechanic opens the throttle valve wider, introducing more air and providing a leaner mixture. This is done by adjusting a separate screw.

The *idle mixture screw* or *valve* shown at the bottom of the top drawing of Figure 10-3 adjusts the idle mixture. So as not to change the mixture proportion no matter how much the screw is rotated, the factory presets the idle mixture screw needle.

During acceleration and cruising, since the throttle is fully open, admitting more air, the mixture is naturally lean for the driving condition; emission is at a minimum. During deceleration, however, the sudden low demand for power causes the mixture to become suddenly rich for the driving condition, thereby, increasing hydrocarbon emissions. Maintaining the distributor at the retarded firing condition will provide little time to burn the fuel in the cylinder. Hence, advancing the spark provides increased burning time. Retarded during idle and advanced during deceleration is then the ignition timing. Also, the throttle opens wide during acceleration and cruising, drawing in more fuel and requiring more burning time. Therefore, the spark also advances during this driving condition. To summarize, retarded at idle, gradually increasing to top dead center as the speed increases, then to advanced timing when accelerating, cruising, and decelerating is, finally, the ignition timing.

In engine design, the anticipated driving conditions set the degree of advance. There are several methods of advancing the spark. The centrifugal governor facilitates the advance, corresponding to the normal increase in speed. As the speed changes, the centrifugal force on the governor also changes proportionately, which, in turn, by some mechanism, makes adjustments to the distributor timing. The other is through the use of a vacuum advance during deceleration. The distributor connects directly through tubing to a port on the underside of the throttle valve. On decelerating, the valve partially closes and, due to the continued suction by the piston, a vacuum is created at the point. The tubing transmits this vacuum to the distributor to make adjustments on the distributor timing for advancing the spark.

The final potential source of hydrocarbon emission is the untuned engine. For example, if the spark plug misfires or simply does not fire, hydrocarbons will simply go out of the tailpipe unburned. A dirty air filter will decrease the amount of air that would normally enter the engine. This causes the mixture to be rich, producing larger emissions. Engine oil not changed regularly results in an increase in the internal friction of engine parts. The engine will become a generally untuned engine and will run poorly, thus producing more emissions. Finally, a malfunctioning PCV valve will not be able to apportion correctly the recirculated blowby gases with the normal fuel feed. The engine becomes out of tune and will emit more emissions.

Control of CO

Incomplete combustion of fuel also results in emissions of carbon monoxide. CO can be found in the blowby but is mostly in the exhaust pipe. As in the case of hydrocarbons, the tailpipe may be controlled by the thermal reactor and the three-way catalyst. The emission of CO also runs parallel to those of hydrocarbons as far as idle and deceleration are concerned. Whenever the mixture is rich, the combustion becomes far from complete and more CO will be emitted at the tailpipe unless the timing is advanced.

Control of NO_x Emissions

Since high temperatures favor the production of NO_x, the emission may be controlled by lowering the combustion temperature through recirculating exhaust gases. Exhaust gases will no longer burn, hence it will no longer produce any heat and, if recirculated, will help to quench the combustion gases. This method is called *exhaust gas recirculation* (EGR).

In the cylinder, the temperature can range from 500°C at the end of the compression stroke (before ignition) to approximately 3000°C right after firing. This temperature value is really a sure way to produce NO_x. To reduce NO_x formation, combustion temperatures must be kept below 1300°C. Thus exhaust gas recirculation may be used.

Another method of lowering temperature is by retarding spark. Peak combustion temperature relates directly to spark timing. If the spark occurs at exactly the right instant, the maximum amount of pressure and heat is obtained and the engine puts out maximum power. Retarding spark slightly more than normally for maximum power would cause the power to fall off but lower the combustion temperature, thus reducing NO_x emissions. As noted under hydrocarbon control above, the three-way catalytic converter may also be used for NO_x control. The chemical reaction was written earlier.

Table 10-1 shows the composition of gasoline along with the boiling points of components, and Figure 10-4c shows the effect of the air/fuel ratio on emissions, power, and economy. In general, emissions are small when the mixture is lean. For NO_x, however, the emission is a maximum at a slightly leaner mixture than stoichiometric and drastically falls off. At a slightly richer mixture than stoichiometric, the power is at a maximum, while at the slightly leaner mixture than stoichiometric, the fuel consumption is at a minimum but the power drops. Beyond a certain value of a lean mixture, the hydrocarbon emissions increase. The reason for this is that beyond this point, because of too much dilution, which results in a much cooler combustion temperature, more hydrocarbon molecules no longer burn.

ATMOSPHERIC POLLUTION

Atmospheric pollution is basically photochemical smog. *Photochemical smog* is a designation given to a mixture of reactants and products that result from the interaction between organics and the oxides of nitrogen. The primary pollutants in smog are *nitric oxide* and *hydrocarbons*. These primary pollutants convert rapidly to the secondary pollutants *ozone, organic nitrates, oxidized hydrocarbons*, and *photochemical aerosols*. Although ozone is a major component of photochemical smog, it performs an important function in the stratosphere by absorbing potentially damaging shortwave ultraviolet radiation that can reach the surface of the earth.

Photochemical Smog

The origin of photochemical smog is the *photolytic decomposition* of nitrogen dioxide. Although NO is the predominant form of NO_x emissions, it can react with O_2 to form

TABLE 10-1 COMPOSITION
OF GASOLINE AND BOILING
POINTS OF COMPONENTS

Component	Boiling point (°F)
Undecane	400
Decane	—
Nonane	300
Octane	—
Heptane	200
Hexane	—
Pentane	100

NO_2. This small amount of NO_2 can trigger subsequent reactions through its decomposition, forming what is called the *NO₂ photolytic cycle*. The cycle starts as follows:

$$NO_2 + h\nu \rightarrow NO + O \tag{10-40}$$

$h\nu$ represents a photon of energy absorbed by nitrogen dioxide, causing its decomposition to NO and O. (Note that the free-radical dot is not used.) The atomic oxygen liberated then reacts with molecular O_2, producing ozone as follows:

$$O + O_2 + M \rightarrow O_3 + M \tag{10-41}$$

M is a third molecule (usually, O_2 or N_2, since they are abundant) that absorbs the excess energy from the reaction. Without M, O_3 will possess too much energy for it to be stable and will simply dissociate back to O and O_2. Ozone subsequently reacts with NO from either reaction (10-40) or from the original atmospheric NO to regenerate NO_2 and molecular O_2, thus completing the cycle.

$$O_3 + NO \rightarrow NO_2 + O_2 \tag{10-42}$$

NO is relatively inert and only moderately toxic. Although, like CO, it can impair the blood oxygen-carrying capacity when it combines with hemoglobin, its concentration in the atmosphere is generally less than 1 ppm, which is considered harmless. On the other hand, NO_2 irritates the alveoli of the lungs and O_3 has many undesirable properties. These properties include causing chest constriction, irritation of the human mucous membrane, cracking of rubber products, and damage to vegetation. Prolonged exposures to even relatively low concentrations of NO_x, characteristics of polluted environments, have been linked to increased bronchitis in children.

Deducing from equations (10-40) to (10-42), the NO_2 cycle does not produce a net increase in NO_2 or O_3. Hence if not disturbed and provided that the starting concentrations of the species are low, there will be no undesirable effects. However, human activities emit not only NO_x to the atmosphere but also *carbon monoxide*, the *carbonyl compounds*, and *hydrocarbons*. Carbon monoxide and hydrocarbons, by their reactions with the hydroxyl radical, disturb the normal NO_2 photolytic cycle through the formation of *peroxyl radicals*. These peroxyls prevent the reaction of O_3 with NO, thus terminating the cycle and causing the O_3 to accumulate. In addition, the photolytic decomposition of carbonyl compounds also disturbs this cycle. This is discussed later.

Reactions of the hydroxyl radical, OH·.

Reaction with NO to produce NO_2 and O_2 is the normal path followed by O_3. In some situations, however, because of the presence of H_2O in the atmosphere, OH· appears. Its production proceeds according to the following sequence of reactions:

$$O_3 + h\nu \rightarrow O(^1D) + O_2 \tag{10-43}$$

$$O(^1D) + H_2O \rightarrow 2OH· \tag{10-44}$$

When O_3 absorbs solar photons having wavelengths between 315 and 1200 nm, it dissociates, producing the ground electronic state of atomic oxygen, the normal oxygen atom, O. However, when it absorbs photons in wavelengths of less than 315 nm near the ultraviolet range, O_3 dissociates, producing $O(^1D)$. The latter product is called the *electronically excited oxygen atom*. $O(^1D)$ cannot transform directly into its ground-state form but needs another species to react with to release its energy[7]—hence reaction (10-44), producing 2 mol of OH·.

The formation of OH· is the beginning of a complex series of reactions in the atmosphere. These reactions disturb the normal NO_2 photolytic cycle, leading to the formation of the components of the more complex photochemical smog. OH·, once formed, can react with the three compounds: CO, hydrocarbons, and carbonyl compounds. Thus if only these three compounds are not present, the photolytic cycle will not be disturbed and there will be an absence of photochemical smog. This, however, is not the case when the atmosphere is polluted. As far as the phenomenon of photochemical smog is concerned, the end products of the reaction of OH·, namely, the peroxyl radicals, are the most important. These radicals, rather than O_3, are the ones that react with NO to regenerate NO_2, thus terminating the normal NO_2 cycle, leaving O_3 undisturbed and allowing it to accumulate. The following reactions of OH· with CO, the carbonyl compounds, and hydrocarbons explain the chemical basis of this phenomenon.

1. *Reaction with carbon monoxide.* The incomplete combustion of the carbon in fuels results in the formation of CO. In the presence of the hydroxyl radical, CO reacts according to the reaction below, producing carbon dioxide and a reactive hydrogen free radical H·. The hydrogen radical then quickly reacts with atmospheric O_2 to produce the hydroperoxyl radical, $HO_2·$.

$$CO + OH· \rightarrow CO_2 + H· \tag{10-45}$$

$$H· + O_2 + M \rightarrow HO_2· + M \tag{10-46}$$

Again, the third species, M, absorbs some of the energy released in the reaction. $HO_2·$ then reacts with NO to regenerate NO_2 and, to regenerate OH· , as well. This is shown as

$$HO_2· + NO \rightarrow NO_2 + OH· \tag{10-47}$$

---

[7]Seinfield, *Atmospheric Chemistry and Physics of Air Pollution.*

Now OH· can hunt for another CO and do its pernicious job again. In turn, NO$_2$ absorbs energy from sunlight, repeating the cycle but in another way.

The reaction sequence of the disturbance under the influence of CO may now be summarized as follows: NO$_2$ photolyzes, producing photolysis NO and O; the O produced reacts with molecular O$_2$, producing O$_3$. Photolysis NO reacts with the HO$_2$· produced from CO, producing back NO$_2$. NO$_2$ photolizes again, producing NO and O and, subsequently, the O$_3$. Hence OH· and CO keeps on recovering NO$_2$ without destroying O$_3$ but, instead, producing it. The sequence therefore has become a "mill" for producing O$_3$. Theoretically, it can go on forever to build up the concentration of ozone.

Fortunately, the buildup of O$_3$ cannot go on forever. The disturbed and undisturbed photolytic cycle described above is a function of cloud shading and the position of the sun in the sky. It is also a function of whether it is nighttime or daytime. Heavy clouds and rain impede sunlight from reaching the surface of the earth, suppressing the photon absorption by NO$_2$. The position of the sun also affects the photolysis of NO$_2$. The farther away it is from the earth, the less $h\nu$ will be absorbed by nitrogen dioxide.

In addition to producing more O$_3$, OH· may also form HNO$_3$ when it reacts with NO$_2$:

$$OH· + NO_2 \rightarrow HNO_3 \tag{10-48}$$

Now, upon considering all the reactions above, the inventory of the composition of photochemical smog up to this point is CO, NO$_2$, NO, O$_3$, and HNO$_3$. O$_3$ and HNO$_3$, which are secondary pollutants formed from the primary pollutants, CO, NO, and NO$_2$, are called *photochemical oxidants*. Also, HNO$_3$ is a precursor of acid rain.

2. *Reactions with carbonyl compounds.* The presence of the functional group

$-\overset{|}{C}=O$, called the *carbonyl group*, characterizes the carbonyl compounds. There are two types of carbonyl compounds: aldehydes and ketones. *Aldehydes* are oxidation products of primary alcohols; *ketones* are oxidation products of secondary alcohols. The general formulas RCHO and RCOR′, respectively, represent these compounds, where R and R′ are hydrocarbon portions of the respective molecules.

OH· reacts with aldehydes, abstracting the H atom from the carbonyl group forming water:

$$RCHO + OH· \rightarrow RC(O)· + H_2O \tag{10-49}$$

The parent aldehyde from where the H was abstracted forms an *aldocarbonyl radical*, RC(O)·, and becomes very reactive. RC(O)· is also called an *acyl radical*. The radical, produced rapidly, reacts with atmospheric O$_2$, forming the peroxyl radical, RC(O)O$_2$·. RC(O)O$_2$· is an *aldocarbonylperoxyl* or *acylperoxyl radical*. The reaction for its formation is

$$RC(O)· + O_2 \rightarrow RC(O)O_2· \tag{10-50}$$

As in CO, the peroxyl radical disturbs the normal photolytic NO$_2$ cycle. The following reaction shows this disturbance:

$$RC(O)O_2· + NO \rightarrow NO_2 + RC(O)O· \tag{10-51}$$

As indicated, the reaction regenerates the NO_2 not by the normal reaction with O_3 but by the aldocarbonylperoxyl radical. Hence, as in CO, ozone will again be allowed to accumulate. In addition, the reaction generates another radical, $RC(O)O\cdot$. This radical is called an *aldocarbonyloxy* or *acyloxy radical*. Since $RC(O)O\cdot$ is a radical, it is very reactive; in fact, it rapidly reacts with molecular O_2 forming yet another peroxyl radical, $RO_2\cdot$, an *alkylperoxyl radical*, and carbon dioxide:

$$RC(O)O\cdot + O_2 \rightarrow RO_2\cdot + CO_2 \tag{10-52}$$

The reactions of $RC(O)\cdot$ and $RC(O)O\cdot$ with O_2 may be compared. Whereas O_2 simply attaches to $RC(O)\cdot$, forming $RC(O)O_2\cdot$, in the case of $RC(O)O\cdot$, COO splits off from the radical. One important point to remember in these reactions is that since the makeup of $RC(O)O\cdot$ already contains CO_2, CO_2 has the tendency to separate out as a molecule. It is equally logical to think that CO could have separated from $RC(O)\cdot$ also. However, this chemical property is clearly absent; CO bonds strongly to R and remains there when the $RC(O)O_2$ peroxyl radical forms.

As with the other peroxyl radicals, $RO_2\cdot$ now disturbs the normal NO_2 photolytic cycle, producing NO_2 and yet another radical $RO\cdot$, an *alkyloxy* or *alkoxy radical*:

$$RO_2\cdot + NO \rightarrow NO_2 + RO\cdot \tag{10-53}$$

Again, the reaction regenerates the NO_2 not by the normal reaction with ozone but by an entirely different route. This bypassing allows O_3 to accumulate still further. This is now the second time that a peroxyl has disturbed the cycle starting from one molecule of RCHO. As will now be shown, there will be a third, a fourth, and a fifth up to the *n*th, depending on the length of the R radical. $RO\cdot$ reacts further with O_2 as follows:

$$RO\cdot + O_2 \rightarrow R'CHO + HO_2\cdot \tag{10-54}$$

In reaction (10-54) R' is an alkyl or some hydrocarbon radical one carbon shorter than R. Also, here again, is the hydroperoxyl radical encountered in connection with the discussion on CO. *From this reaction, deduce, then, that when an alkoxy radical reacts with O_2, the hydroperoxyl free radical is produced accompanied by the production of a carbonyl.* Clearly, $HO_2\cdot$ will go on to react, disturbing the cycle again. $R'CHO$ will also proceed to react as an aldehyde. It will follow the same steps that RCHO followed until it becomes $R''CHO$. $R''CHO$ will become $R'''CHO$ and $R'''CHO$ will become $R''''CHO$ until the simplest aldehyde is reached, HCHO. Assuming that the R's are alkyl radicals, in each step decrease (R to R', or R' to R'', and so on) the NO_2 cycle is disturbed three times. Hence if the alkyl radical is originally *n* carbons long, the normal cycle will be disturbed $3(n - 1)$ times before reaching HCHO.

HCHO further reacts with $OH\cdot$ as follows:

$$HCHO + OH\cdot \rightarrow HC(O)\cdot + H_2O \tag{10-55}$$

Further reaction of $HC(O)\cdot$ with O_2 results in

$$HC(O)\cdot + O_2 \rightarrow CO + HO_2\cdot \tag{10-56}$$

Here, instead of forming $HC(O)O_2\cdot$ as it would be with the longer-chain molecules, $HC(O)O_2\cdot$ splits up, forming CO and the hydroperoxyl radical, $HO_2\cdot$. This newly formed

$HO_2\cdot$ will, once again, disturb the cycle regenerating NO_2 in a different way and, also, again regenerating $OH\cdot$. This regenerated $OH\cdot$ will now, once again, attack yet another aldehyde molecule. All in all, the total disturbance up to HCHO is $3n - 2$. Since every disturbance spares one O_3, the total number of O_3 spared per molecule of RCHO is $3n - 2$.

If $OH\cdot$ cannot repeat the cycle with another molecule of RCHO, it goes through a terminating reaction forming nitric acid, HNO_3. The aldocarbonylperoxyl radical may also undergo a terminating reaction with NO_2 producing a nitrate,

$$RC(O)O_2\cdot + NO_2 \rightarrow RC(O)O_2NO_2 \tag{10-57}$$

If R is a methyl radical, $RC(O)O_2NO_2$ is $CH_3C(O)O_2NO_2$ and is called *acetylperoxylnitrate* or *peroxyacetylnitrate* (PAN). PAN is an eye irritant and is ubiquitous throughout the troposphere.

From the discussions above, two more general constituents of photochemical smog can be added: the aldocarbonylperoxylnitrates and the various photochemical oxidants produced in the foregoing reactions. Up to this point, the inventory of the constituents of photochemical smog constituents is CO, NO NO_2, O_3, HNO_3, aldocarbonylperoxyl-nitrates, and the various photochemical oxidants.

The reactions of ketones are very similar to those of aldehydes. The following reactions show the sequence (R″ is R′ with one H removed):

$$RC(O)R' + OH\cdot \rightarrow RC(O)CH(\cdot)R''\cdot + H_2O \tag{10-58}$$
$$\text{ketocarbonyl radical}$$

$$RC(O)CH(\cdot)R'' + O_2 \rightarrow RC(O)CH(O_2\cdot)R'' \tag{10-59}$$
$$\text{ketocarbonylperoxyl radical}$$

$$RC(O)CH(O_2\cdot)R''\cdot + NO \rightarrow RC(O)CH(O\cdot)R''\cdot + NO_2 \tag{10-60}$$
$$\text{ketocarbonyloxy radical}$$

$$RC(O)CH(O\cdot)R'' + O_2 \rightarrow RC(O)O_2\cdot + R''\overset{O}{\overset{\|}{C}}H \tag{10-61}$$
$$\text{alkylperoxy radical}$$

$$RC(O)O_2\cdot + NO \rightarrow NO_2 + RC(O)O\cdot \tag{10-62}$$
$$\xrightarrow{\text{etc.}}$$

3. *Reactions with hydrocarbons.* Hydrocarbons are compounds of hydrogen and carbon. Hydrocarbons may be saturated or unsaturated. The saturated hydrocarbons are those in which the carbon atoms are joined by a single bond. They are called *alkanes*. Those hydrocarbons in which the carbons are joined by a double bond or a triple bond are said to be *unsaturated*. If there is only one double bond in the molecule,

the hydrocarbon is called an *alkene;* if two double bonds are present, it is called an *alkadiene;* if there are three present, it is called an *alkatriene;* and so on. Similar naming conventions apply to hydrocarbons with triple bonds except that the *-ene* is changed to *-yne.* Thus, for a one triple bond, it is called an alkyne; for two triple bonds, it is called an alkadiyne; for three triple bonds, it is called an alkatriyne; and so on.

Examples of an alkane, alkene, and an alkyne are shown in the following structural formulas:

$$
\underset{\text{Ethane}}{
\begin{array}{c}
\text{H} \quad \text{H} \\
| \quad\; | \\
\text{H}-\text{C}-\text{C}-\text{H} \\
| \quad\; | \\
\text{H} \quad \text{H}
\end{array}}
\qquad
\underset{\text{Ethene}}{
\begin{array}{c}
\text{H} \quad \text{H} \\
| \quad\; | \\
\text{H}-\text{C}=\text{C}-\text{H} \\
| \quad\; | \\
\text{H} \quad \text{H}
\end{array}}
\qquad
\underset{\text{Ethyne}}{
\begin{array}{c}
\\
\text{H}-\text{C}\equiv\text{C}-\text{H} \\
\\
\end{array}}
$$

As shown above, a single bond joins two carbons of ethane; hence ethane is saturated. In ethene and ethyne, on the other hand, double bonds and triple bonds, respectively, join the carbons; they are said to be *unsaturated.*

Now use the symbol RH for an alkane. As in the case of the aldehydes and ketones, the hydroxyl free radical grabs a hydrogen atom when it reacts with an alkane to form H_2O and a new free radical (an alkyl radical),

$$RH + OH\cdot \rightarrow R\cdot + H_2O \tag{10-63}$$

R· then reacts with molecular O_2 to form the alkylperoxyl free radical,

$$R\cdot + O_2 \rightarrow RO_2\cdot \tag{10-64}$$

RO_2· then proceeds to react as a peroxyl does, disturbing the normal NO_2 photolytic cycle. Subsequent reactions are exactly the same as before.

In the case of the unsaturated hydrocarbon, instead of OH· abstracting a hydrogen to form water, the free radical adds into the unsaturation, a normal characteristic reaction of unsaturated hydrocarbons. The symbols for the unsaturated double-bond and unsaturated triple-bond hydrocarbons are, respectively, RHC=CHR′ and RC≡CR′. The reaction of RHC=CHR′ with OH· is

$$RHC=CHR' + OH\cdot \rightarrow \underset{\substack{\text{hydroxyalkyl}\\\text{radical}}}{RHCOH-CHR'\cdot} \tag{10-65}$$

The reaction of RC≡CR′ will be similar.

If RHC=CHR′ is ethene, OH· adds to the unsaturation producing a saturated species:

$$
\begin{array}{c}
\text{H} \quad \text{H} \\
| \quad\; | \\
\text{C}=\text{C} \\
| \quad\; | \\
\text{H} \quad \text{H}
\end{array}
+ OH\cdot \rightarrow
\underset{\substack{\text{hydroxyethyl}\\\text{radical}}}{
\begin{array}{c}
\text{H} \quad \text{H} \\
| \quad\; | \\
\text{HO}-\text{C}-\text{C}\cdot \\
| \quad\; | \\
\text{H} \quad \text{H}
\end{array}}
\tag{10-66}
$$

Subsequent reactions follow:

$$RHC(OH)CHR' \cdot + O_2 \rightarrow RC(OH)CHO_2R' \cdot \qquad (10\text{-}67)$$
hydroxyalkylperoxy
radical

$$RHC(OH)CHO_2R' \cdot + NO \rightarrow RHC(OH)CHOR' \cdot + NO_2 \qquad (10\text{-}68)$$
hydroxyalkyloxy
radical

$$RHC(OH)CHOR' \cdot + O_2 \rightarrow RHC(OH)C(O)R' + HO_2 \cdot \qquad (10\text{-}69)$$

$HO_2 \cdot$ proceeds to disturb the normal NO_2 photolytic cycle; $RHC(OH)C(O)R'$ proceeds according to the reactions of a ketone,

$$RHC(OH)C(O)R' + OH \cdot \rightarrow RHC(O \cdot)C(O)R'' + H_2O \qquad (10\text{-}70)$$

and so on.

Disturbance by photolysis products of carbonyl compounds. The discussions above addressed the reactions of carbonyl compounds with the hydroperoxyl radical. The discussion below focuses on a special reaction of the carbonyl compounds by photolysis and its effect on the NO_2 cycle. The photolysis of aldehydes is

$$RCHO + h\nu \rightarrow R \cdot + HC(O) \cdot \qquad (10\text{-}71)$$

As shown, the reaction produces a hydrocarbon photochemical oxidant product $R \cdot$, as well as an oxidant product of the photochemical reaction of HCHO. Henceforth, reactions similar to those discussed before will follow. Further reactions will ultimately lead to the disturbance of the normal NO_2 cycle and more formation of ozone.

The photolysis of ketones may be represented as

$$RC(O)R' + h\nu \rightarrow R \cdot + R'C(O) \cdot \qquad (10\text{-}72)$$

The reactions of $R \cdot$ and $R'C(O) \cdot$ have already been discussed.

The peroxyls (alkyl and acyl peroxyls) formed during a photochemical reaction may not disturb the normal NO_2 photolytic cycle. They may end up, instead, in a terminal reaction with NO and NO_2, respectively, producing organic nitrates. Letting $RO_2 \cdot$ and $RC(O)O_2 \cdot$ represent the general formula for the peroxyls, the termination reaction may be shown as

$$RO_2 \cdot + NO \rightarrow RNO_3 \qquad (10\text{-}73)$$

$$RC(O)O_2 \cdot + NO_2 \rightarrow RC(O)O_2NO_2 \qquad (10\text{-}74)$$

The products in these reactions are organic nitrates.

Up to this point, observe that the species in photochemical smog are composed of NO, NO_2, HNO_3, CO, organic nitrates, ozone, and photochemical oxidants. In addition, given the right conditions, the species participating in the photochemical smog formation may nucleate, forming aerosols. The last component of a photochemical smog may therefore be called *photochemical aerosol.*

Important photochemical smog reactions to remember.   For future reference, it is helpful to summarize all pertinent photochemical reactions. They are

Normal NO_2 photolytic cycle:

$$NO_2 + h\nu \rightarrow NO + O$$

$$O + O_2 + M \rightarrow O_3 + M$$

$$O_3 + NO \rightarrow NO_2$$

Chain and disturbance reactions:

$$O_3 + h\nu \rightarrow O(^1D) + O_2$$

$$O(^1D) + H_2O \rightarrow 2OH\cdot$$

$$OH\cdot + CO \rightarrow CO_2 + H\cdot$$

$$H\cdot + O_2 \rightarrow HO_2\cdot$$

$$HO_2\cdot + NO \rightarrow NO_2 + OH\cdot$$

$$RCHO + OH\cdot \rightarrow RC(O)\cdot + H_2O$$

$$RC(O)\cdot + O_2 \rightarrow RC(O)O_2\cdot$$

$$RC(O)O_2\cdot + NO \rightarrow NO_2 + RC(O)O\cdot$$

$$RC(O)O\cdot + O_2 \rightarrow RO_2\cdot + CO_2$$

$$RO_2\cdot + NO \rightarrow NO_2 + RO\cdot$$

$$RO\cdot + O_2 \rightarrow R'CHO + HO_2\cdot$$

$$RH + OH\cdot \rightarrow R\cdot + H_2O$$

$$R\cdot + O_2 \rightarrow RO_2\cdot$$

Terminating reactions:

$$OH\cdot + NO_2 \rightarrow HNO_3$$

$$RO_2\cdot + NO \rightarrow RNO_3$$

$$RC(O)O_2\cdot + NO_2 \rightarrow RC(O)O_2NO_2$$

Example 10-8

Assuming the NO_2 photolytic cycle is in progress, write all the chemical reactions that ethane and its reaction products will have in the formation of photochemical smog.

Solution

$$C_2H_6 + OH\cdot \rightarrow C_2H_5\cdot + H_2O$$

$$C_2H_5\cdot + O_2 \rightarrow C_2H_5O_2\cdot$$

$$C_2H_5O_2\cdot + NO \rightarrow NO_2 + C_2H_5O\cdot$$

$$C_2H_5O\cdot + O_2 \rightarrow HO_2\cdot + CH_3CHO$$

$$HO_2\cdot + NO \rightarrow NO_2 + OH\cdot$$

$$CH_3CHO + OH\cdot \rightarrow H_2O + CH_3C(O)\cdot$$

$$CH_3C(O)\cdot + O_2 \rightarrow CH_3C(O)O_2\cdot$$

$$CH_3C(O)O_2\cdot + NO \rightarrow NO_2 + CH_3C(O)O\cdot$$

$$CH_3C(O)O\cdot + O_2 \rightarrow CH_3O_2\cdot + CO_2$$

$$CH_3O_2\cdot + NO \rightarrow NO_2 + CH_3O\cdot$$

$$CH_3O\cdot + O_2 \rightarrow HO_2\cdot + HCHO$$

$$HO_2\cdot + NO \rightarrow NO_2 + OH\cdot$$

$$HCHO + OH\cdot \rightarrow H_2O + HC(O)\cdot$$

$$HC(O)\cdot + O_2 \rightarrow HO_2\cdot + CO$$

$$HO_2\cdot + NO \rightarrow NO_2 + OH\cdot$$

$$CO + OH\cdot \rightarrow CO_2 + H\cdot$$

$$H\cdot + O_2 \rightarrow HO_2\cdot$$

$$HO_2\cdot + NO \rightarrow NO_2 + OH\cdot$$

Terminating reactions:

$$OH\cdot + NO_2 \rightarrow HNO_3 \Rightarrow \text{ nitric acid}$$

$$CH_3O_2\cdot + NO \rightarrow CH_3NO_3 \Rightarrow \text{ methyl nitrate}$$

$$CH_3C(O)O_2\cdot + NO_2 \rightarrow CH_3C(O)O_2NO_2 \Rightarrow \text{ peroxyacetyl nitrate (PAN)}$$

$$C_2H_5O_2\cdot + NO \rightarrow C_2H_5NO_3 \Rightarrow \text{ ethyl nitrate } \textbf{Answer}$$

ACID RAIN

The basic components of acid rain are nitric acid and sulfuric acid. Nitric acid is a component product of the NO_2 photolytic cycle. Sulfuric acid results when SO_2 discharged from combustion processes converts to SO_3, which, in turn, reacts with the moisture in the air, producing the acid.

The fundamental question is: When is precipitation acid rain? To answer this question, let us investigate the normal environmental pH. Absolutely neutral precipitation would have a pH of 7 at 25°C. But one would not expect to observe this value because of other species that can depress the pH below 7. A major naturally occurring species that can depress the pH below neutral value is carbon dioxide. From Henry's law, CO_2 may dissolve in atmospheric moisture or on surface waters. Now, find the corresponding pH for this naturally occurring CO_2. Assume that the temperature and atmospheric pressure are 25°C and 1 atm, respectively. The equilibrium expressions are

$$K_w = \{OH^-\}\{H^+\} = (10^{-14}) \tag{10-75}$$

$$K_{HCO_3} = \frac{\{H^+\}\{CO_3^-\}}{\{HCO_3\}} = 4.69(10^{-11}) \tag{10-76}$$

$$K_{H_2CO_3} = \frac{\{H^+\}\{HCO_3^-\}}{\{CO_2\}} = 4.45(10^{-7}) \tag{10-77}$$

$$K_{CO_2} = \frac{f_g P_{atm}}{\{CO_{2(aq)}\}} = 0.164(10^4) \text{ atm/mol fraction} \tag{10-78}$$

where the K symbols represent the equilibrium constant for the substances used as subscripts, and $\{\cdot\}$ is read as "the activity of."

The partial pressure of CO_2 at normal atmospheric conditions is approximately $10^{-3.5}$ atm, and the fugacity coefficient, f_g, may be assumed equal to 1. Solving for the unknowns, the resulting equation for the hydrogen ion is

$$\{H^+\}^3 - 4.76(10^{-12})\{H^+\} - 4.46(10^{-22}) = 0 \tag{10-79}$$

Solving this equation produces the $\{H^+\}$ and the corresponding pH,

$$\{H^+\} = 10^{-5.7} \text{ mol/L} \Rightarrow \text{pH} = 5.7 \tag{10-80}$$

Thus, from the result above and applying some factor of safety, it may be concluded that when the environmental pH of a stream water falls below 5.0 (< 5.7), anthropogenic contributions of acidity may be the reason.

The following equations summarize the reactions for sulfuric acid formation:

$$SO_2 + OH\cdot \rightarrow HOSO_2\cdot \tag{10-81}$$

$$HOSO_2\cdot + O_2 \rightarrow SO_3 + HO_2\cdot \tag{10-82}$$

$$SO_3 + H_2O \rightarrow H_2SO_4 \tag{10-83}$$

$$HO_2 + NO \rightarrow NO_2 + OH\cdot \tag{10-84}$$

Referring to the equations above, SO_2 emitted by various processes, notably combustion processes, reacts with the hydroxyl free radical, forming the hydrosulfurdioxide free radical, $HOSO_2\cdot$. $HOSO_2\cdot$ then reacts with molecular O_2, producing SO_3 and the hydroperoxyl free radical, HO_2. SO_3 finally reacts with moisture, producing H_2SO_4, the agent of acid rain. The hydroperoxy free radical may also react with NO, emitted from pollution sources or from the NO produced in the NO_2 photolytic cycle, again liberating the hydroxyl radical. The latter reaction is the disturbance of the normal NO_2 photolytic cycle.

The HNO_3 is formed from the reaction of the hydroxyl radical with nitrogen dioxide:

$$OH\cdot + NO_2 \rightarrow HNO_3 \tag{10-85}$$

Once the acid-rain precursors are formed, essentially two mechanisms transport them toward the surface of the earth: dry deposition and wet deposition. *Dry deposition* is a simple transport of the precursors toward the earth's surface through settling and other physical means without the benefit of dissolving into the water in precipitation. In *wet deposition*, on the other hand, rain scrubs out the acids from the atmosphere, thus dissolving them into water. As the precursors reach the earth, they acidify natural waters: streams, rivers, lakes, and reservoirs. Some occurrences of acidification are believed to have already taken place in the Adirondack Mountains of New York State and portions of Ontario, Quebec, Nova Scotia, and Newfoundland in Canada.

DESTRUCTION OF THE OZONE LAYER

As illustrated in Figure 10-5, the earth's atmosphere may be divided into four layers: the troposphere, stratosphere, mesosphere, and thermosphere. The figure shows that the troposphere occupies the layer closest to the surface starting from zero to about 16 km above the ground. This thickness, however, is not constant throughout the globe. In the midlatitudes, it can range from 10 to 12 km (about the altitude of a typical airplane flight).

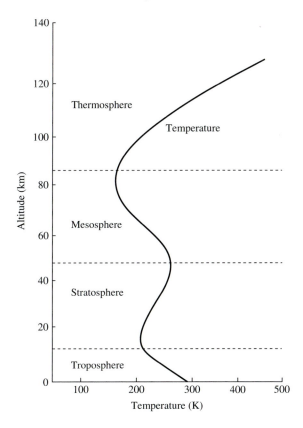

Figure 10-5 Four major layers of atmospheric temperature profile.

In the poles, due to a heavier density, it can range from 5 to 6 km, and in the equator due to lighter density, the thickness of the troposphere layer may go as high as 18 km. The troposphere has a dry adiabatic lapse rate of $-9.57°C/km$.

The troposphere is the layer where clouds form. It is normally turbulent, due to the presence of a favorable lapse rate, enabling the pollutant to mix relatively fast. In contrast, above the troposphere is a calm, stable layer called the stratosphere. From the figure, the layer stretches from an altitude of approximately 16 km to about 49 km above the surface of the earth. The reason for the stability of the stratosphere is the inverted lapse rate that occurs in this layer. The temperature increases as the altitude increases.

Figure 10-6b shows the atmospheric absorption of radiation on a clear day. As indicated ozone and molecular O_2 effectively absorbs the sun's ultraviolet radiation in the range below 0.30 μm. This absorption causes a series of photochemical reaction to occur, liberating heat. This heat liberation is the cause of the temperature rise in the stratosphere layer.

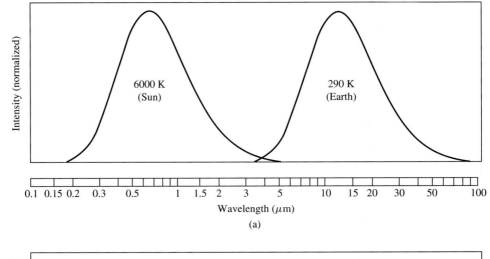

(a)

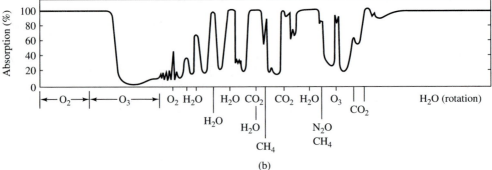

(b)

Figure 10-6 (a) Normalized blackbody radiation curves for the sun and earth. (b) Atmospheric absorption on a clear day.

Because of the temperature inversion, the stratosphere is very stable. Pollutants that come into this layer from the troposphere may become trapped here for years. There is no mechanism for scavenging any pollutant by rain since this layer is very dry and the limit of cloud formation is the troposphere. The only way that a pollutant may decay is through photolytic decomposition and by a chance-drifting back to the troposphere. Once returned to the troposphere, it can be removed by rainfall scavenging and dissolution. The troposphere and the stratosphere comprise about 99.9% of the mass of the atmosphere.

Above the stratosphere is the mesosphere which extends from approximately 49 km to about 86 km above the surface of the earth. With respect to temperature variation, it is similar to the troposphere. As the air expands, it cools. If as the air expands, however, a reaction releases heat, the air may heat up or stay cool, depending on the amount of heat released. In the stratosphere, the heat from the photochemical reaction exceeds the heat required for expansion; hence the layer heats up. In the mesophere, however, the situation is different: The heat requirement for expansion exceeds the heat released by reaction and the air cools. Hence the temperature decreases with altitude, as in the troposphere.

Above the mesosphere is the thermosphere. Figure 10-5 shows a temperature rise. The reason for this rise is the absorption of solar energy by atomic oxygen. The heating effect caused by this absorption exceeds the heat requirement for expansion, heating the atmosphere. (Within the thermosphere is a relatively dense layer of charged particles. This layer is called the *ionosphere*.)

As learned in the photolytic NO_2 cycle, ozone is an undesirable species; however, referring to Figure 10-6, ozone effectively absorbs the short-wavelength ultraviolet (UV) spectrum of less than 0.30 μm. This process occurs in the stratosphere. There is good evidence that a UV-induced alteration of deoxyribonucleic acid (DNA) causes skin cancer. The most useful function of ozone is therefore in the stratosphere.

The reactions below illustrate the formation of ozone in the stratosphere. As shown in Figure 10-6, O_2 absorbs the sun's short-UV wavelength of 0.24 μm and below;

$$O_2 + h\nu \rightarrow O + O \qquad (10\text{-}86)$$

The atomic oxygen, in turn, rapidly reacts with molecular O_2 to form ozone.

$$O + O_2 + M \rightarrow O_3 + M \qquad (10\text{-}87)$$

Radiation on the order of 0.30 μm (0.300 to 0.315 μm) has been linked to skin cancer: nonmelanoma and the more life-threatening malignant melanoma. Absorption in this range of wavelength therefore protects life on earth from the damaging biological effects of radiation. The damaging effect of radiation includes, in addition to skin cancer, eye cataracts and retinal degeneration and immune system suppression. A milder form of biological damage is human sunburn. The equation below portrays the absorption of UV radiation in the range 0.300 to 0.315 μm.

$$O_3 + h\nu \rightarrow O_2 + O \qquad (10\text{-}88)$$

Scientists first expressed concern over the possible destruction of the ozone layer in the 1970s.[8] They found that species such as Cl, Br, OH·, and NO react with ozone, destroying it. Then, in 1985, an announcement was made of the creation of a hole in the ozone layer over Antarctica the size of the United States.[9] This aroused the world to begin acknowledging the seriousness of the problem. Using X to represent the species Cl, Br, OH·, and NO, following are the reactions of ozone destruction:

$$X + O_3 \rightarrow O_2 + XO \qquad (10\text{-}89)$$

$$XO + O \rightarrow X + O_2 \qquad (10\text{-}90)$$

As seen from these reactions, the species X has been regenerated. Hence this catalyst can go on destroying thousands more O_3 molecules. The only way that the destruction can cease is when a species appears that is capable of undergoing a terminating reaction with X, or for X to return to the troposphere, where it may be removed by rain. It must also be remembered that X is composed of Cl, Br, OH·, and NO; thus the reactions above can be compounded several times over.

The dominant sources of the chlorine atom are CH_3Cl, the chlorofluorocarbons CFC-11 and CFC-12, and carbon tetrachloride. The CH_3Cl is mostly of natural origin, whereas the chlorofluorocarbons are man-made. Carbon tetrachloride is both of natural and man-made origin. Anthropogenic sources of less importance are the cleaning solvents trichloroethylene ($CCl_2 = CHCl$), methyl chloroform, and CFC-113. CFC-113 is used in the electronic industry in various critical cleaning and degreasing operations. The key chlorine species for ozone destruction are CFC-11 and CFC-12. A source of bromine atoms are Halons, fluorocarbons that contain bromine atoms and used in fire extinguishers. The sources of OH· and NO have already been discussed. Although they are in the troposphere, they can diffuse into the stratosphere to cause ozone damage.

CFCs are molecules that contain chlorine, fluorine, and carbon. They are very important commercially. Although banned by the U.S. Environmental Protection Agency since 1979, they have been used extensively in the past. CFC-11 and CFC-12 are believed to have lifetimes of 60 and 110 years, respectively; hence, they do not just disappear from the atmosphere.

To determine the chemical formula for a CFC, simply add 90 to its suffix number and interpret the resulting three-digit number as follows: the leftmost digit is the number of C atoms, the middle digit is the number of hydrogen atoms, and the rightmost digit is the number of fluorine atoms. To determine the number of chlorine atoms, subtract the number of hydrogen and fluorine atoms plus the number of single carbon-to-carbon bonds from the total possible number of single bonds that the carbon atoms can have. Thus, for CFC-12: $90 + 12 = 102$. Therefore, there are one C, zero H, and two F atoms. The H and F are two, and since there is only one C, there are no carbon-to-carbon bonds. The total possible number of single bonds of C is four. Therefore, the total number of chlorine atoms is $4 - (2 + 0) = 2$. Hence the formula is CCl_2F_2.

---

[8]M. J. Molina and F. S. Rowland (1974). "Stratospheric Sink for Chlorofluoromethanes: Chlorine Atom Catalyzed Destruction of Ozone." *Nature*, 249, pp. 810–812.

[9]G. M. Masters (1991). *Introduction to Environmental Engineering and Science*. Prentice Hall, Englewood Cliffs, N.J.

Example 10-9

What is the chemical composition of CFC-113?

Solution $90 + 113 = 203$. Therefore, $C = 2$, $H = 0$, and $F = 3$. $H + F = 0 + 3 = 3$. Since there are two C, total C-to-C bonds $= 1$, and the total possible single C bonds $= 7$. Thus the number of Cl atoms $= 7 - (3 + 1) = 3$, and the formula is $C_2Cl_3F_3$.   **Answer**

CFCs are used as aerosol propellants, refrigerants, and solvents. They are also used in foamed plastics. After CFCs were banned in 1979, replacement propellants such as carbon dioxide, isobutane, and propane were used. However, CO_2 is an important agent of the greenhouse effect. The hydrocarbons isobutane and propane are constituent species in photochemical smog reactions. Daily use of these aerosols in the home may also result in substantial problems of indoor air pollution. Presently, simple pumps have been used to propel contents instead of aerosols.

Foamed plastics are of two types: the rigid and flexible. Rigid sheets of foamed urethane plastics are used as insulation in the construction industry and in refrigeration equipment. The CFC trapped in the holes of the foam serves as an insulator and blocks heat transfer. Nonurethane foams such as extruded polystyrene (such as Dow's Styrofoam) are used extensively for egg cartons and food service trays. Expanded polystyrene foams are also used for drinking cups. Flexible foams are used in furnitures and automobile seats.

In the past, ammonia, carbon dioxide, isobutane, methyl chloride, methylene chloride, and sulfur dioxide had been used as refrigerants. These refrigerants are either toxic, noxious, highly flammable, or require high operating pressure that necessitates the use of heavy equipment. CFCs have replaced all these refrigerants with the attendant potential harmful effects to the environment. Hydrogenated CFCs (HCFCs) have, however, been found to be less damaging than the fully halogenated CFCs and may therefore be suitable replacements. In addition, HCFCs, because of the hydrogen bonds, are less stable than CFCs in the environment. They are therefore less likely to reach the stratosphere. Also, its ozone-depleting potential is only 10 to 15% of the regular fully halogenated CFCs.[10]

The creation of the ozone hole over Antarctica may be explained as follows: CFCs are very stable molecules that are not normally degraded in the usual processes taking place in the troposphere. When found in the stratosphere, however, they are broken down by UV, releasing the chlorine atom. Using CFC-12, the reaction is

$$CCl_2F_2 + h\nu \rightarrow Cl + CClF_2 \tag{10-91}$$

Cl then reacts according to reaction (10-89) destroying ozone and producing ClO. The product ClO reacts further with NO_2, forming chlorine nitrate:

$$ClO + NO_2 \rightarrow ClONO_2 \tag{10-92}$$

[10]Ibid.

At this stage Cl is effectively trapped in $ClONO_2$, an inert compound that can do no damage to the ozone. Over Antarctica, however, a phenomenon of atmospheric circulation called the *circumpolar* or *polar vortex* forms.[11] The formation of the polar vortex blocks the warmer midlatitude air from mixing with the air above the pole. Thus the polar air is trapped with no connections to the outside warmer air. This condition cools the air in the stratosphere, which can go down to $-90°C$. Even though stratospheric air is very dry, ice crystals can form at this very low temperature, providing reaction surfaces for chlorine nitrate to react with water to form HOCl and HNO_3:

$$ClONO_2 + H_2O \rightarrow HOCl + HNO_3 \tag{10-93}$$

As long as the polar vortex exists, reaction (10-93) continues to operate, accumulating HOCl; this accumulation is simply waiting for the Antarctic spring.

As the sun first rises in the Antarctic spring of August or September, HOCl photolyzes, forming Cl and the hydroxyperoxyl radical, which destroys the ozone. The sequence of reactions are as follows:

$$HOCL + h\nu \rightarrow Cl + OH \cdot \tag{10-94}$$

$$Cl + O_3 \rightarrow ClO + O_2 \tag{10-95}$$

$$OH \cdot + O_3 \rightarrow O_2 + HO_2 \cdot \tag{10-96}$$

$$ClO + HO_2 \cdot \rightarrow HOCl + O_2 \tag{10-97}$$

With this formation of HOCl, the cycle starts all over again. It is clear that even one molecule of CFC-12 can destroy thousands of molecules of ozone.

GREENHOUSE EFFECT

The sun emits solar radiation in all directions equal to 1372 W/m^2. This is called the *solar constant S*. Designating R as the radius of the earth, πR^2 is its projected area. The amount of energy that the earth absorbs is therefore equal to $S(1 - \alpha)\pi R^2$, where α is the earth's albedo. This absorption of energy heats the earth.

The *Stefan–Boltzmann law* of blackbody radiation is given by

$$W_b = \sigma A T^4 \tag{10-98}$$

where W_b is the total blackbody radiation rate, W; σ is the Stefan–Boltzmann constant, $5.67(10^{-8})$W/m^2-K^4; A the surface area of the object, m^2; and T temperature of the radiating surface, K. As the earth is heated, its temperature rises. This makes the earth capable of reradiating back energy into space according to the Stefan–Boltzmann law. At steady state, equating the rate of absorption of the energy coming from the sun to the rate of reradiation of energy by the earth into space and assuming an albedo of 30%, the predicted overall, global annual mean temperature of the air close to the ground is around

---

[11]H. J. Lugt (1983). *Vortex Flow in Nature and Technology*. Wiley, New York.

255 K.[12] However, 288 K is normally considered as the average value. The difference of 33 K would therefore represent a perturbation due to some factors. This factor is considered to be that of what is called the *greenhouse effect*.

From Figure 10-6, radiations from the sun range from a wavelength of about 0.15 μm to about 4.8 μm. The radiations from the earth, on the other hand, range from about 2.8 to about 85 μm. Hence since those coming from the sun are shorter, they are called shortwave radiations; those coming from the earth are called long-wavelength radiations. The value of 290 K for the earth's radiation curve shown in the figure was produced by adding 2 degrees to 288, to obtain the surface temperature of the earth (288 K is the average temperature of the air, not of the ground). The distributions in the figure may be derived using *Planck's law,*

$$W_{b,\gamma} = \frac{C_1 \gamma^{-5}}{e^{C_2/\gamma T} - 1}$$

(10-99)

where $W_{b,\gamma}$ is the monochromatic radiating power of a blackbody, Btu/ft^2(h)(μm); $T =$ absolute temperature, °R; $C_1 = 1.187(10^8)$ Btu/ft^2(h)(μm^4); $C_2 = 2.590(10^4)$ μm°R; and $\gamma =$ wavelength of radiation, μm.

The phenomenon of the *greenhouse effect* may be explained as follows: The short-wave radiation of the sun is absorbed by the earth, heating it to a temperature of about 255°C. Because of the heat it absorbs, the earth reradiates energy back into space in the form of long-wave radiation. In the atmosphere, however, are molecules that can absorb this reradiated energy, converting it to heat. Being heated, the molecules can now also reradiate the heat they absorb back into earth, producing the perturbation differential temperature of 33 K. This temperature differential is due to the greenhouse effect. The aforementioned molecules that produce this effect are called *greenhouse gases*. As indicated in Figure 10-6, these gases are CO_2, H_2O, N_2O, and CH_4.

It is fortunate that because of the greenhouse effect, the overall average annual temperature has been raised to 15°C (288K − 273 = 15°C). The temperature would have been −18°C only (255 K − 273 = −18°C), a temperature that can make living uncomfortable. On the other extreme, the temperature should not be so high. Scientists have discovered that the CO_2 concentration in the atmosphere has been steadily increasing, as shown in Figure 10-7. From this figure the concentration of CO_2 at the beginning of the nineteenth century was about 280 ppm. Presently, it is about 350 ppm. This is alarming, since continued rise of the concentration may increase global temperatures, which can upset the balance of nature.

It is known that carbon dioxide is a product of complete combustion, and society is very much dependent on fossil fuels. Argument has even been made as to which one to use: coal or natural gas. Advocates say that natural gas is preferrable to coal. But either way, CO_2 is still produced. In addition, burning of forests has compounded the problem. In underdeveloped countries, forest burning is the easiest way to make farms, and this is done every day. One way to solve this problem is to use sources of energy that do not produce CO_2: wind power, hydroelectric, solar, and nuclear. It is

[12]Masters, *Introduction to Environmental Engineering and Science.*

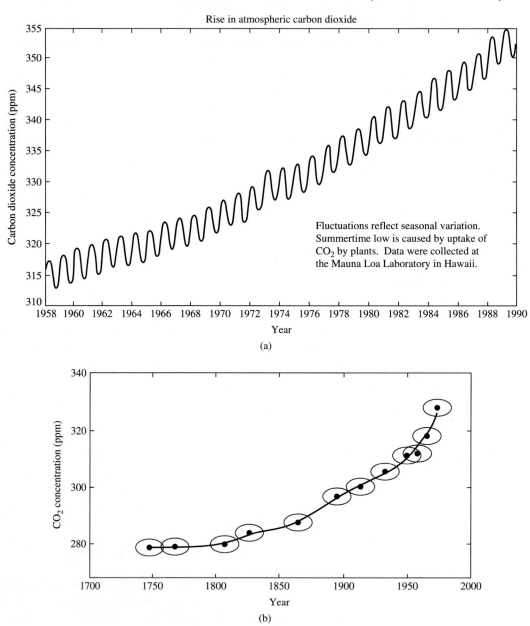

Figure 10-7   (a) Concentrations of atmospheric carbon dioxide at Mauna Loa Laboratory, Hawaii. (b) Atmospheric carbon dioxide concentrations inferred from measurements of glacial ice formed during the last 200 years. (Reprinted with permission from *Nature*, Macmillan Magazine.)

probably more desirable to perfect the method of disposal of nuclear wastes and improve the safety of nuclear plants and use them rather than to continue reliance on fossil fuels.

Added to the list of greenhouse gases are CFC-11 and CFC-12. Emprical equations have been derived to predict the equilibrium-incremental increase in temperature ΔT caused by these pollutants. The formula is[13]

$$\Delta T = \frac{\Delta T_d}{\ln 2} \ln \frac{\{CO_2\}}{\{CO_2\}_0} + 0.057(\{N_2O\}^{0.5} - \{N_2O\}_0^{0.5}) + 0.019(\{CH_4\}^{0.5} - \{CH_4\}_0^{0.5})$$

$$+ 0.14(\{CFC\text{-}11\} - \{CFC\text{-}11\}_0) + 0.16(\{CFC\text{-}12\} - \{CFC\text{-}12\}_0) \qquad (10\text{-}100)$$

where ΔT_d is the equilibrium-incremental change in temperature for a doubling in CO_2 concentration and the zero subscripts refer to the initial value of the parameter affected. The temperature is in degrees Celsius and the concentrations are in ppb.

Example 10-10

In 1850 the concentrations of the greenhouse gases were as follows: $CO_2 = 280$ ppm, $CH_4 = 1150$ ppb, $N_2O = 285$ ppb, CFC-11 = 0, and CFC-12 = 0. In 1985 they were $CO_2 = 345$ ppm, $CH_4 = 1790$ ppb, $N_2O = 305$ ppb, CFC-11 = 0.24 ppb, and CFC-12 = 0.4 ppb. By 2075[14] they are estimated to be $CO_2 = 576$ ppm, $CH_4 = 4402$ ppb, $N_2O = 478$ ppb, CFC-11 = 2.28 ppb, and CFC-12 = 3.80 ppb. (a) Calculate the equilibrium-incremental increase in temperature from 1850 to 2075. (b) Calculate the equilibrium-incremental increase in temperature from 1985 to 2075. Assume that $\Delta T_d = 3°C$.

Solution

$$\Delta T = \frac{\Delta T_d}{\ln 2} \ln \frac{\{CO_2\}}{\{CO_2\}_0} + 0.057(\{N_2O\}^{0.5} - \{N_2O\}_0^{0.5}) + 0.019\left(\{CH_4\}^{0.5} - \{CH_4\}_0^{0.5}\right)$$

$$+ 0.14(\{CFC-11\} - \{CFC-11\}_0) + 0.16(\{CFC-12\} - \{CFC-12\}_0)$$

$$\Delta T_{1850-2075} = \frac{3.0}{\ln 2} \ln \frac{576}{280} + 0.057\left(\{478\}^{0.5} - \{285\}^{0.5}\right) + 0.019\left(\{4402\}^{0.5} - \{1150\}^{0.5}\right)$$

$$+ 0.14(2.28 - 0) + 0.16(3.80 - 0)$$

$$= 3.12 + 0.284 + 0.62 + 0.32 + 0.61 = 4.95°C \quad \textbf{Answer}$$

$$\Delta T_{1985-2075} = \frac{3.0}{\ln 2} \ln \frac{576}{345} + 0.057\left(\{478\}^{0.5} - \{305\}^{0.5}\right) + 0.019\left(\{4402\}^{0.5} - \{1790\}^{0.5}\right)$$

$$+ 0.14(2.28 - 0.24) + 0.16(3.80 - 0.40)$$

$$= 2.22 + 0.25 + 0.46 + 0.29 + 0.54 = 3.76°C \quad \textbf{Answer}$$

[13] Ibid.
[14] Ibid.

GLOSSARY

Acid rain. The type of precipitation that results when oxides of nitrogen and sulfur are emitted into the atmosphere.

Bottom dead center. The limit of travel of the piston in the power and intake strokes.

Carburetion. The mixing of atomized gasoline and air.

Carburetor. The unit above an engine of the automobile where carburetion is carried out.

Catalytic incineration. A method of incineration that employs a catalyst.

Chlorofluorocarbons (CFCs). Molecules containing chlorine, fluorine, and carbon.

Combustion. Reaction of a substance with molecular oxygen to produce energy.

Compression stroke. The piston movement where the mixture of fuel and air is compressed before firing.

Equilibrium reaction. When the rate of forward reaction is equal to the rate of backward reaction, the reaction is in equilibrium.

Equivalence ratio. The reciprocal of the stoichiometric ratio.

Equivalent weight (or equivalent mass). The mass of a substance participating in a chemical reaction per unit mole of the reference species.

Exhaust gas recirculation. A process of controlling NO_x where exhaust gases are recirculated to lower the temperature of the combustion gases.

Exhaust manifold. The manifold that accepts combustion gases from an engine.

Free radical. A radical that contains one or more unpaired electrons or an atomic species formed containing one or more such unpaired electrons.

Greenhouse effect. The absorption of long-wave radiation from the earth, reradiating heat back to earth, causing an increase in temperature.

Hot soak. Emissions after the engine is shut off.

Hydroxyl free radical. The free radical formed from water when an electron bond is broken in the water molecule forming unpaired electrons.

Idling. The condition exemplified by an engine running without movement of a vehicle.

Intake manifold. The manifold that accepts the supply of fuel and air in an engine.

Intake stroke. The suction stroke of a piston.

Ionosphere. A layer of atmosphere in the thermosphere.

Lapse rate. The rate of change of temperature in the atmosphere.

Mesosphere. The atmosphere above the stratosphere; it extends from an altitude of approximately 49 km to approximately 86 km.

NO_2 photolytic cycle. The cycle of NO_2 decomposition and reformation in the atmosphere using ozone for the reformation.

Oxidation. The loss of electrons from a substance.

Peroxyl radicals. Free radicals formed from peroxides.

Photochemical smog. A mixture of reactants and products resulting from the interaction of organics and the oxides of nitrogen.

Power stroke. The movement of a piston where power is abstracted from the rapidly expanding combustion gases after firing.

Primary pollutants. Pollutants formed directly at the source.

Rapid combustion. Combustion accompanied by intense heat and light.

Reference species. Moles of electrons or the moles of positive (or alternatively, negative) oxidation states participating in a given chemical reaction.

Slow or biological combustion. Combustion mediated by organisms in which the release of heat is inconsequential compared to that of rapid combustion.

Stoichiometric ratio. The ratio of the theoretical fuel/air ratio to the actual fuel/air ratio.

Stratosphere. The limit of cloud formation; it extends from an altitude of approximately 16 km to approximately 49 km. This is where the protective ozone layer exists.

Thermosphere. The atmosphere above the mesosphere.

Three-way catalytic converter. A catalytic reactor that converts HC, CO, and NO_x to H_2O, CO_2, and N_2 using the catalysts platinum and rhodium.

Top dead center. The limit of travel of the piston in the compression and exhaust strokes.

Trophosphere. That region of the atmosphere where cloud formation occurs; it extends from the surface of the earth up to an altitude of approximately 16 km.

SYMBOLS

CFC	chlorofluorocarbon
$h\nu$	a photon of energy
k_b	backward rate coefficient
k_f	forward rate coefficient
K	equilibrium constant
K_H	Henry's law constant
R	universal gas constant

PROBLEMS

10-1. Ultimate analysis of a municipal solid waste yields the following results: H = 8.2%, C = 27.2%, N = 0.7%, O = 56.8%, S = 0.1%, ash = 7.0%, and moisture = 43.3%. This solid waste is used to fire a plant that burns 100 tonnes/day of refuse and produces 20 tonnes/day of residue. If the burning is under insufficient air, how much CO is produced per day?

10-2. In Problem 10-1, if the solid waste burns at 20% excess air, how much CO_2 is produced per day assuming negligible NO formation?

10-3. In Problem 10-1, if the solid waste burns at 20% excess air, what are the stoichiometric and equivalence ratios?

10-4. A local village in the Middle East uses cattle manure for fuel. Ultimate analysis of the dried manure shows that the percent S is approximately 0.3%. Assuming all the sulfur is converted to SO_3 and assuming that enough moisture is present in the atmosphere to react with the trioxide, calculate the amount of acid produced per ton of manure.

10-5. Calculate the amount of moisture needed to react with the SO_3 in Problem 10-4.

10-6. Assume in Problem 10-4 that the inhabitants of the village cook food three times a day using 2 kg of manure per cooking. If there are 60 houses in the village, how much sulfuric acid is produced in 1 year?

10-7. A small stream discharging 0.002 cfs runs through the village in Problem 10-4: The stream empties into a lake 5 ha in surface area and 7 m in average depth. The annual precipitation averages 127 cm. Can you predict the resulting pH in the lake as a result of the acid pollution coming from the cooking? If so, what is the resulting pH? Assume the topography of the village is such that all precipitations drain into the lake and that the headwaters of the stream are within the village boundary.

10-8. How will you confirm the accuracy of the prediction of pH in Problem 10-7?

10-9. A flue gas at a temperature of 200°C and 1 atm pressure is analyzed and found to contain 85% N_2, 5% O_2, and 10% CO_2 on the dry basis. If the humidity of the gas is 0.016 mol H_2O per mol of gas, calculate the equilibrium concentrations of CO, NO, and NO_2.

10-10. Recalculate Problem 10-9 assuming that the flue gas temperature is 1600°C.

10-11. Ultimate analysis of a municipal refuse used to fire a solid-waste-to-energy recovery plant shows the following results: H = 7.2%, C 47.6%, N = 2.0%, O = 35.7%, S = 0.3%, ash = 7.2%, and moisture = 4.9%. The plant burns 100 tonnes/day of refuse, producing 20 tonnes/day of residue containing 5% C. Calculate the combustion products for an incomplete combustion. The humidity of the air is 0.016 mol of H_2O per mole of dry air.

10-12. Calculate the percent carbon in the refuse that is gasified in Problem 10-11.

10-13. How much atmospheric oxygen and air were used in Problem 10-11?

10-14. Ultimate analysis of a municipal refuse used to fire a solid-waste-to-energy recovery plant shows the following results: H = 7.2%, C 47.6%, N = 2.0%, O = 35.7%, S = 0.3%, ash = 7.2%, and moisture = 4.9%. The plant burns 100 tonnes/day of refuse, producing 20 tonnes/day of residue containing 5% C. Calculate the combustion products at the combustion zone for complete combustion using 20% excess air and neglecting thermal NO. Use any chart that will give the percent conversion of fuel NO and assume an incoming air humidity of 0.016 mol of H_2O per mole of dry air.

10-15. Using the same data as in Problem 10-14, calculate the ideal temperature of combustion and the corresponding products of combustion.

10-16. Calculate the percent carbon in the refuse that is gasified in Problem 10-14.

10-17. How much atmospheric oxygen and air were used in Problem 10-14?

10-18. What three important emissions can you expect from operation of an automobile?

10-19. How does acceleration of the car affect emission of CO and hydrocarbons? Why?

10-20. Does deceleration decrease emissions? Why?

10-21. Why is the carburetor float used in the carburetor? Does the float tank prevent emission? Why?

10-22. If the throttle valve is closed, is it true that there is no vacuum on the underside?

10-23. Is it true that the engine piston and cylinder assembly does not create a vacuum? If it does create a vacuum, explain why.

10-24. Does delaying the spark burn the fuel almost to completion? Explain. Answer this problem if the spark is advanced.

10-25. Do emissions decrease right after engine is turned off? Why? What is the emission during this time called?

10-26. Why is the three-way catalytic converter called a three-way catalytic converter?

10-27. Write the pertinent reactions for the NO_2 photolytic cycle. What are the general classes of compounds that disturb the NO_2 photolytic cycle?

10-28. What is $O(^1D)$?

10-29. What is photochemical smog? Write the important photochemical smog reactions to remember.

10-30. Write the reactions that lead to formation of the hydroxyl radical.

10-31. What are the three compounds that react with the hydroxyl radical to initiate the complex series of photochemical reactions?

10-32. Write the chemical reactions that illustrate the disturbance of the NO_2 photolytic cycle by CO.

10-33. How many moles of ozone is produced in Problem 10-32?

10-34. How many times is the cycle being disturbed in Problem 10-32?

10-35. Write the chemical reactions that illustrate the disturbance of the NO_2 photolytic cycle by acetaldehyde.

10-36. How many moles of ozone are produced in Problem 10-35?

10-37. How many times is the cycle being disturbed in Problem 10-35?

10-38. Write the chemical reactions that illustrate the disturbance of the NO_2 photolytic cycle by butane.

10-39. How many moles of ozone are produced in Problem 10-38?

10-40. How many times is the cycle being disturbed in Problem 10-38?

10-41. Starting with propionaldehyde, how many times will the normal NO_2 photolytic cycle be disturbed before reaching formaldehyde if propionaldehyde undergoes photochemical reactions starting with $OH\cdot$?

10-42. Assuming that the NO_2 photolytic cycle is in progress, write all the chemical reactions that propane and its reaction products will have in the formation of photochemical smog.

10-43. What are the key chlorine species for ozone destruction?

10-44. Write the chemical reactions for ozone destruction in the stratosphere.

10-45. Explain the positive lapse rate in the stratosphere and the thermosphere.

10-46. Explain the negative lapse rate in the troposhere and the mesosphere.

10-47. What is the normal altitude for airplane flight?

10-48. Write the chemical reactions for acid rain formation.

10-49. What is the chemical formula of CFC-11?

10-50. Name five important compounds responsible for global warming.

10-51. In 1985 the concentrations of the greenhouse gases were as follows: $CO_2 = 280$ ppm, $CH_4 = 1150$ ppb, $N_2O = 285$ ppb, CFC-11 = 0, CFC-12 = 0. In 1985, they were $CO_2 = 345$ ppm, $CH_4 = 1790$ ppb, $N_2O = 305$ ppb, CFC-11 = 0.24 ppb, and CFC-12 = 0.40 ppb. Assuming that $\Delta T_d = 3°C$, what was the average global temperature in 1985?

10-52. A Maryland coal is burned at the rate of 5.50 tonnes/h. If the coal contains 3.0% S, what is the annual rate of emission of SO_2?

BIBLIOGRAPHY

AMMANN, P. R., and R. S. TIMMINS (1966). "Chemical Reactions during Thermal Quenching of Oxygen–Nitrogen Mixtures from Very High Temperatures." *AlChE Journal* 12, pp. 956–963.

COOPER, C. D., and F. C. ALLEY (1986). *Air Pollution Control: A Design Approach.* Waveland Press, Prospect Heights, Ill.

LUGT, H. J. (1983). *Vortex Flow in Nature and Technology.* Wiley, New York.

MASTERS, G. M. (1991). *Introduction to Environmental Engineering and Science.* Prentice Hall, Englewood Cliffs, N.J.

SEINFIELD, J. H. (1986). *Atmospheric Chemistry and Physics of Air Pollution.* Wiley, New York.

TOBOLDT, W. K. and L. JOHNSON (1983). *Automotive Encyclopedia.* Goodheart-Willcox, South Holland, Ill.

VESILIND, P. A., J. J. PEIRCE, and R. F. WEINER (1988). *Environmental Engineering.* Butterworth, Boston.

CHAPTER 11

Solid Waste Management

Imagine garbage left uncollected at one corner of your home for 3 days. Imagine also that the container is uncovered. Can you imagine the odor that results throughout your house? Now assume that the community refuse collection crew is on strike. You put your refuse out, along with your neighbors, at the curbside and it stays there for the duration of the strike. How high can the waste build up, and can you imagine the stench that results throughout the community? From this scenario you should be able to appreciate the importance of and problems associated with solid waste management. *Solid waste management* may be defined as the application of techniques that will ensure the orderly execution of the functions of collection, processing, and disposal of solid waste. These functions are called the *three basic functional elements of solid waste management. Collection* refers to the gathering of solid wastes from places such as residences, commercial, institutional and industrial establishments, and public places. *Processing* refers to the activity applied to solid waste to prepare it for subsequent operation. *Disposal* refers to the placing of solid waste in its ultimate resting place.

SOLID WASTE TERMINOLOGY

There are three states of matter: gas, liquid, and solid. As a result, there are also three types of wastes. A gas that is wasted is a *gas waste,* such as polluted air from a process that is vented into the atmosphere. A liquid that is wasted, is called such as the polluted water of wastewater, *a liquid waste.* Finally, a solid that is wasted is a *solid waste.* To become a waste, the material must be polluted. In some instances, however, a waste may not be a waste. The effluent of the Back River Sewage Treatment Plant in Baltimore, Maryland is a wastewater as far as the city is concerned. However, for Bethlehem Steel Corporation, which uses some of this effluent in its coke oven plants at Sparrows Point, Maryland, this effluent is a resource. Similarly with solid wastes—a sofa discarded by one person may be fixed and used by another.

Solid waste is synonymous with the word *refuse*. Of course, anything rejected is refuse, be it a liquid or a solid or a gas, but in solid waste jargon, refuse is solid waste. There are two components of solid waste: garbage and rubbish. *Garbage* or *food waste* is the animal and plant residue produced as a result of the handling, preparation, cooking, and eating of foods. *Rubbish* is the combustible and noncombustible portion of solid waste, excluding food waste. *Trash* is the combustible portion of rubbish. *White goods* are the bulky portions of rubbish, such as refrigerators, air conditioners, and the like. Construction and demolition debris, rock fragments, and the like are specifically called *rubble*.

SOLID WASTE CHARACTERISTICS

To effectively implement the three fundamental functions of collection, processing, and disposal, solid wastes must be characterized. Solid wastes may be characterized as to their rates of generation, as well as to their physical and chemical characteristics. Physical characterizations useful in solid waste management are itemization of individual components, moisture content, and density. On the other hand, useful chemical characterizations include proximate and ultimate analyses (chemical analysis of component atomic elements) and heating value.

Generation Rates

The overall generation rates of solid waste for the entire community, generation rates of residential as well as industrial, commercial, and institutional establishments and public places, and generation rates of individual solid waste components are required in the design of a solid waste management system. Table 11-1 shows quantities and overall generation rates of municipal solid wastes (MSW) in a study of 14 communities.

From this table, the average residential solid waste generation for rural areas is 0.97 kg/capita-day; and for suburbs, excluding Haddonfield, NJ, it is 1.67 kg/capita-day. The average for Seattle, WA and Babylon, NY, which are large cities, is 1.50 kg/capita-day. The only small city in this study is Cherry Hill, NJ with a solid waste generation of 1.80 kg/capita-day. Without going into so much statistical analysis, this study seems to indicate that there is not much difference in waste production among suburbs and cities, which averages 1.60 kg/capita-day, excluding Haddonfield, NJ. However, the rural areas show much lower production rates, by comparison.

Generation rates of industrial solid wastes in the United States range from 0.4 to 1.60 kg/capita-day. Generation of demolition and construction wastes ranges from 0.05 to 0.40 kg/capita-day. Typical generation rates of some industrial and commercial establishments are shown in Table 11-2.

TABLE 11-1 SAMPLE SOLID WASTES GENERATIONS

Community	Population	Per capita residential generation (kg/day)
Seattle, WA (large city)	497,000	1.4
Cherry Hill, NJ (small city)	73,000	1.8
Babylon, NY (large city)	213,000	1.6
Park Ridge, NJ (suburb)	8,500	2.0
Fennimore, WI (rural)	2,400	0.9
Woodbury, NJ (suburb)	10,500	2.3
Berlin, NJ (suburb)	5,600	2.3
Longmeadow, MA (suburb)	16,300	1.4
Haddonfield, NJ (suburb)	12,000	2.7
Parkways, PA (suburb)	7,000	1.0
Roman, New York (rural)	850	1.0
Lincoln Park, NJ (suburb)	11,000	1.3
Hamburg, NY (suburb)	11,000	1.4
Wilson, WI (rural)	470	1.0

TABLE 11-2 TYPICAL COMMERCIAL AND INDUSTRIAL SOLID WASTE GENERATION RATES

Source	Rate
Canned and frozen foods (tonnes/tonne of raw material)	0.05
Printing and publishing (tonnes/tonne of raw paper)	0.09
Office buildings (kg/employee-day)	1
Restaurants (kg/customer-day)	0.5
Automotive (tonnes/vehicle produced)	0.7
Petroleum refining (tonnes/employee-day)	0.05
Rubber (tonnes/tonne of raw rubber)	0.2

Solid Waste Components

The characterization of solid wastes as to physical components must incorporate those items that are readily identifiable. Table 11-3 shows generations and percentages of the various components identified from four solid waste sources.

Table 11-3 shows that the most abundant solid waste component is paper, which ranges from 11% in the rural to 50% in a large city. Yard wastes and cardboard rank next in abundance, ranging from 11% to 37% and 9% to 40% respectively. It is interesting to note that food waste is among the lowest component. Glass ranks third in abundance after yard wastes and cardboard.

TABLE 11-3 GENERATION AND PERCENTAGES OF VARIOUS SOLID WASTE COMPONENTS

Component	Lincoln Park, NJ [tonnes (%)]	Woodbury, NJ [tonnes (%)]	Seattle, WA, [tonnes (%)]	Wilson, WI [tonnes (%)]
Yard wastes	725(16)	1,800(37)	36,280(16)	9(11)
Food wastes	73(1.6)	27(0.6)	—	—
Cardboard	1,800(40)	454(9)	45,350(21)	18(22)
Newspaper	635(14)	—	45,350(21)	9(11)
Mixed papers	—	907(19)	—	—
Other papers	907(20)	—	63,490(29)	—
Wood waste	—	363(8)	—	—
Plastics	18(0.4)	8(0.2)	454(0.2)	14(17)
Glass	180(4)	363(8)	14,512(7)	18(22)
Aluminum	5(0.1)	5(0.1)	—	—
Ferrous cans and appliances	—	363(8)	—	—
Aluminum and ferrous	—	—	6,349(3)	—
Ferrous	45(1)	—	—	14(17)
Nonferrous	14(0.3)	—	—	—
Appliances	—	—	272(0.1)	—
Batteries	3(0.06)	—	—	—
Tires	18(0.4)	—	—	—
Motor oil	36(0.8)	5(0.1)	9,070(4)	—
Misc.	10(0.2)	544(11)	27(0.01)	—

Additionally, Table 11-4 shows a further percentage breakdown of the various components of solid waste. This table shows, again, that paper is the most abundant component.

TABLE 11-4 RANGE OF COMPOSITIONS OF MUNICIPAL SOLID WASTE

Component	Range (%)	Typical (%)
Yard wastes	0–10	10
Wood	1–3	2
Food wastes	5–25	17
Paper	10–40	33
Cardboard	3–10	8
Plastics	2–8	5
Textiles	0–3	2
Rubber	0–1	0.5
Leather	0–2	0.5
Misc. organics	0–4	2
Glass	4–15	5
Tin cans	1–7	5
Nonferrous	0–1	1
Ferrous metals	1–3	1
Dirt, ashes, etc.	0–10	8

Moisture Content of Solid Waste

Knowledge of the moisture content of solid waste is important when it is used in boilers to produce steam and electricity. It is also important when solid waste is composted or subjected to anaerobic decomposition in sanitary landfills. To subject an organic matter to composting and anaerobic digestion, the water content must be put at the optimum.

Sanitary landfilling is an engineered burial of refuse. Essentially, it consists of spreading waste on the ground, compacting it, and covering it with soil at the end of the working day. Landfill as a method of disposal is discussed at the end of this chapter.

The moisture content is expressed in terms of two bases: wet and dry. The wet percentage moisture of solid waste is equal to the mass of moisture divided by the total wet mass of the solid, and the dry percentage moisture of solid waste is equal to the mass of moisture divided by the dry mass of the solid. The wet and dry percentages P_w and P_d, respectively, are computed using the respective equations.

$$P_w = \begin{cases} \dfrac{W}{S_w}(100) & \text{(11-1)} \\[2mm] \dfrac{W}{S_d}(100) & \text{(11-2)} \end{cases}$$

where W is the mass of moisture, S_w the mass of wet solids, and S_d the mass of dry solids. During analysis, the sample of solid waste is dried at 77°C to drive off the moisture. W is obtained by difference of S_w and S_d.

Table 11-5 shows typical data on moisture content of municipal solid waste components.

TABLE 11-5 TYPICAL DATA ON MOISTURE
CONTENT OF MUNICIPAL SOLID WASTE
COMPONENTS

Component	Range (%)	Typical (%)
Yard wastes	45–85	60
Wood	15–40	25
Food wastes	45–85	60
Paper	3–8	5
Cardboard	3–8	5
Plastics	1–3	2
Textiles	5–15	10
Rubber	2–4	2
Leather	8–10	9
Misc. organics	10–60	25
Glass	0.5–1	0.5
Tin cans	0.5–1	0.5
Nonferrous	0.5–1	0.5
Ferrous metals	0.5–1	0.5
Dirt, ashes, etc.	6–12	8

Example 11-1

Estimate the moisture content (wet and dry bases) of the following solid waste:

Component	Mass (%)	Moisture (%), from Table 11-5	Moisture (kg)
Newspaper	15	6	0.9
Other papers	24	6	1.44
Cardboard	33	5	1.65
Glass	4.2	0.5	0.021
Plastics	0.49	2	0.0098
Aluminum	0.13	0.5, assumed the same as ferrous	0.0007
Ferrous	1.18	0.5	0.006
Nonferrous	0.35	0.5	0.002
Yard wastes	17.97	60	10.78
Food wastes	1.67	60	1.0
Dirt	2.01	8	0.16
Σ	100.0		16.30

Solution   Basis: 100 kg.

$$P_w = \frac{16.30}{100.0}(100) = 16.30\% \quad \textbf{Answer}$$

$$P_d = \frac{16.30}{100 - 16.30}(100) = 19.50\% \quad \textbf{Answer}$$

Density of Solid Wastes

The most important use of the knowledge of the density of solid waste is the determination of its compacted volume. Arguably, why measure density when volume can be measured? Unfortunately or fortunately, for better control of operation, it is easier just to weigh things than to measure volumes. For example, it is so easy to measure the weight of a truck full of solid wastes that arrives at the landfill by driving it to the top of a scale. By knowledge of the compacted density, the volume of landfill space requirement (a compacted volume) can then be calculated easily. Compacted volume is also required to size vehicles used to collect solid wastes.

Densities of solid wastes may be expressed on an *as-compacted* or *as-discarded* basis. The ratio of the as-compacted density ρ_c to the as-discarded density ρ_d is called the *compaction ratio r*, or

$$r = \frac{\rho_c}{\rho_d} \tag{11-3}$$

There are two compaction ratios: the final disposal compaction ratio, such as in landfills, and the compactor machine compaction ratio. Compactor machines are used to reduce the volume of the solid waste before final disposal. Compactor machines compaction ratios can vary from 2 to 4. Landfill as-compacted densities can vary from 297 to 891 kg/m³. Generally, 475 to 594 kg/m³ can be achieved in landfills with a moderate compaction

TABLE 11-6 TYPICAL DENSITIES OF MUNICIPAL
SOLID WASTE COMPONENTS (as discarded)

Component	Range kg/m^3	Typical kg/m^3
Yard wastes	60–200	100
Wood	130–340	230
Food wastes	130–500	300
Paper	40–140	90
Cardboard	40–80	50
Plastics	40–130	60
Textiles	40–100	60
Rubber	80–200	130
Leather	100–260	150
Misc. organics	100–350	150
Glass	150–500	200
Tin cans	50–160	90
Nonferrous	50–240	160
Ferrous metals	150–1200	350
Dirt, ashes, etc.	320–960	480

effort. A poorly compacted landfill can achieve only about 297 kg/m^3 of compacted density. With the knowledge of the as-discarded density along with these figures, the final disposal compaction ratio may be determined. Table 11-6 shows typical as-discarded densities of solid waste components. Municipal solid waste as-discarded densities may vary from 90 to 180 kg/m^3, with a typical value of 130 kg/m^3.

Example 11-2

Estimate the as-discarded density of the following solid waste. If the compaction ratio is 2.5, what size collection vehicle is needed per 1000 kg of waste?

Solution   Basis: 100 kg.

Component	Mass (%)	Density (kg/m^3), from Table 11-6	Volume (m^3)
Newspaper	15	85	0.176
Other papers	24	85	0.28
Cardboard	33	50	0.66
Glass	4.2	195	0.022
Plastics	0.49	65	0.0075
Aluminum	0.13	160	0.0008
Ferrous	1.18	320	0.0037
Nonferrous	0.35	160	0.0022
Yard wastes	17.97	105	0.17
Food wastes	1.67	290	0.0058
Dirt	2.01	480	0.0042
$\sum$	100.0		1.332

$$\rho_d = \frac{100.0}{1.332} = 75.08 \ \frac{\text{kg}}{\text{m}^3} \quad \textbf{Answer}$$

$$\rho_c = 2.5(75.08) = 187.68 \ \frac{\text{kg}}{\text{m}^3}$$

$$\text{Volume} = \frac{1000}{187.68} = 5.32 \ \text{m}^3 \quad \textbf{Answer}$$

Proximate and Ultimate Analyses and Energy Content of Solid Wastes

Proximate analysis is a chemical characterization that determines the amounts of some surrogate parameters in place of the true chemical content. Surrogate parameters normally determined in proximate analysis are moisture (loss at 105°C for 1 h), volatile matter (ignition at 600 to 950°C), fixed carbon (the carbon not burned), and ash. *Ultimate analysis*, on the other hand, is determination of the chemical elements that compose the substance.

The energy content of solid wastes is the *heat of combustion* released when the waste is burned. There are two types of heats of combustion: the *higher heat of combustion* and the *lower heat of combustion*. The higher heat of combustion includes the heat of vaporization of water, while the lower heat of combustion does not include the heat of vaporization of water. In practical applications, the lower heat of combustion represents the net heat available in the combustion reaction. Higher heats of combustion are determined by the bomb calorimeter. If energy values are not available, values of the higher heat of combustion may be obtained as follows: The heating values of carbon and sulfur are 32,851.465 kJ/kg and 9263.37 kJ/kg, respectively. The higher heating value of hydrogen is 141,989.04 kJ/kg. In any given fuel, only the net hydrogen has the heating value. Hence to obtain the heating value due to hydrogen, the combined hydrogen must be subtracted from the total hydrogen. Let H be the fraction of total hydrogen (excluding moisture) and O be the fraction of oxygen in the fuel. Since in H_2O, one unit mass of hydrogen is equal to $\frac{1}{8}$ of a unit mass of oxygen, the fraction of combined hydrogen is then equal to O/8. Hence the fraction of net hydrogen is equal to H − O/8. Considering the heating values of carbon and sulfur, the higher heating value of the fuel is then

$$H_h = 32,851\text{C} + 141,989 \left(\text{H} - \frac{\text{O}}{8} \right) + 9263 \ \text{S} \qquad (11\text{-}4)$$

where H_h is the higher heating value (kJ/kg), C the fraction of carbon, and S is the fraction of sulfur. This equation is called *Dulong's formula*.

Table 11-7 shows typical proximate and ultimate analyses of municipal solid wastes, and Table 11-8 shows typical heats of combustion for components on an as-discarded basis, as well as inert residues resulting from the combustion of these components. The heat of combustion of municipal solid wastes ranges from 9300 to 12,800 kJ/kg, with a typical value of 10,500 kJ/kg.

TABLE 11-7 PROXIMATE AND ULTIMATE ANALYSES OF MUNICIPAL SOLID WASTES

Proximate analysis	Percent	
	Range (kg/m^3)	Typical (kg/m^3)
Volatile matter	30–60	50
Fixed carbon	5–10	8
Moisture	10–45	25
Ash	10–30	25

Ultimate analysis	Percent by mass (dry basis)					
	C	H	O	N	S	Ash
Yard wastes	48	6	38	3	0.3	4.7
Wood	50	6	43	0.2	0.1	0.7
Food wastes	50	6	38	3	0.4	2.6
Paper	44	6	44	0.3	0.2	5.5
Cardboard	44	6	44	0.3	0.2	5.5
Plastics	60	7	23	—	—	10
Textiles	56	7	30	5	0.2	1.8
Rubber	76	10	—	2	—	12
Leather	60	9	12	10	0.4	8.6
Misc. organics	49	6	38	2	0.3	4.7
Dirt, ashes, etc.	25	3	1	0.5	0.2	70.3

TABLE 11-8 TYPICAL DATA ON INERT RESIDUES AND HEATS OF COMBUSTIONS OF MUNICIPAL SOLID WASTE COMPONENTS

Component	Inerts (%)		Heat of combustion (kJ/g)	
	Range	Typical	Range	Typical
Yard wastes	2–5	4	2,000–19,0000	7,000
Wood	0.5–2	2	17,000–20,000	19,000
Food wastes	1–7	6	3,000–6,000	5,000
Paper	3–8	6	12,000–19,000	17,000
Cardboard	3–8	6	12,000–19,000	17,000
Plastics	5–20	10	30,000–37,000	33,000
Textiles	2–4	3	15,000–19,000	17,000
Rubber	5–20	10	20,000–28,000	23,000
Leather	8–20	10	15,000–20,000	17,000
Misc. organics	2–8	6	11,000–26,000	18,000
Glass	96–99	98	100–250	150
Tin cans	96–99	98	250–1,200	700
Nonferrous	90–99	96	—	—
Ferrous metals	94–99	98	250–1,200	700
Dirt, ashes, etc.	60–80	70	2,000–11,600	7,000

Example 11-3

(a) Derive an empirical organic formula for the solid waste shown below. (b) Using the data in Table 11-8, estimate the higher and lower heats of combustion of the organics on as-discarded basis and on a dry basis and compare the result of the higher heating value on a dry basis with the result using Dulong's formula.

Component	Mass %
Newspaper	15
Other papers	24
Cardboard	33
Glass	4.2
Plastics	0.49
Aluminum	0.13
Ferrous	1.18
Nonferrous	0.35
Yard wastes	17.97
Food wastes	1.67
Dirt	2.01
Σ	100.0

Solution   (a) Basis: 100 kg. Using the percentage compositions in Table 11-7 and the typical moisture contents of Table 11-5, the following tables are obtained:

Component	Mass (%)	Moisture (%), from Table 11-5	Dry mass (kg)
Newspaper	15	6	14.1
Other papers	24	6	22.56
Cardboard	33	5	31.35
Glass	4.2	0.5	4.18
Plastics	0.49	2	0.48
Aluminum	0.13	0.5, assumed the same as ferrous	0.13
Ferrous	1.18	0.5	1.17
Nonferrous	0.35	0.5	0.35
Yard wastes	17.97	60	7.19
Food wastes	1.67	60	0.67
Dirt	2.01	8	1.85
Σ	100.0		84.03

| Component | Mass of respective elements | | | | | |
	C	H	O	N	S	Ash
Newspaper	6.2	0.80	6.20	0.04	0.028	0.78
Other papers	9.93	1.35	9.93	0.068	0.045	1.24
Cardboard	13.79	1.88	13.98	0.094	0.063	1.72
Plastics	0.288	0.034	0.109	—	—	0.048
Yard wastes	3.45	0.43	2.73	0.22	0.02	0.34
Food wastes	0.34	0.04	0.188	0.02	0.003	0.017
$\sum$ mass	34.0	4.53	33.14	0.44	0.159	4.15	$\Rightarrow$ 76.29 kg
%	44.57	5.94	43.51	0.58	0.21	5.43
$\sum$ moles	2.83	4.49	2.07	0.031	0.005	—
$\sum$ reduced moles	566	898	414	6.2	1	—

Empirical formula $= C_{566}H_{898}O_{412}N_{6.2}S$.   **Answer**

(b) Basis: 100 kg.

Component	Mass	Heating value (kJ/kg)	Total energy (kJ)
Newspaper	15	16,750	251,250
Other papers	24	16,750	402,000
Cardboard	33	16,300	537,900
Plastics	0.49	32,600	15,974
Yard wastes	17.97	6,500	116,805
Food wastes	1.67	4,650	7,766
$\sum$	92.13		1,331,695

Higher heating value on as-discarded basis $= \dfrac{1,331,695}{92.13} = 14,454.52$ kJ/kg   **Answer**

$$H_2O \text{ from net hydrogen} = \left(4.53 - \frac{33.01}{8}\right)\left(\frac{18}{2.106}\right) = 3.45 \text{ kg}$$

Latent heat of vaporization of water $= 2420$ kJ/kg

Lower heating value on as discarded basis $= \dfrac{1,331,695 - 3.45(2420)}{92.13} = 14,364$ kJ/kg   **Answer**

Higher heating value on dry basis $= \dfrac{1,331,695}{76.29} = 17,456$ kJ/kg   **Answer**

$$\text{Lower heating value on dry basis} = \frac{1{,}331{,}695 - 3.45(2420)}{76.29} = 17{,}346 \text{ kJ/kg} \quad \textbf{Answer}$$

$$H_h = 32{,}851 \text{ C} + 141{,}989 \left(\text{H} - \frac{\text{O}}{8} \right) + 9263 \text{ S}$$

$$= 32{,}851(0.45) + 141{,}989 \left(0.0594 - \frac{0.433}{8} \right) + 9263(0.0021) = 15{,}551 \text{ kJ/kg}$$

higher heating value on dry basis $< 17{,}456$ kJ/kg   **Answer**

Note: The reason for the difference between the higher heating value obtained using Dulong's formula and the higher heating value obtained using the table is due to the uncertainty of the typical heating value chosen from the table.

SOLID WASTE COLLECTION AND TRANSPORTATION

Collection is the first fundamental function of solid waste management. *Solid waste collection* refers to the gathering of solid wastes from places such as residential, commercial, institutional, and industrial areas, as well as public parks. There are, generally, two methods of collection: hauled-container system and stationary-container system. In the *hauled-container system,* the container is hauled from the collection point to the final point of disposal, processing facility, or transfer station. In the *stationary-container system,* the container is emptied into collection vehicles at the point of collection. *Transportation* refers to the hauling of solid wastes to relatively far distances from the collection areas or transfer station. The distance traveled may be to a final point of disposal or processing facility.

Hauled-Container System

Figure 11-1a shows the definition sketch of a hauled-container system. The truck (collection vehicle) spends a time t_1 in driving from the dispatch station to the first collection point, designated as a "container location." In the hauled-container system, the container is so large that it is not emptied at the collection point but is mounted into a collection vehicle and hauled to a transfer station, processing station, or disposal site, called collectively a *destination point*. Let the time taken to mount the container into the collection vehicle be m, and let s be the time spent to unload the cargo of solid waste at the destination point. From the destination point, the vehicle then goes back to the container location, and since the container is very large, it will also take time to unload this container. Let the time for this unloading be u. Also, let h be the haul time, which is the time it takes to drive from the collection point to the destination point and back. If more than one collection point or container location is serviced, time will also be spent in driving from one collection point to another. Let d_ℓ be the average between-collection-point driving time per collection point. The activities involved—mounting the container, driving to the destination point, unloading the cargo, driving back to the collection point,

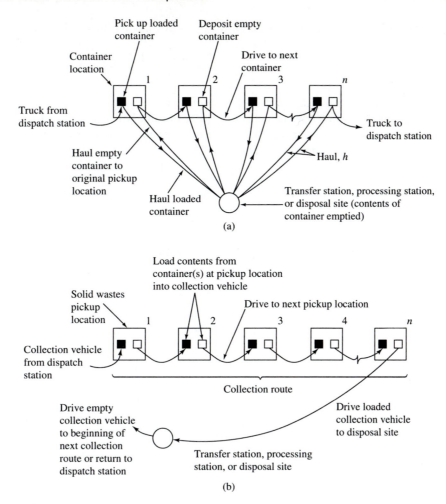

Figure 11-1 Methods of collection system: (a) Hauled-container system; (b) stationary-container system.

unloading the container, and driving between collection points—constitute a *trip*. If t_{net} is the on-the-job time used per trip, then

$$t_{net} = m + u + d_\ell + s + h \qquad (11\text{-}5)$$

Let H be the total allotted time during a workday. Within this allotted time, there will always be some time wasted for non-job-related activities such as taking an unauthorized break on the job, going to stores, talking to friends, and all other off-job activities. Also, include the unavoidable delay due to traffic congestion in this category. Let W be the fraction of H spent on these non-job-related activities. At the end of the working day, the vehicle will also have to spend time t_2 to drive from the last collection point to the dispatch station. The net time spent for collection activities is then only

$(1 - W)H - t_1 - t_2$. The number of trips, N_t, possible during a collection day is therefore

$$N_t = \frac{(1 - W)H - t_1 - t_2}{t_{net}} \tag{11-6}$$

An example length of time to mount and unload a container is 0.40 h, and an example length of time to unload cargo at the site is 0.127 to 0.133 h. The haul time would depend on the speed limit of the route taken; t_1 and t_2, as well as d_ℓ, will have to be determined by an on-site study.

Example 11-4

A contractor agreed to haul the solid wastes from an industrial district of a city. The industries agreed to store their wastes on large containers located at strategic points. Due to the sizes of the containers, the hauled-container system of collection is to be used. Based on a traffic study, t_1, t_2, and d_ℓ were found to be 20, 25, and 8 minutes, respectively. If the round-trip haul distance averaged 60 km at a speed limit of 55 mph, how many containers can be serviced on a collection day of 8 h?

Solution

$$t_{net} = m + u + d_\ell + s + h$$
$$N_t = \frac{(1 - W)H - t_1 - t_2}{t_{net}}$$

Assume that $W = 0.15$, $m + u = 0.4$ h/trip and $s = 0.133$ h/trip. Then

$$t_{net} = 0.4 + \frac{8}{60} + 0.133 + \frac{60(0.62)}{55} = 1.34 \text{ h}$$

$$N_t = \frac{(1 - 0.15)(8) - 20/60 - 25/60}{1.34} = 4.5 \text{ or 4 trips} \quad \textbf{Answer}$$

Note: Actual hours spent on collection day $= 4(1.34) + \dfrac{20 + 25}{60} + 0.15(8) = 7.31$ h.

Stationary-Container System

In this system, containers are emptied at the collection point or container location. There are, generally, two types: one in which the containers are large and must be emptied by mechanical means, and one in which the containers are small and can simply be emptied manually. Examples of the first case are those in which solid wastes are collected from apartment buildings, large commercial establishments, and public parks. In these instances, wastes are stored in large containers strategically located in the premises. The second case refers to the collection in residential areas. Figure 11-1b shows the schematic of a stationary-container system.

There are generally two methods of collecting solid wastes in residential areas: the curbside or alley method and the backyard collection method. In the *curbside* or *alley method,* residences are required to place the solid wastes in containers at the curb or alley during designated collection days. The number of collection days per week varies with localities. In Annapolis, Maryland, for example, presently, regular collection is on Mondays and Thursdays. Collection of bulk items is every other 3 months. There are also regular intervals for collection of yard wastes and recyclable bottles and cans. In the *backyard method*, no collection days need to be specified. The collection crew simply enter residential backyards and empty the owner's solid waste storage containers into tote barrels for loading into the waiting collection vehicle on the street.

One major difference between the hauled-container system and the stationary-container system is that in the former the collection trip empties just one container; in the latter system, the collection trip empties more than just one container. In other words, in the hauled-container system, the number of trips is equal to the number of container locations. Let the number of container locations in the stationary-container system be C_ℓ. Also, there is no m or u in the stationary-container system; instead, there is the time to pick up and load the solid wastes from a container location to the collection vehicle. Call this time p. Using the same meaning for d_ℓ as in the hauled-container system, t_{net} is

$$t_{net} = (p + d_\ell)C_\ell + s + h \qquad (11\text{-}7)$$

As mentioned, d_ℓ is the average between-container location driving time *per container location*. Call the average driving time between container location *dbc*. If there are C_ℓ container locations, the number of driving distances in a collection route is $C_\ell - 1$, and the total amount of time spent driving (in a collection route) is $dbc(C_\ell - 1)$. Hence d_ℓ is

$$d_\ell = \frac{dbc(C_\ell - 1)}{C_\ell} \qquad (11\text{-}8)$$

The number of trips, N_t, is then

$$N_t = \frac{(1 - W)H - t_1 - t_2}{t_{net}} \qquad (11\text{-}9)$$

The values of p and d_ℓ would vary depending on whether the system is mechanically or manually loaded. In mechanically loaded systems p and d_ℓ could be as much as 0.2 h and 0.1 h, respectively. In manually loaded systems, on the other hand, p could be as low as 30 s. dbc, the parameter used to calculate d_ℓ, is simply the driving time between houses; hence it can be very low, on the order of half a minute. At any rate, on-site time study must be performed to determine these durations. Also, if it takes 1 minute for one person to empty a location, it does not follow that it will take only 0.5 minute for two persons to empty the same location. Hence the rate of emptying for a two-man crew, is unique for this size of crew much as the rate of emptying of a one-man crew is unique for this one-man crew. Hence once the size of crew has been determined, a time study should be performed in another collection system having the same size crew. Normally, a two-man crew with one driver is utilized.

Another factor to be considered in a stationary-container system is the determination of the size of the collection vehicle. Clearly, the size of the vehicle must be sufficient to hold all the solid wastes in one trip. If V_d is the volume of the refuse as discarded in one trip or collection route and r is the compaction ratio, the required volume of the

vehicle V_h is

$$V_h = \frac{V_d}{r} \tag{11-10}$$

If the total discarded volume of solid wastes on collection day is V_{dc}, the total number of trips is also given by

$$N_t = \frac{V_{dc}}{V_d} = \frac{V_{dc}}{V_h r} \tag{11-11}$$

Figure 11-2 shows a vehicle of the type used for residential neighborhood collection. This particular vehicle makes 250 stops per day.

Figure 11-2 Solid waste collection vehicle in James Island, South Carolina.

Example 11-5

In an on-site time study of a mechanically loaded stationary-container system, the time it takes to empty all locations in a collection route is 2.32 h. The average driving time between container locations is 0.1 h. If there are 11 container locations, calculate p and d_ℓ.

Solution

$$p = \frac{2.32}{11} = 0.21 \text{ h} \quad \textbf{Answer}$$

$$d_\ell = \frac{dbc(C_\ell - 1)}{C_\ell} = \frac{0.1(11 - 1)}{11} = 0.09 \text{ h} \quad \textbf{Answer}$$

Example 11-6

In an on-site time study of a manually loaded stationary-container system, the time it takes to empty all locations in a collection route is 2.32 h. The average driving time between container locations is 15 s. If there are 270 residences serviced in one collection route, calculate p and d_ℓ.

Solution

$$\text{Number of collection locations} = \text{number of residences}$$

$$p = \frac{2.32}{270}(60)(60) = 31 \text{ s} \quad \textbf{Answer}$$

$$d_\ell = \frac{dbc(C_\ell - 1)}{C_\ell} = \frac{15(270 - 1)}{270} = 14.94 \text{ s} \quad \textbf{Answer}$$

Example 11-7

A total of 11 apartment complexes and commercial establishments have entered into a contract with a solid waste collection firm to collect their solid wastes. The discarded volume of solid wastes to be collected is 96 m^3 per collection day. Determine the number of trips required on collection day, the size of the collection vehicle, and the size of the container to be provided. Assume that $r = 2.5$ and, for simplicity, assume that each of the 11 sources contributes the same amount of solid wastes. Make any additional assumptions necessary.

Solution

$$N_t = \frac{(1 - W)H - t_1 - t_2}{t_{\text{net}}}$$

$$t_{\text{net}} = (p + d_\ell)C_\ell + s + h$$

$$d_\ell = \frac{dbc(C_\ell - 1)}{C_\ell}$$

Assumptions:

$$dbc = 0.1 \text{ h}$$

$$p = 0.2 \text{ h}$$

$$s = 0.1 \text{ h}$$

$$\text{Round-trip distance} = 60 \text{ km at speed limit of 55 mph}$$

$$W = 0.15$$

$$H = 8 \text{ h}$$

$$t_1 = 20 \text{ min}$$

$$t_2 = 25 \text{ min}$$

$$d_\ell = \frac{0.1(11 - 1)}{11} = 0.09 \text{ h}$$

$$t_{\text{net}} = (0.2 + 0.09)11 + 0.1 + \frac{0.62(60)}{55} = 3.97 \text{ h}$$

$$N_t = \frac{(1 - 0.15)8 - 20/60 - 25/60}{3.97} = 1.52$$

If only one trip is planned, the crew will have to go home early. Hence let N_t be equal to 2. Then, in this case, the crew size must be increased to decrease p. For $N_t = 2$, solve for t_{net}.

$$t_{\text{net}} = \frac{(1-W)H - t_1 - t_2}{N_t} = \frac{(1-0.15)8 - 20/60 - 25/60}{2} = 3.03 \text{ h}$$

Solving for p yields

$$t_{\text{net}} = (p + 0.09)11 + 0.1 + \frac{0.62(60)}{55} = 3.03$$

$p = 0.12$ h; hence the crew size must be increased to empty one container location in 0.12 h.

$$N_t = 2 \quad \textbf{Answer}$$

$$V_h = \frac{96/2}{2.5} = 19.2 \text{ m}^3 \text{or next-larger available vehicle size} \quad \textbf{Answer}$$

$$\text{Volume of container} = \frac{96/2}{11} = 4.36 \text{ m}^3 \quad \textbf{Answer}$$

Layout of Collection Routes

To manage a solid waste collection system effectively, the community, industrial park, and the like should be divided into collection districts (or boroughs) in which a number of collection routes must be laid out. Laying out of routes should follow a minimum of guidelines, among which could be the following:

1. Routes should not overlap and should be continuous.
2. On-the-job time, t_{net}, of all collection routes should be made equal as possible.
3. Congested areas should not be collected during rush hours.
4. To service a dead-end street, collect it when it is on the right of a main street, to avoid making a left turn.
5. As practicable as possible, collect a street on both sides.
6. As practicable as possible, select a consistent collection pattern.
7. For collection on steep streets, it is best for the vehicle to be collecting while it is moving downslope rather than upslope. This is safer and will save energy on gasoline.

Laying of collection routes is a four-step process. First, a relatively large-scale map for the area to be serviced should be prepared. On this map, indicate the collection points. Second, estimate the total volume of waste to be collected in the entire collection area and apportion equally to each collection point. Third, decide on the number of routes to collected. This will determine the size of the collection vehicle. Fourth, taking into consideration guidelines 1 to 7 above, layout the collection routes by trial and error such that, most important, each route should have about the same number of collection points and about the same on-the-job time. Figure 11-3 shows a layout of collection routes addressed in the next example.

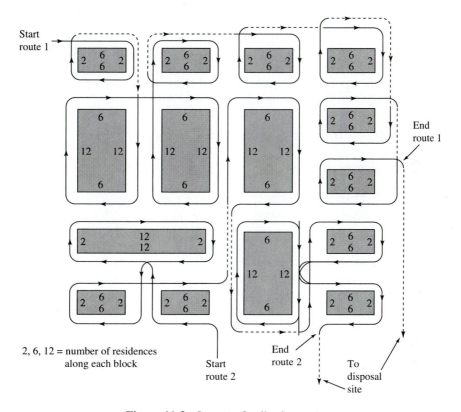

Start route 1

End route 1

2, 6, 12 = number of residences along each block

Start route 2

End route 2

To disposal site

Figure 11-3   Layout of collection routes.

Example 11-8

Figure 11-3 shows blocks of residences. The numbers written on the blocks represent the number of residences along the blocks. The average occupancy per residence is 3.5 and the average solid waste production is 1.6 kg/capita-day. Assuming a collection frequency of once per week, determine the number of trips to be made on a collection day and calculate the volume of the collection vehicle needed. Assume that the compacted volume of solid wastes in the collection vehicle is 325 kg/m^3. Lay out the collection routes.

Solution

Let the number of trips, N_t, be equal to 2 = number of collection routes.   **Answer**

Number of residences = 10(16) + 1(28) + 4(36) = 332

$$\text{Compacted volume of solid wastes collected on collection day} = \frac{332(3.5)(1.6)(7)}{325}$$

$$= 40.0 \text{ m}^3$$

$$V_h = \frac{40}{2} = 20 \text{ m}^3 \quad \textbf{Answer}$$

To lay out the collection route, the number of residences along one collection route should be determined. This is,

$$\text{number of residences per collection route} = \frac{332}{2} = 166$$

Hence the route should be laid out such that 166 residences are collected per trip. The laid-out routes are shown in Figure 11-3.   **Answer**

Solid Waste Transfer Stations and Transportation

As the distance from the collection system to the processing facility or disposal site (collectively called destination point) increases, the cost of hauling or transportation also increases. There will eventually be a certain transport distance, where management must decide whether or not a transfer station is to be built. A *transfer station* is a facility where the wastes collected may be stored temporarily or *transferred* from the smaller collection vehicles to bigger transport vehicles for transportation to the destination point. There are two general types of transfer stations: direct-discharge transfer and storage transfer station. In the former, collection vehicles dump their loads directly into the larger transportation vehicles; in the latter, the solid wastes are emptied into storage pits or platforms. The wastes are then later loaded into big transport vehicles for hauling to the destination point.

Figure 11-4 shows the Western Acceptance Solid Waste Transfer Station in Baltimore County, Maryland. This facility is a direct-discharge transfer station. Notice the three transport trailers inserted into the bays located at the back of the building. In front of this building are large hoppers through which collection vehicles dump their loads into these trailers.

Break-even point analysis.   Let the cost of direct hauling to the destination point from the collection system be K_d dollars per unit mass of solid waste. If the unit cost of this direct haul is k_{hd} dollars per unit mass of solid waste per time t,

$$K_d = k_{hd}(t) \tag{11-12}$$

The costs involved when a transfer station is constructed are operating cost, amortization cost, and hauling cost. Let the sum of the unit cost of operating and amortization be K_{ft} dollars per unit mass of solid waste and the cost of hauling be k_{ht} dollars per unit mass of solid waste per unit time. If the hauling cost using a transfer station is K_t dollars per unit mass of solid waste, then

$$K_t = K_{ft} + k_{ht}(t) \tag{11-13}$$

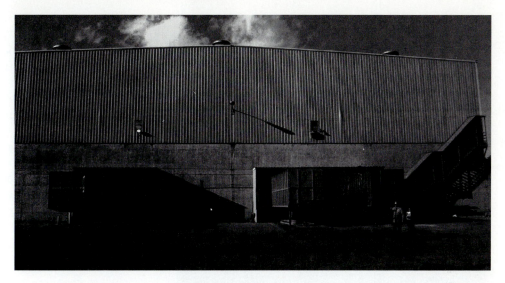

Figure 11-4 Western acceptance solid waste transfer station in Baltimore County, Maryland.

Since using a transfer station will afford the operator the opportunity to utilize a larger transport vehicle, k_{ht} is expected to be less than k_{hd}. Hence there will be a length of haul time that the plot of equations (11-12) and (11-13) will intersect. This point of intersection is called the *break-even point*. Beyond this point, direct haul to the destination point will no longer be economical compared to when a transfer station is constructed. Equating equation (11-12) to equation (11-13), the break-even time t_{even} may be solved; thus

$$t_{even} = \frac{K_{ft}}{k_{hd} - k_{ht}} \qquad (11\text{-}14)$$

Example 11-9

The unit cost of amortization and operation of a transfer station is \$3.00 per tonne of solid waste. The cost of hauling solid waste from the transfer station to a sanitary landfill is \$0.020 per tonne-min, while the cost of direct haul from the collection system to the same landfill is \$0.050 per tonne-min. If the speed limit is 55 mph, compute the distance beyond which direct haul is no longer economical.

Solution

$$t_{even} = \frac{K_{ft}}{k_{hd} - k_{ht}} = \frac{3.0}{0.050 - 0.020} = 100 \text{ min}$$

$$\text{distance} = 55 \left(\frac{100}{60} \right) = 91.7 \text{ miles} \quad \textbf{Answer}$$

SOLID WASTE PROCESSING AND RECOVERY

Processing is the second fundamental function of solid waste management. Processing improves the efficiency of solid waste disposal and prepares solid waste for subsequent recovery of materials and energy. In the not-too-distant past, disposal of solid wastes included open dumps, sanitary landfills, and disposal at sea. Because of environmental problems associated with open dumps and sea disposal, the only acceptable method of solid waste disposal at present is sanitary landfilling. Some of the materials in solid waste can also be recovered and recycled for manufacture of the same or new products. For example, instead of mining silica for glass manufacture, recovered glass may be remelted. Also, recovered paper can be reprocessed to produce new paper. The organics from solid wastes can be processed biologically to produce compost. Electricity can also be recovered from solid waste. Figure 11-5 shows a photograph of the processing facility of the Phoenix Solid Waste Management Center.

Figure 11-5   Processing facility of the Phoenix Solid Waste Management Center.

Topics on processing to be discussed in this section include processing for recovery of materials for recycling, processing for recovery of materials for direct manufacture of solid waste products, and processing for recovery of energy and incineration. The recovery for energy to be discussed pertains to the production of electrical energy from municipal solid wastes. The process of incineration is discussed after energy recovery. Recycling is discussed first.

Recycling

The recovery of solid waste components for possible use as raw materials is called *recycling* or *salvaging*. The future of recycling is either certain or uncertain. Inasmuch as there is resistance on the part of industries to use recycled materials in product manufacture, it is clear that processing virgin raw materials is more economical than processing recycled materials. In a democracy, everything revolves around money. If a certain venture is to be undertaken, the first question is: "How much profit do I get? If I am required to use solid waste components for my raw material, I must ask you: 'Can I make a profit? How about my competitor?'" If recycling is to be successful, the government should listen to these questions. If the government can spend billions of dollars for defense, it should, at least, be able to spend some few millions for recycling. Here is why.

If Table 11-3 is analyzed, the percentage of paper plus cardboard collected are as follows: Lincoln Park, 74%; Woodbury, 28%; Seattle, 71%; and Wilson, 33%. The range is therefore from about 30 to 70%. From these limited data, a recycling rate of 50% is probably possible. But the question is: Now that all these papers have been collected, is there any industry that will take them? If the answer is no, these papers and cardboards will simply pile up and would simply become new solid wastes, themselves, to be disposed of. The recycling program is then a failure. This is where the government should come in and lead the way by providing funds for demonstrations that will show that these recovered materials can be used. Of course, the recovered papers can be used as raw materials!

Going down the list in Table 11-3, yard wastes, wood, and food wastes can all be recovered. The corresponding percentages are Lincoln Park, 18%; Woodbury, 46%; Seattle, 16%; and Wilson, 11%. Hence the range is about 10 to 40% and an average would probably be 20%. Considering the paper and cardboard, the total rate now adds up to 50 plus 20 = 70%. This looks high, but if it is, so be it. Therefore, assuming that these figures are correct, the potential of recovering solid wastes in this country is 70%—just for the organic components alone. If this can be realized, our equivalent solid waste production would indeed be very small. What is needed is government money to fund demonstration projects to show that recycled solid waste components can be a resource for needed raw materials. If this is done, the future of solid waste recycling is not uncertain.

Processing for Recovery of Materials for Recycling

Processing to segregate solid waste components may be done at the point of generation (on-site processing) or at a central processing facility. On-site processing needs the cooperation of the waste producer: homes, commercial establishments, industries, and the like. At this writing, news abounds that the citizens of this country are cooperating in the nationwide effort of recycling (i.e., they are willing to do the processing that is required in recycling or salvaging). In on-site processing, wastes are segregated into types at the point of generation. For example, paper is put into one container, cans in another, and so on. The collection crew would correspondingly have separate containers. In Annapolis, Maryland, special days are set aside for collection of segregated items.

If on-site processing is not done, segregation into components may be done at a central facility. Unit operations in a central facility involve screening, air classifying, and magnetic separations. Size reduction using shredders, although not a segregation or separation process, is also used to produce a more uniformly sized product. Magnetic separation involves the use of electromagnets. This unit operation is used only to separate ferrous materials from the rest of the solid waste. Shown in Figure 11-6 are (a) the schematic of a trommel screen, (b) a hammermill shredder, and (c), a schematic flowsheet of a typical solid waste separation facility.

Screening.   Screening is a unit operation of separating a feed into oversize and undersize products. *Oversize products* are those that do not pass the openings of the screen; *undersize products* are those that do pass the openings of the screen. Screens may be classified as primary, secondary, and tertiary screens, depending on where in the flowsheet the unit is located. The screen with the largest size opening is the *primary screen*. The trommel is a primary screen. They are put ahead of all separation units in a separation facility.

Feed should not be allowed simply to cascade down the length of a trommel screen, nor should it be allowed simply to centrifuge on the side. *Cascading* is due to a slow rotation, while *centrifuging* is due to a fast rotation. The ideal operation would be to allow the feed to climb up the sides of the screen and to drop when it reaches the summit of the rotation. The ideal speed of rotation may be derived as follows: referring to Figure 11-6. Let the screen be inclined at an angle γ and let the tangential velocity of rotation be v_t. At the summit of rotation while the solid waste is on the verge of falling, the component of the weight in the central direction is the cause of the centripetal motion (of the solid). Hence applying Newton's law at this instant gives

$$Mg \, \cos \gamma = M \frac{v_t^2}{r} \qquad (11\text{-}15)$$

where M is the mass of solid waste, g the acceleration due to gravity, and r is the radius of the trommel screen. Solving for v_t,

$$v_t = \sqrt{gr \, \cos \gamma} \qquad (11\text{-}16)$$

Converting equation (11-16) to angular speed ω (in radians per unit time) using the relation $v_t = r\omega$, we obtain

$$\omega = \sqrt{\frac{g \cos \gamma}{r}} \qquad (11\text{-}17)$$

The angular speed ω is called the *critical speed*. A speed below ω but not so small that the solids are simply cascading is called the *cataracting speed*.

Example 11-10

Calculate the critical speed of a 3-m-diameter trommel inclined at an angle of 2°. Express this critical speed in rpm.

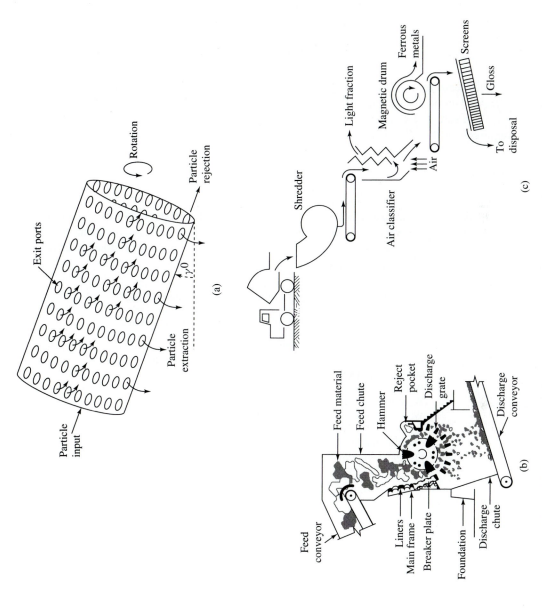

Figure 11-6 (a) Trommel screen. (b) Hammermill shredder. (c) schematic flowsheet of a typical solid waste separation facility.

Solution

$$\omega = \sqrt{\frac{g \, \cos \gamma}{r}} = \sqrt{\frac{9.81 \cos 2°}{1.5}} = 2.56 \text{ rad/s} = \frac{2.56}{2\pi} \text{ rps}$$

$$= \frac{2.56}{2\pi}(60) = 24.45 \text{ rpm} \quad \textbf{Answer}$$

Shredding.   The purpose of shredding is to produce a more uniform product. Figure 11-6b is a sketch of a shredder.

An essential element in the design of shredders is the determination of the power requirement W. This may be calculated by using an empirical equation developed by Bond:[1]

$$W = 10W_i \left(\frac{1}{\sqrt{L_p}} - \frac{1}{\sqrt{L_f}} \right) \tag{11-18}$$

where W_i is called the *Bond work index*, L_p the size of the product (in μm) that is 80% finer (also called *80th centile* finer) and L_f the size of the feed (in μm) that is also 80% finer. The units of W are kWh/ton. Table 11-9 shows values of Bond indices of shredders located in various places.

TABLE 11-9   TYPICAL BOND WORK INDICES OF SHREDDERS

Location	Material shredded	W_i (kWh/ton)
Charleston, SC	MSW	400
St. Louis, MO	MSW	400
Washington, DC	Aluminum cans	650
Washington, DC	Glass	8
Washington, DC	MSW	470
Wilmington, DE	MSW	450
Washington, DC	Paper	200
Washington, DC	Steel cans	260
Pompano Beach, FL	MSW	400

Example 11-11

Determine the power requirement to shred 100 tonnes/day of solid waste having an 80th percentile size of 30 cm to a product having an 80th percentile size of 1.7 cm. Assume that $W_i = 434$.

Solution

$$W = 10W_i \left(\frac{1}{\sqrt{L_p}} - \frac{1}{\sqrt{L_f}} \right)$$

[1]P. A. Vesilind, J. J. Peirce, and R. F. Weiner (1988). *Environmental Engineering*. Butterworth, Boston. p. 284.

$$100 \text{ tonnes/day} = \frac{100(10^3)(2.2)}{2000} = 110 \text{ tons/day}$$

$$W = 10(434)\left[\frac{1}{\sqrt{0.017(10^6)}} - \frac{1}{\sqrt{0.30(10^6)}}\right] = 25.36 \text{ kWh/ton}$$

$$\text{Power} = 25.36(110) = 2789.6 \text{ kWh/day} = 116.23 \text{ kW}$$
$$= 116.23(1.341) = 155.86 \text{ hp} \quad \textbf{Answer}$$

Air classifying. Air classifying uses the same principle as sand filter backwashing. Whereas sand is expanded or fluidized in sand filters, in air classifying, solid waste is expanded or fluidized instead. Also, whereas water is used in filter backwashing, air is used in air classifying. As shown in the schematic of Figure 11-6c, air is introduced at the bottom of the unit. To separate a component, the air velocity must be adjusted so that it is equal to or greater than the terminal settling velocity of the component. Settling velocities of steel fractions could range from 280 to 500 cm/s, whereas settling velocities of paper fractions could range from 10 to 180 cm/s. Hence paper can be separated easily from steel if the air velocity in the air classifier is maintained greater than 10 cm/s but less than 280 cm/s. On the other hand, settling velocities of plastics could range from 30 to 280 cm/s, whereas settling velocities of aluminum fractions could range from 140 to 290 cm/s. From this, a mix of larger size plastics and aluminum cannot be separated.

Processing for Recovery of Materials for Direct Manufacture of Solid Waste Products

After the components have been separated, the organic fractions may be further processed to produce desired products. One of these is refuse-derived fuels (RDFs) and compost. RDFs may be in powdered form produced by grinding using ball mills or they may be in cubed form produced by pelletizing. Figure 11-7 shows a flowsheet for the manufacture of RDFs.

Composting may also be applied to the organic fraction of solid waste. Composting is an enhanced process of rapidly oxidizing a solid organic material with oxygen. Some important considerations in the design of composting facilities include particle size, oxygen requirement, seeding, moisture content, carbon/nitrogen ratio, temperature, pH, and odor control. The optimum size varies from 20 to 75 mm. Oxygen should reach all points in the mass. In the windrow method, this will necessitate frequent turning of the piles in 4- to 5-day intervals. To reduce composting time, seeding with partially decomposed solid waste is important. Moisture content should be controlled so as to allow free passage of air to the mass and to provide the requisite amounts for biological reactions. The optimum appears to be 55%. The carbon/nitrogen ratio should be maintained from 30 to 50% by mass. For other wastes, such as bark, phosphorus may also be added. For best results the temperature should be maintained at 55 to 60°C. If the temperature exceeds 66°C, the biological reaction is reduced significantly; if this happens, the heat may be released by turning up the windrow or, in the case of the static-pile method, by

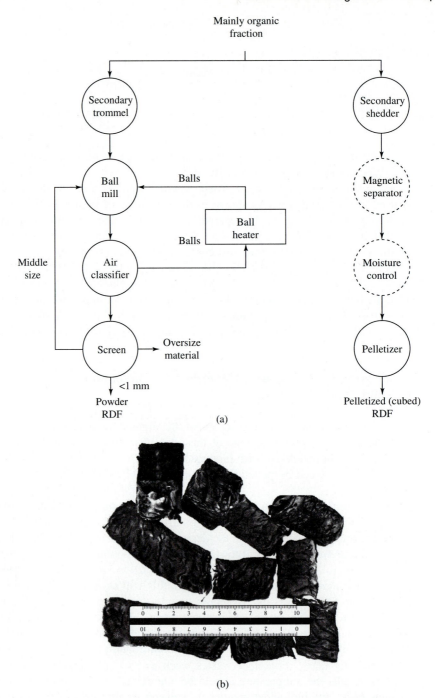

Figure 11-7 (a) Flowsheets for manufacturing RDFs. (b) Densified RDF.

adjusting the blower. The temperature of the pile can reach 80°C. The pH approaches neutrality after an initial drop due to the production of acid intermediates. If the pH goes above 8.5, there will be a loss of nitrogen, which can upset the carbon/nitrogen ratio.

In the profession, NIMBY ("not in my backyard") is often quoted, especially for siting of landfills. One of the major reasons for this is the production of odors. However, a composting plant has been constructed in a 20-acre facility in Prairieland, Minnesota utilizing a process called SILODA.[2] This process is all completely enclosed. The solid waste simply goes to one entry point, is subjected to the composting process inside the totally enclosed composting building, and goes out at the exit point already composted. This could change the word NIMBY to YIMBY ("yes in my backyard"). Figure 11-8 shows schematics of the two basic methods of composting: the static pile and windrow methods.

Electrical Energy Recovery

MSW (municipal solid wastes) has a heating value ranging from 9300 to 12,800 kJ/kg. It is possible to recover this energy by using MSW to fire boilers in order to produce steam that can be used to drive a steam turbine. The turbine then turns a generator producing electricity. (See Figure 11-8c.)

Before electricity is produced, there are efficiencies and losses that must be accounted for. Heat losses include heat losses due to the sensible heat content of the ash and the unburned carbon remaining in the ash, heat loss due to radiation, and water losses. Since the actual furnace combustion suspends all the combustion water in gaseous form, energy of vaporization is needed for this suspension. This wasted energy is the water loss. Water losses include loss due to the moisture content, loss due to combined water, and loss due to the net hydrogen water. The latent heat of vaporization of water is equal to 2420 kJ/kg.

For the sensible heat of the ash, the specific heat is normally taken as 1047 J/kg-C°. The heating value of carbon is 32,851 kJ/kg. Radiation losses range from 0.003 to 0.005 kJ/kg of fuel. After all the losses are accounted for, these are subtracted from the higher heating value of MSW to obtain the sensible heat content of the stack gases. These gases are then passed through boiler tubes. As the gases travel through the tubes, the sensible heat contents are given up heating the water in the boiler to steam. It is this steam introduced to the steam turbine that drives the generator to produce electricity. For steam-turbine generator systems of less than 12.5 MW capacity, thermal efficiencies range from 24 to 40%, with a typical value of 29%, excluding the boiler; for systems over 12.5 MW capacity, thermal efficiencies range from 28 to 32%, with a typical value of 31.6%, also excluding the boiler. MSW boilers have thermal efficiencies of about 70%. The power plant itself uses electricity. Approximately 6% of gross plant electrical output is allotted for this service. In addition, an unaccounted for plant loss of 5% is also provided.[3] A schematic of an energy recovery plant is in Figure 11-8c.

[2]R. Woods (1991). "A French Revolution Comes to Minnesota." *Waste Age*, 2(1), pp. 58–62.

[3]H. S. Peavy, D. R. Rowe, and G. Tchobanoglous (1985). *Environmental Engineering*. McGraw-Hill, New York. p. 674.

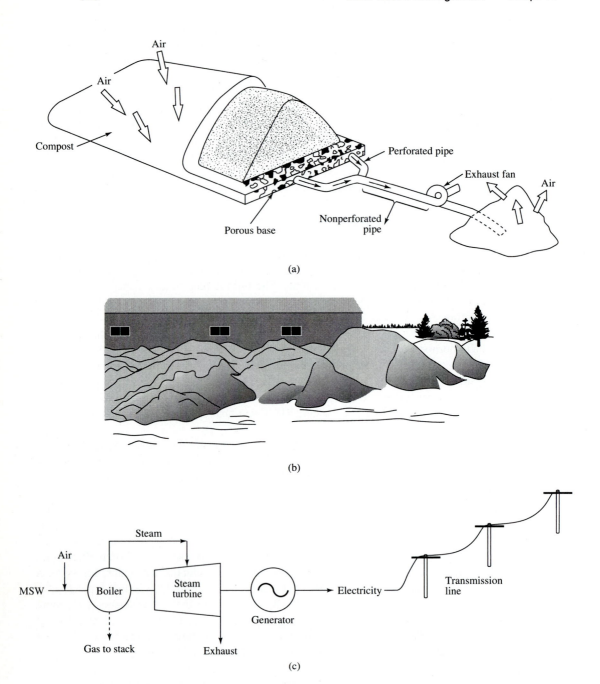

Figure 11-8 (a) Static pile composting method. (b) Windrow composting method. (c) Electrical energy recovery.

Example 11-12

MSW, at a rate of 100 tonnes/day with an empirical formula of $C_{562}H_{900}O_{414}N_{6.6}S$, is used to fire an MSW energy recovery plant. The MSW contains 20% moisture and 20% inert. The ash contains 5% carbon and the higher heating value of the MSW is 12,000 kJ/kg. Also, assume that the fuel enters at 25°C and that the ash is at 420°C. How much steam energy is available from the boiler?

Solution Basis: 100 tonnes MSW/day.

$$\text{Total heat input} = 100(10^3)(12,000) = 1.2(10^9) \text{ kJ/day}$$

$$\text{Heat loss in the ash} = \text{loss due to sensible heat} + \text{heat of combustion of C}$$

$$\text{Inerts} = 0.20(100)(10^3) = 20,000 \text{ kg/day}; \qquad \text{let } x = \text{mass of C}$$

$$\frac{x}{20,000 + x} = 0.05 \qquad x = 1053 \text{ kg/day}$$

$$\text{Ash} = 20,000 + 1053 = 21,053 \text{ kg/day}$$

$$\text{Heat loss in ash} = 21,053(1.047)(420 - 25) + 1053(32,851) = 4.33(10^7) \text{ kJ/day},$$

where 25°C is the temperature for sensible heat calculations.

$$\text{Water losses} = \text{loss due to moisture} + \text{loss due to combined water}$$

$$+ \text{ heat loss due to net hydrogen water}$$

$$\text{Moisture} = 0.20(100)(10^3) = 20,000 \text{ kg/day}$$

$$C_{562}H_{900}O_{414}N_{6.6}S = 0.60(100)(10^3) = 60,000 \text{ kg/day}$$

$$C_{562}H_{900}O_{414}N_{6.6}S = 562(12) + 900(1.008) + 414(16) + 6.6(14) + 32 = 14,399.6$$

$$\text{Combined water} = \frac{414(16)}{14,399.6}(60,000)\left(\frac{18}{16}\right) = 27,600\left(\frac{18}{16}\right) = 31,051 \text{ kg/day}$$

$$\text{Net hydrogen water} = \left[\frac{900(1.008)}{14,399.6}(60,000) - \frac{2.016}{16}(27,600)\right]\left(\frac{18}{2.016}\right) = 2,701 \text{ kg/day}$$

$$\text{Total water} = 20,000 + 31,051 + 2710 = 53,761 \text{ kg/day}$$

$$\text{Water losses} = 53,761(2420) = 1.301(10^8) \text{ kJ/day}$$

Assume that radiation losses = 0.004 kJ/kg MSW; then

$$\text{radiation losses} = 0.004(100)(10^3) = 400 \text{ kJ/day}$$

$$\text{total losses} = 4.33(10^7) + 1.301(10^8) + 400 = 1.734(10^8) \text{ kJ/day}$$

$$\text{sensible heat in stack gas} = 1.2(10^9) - 1.734(10^8) = 1.0266(10^9) \text{ kJ/day, based on } 25°C.$$

(For simplicity of calculation, flue gas has been assumed to cool down to the base temperature of 25°C.)

Assuming a boiler efficiency of 70%,

$$\text{steam energy available} = 0.7(1.0266\}(10^9) = 7.18(10^8) \text{ kJ/day} \quad \textbf{Answer}$$

Example 11-13

If the steam in Example 11-12 is used to drive a steam-turbine generator plant, calculate the plant export of electrical energy. Calculate the overall efficiency of the plant.

Solution   Assumptions: thermal efficiency = 31.6%; plant service allowance = 6%; unaccounted for losses = 5%. Then

$$\text{net electrical export} = [0.316(7.18)(10^8)][1 - (0.06 + 0.05)]$$

$$= 2.02(10^8) \text{ kJ/day} = \frac{2.02(10^8)}{3600(24)} = 2339 \text{ kW} \quad \textbf{Answer}$$

$$\text{overall efficiency} = \frac{2.02(10^8)}{1.2(10^9)}(100) = 16.8\% \quad \textbf{Answer}$$

Incineration

The incineration plant is very similar to that of the energy recovery plant. They differ only in purpose. While the energy recovery plant aims to recover electricity, the incineration plant aims to reduce the volume of waste to be disposed of. Volume reduction is approximately 90%. The heat balance calculation is the same as that for the energy recovery plant. Figure 11-9a shows the basic components of incinerator design; the bottom part is a schematic of the Krefeld incinerator with energy recovery.

DISPOSAL OF SOLID WASTES

Disposal is the third fundamental function of solid waste management. In the not-too-distant past, dumps and disposal at sea were practiced. At present, however, because of inherent environmental problems associated with these methods, they are no longer allowed. The only method of disposal currently allowed and permitted is the use of sanitary landfill. The sanitary landfill method is an engineered burial of refuse. It consists essentially of spreading waste on the ground, compacting it, and covering it with soil at the end of the working day or other suitable intervals for wastes other than sanitary solid wastes.

Landfilling Methods

There are generally two methods of sanitary landfilling: the *area method* and the *trench method*. The area method is used when it is impossible to excavate, especially when the groundwater is high. In this method, a berm is constructed and the solid wastes are simply dumped on the ground, spread in layers of about 0.5 m, and compacted. Another layer of 0.5 m is then placed on top of the previous layer and also compacted. Layering and compacting are repeated until a height of 2 to 3 m is reached. At this point and at the end of a working day a cover of 150 to 300 mm of earth is compacted on the top of the height. This cover is called a *daily cover*.

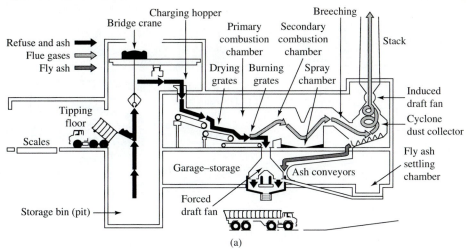

(a)

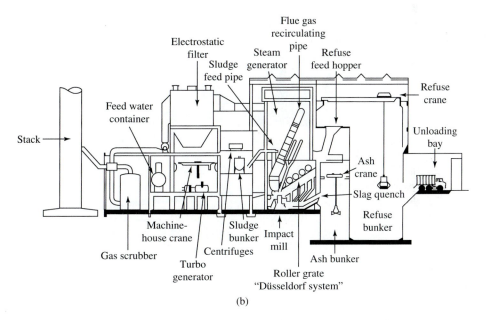

(b)

Figure 11-9   (a) Basic components of incinerator design. (b) Krefeld incinerator with energy recovery.

   Locate the term "earthen levee or berm" in Figure 11-10a. To the right of this berm is a completed unit of compacted solid waste or fill shown shaded. This unit is called a *cell*. Another cell to the right of the previous cell is then filled and compacted in the same way. The cell filling then proceeds to the right until the entire horizontal span is filled. In the figure a total of five horizontal cells are filled in the span. Cells are then further constructed on top of the previous cells to complete another horizontal span of

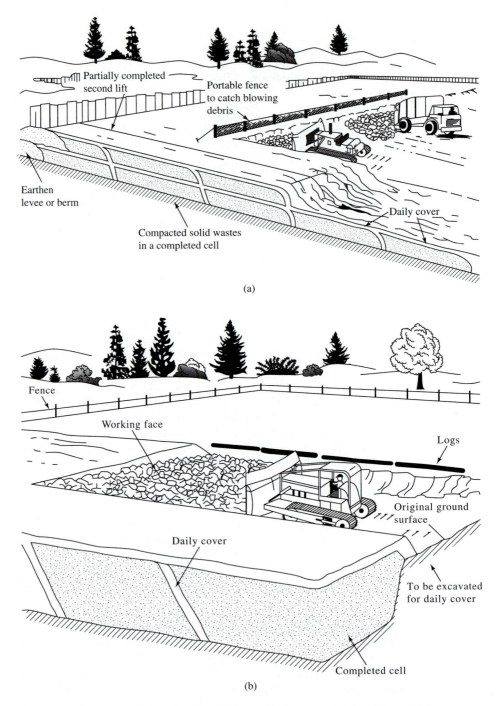

(a)

(b)

Figure 11-10 Methods of landfilling: (a) area method; (b) trench method.

cells. After a horizontal span is completely filled, another span of horizontal cells may then be compacted on top of the previous span of horizontal cells. The span of cover between the upper and lower span of cells is called the *intermediate cover*. The vertical dimension of the entire span of a series of horizontal cells, including the intermediate cover as indicated in the figure, is called a *lift*. A partially filled second lift is shown in the figure.

There is no limit to the number of lifts that can be filled as long as the side slope of the resulting heap does not exceed a certain criterion. This slope is normally limited to a maximum of 1 vertical to 4 horizontal. When the limit of the filling is reached, a final compacted cover of about 0.6 m over the entire heap of fill is installed. A cap system is then also installed on top of the final cover.

The area method is used when it is impossible to excavate, especially when the groundwater is high. The cover materials, however, may have to be imported off site. When it is possible to excavate, the trench method is used. This method has the benefit of having the cover material right at the site from the earth excavated from the trench. The method begins by excavating a trench as shown in Figure 11-10b. The width of the trench must allow free and easy movement of the compacting equipment and vehicles. Other than the addition of excavation, the method of filling, compacting, and covering are exactly the same as in the area method.

Basic Aspects in Landfill Implementation

Implementing the landfill method of disposing of solid wastes should consider five basic aspects: site selection, leachate control, gas control, operation plan, and permit application. Although permit application is put last, it strongly influences the other four aspects. It behooves the applicant for an operation of a landfill to involve the permit engineer of a state or local permitting agency when considering these aspects. *Leachate* is the polluted water that results from rain percolating through the mass of solid wastes (and leaching out any pollutant that the waste may contain).

Site selection.   Factors to be considered in selecting a site include location of the site, haul distance, soil types and geology, surface hydrology, and hydrogeologic conditions. For political reasons, *site location* is probably the most important factor in selecting a site. The following two examples illustrate the importance of this factor: the first occurred on the western shore of Maryland and the second on the eastern shore.

A hearing was held for the reissuance of an existing landfill permit for the one on the western shore. The hearing room was full of people, including a state delegate. Almost all these people, including the delegate, spoke against the reissuance of the permit because big transportation vehicles and trailer trucks hauling solid wastes destined for the landfill coming not only from Maryland but also from other states passed through the main road of the community. The traffic not only caused dust to be generated but also posed a great danger to the community. The applicant closed the landfill down.

The second case also involved issuance of a permit. However, the operation plan called for minimum disruption of the community, and a good public relations campaign was launched explaining to the community how a landfill operates, and that if operated properly, it not only would dispose of their solid waste but would also be a good neighbor. Whether good neighbor or not, the fact is that the hearing was also full of people, but this time, almost all of them spoke for issuance of the permit. Hence the location of the landfill should not cause traffic to pass though neighborhoods; access to the site should be on back roads. If this cannot be done, a landfill should not be located on the site.

Shorter *haul* distances to the landfill would obviously be preferred to longer *haul* distances. This was illustrated in the previous example on break-even point analysis. *Soil types* and *geology* are also very important factors. These determine whether or not cover soil will be available at the site or be imported off site, making a big difference in operational cost. Soil borings should be made at the site to determine the types of soil available. A sample of soil may be characterized according to some form of general description; for a more detailed characterization, some method of soil classification may be used. To understand fully the factor of soil as it affects site selection or its use as a fill, general soil description and classification are discussed briefly below.

General Soil Description.   The general description of soil may include its texture, structure, and consistency. The *texture* of a sample refers to the grain size distribution of the individual component sizes. Components may include the following:

Boulders:	larger than 15 cm	Cobbles:	5 to 15 cm
Gravel:	2 mm to 5 cm	Sand:	0.05 to 2 mm
Silt:	0.002 to 0.05 mm	Clay:	smaller than 2 μm

A soil composed of sand is *coarse-textured*; a soil composed of clay is *fine-textured*. If the percentages of the components are evenly represented in the sample, the soil is said to be *well graded*; if the sample is composed of practically the same size, the soil is said to be *poorly graded*. Soil *structure* is a description of the shape and geometry of clusters of individual soil particles making up the soil mass. These clusters are separated from each other by *surfaces of weakness*. A "cluster" composed of a single grain such as a grain of sand is said to be single-grain structured. Single-grain structured soils lack cohesion. Soil *consistency* refers to the ability of soil to hold together when pressure or stress is applied to the mass. Typical descriptions of consistency of soils are as follows:

Loose	noncoherent even when pressed
Friable	soil material crushes easily under gentle to moderate pressure between thumb and forefinger
Firm	soil material crushes under moderate pressure between thumb and finger, but resistance is distinctly noticeable
Hard	soil material cannot be crushed
Stiff	soil material deforms without crushing under moderate pressure

Plastic   soil material behaves like modeling clay
Soft   soil material is easily deformed
Sticky   soil material sticks to the hand

Soil Classification.   A number of systems has been used to classify soils, including the *CSSS* (Comprehensive Soil Survey System), the *AASHTO* (American Association of State Highway and Transportation Officials) *system*, the *FAA* (Federal Aviation Administration) *system*, and the *USCS* (Unified Soil Classification System). The CSSS is similar to the biological method of classifying living things, which is composed of *kingdom, phylum, class, order, family, genus*, and *species*. The corresponding classification in the CSSS are *order, suborder, great group, subgroup, family*, and *series*. The AASHTO and the FAA systems are geared toward the performance of soils on roads and runways. During World War II, Casagrande developed the USCS for use in the construction of airfields. His method became very popular and, along with the AASHTO classification, is used in landfill technology.

Casagrande divided all soils into coarse- and fine-grained soils. If more than 50% of the sample does not pass a 200-mesh screen (200 openings to the inch), the soil is classified as coarse-grained; if 50% or more passes the screen, the soil is classified as fine-grained. The coarse-grained soil is further subdivided into sand and gravel. If more than 50% does not pass a 4-mesh screen (4 openings to the inch), the soil is classified as gravel, G; if 50% or more passes the screen, the soil is classified as sand, S. G is suffixed with W if it is well graded, and with P if its poorly graded. It is suffixed with M if it is silty; and with C if it is clayey. The same suffixes are also used for sand (i.e., SW, SP, SM, and SC). Summarizing the classification for sand and gravel gives us

GW   well-graded gravels or well-graded gravel-sand mixtures, with little or no fines

GP   poorly graded gravels or poorly graded gravel–sand mixtures with little or no fines

GM   silty gravels or silty gravel–sand mixtures

GC   clayey gravels or clayey gravel–silt mixtures

SW   well-graded sands or gravelly sands with little or no fines

SP   poorly graded sands or poorly graded gravelly sands with little or no fines

SM   silty sands or silt–sand mixtures

SC   clayey sands or clayey silt–sand mixtures

Fine-grained soils are classified as silt (M) or clay (C), modified by the *Atterberg limits* for fine-grained or cohesive soils. The Atterberg limits are *shrinkage limit*, plastic limit, liquid limit, and *flocculating limit*. A soil sample is at the *shrinkage limit* when it is saturated and is at its smallest volume. The water content when the soil truly behaves like a plastic material is called the *plastic limit*; the water content at which the soil losses its plastic characteristics and starts to behave like water is called the *liquid limit*; and the water content at which the soil behaves like true water is called the *flocculation limit*. The flocculation limit may also be considered as the point at which, because of the increasing concentration of soil, the water departs from its true nature as water and starts to become viscous. Hence the range between the flocculation limit and the liquid limit is called the *viscous liquid range*. Going from the liquid limit to the plastic limit, the soil increases its plasticity, hence this range is called the *plastic range* and the liquid content is called the *plasticity index*. A material is a *high plastic* if the plasticity index is high; it is a *low plastic* if the plasticity index is low. A soil with a liquid limit of less than 50 (mass of water divided mass of dry soil times 100) is a low-plasticity soil (L); those with equal to or greater than 50 is a high-plasticity soil (H). Also, going from the plastic limit to the shrinkage limit, the soil increases its "hardness"; hence the range is called the *semisolid range*.

In Casagrande's USCS classification, M and C are suffixed by L and H. Also, if the silt or clay is organic, M is replaced by O and suffixed by either L or H, as the case may be. Following is a summary of the classification for silt and clay:

ML   low-plasticity silts, very fine sands, silty or clayey fine sands, and clayey silts

CL   low- to medium-plasticity clays, gravelly clays, sandy clays, silty clays, and lean clays

OL   low-plasticity organic silts and organic clays

MH   high-plasticity silts, micaceous or diatomaceous fine sandy or silty soils, and elastic silts

CH   high-plasticity clays and fat clays

OH   high-plasticity organic clays and organic silts

The ratio of the solid waste volume to cover volume normally varies from 3:1 to 4:1. Table 11-10 shows the suitability of general soil types as cover materials. Table 11-11 shows some characteristics of soils pertinent to sanitary landfills.

TABLE 11-10 SUITABILITY OF GENERAL SOIL TYPES AS COVER MATERIALS[a]

Function	Gravel	Clayey–silty gravel	Sand	Clayey–silty sand	Silt	Clay
Prevent rodents from burrowing	G	F–G	G	P	P	P
Keep flies from emerging	P	F	P	G	G	E[b]
Minimize moisture entering fill	P	F–G	P	G–E	G–E	E[b]
Minimize gas venting through cover	P	F–G	P	G–E	G–E	E[b]
Supports vegetation	P	G	P–F	E	G–E	F–G

[a]E, excellent; G, good; F, fair; P, poor.

[b]Except when cracks extends through cover.

TABLE 11-11 SOIL CHARACTERISTICS PERTINENT TO SANITARY LANDFILLS [a]

Soil	Potential frost action	Drainage characteristics	Permeability (cm/s)	Compaction characteristics
GW	None to very slight	Excellent	$> 10^{-2}$	Good: tractor, rubber-tired and steel-wheeled roller
GP	None to very slight	Excellent	$> 10^{-2}$	Good: tractor, rubber-tired and steel-wheeled roller
GM	Slight to medium	Fair to poor; poor to practically impervious	10^{-3}–10^{-6}	Good with close control: rubber-tired and sheepsfoot
GC	Slight to medium	Poor to practically impervious	10^{-6}–10^{-8}	Fair: rubber-tired and sheepsfoot roller
SW	None to very slight	Excellent	$> 10^{-3}$	Good: tractor
SP	None to very slight	Excellent	$> 10^{-3}$	Good: tractor
SM	Slight to high	Fair to poor; poor to practically impervious	10^{-3}–10^{-6}	Good with close control: rubber-tired and sheepsfoot roller
SC	Slight to high	Poor to practically impervious	10^{-6}–10^{-8}	Fair: sheepsfoot and rubber-tired roller
ML	Medium to very high	Fair to poor	10^{-3}–10^{-6}	Good to poor, close control essential: rubber-tired and sheepsfoot roller
CL	Medium to high	Practically impervious	10^{-6}–10^{-8}	Fair to good: sheepsfoot and rubber-tired roller
OL	Medium to high	Poor	10^{-4}–10^{-6}	Fair to good: sheepsfoot roller
MH	Medium to very high	Fair to poor	10^{-4}–10^{-6}	Poor to very poor: sheepsfoot roller
CH	Medium	Practically impervious	10^{-6}–10^{-8}	Fair to poor: sheepsfoot roller
OH	Medium	Practically impervious	10^{-6}–10^{-8}	Poor to very poor: sheepsfoot roller

[a] Peat and highly organic soils are not recommended for landfills.

The proposed site should be investigated for its *hydrologic characteristics* (e.g., its runoff potential). The impact of precipitation and the resulting infiltration into the landfill may be minimized by installing suitable covers (to be discussed later). Surface drainage from precipitation must be incorporated in a stormwater management design. This will entail using runoff prediction formulas such as the rational method or the curve number method of the Soil Conservation Service. No further examples will be introduced, as prediction of runoff has been discussed elsewhere in this book. The *hydrogeologic condition* of the site should also be investigated. Water table fluctuations should be studied to determine the highest elevation of the groundwater surface. There should be a safe distance between the highest water table elevation and the bottom of the proposed landfill. Maryland regulations require a distance of about 900 mm.

Example 11-14

A town with a population of 10,000 produces solid wastes at a rate of 5 lb/capita-day. If the waste is compacted to 1000 lb/yd^3, how many acre-feet of landfill is needed in a year? Assuming that the ratio of solid waste to cover is 4:1, what volume of cover soil is needed in a year, and what type of soil would you recommend as a cover?

Solution

$$\text{Volume of landfill needed} = \frac{5(10,000)}{1,000}(365)(27)\left(\frac{1}{43,560}\right) = 11.31 \text{ acre-feet} \quad \textbf{Answer}$$

$$\text{Volume of cover} = \frac{11.31}{4} = 2.83 \text{ acre-feet} \quad \textbf{Answer}$$

Normally, in practice, any soil available is used for cover.   **Answer**

Leachate control.   Leachate control is the second of the five basic aspects in implementing the sanitary landfill method of solid waste disposal. Leachate is given special consideration since it had been an important source of groundwater contamination in the past, especially in connection with the use of open dumps. To control the production of leachate, a *cap system* is installed above the final cover and a *liner system* is installed at the bottom of the landfill. Basically, then, with the installation of the top cap and the bottom liner, precipitation is practically excluded. This arrangement is desirable provided that enough moisture is provided in the solid waste heap to effect the necessary biological decomposition. (Read Chapter 2 for chemistry of water requirements in anaerobic decomposition.)

The most important part of the cap system is the *cap liner* or, simply, the *cap*. This cap may be made of synthetic material or 450 mm of compacted clay having a permeability of about 10^{-5} cm/s. Above the cap is a drainage layer about 150 mm thick, preferably composed of sand or soil that has a permeability of 10^{-2} cm/s or greater to allow fast drainage. Above the drainage layer is a geotextile-filter membrane to protect the drainage layer from being clogged. Finally, about 300 mm of soil suitable for plant growth is put on top of the drainage layer. This soil is called the *vegetative layer*. The purpose of the cap is to prevent rain from permeating the solid waste, hence the installa-

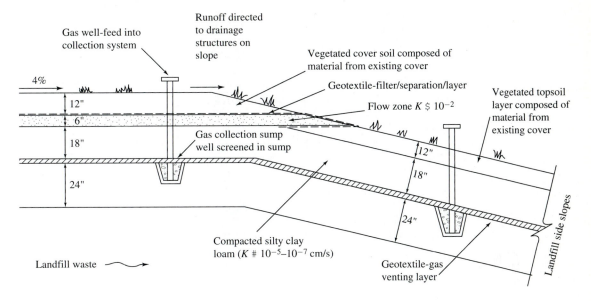

Gas well-feed into
collection system

Runoff directed
to drainage
structures on
slope

Vegetated cover soil composed of
material from existing cover

Geotextile-filter/separation/layer

Flow zone $K \gtrsim 10^{-2}$

Vegetated topsoil
layer composed of
material from
existing cover

4%

12"

6"

18"

24"

Gas collection sump
well screened in sump

12"

18"

24"

Compacted silty clay
loam ($K \# 10^{-5}–10^{-7}$ cm/s)

Landfill waste

Geotextile-gas
venting layer

Landfill side slopes

Figure 11-11   Cap system equipped with gas wells.

tion of the drainage layer above the cap. Allowing rain will leach out a large amount of pollutants from the solid waste, resulting in a very concentrated leach water or leachate. The cap liner, drainage layer, and the vegetative cover are also collectively called simply a *cap*.

Figure 11-11 shows a cap system used in one of the landfills in Maryland. This particular system is equipped with gas wells used for collecting gases. Note that the wells are located at the final cover and connected to the geotextile-gas venting layer.

Figure 11-12 shows the bottom lining of a sanitary landfill with leachate pipe. In general, the bottom lining system is composed of the *subgrade, subbase* of the liner, the *liner, drainage layer*, and a *protective drainage cover*. All subgrade soils should be free of organics and other deleterious materials and should not contain particles larger than 15 cm or so in diameter. Soils classified as SC, SM, SW, SP, ML, or CL as determined by ASTM D422-D421 for grain size determination are suitable for subgrade construction. Subgrades are constructed in lifts 12 in. thick and compacted to at least 90% of the maximum attainable dry density and to a moisture content of $\pm$ 2% of optimum moisture content as determined by a suitable laboratory procedure. Normally, the modified Proctor compaction test (ASTM D-1557) is used.

The subbase is normally 0.6 m in thickness. It is constructed in lifts of 20 cm and compacted to at least 90% of the maximum attainable dry density, as in the case of the subgrade. The moisture content should be approximately 2 to 3% wet of the soil's optimum moisture content. The permeability of the subbase should be equal to or less than 10^{-5} cm/s. In the determination of permeability (Triaxial or Falling Head Methods, EPA 9100), the sample should first be compacted to 90% of the maximum dry density

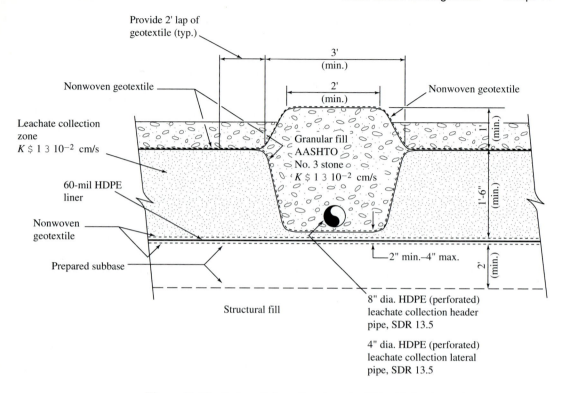

Figure 11-12   Bottom lining of sanitary landfill with leachate pipe.

at water content that is approximately 2% wet of the optimum. Soils classified as ML or CL are suitable for subbase construction.

Above the subbase is installed the *liner*. Immediately above and below the liner [shown as 60-mil HDPE *(high-density polyethylene)*], are *geotextiles* (*geo* means earth; *textile* relates to clothing and is a misnomer for this material). The bottom and top geotextiles are supposed to protect the liner from any sharp object in the subbase and from the overlying drainage zone, respectively. Above the top geotextile is about 1.5 ft minimum of drainage materials of permeability greater than 10^{-2} cm/s. Sand is an excellent drainage material. Above this zone is a protective cover of gravel materials of at least 0.5 ft. A *geonet* is sometimes used as a drainage channel in addition to the drainage zone of permeable materials. (Again, *geo* means earth and *net* means net; that is, geonet is a plastic material constructed of a matrix of nets.) The geonet is placed directly above the liner and serves as a channel toward the leachate pipe.

A leachate pipe header is shown in the figure. This accepts all drainage leachate from a number of leachate pipe laterals, as well as any leachate passing directly through its perforations. Laterals collect leachate from the fill by allowing passage through their perforations. As shown, the header is buried in a trench filled with granular materials (indicated as AASHTO No. 3). Completely wrapping the fill is a nonwoven geotextile. The purpose of this wrapping is to filter out any fines in order to protect the granular fill and the header from clogging.

Spacing of laterals. Referring to Figure 11-13, the spacing between laterals L and the flow q to a unit length of lateral will now be derived as follows: Since vertical percolation is larger than the horizontal percolation, the leachate over the liner will soon

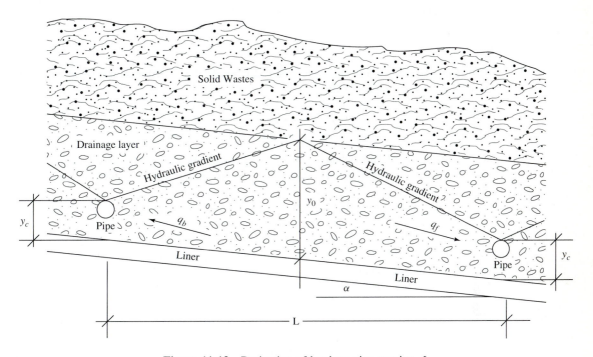

Figure 11-13 Derivation of leachate pipe spacing, L.

build up. The bottom slope of a landfill is normally 2%. Since the slope is very small, the hydraulic gradient will be practically horizontal; the hydraulic gradient can therefore be approximated accurately by a straight line. This approximation is shown in the figure. Two flow directions of leachate will result: q_f, in the direction of the slope, and q_b, in the opposite direction, toward the upstream leachate pipe. q_f and q_b have units of rate of flow per unit length of leachate pipe. The average hydraulic gradient for q_f is $[(L/2) \tan \alpha + y_0 - y_c]/(L/2)$. Considering a unit width perpendicular to the direction of flow, the average flow area is $(\frac{1}{2})(y_0 + y_c)$. With the hydraulic conductivity K, q_f is given, using Darcy's law, by

$$q_f = \frac{y_0 + y_c}{2} K \frac{(L/2) \tan \alpha + y_0 - y_c}{L/2} \qquad (11\text{-}19)$$

Using analysis similar to q_b gives

$$q_b = \frac{y_0 + y_c}{2} K \frac{y_0 - (L/2) \tan \alpha - y_c}{L/2} \tag{11-20}$$

The total flow q to a particular leachate pipe is the sum of q_f and q_b. Hence, combining the equations above and simplifying yields

$$q = \frac{2K(y_0 + y_c)}{L}(y_0^2 - y_c^2) \tag{11-21}$$

Equation (11-21) may be solved for L, producing

$$L = \frac{2K(y_0^2 - y_c^2)}{q} \tag{11-22}$$

This formula shows that the liner need not be sloped. This may be compared with Moore's formula, which reads

$$L = \frac{2\eta h}{\sqrt{(e/K) + \text{slope}^2} - \text{slope}}$$

where h is the depth of leachate, η the porosity, e the percolation rate, K the permeability, and slope the slope of the liner. Moores's formula requires that a slope to the liner be provided.

At the start of rainfall, water begins to infiltrate the surface and to percolate downward. This vertical percolation creates an advancing front. Initially, the head above this front is equal to the thin sheet of rain over the surface of the ground plus the length of the front. When the rain stops, the sheet of rain on the ground surface and over the front disappears, and the head above the front simply reduces to the length of the front itself. Since the thickness of the sheet of rain is much smaller than the length of the advancing front, it can be assumed that the hydraulic gradient driving force for vertical percolation is equal to unity. Therefore, considering $dh/d\ell$ to be equal to 1, the vertical percolation velocity is $K(1)$, and the vertical flow q_u per unit of horizontal area is then equal to K.

If the flow is percolating downward across several layers, q_u for the top layer may be written as

$$q_u = K_1 \frac{\Delta h_1}{y_1} \tag{11-23}$$

where K_1 is the vertical hydraulic conductivity of the top or first layer, Δh_1 the head loss of the layer, and y_1 the thickness of the layer. The flows through each successive layer are all equal to q_u. Similar expressions can be written for all the other layers up to the nth. With equal q_u's, the head loss across each layer may be solved from the expressions and the total head loss Δh across all the layers can then be added, obtaining

$$\Delta h = \left(\frac{y_1}{K_1} + \frac{y_2}{K_2} + \cdots + \frac{y_n}{K_n}\right) q_u \tag{11-24}$$

where the subscripts denote the respective layers. The terms inside the parentheses may be written as y/K, where y is the sum thickness of all layers and K is the overall hydraulic conductivity. Hence

$$K = \frac{y}{y_1/K_1 + y_2/K_2 + \cdots + y_n/K_n}$$ (11-25)

Example 11-15

Design the spacing of laterals for an uncapped landfill for a required maximum leachate head of 30.5 cm. Also, determine the flow q to each lateral and the diameter of the lateral, assuming that its length is 100 m. Assume that the drainage layer is composed of gravel with a hydraulic conductivity of 100 m/day and a drainage layer slope of 1%. Assume that the overall vertical hydraulic conductivity through the solid wastes is 10^{-3} cm/s.

Solution

$$L = \frac{2K(y_0^2 - y_c^2)}{q}$$

Assume that $y_c = 0$. $q_u = K = 10^{-3}\text{cm}^3/\text{s-cm}^2 = 10^{-5}\text{m}^3/\text{s-m}^2$. Since the length of lateral is 100 m, downward percolating water $= 10^{-5}(L)(100) = 0.001L$ m^3/s. Thus, since there are two laterals collecting a draining area,

$$q = \frac{0.001L}{2(100)} = 5(10^{-6})L \ \frac{\text{m}^3}{\text{m-s}}$$

$$L = \frac{2(100)(0.305^2 - 0^2)}{[5(10^{-6})L](60)(60)(24)} = \frac{43.067}{L}$$

$$= 6.56 \text{ m} \quad \textbf{Answer}$$

$$q = 5(10^{-6})(6.56) = 3.28(10^{-5})\text{m}^3/\text{m-s} \quad \textbf{Answer}$$

Assuming a self-cleaning velocity of 0.77 m/s,

$$\text{cross-sectional area at lower end of lateral} = \frac{3.28(10^{-5})(100)}{0.77} = 0.0042 \text{ m}^2;$$

$$d = 0.074 \text{ m} = 7.5 \text{ cm} \quad \textbf{Answer}$$

Example 11-16

Solve Example 11-15 assuming that the landfill is capped using the cap system shown in Figure 11-11 and that the height of the solid waste mass is 25 m. Use the following K's: 12-in. layer $= 10^{-2}$ cm/s, 6-in. layer $= 10^{-2}$ cm/s, 18-in. layer $= 10^{-6}$ cm/s, 24-in. layer $= 10^{-2}$ cm/s, and solid waste mass $= 10^{-3}$ cm/s. Assume that the geotextile-gas venting layer does not contribute resistance to flow.

Solution

$$K = \frac{y}{y_1/K_1 + y_2/K_2 + \cdots + y_n/K_n}$$

$$= \frac{2.54(12 + 6 + 18 + 24) + 2500}{2.54(12)/10^{-2} + 2.54(6)/10^{-2} + 2.54(18)/10^{-6} + 2.54(24)/10^{-2} + 2500/10^{-3}}$$

$$= 5.5(10^{-5}) \text{ cm/s}$$

$$q_u = K = 5.5(10^{-5}) \text{ cm}^3/\text{s-cm}^2 = 5.5(10^{-7}) \text{ m}^3/\text{s-m}^2$$

$$L = \frac{2K(y_0^2 - y_c^2)}{q} \qquad \text{assume that } y_c = 0$$

Since the length of lateral is 100 m, downward percolating water $= 5.5(10^{-7})(L)(100) = 5.5(10^{-5})L$ m^3/s. Thus since there are two laterals collecting a draining area,

$$q = \frac{5.5(10^{-5})L}{2(100)} = 2.75(10^{-7})L \; \frac{\text{m}^3}{\text{m-s}}$$

$$L = \frac{2(100)(0.305^2 - 0^2)}{[2.75(10^{-7})L](60)(60)(24)} = \frac{783.04}{L}$$

$$= 27.98\text{m} \quad \textbf{Answer}$$

$$q = 2.75(10^{-7})(27.98) = 7.69(10^{-6}) \text{ m}^3/\text{m-s} \quad \textbf{Answer}$$

$$\text{Cross-sectional area at lower end of lateral} = \frac{7.69(10^{-6})(100)}{0.77} = 0.0011 \text{ m}^2$$

$$d = 0.038 \text{ m} = 3.8 \text{ cm} \quad \textbf{Answer}$$

Example 11-17

Determine the spacing of the laterals of Example 11-16 if the liner grade is increased to 4%.

Solution $L = 27.98$ m.   **Answer**

Loads on Leachate Pipes.   When a pipe is buried, it must be able to support the weight of the fill and any superimposed load above the fill. The superimposed load may be due to a static load on top of the fill or to any moving load. The method used to compute the fill load on buried pipes has been, up to this point, empirical and is based on the work done at Iowa State College under the direction of Dean Marston.[4]

There are two types of pipes: rigid and flexible. Rigid pipes are those whose every element material making up the pipe does not yield under stress, while flexible pipes are those whose element materials do yield under stress. For leachate pipes in landfills, PVC and vitrified clay pipes are used. PVC is an example of a flexible pipe, and vitrified clay pipe is an example of a rigid pipe.

For rigid pipes, Marston's formula is

$$w = C\gamma B^2 \tag{11-26}$$

and for flexible pipes his formula is

$$w = C\gamma BD \tag{11-27}$$

---

[4]M. G. Sprangler (1948). "Underground Conduits—an Appraisal of Modern Research." *Transactions of ASCE, 113*, pp. 316–374.

where w is the force or load per unit length of pipe, γ the specific weight of the fill, B the trench width, and D the outside diameter of the pipe. Equations (11-26) and (11-27) apply to pipes buried in trenches with undisturbed sides. For a given type of fill, the value of C depends on the ratio of the depth of cover H, measured from the top of the pipe in the trench to the surface of the trench, and the width of the trench B. As the overburden settles to impose a load on the pipe, the undisturbed sides of the trench help to support the load. Table 11-12 shows values of C to be used with equations (11-26) and (11-27).

TABLE 11-12 VALUES OF C FOR USE IN EQUATIONS (11-26) AND (11-27)

H/B	Sand and gravel ($\gamma = 15.7$ kN/m^3)	Saturated top soil ($\gamma = 15.7$ kN/m^3)	Clay ($\gamma = 18.9$ kN/m^3)	Saturated clay ($\gamma = 20.4$ kN/m^3)
1.0	0.84	0.86	0.88	0.90
2.0	1.45	1.50	1.55	1.62
3.0	1.90	2.00	2.10	2.20
4.0	2.22	2.33	2.49	2.65
5.0	2.45	2.60	2.80	3.03
6.0	2.60	2.78	3.04	3.33
7.0	2.75	2.95	3.23	3.57
8.0	2.80	3.03	3.37	3.76
9.0	2.88	3.11	3.48	3.92
10.0	2.92	3.17	3.56	4.04
12.0	2.97	3.24	3.68	4.22
14.0	3.00	3.28	3.75	4.34

The value of H used to determine C from Table 11-12 should be measured from the top of the pipe to the top of the trench. Above the trench would be another load from the compacted solid waste. To incorporate this additional load, γ in equations (11-26) and (11-27) should be computed as the volume average of the γ of the solid waste and the γ of the trench fill.

For headers and lateral pipes that are simply laid on top of the liner, the loads imposed on the pipes are under *embankment* or *broad-fill load conditions* and equations (11-26) and (11-27) do not apply. Under embankment conditions, the overburden directly above the pipe is dragged downward by the overburden not directly above the pipe. This condition causes the transmission of load to the pipe from the side overburden, in addition to the load directly above the pipe. Embankment conditions therefore impose more loads to the pipe compared to those under trench conditions. For embankment load conditions, Marston's formula is

$$w = C_p \gamma D^2 \tag{11-28}$$

Table 11-13 shows some values of C_p for use with equation (11-28).

TABLE 11-13 VALUES OF C_p FOR
USE WITH EQUATION (11-28)

H/D	Rigid pipe	Flexible pipe
1.0	1.2	1.1
2.0	2.8	2.6
3.0	4.7	4.0
4.0	6.7	5.4
6.0	11.0	8.2
8.0	16.0	11.0

Another load that must be considered for buried pipes is the superimposed loads by machineries used to operate the landfill and the trucks used to unload the solid waste into the fill. This load may be computed using *Boussinesq's equation*:

$$p = \frac{3d^3 P}{2\pi Z^5} \tag{11-29}$$

where p is the variable pressure at any point in the fill, d the depth of the point below the surface, and Z the slant distance of the point from the superimposed load P at the surface.

The total force due to the superimposed load is calculated by subdividing the horizontal projected area of the pipe into small subareas. The pressure p in each subarea is then calculated using equation (11-29) and the corresponding force calculated. The total force on the pipe is equal to the sum of the forces in each individual subarea. The impact force can be twice the force calculated. Therefore, if the force is due to a moving load such as a moving crawler that can drop into a cavity because of unevenness of the fill, the calculated force should be multiplied by a factor such as 2. If the crawler drops into a cavity and a leachate pipe happens to be underneath it, the impact force will be transmitted to the pipe.

Another factor that must considered in design of buried pipes is the effect of the load-carrying capacity of the pipes when subjected to different types of bedding. Bedding improves the ability of the pipe to support loads. The crushing strength of a pipe is normally determined by the three-edge bearing test (ASTM C14; see Figure 11-14). The result of this test must be multiplied by a factor that depends on the type of bedding used to determine the *field supporting strength*. The field supporting strength must then be divided by a factor of safety to obtain the *design strength*. The factor of safety is usually 1.5 or greater. Figure 11-14 shows some types of bedding along with the corresponding factors to be used to obtain the field supporting strength. Table 11-14 shows some values of γ.

Example 11-18

Calculate the load on the leachate header pipe in Figure 11-12. Assume that the height of solid waste heap is 50 m and the compacted density is 600 kg/m³. Assume that a crawler dozer weighing 50,000 lb is used to compact the waste. The distance between wheels of the crawler is 5.0 ft.

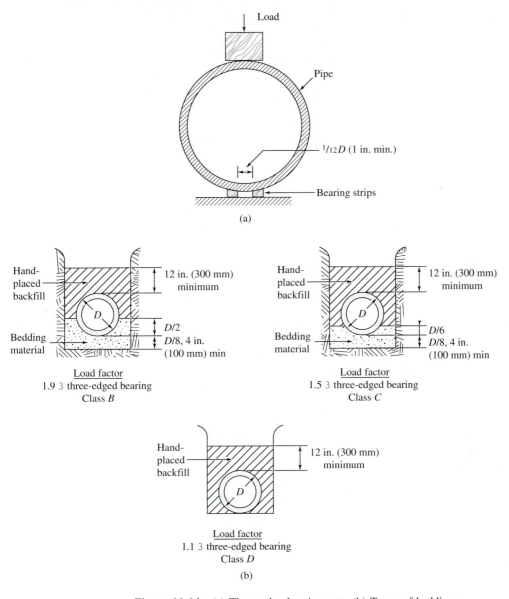

Figure 11-14 (a) Three-edge bearing test. (b) Types of beddings.

Solution In Figure 11-12, although the header is shown as buried in a trench, the sides are disturbed; the condition shown is under an embankment condition. Depth of drainage layer (composed of the regular layer, 1.5 ft + protective layer, 0.5 ft) = 2 ft. Based on Table 11-14, assume that γ of the layer = 130 lb/ft^3 = 2085 kg/m^3, dry. To compute average γ, moisture must be included. Assume 1% moisture.

TABLE 11-14 VALUES OF γ

Soil class	γ (lb/ft^2 dry)	Soil class	γ (lb/ft^3 dry)
GW	125–135	GP	115–125
GM	120–135	GC	115–130
SW	110–130	SP	100–120
SM	110–125	SC	105–125
ML	95–120	CL	95–120
OL	80–100	MH	70–95
CH	75–105	OH	65–100

$$\text{Average density of load over pipe} = \frac{[2085(1.01)][(1)(1)(2/3.281)] + 600(1)(1)(50)}{(1)(1)(2/3.281) + (1)(1)(50)}$$

$$= 617.89 \text{kg/m}^3 = 6061.46 \text{ N/m}^3$$

$$w = C_p \gamma D^2 \qquad H/D = \frac{50 + [2 - (8/12)/3.281]}{0.8(2.54)(10^{-2})} = 2480$$

Since the pipe is HDPE (a plastic pipe), C_p must be for flexible pipes. In Table 11-13, however, the highest H/D value is 8.0 $\ll$ 2480. To use the table, assume an equivalent depth of overburden of higher density. Letting $H/D = 8$, the equivalent depth is $(8)(8) = 64$ in (where 8 is the diameter of the pipe). Hence $C_p = 11.0$.

$$\gamma_{eq} = \frac{6,061.46 \{50 + [2 - (8/12)]/3.281\} (1)(1)}{64(2.54)(10^{-2})(1)(1)} = 187,953 \text{ N/m}^3$$

$$w = 11.0(187,953)[8(2.54(10^{-2})]^2 = 85,367 \text{ N/m}$$

$$p = \frac{3d^3 P}{2\pi Z^5}$$

Assume that the crawler is directly on top of a pipe. At this condition, compute the pressure on the pipe midway between the wheels of crawler.

$$\text{Load from each wheel} = \frac{50,000(0.454)}{2} = 11,350 \text{ kg}$$

$$Z = \sqrt{\left[50 + \frac{2 - (8/12)}{3.281}\right]^2 + \left[\frac{5}{2(3.281)}\right]^2} = 50.4 \text{ m}$$

$$p = \frac{3\left[50 + \frac{2-(8/12)}{3.281}\right]^3}{2(\pi)(50.4^5)} = 20.94 \text{ N/m}^2 \text{ due to one wheel}$$

The projected area of the pipe may be subdivided into small subareas to obtain the total force per unit meter; however, since the pipe is very small, this procedure is not warranted. Assume that total superimposed pressure, including impact pressure $= 2p = 2(20.94) = 41.87 \text{ N/m}^2$.

Considering the other wheel,

$$\text{total load per meter} = 2(41.87[(8)(2.54)(10^{-2})][1] = 17 \text{ N/m}$$

$$\text{total load on pipe} = 85.4 + 0.017 = 85.42 \text{ kN/m} \quad \textbf{Answer}$$

Example 11-19

In Example 11-18, compute the load on the pipe during initial filling of the landfill.

Solution   Depth of drainage layer (composed of the regular layer, 1.5 ft + protective layer, 0.5 ft) = 2 ft. Assume that γ of the layer = 130 lb/ft^3 = 2085 kg/m^3. During initial filling, depth of compacted fill = 0.

$$w = C_p \gamma D^2 \qquad H/D = \frac{[2 - (8/12)]/3.281}{8(2.54)(10^{-2})} = 2.0$$

From Table 11-13, $C_p = 2$. Thus

$$w = [2(2,085)(1.01)]19.81[(8)(2.54)(10^{-2})]^2 = 1706 \text{ N/m}$$

$$p = \frac{3d^3 P}{2\pi Z^5}$$

Assume that the crawler is directly on top of a pipe. At this condition, compute the pressure on the pipe midway between the wheels of crawler.

$$\text{Load from each wheel} = \frac{50,000(0.454)}{2} = 11,350 \text{ kg}$$

$$Z = \sqrt{\left[\frac{2 - (8/12)}{3.281}\right]^2 + \left[\frac{5}{2(3.281)}\right]^2} = 0.86 \text{ m}$$

$$d = \frac{2 - (0.8/12)}{3.281} = 0.41 \text{ m}$$

$$p = \frac{3(0.41^3)(11,350)(9.81)}{2(\pi)(0.86^5)} = 7789 \text{ N/m}^2 \text{ due to one wheel}$$

Assume total superimposed pressure including impact pressure $= 2p = 2(7789) = 15,577$ N/m^2. Considering the other wheel, total load per meter $= 2(15,577)[(8)(2.54)(10^{-2})](1) = 3,165$ N/m. Thus

$$\text{total load on pipe} = 1.71 + 3.17 = 4.88 \text{ kN/m} \quad \textbf{Answer}$$

Example 11-20

In Example 11-19, what three-edge bearing load would you specify if the bedding used were class D? Assume a safety factor of 2.0.

Solution   From the previous examples, the actual load equals 4.88 kN/m = design strength. Hence field supporting strength = 4.88(2) = 9.76. From Figure 11-14, bedding factor = 1.1. Thus

$$\text{three-edge bearing load} = \frac{9.76}{1.1} = 8.87 \text{ kN/m} \quad \textbf{Answer}$$

Gas control.   Gas control is the third of the five basic aspects in implementing the sanitary landfill method of solid waste disposal. Landfill gases are composed of methane, carbon dioxide, nitrogen, hydrogen sulfide, and ammonia. As discussed in Chapter 2, these gases are produced by the anaerobic decomposition of the organic portions of solid waste. The percentage composition of these gases can vary as the decomposition progresses. For example, the following changes in composition may be observed: From 0 to 3 months, $N_2 = 5\%$, $CO_2 = 90\%$, and $CH_4 = 5\%$. However, at the end of 40 to 50 months, $N_2 = 0.5\%$, $CO_2 = 50\%$, and $CH_4 = 50\%$. The rest of the percentages are for the other components. Landfill gases are composed mostly of carbon dioxide and methane.

If not properly vented, methane can accumulate in enclosed spaces and develop an explosive mixture with air. The range of flammability of methane in air is from 5 to 15% by volume of the gas. Concentrations at 40% for both carbon dioxide and methane have been found at lateral distances of up to 120 m from the edges of landfills. Hence control of lateral movement of landfill gases is very important. Buffer zones of about 100 to 150 m from residences should therefore be provided to prevent the possible accumulations of these gases at basements of homes.

Some methods of gas control are shown in Figure 11-15: cell vent, trench vent, perforated-pipe vent, and barrier vent. The cell vent, composed of gravel, is put directly above the daily cell cover. In the trench vent, a trench is dug as deep as the solid wastes and filled with gravel. In the perforated-pipe vent, perforated pipes are used to collect and convey the gases to riser pipes for gas venting. If gases are recovered for the heating value of methane, the riser pipes may be connected to a main for pumping to a processing plant. In the barrier method, an impermeable lining of clay is installed to intercept the moving gases, as indicated at the bottom of the figure.

Operation plan.   The operation plan is the fourth of the five basic aspects in implementing the sanitary landfill method of solid waste disposal. The use of the landfill from the initial filling to closure of the site is basically a construction operation. To implement the various stages of construction, a good operation plan must be in place. Factors considered in the design of a good operational plan may include access to the site and to various points on the site, filling sequence, source of cover material, drainage, equipment requirement and maintenance, provision for wetting the landfill heap, fire prevention, litter control, communication and employee facilities, days and hours of operation, and closure plan. *Access to the site* should not be through populated areas and should be via back roads. If this cannot be done, a landfill should not be located at the site. Access roads to the site may be lined with trees and the entire landfill site may be surrounded with trees to keep it away from public sight and to serve as *sound breaks* from the equipment noises used in the construction. Any construction is always noisy, and no residence would want to be near it. *Access to various points in the site* must also be carefully delineated in an *operation map*. This map will indicate at a glance where the stage of construction or *filling* is at any point in time.

For proper control of the *filling sequence*, the entire landfill site should be sub-divided into subareas called *cells*. The term *cell* is not to be confused with the same

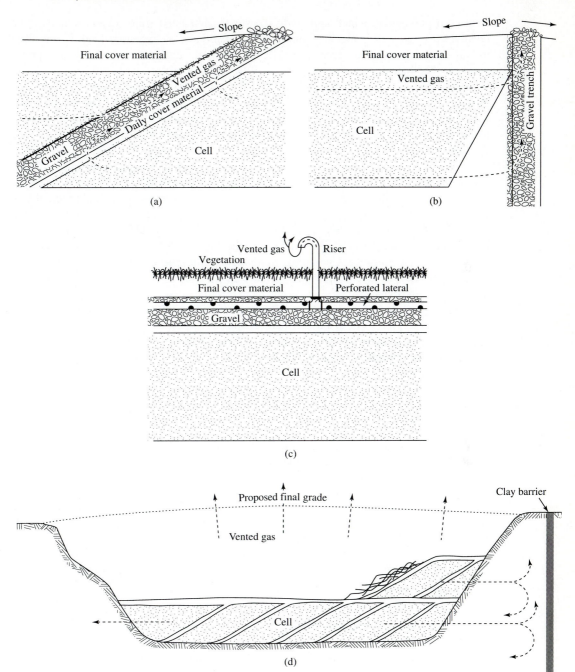

Figure 11-15   Methods of gas control: (a) cell vent; (b) trench vent; (c) perforated-pipe vent; (d) barrier vent.

term used previously, which applies to the small compacted solid waste volumes. The *compacted single cell* is a cell within the *subdivision cell* of the entire landfill site. This use of the same term to mean two different things is unfortunate but reflects current practice. The meaning of the word *cell* will be understood by its context. Figure 11-16 is an example of a landfill site subdivided into a number of subdivision cells or cells. Also, notice the plan view schematic of the leachate collection system. The header is the pipe running diagonally across the cells, and the laterals are the several pipes connecting to the header. The headers are connected to the main located at the north end (top) of the drawing.

The source of *cover* would have been determined in the planning stage of site selection. Covers may be obtained at the site or imported off-site, but wherever the source, the method of securing them so that cover will always be available should be elaborated in the operation plan. The question of *drainage* is also a consideration in site selection. As part of the operational plan, drainage ditches should be installed to divert surface runoff from the site. The surface of the finished fill should be maintained at a grade of 1 to 2% to prevent ponding. *Equipment requirements and maintenance* should also be a part of the operational plan. Figure 11-17 shows some of the equipment used in landfill operations.

During filling and compaction, heavy equipment could be standing directly on top of the leachate pipe. As discussed before, the design of the pipe must consider this situation; the pipe must be able to withstand the crushing strength imposed on the pipe by such heavy equipment. Table 11-15 gives the weights of some equipment used in sanitary landfills.

TABLE 11-15   WEIGHT OF SOME EQUIPMENT USED IN SANITARY LANDFILLS

Equipment type	Flywheel horsepower	Unequipped weight (lb)	Equipped weight (lb)
Crawler dozer	<80	<15,000	19,000
	110–130	20,000–25,000	32,000
	250–280	47,500–52,000	67,000
Crawler loader	<70	<20,000	23,000
	100–130	25,000–32,500	31,000–32,000
	150–190	32,500–45,000	45,000–47,000
Rubber-tired loader	<100	<20,000	17,000–18,000
	120–150	22,500–27,500	23,000–26,000

As discussed in Chapter 2, water is needed for anaerobic decomposition to occur. Although initially, there is air in the landfill, it will soon be consumed and the entire process will become anaerobic. The amount of water needed can be determined by following the steps discussed in Chapter 2.

Since the landfill would have been capped, only a theoretically small amount of rain can get into the heap. Therefore, if the heap is organic, *provisions for wetting the*

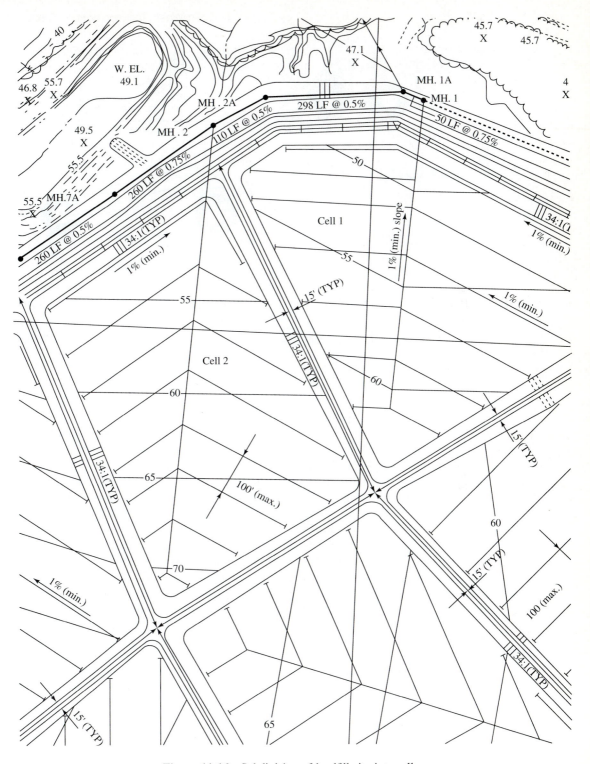

Figure 11-16 Subdivision of landfill site into cells.

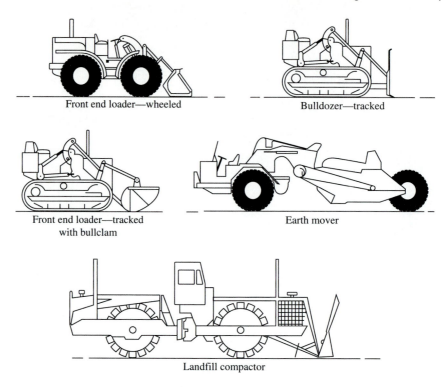

Front end loader—wheeled

Bulldozer—tracked

Front end loader—tracked
with bullclam

Earth mover

Landfill compactor

Figure 11-17   Sanitary landfill equipment.

landfill heap should be accounted for in the design of the operation plan. The wetting can be accomplished by recycling leachate to the heap. This will entail calculations of pump horsepower and piping size. These have been discussed elsewhere in the book and are not repeated here. During decomposition, periodic sampling of the heap should be made to ascertain that adequate moisture is present to effect the decomposition. The result of moisture analysis will regulate the recirculation of leachate.

Since one of the products of decomposition is methane gas, it must be expected that a fire will occur from time to time. In the design of the operational plan, *firefighting equipment* should be included. There should be adequate water on site to fight fire; and if the water is nonpotable, this should clearly be marked. In the design of subdivision cells, proper separation by appropriate barriers should be provided to prevent continuous burn-through in case of fire. Figure 11-16 shows how cells are separated.

At the beginning of this chapter it was noted that paper, which can easily be blown away, is the most abundant component of solid waste. Good *litter control* must therefore be provided in design of the operation plan. This may done by erecting fences around an active cell.

Communication and employee facilities such as telephone installation, rest rooms, and drinking water should be provided. The *days and hours of operation* should be posted at the landfill gate. This is good public relations. Finally, a very important part of an operation plan is the *closure plan*.

Closure plan refers to the activities performed after a landfill cell is filled. The activities involve the construction of the *cap system* and *post-fill monitoring*. Post-fill monitoring is required to ascertain that the groundwater is not contaminated. This is done by installing *monitoring wells*. Figure 11-18 shows cross sections of landfill heaps with monitoring wells installed. The wells should be installed downgradient of groundwater flow. Wells may also be installed upgradient to the direction of groundwater flow to account for the uncertainty of the direction of groundwater and for a possible comparison of contaminated and uncontaminated groundwater. As noted in the figure, a double liner has been installed for the hazardous waste landfill.

Permit application. The application for a permit is the last of the five basic aspects in implementing the sanitary landfill method of solid waste disposal. No landfill can be operated legally without first obtaining a permit. States vary in requirements, but in Maryland, the process is divided into four phases. Phase I is the permit application itself. Applicants fill out a form. The second phase is submission of a hydrogeologic and geologic report. This report is reviewed by the state, and after it is in an acceptable form and content, the third phase can begin. The third phase involves the submission of engineering designs and operation plan. Engineering designs pertain to the design of the control of leachate and gases and the design of drainage for stormwater control.

The last phase is the hearing. At the hearing, the applicant presents his side, normally with the help of a consultant. Members of the public may also voice their objection or approval. Based on this hearing, the permit may or may not be issued to the applicant. Following an unfavorable decision, the applicant may request that the hearing be adjudicated before a hearing officer, who is usually designated by the state. If, on adjudication, the decision is still unfavorable, the applicant may move the case to the courts. The public may do likewise if the decision is unfavorable to them.

GLOSSARY

Area method. A method of sanitary landfilling used when it is impossible to excavate.

Break-even point. The point where the cost of transporting solid waste directly to the ultimate point of disposal is equal to the cost when transportation to the point of disposal is facilitated using a transfer station.

Cap. A system of construction over a landfill heap provided to prevent precipitation from leaching through the heap.

Cataracting speed. The speed of rotation of a rotating screen equal to or less than the critical speed but not so fast that the solids are simply centrifuging and not so slow that the solids are simply cascading.

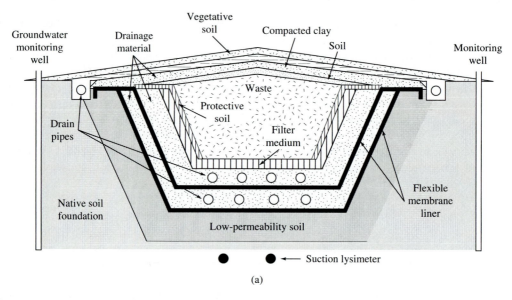

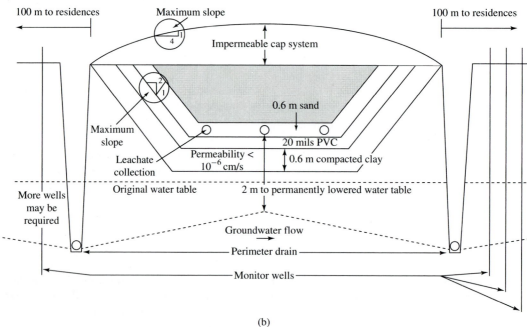

Figure 11-18 Generalized cross sections of landfill heaps: (a) hazardous waste landfill; (b) municipal landfill.

Cell.   Daily cell: a working compartment where the solid waste is compacted in the course of a working day. Subdivision cell: a major subdivision of a landfill site.

Centrifuging.   The mode of screening where the solids simply cling to the sides as the screen is rotating.

Closure plan.   The activities performed after a landfill subdivision cell is filled.

Coarse-textured soil.   Soil the major components of which are large.

Collection.   The gathering of solid wastes from places such as residences, commercial, industrial and institutional establishments, and public places.

Compaction ratio.   The ratio of the as-compacted density to the as-discarded density.

Critical speed.   The rotational speed of a rotating screen where the solids would drop off as they reach the summit of rotation.

Crushing strength.   Force supported by a pipe before it fails.

Daily cover.   Soil that is used to cover compacted solid waste at the end of an operating day.

Design strength.   Field-supporting strength divided by the factor of safety.

Destination point.   The processing facility, transfer station, or disposal site.

Disposal.   The placing of solid waste in its final resting place.

Field-supporting strength.   The three-edged bearing crushing strength of pipe multiplied by a factor that depends on the type of bedding used to bury the pipe.

Fine-textured soil.   Soil the major components of which are fine.

Garbage or food waste.   The animal and plant residue produced as a result of the handling, preparing, and eating of food.

Hauled-container system.   A system of collecting solid waste where the container containing the waste is hauled to a destination point.

Heat of combustion.   The energy released when a substance is burned.

Higher heat of combustion.   The heat of combustion, which includes the heat of vaporization of water.

Intermediate cover.   The horizontal layer of cover between cells in a landfill heap.

Leachate.   Precipitation that has percolated through a landfill heap, leaching out any contaminant.

Lift.   The cell height plus the thickness of the intermediate cover.

Liner system.   A construction at the bottom of a sanitary landfill provided to prevent leachate to percolate toward the groundwater.

Lower heat of combustion.   The net heat of combustion (i.e., excluding the heat of vaporization of water).

Oversize products.   Products that did not pass through the openings of the screen.

Poorly graded soil.   Soil composed of components of practically the same size.

Processing.   The activity applied to solid waste to prepare it for subsequent operation.

Proximate analysis.   A chemical characterization that determines the concentrations of surrogate parameters in place of the true chemical content.

Recycling or salvaging.   The recovery of solid waste components for possible reuse.

Refuse.   Solid waste.

Rubbish.   Combustible and noncombustible portions of solid waste.

Rubble.   Construction and demolition debris, rock fragments, and the like.

Sanitary landfilling. Engineered burial of refuse.

Screening. The unit operation of separating a feed into oversize and undersize products.

Soil consistency. The ability of the soil to hold together when pressure is applied to it and released.

Soil structure. The shape and geometry of clusters of particles that make up the soil mass.

Solid waste management. The application of techniques that will ensure orderly execution of the functions of collection, processing, and disposal of solid waste.

Solid wastes. Solid objects being discarded.

Stationary-container system. A system of solid waste collection where the waste is emptied directly into the collection vehicle at the point of collection.

Texture of soil. Refers to the predominant size fraction of components in soil.

Transfer station. A facility where collected solid wastes are transferred to a larger transport vehicle or stored temporarily for transfer to the larger transport vehicle for transportation to a destination point.

Trash. The combustible portion of rubbish.

Trench method. A method of sanitary landfilling where a trench is constructed to accept the solid waste.

Ultimate analysis. Analysis of the elements composing the substance.

Undersize products. Products that pass through the openings of the screen.

Vegetative layer. A layer in the cap system provided for plants to grow.

Well-graded soil. Soil composed of different-size fractions.

White goods. The bulky portion of solid waste, such as refrigerators, air conditioners, and the like.

SYMBOLS

C_ℓ	number of container locations
d_ℓ	between-collection-point driving time
dbc	average driving time between container locations
D	diameter of leachate drain pipe
h	haul time; piezometric height
H	total available time in a collection day
H_h	higher heating value
k_{ft}	transfer station unit cost of operating and amortization per unit mass of solid waste from transfer station to destination point
k_{hd}	cost of direct haul per unit mass of solid waste per unit time from collection system to destination point
k_{ht}	transfer station hauling cost per unit mass of solid waste per unit time from transfer station to destination point

K	hydraulic conductivity
K_d	cost of direct haul per unit mass of solid waste from collection system to destination point
K_t	transfer station cost per unit mass of solid waste
L_f	size of shredder feed which is 80% finer
L_p	size of shredder product which is 80% finer
m	time spent to mount container to collection vehicle
M	mass of solid waste in trommel screen
N_t	total number of trips in a collection day
p	time spent to pick up and load solid waste in a collection point to collection vehicle
P_d	percent moisture in solid waste, dry basis
P_w	percent moisture in solid waste, wet basis
q_b	percolation per unit length of drain pipe upslope of liner
q_f	percolation per unit length of drain pipe downslope of liner
q_u	vertical percolation per unit of horizontal area
r	compaction ratio
s	time spent at site
S_d	mass of dry solid waste
S_w	mass of wet solid waste
t_1	driving time from dispatch station to first collection point
t_2	driving time from last collection point or destination point to dispatch station
t_{even}	break-even time
t_{net}	on-the-job time used per trip
u	time spent to unload container from collection vehicle
V_d	as-discarded volume of solid waste per trip
V_{dc}	as-discarded volume of solid waste in a collection day
V_h	volume of collection vehicle
w	load per unit length of buried pipe
W	percent of total time wasted in a collection day; shredder power requirement
W_i	Bond index
y_c	depth of percolate at drain pipe
y_0	depth of percolate midway between drain pipes
α	liner slope
γ	angle of inclination of trommel screen; specific weight of soil
ρ_c	as-compacted density
ρ_d	as-discarded density
ω	critical speed of trommel screen

PROBLEMS

11-1. Estimate the moisture content (wet and dry bases) of the following solid waste.

Component	Mass (%)
Newspaper	15
Other papers	24
Cardboard	33
Glass	4.2
Plastics	0.49
Yard wastes	17.97
Food wastes	1.67
Dirt	2.01

11-2. Has recovery of materials been practiced in your community?

11-3. Estimate the as-discarded density of the following solid waste. If the solid waste is compacted to a density of 600 kg/m^3 at the landfill, calculate the at-site compaction ratio.

Component	Mass (%)
Newspaper	15
Other papers	24
Cardboard	33
Glass	4.2
Plastics	0.49
Yard wastes	17.97
Food wastes	1.67
Dirt	2.01

11-4. What efforts, if any, have been made on your campus to reduce the volume of solid waste?

11-5. Derive an empirical organic formula for the solid waste shown in the table below.

Component	Mass (%)
Newspaper	15
Other papers	24
Cardboard	33
Glass	4.2
Plastics	0.49
Yard wastes	17.97
Food wastes	1.67
Dirt	2.01

11-6. In Problem 11-5, estimate the higher and lower heats of combustion of the organics on as-discarded basis and on dry basis and compare result of the higher heating value on dry basis with the result using Dulong's formula.

11-7. When you go home tonight, sort out the components in your solid waste bin.

11-8. A traffic study for an industrial district produced the following results: t_1, t_2, and $d_\ell = 20$, 25, and 8 min, respectively. The round-trip haul distance averaged 60 km at a speed limit of 55 mph in the 50-km stretch of this distance and 25 mph in the rest of the distance. A contractor agreed to haul the solid wastes from this district. The industries agreed to store their wastes on large containers located at strategic points. Due to the sizes of the containers, the hauled-container system of collection is to be used. How many containers can be serviced on a collection day of 8 h?

11-9. Call the department of public works in your community and inquire as to the rate of solid waste production.

11-10. The average between-container-location driving time per container location is 0.09 h. If the average time to "clean up" a location is 0.21 h and the average driving time between container locations is 0.1 h, there are how many container locations in this collection route? How long does it take to collect all the solid wastes in the route?

11-11. How many times a week is domestic solid waste collected in your community? The yard waste and recyclables?

11-12. The average between-container-location driving time per container location or residence is 14.94 s. If the average time to "clean up" a location is 31 s and the average driving time between location or residence is 15 s, there are how many residences in this collection route? How long does it take to collect all the solid wastes in the route?

11-13. Calculate the mass and volume of solid waste generated per week for a family of four.

11-14. If there are 166 residences in the collection route, what fraction of the volume of the collection vehicle is occupied by the solid waste of the family in Problem 11-13 if collection is done twice a week? Assume that the family of four is typical in your route.

11-15. A total of 11 apartment complexes and commercial establishments have entered into a contract with a solid waste collection firm to collect their solid wastes. A 20-m^3 vehicle is available for collection. The discarded volume of solid wastes to be collected is 96 m^3/week. If it is decided to have two collections per week, how many trips are to be made on a collection day? Also, determine the size of the container to be provided. Assume that $r = 2.5$ and, for simplicity, assume that each of the 11 sources contributes the same amount of solid wastes. Make any additional assumptions necessary.

11-16. Lay out the collection route in your area for Problem 11-15.

11-17. In a collection route of your choice, assume that the average occupancy per residence is 3.5 and the average solid waste production is 1.6 kg/capita-day. Assuming a collection frequency of two times per week, determine the number of trips to be made on collection day and calculate the volume of the collection vehicle needed. Assume that the compacted volume of solid wastes in the collection vehicle is 325 kg/m^3.

11-18. What is the daily production of wastepaper in town of 200,000?

11-19. The cost of amortization and operation of a transfer station handling 1200 tonnes/day of solid waste is $3600.00 per day. The cost of hauling solid waste from the transfer station to a sanitary landfill is $24.00 per minute of driving time, while the cost of direct haul from the collection system to the same landfill is $60.00 per minute of driving time. If the speed limit is 55 mph, compute the distance beyond which direct haul is no longer economical.

11-20. A plant burns 100 tonnes/h of solid waste represented by the following formula: $C_{562}H_{900}O_{414}N_6S$. How much air is needed if 50% excess air is used?

11-21. Calculate the critical speed of a horizontal trommel screen.

11-22. Estimate the heating value of the solid waste in Problem 11-1.

11-23. A municipal solid waste has the following empirical formula: $C_{562}H_{900}O_{414}N_{6.6}S$. What is its heating value?

11-24. Compare the power requirements between the shredder in Wilmington and the shredder in Charleston in shredding 100 tonnes/day of solid waste having an 80th percentile size of 30 cm to a product having an 80th percentile size of 1.7 cm.

11-25. A magnetic separator is used to separate ferrous materials from a municipal solid waste. The feed rate to the unit is 1500 kg/h, and it contains 5.5% of ferrous materials. Of the 50 kg in the product stream, 40 kg is ferrous. What is the percent recovery?

11-26. If C is burned completely, what is the product of combustion? How much heat is liberated per kilogram? If all the heat is utilized, how much steam can be produced?

11-27. MSW, at a rate of 100 tonnes/day, with an empirical formula of $C_{562}H_{900}O_{414}N_{6.6}S$, is used to fire an MSW energy recovery plant. The MSW contains 20% moisture and 20% inert. The ash contains 5% carbon. Also, assume that the fuel enters at 25°C and that the ash is at 420°C. How much steam energy is available from the boiler? *Hint:* Use Dulong's formula.

11-28. A solid waste has the following formula: $(CH_2O)_{106}$. How much heat will be available if this solid waste is used for generating steam?

11-29. Repeat Problem 11-28 for the formula $(CH_4O)_{106}$.

11-30. How much heat is available from the hydrogen portion of the solid waste in Problem 11-29?

11-31. If the steam in Problem 11-29 is used to drive a steam-turbine generator plant, calculate the plant export of electrical energy. Calculate the overall efficiency of the plant.

11-32. A town having a population of 10,000 produces solid wastes at rate of 5 lb/capita-day. If the waste is compacted to 1000 lb/yard^3 and 6.0 acre of landfill space is available, how high will the landfill be in a year using the area method of landfilling? Assume side slopes of 4H to 1V.

11-33. Do you know how much gold can be recovered from the solid waste of Problem 11-32?

11-34. The average precipitation in a landfill area is 12 in./month. Design the spacing of the laterals for a required maximum leachate head of 30.5 cm if the landfill is uncapped. Assume that the drainage layer is composed of gravel with a hydraulic conductivity of 100 m/day and a drainage layer slope of 1%. Assume that the overall vertical hydraulic conductivity through the solid wastes is 10^{-3} cm/s.

11-35. Is the average drainage layer slope of 1% in Problem 11-34 necessary to the solution of the problem? Why?

11-36. Solve Problem 11-34 assuming that the landfill is capped using the cap system discussed in this chapter and that the height of the solid waste mass is 25 m. Use the following K's: 12-in. layer = 10^{-2} cm/s, 6-in. layer = 10^{-2} cm/s, 18-in. layer = 10^{-6} cm/s, 24-in. layer = 10^{-2} cm/s, and solid waste mass = 10^{-3} cm/s.

11-37. In Problem 11-36, if the spacing of the laterals is 100 ft, will the system be able to drain the percolate? If not, what depth of drainage layer should be provided to drain out the percolate?

11-38. Calculate the load on a header laid in an undisturbed trench under a landfill. Assume that the height of solid waste heap is 50 m and the compacted density is 600 kg/m^3. The trench is 2 ft deep and the diameter of the header is 8 in. Assume that a crawler dozer weighing 50,000 lb is used to compact the waste. The distance between wheels of the crawler is 5.0 ft.

11-39. In Problem 11-38, compute the load on the pipe during initial filling of the landfill.

11-40. In Problem 11-39, what three-edge bearing load would you specify if the bedding used is class D? Assume a safety factor of 2.0.

11-41. A 10-ft-diameter rigid concrete pipe rests on an unyielding ground. The trench is then backfilled with sand with a specific weight of 100 lb/ft^3 to a depth of 8 ft. Directly above the pipe is a concentrated load of 20,000 lb. Calculate the vertical load on the pipe.

11-42. Calculate the vertical load in Problem 11-41 if the pipe is flexible.

11-43. What is a closure plan?

11-44. Discuss the steps in obtaining a landfill permit.

BIBLIOGRAPHY

BRUNNER, D. R., and D. J. KELLER (1972). *Sanitary Landfill Design and Operation.* SW-65ts. U.S. Environmental Protection Agency, Washington, D.C.

CONNOR, K. R., R. C. BROCKWAY, and B. R. HENNING (1994). "Taking Trash Out of Hiding." *Civil Engineering,* February.

NELS, C. (1983). "Thermal Treatment of Household Wastes," in G. Tharun, N. C. Thanh, and R. Bidwell (eds.), *Environmental Management for Developing Countries.* Asian Institute of Technology, Bangkok, Thailand.

PEAVY, H. S., D. R. ROWE, and G. TCHOBANOGLOUS (1985). *Environmental Engineering.* McGraw-Hill, New York.

PRATT, B., C. DOHERTY, A. C. BROUGHTON, and D. MORRIS (1991). *Beyond 40 Percent.* Island Press, Washington, D.C.

SPRANGLER, M. G. (1948). "Underground Conduits—an Appraisal of Modern Research." *Transaction of ASCE,* 113, pp. 316–374.

STRATTON, F. E., and H. ALTER (1978). "An Application of Bond Theory to Solid Waste Shredding." *Journal of the Environmental Engineering Division, ASCE,* 104(EE1).

VESILIND, P. A., J. J. PEIRCE, and R. F. WEINER (1988). *Environmental Engineering.* Butterworth, Boston.

WOODS, R. (1991). "A French Revolution Comes to Minnesota." *Waste Age,* 2(1), pp. 58–62.

WOODS, R. (1993). *Public Works,* 124(12).

CHAPTER 12

Air Pollution Control

Air pollution may be defined as the presence of substances in concentrations that make the air harmful or dangerous to breathe or to cause damage to plants, animals, and properties. This problem is not a recent phenomenon. In 1881, Chicago and Cincinnati passed antismoke ordinances to combat air pollution. In the Middle Ages, King Edward I banned the burning of coal in lime kilns when in 1307 the smoke resulting from their operation became a serious problem. The control of air pollution runs counter to industrial development; hence despite the effect that pollution has on the population, little effort was made to curb it, especially in the industrially developing world. However, after World War II, two major air pollution episodes, in part, prompted the world to take action: one was in Donora, Pennsylvania, and the other, in London, England. These episodes were discussed in Chapter 1.

TYPES OF POLLUTANTS

In general, there are only two types of pollutants: particulates and gaseous. *Particulates* are finely divided solids and liquids, such as dust, fumes, smoke, fly ash, mist, and spray. Gaseous pollutants are, by definition, pollutants in gaseous form. Gaseous pollutants have the property of filling any available space until their concentrations fall to equilibrium by diffusion. If the space is large, the resulting concentration may be negligible; if the space is small, the resulting concentration may not be negligible as in concentrations of

582

carbon monoxide formed from the continuous running of a motor vehicle in a closed garage. This concentration can cause asphyxiation.

Dusts (1.0 to 1000 μm) are small particles of solids created from the breakup of larger particles by operations such as crushing, grinding, and blasting. *Fumes* (0.03 to 0.3 μm) are fine solid particles that condense from vapors of solid materials. *Smoke* (0.5 to 1.0 μm) is unburned carbon that results from the incomplete combustion of carbon-containing substances. *Fly ash* (1.0 to 1000 μm) is the noncombustible particle admixed with combustion gases in the burning of coal. *Mists* (0.07 to 10 μm) are particles formed from the condensation of liquid vapors. *Mists* concentrate into *fogs*. *Sprays* (10 to 1000 μm) are particles formed from the atomization of liquids.

All pollutants may be organic or inorganic. *Organic pollutants* are those of organic origin. Gasoline is an organic pollutant. If a pollutant is not organic, it must be inorganic. Carbon monoxide, carbon dioxide, and sulfur dioxide are examples of inorganic pollutants. Organic pollutants, in turn, may be natural or synthetic. *Natural* organic pollutants are those of biological origin. Forms of natural organic air pollutants that are microscopic or "macro" microscopic such as pollens, microscopic algae, spores, bacteria, viruses, and the like that are *alive* or *have just recently died* are specifically called *biological air pollutants*. *Synthetic* organic pollutants are pollutants from manufactured products of non-biological origin.

Pollutants may also be classified as primary or secondary. *Primary pollutants* are those directly emitted from the source. *Secondary* pollutants are those formed in the atmosphere after they have been emitted from the source. For example, ozone is formed from the NO_2 photolytic cycle—not directly from a source, hence, it is a secondary pollutant. On the other hand, NO that produces NO_2 is formed directly from the combustion of fuels; it is a primary pollutant.

The U.S. Environmental Protection Agency established National Ambient Air Quality Standards (NAAQS) for two levels of pollutants. These also have the names of primary and secondary, not to be confused with the types of pollutants. *Primary* NAAQS are designed to protect public health with an adequate margin of safety, while *secondary* NAAQS are designed to protect public welfare (plants, animals, and properties). NAAQS now exists for six pollutants. These pollutants are called *criteria* pollutants, and they include carbon monoxide, nitrogen dioxide, ozone, sulfur dioxide, PM-10 (particulate) and lead. PM-10 are particulates of equivalent spherical diameter of 10 μm or less. It has been found that PM-10 are the ones that directly affect health and not the whole range of sizes of particulates. Hence, sizes greater than 10 μm are not set ambient standards. The National Ambient Air Quality Standards for the criteria pollutants are shown in Table 12-1. It is important to note the health effect of carbon monoxide. The toxic effects of this compound are mainly due to its reaction with hemoglobin in the blood.

When carbon monoxide and oxygen are introduced to the blood through the alveolar sacs of the lungs, the former is absorbed 210 times in terms of the equilibrium absorption coefficient of oxygen.[1] Thus, if [HbCO] is the concentration of carboxyhemoglobin and

---

[1] C. D. Cooper and F. C. Alley (1986). *Air Pollution Control: A Design Approach*. Waveland Press, Prospect Heights, Ill., p. 38.

TABLE 12-1 NATIONAL AMBIENT AIR QUALITY STANDARDS

Pollutant	Average time	Primary	Secondary	Objective
Carbon monoxide	8 h	10 mg/m^3 (9 ppm)	None	To limit carboxy-hemoglobin
	1 h	40 mg/m^3 (35 ppm)	23 mg/m^3	
Nitrogen dioxide	Annual	100 μg/m^3 (0.053 ppm)	Same	To prevent health risks and improve visibility
Ozone	1 h	235 μg/m^3 (0.12 ppm)	Same	To prevent eye irritation and breathing difficulties
Sulfur dioxide	Annual	80 μg/m^3 (0.03 ppm)	None	To prevent increase in respiratory disease, plant damage, and damage to structures
	24 h	365 μg/m^3 (0.14 ppm)		
PM-10	Annual	50 μg/m^3	Same	To improve visibility and to prevent health effects
	24 h	150 μg/m^3	Same	
Lead	3 month	1.5 μg/m^3	Same	To prevent health problems

[HbO$_2$] is the concentration of oxyhemoglobin, their equilibrium concentration ratio is

$$\frac{[\text{HbCO}]}{[\text{HbO}_2]} = 210 \left(\frac{\overline{P}_{\text{CO}}}{\overline{P}_{\text{O}_2}} \right) \tag{12-1}$$

where $\overline{P}_{\text{CO}}$ and $\overline{P}_{\text{O}_2}$ are the partial pressures of carbon monoxide and oxygen in the lungs, respectively. The concentration of carbon monoxide in the blood, however, is expressed empirically in terms of the percent of saturation, % HbCO,[2]

$$\%\text{HbCO} = 0.005[\text{CO}]^{0.85}(\alpha t)^{0.63} \tag{12-2}$$

where [CO] is the concentration of CO in ppm, α a physical activity level (1 for sedentary activity up to 3 for heavy work), and t the duration of activity in minutes. Table 12-2 shows the health effects of HbCO at various levels in the blood.

Example 12-1

The concentration of CO in a street intersection reaches the federal ambient standard of 35 ppm. Crewmen from the department of public works are repairing a break in the water line. Estimate the CO concentration in their blood after 1 h of work and make conclusions as to their work performance.

[2]G. M. Masters (1991). *Introduction to Environmental Engineering and Science.* Prentice Hall, Englewood Cliffs, N.J. p. 282.

TABLE 12-2 HEALTH EFFECTS OF HbCO AT VARIOUS LEVELS IN THE BLOOD

HbCO level (%)	Demonstrated effect
Less than 1.0	No apparent effect
1.0–1.5	Some evidence of effect on behavior is noticeable
1.5–4.5	Time interval discrimination, visual acuity, brightness determination, and certain other psychomotor functions are impaired
> 4.5	Cardiac and pulmonary functional changes
4.5–80.0	Headache, fatigue, drowsiness, coma, respiratory failure, and death

Solution

$$\%HbCO = 0.005[CO]^{0.85}(\alpha t)^{0.63}$$

Let $\alpha = 2$. Then $\% \ HbCO = 0.005[35]^{0.85}[2(60)])^{0.63} = 2.1\%$; hence, from Table 12-2 their work performance is impaired. **Answer**

SOURCES OF POLLUTANTS

All sources of pollutants may be categorized under the general headings of transportation, stationary source fuel combustion, industrial processes, solid waste disposal, and miscellaneous. Miscellaneous sources include forest fires and agricultural burning, for example. Table 12-3 shows percentage estimates of the criteria pollutants coming from various sources.[3] Volatile organic compounds (VOC) substitute for ozone. Since VOCs

TABLE 12-3 PERCENTAGE ESTIMATES OF CRITERIA POLLUTANTS FROM VARIOUS SOURCES

Source	Particulate	SO_x	NO_x	VOC	CO	Lead
Transportation	21.0	4.0	44.0	33.0	70.0	41.5
Stationary source fuel combustion	26.0	81.0	52.0	12.0	12.0	6.0
Industrial processes	37.0	15.0	3.0	41.0	7.0	22.0
Solid waste disposal	4.0	0	0.5	3.0	3.0	31.0
Miscellaneous	12.0	0	0.5	11.0	8.0	0.0

are a precursor of ozone, their estimate is a surrogate measure of the ozone concentration. The particulates in the table include all the range of sizes, not only PM-10. SO_x's are the oxides of sulfur but are composed mostly of SO_2. NO_x's are the oxides of nitrogen. As shown in the table, combustion is the major source of NO_x, which accounts for 52% from stationary sources and 44% from transportation. The major source of NO_x in transportation is the combustion in internal combustion engines. NO is favored over

[3]USEPA (1988). *National Air Pollutant Emission Estimates, 1940–1986.* U.S. Environmental Protection Agency, Washington, D.C.

NO_2 in high-temperature combustion, so that NO_2 is negligibly small. In general, NO_x are formed at high temperatures of combustion. It is interesting to note that solid waste disposal accounts for 31% of the lead emitted into the atmosphere. Lead in gasoline mostly accounts for the 41% contribution of lead from transportation.

TYPES OF CONTROL

Air pollution control may be defined as the various measures taken to meet certain emission standards. Since the standards are met, the pollutants are said to be controlled. These measures may include changes in processes or raw materials, modification of equipment, and installation of devices at the end of process equipment to treat the ensuing effluent. These devices are called *air pollution control equipment.* Another term related to control is *abatement.* Whereas control is aimed at meeting a certain standard, *abatement* simply refers to measures used to reduce the quantity of emissions regardless of the standard. When the abatement measure succeeds in meeting the standard, abatement becomes a control.

Corresponding to the two general types of pollutants are also two general types of pollution control: particulate control and the control of gaseous pollutants. In this chapter we discuss the various types of control equipment used in air pollution management. In addition, special discussion is devoted to indoor air pollution. Indoor air pollution is a concern because studies have shown that radon is emitted from rocky soils and eventually seeps into houses. Radon causes lung cancer. Figure 12-1 shows an industrial facility whose process equipment are controlled. The emissions coming out of the stacks are harmless steam.

Particulate Control

There are three general types of particulate control equipment: force-field settlers, baghouse filters, and scrubbers. *Force-field settlers* are those equipment that relies on a field of force to pull down particulates to a collection point or surface. There are three types of fields of force: gravitational, centrifugal, and electrical. The type of settling that relies on gravitation is called *gravitational settling.* This is exemplified by the sedimentation basins used in water and wastewater treatment. Gravitational settling is also used in waste air treatment. Settlers that utilize centrifugal forces are *cyclones* or *centrifugal settlers.* Settlers that utilize an electric field of force to collect particulates are called *electrostatic precipitators* (ESPs). Filters operate in much the same way as the ordinary vacuum cleaner. Waste air is introduced into the unit and the particulates are filtered out. In scrubbers, particulates are *scrubbed out* from the waste air by using water droplets of various sizes depending on the type of scrubber used. The particulates are removed by impaction and interception. In *impaction,* the center of mass of the particulate collides directly with the center of mass of the water droplet—a direct impact; in *interception,* the center of masses of the particulate and the water droplet are not in direct collision, but oblique to each other—an indirect impact. Upon impact, the particulate is wetted and carried by the water droplet, thus, effecting removal.

Figure 12-1 Industrial facility: emissions from stacks are harmless steam. (Used by permission of SCM Chemicals)

Gravitational settling.   The practical limit of particulate size to be removed by gravitational settling is 50 μm, the size of about one-half the diameter of a human hair. Below this limit would require a large settling chamber and a long detention time. The principles of design are similar to that in water and wastewater treatment, except that in the present case, air is involved instead of water. As in grit chambers, the flow-through velocity along the chamber should be maintained at less than 0.3 m/s. The terminal velocity of a settling particle has been derived in Chapter 6. For convenience, it is reproduced below.

$$v = \sqrt{\frac{4}{3}g\frac{(\rho_p - \rho_a)d}{C_D\rho_a}} \tag{12-3}$$

where v is the terminal settling velocity, g the gravitational acceleration, ρ_p the density of the particulate, ρ_a the density of air, d the diameter of the spherical particle, and C_D the drag coefficient. Since the density of air is very much smaller than that of the particulate, equation (12-3) may be written as

$$v = \sqrt{\frac{4}{3}g\frac{\rho_p d}{C_D\rho_a}} \tag{12-4}$$

With L equal to length, W to width, and Q to rate of flow, v is equal to Q/WL. Substituting in equation (12-4) and solving for d gives

$$d = \frac{0.75}{g} \frac{C_D Q^2 \rho_a}{L^2 W^2 \rho_p} \tag{12-5}$$

d is now the diameter of the spherical particle that can be removed 100%. This diameter corresponds to v_0, the settling velocity for 100% removal. Particles of other diameters less than d will be removed in the ratio of their terminal settling velocities. Expressing Q in terms of the flow-through velocity v_h, depth H, and W yield

$$d = \frac{0.75}{g} \frac{C_D v_h^2 H^2 \rho_a}{L^2 \rho_p} \tag{12-6}$$

From Chapter 6, the fractional removal of solids R is

$$R = 1 - x_0 + \int_0^{x_0} \frac{v_p}{v_0} dx \tag{12-7}$$

where x_0 is the fraction of particulates of settling velocity equal to or less than v_0, and v_p is the settling velocity of the particle of diameter less than d. The formulas above apply to spherical particles only. For nonspherical particles, the volume shape factor β should be used.

A procedure for determining β is detailed as follows: Soak a sample of particulates to saturation with water and record the resulting volume and weight. Evaporate the water to dryness and dry to constant weight. The loss in weight represents the volume of water in the voids; hence the average volume of a particulate V_p is equal to the volume of the soaked sample minus the volume of water evaporated (represented by the loss in weight) divided by the number of particles. The number of particles must have to be counted under a microscope using a particulate counter. Since the volume of the particle is also equal to the volume of the equivalent spherical particle,

$$V_p = \frac{4\pi}{3} \left(\frac{d}{2}\right)^3 = \beta d_p^3 \tag{12-8}$$

where d is the average diameter of the equivalent spherical particle and d_p is the average "diameter" of the particle. From equation (12-8), β may be obtained.

Example 12-2

A sample of particulates of average diameter 60 μm is soaked with water in a graduated cylinder to the 1-mL mark. The loss in weight after evaporating to dryness is 0.01 g. Determine β if the number of particles in the sample is $5.0(10^6)$.

Solution

$$0.01 \text{ g} = 0.01 \text{ cc}$$

$$V_p = \frac{1.0 - 0.01}{5.0(10^6)} = 2.0(10^{-7}) \text{ cm}^3$$

$$\beta = \frac{2.0(10^{-7})}{[60(10^{-6})(10^2)]^3} = 0.93 \quad \textbf{Answer}$$

Example 12-3

Measurement of the dust distribution of a certain industrial operation yields the results shown in the table below. These results are to be used to design a settling chamber. The horizontal velocity is to be 0.3 m/s. The temperature is 77°C, the specific gravity of the particle is 2.0, and the chamber length and depth equal 7.5 m and 1.5 m, respectively. (a) What is the terminal settling velocity of the particle that is removed 100%? (b) Determine the expected percent removal of the particles. Assume that β is 0.90.

Particle size (μm)	Wt %
0–10	8
10–20	10
20–30	12
30–40	15
40–50	19
50–60	14
60–70	13
70–80	9

Solution

(a) $$d = 1.24\beta^{0.333} d_p = \frac{0.75}{g} \frac{C_D v_h^2 H^2 \rho_a}{L^2 \rho_p}$$

$$d_p = \frac{1}{1.24(0.90^{0.333})} \left(\frac{0.75}{9.81}\right) \left[\frac{C_D(0.3^2)(1.5^2)(1.2)}{7.5^2(2.0)(1000)}\right] = 1.37(10^{-7})C_D$$

From Appendix 19, $\rho_a = 1.2$ kg/m^3 and $\mu = 2.1(10^{-5})$ kg/m-s at 77°C.

$$v = \sqrt{\frac{4}{3} g \frac{\rho_p d}{C_D \rho_a}} = \sqrt{\frac{4}{3}(9.81) \frac{(2.0(1000)(1.24)(0.90^{0.333})d_p}{C_D(1.2)}} = 161.56\sqrt{\frac{d_p}{C_D}}$$

$$v_0 = 161.56\sqrt{1.37(10^{-7})} = 0.060 \text{ m/s} \quad \textbf{Answer}$$

(b) $$\text{Re} = \frac{dv\rho_a}{\mu} = \frac{1.24(0.90^{0.333})d_p v(1.2)}{2.1(10^{-5})} = 6.84(10^4)d_p v$$

$$C_D = \begin{cases} \dfrac{24}{\text{Re}} & \text{(for Re} < 1) \\[2ex] \dfrac{24}{\text{Re}} + \dfrac{3}{\text{Re}^{1/2}} + 0.34 & \text{(for } 1 \le \text{Re} \le 10^4) \\[2ex] 0.4 & \text{(for Re} > 10^4) \end{cases}$$

By trial and error:

Particle size (μm)	Wt %	d_{pi} (10^{-6} m)	C_D	v_i (m/s)	Re_i
0–10	8	5.0	37,844	0.00186	0.00064
10–20	10	15	1,401	0.0167	0.0172
20–30	12	25	305	0.046	0.079
30–40	15	35	111	0.091	0.22
40–50	19	45	52	0.15	0.46
50–60	14	55	29	0.22	0.84
60–70	13	65	17	0.32	1.40
70–80	9	75	11	0.42	2.15

$$R = 1 - x_0 + \int_0^{x_0} \frac{v_p}{v_0} dx$$

Solve for x_0:

$$x_1 = 8 + 10 + 12 = 30 \qquad x_2 = 30 + 15 = 45$$

$$30 \Rightarrow 0.046 \qquad \frac{x_0 - 30}{45 - 30} = \frac{0.060 - 0.046}{0.091 - 0.046}$$

$$x_0 \Rightarrow 0.060$$

$$45 \Rightarrow 0.091 \qquad x_0 = 34.67\%$$

Wt %	v_i (m/s)	$\sum$ Wt %
8	0.00186	8
10	0.0167	18
12	0.046	30
	0.060	x_0
15	0.091	45
19	0.15	64
14	0.22	78
13	0.32	91
9	0.42	100

$$R = 1 - 0.3467 + \frac{1}{0.060}[0.00186(8 - 0) + 0.0167(18 - 8) + 0.046(30 - 18)$$

$$+ 0.060(34.67 - 30)]\left(\frac{1}{100}\right) = 0.6533 + 0.169 = 0.8223 \quad \text{or} \quad 82.23\% \quad \textbf{Answer}$$

Centrifugal settling. The dimensions of standard centrifugal settlers or cyclones are shown in Figure 12-2. There are three general types of cyclones: high-throughput, conventional, and high-efficiency. High-throughput cyclones are those that process high

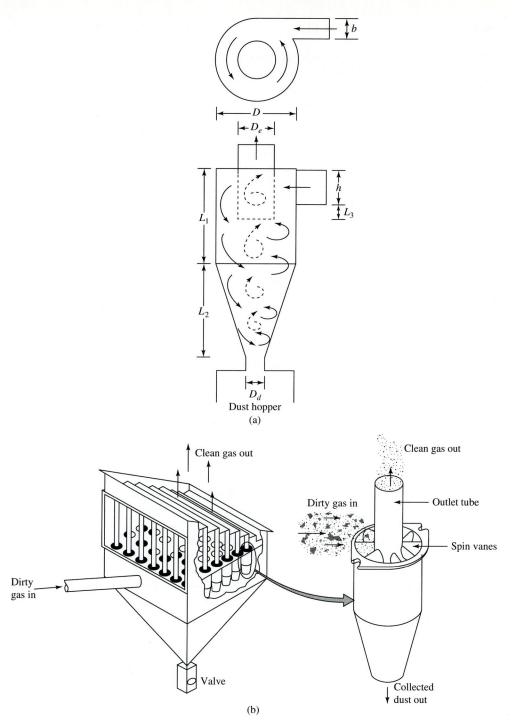

Figure 12-2 (a) Dimensions of standard cyclone. (b) Small cyclones arranged in parallel in housing.

volumes of waste air input but are operating at low efficiencies. The conventional cyclones are in between high-throughput and high-efficiency cyclones. Cyclones can also be arranged to operate in multiples to produce higher efficiencies. The efficiency of removal is, of course, not only a function of the size of the particle but of other variables such as the airflow. Table 12-4 shows the definition of various types of standard cyclones.

TABLE 12-4   STANDARD CYCLONE DESIGN DIMENSIONS

Dimensions	Conventional	High-throughput	High efficiency
Cylinder diameter, D	D	D	D
Cylinder length, L_1	$2D$	$1.5D$	$1.5D$
Cone length, L_2	$2D$	$2.5D$	$2.5D$
Total height, $L_1 + L_2$	$4D$	4	$4D$
Outlet length, $h + L_3$	$0.675D$	$0.875D$	$0.5D$
Inlet height, h	$0.5D$	$0.75D$	$0.5D$
Inlet width, b	$0.25D$	$0.375D$	$0.2D$
Outlet diameter, D_e	$0.5D$	$0.75D$	$0.5D$
Dust exit diameter, D_d	$0.25D$	0.375	0.375

From Example 12-3, particulates of diameters up to 60 to 70 μm settle at Reynolds numbers equal to or less than 1, the Stokes law range. In centrifugal settling, particulates also settle at the Stokes law range, and $C_D = 24/\text{Re}$ may be substituted into the equations derived for the terminal settling velocity.

The g in the equations above should be substituted by the g in a centrifugal field. In the centrifugal field, two forces are acting on a particle: the *inertial force*, which keeps the particle moving in a straight line as soon as it enters the body of the cyclone, and the *sweeping force* of the air, which tends to move the particle in the direction of flow. The vector sum of these forces produces a curved trajectory for the particle, which eventually bumps the side of the cyclone. The radial component of this "trajectory" force is the centrifugal field of force, and the body force acting on the particle in radial motion is the centrifugal force mg, where g is the centrifugal acceleration.

To find the value of the centrifugal acceleration, consider a particle and its component motion in the radial direction. Since the particle is forced to undergo a radial motion, by Newton's law there is an unbalanced force in this direction. This unbalanced force is the centripetal force that tends to force the particle toward the center of rotation and is simply v_t^2/r multiplied by the mass m of the particle (mv_t^2/r), where v_t is the tangential velocity of the particle and r is the radius at the location of the particle. mv_t^2/r is also equal to mg. The g in this expression is the centrifugal acceleration. Hence solving for g, we have

$$g = \frac{v_t^2}{r} \tag{12-9}$$

Assuming no slippage between particle and air, v_t may be assumed equal to the velocity of the waste air revolving around the cyclone interior.

As the airflow is introduced from the inlet into the cyclone chamber, it is subjected to a sudden increase in the cross-sectional area of flow. As the air negotiates the expanding cross section and, along with the curvature of the cyclone, an outer vortex flow is formed. It is in this outer vortex that removal of particulates occurs. This vortex acts as the settling chamber or the settling zone. The dimensions of this settling zone are: depth $= (D/2 - D_e/2)$, length $= [\pi(D + D_e)/2]$, and width $= (L_1 + L_2/2)$. The average cross-sectional area of flow A_h as the air spirals through the outer vortex is

$$A_h = \left(\frac{D}{2} - \frac{D_e}{2}\right)\left(L_1 + \frac{L_2}{2}\right) \tag{12-10}$$

Calling the inlet flow as Q, v_t can therefore be calculated. Substituting the result in equation (12-9) and simplifying yields

$$g = \frac{\{4Q/[(D - D_e)(2L_1 + L_2)]\}^2}{r} \tag{12-11}$$

Since the air has to go out, an inner vortex is also formed. This vortex is the outlet zone of this sedimentation basin.

The overflow area of the settling zone may be considered as the average of the outside and the inside lateral areas of the outer-vortex settling chamber. This area is

$$A_s = \frac{\pi}{2}\left(L_1 + \frac{L_2}{2}\right)(D + D_e) \tag{12-12}$$

and hence the overflow velocity is

$$v_0 = \frac{Q}{(\pi/2)(L_1 + L_2/2)(D + D_e)} \tag{12-13}$$

Equating the overflow velocity to the terminal settling velocity [equation (12-4)], substituting $C_D = 24/\text{Re}$, and solving for d, we obtain

$$d = \sqrt{\frac{72\mu}{\pi g \rho_p (2L_1 + L_2)(D + D_e)}} \tag{12-14}$$

The other consideration in the design of cyclones is the pressure drop $-\Delta P$ across the unit. One empirically derived formula is given below.[4] This pressure drop is used to size the blower used for forcing the waste air into the unit.

$$-\Delta P = \frac{0.0027QF}{C D_e^2 bh (L_1/D)^{1/3} (L_2/D)^{1/3}} \frac{1}{0.01C_i + 1} \tag{12-15}$$

where $-\Delta P$ is the pressure drop in inches of water ($-\Delta P = P_1 - P_2$); Q the waste air flow in cfm; F is 0.8 for high-throughput cyclones, 1.0 for conventional cyclones, and 2.2 for high-efficiency cyclones; $C = 0.5$ with no guide vanes from the inlet to the cyclone chamber, $= 1.0$ with guide vanes straight from the inlet to the chamber, and $= 2.0$ with guide vanes from the inlet to the inner edge of the outer vortex; and C_i is the inlet dust loading in grains/ft^3. The units of all dimensions are in feet.

[4]H. E. Hesketh (1974). *Understanding and Controlling Air Pollution.* Ann Arbor Science, Ann Arbor, Mich., p. 312.

Example 12-4

Measurement of the dust distribution of a certain industrial operation yields the results shown in the table below. These results are to be used to design a standard conventional cyclone. The airstream is 21.0 m³/s and the diameter of the cyclone is 2.0 m. The temperature is 650°C and the specific gravity of the particle is 2.0. (a) What are the diameter and the terminal settling velocity of the particle that is removed 100%? (b) Determine the expected percent removal of the particles. Assume that β equals 0.90.

Particle size (μm)	Wt %
0–10	8
10–20	10
20–30	12
30–40	15
40–50	19
50–60	14
60–70	13
70–80	9

Solution

(a) $$d = 1.24\beta^{0.333} d_p = \sqrt{\frac{72Q\mu}{\pi g \rho_p (2L_1 + L_2)(D + D_e)}}$$

$$d_p = \left[\frac{1}{1.24(0.90^{0.333})}\right] \sqrt{\frac{72Q\mu}{\pi g \rho_p (2L_1 + L_2)(D + D_e)}} = 4.0\sqrt{\frac{Q\mu}{g\rho_p(2L_1 + L_2)(D + D_e)}}$$

From Appendix 19, $\rho_a = 0.59$ kg/m³ and $\mu = 4.0(10^{-5})$ kg/m-s at 650°C. From Table 12-4, $L_1 = 2D = 2(2) = 4$ m; $L_2 = 2D = 4$ m; $D_e = 0.5D = 1.0$ m; $r = 0.5 + (1.0 - 0.5)/2 = 0.75$ m.

$$g = \frac{\left[\frac{4Q}{(D-D_e)(2L_1+L_2)}\right]^2}{r} = \frac{\left[\frac{4(21)}{(2.0-1.0)[2(4)+4]}\right]^2}{0.75} = 65.33 \text{ m/s}^2$$

$$d_p = 4.0\sqrt{\frac{21(4.0)(10^{-5})}{65.33(2.0)(1000)[2(4)+4](2+1)}} = 53.5(10^{-6}) \text{ m} \quad \textbf{Answer}$$

$$v_0 = \frac{Q}{(\pi/2)(L_1 + L_2/2)(D + D_e)} = \frac{21}{(\pi/2)(4+2)(2+1)} = 0.74 \text{ m/s} \quad \textbf{Answer}$$

(b) $$v = \sqrt{\frac{4}{3}g\frac{\rho_p d}{C_D \rho_a}} = \sqrt{\frac{4}{3}(65.33)\frac{2.0(1000)(1.24)(0.90^{0.333})d_p}{C_D(0.59)}}$$

$$= 594.58\sqrt{\frac{d_p}{C_D}}$$

$$\text{Re} = \frac{dv\rho_a}{\mu} = \frac{1.24(0.9^{0.333})d_p v(0.59)}{4.0(10^{-5})} = 1.77(10^4)d_p v$$

$$C_D = \frac{24}{\text{Re}} = \frac{24}{1.77(10^4)d_p v} = \frac{1.36(10^{-3})}{d_p v}$$

$$v = 2.60(10^8)d_p^2$$

Particle size (μm)	Wt %	d_{pi} (10^{-6} m)	v_i (m/s)
0–10	8	5.0	0.0065
10–20	10	15	0.06
20–30	12	25	0.16
30–40	15	35	0.32
40–50	19	45	0.53
50–60	14	55	0.79
60–70	13	65	1.10
70–80	9	75	1.46

$$R = 1 - x_0 + \int_0^{x_0} \frac{v_p}{v_0}\,dx$$

Solve for x_0:

$$x_1 = 8 + 10 + 12 + 15 + 19 = 64 \qquad x_2 = 64 + 14 = 78$$

$$64 \Rightarrow 0.53 \qquad \frac{x_0 - 64}{78 - 64} = \frac{0.74 - 0.53}{0.79 - 0.53}$$

$$x_0 \Rightarrow 0.74$$

$$78 \Rightarrow 0.79 \qquad x_0 = 75.30\%$$

Wt %	v_i (m/s)	$\sum$ Wt %
8	0.0065	8
10	0.06	18
12	0.16	30
15	0.32	45
19	0.53	64
	0.74	x_0
14	0.12	78
13	0.17	91
9	0.22	100

$$R = 1 - 0.753 + \frac{1}{0.74}[0.0065(8 - 0) + (0.06)(18 - 8) + 0.16(30 - 18)$$

$$+ 0.32(45 - 30) + 0.53(64 - 45) + 0.74(75.3 - 64)]\left(\frac{1}{100}\right) = 0.247 + 0.348$$

$$= 0.595 \quad \text{or} \quad 59.5\% \quad \textbf{Answer}$$

Example 12-5

In Example 12-4, what is the pressure drop across the unit? Assume that $C_i = 1.0$ grain/ft^3 and that the cyclone has no vanes.

Solution

$$-\Delta P = \frac{0.0027QF}{CD_e^2 bh(L_1/D)^{1/3}(L_2/D)^{1/3}} \frac{1}{0.01C_i + 1}$$

$Q = 21(3.281^3)(60) = 44{,}503$ cfm; $F = 1.0; C = 0.5; D_e = 1(3.281) = 3.281$ ft; $D = 2(3.281) = 6.562$ ft $b = 0.25(2)(3.281) = 1.64$ ft; h = $0.5(2)(3.281) = 3.281$ ft; $L_1 = 4(3.281) = 13.124$ ft; $L_2 = 13.124$ ft. Substitution yields

$$-\Delta P = \left[\frac{0.0027(44{,}503)(1)}{0.5(3.281^2)(1.64)(3.281)(13.124/6.562)^{0.333}(13.124/6.562)^{0.333}}\right]\left[\frac{1}{0.01(1) + 1}\right]$$

$$= 2.59 \text{ in. of water} \quad \textbf{Answer}$$

Example 12-6

Solve Example 12-4 if a bank of 64 standard conventional cyclones with diameters of 24 cm is used instead of a single large unit.

Solution

$$\text{(a)} \qquad d_0 = 1.24\beta^{0.333} d_p = \sqrt{\frac{72Q\mu}{\pi g \rho_p (2L_1 + L_2)(D + D_e)}}$$

$$d_p = \left[\frac{1}{1.24(0.90^{0.333})}\right]\sqrt{\frac{72Q\mu}{\pi g \rho_p (2L_1 + L_2)(D + D_e)}} = 4.0\sqrt{\frac{Q\mu}{g \rho_p (2L_1 + L_2)(D + D_e)}}$$

From Appendix 19, $\rho_a = 0.59$ kg/m^3 and $\mu = 4.0(10^{-5})$ kg/m-s at 650°C. From Table 12-4, $L_1 = 2D = 2(0.24) = 0.48$ m; $L_2 = 2D = 0.48$ m; $D_e = 0.5D = 0.12$ m; $Q = 21/64 = 0.328$ m^3/s; $r = D_e/2 + (1/2)(D/2 - D_e/2) = 0.12/2 + (1/2)(0.24/2 - 0.12/2) = 0.09$ m.

$$g = \frac{\left[\frac{4Q}{(D - D_e)(2L_1 + L_2)}\right]^2}{r} = \frac{\left[\frac{4(0.328)}{(0.24 - 0.12)(2(0.48) + 0.48)}\right]^2}{0.09} = 640.53 \text{ m/s}^2$$

$$d_p = 4.0\sqrt{\frac{0.328(4.0)(10^{-5})}{640.53(2.0)(1000)[2(0.48) + 0.48](0.24 + 0.12)}} = 17.8(10^{-6}) \text{ m} \quad \textbf{Answer}$$

$$v_0 = \frac{Q}{(\pi/2)(L_1 + L_2/2)(D + D_e)} = \frac{0.328}{(\pi/2)(0.48 + 0.24)(0.24 + 0.12)}$$

$$= 0.81 \text{ m/s} \quad \textbf{Answer}$$

(b) $\quad v = \sqrt{\dfrac{4}{3}g\dfrac{\rho_p d}{C_D \rho_a}} = \sqrt{\dfrac{4}{3}(640.53)\dfrac{2.0(1000)(1.24)(0.9^{0.333})d_p}{C_D(0.59)}} = 1862\sqrt{\dfrac{d_p}{C_D}}$

$$\text{Re} = \frac{dv\rho_a}{\mu} = \frac{1.24(0.9^{0.333})d_p v(0.59)}{4.0(10^{-5})} = 1.77(10^4)d_p v$$

$$C_D = \frac{24}{\text{Re}} = \frac{24}{1.77(10^4)d_p v} = \frac{1.36(10^{-3})}{d_p v}$$

$$v = 2.55(10^9)d_p^2$$

Particle size (μm)	Wt %	d_{pi} (10^{-6} m)	v_i (m/s)
0–10	8	5.0	0.064
10–20	10	15	0.57
20–30	12	25	1.59
30–40	15	35	3.12
40–50	19	45	5.16
50–60	14	55	7.71
60–70	13	65	10.77
70–80	9	75	14.34

$$R = 1 - x_0 + \int_0^{x_0} \frac{v_p}{v_0}dx$$

Solve for x_0:

$$x_1 = 8 + 10 = 18 \qquad x_2 = 18 + 12 = 30$$

$$18 \Rightarrow 0.57 \qquad \frac{x_0 - 18}{30 - 18} = \frac{0.81 - 0.57}{1.59 - 0.57}$$

$$x_0 \Rightarrow 0.81$$

$$30 \Rightarrow 1.59 \qquad x_0 = 20.82\%$$

Wt %	v_i (m/s)	$\sum$ Wt %
8	0.064	8
10	0.57	18
	0.81	x_0
12	1.59	30

$$R = 1 - 0.21 + \frac{1}{0.81}[0.064(8 - 0) + (0.57)(18 - 8)]$$

$$+ 0.81(20.82 - 18)] \left(\frac{1}{100}\right) = 0.79 + 0.10 = 0.89 \quad \text{or} \quad 89\% \quad \textbf{Answer}$$

Electrostatic precipitators. ESPs are one of the force-field settlers. Whereas the previous settlers discussed use gravitational and centrifugal fields, ESPs utilize the electric field of force. The principles involved are the same except that the acceleration is the electric field acceleration. There are generally two types of electrostatic precipitators: plate and tube type. Figure 12-3 shows schematics of these types of precipitators.

Referring to the bottom of the figure, a high voltage (could be as high as 100,000 V) is impressed between the two electrodes. The grounded positive electrode is called the *collector* and the negative electrode is called the *discharge electrode*. The discharge electrode creates a shower of electrons that glow in the dark called a *corona discharge*. Because of their negative charges, the electrons migrate toward the positive collector plate. In the process they bump into the constituent of air and particulate matter. The air becomes negatively charged ions, and those electrons that are retained by the particulate render the particulate negative. In addition, the negatively charged air ions attached themselves to particulates. Having thus obtained a negative charge, the particulates travel toward the positive electrode and become neutralized and are collected there. They are removed from the collector electrode by gravitational forces, by rapping or by flushing the plates with liquids. Actual ESPs may have hundreds of parallel plates and tube collectors, with total collection areas measured in tens of thousands of square meters.

Derivation of the Electric Field Acceleration and Other Parameters. From physics, the force of attraction **F** between two charged particles is given by *Coulomb's law*,

$$\mathbf{F} = \frac{1}{4\pi\epsilon_0} \frac{q'q}{r^2}\hat{e}_r \tag{12-16}$$

where q' and q are the charges of the particles in coulombs (C), ϵ_0 the *permittivity constant* [equals $8.85(10^{-12})C^2/\text{N-m}^2$], r the distance between the charges, and $\hat{e}_r$ a unit vector. Therefore,

$$F = \frac{1}{4\pi\epsilon_0} \frac{q'q}{r^2} \tag{12-17}$$

By definition, the electric field **E** is the force **F** divided by the charge. Hence, from equation (12-16), the electric field at the location of q', a distance r from q, is

$$\mathbf{E} = \frac{1}{4\pi\epsilon_0} \frac{q}{r^2}\hat{e}_r \tag{12-18}$$

Consider a sphere with a total charge of q_s. If the radius of the sphere is R, using equation (12-18), the electric field at its periphery is

$$E = \frac{1}{4\pi\epsilon_0} \frac{q_s}{R^2} \tag{12-19}$$

Now consider this charged particulate migrating from the discharge electrode toward the collector plate. The voltage difference between the electrodes will create an electric

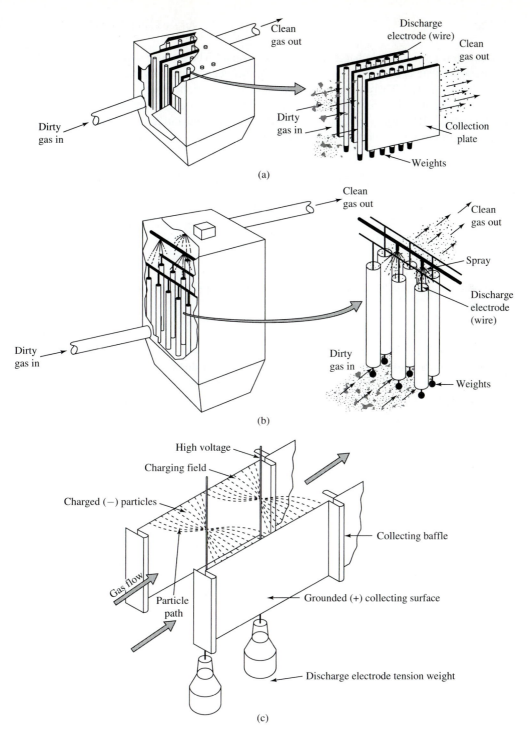

Figure 12-3 Schematics of electrostatic precipitators: (a) plate type; (b) tube type; (c) detailed view of plate-type precipitator element.

field, thus inducing the charge q_s on the particulate. The electric field at the periphery of this particulate is E. However, the charge q_s has been induced by the potential difference of the electrodes. Therefore, the electric field at the periphery due to q_s from the particle is equal to the electric field at the same point due to the potential difference of the electrodes. Therefore, by equation (12-19), the charged induced by the voltage across the electrodes on the particulate of diameter d is

$$q_s = \pi E \epsilon_0 d^2 \qquad (12\text{-}20)$$

From Newton's second law and from the definition of an electric field,

$$F = ma = E q_s \qquad (12\text{-}21)$$

and

$$a = g = \frac{\pi E^2 \epsilon_0 d^2}{m} \qquad (12\text{-}22)$$

Since m equals $\pi d^3 \rho_p / 6$, substituting in equation (12-22) yields

$$a = g = \frac{6\epsilon_0 E^2}{d \rho_p} \qquad (12\text{-}23)$$

Equation (12-23) may be substituted in equation (12-4) to produce the terminal settling velocity of the particulate. This velocity is also called the *drift or electric wind* velocity.

The volume of space from the discharge electrode to the collector electrode is a sedimentation chamber. For the tube-type ESP, the overflow area of this chamber may be considered as the average of the outside and the inside (equals zero) lateral areas of the tube. If the tube diameter is D and the length is L, this area is

$$A_s = \frac{\pi D L}{2} \qquad (12\text{-}24)$$

and the overflow velocity is

$$v_0 = \frac{Q}{\pi D L / 2} \qquad (12\text{-}25)$$

Equating the overflow velocity to the drift velocity, substituting $C_D = 24/\text{Re}$, and solving for d (the diameter of the particle that can be removed 100%), we have

$$d = \frac{6\mu}{\pi \epsilon_0} \frac{Q}{E^2 D L} \qquad (12\text{-}26)$$

For the plate-type ESP, the overflow area of the chamber is LW, where L is the length and W is the width. Hence the overflow velocity is

$$v_0 = \frac{Q}{LW} \qquad (12\text{-}27)$$

As in the case of the tube type, equating the overflow velocity to the drift velocity, substituting $C_D = 24/\text{Re}$, and solving for d gives us

$$d = \frac{3\mu}{\epsilon_0} \frac{Q}{E^2 LW} \qquad (12\text{-}28)$$

E is force/charge. If the force is in newtons and the charge is in coulombs and if both the numerator and denominator are multiplied by meters, the result would be newton-meter per coulomb-meter or volt per meter. Hence E has also the unit of volts per meter. Therefore, if the voltage drop between electrodes is V volts and the distance between the discharge and the collector electrodes is Z_0 meters, the average E between electrodes is

$$E = \frac{V}{Z_0} \tag{12-29}$$

The other consideration in the design of ESPs is the power requirement. Power is consumed whenever electric current is flowing through a wire; in the absence of current flow, no power is consumed. In ESPs, current flow is required to produce the corona; the greater the corona discharge, the greater is the current flow. In addition, the ac current from the power plant is converted to dc by a rectifier. Hence, from electricity, the power consumption P_c is

$$P_c = I_c V \tag{12-30}$$

where I_c is the corona current.

Example 12-7

Measurement of the dust distribution of a certain industrial operation yields the results in the table below. These results are to be used to design a plate-type ESP. The airstream is 330 m³/s and the plates are spaced 30 cm apart. The voltage drop between discharge and collector plates is 70,000 V. The aspect ratio, which is the ratio of the length of the plate to its width, is 1.2. The plate width is 10 m and there are a total of 80 channels. The temperature is 650°C and the specific gravity of the particles is 2.0. (a) What are the diameter and terminal settling velocity of particles that are removed 100%? (b) Determine the expected percent removal of the particles. (c) What is the power requirement assuming that the corona current is 2.3 A? Assume that β is 0.90.

Particle size (μm)	Wt %
0–10	8
10–20	10
20–30	12
30–40	15
40–50	19
50–60	14
60–70	13
70–80	9

Solution

(a)
$$d = 1.24\beta^{0.333} d_p = \frac{3\mu}{\epsilon_0} \frac{Q}{E^2 LW}$$

$$d_p = \left(\frac{1}{1.24(0.90^{0.333})} \right) \left(\frac{3\mu}{\epsilon_0} \frac{Q}{E^2 LW} \right) = 0.835 \left(\frac{3\mu}{\epsilon_0} \frac{Q}{E^2 LW} \right)$$

From Appendix 19, $\mu = 4.0(10^{-5})$ kg/m-s and $\rho_a = 0.59$ kg/m^3 at 650°C.

$$Q = \frac{330}{80(2)} = 2.06 \text{ m}^3/\text{s} \qquad \epsilon_0 = 8.85(10^{-12}) \text{ C}^2/\text{N-m}^2) \qquad W = 10 \text{ m}$$

$$L = 1.2(10) = 12 \text{ m} \qquad E = \frac{70,000}{0.30/2} = 4.67(10^5) \text{ V/m}$$

$$d_p = 0.835 \left(\frac{3\mu}{\epsilon_0} \frac{Q}{E^2 LW} \right) = 0.835 \left[\frac{3(4)(10^{-5})(2.06)}{8.85(10^{-12})[4.67(10^5)]^2(12)(10)} \right]$$

$$= 8.91(10^{-7}) \text{ m} \quad \textbf{Answer}$$

$$v_0 = \frac{Q}{LW} = \frac{2.06}{12(10)} = 0.017 \text{ m/s} \quad \textbf{Answer}$$

(b) $\qquad g = \dfrac{6\epsilon_0 E^2}{d\rho_p} = \dfrac{6(8.85)(10^{-12})[(4.67)(10^5)]^2}{1.24(0.9^{0.333})d_p(2.0)(1000)} = \dfrac{4.83(10^{-3})}{d_p}$

$$v = \sqrt{\frac{4}{3}g \frac{\rho_p d}{C_D \rho_a}} = \sqrt{\frac{4}{3} \frac{4.83(10^{-3})}{d_p} \frac{2.0(1000)(1.24)(0.9^{0.333})d_p}{C_D(0.59)}} = \frac{5.11}{\sqrt{C_D}}$$

$$\text{Re} = \frac{dv\rho_a}{\mu} = \frac{1.24(0.9^{0.333})d_p v(0.59)}{4.0(10^{-5})} = 1.77(10^4)d_p v$$

$$C_D = \frac{24}{\text{Re}} = \frac{24}{1.77(10^4)d_p v} = \frac{1.36(10^{-3})}{d_p v}$$

$$v = 1.92(10^4)d_p$$

Particle size (μm)	Wt %	d_{pi} $(10^{-6}$ m)	v_i (m/s)
0–10	8	5.0	0.096
10–20	10	15	0.288
20–30	12	25	0.48
30–40	15	35	0.672
40–50	19	45	0.864
50–60	14	55	1.056
60–70	13	65	1.248
70–80	9	75	1.44

$$R = 1 - x_0 + \int_0^{x_0} \frac{v_p}{v_0} dx$$

Solve for x_0:

$$x_1 = 8 \qquad x_2 = 8 + 10 = 18$$

$$x_0 \Rightarrow 0.017 \qquad \frac{x_0 - 8}{8 - 18} = \frac{0.017 - 0.096}{0.096 - 0.288}$$

$$8 \Rightarrow 0.096$$

$$18 \Rightarrow 0.288 \qquad x_0 = 3.89\%$$

Particle size (μm)	Wt %	d_{pi} (10^{-6} m)	v_i (m/s)	$\sum$ Wt %
			0.017	3.89
0–10	8	5.0	0.096	8
10–20	10	15	0.288	18
20–30	12	25	0.481	30
30–40	15	35	0.672	
40–50	19	45	0.864	
50–60	14	55	1.056	
60–70	13	65	1.248	
70–80	9	75	1.44	

$$R = 1 - 0.0389 + \frac{1}{0.017}[0.017(3.89 - 0)]\left(\frac{1}{100}\right) \approx 1.0 \quad \text{or} \quad 100\% \quad \textbf{Answer}$$

(c) $\qquad P_c = I_c V = 2.3(70,000) = 161,000$ W or 161 kW $\quad$ **Answer**

Example 12-8

Measurement of the dust distribution of a certain industrial operation yields the results shown in the table below. These results are to be used to design a tube-type ESP. The airstream is 330 m^3/s and there are 36,000 collector tubes of 3 in. ID. The voltage drop between discharge and collector tubes is 70,000 V. The tube is 10 m long. The temperature is 650°C and the specific gravity of the particles is 2.0. (a) What are the diameter and terminal settling velocity of particles that are removed 100%? (b) Determine the expected percent removal of the particles. (c) What is the power requirement assuming that the corona current is 2.3 A? Assume that β equals 0.90.

Particle size (μm)	Wt %
0–10	8
10–20	10
20–30	12
30–40	15
40–50	19
50–60	14
60–70	13
70–80	9

Solution

(a)
$$d = 1.24\beta^{0.333} d_p = \frac{6\mu}{\pi\epsilon_0} \frac{Q}{E^2 DL}$$

$$d_p = \left[\frac{1}{1.24(0.90^{0.333})}\right]\left(\frac{6\mu}{\pi\epsilon_0}\frac{Q}{E^2 DL}\right) = 0.835\left(\frac{6\mu}{\pi\epsilon_0}\frac{Q}{E^2 DL}\right)$$

From Appendix 19, $\mu = 4.0(10^{-5})$ kg/m-s and $\rho_a = 0.59$ kg/m³ at 650°C.

$$Q = \frac{330}{36,000} = 0.00917 \text{ m}^3/\text{s} \qquad \epsilon_0 = 8.85(10^{-12}) \text{ C}^2/\text{N-m}^2$$

$$D = \frac{3}{12(3.281)} = 0.0762 \text{ m} \qquad L = 10 \text{ m} \qquad E = \frac{70,000}{3/[2(12)(3.281)]} = 1.84(10^6) \text{ V/m}$$

$$d_p = 0.835\left(\frac{6\mu}{\pi\epsilon_0}\frac{Q}{E^2 DL}\right) = 0.835\left[\frac{6(4)(10^{-5})(0.00917)}{\pi(8.85)(10^{-12})[1.84(10^6)]^2(0.0762)(10)}\right]$$

$$= 2.5(10^{-8}) \text{ m} \quad \textbf{Answer}$$

$$v_0 = \frac{Q}{\pi DL/2} = \frac{0.00917}{[\pi(0.0762)(10)]/2} = 0.0077 \text{ m/s} \quad \textbf{Answer}$$

(b)
$$g = \frac{6\epsilon_0 E^2}{d\rho_p} = \frac{6(8.85)(10^{-12})[(1.84)(10^6)]^2}{1.24(0.90^{0.333})d_p(2.0)(1000)} = \frac{7.5(10^{-2})}{d_p}$$

$$v = \sqrt{\frac{4}{3}g\frac{\rho_p d}{C_D \rho_a}} = \sqrt{\frac{4}{3}\frac{7.5(10^{-2})}{d_p}\frac{2.0(1000)(1.24)(0.90^{0.333})d_p}{C_D(0.59)}} = \frac{20.15}{\sqrt{C_D}}$$

$$\text{Re} = \frac{dv\rho_a}{\mu} = \frac{1.24(0.90^{0.333})d_p v(0.59)}{4.0(10^{-5})} = 1.77(10^4)d_p v$$

$$C_D = \frac{24}{\text{Re}} = \frac{24}{1.77(10^4)d_p v} = \frac{1.36(10^{-3})}{d_p v}$$

$$v = 2.98(10^5)d_p$$

Particle size (μm)	Wt %	d_{pi} (10^{-6} m)	v_i (m/s)
0–10	8	5.0	1.49
10–20	10	15	4.47
20–30	12	25	7.45
70–80	9	75	22.35

$$R = 1 - x_0 + \int_0^{x_0} \frac{v_p}{v_0} dx$$

Solve for x_0:

$$x_1 = 8 \qquad x_2 = 8 + 10 = 18$$

$$x_0 \Rightarrow 0.0077 \qquad \frac{x_0 - 8}{8 - 18} = \frac{0.0077 - 1.49}{1.49 - 4.47}$$

$$8 \Rightarrow 1.49$$

$$18 \Rightarrow 4.47 \qquad x_0 = 3.03\%$$

Particle size (μm)	Wt %	d_{pi} (10^{-6} m)	v_i (m/s)	$\sum$ Wt %
			0.0077	3.03
0–10	8	5.0	1.49	8
10–20	10	15	4.47	18
20–30	12	25		30

$$R = 1 - 0.0303 + \frac{1}{0.0077}[0.0077(3.03 - 0)]\left(\frac{1}{100}\right) \approx 1.0 \quad \text{or} \quad 100\% \quad \textbf{Answer}$$

(c) $\qquad P_c = I_c V = 2.3(70{,}000) = 161{,}000$ W or 161 kW **Answer**

Baghouse filters. In fabric filtration, the dirty air is passed through a woven or felted fabric that filters out the particulate matter. The initial process of removal when the fabric is still clean is by direct interception, impaction, diffusion, and electrostatic attraction. After a dust mat or cake has been formed, submicron particles are removed by a sieving action. The principle of operation is similar to that of an ordinary vacuum cleaner used in homes.

The filtering element is usually shaped in tubular form as shown in Figure 12-4, called *filter bags*. Filter bags may range from 1.8 to 9 m in length and may have diameters of around 20 cm. Several of these bags are grouped together and put into a compartment, and several of these compartments are put together and assembled inside a structure called a *baghouse*. The capacity of a compartment is determined by the area of the fabric filter. Table 12-5 shows some design guidelines to size compartments.

Classified according to the methods of cleaning, there are three types of baghouses: shaker, reverse-air, and pulse-jet baghouses. In the reverse-air and shaker baghouses, the bags in each compartment are bundled in bag headers at the bottom and the individual tube bags are suspended from the ends at the top. Dirty air is introduced at the bottom and flows through the fabric from the inside of the tubes. The cakes are therefore collected at the inside of the bags. When it is time for the filters to be cleaned, one compartment is isolated from the rest. In the reverse-air system, air is blown through the bag in the reverse

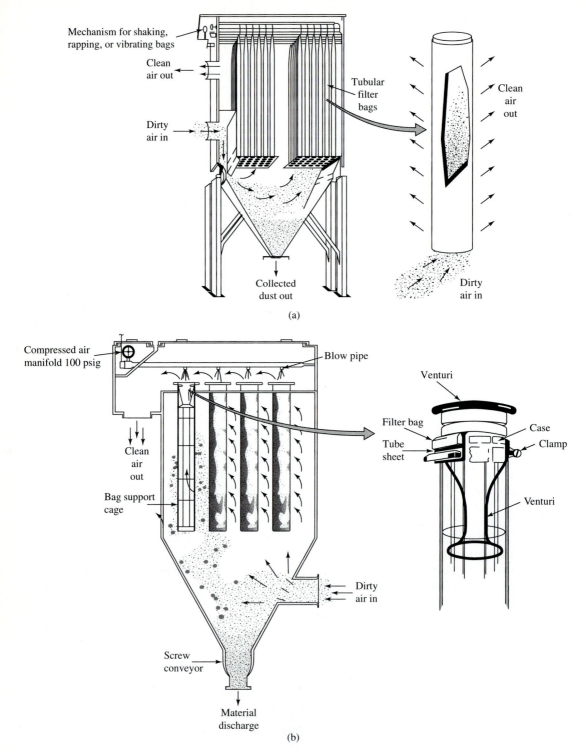

Figure 12-4 (a) Shaker baghouse. (b) Pulse-jet baghouse.

TABLE 12-5 FABRIC FILTER AREA VERSUS
NUMBER OF COMPARTMENTS

Filter area (m^2)	Number of compartments
1–200	1–2
201–600	3–4
601–1000	5–7
1001–1600	8–10
1601–2200	11–13
2201–3000	14–16
3001–4100	17–20
>4100	>20

direction from the outside toward the inside. In the shaker system, the bags are shaken. In both systems, during cleaning, chunks of the previously collected dust fall down to the bottom of the hopper below. As soon as the compartment has been cleaned, it is put back to operation, and the next compartment in-line for cleaning is now, in turn, isolated, and another compartment cleaning begins. The cleaning proceeds around a cycle.

In the pulse-jet baghouses, bags are suspended on tube-bag headers at the top and the bags are mounted on cage supports. The cage is necessary to prevent bag collapse. Dirty air is introduced so as to flow from the outside toward the inside of the bag, thus the necessity of the cage support. Clean, filtered air then flows up through the tube header, thence into the outlet. The bags are cleaned by pulsing a blast of high-pressure air from a blow pipe at 90 to 100 psi at intervals of every few minutes. There is no need to isolate a compartment, as cleaning is done at the same time that filtration is going on. In fact, no compartment is needed.

Variables to be established in the design of baghouses are filtering area, cleaning cycle, pressure drop, and compressor requirement for pulse-jet baghouses. Of course, pumping station design for the system has already been discussed elsewhere in the book. The filtering area is calculated using *filtration velocities* or *air/cloth ratios* that must be determined experimentally for a particular type of dust. Table 12-6 shows some air to cloth ratios for some types of dust, and Table 12-7 shows some types of filter fabric that have been used along with recommendations for maximum operating temperatures, as well as an indication of chemical resistances. Note that polypropylene and Teflon have excellent chemical resistances to acids and alkalis. In addition, Teflon has a high temperature of operation of 400°F. Glass has the highest temperature of operation (550°F) but has a poor resistance toward alkali, although it has a high resistance toward acids.

Example 12-9

A baghouse is to be constructed to control emissions from a grain elevator. The filter bags to be used measure 0.3 m in diameter and 6.0 m in length. The system is used to control 21 m^3/s of waste-air flow. Determine the number of bags required.

Solution   From Table 12-6, use an air/cloth ratio of 3.0 cfm/ft^2 = 0.0152 m/s. Then

$$\text{no. of bags} = \frac{21}{0.0152(\pi)(0.3)(6)} = 244.3, \text{ say 245 bags} \textbf{Answer}$$

TABLE 12-6 AIR/CLOTH RATIOS FOR VARIOUS TYPES OF DUST

Dust	Maximum air/cloth ratio (cfm/ft^2)
Leather, paper, tobacco, wood	3
Alumina, clay, coke, charcoal, cocoa, lead oxide, mica, soap, sugar, talc	2
Aluminum oxide, carbon, fertilizer, graphite, iron ore, lime, paint pigments	2
Asbestos, limestone, quartz, silica	2
Cork, feeds and grain, marble, oyster shell	3
Bauxite, ceramics, chrome ore, feldspar, flour, flint, glass, gypsum, plastics	2

TABLE 12-7 BAGFILTER FABRICS

Fabric	Recommended maximum temperature (°F)	Chemical resistance	
		Acid	Alkali
Polypropylene	195	Excellent	Excellent
Orlon	255	Good	Fair
Nomex	395	Fair	Good
Teflon	395	Excellent	Excellent
Glass	545	Excellent	Poor
Dacron	270	Good	Fair
Dynel	155	Good	Good
Cotton	175	Poor	Good
Wool	195	Good	Poor
Nylon	195	Poor	Good

Design of Cleaning Cycle. For the shaker and reverse-air baghouses, the compartments are cleaned around a cycle. Imagine that compartment 1 has just been cleaned and immediately put back to service. Then a second compartment is now in line for the next isolation and cleaning. Let t_r, the *run time*, be the time from when the first compartment is put into service and the time when the second compartment is isolated for cleaning. Also, let t_c be the time it takes to clean a compartment. Call $t_r + t_c$ the *run–clean* time of the cycle. If there are N compartments in a cleaning cycle, there will be N of these times; and the total length of run–clean times is $N(t_r + t_c)$ [i.e., $N(t_r + t_c)$ is the *cycle time*]. Let t_f be the total filtration time of compartment 1 before it is cleaned again. During the $N(t_r + t_c)$ run–clean times around the cycle, compartment 1 is running, except for its cleaning time. Therefore, its filtration time t_f, which is equal to the filtration time of any compartment, is

$$t_f = N(t_r + t_c) - t_c \tag{12-31}$$

The value of t_c can range from 1 to 5 min.

From the discussion above it can be deduced that the air/cloth ratio is highest during cleaning. Hence, if the design value is not to be exceeded, the air/cloth ratio should be based on $N - 1$ compartments running any time.

Example 12-10

In Example 12-9, determine the number of compartments and the number of bags in each compartment to be used. Design the cleaning cycle.

Solution   Total filter area $= 245(\pi)(0.3)(6) = 1385$ m^2; hence, from Table 12-5, number of compartments $= 9$.   **Answer**
To maintain design conditions, when one compartment is isolated for cleaning, there should be 245 bags in use. Hence

$$\text{number of bags/compartment} = \frac{245}{9-1} = 30.63, \text{ say 30 bags} \quad \textbf{Answer}$$

Assume that a study on a similar baghouse treating a similar waste has established the following values: $t_r = 9$ min and $t_c = 2.5$ min. Then

$$t_f = N(t_r + t_c) - t_c = 9(9 + 2.5) - 2.5 = 101 \text{ min} \quad \textbf{Answer}$$

Determination of Pressure Drop.   As filtration proceeds, dust continues to build up on the filter, forming a cake. This process forms two resistances in the path of airflow: the filter cloth and the cake. The theory involved is the same as in the filtration through a sludge cake discussed in Chapter 8. The equation for the pressure drop $-\Delta P$ is as follows: ($-\Delta P$ is $P_1 - P_2$ or ΔP is $P_2 - P_1$):

$$-\Delta P = \frac{\mu c \bar{\alpha}}{2A^2 t} V^2 + \frac{\mu R_m}{At} V \tag{12-32}$$

where μ is the dynamic viscosity of air, c the mass of cake collected per unit volume of air filtered, $\bar{\alpha}$ the average specific cake resistance, V the volume of air filtered, A the filter cloth area, t the filtration time, and R_m the resistance of the filter cloth. But $V/At = V$, the filtration velocity or air/cloth ratio. Substituting in equation (12-32) and dividing all throughout by V,

$$\frac{-\Delta P}{V} = \frac{\mu c \bar{\alpha} V t}{2} + \mu R_m \tag{12-33}$$

But $c = (L_i Q_0 - L_e Q_0)/Q_0 = L_i - L_e$, where L_i and L_e are the inlet and exit dust concentrations, respectively, and Q_0 is the waste airflow. Substituting in equation (12-33), yields

$$\frac{-\Delta P}{V} = \frac{\mu \bar{\alpha}}{2} V(L_i - -L_e)t + \mu R_m \tag{12-34}$$

In equation (12-34), the plot of $V(L_i - L_e)t$ and $-\Delta P/V$ is a straight line with a slope of $\mu \bar{\alpha}/2$ and intercept of μR_m; the equation requires only two experimental points to determine the parameters. Also, from Chapter 8, $\bar{\alpha}$ is $\bar{\alpha}_0(\Delta P)^s$, where $\bar{\alpha}_0$ is a proportionality constant and s is a compressibility coefficient.

Example 12-11

From the test results shown below, determine the resistance of the filter medium, the specific cake resistance $\bar{\alpha}_0$, the compressibility coefficient s, and the pressure drop (in cm H$_2$O) after 101 min of operation. The air/cloth ratio is 0.9 m/min, the inlet dust concentration is 5.0 g/m^3, and exit dust concentration is practically zero. Assume that the temperature is 45°C.

Time (min)	$-\Delta P (\text{N/m}^2)$
0	151
5	379
10	506
20	611
30	694
60	992

Solution

$$\frac{-\Delta P}{V} = \frac{\mu \bar{\alpha}}{2} V(L_i - L_e)t + \mu R_m$$

Time (min)	$-\Delta P (\text{N/m}^2)$	$-\Delta P/V (\text{N-min/m}^3)$	$V(L_i - L_e)t \ (\text{kg/m}^2)$
0	151	167.78	0
5	379	421.11	0.0225
10	506	562.22	0.045
20	611	678.89	0.09
30	694	771.11	0.135
60	992	1102.22	0.27

$$\left(\frac{-\Delta P}{V}\right)_1 = \frac{167.78 + 421.11 + 562.22}{3} = 383.70$$

$$[V(L_i + L_e)t]_1 = \frac{0 + 0.0225 + 0.045}{3} = 0.0225$$

$$\left(\frac{-\Delta P}{V}\right)_2 = \frac{678.89 + 771.11 + 1102.22}{3} = 850.74$$

$$[V(L_i + L_e)t]_2 = \frac{0.09 + 0.135 + 0.27}{3} = 0.165$$

From Appendix 19, μ of air at 45°C = 1.8(10^{-5}) kg/m-s = 1.08(10^{-3}) kg/m-min. Thus

$$383.70 = \frac{\mu \bar{\alpha}}{2}(0.0225) + \mu R_m \tag{a}$$

$$850.74 = \frac{\mu \bar{\alpha}}{2}(0.165) + \mu R_m \tag{b}$$

Solving, $\mu R_m = 298.18$; $\bar{\alpha}$ = average $\bar{\alpha}$ = 6.07(10^6). ρ_{water} = 990 kg/m^3 at 45°C.

$$\text{Resistance of filter medium} = \mu R_m V = 298.18(0.9) = 268.36 \frac{\text{N}}{\text{m}^2} = \frac{268.36}{990(9.81)}$$

$$= 0.0276 \text{ m } H_2O = 2.76 \text{ cm } H_2O \quad \textbf{Answer}$$

$$\frac{-\Delta P}{V} = \frac{\mu\bar{\alpha}}{2} V(L_i - L_e)t + \mu R_m = \frac{\mu\bar{\alpha}}{2} V(L_i - L_e)t + 298.81$$

Solving for $\bar{\alpha}$ gives

$$\bar{\alpha} = \frac{2(-\Delta P/V - 298.81)}{\mu V(L_i - L_e)t}$$

Hence

Time (min)	$-\Delta P/V$ (N-min/m^3)	$V(L_i - L_e)t$ (kg/m^2)	$\bar{\alpha}$
0	167.78	0	—
5	421.11	0.0225	$1.01(10^7)$
10	562.22	0.045	$1.08(10^7)$
20	678.89	0.09	$0.78(10^7)$
30	771.11	0.135	$0.65(10^7)$
60	1102.22	0.27	$0.55(10^7)$

$$\bar{\alpha} = \bar{\alpha}_0(-\Delta P)^s \qquad \ln\bar{\alpha} = \ln\bar{\alpha}_0 + s\ln(-\Delta P)$$

$$\bar{\alpha}_1 = \frac{(1.01 + 1.08 + 0.78)10^7}{3} = 0.96(10^7)$$

$$-\Delta P_1 = \frac{421.11 + 562.22 + 678.89}{3}(0.9) = 498.67$$

$$\bar{\alpha}_2 = \frac{(0.78 + 0.65 + 0.55)10^7}{3} = 0.66(10^7)$$

$$-\Delta P_2 = \frac{678.89 + 771.11 + 1102.22}{3}(0.9) = 850.74$$

$$\ln 0.96(10^7) = \ln\bar{\alpha}_0 + s\ln 498.67 \qquad\qquad (c)$$

$$\ln 0.66(10^7) = \ln\bar{\alpha}_0 + s\ln 850.74 \qquad\qquad (d)$$

Solving, $\bar{\alpha}_0 = 7.53(10^8)$; $s = -0.703$ **Answer**

$$\frac{-\Delta P}{V} = \frac{\mu\bar{\alpha}}{2} V(L_i - L_e)t + 298.18 = \frac{\mu\bar{\alpha}_0(-\Delta P)^s}{2} V(L_i - L_e)t + 298.18$$

$$\frac{1}{V}(-\Delta P)^{1+0.703} = \frac{1.8(10^{-3})(7.53)(10^8)}{2}(0.9)(0.005)(101) + \frac{298.18}{-\Delta P^{-0.703}}$$

$$(-\Delta P)^{1.703} = 3.08(10^5) + 298.18(-\Delta P)^{0.703} \Rightarrow Y = (-\Delta P)^{1.703} - 298.18(-\Delta P)^{0.703}$$

$$= 3.08(10^5)$$

$-\Delta P(\mathrm{N/m^2})$	$\dfrac{(-\Delta P)^{1.703}}{V}$	$298.18(-\Delta P)^{0.703}$	Y
1500	$2.85(10^5)$	$5.10(10^4)$	$2.33(10^5)$
2500	$6.80(10^5)$	$7.30(10^4)$	$6.07(10^5)$

$$
\begin{array}{lll}
1500 & \Rightarrow & 2.33 \\
x & \Rightarrow & 3.08 \\
2500 & \Rightarrow & 6.07
\end{array}
\qquad
\dfrac{x-1500}{2500-1500} = \dfrac{3.08-2.33}{6.07-2.33}
$$

$$x = 1700$$

$$-\Delta P = 1700\,\frac{\mathrm{N}}{\mathrm{m^2}} \Rightarrow \mathrm{m\ H_2O} = \frac{1700}{990(9.81)} = 0.176 = 17.6 \text{ cm water} \quad \textbf{Answer}$$

Compressor Power.   In addition to the power needed to force the waste air through the baghouse, compressed air is also needed for pulse-jet baghouses. In pulse-jet baghouses, the bags are cleaned by pulsing a blast of high-pressure air from a blow pipe at 90 to 100 psi at intervals of every few minutes. Typically, the flow rate of compressed air is about 0.5 to 1.0% that of the filtered air when both are corrected to the same temperature and pressure. This high-pressure air is provided by a compressor. The air is drawn from the surroundings into the intake valves of the compressor, is compressed inside the unit, and expelled at the discharge port at a much higher pressure into the blow pipe. In general, as in the case of pumps, compressors can be of the centrifugal or positive-displacement type. Figure 12-5 shows sectional views of a centrifugal compressor and two positive-displacement compressors. Figure 12-5b and c are of the positive-displacement type. The air is compressed by means of a piston.

The centrifugal compressor is shown in Figure 12-5a. Notice that there are four impellers in the unit; hence this compressor is a four-stage compressor. The air flows through the four stages in succession starting from the inlet and out in the discharge port. Every time the air exits from a stage, its pressure is compressed to a higher pressure than that at the inlet to the stage.

The other types of positive-displacement machines use a plunger, which is simply a rod that moves back and forth inside a very narrow cylinder. The positive-displacement compressor at the top of the drawing is a two-stage, single-acting unit. The bottom is a two-stage, double-acting unit. *Single acting* means that only one side of the piston compresses the gas, while *double acting* means both sides compress the gas.

Air compression is of three different types: isothermal, adiabatic, and polytropic. In the process of compression, the air is heated and the resulting temperature may become excessive. For this reason the compressed air is cooled by aftercoolers and intercoolers. In the design of the machine, channels are provided in the body to enable coolant such as water to be circulated. Units that provide coolant between stages are called *intercoolers*, and those that provide coolant at the final discharge port are called *aftercoolers*. The cooling process can become so effective that the air exit temperature may approach that of the inlet. The air is then said to be subjected to an *isothermal* process.

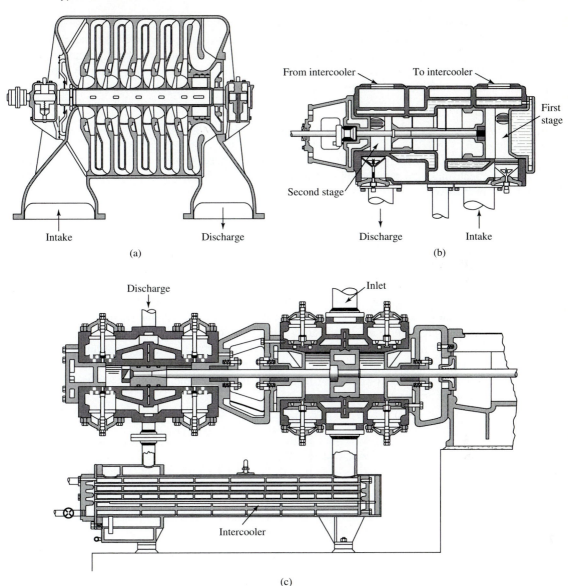

Figure 12-5 Sectional views of compressors: (a) centrifugal; (b) two-stage, single-acting positive-displacement type; (c) two-stage, double-acting positive-displacement type.

For a given parcel of air, the compression process happens so fast that heat exchange with the surroundings may be neglected. Processes that do not exchange heat with the surroundings are called *adiabatic*. By definition, since no heat is exchanged with the surroundings, adiabatic processes are isentropic processes. For large compressors, however, isothermal and adiabatic processes are not maintained. Under these conditions, the process is called *polytropic*. Polytropic processes may be viewed as general processes between isothermal and adiabatic, where the adiabatic and isothermal processes may be considered special cases. From fluid mechanics, applying the energy equation between the inlet and the outlet of a given stage, the ideal work of compression per unit mass of air is

$$w_{\text{comp}} = -\frac{dP}{\rho_a} \quad \text{(for a unit mass of gas)} \tag{12-35}$$

where dP is the pressure drop and ρ_a is the mass density of air. Again, from fluid mechanics, the relationships of P and ρ_a for isothermal and polytropic processes are, respectively,

$$\frac{P}{\rho_a} = \frac{P_1}{\rho_{a1}} \tag{12-36}$$

$$\frac{P}{\rho_a^n} = \frac{P_1}{\rho_{a1}^n} \tag{12-37}$$

where P is pressure, n the exponent for the polytropic process, and the subscript 1 designates inlet conditions. For adiabatic processes, $n = \gamma$, the ratio of specific heat at constant pressure to specific heat at constant volume, c_p/c_v. For air, $\gamma = 1.4$.

Solving equations (12-36) and (12-37), respectively, for ρ_a, substituting the results, in turns, in equation (12-35), and integrating from inlet to outlet (points 1 to 2), the compression work formulas for isothermal and polytropic processes per unit mass of air per stage are, respectively,

$$w_{\text{comp}} = \begin{cases} \dfrac{P_1}{\rho_{a1}} \ln \dfrac{P_2}{P_1} & (12\text{-}38) \\[4mm] \dfrac{P_1 n}{(n-1)\rho_{a1}} \left[\left(\dfrac{P_2}{P_1} \right)^{1-1/n} - 1 \right] & (12\text{-}39) \end{cases}$$

Assuming that ideal gas laws apply, ρ_{a1} may be eliminated in favor of temperature T_1 and the equations above become, respectively,

$$w_{\text{comp}} = \begin{cases} \dfrac{RT_1}{M} \ln \dfrac{P_2}{P_1} & (12\text{-}40) \\[4mm] \dfrac{RT_1 n}{(n-1)M} \left[\left(\dfrac{P_2}{P_1} \right)^{1-1/n} - 1 \right] & (12\text{-}41) \end{cases}$$

where M is the molecular weight of air ($= 28.84$). P_2/P_1 is called the *compression ratio*.

w_{comp} is a work per unit mass of gas. If a rate of flow Q of air is compressed, multiplication of $\rho_{a1}Q$ by w_{comp} will produce the fluid power $\mathcal{P}_f$, the power received by the air. Dividing $\mathcal{P}_f$ by the efficiency of the compressor η produces the brake power $\mathcal{P}_s$. In turn, dividing $\mathcal{P}_s$ by the electrical efficiency $\mathcal{M}$ produces the electrical power input $\mathcal{P}_i$.

In the design of multistage compressors, the work in each stage is made equal. To find the compression ratio in each stage, write the following equation:

$$\frac{P_2}{P_1}\frac{P_3}{P_2}\frac{P_4}{P_3}\cdots\frac{P_{s+1}}{P_s}\cdots\frac{P_{n_s+1}}{P_{n_s}} = \frac{P_{n_s+1}}{P_1} \tag{12-42}$$

where the ratios in parentheses are the compression ratios in each stage and n_s is the number of stages. Equal work per stage means equal compression ratios in all stages. Hence

$$\left(\frac{P_{s+1}}{P_s}\right)^{n_s} = \frac{P_{n_s+1}}{P_1} \tag{12-43}$$

and the compression ratio in each stage is

$$\frac{P_{s+1}}{P_s} = \left(\frac{P_{n_s+1}}{P_1}\right)^{1/n_s} \tag{12-44}$$

Under polytropic conditions using the ideal gas laws, equation (12-37) may be written in terms of temperatures as

$$\frac{T}{T_1} = \left(\frac{P}{P_1}\right)^{(n-1/n)} \tag{12-45}$$

Example 12-12

Ambient air at 26°C at a barometric pressure of 29.90 in. Hg is to be compressed to 100 psig. A three-stage reciprocating compressor is to be used. (a) What is the brake horsepower if the mechanical efficiency is 80%? (b) What is the discharge temperature from the first stage? (c) If intercoolers and aftercoolers are to be used and the temperature of the cooling water is to rise by 20°F, how much water is needed for the compressed air to leave at 26°C? Assume that 1.4 m³/s of ambient air is to be compressed. c_p for air is 0.25 Btu/lb-°F for a temperature range of 32 to 2550°F.

Solution

(a) $$\mathcal{P}_f = n_s(\rho_{a1}Q)\frac{RT_1 n}{(n-1)M}\left[\left(\frac{P_2}{P_1}\right)^{1-1/n} - 1\right]$$

assuming that compression is adiabatic, $n = \gamma = 1.4$. $\rho_{a1} = 1.3$ kg/m³.

$$\frac{P_2}{P_1} = \left(\frac{100 + 14.7}{14.7}\right)^{1/3} = 1.98$$

$$R = 82.05\frac{\text{atm-cm}^3}{(\text{g-mol})(\text{K})} = 82.05\frac{(101,330 \text{ N/m}^2)(10^{-6} \text{ m}^3)}{(10^{-3} \text{ kg-mol})(\text{K})} = 8314.13\frac{\text{N-m}}{(\text{kg-mol})(\text{K})}$$

$$\mathcal{P}_f = 3(1.3)(1.4)\frac{8314.13(26 + 273.16)(1.4)}{(1.4 - 1)(28.84)}\left[(1.98)^{1-1/1.4} - 1\right] = 355,177.88 \text{ N-m/s}$$

$$= 355.18 \text{ kW} = 355.18(1.341) = 476.30 \text{ hp}$$

$$\mathcal{P}_s = \frac{476.30}{0.80} = 597.38 \text{ hp} \quad \textbf{Answer}$$

(b)
$$\frac{T}{T_1} = \left(\frac{P}{P_1}\right)^{(n-1)/n}$$

$$T = (26 + 273.16)(1.98)^{(1.4-1)/1.4} = 363.63 \text{ K} = 90.47°C \quad \textbf{Answer}$$

(c) $0.25 \text{ Btu/lb-}°F = \dfrac{0.25(252.2 \text{ cal/Btu})(4.184 \text{ N-m/cal})}{(0.454 \text{ kg})[\frac{5}{9}(°C/°F)]} = 1045.91\dfrac{\text{N-m}}{\text{kg-}°C}$

$$1.4 \text{ m}^3/\text{s of air} = 1.4\rho = 1.4(1.3) = 1.82 \text{ kg/s}$$

$$\text{Heat load to each cooler} = 1.82(1045.91)(90.47 - 26.0) = 122{,}722.27 \text{ N-m/s}$$

$$= 29{,}331.33 \text{ cal/s}$$

$$\text{Total heat load} = 29{,}331.33(3) = 87{,}993.98 \text{ cal/s}$$

$$20°F = 20\left(\frac{5}{9}\right) = 11.11°C$$

$$\text{H}_2\text{O needed} = \frac{87{,}993.98}{11.11(1.0)} = 7920.25 \text{ g/s} = 7.92 \text{ kg/s} \quad \textbf{Answer}$$

Scrubbers.   Figure 12-6 shows various types of scrubbers. Parts (a) and (b) show ordinary spray chambers. On the left of these drawings, dirty gas is introduced at the bottom side of the unit and flows upward countercurrent to the settling of atomized liquid droplets; on the right, the dirty gas flows cross-flow to the settling of the atomized spray water droplets. Although two sets of sprays atomize the water in horizontal directions, the settling of the resulting droplets is still downward, cross-flow to the direction of the gas.

The scrubber shown in Figure 12-6d is a venturi attached to the larger droplet separator and mist eliminator. At the bottom right, the scrubber is a combination of an ordinary spray and cyclone. Hence this is called a *cyclone spray chamber.* In this type of spray chamber, the cyclonic driving force will not form a true vortex, since the path that would have led to its formation will simply be broken by the sprayed liquid. What the configuration does is delay the exit of the gas providing longer contact time with the liquid droplets. Overall, however, the settling of droplets is downward, while the gas has a net upward movement. Hence the mechanism is a countercurrent flow between droplets and particulates, similar to the countercurrent spray chamber. Of the above-mentioned scrubbers, the venturi can be designed to have efficiencies ranging from very low to very high. Venturi efficiencies can go as high as 99%.

In scrubbers, particulates are *scrubbed out* from the waste air by using water droplets of various sizes, depending on the type of scrubber used. The particulates are removed

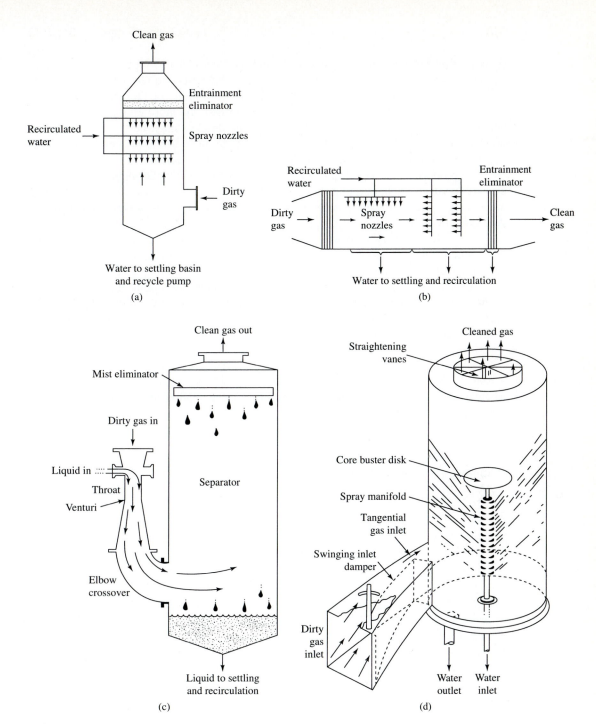

Figure 12-6 Types of scrubbers: (a) countercurrent spray chamber; (b) cross-flow spray chamber; (c) venturi scrubber; (d) cyclone spray chamber.

by impaction and interception between droplets and particulates. Upon impact, the particulate is wetted and carried by the water droplet, thus effecting removal.

To provide impact on millions of particulates, millions of droplets must also be formed. This means that the more droplets are formed, the more efficient the unit will be, and conversely. Therefore, to be effective, the droplets must be small. Ideally, the droplet should be large, in order for the particulate to hit it easily. However, large sizes will produce few droplets—and more droplets are what is needed.

To produce small particles, energy must be expended. Smaller-diameter spray nozzles will produce smaller droplets but will also result in higher-pressure drops consuming more energy. Since efficiency increases as the droplet size decreases, efficiency increases with pressure drop.

Let a flow of dirty air with a concentration C_0 of particulate be scrubbed. Assume that, initially, there is no pressure drop. At this initial condition, the concentration of the particulate at the effluent of the unit will be also C_0. Setting a pressure drop and increasing it will cause a decrease of the particulate concentration at the effluent. This means that the effluent concentration varies inversely as the pressure drop. This behavior may be modeled as a first-order process. Calling C the effluent concentration and k the first-order process coefficient, the model is

$$\frac{dC}{d(-\Delta P)} = -kC \tag{12-46}$$

where a minus sign is prefixed to account for the inverse relationship. Integrating between limits of $C = C_0$ when $(-\Delta P) = 0$ and $C = C$ when $(-\Delta P) = (-\Delta P)$ yields

$$\frac{C}{C_0} = e^{-k(-\Delta P)} \tag{12-47}$$

Therefore, the efficiency η of the unit is

$$\eta = 1 - e^{-k(-\Delta P)} \tag{12-48}$$

For a countercurrent, vertical flow spray chamber, $k(-\Delta P)$ has been determined empirically as[5]

$$k(-\Delta P) = \frac{3Q_\ell v_d Z_0 \eta_d}{2Q_g d_d (v_d - v_g)} \tag{12-49}$$

where Q_ℓ is the volumetric liquid flow rate converted to droplets, v_d the terminal settling velocity of the droplet, Z_0 the length of the scrubber contact zone, η_d the fractional target efficiency for a single droplet, Q_g the volumetric gas flow rate, d_d the diameter of the droplet, and v_g the superficial gas velocity. Not all of Q_ℓ is converted into droplets; how much is converted should be estimated before equation (12-49) can be used.

$v_d - v_g$ is the net settling velocity of the droplet in the countercurrent, vertical flow chamber. In the cross-flow, v_g is not acting in the direction opposite to v_d, hence v_d is

[5]S. Calvert (1977). "Scrubbing," in A. C. Stern (ed.), *Air Pollution*, Vol. IV. Academic Press, New York.

not diminished by v_g and the net settling velocity is simply v_d. Therefore, v_g may be neglected and $k(-\Delta P)$ becomes, for cross-flow chambers,

$$k(-\Delta P) = \frac{3Q_\ell Z_0 \eta_d}{2Q_g d_d} \qquad (12\text{-}50)$$

η_d has also been determined empirically as

$$\eta_d = \left(\frac{C\rho_p d_p^2 v_p / 9\mu d_d}{C\rho_p d_p^2 v_p / 9\mu d_d + 0.7} \right)^2 \qquad (12\text{-}51)$$

where C is called the dimensionless *Cunningham correction factor*, ρ_p the density of the particle, d_p the diameter of the particle, v_p the velocity of the particle (practically equal to v_g), μ the absolute viscosity of the air, and d_d is the diameter of the droplet. Table 12-8 shows values of the Cunningham factor for several size particulates. d_d for spray chambers normally ranges from 500 to 1000 μm. However, in an actual design, the manufacturer's literature should be consulted for the correct droplet size for a particular spray nozzle pressure drop.

TABLE 12-8 CUNNINGHAM CORRECTION FACTOR, C, AT 1 ATM AND 25°C

d_p (μm)	C
10	1.0
2.0	1.1
1.0	1.2
0.50	1.3
0.05	5.0
0.01	23.0

For venturi scrubbers, $k(-\Delta P)$ has been determined empirically as[6]

$$k(-\Delta P) = -\left[\frac{Q_\ell v_g \rho_\ell d_d}{55 Q_g \mu} \left(-0.7 - K_p f + 1.4 \ln \frac{K_p f + 0.7}{0.7} + \frac{0.49}{0.7 + K_p f} \right) \frac{1}{K_p} \right] \qquad (12\text{-}52)$$

f is an empirical factor that is equal to 0.25 for hydrophobic particles and 0.50 for hydrophylics. The empirical factor K_p is

$$K_p = \frac{C\rho_p d_p^2 v_p}{9\mu d_d} \qquad (12\text{-}53)$$

[6]S. Calvert, J. Goldschmid, D. Leith, and D. Mehta (1972). "Wet Scrubber System Study," in *Scrubber Handbook*, Vol. I. NTIS, PB-213016. U. S. Department of Commerce, Washington, D.C.

In venturis, gas is atomized at the throat. An empirical equation for d_d (in μm) is[7]

$$d_d = \frac{58,600}{v_g} \left(\frac{\sigma_\ell}{\rho_\ell} \right)^{0.5} + 597 \left[\frac{\mu_\ell}{(\sigma_\ell \rho_\ell)^{0.5}} \right]^{0.45} \left(1000 \frac{Q_\ell}{Q_g} \right)^{1.5} \qquad (12\text{-}54)$$

where σ_ℓ is the liquid surface tension (dyn/cm), ρ_ℓ the density of the liquid(g/cm^3), μ_ℓ the absolute viscosity of the liquid (poise, dyn-s/cm^2 or g/cm-s), v_g the gas velocity (cm/s), and the remaining symbols are as defined earlier. This equation is dimensional and the units to be used must be as specified. Q_ℓ and Q_g should have the same units.

The other concern in the design of scrubbers is the pressure drop required for atomization. For nozzles, the pressure drops from the inlet value to atmospheric pressure in the chamber. There are usually several nozzles operated in parallel. For venturis, the pressure drop has been determined empirically as[8]

$$-\Delta P = 2\rho_\ell v_g^2 \frac{Q_\ell}{Q_g} \left(1 - x^2 + \sqrt{x^4 - x^2} \right) \qquad (12\text{-}55)$$

where the terms are as defined earlier and x is given by

$$x = \frac{3\ell_t C_D \rho_g}{16 d_a \rho_\ell} + 1 \qquad (12\text{-}56)$$

Also, ℓ_t is the length of the venturi throat, C_D is the drag coefficient, $\rho_g = \rho_a$ is the density of the air, and all other symbols are as defined earlier.

The procedure for design is similar to that used in cyclones, gravitational settlers, EXPs, and baghouses in that size distributions of the particulates must first be determined. This is necessary since in applications the particulates are a mixture of sizes. Relevant formulas derived above will then apply to each size fraction. As seen in previous examples, considering the distribution of sizes involves complicated calculations in design. In an actual design, the formulas should be programmed into a computer. The following examples will no longer consider size distribution but simply work on a single size of particulate.

Example 12-13

Measurement of the dust distribution of a certain industrial operation yields the results in the table below. These results are to be used to design a countercurrent spray chamber. $Q_\ell/Q_g = 0.02(10^{-3})$, $v_g = 25$ cm/s, and $Z_0 = 3.5$ m. The pressure is atmospheric and the temperature is 77°C. (a) What is the efficiency of removal for the 0- to 10-μm size fraction? (b) If Q is 21 m^3/s, what is the cross-sectional area of the chamber? (c) What is the pressure drop through the spray nozzle if the pressure at the nozzle inlet is 20 in. of

[7]S. Nukiyama and Y. Tanasawa (1938). "An Experiment on the Atomization of Liquid by Means of an Air Stream." *Transactions of the Society of Mechanical Engineers*, 4(14), Japan, p. 86.

[8]C. S. Yung, H. F. Barbarika, and S. Calvert (1977). "Pressure Loss in Venturi Scrubbers." *Journal of the Air Pollution Control Association,* 27, pp. 348–351.

water? Assume that the pressure drop produces droplets of $d_d = 0.030$ cm and that 20% of the sprayed water is converted to suspended droplets. Also, assume that the specific gravity of a particulate equals 2.0.

Particle size (μm)	Wt. %
0–10	8
10–20	10
20–30	12
30–40	15
40–50	19
50–60	14
60–70	13
70–80	9

Solution

(a)
$$\eta = 1 - e^{-k(-\Delta P)}$$

$$k(-\Delta P) = \frac{3 Q_\ell v_d Z_0 \eta_d}{2 Q_g d_d (v_d - v_g)}$$

$$\eta_d = \left(\frac{C \rho_p d_p^2 v_p / 9 \mu d_d}{C \rho_p d_p^2 v_p / 9 \mu d_d + 0.7} \right)^2$$

$d_p = 5$ μm; $v_p = 0.25$ m/s. $C = 1$ from Table 12-8. From Appendix 19, $\rho_a = 1.2$ kg/m^3 and $\mu = 2.1(10^{-5})$ kg/m-s at 77°C.

$$\eta_d = \left(\frac{1(2000)[(5)(10^{-6})]^2 (0.25)/9(2.1)(10^{-5})(0.030)(10^{-2})}{\left\{ 1(2000)[(5)(10^{-6})]^2 (0.25)/9(2.1)(10^{-5})(0.030)(10^{-2}) \right\} + 0.7} \right)^2$$

$$= \left(\frac{0.22}{0.22 + 0.7} \right)^2 = 0.057$$

$$v_d = \sqrt{\frac{4}{3} g \frac{\rho_d d_d}{C_D \rho_a}} = \sqrt{\frac{4}{3}(9.81) \frac{(974)(0.030)(10^{-2})}{C_D (1.2)}} = 1.78 \sqrt{\frac{1}{C_D}}$$

$$\text{Re} = \frac{d v_d \rho_a}{\mu} = \frac{(0.030)(10^{-2}) v_d (1.2)}{2.1(10^{-5})} = 17.1 v_d$$

$$C_D = \begin{cases} \dfrac{24}{\text{Re}} & \text{(for Re} < 1) \\[2ex] \dfrac{24}{\text{Re}} + \dfrac{3}{\text{Re}^{1/2}} + 0.34 & \text{(for } 1 \le \text{Re} \le 10^4) \\[2ex] 0.4 & \text{(for Re} > 10^4) \end{cases}$$

By trial and error:

C_D	v_d (m/s)	Re
2.12	1.22	20.90
2.14	1.22	20.80

$$k(-\Delta P) = \frac{3[0.20(10^{-3})](1.22)(3.5)(0.057)}{2(0.03)(10^{-2})(1.22 - 0.25)} = 0.25$$

$$\eta = 1 - e^{-k(-\Delta P)} = 1 - e^{-0.25} = 0.22 \quad \text{or} \quad 22\% \quad \textbf{Answer}$$

(b) Cross-sectional area $= \dfrac{21}{0.25} = 84 \text{ m}^2.$ **Answer**

(c) $(-\Delta P) = 20 - 0 = 20$ in. of water.   **Answer**

Example 12-14

Solve Example 12-13 for a cross-flow spray chamber.

Solution

(a) $$\eta = 1 - e^{-k(-\Delta P)}$$

$$k(-\Delta P) = \frac{3Q_\ell Z_0 \eta_d}{2Q_g d_d}$$

$$\eta_d = \left(\frac{C\rho_p d_p^2 v_p/9\mu d_d}{C\rho_p d_p^2 v_p/9\mu d_d + 0.7} \right)^2$$

$d_p = 5$ μm; $v_p = 0.25$ m/s. $C = 1$ from Table 12-8. From Appendix 19, $\rho_a = 1.2$ kg/m^3 and $\mu = 2.1(10^{-5})$ kg/m-s at 77°C.

$$\eta_d = \left(\frac{1(2000)[(5)(10^{-6})]^2(0.25)/9(2.1)(10^{-5})(0.030)(10^{-2})}{\{1(2000)[(5)(10^{-6})]^2(0.25)/9(2.1)(10^{-5})(0.030)(10^{-2})\} + 0.7} \right)^2$$

$$= \left(\frac{0.22}{0.22 + 0.7} \right)^2 = 0.057$$

$$k(-\Delta P) = \frac{3Q_\ell Z_0 \eta_d}{2Q_g d_d}$$

$$k(-\Delta P) = \frac{3[0.20(10^{-3})](3.5)(0.057)}{2(0.03)(10^{-2})} = 0.20$$

$$\eta = 1 - e^{-k(-\Delta P)} = 1 - e^{-0.20} = 0.18 \quad \text{or} \quad 18\% \quad \textbf{Answer}$$

(b) Cross-sectional area $= \dfrac{21}{0.25} = 84 \text{ m}^2.$ **Answer**

(c) $-\Delta P = 20 - 0 = 20$ in. water. **Answer**

Example 12-15

Measurement of the dust distribution of a certain industrial operation yields the results in the table below. These results are to be used to design a venturi scrubber. $Q_\ell/Q_g = 10^{-3}$ and the throat length is 30 cm. The pressure is atmospheric and the temperature is 77°C. (a) What is the velocity v_g at the throat for a pressure drop of 55 in. water? (b) What is the efficiency of removal for the 0- to 10-μm size fraction? At 1 atm and 77°C, σ equals 63 dyn/cm. Assume that $f = 0.5$.

Particle size (μm)	Wt %
0–10	8
10–20	10
20–30	12
30–40	15
40–50	19
50–60	14
60–70	13
70–80	9

Solution

(a)
$$-\Delta P = 2\rho_\ell v_g^2 \frac{Q_\ell}{Q_g}\left(1 - x^2 + \sqrt{x^4 - x^2}\right)$$

$\rho_g = 1.2$ kg/m^3, $\mu_\ell = 3.5(10^{-4})$ kg/m-s $= 3.5(10^{-3})$ g/cm-s, $\mu = 2.1(10^{-5})$ kg/m-s, and $\rho_\ell = 0.974$ kg/cm^3 at 77°C.

$$55 \text{ in. water} = \frac{55}{12(3.281)}(974)(9.81) = 13{,}347.6 \text{ N/m}^2$$

$$x = \frac{3\ell_t C_D \rho_g}{16 d_d \rho_\ell} + 1 = \frac{3(0.30)C_D(1.2)}{16(d_d)(974)} = 6.93(10^{-5})\frac{C_D}{d_d} + 1$$

$$d_d = \frac{58{,}600}{v_g}\left(\frac{\sigma_\ell}{\rho_\ell}\right)^{0.5} + 597\left(\frac{\mu_\ell}{\sigma_\ell \rho_\ell}\right)^{0.45}\left(1000\frac{Q_\ell}{Q_g}\right)^{1.5}$$

$$= \frac{58{,}600}{v_g}\left(\frac{63}{0.974}\right)^{0.5} + 597\left\{\frac{3.5(10^{-3})}{[63(0.974)]^{0.5}}\right\}^{0.45}[(1000(10^{-3})]^{1.5} = \frac{4.7(10^5)}{v_g} + 18.56$$

$$= \frac{4.7(10^5)}{v_g} + 18.56$$

$$= \frac{4.7(10^5)}{v_g} + 18.56 \qquad\qquad (a)$$

[Note that in equation (a) only, v_g and d_d have units of cm/s and μm, respectively.]

$$C_D = \begin{cases} \dfrac{24}{\text{Re}} & \text{(for Re < 1)} \\[2mm] \dfrac{24}{\text{Re}} + \dfrac{3}{\text{Re}^{1/2}} + 0.34 & \text{(for } 1 \leq \text{Re} \leq 10^4) \\[2mm] 0.4 & \text{(for Re > } 10^4) \end{cases} \qquad (b)$$

where $\text{Re} = d_d v_g \rho_g / \mu$.

$$x = 6.93(10^{-5})\frac{C_D}{d_d} + 1 \qquad (c)$$

$$13{,}347.6 = 2(974)v_g^2\left(10^{-3}\right)\left(1 - x^2 + \sqrt{x^4 - x^2}\right)$$

$$6851.95 = v_g^2\left(1 - x^2 + \sqrt{x^4 - x^2}\right) = Y \qquad (d)$$

By trial and error of (a) to (d):

v_g (m/s)	d_d (μm)	Re	C_D	x	Y
132.0	55.17	416.11	0.52	1.66	7734.24
120.0	58.73	402.70	0.55	1.65	6379.59

$$
\begin{array}{lll}
132.0 & \Rightarrow & 7734.24 \\
y & \Rightarrow & 6851.95 \\
120.0 & \Rightarrow & 6379.59
\end{array}
\qquad
\frac{y - 132.0}{120.0 - 132.0} = \frac{6851.95 - 7734.24}{6379.59 - 7734.24}
$$

$$y = v_g = 124.18 \text{ m/s} \quad \textbf{Answer}$$

(b) $k(-\Delta P) =$

$$-\left[\frac{Q_\ell v_g \rho_\ell d_d}{55 Q_g \mu}\left(-0.7 - K_p f + 1.4\ln\frac{K_p f + 0.7}{0.7} + \frac{0.49}{0.7 + K_p f}\right)\frac{1}{K_p}\right]$$

$$K_p = \frac{C \rho_p d_p^2 v_p}{9 \mu d_d} = \frac{1.0(2)(1000)[5(10^{-6})]^2(124.18)}{9(2.1)(10^{-5})(57)(10^{-6})} = 576.35$$

$$k(-\Delta P) = -\left\{\frac{10^{-3}(124.18)(974)(57)(10^{-6})}{55(2.1)(10^{-5})}\left[-0.7 - (576.35)(0.5)\right.\right.$$

$$+1.4\ln\frac{576.35(0.5) + 0.7}{0.7} + \left.\frac{0.49}{0.7 + 576.35(0.5)}\right]\frac{1}{576.35}\right\} = -\left[5.97(-280.44)\frac{1}{576.35}\right] = 2.90$$

$$\eta = 1.0 - e^{-k(-\Delta P)} = 1 - e^{-2.90} = 0.945 \quad \text{or} \quad 94.5\% \quad \textbf{Answer}$$

Summary of sizes removed.   Table 12-9 shows the summary of particle sizes that are removed by the various particulate control devices. It appears that bagfilters are efficient in removing particulates less than 1.0 μm in diameter, although it cannot with-

TABLE 12-9 SUMMARY OF SIZES OF PARTICLES REMOVED BY
VARIOUS PARTICULATE CONTROL DEVICES

Device	Minimum particle size (μm)	Efficiency (%)
Venturi	>0.5	<99
ESP	>1	<99
Bagfilter	<1	>99
Spray chamber	>10	<80
Cyclonic spray chamber	>3	<80
Impingement scrubber	>3	<80
Centrifugal settler	>5	<80
Gravitational settler	>50	<50

stand much higher temperatures. For operation at high temperatures with the capability
to remove small particles, ESPs are the choice.

Control of Gaseous Pollutants

The principal gaseous pollutants of concern in air pollution are SO_x, NO_x, CO, the
hydrocarbons, and other organic and inorganic gases. In general, these emissions may
be controlled by absorption, adsorption, and incineration. For example, SO_2 may be
absorbed using a solution of lime in water; hydrocarbons may be adsorbed in activated
carbon or incinerated; and CO may be completely burned to CO_2. Before an absorption
or adsorption unit is designed there should be a match between the absorbent or adsorbent
material and the solute to be removed.

Some basic principles of adsorption and design techniques were discussed in Chapter 9, and some fundamentals of incineration in Chapters 10 and 11. Absorption was also
discussed in Chapter 9 in connection with the removal of ammonia. Actually, the discussion was on stripping, the reverse of absorption. The discussion below will continue
on topics not covered before. Figure 12-7 shows a flowsheet for an activated carbon
fixed-bed adsorption, a packed tower absorber, and a thermal incinerator.

Pressure drop across fixed adsorption beds and expansion of fluidized beds.
Since dirty air has to be forced through the bed, pressure drop determines the size
and power requirement of the blower. To provide intimate contact between solute and
adsorbent, the bed may also be fluidized or expanded. These topics have been discussed
in Chapter 6. For convenience the formulas for head losses and the formula for bed
expansion are reproduced below.

$$h = \frac{\ell(1 - \eta)V_s^2}{\eta^3 g} \sum \frac{f_i' x_i}{d_{pi}} \qquad \text{head loss on stratified bed} \qquad (12\text{-}57)$$

$$h_{fb} = \frac{\ell_e(\gamma_p - \gamma_a)}{\gamma_a} \sum x_i(1 - \eta_e) \qquad \text{head loss on expanded bed} \qquad (12\text{-}58)$$

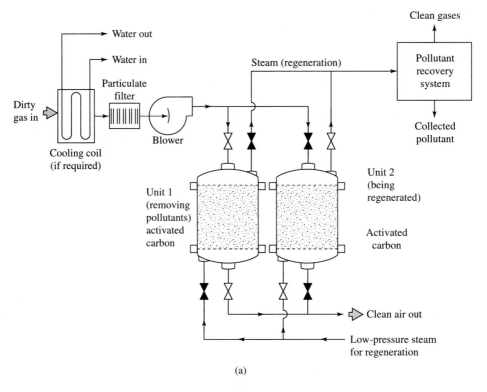

(a)

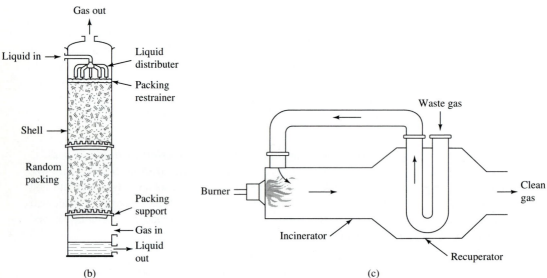

(b)   (c)

Figure 12-7 (a) Flowsheet for activated carbon adsorption. (b) Packed tower absorber. (c) Thermal incinerator.

$$\frac{\ell_e}{\ell} = (1 - \eta) \sum \frac{x_i}{1 - \eta_{e,i}} \qquad \text{bed expansion} \qquad (12\text{-}59)$$

$$f_i' = 150\left(\frac{1 - \eta}{\text{Re}}\right) + 1.75$$

$$\eta_e = \left(\frac{V_B}{v}\right)^{0.22}$$

where h is the head loss in an unexpanded bed, h_{fb} the head loss in bed expansion, ℓ the length of unexpanded bed, η the porosity of unexpanded bed, V_s the superficial velocity of air, g the gravity constant, x_i the mass fraction of particles of size d_{pi} or of particles in layer i, ℓ_e the length of expanded bed, γ_p the specific weight of particle, γ_a the specific weight of air, η_e the porosity of expanded bed, V_B the upflow superficial velocity of air, and v the terminal settling velocity of particles.

Example 12-16

An activated carbon adsorber measuring 4 by 2 m in cross section and 0.7 m deep is used to treat a polluted air containing 5000 ppm of C_6H_6 at 1 atm and 37°C. The X/M value of the bed is 5.0 kg of C_6H_6 per 50 kg of C. The bed has a bulk or packed density of 480 kg/m³, η of 0.4, and a particle size of 4 by 10 mesh. Assuming that the unit absorbs 3.44 kg-mol of C_6H_6 per hour, determine the pressure drop. Assume that $\beta = 0.52$.

Solution

$$h = \frac{\ell(1 - \eta)V_s^2}{\eta^3 g} \sum \frac{f_i' x_i}{d_{pi}}$$

$$f_i' = 150\frac{(1 - \eta)}{\text{Re}} + 1.75$$

$$\text{Total moles of polluted air} = \frac{10^6}{5000}(3.44) = 688 \text{ kg-mol/h}$$

$$= 688(10^3)(22.4)\left(\frac{1}{10^3}\right)\left(\frac{37 + 273}{273}\right) = 17,500 \text{ m}^3/\text{h}$$

$$V_s = \frac{17,500}{2(4)(60)(60)} = 0.61 \text{ m/s}$$

No. 4 mesh = 4.699 mm opening; no. 10 mesh = 1.651 mm opening.

$$4 \text{ by } 10 \text{ mesh} = \frac{4.699 + 1.651}{2} = 3.175 \text{ mm} = d_p$$

$$\text{Re} = \frac{dV_s\rho_a}{\mu} \qquad \rho = 1.27 \text{ kg/m}^3 \qquad \mu = 1.8(10^{-5}) \text{ kg/m-s}$$

$$d = 1.24\beta^{0.333}d_p = 1.24(0.52^{0.333})(0.003175) = 0.0032 \text{ m}$$

$$\text{Re} = \frac{0.0032(0.61)(1.27)}{1.8(10^{-5})} = 137.72$$

$$f_i' = 150 \left(\frac{[1 - 0.4]}{137.72} \right) + 1.75 = 2.40$$

$$h = \frac{0.7(1 - 0.4)(0.61^2)}{0.4^3(9.81)} \left[\frac{2.40(1)}{0.003175} \right] = 188.16 \text{ m of air}$$

$$188.16(1.27)(9.81) = h_{water}(\rho_{water} \text{ at } 37^\circ\text{C})(9.81) = h_{water}(994)(9.81)$$

$$h_{water} = 0.24 \text{ m} = 24 \text{ cm} = 9.44 \text{ in.} \quad \textbf{Answer}$$

Example 12-17

The dirty air in Example 12-16 is to be treated in a fluidized-bed activated carbon adsorber using the same type of activated carbon. It is desired to expand the bed to 150% from an initial depth of 0.7 m. Calculate the head loss. Also, determine the settling velocity of the carbon particles. The specific gravity of activated carbon is 1.8 and the flow of dirty air is 17,500 m^3/h.

Solution

$$h_{fb} = \frac{\ell_e(\gamma_p - \gamma_a)}{\gamma_a} \sum x_i(1 - \eta_e)$$

$$\frac{\ell_e}{\ell} = (1 - \eta) \sum \frac{x_i}{1 - \eta_{e,i}}$$

$$1.5 = (1 - 0.4)\frac{1}{1 - \eta_e}$$

Solving, $\eta_e = 0.6$.

$$h_{fb} = \frac{1.5(0.7)[1.8(1000) - 1.27](9.81)}{1.27(9.81)}(1 - 0.6) = 594.86 \text{ m of air}$$

$$= 0.76 \text{ m of water} \quad \textbf{Answer}$$

$$\eta_e = \left(\frac{V_B}{v} \right)^{0.22}$$

$$V_B = \frac{17,500}{2(4)(60)(60)} = 0.61 \text{ m/s}$$

$$0.6 = \left(\frac{0.61}{v} \right)^{0.22} \qquad v = 6.3 \text{ m/s} \quad \textbf{Answer}$$

Experimental determination of adsorption isotherms and design methods.   As in the case of wastewater treatment, adsorption isotherms need to be determined. The procedure is very similar. In one procedure, a small bucket containing a few milligrams of the adsorbent is suspended inside a cell containing the adsorbate.[9] An inert gas (typically helium) is used to dilute the adsorbate. The bucket containing the adsorbent is suspended

---

[9]E. D. Sloan (1974). "Nonideality of Binary Adsorbed Mixtures of Benzene and Freon on Graphitized Carbon," Ph.D thesis, Clemson University, Clemson, S.C.

from a microbalance capable of reading in micrograms. As the adsorbate is adsorbed, the increase in mass is recorded continuously until saturation is reached. The saturation value gives the X/M corresponding to the diluted concentration of adsorbate. This procedure is repeated for each adsorbate dilution. Once the the desired X/M has been determined, the design procedure is similar to that already discussed in Chapter 9.

Table 12-10 shows some of the adsorbents used in both air and water pollution. The term *activated* refers to the increased internal surface area of the adsorbent when subjected to special processes. For example, one way to activate carbon is by using steam to initiate the water gas reaction. This increases the surface area by removal of some atoms on the surface in the form of carbon monoxide. Molecular sieves are different from activated materials in that they are normally polymeric materials that form pores of molecular dimensions on the order of angstrom units. For example, carbon sieves have 5-Å pores. Molecular sieves perform by "sieving" the molecules.

TABLE 12-10   TYPICAL ADSORBENTS AND THEIR PROPERTIES

Adsorbent	Internal porosity (%)	External porosity, η (%)	Bulk density (kg/m^3)	Surface area (m^2/kg $\times$ 10^{-3})
Carbons	55–75	35–40	160–480	600–1400
Fuller's earth	50	40	470–630	120–240
Iron oxide	22	37	1400	20
Magnesia	75	45	400	200
Silica gel	70	40	400	320
Acid-treated clay	30	40	550–870	100–250
Activated alumina and bauxite	35	45	700–870	200–250
Aluminosilicate sieves	50	35	655–695	595–695
Bone char	50	18	640	100

Tower height of an absorber.   Although the subject matter is absorption, the subsequent formulas that will be derived apply equally well to stripping. Stripping is simply the reverse of absorption.

Absorption units are normally constructed of towers with packings dumped randomly or packings stacked by hand. These towers are called *packed tower absorbers.* Packings that have been used consist of Berl saddles, Intalox saddles, Raschig rings, and so on. For a given load of pollutants to be removed at a given condition, the design of absorption towers involves the determination of the cross-sectional areas and heights. The cross-sectional area is, of course, simply equal to the flow rate divided by the superficial velocity. This superficial velocity must be less than the flooding velocity. The design of the height is, however, a bit tricky. Let us now tackle the derivation of this quantity for packed tower absorbers.

Between liquid and gaseous phases, the transfer of mass from one phase to the other must pass through an interfacial boundary surface. Call the concentration of the solute at this surface the y_i mole fraction referred to the gas phase. The corresponding concentration referred to the liquid phase is x_i. x_i and y_i are in equilibrium. If y mole fraction is the concentration in the bulk gas phase, the driving force toward the interfacial boundary is $y - y_i$ and the rate of mass transfer is $k_y(y - y_i)$, where k_y is the coefficient of mass transfer. For this rate of mass transfer to exist, it must be balanced by an equal rate of mass transfer at the liquid phase. The liquid-phase mass transfer rate is $k_x(x_i - x)$, where k_x is the coefficient of mass transfer based on the liquid side and x is the bulk mole fraction concentration of the solute in the liquid phase. Hence

$$k_y(y - y_i) = k_x(x_i - x) \tag{12-60}$$

In this equation the interfacial concentrations x_i and y_i are impossible to determine experimentally. Hence, instead of them, use x^* and y^*. x^* is the equilibrium concentration corresponding to x, and y^* is the equilibrium concentration corresponding to y. The corresponding driving forces are $y - y^*$ and $x^* - x$ and equating the gas-side equations,

$$k_y(y - y_i) = K_y(y - y^*) \tag{12-61}$$

On the liquid side,

$$k_x(x_i - x) = K_x(x^* - x) \tag{12-62}$$

K_y and K_x are called *overall mass transfer coefficients* for the gas and liquid sides, respectively. To differentiate, k_y and k_x are called *individual mass transfer coefficients* for the respective sides.

It is instructive to determine the equation relating the overall and the individual mass transfer coefficients. Equation (12-61) may be rearranged to obtain

$$\frac{1}{K_y} = \frac{y - y^*}{k_y(y - y_i)} = \frac{(y - y_i) + (y_i - y^*)}{k_y(y - y_i)} = \frac{1}{k_y} + \frac{y_i - y^*}{k_y(y - y_i)} \tag{12-63}$$

Replacing $k_y(y - y_i)$ using equation (12-60),

$$\frac{1}{K_y} = \frac{1}{k_y} + \frac{y_i - y^*}{k_x(x_i - x)} \tag{12-64}$$

Letting $(y_i - y^*)/(x_i - x)$ equal m,

$$\frac{1}{K_y} = \frac{1}{k_y} + \frac{m}{k_x} \tag{12-65}$$

A parallel derivation for K_x yields

$$\frac{1}{K_x} = \frac{1}{k_x} + \frac{1}{k_y} = \frac{1}{mK_y} \tag{12-66}$$

The coordinates x_i and y_i are the coordinates of the point (x_i, y_i) on the equilibrium curve. On the other hand, the coordinates x and y are coordinates of point (x, y) along the length of the tower. Since the tower operation is not in equilibrium, (x, y) is not on the equilibrium curve. The various values of (x, y) along the height of the tower plot a line called the *operating line*. Point (x^*, y^*) may or may not be on the equilibrium curve. For the operating point (x, y), the corresponding points on the equilibrium curve are (x, y^*) and (x^*, y). Point (x^*, y^*) can be on the equilibrium curve only if point (x, y) is on the equilibrium curve; otherwise, point (x^*, y^*) does not exist.

As discussed, there are three points on the equilibrium curve: (x, y^*), (x_i, y_i), and (x^*, y). The slope between (x, y^*) and (x_i, y_i) is $(y_i - y^*)/(x_i - x)$, which is equal to m. Hence m is the slope of the equilibrium curve if it is a straight line.

From Chapter 9, the rate of mass transfer in the gas and the liquid phases are $G\,dY$ and $L\,dX$, respectively, where G is the mole flow rate of solute-free carrier gas, Y the moles of solute per mole of solute-free carrier gas, L the mole flow rate of solute-free carrier liquid, and X the moles of solute per mole of solute-free carrier liquid. $G\,dY$ is equal to $k_y(y - y_i)\,dA[= K_y(y - y^*)\,dA]$ and $L\,dX$ is equal to $k_x(x_i - x)\,dA[= K_x(x^* - x)\,dA)]$, where A is the area of mass transfer or the interfacial area of contact for mass transfer. Calling G' the gas flow rate (mixture of carrier gas and solute), G is equal to $G'(1 - y)$; and calling L' the liquid flow rate (mixture of carrier liquid and solute), L is equal to $L'(1 - x)$. Y and X are also, respectively, equal to $y/(1 - y)$ and $x/(1 - x)$. Hence $dY = dy/(1 - y)^2$ and $dX = dx/(1 - x)^2$.

Since the objective is to design the height of the tower, also express dA in terms of height. Call the superficial area S, the height Z, and the interfacial contact area per unit bulk volume of tower a. Hence $dA = aS\,dZ$. Performing the necessary substitutions, the mass transfer expressions become, for the gas and liquid sides, respectively,

$$K_y a(y - y^*)\,dZ = \frac{G'}{S}\frac{dy}{1 - y} = V_{My}\frac{dy}{1 - y} \tag{12-67}$$

$$K_x a(x^* - x)\,dZ = \frac{L'}{S}\frac{dx}{1 - x} = V_{Mx}\frac{dx}{1 - x} \tag{12-68}$$

where V_{My} is called the *gas-side molar mass velocity* and V_{Mx} is called the *liquid-side molar mass velocity*. Integrating the equations and simplifying produces the formulas for tower height based on the gas and liquid, respectively:

$$\int_{y_1}^{y_2} \frac{dy}{(1 - y)(y - y^*)} = \overline{\left(\frac{K_y a}{V_{My}}\right)} Z_T \tag{12-69}$$

$$\int_{x_1}^{x_2} \frac{dx}{(1 - x)(x^* - x)} = \overline{\left(\frac{K_x a}{V_{Mx}}\right)} Z_T \tag{12-70}$$

where Z_T is the tower height and the expressions with overbars are average values between elevations 1 and 2, respectively.

If the Z_T values of the equations above are solved, they will be expressed in terms of the product of the reciprocal of the overbarred factors by the respective integrals. Hence the tower height may be expressed as the product of two factors. Call the first the *height of a mass transfer unit H* and the second the *number of mass transfer units* N_t. Therefore,

$$Z_T = H N_t \tag{12-71}$$

H based on the gas side will be designated as H_y and H on the liquid side will be designated as H_x. The corresponding designations for N_t are N_{ty} and N_{tx}, respectively. Hence

$$H_y = \frac{1}{K_y a / V_{My}} \tag{12-72}$$

$$H_x = \frac{1}{K_x a / V_{Mx}} \tag{12-73}$$

$$N_{ty} = \int_{y_1}^{y_2} \frac{dy}{(1-y)(y-y^*)} \tag{12-74}$$

$$N_{tx} = \int_{x_1}^{x_2} \frac{dx}{(1-x)(x^*-x)} \tag{12-75}$$

Operating line. $G\,dY$ and $L\,dX$ are equal; thus

$$G\,dY = L\,dX \tag{12-76}$$

But

$$dY = \frac{dy}{(1-y)^2} \quad \text{and} \quad dX = \frac{dx}{(1-x)^2} \tag{12-77}$$

Substituting equations (12-77) into equation (12-76) produces

$$G\frac{dy}{(1-y)^2} = L\frac{dx}{(1-x)^2} \tag{12-78}$$

which can be integrated from $x = x_1$ to $x = x$ and from $y = y_1$ to $y = y$ to yield

$$G\left(\frac{1}{1-y} - \frac{1}{1-y_1}\right) = L\left(\frac{1}{1-x} - \frac{1}{1-x_1}\right) \tag{12-79}$$

x_1 and y_1 are the concentrations at the bottom. Note that G and L are constant; G' and L' are not. Equation (12-79) yields the concentration at any elevation in the tower; hence it is called the equation of the *operating line*.

When plots of the operating line and the equilibrium line are far apart, the driving force will be large and the rate of mass transfer will also be large. When the two lines are close to each other, the driving force will be small, and accordingly, the rate of mass transfer will also be small. The limiting condition is reached when the two lines touch or intersect each other. The liquid flow rate corresponding to this condition is the *minimum flow rate*.

Example 12-18

A packed absorption tower is designed to remove SO_2 from a coke oven stack. The stack gas flow rate measured at 1 atm and 30°C is 10 m³/s, and the SO_2 content is 3.0%. Using an initially pure water, 90% removal is desired. The equilibrium curve of SO_2 in water may be approximated by $y_i = 30x_i$. Determine the water requirement if 150% of the minimum flow rate is deemed adequate. Calculate the height of the tower. Assume that $K_y a = 5.35$ lb-mol/ft³-h-mole fraction and $K_x a = 117$ lb-mol/ft³-h-mole fraction. Assume that the total cross-sectional area of the tower is 11.0 m².

Solution

$$G\left(\frac{1}{1-y} - \frac{1}{1-y_1}\right) = L\left(\frac{1}{1-x} - \frac{1}{1-x_1}\right)$$

$$y_1 = 0.03$$

$$y_2 = \frac{0.1(0.03)(1)}{1 - 0.03(1) + 0.1(0.03)(1)} = 0.0031$$

$$x_2, y_2 = 0.0, 0.0031$$

$$x_1, y_1 = x_1, 0.03$$

At the bottom of the tower, assume that the operating line is at equilibrium point; hence

$$y_i = 30x_i \qquad 0.03 = 30x_i \qquad x_i = x_1 = 0.001$$

$$G\left(\frac{1}{1-0.0031} - \frac{1}{1-0.03}\right) = L\left(\frac{1}{1-0} - \frac{1}{1-0.001}\right)$$

$$\frac{L}{G} = 28$$

$$G = 10(0.97)\left(\frac{273}{30+273}\right)(10^3)\left(\frac{1}{22.4}\right)(10^{-3}) = 0.39 \text{ kg-mol/s}$$

$$L = (1.5)(28)(0.39)(18) = 295 \text{ kg/s} = \frac{295}{\rho_{water}} = \frac{295}{980} = 0.30 \text{ m}^3/\text{s} \quad \textbf{Answer}$$

$$H_y = \frac{1}{K_ya/V_{My}}$$

$$N_{ty} = \int_{y_1}^{y_2} \frac{dy}{(1-y)(y-y^*)}$$

$$\text{Average } y = \frac{0.03 + 0.0031}{2} = 0.017$$

$$G' = \frac{0.39}{1-0.017} = 0.40 \text{ kg-mol/s}$$

$$V_{My} = \frac{0.40}{11} = 0.036 \text{ kg-mol/m}^2\text{-s}$$

$$K_ya = 5.35 \text{ lb-mol/ft}^3\text{-h-mol fraction} = \frac{5.35(0.454)}{(1/3.281)^3(60)(60)} = 0.024\frac{\text{kg-mol}}{\text{m}^3\text{-s-mol fraction}}$$

$$H_y = \frac{1}{0.024/0.036} = 1.5 \text{ m per transfer unit}$$

$$0.39\left(\frac{1}{1-y} - \frac{1}{1-0.03}\right) = (1.5)28(0.39)\left(\frac{1}{1-x} - \frac{1}{1-0.001}\right) \qquad x = \frac{0.39y}{16.38 + 15.99y}$$

From $y_i = 30x_i$, $y^* = 30x$.

y	x	y^*	$\dfrac{1}{y-y^*}$	$\dfrac{1}{1-y}$	Δy	$\dfrac{\Delta y}{(1-y)(y-y^*)}$
0.0031	0.0000	0.0000	—	—	—	—
0.0050	0.00012	0.0036	714.3	1.0050	0.002	1.44
0.0080	0.00019	0.0057	429.19	1.0081	0.003	1.38
0.0110	0.00026	0.0078	309.9	1.0111	0.003	0.94
0.0130	0.00031	0.0092	261	1.0132	0.002	0.53
0.0160	0.00038	0.011	210.7	1.0163	0.003	0.64
0.0190	0.00044	0.013	172	1.0194	0.003	0.53
0.0210	0.00049	0.0147	159	1.0215	0.002	0.32
0.0240	0.00056	0.0168	138	1.0245	0.003	0.28
0.0270	0.00063	0.0188	122	1.0277	0.003	0.38
0.03	0.00060	0.021	109	1.0309	0.003	0.34
	6.78	10.20			Σ	6.78

$$Z_T = 1.5(6.78) = 10.20 \text{ m} \quad \textbf{Answer}$$

Incineration.   Combustible wastes may also be destroyed by incineration. Incineration connotes destruction of a material by burning or combustion. Incineration is always combustion, but combustion may not always be incineration. (For example, biological combustion is not incineration.) Incineration is a rapid combustion process. Since incineration of solid wastes has been discussed elsewhere in the book, the concern in this section is incineration of gaseous wastes, such as volatile organic compounds (VOCs). This class of wastes includes not only hydrocarbons but organic acids, aldehydes, and ketones as well as organics containing chlorine, sulfur, and nitrogen.

Incineration may be direct-flame, thermal, or catalytic. In *direct-flame incineration*, the contaminant stream has a high heating value that it will just burn by itself in a combustor. By contrast, the contaminant stream in *thermal incineration* has such a low heating value that it needs supplementary fuel for it to burn. Figure 12-7c shows a thermal incinerator, so called because the influent waste stream is preheated (thermal) by a heat exchanger or recuperator which gets its heat from the burning of the waste stream itself. Because of the indirect nature of the combustion, thermal incinerators are also called *afterburners*. *Catalytic incineration* employs the use of catalysts, which are substances that speed up the reaction but are not themselves consumed. The catalyst draws the reactants to its surface by diffusion and adsorption. At the surface, the reaction occurs at a much faster rate than it would without the surface. Catalysts also assure that the reaction occurs at a much lower temperature.

There are two types of catalysts: homogeneous and heterogeneous. *Homogeneous catalysts* function in the same phase as the reactants, while *heterogeneous catalysts* are a separate phase from the reactants. For example, the third body, M, in the formation of ozone is in the same phase as O and O_2; hence it is a homogeneous catalyst. The catalysts used in incineration are the heterogeneous ones. Examples of heterogeneous catalysts that have been used are Pd, Pt, Cr, Mn, Cu, Co, Au, W, Os, and Mo.

Parameters of interest in the design of incinerators are residence time, temperature, the amount of supplementary fuel, and air. There should be sufficient time to effect the desired degree of completeness of combustion. The temperature must be such as to minimize the formation of NO_x. From experience, residence times are normally set at 0.3 to 1.0 s for thermal incinerators and 0.03 to 0.1 s for catalytic incinerators at flow velocities of 6.0 to 12.0 m/s. This range of velocities is sufficient to provide turbulence. The temperature is set in accordance with Arrhenius equation. How this is done is shown in the following development.

If the rate of destruction of pollutants is modeled as a first-order process, $dC/dt = -kC$, where C is the concentration of the pollutant, k the first-order decay rate or reaction velocity constant, and t is the reaction time. Integration and simplification will ultimately result in the equation for efficiency η written as follows:

$$\eta = 1 - e^{-kt} \qquad (12\text{-}80)$$

If the reaction order is other than first, the use of equation (12-80) for efficiency requires that the reaction velocity be converted to a pseudo-first-order velocity rate by manipulating with concentrations to have k in units of per time.

In design, the efficiency desired is known; hence, setting the residence time in equation (12-80) will solve k, which is also given by the Arrhenius equation as

$$k = A \exp\left(-\frac{E}{RT}\right) \qquad (12\text{-}81)$$

where A is called a *preexponential factor*, E the reaction activation energy, R the universal gas constant, and T the absolute temperature. To use equation (12-81), the constants A and E must be determined experimentally. Once they are known, T can be calculated using the k value obtained in equation (12-80).

The other design parameters that must be determined are the amount of supplementary fuel and burner air needed for a given amount of waste or polluted air. Of course, if the waste is self-burning, the amount of supplementary fuel in the equation that will be derived below can just be set equal to zero. The amount of burner air can be established by stoichiometry. The quantity of supplementary fuel may be calculated by performing material and enthalpy balances. An enthalpy balance is a heat balance at constant pressure, which is the condition of operation in an incinerator.

Let $\dot{M}_G$ be the rate of supplementary fuel flow, $\dot{M}_{PA}$ the rate of waste or polluted airflow, $\dot{M}_{BA}$ the rate of burner airflow, and $\dot{M}_E$ the rate of exhaust gases flow. By material balance, the following equation is obtained:

$$\dot{M}_G + \dot{M}_{PA} + \dot{M}_{BA} = \dot{M}_E \qquad (12\text{-}82)$$

Designating enthalpies as h and the lower heating value as H_ℓ, the corresponding enthalpy balance is

$$\dot{M}_G H_{\ell G}(1-f_L)+\dot{M}_G h_G+\dot{M}_{PA}h_{PA}+\dot{M}_{PA}y_w X H_{\ell PA}(1-f_L)+\dot{M}_{BA}h_{BA} = \dot{M}_E h_E$$

(12-83)

where y_w is the weight fraction of burnable pollutant in polluted air, f_L the fractional loss of input heating value, and X the fractional destruction of the pollutant. The meanings of the subscripts for h and H_ℓ should be clear. $H_{\ell PA}$ is the lower heating value of the pollutant in the polluted air. Substituting the $\dot{M}_E$ in equation (12-82) for the $\dot{M}_E$ in equation (12-83) and solving for $\dot{M}_G$ yields

$$\dot{M}_G = \frac{\dot{M}_{PA}(h_E - h_{PA}) + \dot{M}_{BA}(h_E - h_{BA}) - \dot{M}_{PA}y_w X H_{\ell PA}(1 - f_L)}{H_{\ell G}(1 - f_L) - (h_E - h_G)}$$

(12-84)

The burner is often operated with a preset ratio R of burner air to gas, $\dot{M}_{BA}/\dot{M}_G$. Using this ratio gives

$$\dot{M}_G = \frac{\dot{M}_{PA}(h_E - h_{PA}) - \dot{M}_{PA}y_w X H_{\ell PA}(1 - f_L)}{H_{\ell G}(1 - f_L) - (R + 1)(h_E) + h_G + Rh_{BA}}$$

(12-85)

In the equations above, if the waste gas is not admixed with air, y_w is equal to 1. Also, $\dot{M}_G$ should refer to the supplementary fuel that is completely burnable to carbon dioxide and water. Moreover, since the quantities of pollutants in the polluted air are parts per million or parts per billion in concentration, the h's in the equations above, except h_G, may be considered equal to that of air. Obtaining the values from handbooks, the enthalpy of air may be derived. The result, referred to 0°C as datum, is,

$$h = 1.08C$$

(12-86)

where h is in kJ/kg and C is in °C.

Example 12-19

The variation of the rate constant with temperature for the decomposition of HI(g) to $H_2(g)$ and $I_2(g)$ is shown in the following table. Calculate the preexponential factor and the activation energy. How would k in the table be manipulated to have units of s^{-1}?

T (K)	k (L/gmol-s)
555	$3.52(10^{-7})$
575	$1.22(10^{-6})$
645	$8.59(10^{-5})$
700	$1.16(10^{-3})$
781	$3.95(10^{-2})$

Solution

$$k = A \exp\left(-\frac{E}{RT}\right)$$

$$\ln k = \ln A - \frac{E}{RT}$$

T (K)	k (L/gmol-s)	$\ln k$	$\dfrac{1}{T}$
555	$3.52(10^{-7})$	-14.86	0.0018
575	$1.22(10^{-6})$	-13.62	0.0017
645	$8.59(10^{-5})$	-9.36	0.0016
700	$1.16(10^{-3})$	-6.76	0.0014
781	$3.95(10^{-2})$	-3.23	0.0013

$$(\ln k)_1 = \frac{-14.86 - 13.62 - 9.36}{3} = -12.61 \qquad \left(\frac{1}{T}\right)_1 = \frac{0.0018 + 0.0017 + 0.0016}{3} = 0.0017$$

$$(\ln k)_2 = \frac{-9.36 - 6.76 - 3.23}{3} = -6.45 \qquad \left(\frac{1}{T}\right)_2 = \frac{0.0016 + 0.0014 + 0.0013}{3} = 0.0014$$

$$-12.61 = \ln A - (0.0017)\left(\frac{E}{R}\right) \tag{a}$$

$$-6.45 = \ln A - (0.0014)\left(\frac{E}{R}\right) \tag{b}$$

Solving simultaneously gives

$$\frac{E}{R} = 20{,}533.333; \; A = 4.82(10^9)\text{L/gmols-s} \quad \textbf{Answer}$$

$$E = 20{,}533.333(8.314) = 1.71(10^5) \text{ J/gmol} \quad \textbf{Answer}$$

For k to have units of s^{-1}, the values of k in the table should be multiplied by the corresponding average concentrations of HI in gmol/L.   **Answer**

Example 12-20

(a) Estimate the temperature of combustion required to destroy 99.9% of an organic compound using a residence time of 0.5 s. E and A were initially determined to be 45.2 kcal/gmol and $9.0(10^{10})$ per second, respectively. (b) Calculate the methane gas required in an afterburner if the polluted air containing the organic compound at 0.003 weight fraction flows at a rate of 228 m³/s at 95°C and 1 atm pressure. The burner air is estimated to be 6.0 m³/s entering at 26°C. Methane enters at a temperature of 24°C. H_ℓ for methane is 21,575 Btu/lbm and that for the pollutant is 18,000 Btu/lbm. Assume that the overall heat loss of input heating value equals 10%. c_p for methane equals 0.55 cal/g-°C. (c) Calculate the volume of the combustion chamber.

Solution

(a)
$$\eta = 1 - e^{-kt}$$

$$0.999 = 1 - e^{-k(0.5)} \qquad k = 13.82 \text{ s}^{-1}$$

$$\ln k = \ln A - \frac{E}{RT}$$

$$T = \frac{E}{R \ \ln(A/k)} = \frac{45{,}200 \text{ cal/gmol}}{1.987 \text{ cal/gmol-K} \ \ln[9.0(10^{10})\text{s}^{-1}/13.82\text{s}^{-1}]} = 1007K$$

$$= 734°C \quad \textbf{Answer}$$

(b) $\quad \dot{M}_G = \dfrac{M_{PA}(h_E - h_{PA}) + \dot{M}_{BA}(h_E - h_{BA}) - \dot{M}_{PA}y_w X H_{\ell PA}(1 - f_L)}{H_{\ell G}(1 - f_L) - (h_E - h_G)}$

$$h = 1.08C$$

$$\dot{M}_{PA} = 228(\rho_{PA} \text{ at } 95°C) = 228(1.18) = 269.04 \text{ kg/s} \qquad \dot{M}_{BA} = 6.0(\rho_{BA} \text{ at } 26°C)$$

$$= 6.0(1.32) = 7.92 \text{ kg/s}$$

$$h_E = 1.08(734) = 792.72 \text{ kJ/kg} \qquad h_{PA} = 1.08(95) = 102.6 \text{ kJ/kg}$$

$$h_{BA} = 1.08(26) = 20.08 \text{ kJ/kg} \qquad h_G = 0.55(1000)(24 - 0)(4.184)(10^{-3}) = 55.23 \text{ kJ/kg}$$

$$y_w = 0.003 \qquad X = 0.999$$

$$H_{\ell G} = 21{,}575 \text{ Btu/lb}_m = 21{,}575 \ \frac{\text{Btu}}{\text{lb}_m} \frac{1.0 \text{ lb}_m}{0.454 \text{ kg}} \frac{252.2\text{cal}}{1.0 \text{ Btu}} \frac{4.184 \text{ J}}{1.0 \text{ cal}} = 21{,}575(2322.57)$$

$$= 5.02(10^7) \ \frac{\text{J}}{\text{kg}}$$

$$H_{\ell PA} = 18{,}000 \text{ Btu/lb}_m = 18{,}000(2322.57) = 4.18(10^7) \text{ J/kg}$$

$$\dot{M}_G = \frac{269.04(792.72 - 102.6) + 7.92(792.72 - 20.08) - 269.04(0.003)(0.999)(4.18)(10^4)(1 - 0.10)}{5.02(10^4)(1 - 0.10) - (792.72 - 55.23)}$$

$$= 3.63 \quad \textbf{Answer}$$

(c)
$$\dot{M}_E = \dot{M}_G + \dot{M}_{PA} + \dot{M}_{BA}$$

$$= 3.63 + 269.04 + 7.92 = 280.59 \text{ kg/s}$$

Assuming practically still air,

$$Q = \frac{280.59}{\rho_a \text{ at } 734°C} = \frac{280.59}{0.55} = 510 \text{ m}^3/s$$

Thus

$$\text{volume of combustion chamber} = 510(0.5) = 255 \text{ m}^3 \quad \textbf{Answer}$$

INDOOR AIR POLLUTION CONTROL

In an underdeveloped country in Asia, most dwellings in the countryside are made of tree branches with roofs and sides made of leaves and floors made of bamboo strips or even twigs. These dwellings are excellent since there are no indoor air pollution and can be comfortably cool all summer and comfortably warm all winter. In advanced countries such as the United States, dwellings can also be made comfortably cool all summer and comfortably warm all winter, but there is a price to pay. Unlike dwellings in Asia, where there are free air interchanges between the inside of the thatch and ambient air through large holes in the dwelling envelope, people in advanced countries make their houses as airtight as possible, to conserve energy. This means that any pollutant released inside the home stays inside the home, practically. People in Asia do not die so much of cancer; in the United States, increasing numbers do. If not malnourished, people in Asia die of old age, except those very, very few rich people who can afford homes like those in the United States; they can also die of cancer.

Combustion occurs in homes daily to cook food and heat water, and in winter to provide space heating. This produces carbon monoxide and NO_x. Some photocopying machines convert oxygen to ozone. Tobacco smoke contains numerous suspected carcinogens, including benzene, hydrazine, benzo[a]anthracene, benzo[a]pyrene, and nickel. Tobacco smoke particles are on the order of 0.2 μm; this size can easily get deep into the lungs. Other indoor pollutants as a result of tobacco smoke are carbon monoxide, nicotine, nitrosoamines, acrolein, and other aldehydes. If pets are kept at home, there would likely be danders and asthmatic persons would be in danger. If the method of heating is forced-air recirculation, feathers can accumulate to an inch thick in air cleaners. Asthma is aggravated by these kinds of houses. Practically, anything that gasifies will accumulate if the house is airtight. For homes in advanced countries, Table 12-11 summarizes indoor air pollutants, their sources, and exposure guidelines. The guideline concentration for radon is given in terms of working levels. The unit of *working level* is defined as 100 pCi/L of radon-222 in equilibrium with its progeny.

A very important indoor air pollutant that has caused so much dreaded attention is radon. Radon is a chemically inert gas that emits alpha radiation. Radon gas and its daughter products such as polonium, lead, and bismuth can attach themselves to inhaled

TABLE 12-11 SOURCES AND EXPOSURE GUIDELINES OF INDOOR AIR POLLUTANTS

Pollutant and indoor sources	Guideline, average concentrations
Asbestos and other fibrous aerosols Friable asbestos: fireproofing, thermal and acoustic insulation, decoration; hard asbestos: vinyl floor and cement products	0.2 fibers/mL for fibers longer than 5 μm
Carbon monoxide Kerosene and gas space heaters, gas stoves, wood stoves, fireplaces, smoking	10 mg/m^3 for 8 h, 40 mg/m^3 for 1 h
Formaldehyde Particleboard, paneling, plywood, carpets, ceiling tile, urea–formaldehyde foam insulation	120 μg/m^3
Inhalable particulate matter Smoking, vacuuming, wood stoves, fireplaces	55–110 μg/m^3 annual, 150–350 μg/m^3 for 24 h
Nitrogen dioxide Kerosene, and gas space heaters, gas stoves	100 μg/m^3 annual
Ozone Photocopying machines, electrostatic air cleaners	235 μg/m^3 once a year
Radon and radon progeny Diffusion from soil, groundwater, building materials	0.01 working levels annual
Sulfur dioxide Kerosene space heaters	80 μg/m^3 annual, 365 μg/m^3 for 24 h
Volatile organics Cooking, smoking, room deodorizers, cleaning sprays, paints, varnishes, solvents, carpets, furniture, draperies	None available

particulates and can become lodged deep in the alveoli. Disease caused by continuous lung irradiation is believed to be the reason for some 5000 to 20,000 lung cancer deaths per year in the United States. Radon seeps out of the soil and can accumulate in houses. It can exist in groundwaters and be released and inhaled in high concentrations when the shower is used. It is one of the intermediate products when uranium decays naturally to the stable lead isotope. Figure 12-8 shows the uranium-238 decay chain and the major entry points of radon into the house. Aside from uranium-238, which decays to the stable lead-206, three other chains that produce radon exist. They are uranium-235, which ends with lead-207; thorium-232, which ends with lead-208; and plutonium-241, which ends with bismuth-209. In Figure 12-8, entries of radon are identified by arrows; for example, A denotes entry of radon into homes from the soil underneath the basement slab and K denotes entry of radon from the groundwater through the shower. The EPA suggests a radon guideline concentration of 4 pCi/L as a level at which residents might consider remedial action and a concentration of 8 pCi/L as the level at which action is recommended.

If the concentration can be diluted to below harmful levels, the problem is mitigated. This is exactly one of the ways of radon control—*dilution*. Ambient air is blown in from outside, thus diluting the concentration of any pollutant that happens to be inside the house, including radon. Warm air tends to rise and this is exactly what happens in homes. This is called the *stack effect*. As the warm air rises, it leaves a void at the bottom. This void is a partial vacuum, causing radon to be sucked into the house from the outside soil. One of the engineering controls is, therefore, to plug holes such as those in the ceiling, electric fixture holes, openings around ducts, basement walls, and so on to eliminate stack effects. A third method is *subslab* suction. This is done by penetrating the basement slab with a pipe and installing a blower to suck the radon from under the slab. Schematics of the dilution and subslab suction method are shown in Figure 12-9. The *plugging-point method* to eliminate stack effect is shown in Figure 12-10. The dilution or ventilation requirement can be specified in terms of *air exchanges per hour*, the number of times that a volume of air equal to the volume of space in the house is exchanged with the outside air per hour.

To estimate resulting concentrations, the Reynolds transport theorem (see Appendix 21) may be applied to the volume of space in the house. The Lagrangian derivative is

$$\frac{D \int_M C \, d\Psi}{Dt} = \dot{S} - kC\Psi \tag{12-87}$$

where $\dot{S}$ is the rate of generation from a source, k a decay parameter, C the concentration in the space, and Ψ the volume of space. The rate of generation has been estimated to range from 0.1 to 100 pCi/m^2-s with a value of 1.0 pCi/m^2-s as typical.

For the Eulerian derivative, the local derivative is

$$\frac{\partial \int_{CV} C \, d\Psi}{\partial t} = \Psi \frac{\partial C}{\partial t} \tag{12-88}$$

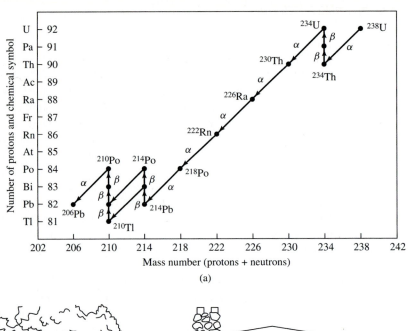

(a)

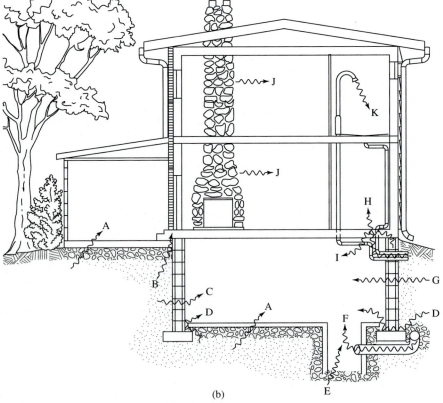

(b)

Figure 12-8 (a) Uranium-238 decay chain. (b) Major points of entry.

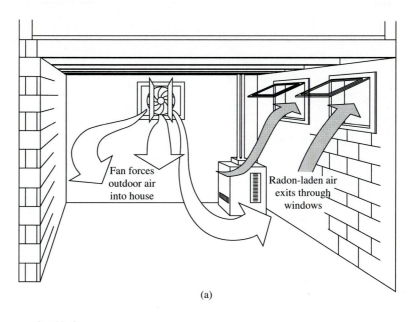

(a)

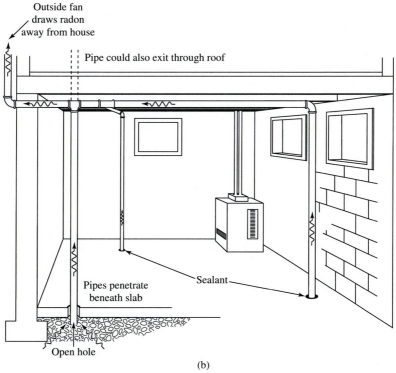

(b)

Figure 12-9 Radon entry control: (a) dilution method; (b) subslab suction method.

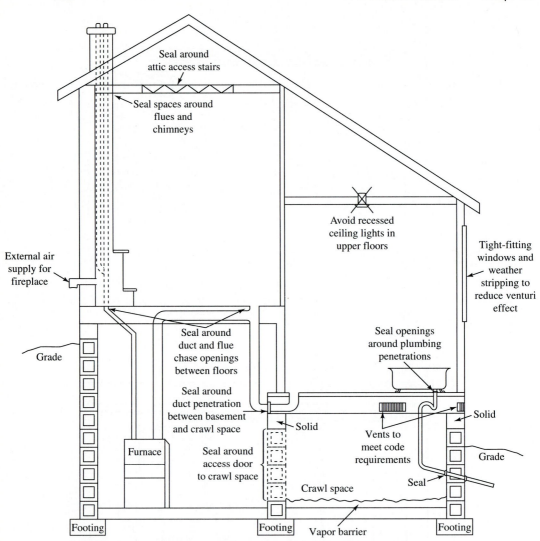

Figure 12-10 Plugging-point method to eliminate stack effect.

and the convective derivative is

$$\oint_A C\mathbf{v} \bullet \hat{n}\, dA = -C_a I \Psi + C I \Psi \tag{12-89}$$

where C_a is the ambient air concentration and I is the air exchanges per hour. Finally, the Reynolds transport theorem becomes

$$\dot{S} - kC\Psi = \Psi\frac{\partial C}{\partial t} - C_a I \Psi + C I \Psi \tag{12-90}$$

Solving for C yields

$$C = \frac{\dot{S}/V + C_a I}{I + k}\{1 - \exp[-(I + k)t]\} + C_0 \exp[-(I + k)t] \qquad (12\text{-}91)$$

Equation (12-91) may be used to calculate the number of air exchanges required to dilute indoor air pollutant concentration to the safe level. Normally, the blower will not be required to run all the time but is operated to cut in and out at regular intervals. Table 12-12 shows some values of k for different types of indoor air pollutants.

TABLE 12-12 k VALUES FOR INDOOR AIR POLLUTANTS

Pollutant	$k \ (h^{-1})$
SO_2	0.20
Particles ($< 0.5\mu$m)	0.5
Radon	$7.5(10^{-3})$
CO	0.0
NO	0.0
NO_x (as N)	0.12
HCHO	0.35

Example 12-21

The basement of a home measures 24 ft by 36 ft by 92 in. (a) Assuming that the soil emits 1.0 pCi/m^2-s of radon, design for the number of air exchanges to reduce the concentration to 0.5 pCi/L. (b) Size the blower required and design its automatic operation. Assume an ambient air temperature of 27°C. Assume that 20% of emissions enters the house; also assume that the available fan has a capacity of 180 cfm.

Solution

(a) $\qquad C = \dfrac{\dot{S}/V + C_a I}{I + k}\{1 - \exp[-(I + k)t]\} + C_0 \exp[-(I + k)t]$

Design should be based on the assumption that steady state has been attained. Hence

$$C = \frac{\dot{S}/V + C_a I}{I + k}$$

$$\text{Area of emission} = \left(\frac{24}{3.281}\right)\left(\frac{36}{3.281}\right) + \left(\frac{92/12}{3.381}\right)\left(\frac{24}{3.281}\right) (2)$$

$$+ \left(\frac{92/12}{3.381}\right)\left(\frac{36}{3.281}\right) (2) = 165.72 \text{ m}^2$$

$$V = \frac{24(36)(92/12)}{(3.281)^3} = 187.54 \text{ m}^3 \qquad C_a = 0$$

$$C = 0.5 \text{ pCi/L} = 0.5(10^3) \text{ pCi/m}^3$$

$$\dot{S} = (1.0 \text{ pCi/m}^2\text{-s})(165.72 \text{ m}^2)(3600 \text{ s/h})(0.20) = 119,320.45 \text{ pCi/-h}$$

$$0.5(10^3) = \frac{119,320.45/187.54 + 0(I)}{I + 7.6(10^{-3})}; \ I = 1.26 \text{ air exchange per hour} \quad \textbf{Answer}$$

(b) $\qquad$ Blower size $= 1.26(187.54) = 236.30 \text{ m}^3/\text{h} = 3.94 \text{ m}^3/\text{m}$

$$= 139.16 \text{ cfm capacity at } 27°C$$

$$\text{Number of fans} = \frac{139.16}{180} = 0.77 \quad \text{use 2.0 fans} \quad \textbf{Answer}$$

Total volume of ambient air required per day $= 139.16(60)(24) = 200,390.4 \text{ ft}^3$

$$\text{Length of time required to fulfill requirement} = \frac{200,390.4}{2(180)(60)} = 9.28 \text{ h, say 10 h}$$

Therefore, operate fan for 10 h/day. Cut-in time = 6.00 P.M.; cutout time = 4:00 A.M. **Answer**

GLOSSARY

Adiabatic compression. A compression process where heat is not allowed to escape to the surroundings during compression.

Afterburner. A thermal incinerator.

Air pollution. The presence of substances in such concentrations that make the air harmful or dangerous to breathe or causes damage to plants, animals, and properties.

Air pollution control. Measures taken to meet certain emission standards.

Air pollution control equipment. A device or system of devices constructed to meet emission standards.

Baghouse. A structure housing bagfilters.

Biological pollutants. Forms of natural organic pollutants that are microscopic or "macro" microscopic, such as pollens, microscopic algae, spores, bacteria, viruses, and the like, that are alive or have just recently died.

Catalytic incineration. A method of incineration that employs a catalyst.

Compression ratio. The ratio of the exit absolute pressure to the inlet absolute pressure in a stage of a compressor.

Corona discharge. A shower of electrons surrounding a discharge electrode.

Criteria pollutants. Pollutants of which primary and secondary standards have been established.

Cyclone or centrifugal settler. A force-field settler that relies on centrifugal field.

Direct-flame incinerator. An incinerator that does not need an auxiliary fuel for firing.

Drift or electric wind velocity. Settling velocity of particles in an electric field.

Dust. Small particles of solids created from the breakup of larger particles.

Electrostatic precipitator (ESP). A force-field settler that relies on an electric field.

Emission standards. Values of parameters to be met at emission points.

Fly ash. Noncombustible fine particles admixed with combustion gases in the burning of coal.

Fog. Concentrated form of mist.

Force-field settlers. Equipment that rely on a field of force to pull down particles to a point of collection.

Fumes. Fine solid particles formed from the condensation of vapors of solid materials.

Gravitational settling. A force-field settling that relies on gravitational field to settle particles.

Heterogeneous catalyst. A catalyst that is not in the same phase as the reactants.

Homogeneous catalyst. A catalyst that exist in the same phase as the reactants.

Isothermal compression. A compression process where the temperature of the gas compressed remains constant.

Mass transfer. The movement of mass due to concentration difference.

Mist. Particles formed from the condensation of liquid vapors.

Molecular sieve. A polymeric material possessing pores of molecular dimension capable of sieving molecules.

Natural organic pollutant. Organic pollutant of biological origin.

Operating line. The locus of the vapor and liquid compositions in an absorber.

Organic pollutant. Pollutant of organic origin.

Particulates. Finely divided solids and liquids.

Polytropic compression. A general process of compression.

Primary NAAQS standards. Air quality standards established to protect health with an adequate margin of safety.

Primary pollutants. Pollutants formed directly from the source.

Run–clean time. The run time plus the cleaning time.

Run time. The time between a first compartment is put into service and when a second compartment is isolated for cleaning.

Scrubbers. Units used for removing particulates by using water droplets.

Secondary NAAQS standard. Air quality standards established to protect public welfare.

Secondary pollutants. Pollutants formed from primary pollutants.

Smoke. A colloid of solid particles dispersed in a gas.

Spray. Particles formed from the atomization of liquids.

Stack effect. The rising of warm air in a building.

Thermal incinerator. An incinerator that requires an auxiliary fuel and preheating of the combustion air to fire.

SYMBOLS

A	preexponential factor
A_s	overflow area of settling zone
c	mass of dust collected per unit volume of air filtered
c_p, c_v	specific heats at constant pressure and volume, respectively

C	Cunningham correction factor; concentration of indoor air pollutant
C_a	concentration of ambient air pollutant
C_D	drag coefficient
d	diameter of particle
d_d	diameter of droplet
d_0	diameter of particle 100% removed
D	diameter of cyclone; diameter of tube-type ESP
E	electric field; activation energy
$\mathbf{E}$	electric field vector
f'	modified Fanning's friction factor
F	force between two charged particles
$\mathbf{F}$	force vector between two charged particles
g	acceleration in a force field
G	carrier gas molar gas flow
G'	molar gas flow
h	head loss; enthalpy per unit mass
h_{fb}	expansion head loss
H	depth of settling zone
H_x, H_y	liquid- and gas-side height of a mass transfer unit, respectively
I	air exchanges per hour
I_c	corona current
k	first order process constant; pollutant decay rate
k_x, k_y	individual mass transfer coefficients based on liquid and gas sides, respectively
K_x, K_y	overall mass transfer coefficients based on liquid and gas sides, respectively
ℓ	length of bed
ℓ_e	length of expanded bed
L	length of settling zone, carrier liquid molar flow
L'	liquid molar flow
L_i, L_e	influent and effluent dust concentrations, respectively
M	molecular weight
$\dot{M}_{BA}$	mass flow rate of burner air
$\dot{M}_E$	mass flow rate of exhaust gases
$\dot{M}_G$	mass flow rate of supplementary fuel
$\dot{M}_{PA}$	mass flow rate of waste or polluted air
n	exponent for a polytropic process
N	number of compartments
N_{tx}, N_{ty}	liquid- and gas-side number of mass transfer units
ΔP	pressure drop

P_c	corona power
P_n	pressure at outlet of last compressor stage
P_s	pressure at outlet of any compressor stage
P_1, P_2	pressures at inlet and outlet of compressor, respectively
q, q'	charges of two attracting particles
q_s	charge of sphere
Q	flow rate
Q_g	volumetric gas flow rate
Q_ℓ	volumetric liquid flow rate converted to droplets
r	radius of motion in a cyclone; distance between two charge particles
R	fractional removal of particles; universal gas constant
R_m	resistance of filter cloth
Re	Reynolds number
$\dot{S}$	rate of generation of indoor air pollutant
t	time
t_c	time to clean a compartment
t_f	filtration time of a given compartment
t_r	run time from putting a compartment into service and shutting down the next compartment in a cleaning cycle
T_1, T_2	temperatures to inlet and outlet of compressor, respectively
v	terminal settling velocity of particle
v_d	terminal settling velocity of droplet
v_h	flow-through velocity
v_p	velocity of particles of diameter d
v_t	tangential velocity of particle
v_0	velocity of particles removed 100%
V	filtration velocity of air/cloth ratio; volts; voltage drop between discharge and collector electrodes; volume of space in a house
Ψ	volume of air filtered
V_{Mx}, V_{My}	liquid- and gas-phase molar gas velocities
V_p	volume of a particle of equivalent diameter, d, or nominal diameter, d_p
V_s	superficial velocity
W	width of settling zone
x	mole of solute in liquid phase
x^*	mole fraction of solute in liquid phase in equilibrium with x
x_i	mole fraction of solute in liquid phase at interface
x_0	mass fraction of particles of settling velocity equal to or less than v_0
y	mole of solute in gas phase
y^*	mole fraction of solute in gas phase in equilibrium with y

y_i	mole fraction of solute in gas phase at interface
y_w	weight fraction of burnable pollutant in polluted air
Z_T	tower height
Z_0	distance between discharge and collector electrodes; length of scrubber contact zone
$\overline{\alpha}, \overline{\alpha}_0$	specific cake resistances
β	volume shape factor
ϵ_0	permittivity constant
η	porosity
η_d	fractional target efficiency of a single droplet
ρ_p	density of particle
σ_ℓ	liquid surface tension

PROBLEMS

12-1. The concentration of CO in a street intersection reaches the federal ambient standard of 35 ppm. Crewmen from the department of public works are repairing a break in the water line. Calculate the length of time before their performance became impaired.

12-2. What are the two levels of pollutants for which the USEPA has national ambient air quality standards?

12-3. What are the six criteria pollutants?

12-4. A sample of 1 m^3 of air was found to contain 35 ppm. What is the concentration in mg/m^3?

12-5. Compare 1 ppm by mass of CO_2 in air to 1 ppm by mass of CO_2 in water.

12-6. A sample of particulates of average diameter of 60 μm is soaked with water in a graduated cylinder to the 1-mL mark. The loss in weight after evaporating to dryness is 0.01 g. If β equals 0.93, calculate the number of particles in the sample.

12-7. A Maryland coal is burned at the rate of 5.50 tonnes/h. If the coal contains 3.0% S, what is the annual rate of emission of SO_2?

12-8. The following results are to be used to design a settling chamber. The horizontal velocity is to be 0.3 m/s. Other pertinent information are temperature equals 650°C, specific gravity of particle equals 2.0, and chamber length and depth equal 7.5 m and 1.5 m, respectively. Assume that β equals 0.90. What is the terminal settling velocity of the particle that is removed 100%?

Size (μm)	Wt %
0–10	8
10–20	10
20–30	12
30–40	15
40–50	19
50–60	14
60–70	13
70–80	9

12-9. In Problem 12-8, determine the expected percent removal of the particles.

12-10. A quantity of 250 m^3/m of polluted air containing particulates is settled using a settling chamber measuring 14 m long, 2 m high, and 2 m wide. If the specific gravity of the particulates is 2.0, calculate the maximum size that can be removed 100%. Assume that the temperature is 80°C.

12-11. The results shown in the table below are to be used to design a standard conventional cyclone. The airstream is 21.0 m^3/s and the diameter of the cyclone is 2.0 m. The temperature is 77°C and the specific gravity of the particles is 2.0. Assume that β equals 0.90. What are the diameter and the terminal settling velocity of the particles that are 100% removed?

Particle size (μm)	Wt %
0–10	8
10–20	10
20–30	12
30–40	15
40–50	19
50–60	14
60–70	13
70–80	9

12-12. In Problem 12-11, determine the expected percent removal of the particles.

12-13. Compare the removal efficiencies for a particulate having a diameter of 2.5 μm and a density of 1200 kg/m^3 using cyclone diameters of 0.30 m and 0.60 m, respectively. The gas flow rate is 3.0 m^3/s and the temperature is 30°C.

12-14. In Problem 12-11, what is the pressure drop across the unit assuming that a vane guides the flow toward the inner side of the outer vortex? Assume that $C_i = 1.0$ grain/ft^3.

12-15. Design a standard cyclone to remove 15-μm particulate at 50% efficiency. The airflow is 7.0 m^3/m at a temperature of 80°C. The specific gravity of the particulate is 2.0.

12-16. In Problem 12-11, determine the expected percent removal of the particles if a bank of 32 standard conventional cyclones with diameters of 24 cm is used instead of the single large unit.

12-17. The drift velocity u in the electric field of an electrostatic precipitator is given by $u = 2.0(10^5)d_p$, where d_p is the diameter of the particle. Determine the plate area required to remove particles of diameter 0.8 μm at 95% efficiency. The polluted airflow rate is 6 m^3/s.

12-18. The data in Problem 12-11 are to be used to design a plate-type ESP. The plates are spaced 30 cm apart and the airstream is 330 m^3/s. The voltage drop between discharge and collector plates is 70,000 V. The aspect ratio, which is the ratio of the length of the plate to its width, is 1.2. The plate width is 10 m and there are a total of 80 channels. The temperature is 77°C and the specific gravity of the particles is 2.0. (**a**) What are the diameter and terminal settling velocity of particles that are 100% removed? (**b**) Determine the expected percent removal of the particles. (**c**) What is the power requirement assuming that the corona current is 2.3 A? Assume that β equals 0.90.

12-19. A tube ESP is used to removed particulates that are 1.0 μm in diameter. The electric field intensity is 100,000 V/m and the flow rate is 0.2 m³/s. The particle charge is 0.25 femtocoulomb (femto = 10^{-15}). Determine the collection efficiency of the ESP.

12-20. The data in Problem 12-11 are to be used to design a tube-type ESP. The airstream is 330 m³/s and there are 36,000 collector tubes of 3-in. ID. The voltage drop between the discharge and collector tubes is 70,000 V. The tube is 10 m long. The temperature is 77°C and the specific gravity of the particles is 2.0. **(a)** What are the diameter and terminal settling velocity of the particles that are 100% removed? **(b)** Determine the expected percent removal of the particles. **(c)** What is the power requirement assuming that the corona current is 2.3 A? Assume that β equals 0.90.

12-21. Under what industrial plant conditions would you recommend the use of a baghouse?

12-22. Using an air/cloth ratio of 3.25 cfm/ft², design a baghouse to control emissions from a grain elevator. The filter bags to be used measure 0.3 m in diameter and 6.0 m in length. The system is used to control 21 m³/s of waste-air flow.

12-23. Determine the number of bags necessary to treat 16 m³/s of polluted air laden with particulates. The baghouse is to be subdivided into eight compartments. The air/cloth ratio is 10 m/min. The bags are 0.25 m in diameter and 7.0 m in length.

12-24. Design the cleaning cycle of Problem 12-23.

12-25. In Problem 12-22, determine the number of compartments and the number of bags in each compartment to be used. Design the cleaning cycle.

12-26. Ambient air at 26°C and at barometric pressure of 29.90 in. Hg is to be compressed to 100 psig. A four-stage compressor is to be used. If the pressure at the end of the third stage is 51 psig, calculate the value of n. If the motor and brake efficiencies are respectively 90% and 80%, what is the power input to the compressor?

12-27. As far as dying of cancer is concerned, what factor of the homes in Asia makes them fortunate? What types of people in Asia could also die of cancer?

12-28. From the test results shown below, determine the resistance of the filter medium, the specific cake resistance $\bar{\alpha}_0$, the compressibility coefficient s, and the pressure drop in cm H_2O after 200 minutes of operation. The air-to-cloth ratio is 0.9 m/min, the inlet dust concentration is 5.0 g/m³, and exit dust concentration is practically zero. Assume that the temperature is 45°C.

Time (min)	$\Delta P(N/m^2)$
0	151
5	379
10	506
20	611
30	694
60	992

12-29. A high-volume sampler draws air at a rate of 80 ft³/min for 24 h. How much residue on the filter is collected if the concentration is 150 μg/m³?

12-30. Give an idea of how large 100 μm is.

12-31. A three-stage compressor is used to compress ambient air at 30°C and a barometric pressure of 29.90 in. Hg to 100 psig. **(a)** What is the brake horsepower if the mechanical efficiency is 80%? **(b)** What is the discharge temperature from the first stage? **(c)** If intercoolers and aftercoolers are to be used and the temperature of the cooling water is to rise by 20°F, how much water is needed for the compressed air to leave at 26°C? Assume that 1.4 m³/s of ambient air is to be compressed. c_p for air is 0.25 Btu/lb-°F for a temperature range of 32 to 2550°F.

12-32. Particles of fly ash with a density of 710 kg/m³ and minimum size of 10 μm are to be removed using a venturi scrubber. The scrubber has a throat area of 1.0 m² and a throat length of 50 cm. The droplet diameter formed is 100 μm. If the liquid and gas flow rates are 0.15 m³/s and 96.0 m³/s, respectively, what is the efficiency of removal of the unit?

12-33. The data in Problem 12-11 are used to design a countercurrent spray chamber. $Q_\ell/Q_g = 0.02(10^{-2})$, $v_g = 25$ cm/s, and $Z_0 = 3.5$ m. The pressure is atmospheric and the temperature is 77°C. **(a)** What is the efficiency of removal for the 0- to 10-μm size fraction? **(b)** If Q is 21 m³/s, what is the cross-sectional area of the chamber? **(c)** What is the pressure drop through the spray nozzle if the pressure at the nozzle inlet is 20 in. water. Assume that the pressure drop produces droplets of $d_d = 0.030$ cm and that 20% of the sprayed water is converted to suspended droplets. Also, assume that the specific gravity of the particulates is 2.0.

12-34. Water at the rate of 0.15 m³/s is atomized in a venturi having a throat area of 1.0 m². The rate of flow of the polluted air through the throat is 100 m³/s. If the temperature of the water is 25°C, calculate the droplet diameter. Assume that 20% of the water introduced at the throat is converted to droplets.

12-35. Solve Problem 12-33 for a cross-flow spray chamber.

12-36. Solve Problem 12-35 by reducing Q_ℓ/Q_g to 10^{-4}.

12-37. Polluted air containing to 3.44 kg-mol of C_6H_6 per hour is treated by an activated carbon adsorber measuring 4 m by 2 m in cross section and 0.7 m deep. The air contains 5000 ppm of C_6H_6 at 1 atm and 37°C. The X/M of the bed is 5.0 kg of C_6H_6 per 50 kg of C. The bed has a bulk or packed density of 480 kg/m³, η of 0.4, and a particle size of 4 by 10 mesh. Determine the pressure drop. Assume that $\beta = 0.70$.

12-38. The dirty air in Problem 12-37 is to be treated in a fluidized-bed activated carbon adsorber using the same type of activated carbon. It is desired to expand the bed to 150% from an initial depth of 0.7 m. Calculate the head loss and superficial area of bed required. The specific gravity of activated carbon is 1.8 and the flow of dirty air is 17,500 m³/h.

12-39. Differentiate between surface and contact condensers.

12-40. Polluted air containing 2.3% SO_2 is to be scrubbed using a packed absorption tower. The stack gas flow rate measured at 1 atm and 30°C is 10 m³/s, and the SO_2 content is 3.0%. Using an initially pure water, 90% removal is desired. The equilibrium curve of SO_2 in water may be approximated by $y_i = 30x_i$. Determine the water requirement if 150% of the minimum flow rate is deemed adequate. Calculate the height of the tower. Assume that $K_ya = 5.35$ lb-mol/ft³-h-mole fraction and $K_xa = 117$ lb-mol/ft³-h-mole fraction. Assume that the total cross-sectional area of tower equals 11.0 m².

12-41. The following data were obtained for the decomposition of HI(g) to H$_2$(g) and I$_2$(g). If the average concentration of HI during the test was 3.0 mol/L, convert the k values to their pseudo-first-order counterparts.

T (K)	k (L/gmol-s)
555	$3.52(10^{-7})$
575	$1.22(10^{-6})$
645	$8.59(10^{-5})$
700	$1.16(10^{-3})$
781	$3.95(10^{-2})$

12-42. Define *adsorption*. What is the difference between chemical and physical adsorption?

12-43. (a) For 99.99% destruction of an organic compound, estimate the temperature of combustion using a residence time of 0.5 s. E and A were initially determined to be 45.2 kcal/gmol and $9.0(10^{10})$ per second, respectively. (b) Calculate the methane gas required in an afterburner if the polluted air containing the organic compound at 0.003 weight fraction flows at a rate of 228 m^3/s at 95°C and 1 atm pressure. The burner air is estimated to be 6.0 m^3/s entering at 26°C. Methane enters at a temperature of 24°C. H_ℓ for methane is 21,575 Btu/lb$_m$ and that for the pollutant is 18,000 Btu/lb$_m$. Assume that the overall heat loss of input heating value equals 10%. The c_p value for methane is 0.55 cal/g-°C. (c) Calculate the volume of the combustion chamber.

12-44. The basement of a home measures 24 ft by 36 ft by 92 in. (a) Assuming that the soil emits 100 pCi/m^2-s of radon, design for the number of air exchanges to reduce concentration to 0.5 pCi/L. (b) Size the blower required and design its automatic operation. Assume an ambient air temperature of 27°C. Assume that 20% of the emissions enters the house; also, assume that the available fan has a capacity of 180 cfm.

12-45. Name four types of adsorption units.

BIBLIOGRAPHY

CALVERT, S. (1977). "Scrubbing," in A. C. Stern (ed.), *Air Pollution*, Vol. IV. Academic Press, New York.

CALVERT, S., J. GOLDSCHMID, D. LEITH, and D. MEHTA (1972). "Wet Scrubber System Study," in *Scrubber Handbook*, Vol. I. NTIS, PB-213016. U. S. Department of Commerce, Washington, D.C.

COOPER, C. D., and F. C. ALLEY (1986). *Air Pollution Control: A Design Approach.* Waveland Press, Prospect Heights, Ill.

DANIELSON, J. A. (1973). *Air Pollution Engineering Manual.* U.S. Environmental Protection Agency, Research Triangle Park, N.C.

HESKETH, H. E. (1974). *Understanding and Controlling Air Pollution.* Ann Arbor Science, Ann Arbor, Mich.

HOLTZCLAW, H. F., Jr. and W. R. ROBINSON (1988). *General Chemistry.* D.C. Heath, Lexington, Mass.

MASTERS, G. M. (1991). *Introduction to Environmental Engineering and Science.* Prentice Hall, Englewood Cliffs, N.J.

NUKIYAMA, S., and Y. TANASAWA (1938). "An Experiment on the Atomization of Liquid by Means of an Air Stream." *Transactions of the Society of Mechanical Engineers*, 4(14), Japan, p. 86.

SEINFELD, J. (1975). *Air Pollution.* McGraw-Hill, New York.

SLOAN, E. D. (1974). *"Normality of Binary Adsorbed Mixtures of Benzene and Freon on Graphitized Carbon."* Ph.D. Thesis. Clemson University, Clemson, S.C.

USEPA (1988). *National Air Pollutant Emission Estimates, 1940–1986.* U.S. Environmental Protection Agency, Washington, D.C.

YUNG, C. S., H. F. BARBARIKA, and S. CALVERT (1977). "Pressure Loss in Venturi Scrubbers." *Journal of the Air Pollution Control Association,* 27, pp. 348–351.

CHAPTER 13

Hazardous Waste Management and Risk Assessment

In the United States, the systematic management of hazardous substances is essentially prompted by three environmental laws: Toxic Substances Control Acts (TSCA), Resource Conservation and Recovery Act (RCRA), and Comprehensive Environmental Response, Compensation, and Liability Act (CERCLA). (See also Table in Appendix for list of environmental legislations.) TSCA was enacted in 1976 in response to the Kepone disaster of Hopewell, Virginia, which contaminated the James River and the Chesapeake Bay portion close to the James River with a large concentration of Kepone in 1975. TSCA controls the production and use of hazardous substances in commerce before they become hazardous wastes. Prior to 1976 the laws and regulations in the United States, such as the Resource Recovery Act of 1970, restricted the disposal of hazardous wastes to the air and water. These restrictions greatly favored disposal into land, creating a large environmental loophole. To close this loophole, RCRA was enacted in 1976 to protect, among other things, the quality of the groundwater and the land from contamination of solid waste. The scope of RCRA was subsequently broadened by the Hazardous and Solid Waste Amendments of 1984 (HSWA). A major theme of this amendment is the added protection to the groundwater by requiring more stringent controls, such as the installation of double liners, leachate collection system, and groundwater monitoring. Aside from the hazardous waste program, which is Subtitle C, RCRA also contains Subtitle D, which relates to the solid waste program, and Subtitle I, which relates to control of underground storage tanks (USTs). Through RCRA, the EPA is able to track hazardous wastes from the point of generation to the ultimate disposal point—the *cradle-to-grave* concept.

656

RCRA is only able to regulate hazardous wastes at active sites. Therefore, new regulations were required for those wastes generated in the past and are simply dangerously scattered all over the country in abandoned sites contaminating the environment. The new law was CERCLA, enacted in 1980. Because of the large sum of money alloted for the cleanup of abandoned sites, CERCLA is also called the Superfund. The superfund of CERCLA was largely augmented, giving rise to the Superfund Amendment and Reauthorization Act of 1986 (SARA).

In the United States, although undoubtedly hazardous, radioactive wastes are not regulated under RCRA but by other laws. These include the Nuclear Waste Policy Act of 1982, the Low-Level Radioactive Waste Policy Act Amendments of 1985, and the Nuclear Waste Policy Act Amendments of 1987.

TYPES OF HAZARDOUS WASTES

Just what is a hazardous waste? *Hazardous waste* may be as any waste or combination of wastes that poses substantial danger to human beings, plants, and animals. Specifically, a waste is hazardous if it possesses one or more of the following characteristics: ignitability, corrosivity, reactivity, toxicity, and radioactivity.

Ignitability

Ignitability is that characteristic of waste that could cause it to catch fire in the process of transport, storage, treatment, and disposal. Examples of this type of waste are oil and solvents and flammable gases such as methane. Specifically, if a representative sample has any of the following properties, a waste is said to exhibit the characteristic of ignitability: it is a liquid with a flash point below 60°C; it is not a liquid but under standard temperature and pressure is capable of causing fire through friction, absorption of moisture, or spontaneous chemical reaction; it is a flammable compressed gas; it is a strong oxidizer capable of causing fire.

Corrosivity

The characteristic of *corrosivity* is given to a waste that has either a very high pH or a very low pH. Substances and wastes of these pH levels are naturally corrosive, hazardous, and dangerous. Specifically a waste of pH equal to or less than 2 or equal to or greater than 12.5 has a hazardous characteristic of corrosivity.

Reactivity

With respect to rate of reaction, a *reactive waste* is synonymous with unstable waste. Due to its extreme rate of reaction, an unstable waste can create an explosive condition at any stage of the management cycle of transport, storage, treatment, and disposal. Hence a reactive waste is a hazardous waste. Specifically, if a sample exhibits any of the following properties, a waste is said to have the characteristic of reactivity: it undergoes a violent change with or without detonating; it reacts violently with water and forms potentially dangerous mixtures with water, whether explosive or toxic; and it is a cyanide- or sulfide-bearing waste capable of releasing toxic gases, vapors, or fumes at the pH range 2 to 12.5 in such a quantity as to cause harm to human health or the environment.

Toxicity

A *toxic* substance is a poisonous substance and a poisonous substance has to be hazardous. Hence a waste that possesses the characteristic of toxicity is a hazardous waste.

To standardize the determination of toxicity, standard laboratory procedures have been developed: the first procedure is called the *extraction procedure (EP) toxicity test;* the second procedure is called the *toxicity characteristics leaching procedure (TCLP).* There are two major differences between these two procedures: TCLP has more parameters than EP and the pH used in the digestion procedure for EP is lower (pH = 2) than that for TCLP (about the pH of acid rain, which is approximately 4.0). TCLP is a replacement for EP.

A list of toxic substances is chosen with their corresponding maximum regulatory concentration for toxicity given. If a particular waste contains any one of the toxic substances listed in concentrations greater than the regulatory maximum, the waste is considered hazardous. Tables 13-1 and 13-2 list the maximum concentrations of contaminants for the EP toxicity test and the TCLP toxicity test, respectively. TCLP has more toxic substances listed.

Example 13-1

A sample of waste is analyzed using the EP toxicity test. It contains the following concentrations: lindane 0.3 mg/L, toxaphene 0.2 mg/L, and 2,4-D 4 mg/L. All the other toxics listed in Table 13-1 are absent. Is the waste a hazardous waste?

Solution No. **Answer**

Radioactivity

Radioactivity may be defined as the spontaneous breakup of the nucleus of an atom. While the characteristic of reactivity of a hazardous waste refers to the instability of the atom or the molecule, the characteristic of radioactivity, on the other hand, refers to the nucleus—not the atom or the molecule. In the dimension of radioactivity, the nucleus of a radioactive atom is too large, making it unstable, and an unstable nucleus breaks up and the atom changes to another one lower in mass. In hazardous waste management, three types of radioactivity or radiation are of concern: alpha particle, beta particle, and gamma radiations.

Alpha radiation. The release of alpha radiation from the nucleus of a radioactive element is equivalent to the release of a positively charged helium nucleus consisting of two neutrons and two protons. This release causes the parent atom to lose four *atomic*

TABLE 13-1 MAXIMUM CONCENTRATIONS OF CONTAMINANTS FOR THE EP
TOXICITY TEST

EPA hazardous waste number	Contaminant	Maximum concentration (mg/L)
D004	Arsenic	5.0
D005	Barium	100
D006	Cadmium	1.0
D007	Chromium	5.0
D008	Lead	5.0
D009	Mercury	0.2
D010	Selenium	1.0
D011	Silver	5.0
D012	Endrin (1,2,3,4,10,10-hexachloro-1,7-epoxy-1,4,4a,5,6,7,8,8a-octahydro-1,4,endo,endo-5,8-dimethanonaphthalene)	0.02
D013	Lindane (1,2,3,4,5,6-hexachlorocyclohexane, γ isomer)	0.4
D014	Methoxychlor [1,1,1-trichloro-2,2-bis(p-methoxyphenyl)ethane]	10.0
D015	Toxaphene ($C_{10}H_{10}Cl_8$, technical chlorinated camphene, 67–69% chlorine)	0.5
D016	2,4-D (2,4-dichlorophenoxyacetic acid)	10.0
D017	2,4,5-TP Silvex (2,4,5-trichlorophenoxypropionic acid	1.0

mass numbers, which are the sum of the number of protons and the number of neutrons in the nucleus. Alpha particle emission occurs mainly in radioisotopes whose atomic number is greater than 82. (*Atomic number* is the number of protons in the nucleus.)

In terms of absolute masses, the mass released from the nucleus may be calculated in terms of the *atomic mass unit* (amu), which is defined as one-twelfth the mass of the C-12 atom. Using this scale, the neutron has 1.0088665 amu, the proton has 1.0088925 amu, and the electron has 0.0005486 amu. The amu itself may be calculated using Avogadro's number of $6.022(10^{23})$ particles per gram-mole. Hence amu $= \frac{1}{12}[12/6.022(10^{23})] = 1.6606(10^{-24})$ g. In this calculation, 12 is the number of grams in a gram-mole of C-12, which is also the atomic weight of C-12.

Since amu is one-twelfth the mass of the C-12 atom, multiplying amu by 12 will give the mass of the C-12 atom: $1.6606(10^{-24})(12) = 1.9927(10^{-23})$ g, mass of C-12 atom. However, in this calculation, 12 is also the atomic weight of C-12. From this it is concluded that the atomic weight of C-12 is the weight or mass of an atom of C-12 expressed in terms of the number of amu. This finding may be extended to any other element. Therefore, the *atomic weight* of any element is the weight or mass of its atom expressed in terms of the number of amu. In a similar manner, the molecular weight of a compound is the weight or mass of the molecule of the compound expressed in terms of the number of amu.

TABLE 13-2 MAXIMUM CONCENTRATIONS OF CONTAMINANTS FOR THE TCLP TOXICITY TEST

EPA hazardous waste number	Contaminant	Maximum concentration (mg/L)
D004	Arsenic	5.0
D005	Barium	100
D006	Cadmium	1.0
D007	Chromium	5.0
D008	Lead	5.0
D009	Mercury	0.2
D010	Selenium	1.0
D011	Silver	5.0
D012	Endrin (1,2,3,4,10,10-hexachloro-1,7-epoxy-1,4,4a,5,6,7,8,8a-octahydro-1,4,endo,endo-5,8-dimethanonaphthalene)	0.02
D013	Lindane (1,2,3,4,5,6-hexachlorocyclohexane, γ isomer)	0.4
D014	Methoxychlor [1,1,1-trichloro-2,2-bis(p-methoxy-phenyl)ethane]	10.0
D015	Toxaphene ($C_{10}H_{10}Cl_8$, technical chlorinated camphene, 67–69% chlorine)	0.5
D016	2,4-D (2,4-dichlorophenoxyacetic acid)	10.0
D017	2,4,5-TP Silvex (2,4,5-trichlorophenoxypropionic acid	1.0
D018	Benzene	0.5
D019	Carbon tetrachloride	0.5
D020	Chlordane	0.03
D021	Chlorobenzene	100.0
D022	Chloroform	6.0
D023	o-Cresol	200.0
D024	m-Cresol	200.0
D025	p-Cresol	200.0
D027	1,4-Dichlorobenzene	7.5
D028	1,2-Dichloroethane	0.5
D029	1,1-Dichloroethylene	0.7
D030	2,4-Dinitrotolulene	0.13
D031	Heptachlor (and its hydroxide)	0.008
D032	Hexachlorobenzene	0.13
D033	Hexachloro-1,3-butadiene	0.5
D034	Hexachloroethane	3.0
D035	Methyl ethyl ketone	200.0
D036	Nitrobenzene	2.0
D037	Pentachlorophenol	100.0
D038	Pyridine	5.0
D039	Tetrachloroethylene	0.7
D040	Trichloroethylene	0.5
D041	2,4,5-Trichlorophenol	400.0
D042	2,4,6-Trichlorophenol	2.0
D043	Vinyl chloride	0.2

Example 13-2

What are the masses of the neutron, proton, and electron expressed in terms of grams and pounds?

Solution

$$\text{Mass of neutron} = 1.6606(10^{-24})(1.0088665) = 1.67532(10^{-24})\text{g}$$

$$= 3.6901(10^{-27}) \text{ lb} \quad \textbf{Answer}$$

$$\text{Mass of proton} = 1.6606(10^{-24})(1.0088925) = 1.67537(10^{-24})\text{g}$$

$$= 3.6902(10^{-27}) \text{ lb} \quad \textbf{Answer}$$

$$\text{Mass of electron} = 1.6606(10^{-24})(0.0005486) = 9.1101(10^{-28})\text{g}$$

$$= 2.0066(10^{-30}) \text{ lb} \quad \textbf{Answer}$$

Example 13-3

What are the average masses of one atom of hydrogen and one molecule of water, respectively?

Solution

$$H = 1.008; \quad \text{hence mass} = 1.008(1.6606)(10^{-24}) = 1.6739(10^{-24}) \text{ g} \quad \textbf{Answer}$$

$$H_2O = 1.008(2) + 16 = 18.016;$$

$$\text{hence mass} = 18.016(1.6606)(10^{-24}) = 2.9917(10^{-23}) \text{ g} \quad \textbf{Answer}$$

Example 13-4

U-238 decays, releasing the α particle. Write the nuclear reaction.

Solution When U-238 releases the alpha particle, the atomic number and the atomic mass number decrease by 2 and 4, respectively. The atomic number of uranium is 92. From the periodic table, the atom with an atomic number of 90 is thorium. Hence

$$^{238}U_{92} \rightarrow {}^{234}Th_{90} + {}^{4}He_2 \quad \textbf{Answer}$$

In the reaction above, the preexponent is the atomic mass number and the subscript is the atomic number.

Beta radiation. Beta radiations are electrons released from a radioactive nucleus as a result of the breakup of a neutron into a proton and an electron. As a result of this breakup, the nucleus gains a positive charge of one proton per neutron that is split. Since the mass of the electron that is split away as the beta radiation is small, the reaction does not change the atomic mass number of the parent atom. However, due to the gain in charge, the parent atom changes to another atom of higher atomic number. Being an electron, beta radiation has a negative charge.

Example 13-5

Sr-90 decays to yttrium by the release of a β particle. Write the nuclear reaction.

Solution

$$^{90}Sr_{38} \rightarrow {}^{90}Y_{39} + \beta \quad \textbf{Answer}$$

Gamma radiation. Gamma radiation or gamma ray, as it is also called, has no charge or mass, being simply an electromagnetic radiation that travels at the speed of light. Being an electromagnetic radiation, gamma radiation may be thought of as a wave or as a photon; it has a very short wavelength, in the range 10^{-3} to 10^{-7} μm. When the electron in an atom moves from a higher energy level (excited state) to the lower level, energy is radiated. An analogous phenomenon, also, occurs in a nuclear reaction; a nucleus in an excited state releases the gamma rays when it transforms to a more stable form. Gamma radiation may accompany either alpha or beta radiation.

Units of radiation. There are a number of radiation units used. The *curie* (Ci), the basic unit of decay rate, is defined as the disintegration of $3.7(10^{10})$ atoms per second. A unit called the *becquerel* (Bq) is related to the curie as follows: 1 Ci $= 3.7(10^{10})$ Bq. The *roentgen* (R) is the dose of gamma or x radiation that produces 1 esu of static electricity in 1 cubic centimeter of air at standard temperature and pressure (STP). The energy absorbed by the air from a roentgen dose of radiation may be calculated as follows: to form 1 esu of electricity in air at STP, $1.61(10^{12})$ ion pairs are produced per gram of air. To produce an ion pair in air, from empirical studies, 34 electron volts is needed. Thus to produce 1 esu, the energy absorbed per gram of air is $34(1.61)(10^{12}) = 5.47(10^{13})$ eV. Using the relation, 1 eV $= 1.602(10^{-12})$ erg, we have 1 R $= 5.47(10^{13})(1.602)(10^{-12}) = 87.63$ ergs/g of air.

The *rad* (radiation absorbed dose) is defined as the absorption of 100 ergs of energy resulting from a radioactive radiation per gram of the absorbing substance. Thus, in air a roentgen is equal to $87.63/100 = 0.88$ rad. The ratio of the absorbed energy from gamma radiation to the absorbed energy from a given radiation that produces the same biological effect as the gamma radiation is called the *relative biological effect* (RBE) of the given radiation. Thus if the absorbed dose of 1 rad of a gamma radiation produces the same biological effect as 0.2 rad of a neutron radiation, the RBE of the neutron radiation is $1/0.2 = 5$. The product of the dose in rads and RBE is called the *roentgen equivalent man* (rem). Although the rad is a convenient unit for expressing the amount of energy absorbed, it does not indicate the relative injury the radiation causes to a biological substance. Expressing it as a product with RBE (the rem) measures this relative biological injury. Hence, in the example above, the neutron radiation causes five times more injury than the simple gamma radiation.

Example 13-6

The rad value of radiation A is 10 rads and that of radiation B is 2 rads. If A has an RBE of 1 while B has an RBE of 7, which is the potentially more damaging of the two radiations?

Solution   To make comparisons, the two radiations must be converted into rems.

$$\text{Number of rems} = \text{rad RBE}$$

$$\text{Number or rems of A} = 10(1) = 10$$

$$\text{Number or rems of B} = 2(7) = 14$$

Thus B is potentially more damaging.   **Answer**

HEALTH EFFECTS

As mentioned before, in the United States, radioactive wastes are not regulated under RCRA. For this reason, two types of hazardous wastes should be distinguished: RCRA wastes (regulated by RCRA) and radioactive wastes. The effect of RCRA wastes on health may be exemplified by the tragedy of Love Canal.[1] This was discussed in Chapter 1.

Radioactive Wastes

Radioactive wastes contain remnants of nuclei that emit alpha, beta, and gamma radiations. As an alpha particle passes through biological tissue, its positive charge attracts orbital electrons. This interaction can knock out the electrons, causing ionization of tissue substances.

Alpha particles are relatively massive and can easily be stopped. The skin is sufficient protection from an external source; however, if the radiation is taken internally it can cause great damage by direct internal bombardment.

As the negatively charged β particles approach an atom electron, Coulomb forces can raise orbital electrons to higher levels, possibly knocking them off the orbit, producing ionization. Although this type of ionization occurs less frequently than that caused by alpha radiation, beta radiations can penetrate deeper into biological tissues. Alpha particles travel less than 100 μm into tissue, while beta particles travel several centimeters. Beta radiation can, however, be stopped by modest shielding; a 1-cm-thick aluminum, for example, is sufficient.

The short wavelengths of gamma radiation means that they are effective at producing ionization. They are thus biologically damaging. Possessing no mass, they are very difficult to contain. They require several centimeters of lead for adequate shielding. Depending on the source of the gamma radiation, thickness could go up to 30 cm of lead.

The source of radiation to a biological tissue may be external or internal. External sources of radiation are those that come from outside the body; internal sources are those that have been taken internally, such as the radon inhaled into the lungs. All these forms of radiation are dangerous to living things. The electron excitations and ionizations cause the molecules to become unstable and break up to form new ones that were nonexistent before. Then the organism responds to the damage at such a slow rate that the effect

[1]C. A. Wentz (1989). *Hazardous Waste Management*. McGraw-Hill, New York, pp. 306–312.

may not become noticed even after several years (20 to 30 years or so). Low-level radiation can cause somatic and/or genetic damage.[2] (Somatic cells are those that are differentiated into different tissues, organs, etc.) Somatic effects include higher risk of cancer, leukemia, sterility, cataracts, and a reduction in lifespan. By altering the DNA in genes and chromosomes, genetic effects affect future generations.

NUCLEAR FISSION

In the case of orbital electrons, when an electron goes from a higher, unstable energy level to a lower, stable energy level, energy is released. As a rough analogy, a given nucleus fissions because it wants to transform to a lower, stable form. As in the orbital electrons, energy is also released by this process. It is this released energy that is tapped in nuclear reactions.

Fission is a breakup of the nucleus; hence at least two fission fragments result from the process. Consider a fission reaction using U-235. In the process, the uranium atom captures a neutron, making the nucleus unstable and thus causing breakup. Since the atomic mass number of a neutron is 1, the sum of the atomic mass numbers of the product species must be $135 + 1 = 136$. An illustration of a fission reaction of U-235 is

$$^{235}U_{92} + {}^1n_0 \rightarrow {}^{95}Mo_{42} + {}^{139}La_{57} + 2\,{}^1n_0 \tag{13-1}$$

Mo and La are the fission fragments. If the atomic mass numbers of the reactants and products of reaction (13-1) are added, they would all be found to be equal to 136. Since they are too small to emit alpha particles, fission fragments are beta or gamma emitters only. Reaction (13-1) yields 204 MeV of energy.

About one fission reaction in 10,000 yields three fission fragments instead of two. This third fragment is tritium (H-3). Since it is a chemical form of hydrogen, it exchanges freely with the nonradioactive form of hydrogen in the cooling water used in nuclear reactors. Containment of this fragment is thus difficult. Tritium has a half-life of 12.3 years.

Table 13-3 lists fission fragments that can be produced in a fission reactor. Cs-137 concentrates in muscles, Sr-90 concentrates in bones, and I-131 concentrates in the thyroid gland.

For a nuclear reaction to proceed, at least one neutron must be able to strike another nucleus for the reaction to continue. For this to happen there must be a minimum amount of nuclear mass present at any time; this mass is called the *critical mass*. Thus the presence of at least the critical mass ensures a continuous reaction called a *fission chain reaction*. In practice, reactors are built to provide excess critical mass to ensure sufficient neutrons. There then exists the potential for a runaway reaction. For this reaction not to happen, excess neutrons are controlled by means of a *moderator*. This is made of materials such as boron, cadmium, or hafnium that have a large cross section

---

[2]G. M. Masters (1991). *Introduction to Environmental Engineering and Science.* Prentice Hall, Englewood Cliffs, N.J., p. 58.

TABLE 13-3 REACTOR FISSION FRAGMENTS

Isotope	Type of emission	Half-life
Ag-111	Beta	8 days
Sn-125	Beta	9 days
Sb-125	Beta	2 yr
Te-127	Gamma	100 days
Te-129	Beta	33 days
I-129	Beta	$2(10^7)$ yr
I-131	Beta	8 days
Xe-131	Beta	10 days
Ce-141	Beta	32 days
Ce-144	Beta	285 days
Pr-143	Beta	13 days
Nd-147	Beta	11 days
Pm-147	Beta	3 yr
Pm-148	Gamma	40 days
Eu-156	Beta	15 days
Kr-85	Beta/gamma	10 yr
Sr-89	Beta	50 days
Sr-90	Beta	29 yr
Y-91	Beta/gamma	55 days
Zr-95	Beta	63 days
Nb-95	Beta/gamma	35 days
Ru-103	Beta	40 days
Ru-106	Beta	1 yr
Xe-133	Beta/gamma	5 days
Cs-134	Beta	2 yr
Cs-137	Beta	30 yr
Ba-140	Beta	12 days

for capturing neutrons. Moderators are made into control rods that are inserted in and out of the reactor to moderate the reaction.

Fission reactions produces a large amount of heat. This must be dissipated away by means of cooling water to prevent mechanical failure of the reactor assembly—a process called *meltdown*. The ultimate uncontrolled reaction is called *atomic explosion*.

CRADLE-TO-GRAVE MANAGEMENT

Under RCRA a cradle-to-grave chain of management has been instituted. The chain is composed of the generator, the transporter, treatment systems, storage facilities, and disposal sites, as well as the EPA or appropriate state agency. The *generator* produces the hazardous waste; the *transporter* receives the waste from the generator and transports it to a destination such as treatment plants or storage facilities. Storage facilities are those structures that are used to store the waste temporarily before treatment and final disposal.

A facility that treats, stores, and disposes the hazardous waste is called a *TSD facility*. The TSD facility receives the waste from the transporter.

Central to this type of management is the *hazardous waste manifest*. The manifest is a multiple-form document and contains information about the waste, and on the waste generator, transporter, and TSD facility. The way that the system works is that the generator initially completes the forms. One copy is retained; another is sent to the USEPA or appropriate USEPA-delegated state agency. The remainder copies are given to the transporter, who retains one copy. At this point the USEPA or the state has been alerted that a hazardous waste is in the shipment process. Upon arrival, the transporter gives the copies of the manifest (minus one copy) to the TSD facility. The facility then retains one copy and returns one copy to the generator and another copy to the appropriate USEPA or state agency. As can be seen, by this system, the whereabouts of the waste is known at all times.

If a certain number of days have elapsed since the waste was given to the transporter and the generator has not received the copy of the manifest, it is the duty of the generator to investigate into the whereabouts of the waste. If, again, after a certain additional number of days, the generator is still unable to locate the waste, the generator must then file an exception with the USEPA. At any rate, the location of the waste must be investigated continually until found.

To qualify for a generator, transporter, or a TSD facility, certain requirements must be met. For example, the generator and the transporter must have an *USEPA identification number*. The generator must refuse to give the waste to a transporter who does not have the number and, vice versa. The USEPA identification number enables the agency to monitor and track all hazardous wastes produced. The generator must comply with all *packaging requirements*. These require the generator to identify and quantify the wastes that are put in a certain container for transport. The transporter must follow all *placarding requirements* while transporting the waste. The placard serves as a warning to the public that a hazardous waste is being transported. The transporter normally follows the transportation regulation developed by the Department of Transportation. The transporter must also comply with all *reporting requirements* in case of an *accident during transport,* which must be reported immediately to local authorities. In addition to the USEPA identification number, the TSD facility must have a *permit* in order to carry out its business. The final disposal facility must be an *approved site*. Figure 13-1 shows a photograph of hazardous waste containers.

TREATMENT METHODS

Upon arrival at the TSD facility, the waste may be treated by a number of ways depending on the type. Treatment methods used for hazardous wastes include biological treatment, neutralization, oxidation-reduction, precipitation, carbon adsorption, ion exchange, electrodialysis, reverse osmosis, fixation/stabilization, and incineration. Depending on the type of waste, any one or combination of these methods may be used. Some of the

Figure 13-1 Hazardous waste containers.

foregoing treatment methods have already been discussed in this book and will therefore not be repeated, except for some that warrant repetition.

Biological waste treatment of hazardous organic wastes is very similar to conventional waste treatment of organic wastes. The only difference is that whereas in conventional treatment, the mean cell residence time may range from 4 to 15 days, in hazardous waste treatment, the mean cell residence time may extend to 3 to 6 months. Most organic hazardous wastes are anthropogenic, and they normally are halogenated. Halogenated compounds are very resistant to biodegradation. These compounds include pesticides, plasticizers, solvents, and trihalomethanes. Although these compounds are toxic, they may be degraded by acclimatization with the attendant penalty of a very long mean cell residence time.

The concept of acclimatization is also manifested in *in situ* biodegradation. In this process, bacteria in the soil or in the environment have been in contact with the waste for a long period of time; hence they have been acclimated to the waste. They have developed the necessary enzymes to degrade the particular type of waste. In practice, what is needed to accelerate the biodegradation process is to add nutrients, nitrogen and phosphorus, to the soil.

The treatment methods enumerated above do not apply automatically to all types of waste. Just what type of method is applicable to a given waste must be determined by some treatability experiment or a search of the literature.

Neutralization

When a hazardous waste has either a high or a low pH, it can be treated by neutralization. For example, sulfuric acid and hydrochloric acid are added to a waste high in caustic, while NaOH or $Ca(OH)_2$ are added to wastes that are acidic. The final pH before discharge should meet regulatory requirements which is 6.5 to 8.5 in Maryland and would be different elsewhere.

Oxidation-Reduction

Whenever a constituent of hazardous waste is amenable to oxidation-reduction (redox) as determined by some treatability experiment or by a literature search, this method may be used. For example, if the waste is a cyanide, it may be oxidized by *alkaline chlorination.* (*Alkaline* here must be emphasized, since acid chlorination produces hydrogen cyanide, which is poisonous.) The process is carried out in two steps. The first step is conducted at pH 10 and the pertinent reaction is as follows:

$$NaCN + 2NaOH + Cl_2 \rightleftharpoons NaCNO + 2NaCl + H_2O \qquad (13\text{-}2)$$

The second step is conducted at pH 8 and the reaction is as follows:

$$2NaCNO + 3Cl_2 + 4NaOH \rightleftharpoons 2CO_2 + N_2 + 6NaCl + 2H_2O \qquad (13\text{-}3)$$

Another example of a redox treatment is the reduction of hexavalent chromium to the trivalent form in electroplating operations using sulfur dioxide. Trivalent chromium is much less toxic and more easily precipitated than hexavalent chromium. The reactions are

$$3SO_2 + 3H_2O \rightarrow 3H_2SO_3 \qquad (13\text{-}4)$$

$$2CrO_3 + 3H_2SO_3 \rightarrow Cr_2(SO_4)_3 + 3H_2O \qquad (13\text{-}5)$$

Precipitation

Precipitation is a process in which an insoluble product is formed called a *precipitate,* which is then settled or removed. Frequently, the precipitation involves the use of $Ca(OH)_2$ or caustic soda to form a precipitate of metal hydroxide. For example, the following reaction shows the removal of a precipitable metal M using lime:

$$M^{2+} + Ca(OH)_2 \rightarrow M(OH)_2 + Ca^{2+} \qquad (13\text{-}6)$$

Solidification and Stabilization

Inorganic hazardous wastes cannot be destroyed, unlike organic wastes, which can be incinerated. A method that has been developed to treat these wastes is solidification and stabilization. In this method, after the waste has been concentrated into ash or sludge, it is mixed in a binding agent to *solidify* and to *stabilize.* After solidification, the waste constituent should not leach out from the solid matrix for it to be considered treated. Portland cement has been used as a solidifying agent.

Incineration

In general, there are three types of radioactive wastes: high-level wastes (HLW), intermediate-level wastes (ILW), and low-level wastes (LLW). *High-level wastes* are those with radioactivities measured in curies per liter; intermediate-level wastes are those with radioactivities measured in millicuries per liter; and low-level wastes are those with radioactivities measured in microcuries per liter. Radioactive wastes containing isotopes higher than uranium in the periodic table are called *transuranic wastes.* For low-level wastes, although an effective way to reduce the volume (by 80 to 90%), incineration should only be used if monitored continuously for hazardous radioactive concentrations in the effluent gas.

The permitting of a RCRA hazardous waste incinerator is a complex process, the most important element of which is the trial burn. A *trial burn*, which has several components and complicated requirements, is a treatability study that determines the suitability of incineration to destroy the waste. The results of the trial then form the basis of the permit that is subsequently issued if it passes a public hearing. The following are determined during the trial burn:

1. Allowable waste analysis procedures
2. Allowable waste feed rate and composition
3. Acceptable operating limits for CO in the stack
4. Combustion temperature and combustion gas flow rate
5. Allowable variations in incinerator design and operating procedures

There are three performance standards that a hazardous waste incinerator must meet: effluent standard on *principal organic hazardous constituents* (POHC), hydrochloric acid, and particulate. The POHC is one or more designated constituents in the effluent upon which the *destruction and removal efficiency* (DRE) of the incinerator is based. The DRE for all designated POHCs is normally specified as 99.99% or higher. For dioxins and dibenzofurans, RCRA requires 99.9999% DRE. The DRE for RCRA wastes is given by

$$DRE = \frac{W_{in} - W_{out}}{W_{in}} \qquad (13\text{-}7)$$

where W_{in} is the mass feed rate of a POHC in the feed stream and W_{out} is the emission rate of the same POHC prior to release to the atmosphere. PCBs are regulated under TSCA.

The combustion efficiency is defined by

$$\text{combustion efficiency} = \frac{[CO_2]}{[CO_2] + [CO]} \qquad (13\text{-}8)$$

If the residence time of liquid PCBs in the furnace is 2 s, TSCA requires the combustion temperature to be at $1200 \pm 100°C$ with 3% excess oxygen; alternatively, if the furnace residence time is 1.5 s, the combustion temperature is to be at $1600 \pm 100°C$ with 2% excess air. The EPA interprets this requirement as equivalent to a DRE value of 99.9999%.

Hazardous wastes normally contain the chlorine atom; hence, when burned, hydrochloric acid is formed. Since the concentration can become hazardous, HCl is also limited by some performance standard, such as a requirement of no greater than the larger of 1.8 kg/h or 1% of the HCl prior to entering any pollution control equipment.

Particulates emitted from an incinerator can be hazardous if the waste incinerated is also hazardous. For this reason, particulates coming out of a hazardous incinerator are also limited by some standard. The standard is set to dry standard conditions corrected to

TABLE 13-4 INCOMPLETE LISTING OF HAZARDOUS SUBSTANCES

Hazardous waste number	Hazardous substance
P023	Acetaldehyde, chloro-
P002	Acetamide, n-(aminothioxomethyl)-
P057	Acetamide, 2-fluoro-
P058	Acetic acid, fluoro-, sodium salt
P066	Acetimidic acid, n-[(methylcarbamoyl)oxy]thio-, methyl ester
.	
.	
.	
U077	Ethylene dichloride
U359	Ethylene glycol monoethyl ether
.	
.	
.	
U043	Vinyl chloride
U248	Warfarin
U239	Xylene
U200	Yohimban-16-carboxylic acid, 11,17-dimethoxy-18-[(3,4,5-trimethoxybenzoyl)oxy]-, methyl ester
U249	Zinc phosphide

some level of percent oxygen in the stack X. Let Y be the actual mole fraction of oxygen in the stack in terms of the stack gas nitrogen and oxygen contents corrected to dry standard conditions. Using 21 as the mole percent of oxygen in dry air, the mole fraction of oxygen used in combustion is proportional to $21 - Y$. Using the correction X, the mole fraction of oxygen used in combustion would also be proportional to $21 - X$ (corrected to dry standard conditions). The amount of particulate matter produced is proportional to the amount of oxygen used in combustion. Hence, letting P_m be the actually measured particulate concentration, P_m is proportional to $21 - Y$. Also, letting P_c represent the corrected particulate concentration at the correction level of X, P_c is proportional to $21 - X$. Formulating the proportionality equation and solving for P_c, the following is obtained:

$$P_c = P_m \frac{21 - X}{21 - Y} \tag{13-9}$$

If X is equal to 7, the numerator of equation (13-9) is equal to 14.[3]

Table 13-4 shows a short and an incomplete listing of hazardous substances that can form the components of hazardous waste. The hazardous waste numbers are USEPA numbers. This table simply gives the reader some idea of what a hazardous substance can be. For an extensive list, the USEPA should be consulted.

[3]M. L. Davis and D. A. Cornwell (1991). *Introduction to Environmental Engineering.* McGraw-Hill, New York, p. 696.

Example 13-7

A trial burn at 1000°C for a hazardous waste mixture designates the following as the POHCs: chlorobenzene, toluene, and xylene. The waste feed and stack emissions are shown below. The stack gas rate of flow is 400 m³/min at dry standard conditions. The permit requires the following to be met: DRE = 99.99%, HCl emissions to be no greater than the larger of 1.8 kg/h or 1% of the HCl prior to entering the control device, particulates = 180 mg/m³ at dry standard conditions corrected to 7% O_2. Is the incinerator in compliance? The mole fraction of O_2 in the stack gas is 14%.

Compound	Inlet (kg/h)	Outlet (kg/h)
Chlorobenzene (C_6H_5Cl)	150	0.010
Toluene (C_7H_8)	420	0.030
Xylene (C_8H_{10})	430	0.065
HCl	—	1.0
Particulates	—	3.50

Solution *For the POHC requirement:*

$$DRE = \frac{W_{in} - W_{out}}{W_{in}}$$

$$DRE_{chlorobenzene} = \frac{150 - 0.010}{150}(100) = 99.993\%$$

$$DRE_{toluene} = \frac{420 - 0.030}{420}(100) = 99.993\%$$

$$DRE_{xylene} = \frac{430 - 0.065}{430}(100) = 99.985\%$$

For the POHC requirement, the incinerator is not in compliance.

For the HCl requirement: Of the three POHC, only chlorobenzene yields HCl,

$$C_6H_5Cl + 7O_2 \rightarrow 6CO_2 + 2H_2O + HCl$$

$$1 \text{ mol chlorobenzene} \rightarrow 1 \text{ mol HCl}$$

$$C_6H_5Cl = 6(12) + 5(1.008) + 35.45 = 112.49$$

$$\text{HCl in gas prior to control equipment} = \frac{150}{112.49} = 1.33 \text{ kg-mol/h} = 1.36(HCl) = 48.61 \text{ kg/h}$$

$$1.0\% \text{ of } 48.61 = 0.4861 \text{ kg/h}$$

1.0 kg/h < 1.8 kg/h; thus for the HCl requirement, the incinerator is in compliance.

For the particulate requirement:

$$P_c = P_m \frac{21 - X}{21 - Y} = \frac{3.50}{400(60)}(1000)(1000)\left(\frac{21 - 7}{21 - 14}\right) = 291.67 \text{ mg/m}^3$$

For the particulate, the incinerator is not in compliance.

Overall, the incinerator is not in compliance.   **Answer**

FINAL DISPOSAL

After various treatment methods have been applied, there will always remain some residue to be disposed of. Land disposal techniques include landfills, surface impoundments, underground injection wells, and waste piles. *Landfill* designs have been discussed in Chapter 11. It must be noted that hazardous waste landfills require at least two liners. Also, wastes should not be mixed, and only compatible wastes should be placed in the same cell. In addition, the location of each cell should be clearly identified in a locator map or similar means. *Surface impoundments* are excavated or diked areas used to store liquid hazardous wastes. This storage is only temporary. As in the case of landfills, impoundments should be double-lined and provided with a leachate collection system. *Injection wells* have been used as a cheap method of disposing of hazardous waste; they should be located well below drinking water aquifers. Regulations regarding construction, operation, and monitoring of injection wells have become more stringent and, as in the case of landfills, reliance on this method of final disposal is strongly discouraged. *Waste piles* are noncontainerized accumulation of solid hazardous waste used for temporary storage.

Since hazardous wastes are simply kept on hold, all these methods of final disposal may be considered temporary. Unless they are rendered innocuous before disposal, these wastes are like time bombs waiting to explode at the right moment. Research must therefore continue to find economical ways that will render hazardous wastes harmless before being put in their final resting place.

Final Disposal of Radioactive Wastes

There is no way that the half-lives of radioactive wastes can be reduced; therefore, the only method of treatment for these types of wastes is time. Wastes are held until the radioactivity has died down to safe levels; at these levels, the wastes can then be discharged into the environment. Holding of the waste can take days, months, or hundreds of years, as can be computed from the half-lives of the fission products as shown in Table 13-3.

High- and low-level wastes are put in different holding depositories. In the United States, the only site for high-level wastes is Yucca Flats, Nevada; the sites for low-level wastes are Maxey Flats, Kentucky; Sheffield, Illinois; and West Valley, New York. High-level wastes are produced from nuclear reactors, which number around 102 in the United States.[4] Low-level wastes are produced from such places as research laboratories and hospitals. Yucca Flats is designated as a monitored, retrievable storage facility, which means that the waste is to be retrieved after a number of years.

---

[4]USEPA (1988). *Nucleus News*. American Nuclear Society.

States entered into compacts to manage their low-level wastes. One of these compacts, the *Midwest Compact,* considers an engineered containment structure to store waste. This structure can be above or below ground and consists of water-free storage vaults into which radioactive wastes in their respective containers can be deposited. Before being placed in containers, radioactive wastes must be reduced in volume. Volume reduction includes evaporation (for liquid wastes), incineration, and compression (for solid wastes).

Reactor wastes contain the radionuclide plutonium, which has a half-life of 24,390 years, posing great problems for disposal. The fissile isotope, U-235, which provides the energy in the nuclear fuel, is only 2 to 3% of the fuel mass; the rest is the nonfissile U-238. During nuclear reaction, U-238 can capture a neutron that leads to the production of plutonium, which because of its very long half-life, is very difficult to dispose of. Plutonium may be separated from the spent fuel to reduce the overall half-life of the resulting waste; however, this isotope is a critical ingredient in the production of atomic bombs. Hence, if separated, the world could be in danger of the proliferation of nuclear weapons if the technology for isotope separation falls into the wrong hands. In France, plutonium is produced for use in nuclear electric power generation. In the overall, disposal of radioactive wastes remains a very difficult problem for environmental engineers.

RISK ASSESSMENT

Chloroform is produced daily in the chlorination of public water supplies; it is also a potential carcinogen. Knowing this potential, should chlorination be stopped? If so, is it justified? Making this decision involves taking some amount of risk; elimination of chlorination would mean that the risk of getting cancer from this cause is zero, and continuing chlorination would mean that the risk of getting cancer always exists. The subject of determining the probability of getting an undesirable health effect from exposure to hazards is called *risk assessment*. Risk assessment has four essential elements: hazard or risk identification, dose–response assessment, hazard or risk exposure assessment, and hazard or risk characterization.

Hazard or *risk identification* refers to determining whether or not a particular chemical is causally linked to an observed health effect such as cancer or a birth defect. This may done by studies on animals using the suspected chemical or chemicals and positive results assumed applicable to humans. Assuming positive results in the hazard identification stage, the dose–response assessment may be performed. *Dose–response assessment* refers to the determination of the probability of an undesirable health response (probability of risk) to a given dose. In other words, a relationship between dose of suspected chemical and probability of risk is developed. *Risk exposure assessment* refers to the determination of the circumstances of exposure of a person to a particular chemical. The circumstances may include the length of time of exposure between the chemical and the person, the concentration of the chemical at the point of exposure, and the volumetric rate of flow of the medium carrying the chemical. As determined from the results of the risk exposure assessment, the dose–response relationship may be modified depending on the actual percentage of exposure the person has had with the chemical. *Risk characterization* is the integration of the results of the foregoing three steps providing estimates of the magnitude of the public health problem. In risk characterization, uncertainties of the statistical methods used and the nature of these uncertainties should be explained. This gives the reader an idea of how much reliance must be placed on a risk assessment report.

In hazardous waste management, risk assessment is normally done on the potential risk of chemicals to cause cancer. Therefore, carcinogenesis will be discussed, first, followed by elaboration on the dose–response and exposure assessment stages of risk assessment.

Carcinogenesis

Cancer is thought to be produced by the mutation of DNA. DNA mutation causes the cell to malfunction, leading in some cases to cell death, cancer, reproductive failure, or abnormal offspring. Carcinogenesis induced by chemicals occurs in two stages: initiation and promotion. *Initiation* is the mutation of DNA as a result of exposure to the chemical; *promotion,* influenced by the same or another chemical, is the stage where the cell no longer recognizes growth constraints and continues to grow abnormally developing *tumors.* Tumors may be benign or malignant. *Benign tumors* are not cancerous; their existence is simply restricted to their own boundaries. On the other hand, *malignant tumors* are those that break apart (undergo *metastasis*) and enter other tissues and spread. Malignant tumors are therefore cancerous.

Chemicals may be categorized as human carcinogens, probable human carcinogens, possible human carcinogens, and noncarcinogens. *Human carcinogens* are those chemicals that are shown by epidemiological studies to cause cancer. *Probable human carcinogens* are those chemicals that are found to cause cancer by *limited* epidemiological studies and to show carcinogenity by an adequate number of laboratory studies. The word *probable* comes from the fact that *limited* epidemiological studies show the chemical to be carcinogenic, but because the human data are only limited, the carcinogenity is only probable. *Possible human carcinogens* are those showing carcinogenity on animals, but no data on humans are available. Since there are no data on humans, the carcinogenity is only *possible. Noncarcinogenic* chemicals are those that do not cause cancer in humans.

Dose–Response Assessment

In the determination of the dose–response characteristic of a carcinogenic substance, it is assumed that any dose will always result in a certain probability of getting an undesirable health effect called *probability of risk.* Hence the graph of probability of risk against dose will pass through the origin.

There are two problems associated with determining the dose–response characteristic of a carcinogen: (1) the experiment cannot be done directly on humans, and (2) the concentration used in the experiment is large. Needless to say, no experiment to determine carcinogenity of chemicals should be done directly on humans. Knowledge of human carcinogens is derived from epidemiological studies. This situation poses a problem in ascertaining the human carcinogenity of a suspected chemical. In practice, aside from epidemiological studies, all that is done is to perform experiments on nonhumans. In addition, dose–response assessment is not determining whether or not a chemical is a carcinogen but determining the *probability* of carcinogenic response to a certain dose.

Despite the inappropriateness and because experiments cannot be done directly on humans, tests are conducted on animals. To apply dose–response data from animal bioassays to humans, a *scaling factor* must be applied. In deriving this factor, assumption is made that the doses are equivalent if the doses per unit body weight are equal. Sometimes dose per unit of body surface area is also used. Strictly, the rates of absorption of the chemical by the animal and the human are different, and this should be taken into account to modify the dose–response characteristic. However, this information is normally lacking, and absorptions are assumed to be equal.

The other problem is the large concentrations or doses used in the animal bioassay. Humans are subjected to carcinogens at only low concentrations. In the animal bioassay, by necessity, the concentrations or doses are large. Animals have very short life spans compared to humans. If they are subjected to very low doses of a suspected carcinogen, because of their short life spans, cancer will not develop during their lifetimes. To induce cancer during the animal's life, it is necessary to use higher doses, resulting in a very unrealistic situation. On this account animal bioassays can be severely criticized. Nevertheless, despite this drawback, dose–response data obtained from animals are used to obtain a factor by extrapolating to concentrations applicable to human exposure. If the probability of risk (dimensionless) is plotted against dose (standard unit is mg/kg-day), the curve at the very low doses would be a straight line. The slope at these low doses is the factor referred to above and is called the *potency factor* (PF). PF is one unit of probability. The units of this factor are $(mg/kg\text{-}day)^{-1}$. From these units, the potency factor may be defined as the probability of risk produced by an exposure of a daily dose of 1.0 mg/kg-day. But the lifetime of the animal tested has been made equivalent to the lifetime of humans (assumed to be 70 years). Therefore, the "human" definition of potency factor is the probability of risk produced by lifetime exposure of an average daily dose, also called *chronic daily intake* (CDI) of 1.0 mg/kg-day absorbed into the bloodstream. If the average daily dose is known, the lifetime risk is

$$\text{lifetime risk } = \text{ CDI (in mg/kg-day)} \times \text{PF} \qquad (13\text{-}10)$$

Table 13-5 shows potential factors of various potential carcinogens.

For noncarcinogenic substances, the key assumption is that there exists an exposure threshold below which the chemical has no adverse effect on the person beyond that caused by the background concentration. The lowest dose of a chemical that results in an adverse response in a person is called the *lowest-observed-effect level* (LOEL). Conversely, the highest dose that does not produce a response is called the *no-observed-effect level* (NOEL). A dose somewhere below NOEL is the *acceptable daily intake* (ADI). ADI is obtained from the NOEL by applying *uncertainty factors*, also called *safety factors*. These factors are obtained as follows: To account for differences in sensitivities between the most sensitive members of the population exposed, such as pregnant women, small children, and the elderly and normal healthy people, a 10-fold factor of safety is used.

TABLE 13-5   POTENCY FACTORS FOR VARIOUS POTENTIAL CARCINOGENS

Chemical	PF, oral route $(\text{mg/kg-day})^{-1}$	PF, inhalation route $(\text{mg/kg-day})^{-1}$
Arsenic	1.75	50
Benzene	$2.0(10^{-2})$	$2.9(10^{-2})$
Benzo[a]pyrene	11.5	6.11
Cadmium	—	6.1
Carbon tetrachloride	0.13	—
Chloroform	$6.1(10^{-3})$	$8.1(10^{-2})$
Chromium VI	—	41
DDT	0.34	—
1,1-Dichloroethylene	0.58	1.16
Dieldrin	30	—
Heptachlor	3.4	—
Hexachloroethane	$1.4(10^{-2})$	—
Methylene chloride	$7.5(10^{-3})$	$1.4(10^{-2})$
Nickel and compounds	—	1.19
PCBs	7.7	—
2,3,7,8-TCDD (dioxin)	$1.56(10^5)$	—
Tetrachloroethylene	$5.1(10^{-2})$	$1-3.3(10^{-3})$
Trichloroethylene	$1.1(10^{-2})$	$1.3(10^{-2})$
Vinyl chloride	2.3	0.295

Another factor of 10 is applied if data are obtained from animal studies and extrapolated to humans (which is normally the case). Finally, another factor of 10 is also used if the animal data available are limited. Hence the overall safety factor to be applied to the NOEL can be as high as 1000. ADI is obtained by dividing NOEL by the overall factor of safety. Table 13-6 shows some ADI values.[8] (ADI, like CDI, is the amount of chemical absorbed into the bloodstream.) Comparing Tables 13-5 and 13-6, a chemical can be considered as both a potential carcinogen and a noncarcinogen. This means that a chemical may be studied for its carcinogenic or noncarcinogenic effects. See the values for cadmium and chloroform, for example. If the ADI is subtracted from CDI, the remainder represents the dose effective in producing a risk. By analogy with carcinogens, the risk formula for noncarcinogens is therefore

$$\text{risk} = (\text{CDI} - \text{ADI}) \times \text{PF} \tag{13-11}$$

The problem with using equation (13-11) is that no PF exists, as yet, for noncarcinogens. The hazard index defined as $\frac{\text{CDI}}{\text{ADI}}$ is used in practice, instead. HIs are added for all the chemicals of concern. Hazardous concentrations are said to exist if the sum is greater than 1 and considered acceptable if the sum is less than 1.

For calculating daily intake requirements, the USEPA recommends standard values. These are shown in Table 13-7.[9]

[8] Source: USEPA IRIS database (1989).

[9] USEPA (1989). *Superfund Public Health Evaluation Manual.* Office of Emergency and Remedial Response, Washington, D.C.

TABLE 13-6 ADIs FOR CHRONIC NONCARCINOGENIC EFFECTS OF VARIOUS CHEMICALS

Chemical	ADI (mg/kg-day)
Acetone	0.1
Cadmium	$5(10^{-4})$
Chloroform	0.01
1,1-Dichloroethylene	0.009
cis-1,2-Dichloroethylene	0.01
Methylene chloride	0.06
Phenol	0.04
PCB	0.0001
Tetrachloroethylene	0.01
Toluene	0.3
1,1,1-Trichloroethane	0.09
1,1,2-Trichloro-1,2,2-trifluoroethane (CFC-113)	30
Xylene	2.0

TABLE 13-7 RECOMMENDED STANDARD VALUES FOR DAILY INTAKE CALCULATIONS

Parameter	Standard value
Average body weight, adult	70 kg
Average body weight, child	10 kg
Amount of water ingested daily, adult	2 L
Amount of water ingested daily, child	1 L
Amount of air breathed daily, adult	20 m^3
Amount of air breathed daily, child	5 m^3
Amount of fish consumed daily, adult	6.5 g
Lifetime exposure	70 years

Example 13-8

The city of Annapolis with a population of 33,178 disinfects its water supply using chlorine. If 50 μg/L of chloroform is being produced (a) estimate the maximum lifetime risk of an Annapolitan to get cancer associated with the drinking of this water. (b) What extra cancer deaths would be expected in Annapolis per year from drinking its water?

Solution (a) From Table 13-7, assume an average man of 70 kg, lifetime exposure of 70 years, and a water intake of 2 L/day. Also, from Table 13-5, chloroform has a potency factor of $6.1(10^{-3})$.

$$\text{CDI} = \frac{50(10^{-6})(1000)(2)}{70} = 0.00143 \text{ mg/kg-day}$$

$$\text{risk} = \text{CDI} \times \text{PF} = 0.00143(6.1)(10^{-3}) = 8.7(10^{-6})$$

or 8.7 deaths per million of population **Answer**

(b) $\quad$ Additional cancer deaths $= \dfrac{8.7}{10^6}(33{,}178)\left(\dfrac{1}{70}\right) = 0.004$ person **Answer**

Example 13-9

Estimate the risk of getting an undesirable health effect due to chloroform from drinking the Annapolis water. Assume that the PF value of Table 13-5 applies. Also determine HI.

Solution

$$\text{Risk} = (\text{CDI} - \text{ADI}) \times \text{PF}$$

From Table 13-6, ADI = 0.01 mg/kg-day and from Example 13-8, CDI = 0.00143 mg/kg-day.

$$\text{Risk} = (0.00143 - 0.01)(6.1)(10^{-3}) = -3.6(10^{-5}) \text{ or zero} \quad \textbf{Answer}$$

$$\text{HI} = \frac{0.00143}{0.01} = 0.143 \quad \textbf{Answer}$$

Risk Exposure Assessment

The dose–response data developed above assume that the person is being exposed to the chemical all the time. However, this may or may not be the case. To evaluate a more accurate response of the person, the exposure circumstances must be delineated and defined. For example, a potential carcinogen volatilized from a landfill may have an effect only if the prevailing wind is blowing toward a residence. At other times, the effect is zero. The dose–response developed above must therefore be corrected by the fraction of time that the wind is not blowing toward the residence. Just how much correction is to be applied on a particular exposure situation should be determined on a case-by-case basis.

One other potentially important exposure is the human consumption of contaminated fish. If the water is contaminated with aldrin, for example, it is relatively easy to determine the ambient concentration. However, just how much of this will end up in the fish that is consumed by humans? Some chemical will simply be excreted by the fish, but some will not. Those that are not excreted or are excreted with difficulty are said to *bioconcentrate* in the fish body. The amount of bioconcentration is measured in terms of the *bioconcentration factor* (BF), which relates the concentration of the chemical in the fish tissue and the concentration in the water. Thus

$$\text{concentration in fish} = \text{concentration in water} \times \text{BF} \quad (13\text{-}12)$$

Table 13-8 shows some bioconcentration factors for several chemicals.

Another point that must be considered in human risk exposure assessment is the possibility that a chemical may degrade before reaching the point of contact with a person. It is possible that at the point of exposure the concentration would already be zero. The amount of degradation of a chemical may be measured in terms of its half-life. Table 13-9 shows some half-lives of various chemicals.

TABLE 13-8 BIOCONCENTRATION FACTORS FOR SEVERAL CHEMICALS

Chemical	Bioconcentration factor (L/kg)
Aldrin	28
Arsenic and compounds	44
Benzene	5.2
Cadmium and compounds	81
Carbon tetrachloride	19
Chlordane	14,000
Chloroform	3.75
Chromium III, VI, and compounds	16
Copper	200
DDE	51,000
DDT	54,000
1,1-Dichloroethylene	5.6
Dieldrin	4,760
Formaldehyde	0
Heptachlor	15,700
Hexachloroethane	87
Nickel and compounds	47
PCBs	100,000
2,3,7,8 TCDD (dioxin)	5,000
Tetrachloroethylene	31
1,1,1-Trichloroethane	5.6
Trichloroethylene	10.6
Vinyl chloride	1.17

TABLE 13-9 HALF-LIVES (DAYS) OF VARIOUS CHEMICALS

Chemical	Air		Surface water	
	Low	High	Low	High
Benzene	6	—	1	6
Benzo[a]pyrene	1	6	0.4	—
Carbon tetrachloride	8030	—	0.3	300
Chlordane	40	—	420	500
Chloroform	80	—	0.3	30
DDT	—	—	56	110
1,1-Dichloroethane	45	—	1	5
Formaldehyde	0.8	—	0.9	3.5
Heptachlor	40	—	0.96	—
Hexachloroethane	7900	—	1.1	9.5
PCBs	58	—	2	12.9
2,3,7,8-TCDD (dioxin)	—	—	365	730
1,1,1-Trichloroethane	803	1752	0.14	7
Trichloroethylene	3.7	—	1	90
Vinyl chloride	1.2	—	1	5

Example 13-10

The PCB chronic water quality standard in the state of Maryland is 0.014 μg/L. Assume that a risk exposure assessment determines a correction factor to the dose-response data of 0.60. Assuming that the water quality standard is met and that, in fact, the concentration of PCB in the ambient water is 0.014 μg/L, what is the risk of a person dying of cancer due to PCB in Maryland?

Solution　　　From the previous tables, the following are obtained for PCB: BF = 100,000 L/kg, PF = 7.7, adult weight = 70 kg, and daily fish consumption = 6.5 g.

$$\text{Concentration in fish} = \text{concentration in water times BF}$$

$$\text{PCB in fish tissue} = 0.014(10^{-6})(1000)(100,000) = 1.4 \text{ mg PCB/kg fish}$$

$$\text{CDI} = \frac{6.5(10^{-3})(1.4)}{70} = 0.00013 \text{ mg/kg-day}$$

$$\text{Risk} = \text{CDI (PF) (correction factor)} = 0.00013(7.7)(0.60)$$

$$= 0.0006 \text{ or 6 persons in 10,000}\quad\textbf{Answer}$$

Example 13-11

A factory discharges PCB into a stream resulting in an instream concentration of 0.20 μg/L just below the outfall. At a distance 100 miles downstream, a town uses this stream as a source of water. Calculate the lifetime risk of getting cancer from drinking this water. Assume that the velocity of the stream is 1.0 mph, and assume that dispersion is negligible.

Solution

$$\text{Time of travel} = \frac{100}{1.0} = 100 \text{ h} = 4.17 \text{ days}$$

From Table 13-9, half-life ranges from 2 to 12.9 days; use 5 days. Letting C be the concentration, we have

$$C/2 = Ce^{-k(5)}\quad k = 0.1386 \text{ per day}$$

$$\text{concentration at point of use} = 0.20e^{-0.138^6(4.1^7)} = 0.112 \ \mu\text{g/L}$$

$$\text{Water ingested daily per person} = 2 \text{ L/day and}$$

$$\text{average weight of person} = 70 \text{ kg (from table)}$$

$$\text{CDI} = \frac{0.112(10^{-6})(1000)(2)}{70} = 3.2(10^{-6}) \text{ mg/kg-day}$$

$$\text{PF} = 7.7$$

$$\text{Risk} = \text{CDI(PF)} = 3.2(10^{-6})(7.7) = 2.46(10^{-5})$$

or 2.46 persons per 100,000 population   **Answer**

GLOSSARY

Acceptable daily intake (ADI). A dose below the NOEL obtained by applying a safety factor to the NOEL.

Alpha radiation. The release of positively charged helium atom.

Atomic explosion.   Uncontrolled nuclear fission.

Atomic mass number. The sum of the number of protons and neutrons in the nucleus of an atom.

Atomic mass unit (amu). The mass equal to one-twelfth the mass of the C-12 atom.

Atomic number. The number of protons in the nucleus of an atom.

Atomic weight.   The weight of an element expressed in terms of atomic mass unit, amu.

Becquerel. One nuclear disintegration per second.

Benign tumors. Tumors that are not cancerous.

Beta radiation. The release of electrons during a breakup of the nucleus of an atom.

Bioconcentration factor. The factor used to multiply the concentration of a bioconcentrating chemical in water to obtain the concentration of the chemical in fish.

Combustion efficiency. The ratio of the concentration of carbon dioxide in the effluent to the sum of the concentrations of carbon dioxide and carbon monoxide in the effluent.

Corrosivity. Characteristic of a material that has high or low pH.

Critical mass.   The minimum amount of nuclear mass to sustain a nuclear reaction.

Curie.   A unit of radiation equal to $3.7(10^{10})$ nuclear disintegration per second.

Dose–response assessment. The determination of the probability of getting an undesirable health effect from a given dose of a substance.

Fission chain reaction. A continuous nuclear reaction.

Fission reaction. The breakup of a radioactive nucleus.

Gamma radiation. The electromagnetic radiation released during the breakup of the nucleus of an atom.

Generator. The facility that produces the hazardous waste.

Hazard or risk identification. A determination of whether or not a particular chemical is causally linked to an observed health effect.

Hazardous waste. Any waste or combination of wastes that poses substantial danger to humans, plants, and animals.

Hazardous waste manifest. A multiple-form document that contains information about the waste, the generator, transporter, and the TSD facility.

High-level waste (HLW). Radioactive waste that contains radioactivity measured in curies per liter.

Ignitability. A characteristic that causes a material to catch fire.

Initiation. The mutation of DNA due to exposure to a chemical.

Intermediate-level waste (ILW). Radioactive waste that contains radioactivity measured in millicuries per liter.

Lowest-observed-effect level (LOEL). The lowest dose of a chemical that results in an adverse effect in a person.

Low-level waste (LLW). Radioactive waste that contains radioactivity measured in microcuries per liter.

Malignant tumor. A cancerous tumor.

Meltdown. A mechanical failure of a nuclear reactor.

Moderators. Control rods used to prevent runaway reactions in a nuclear reactor.

Molecular weight. The weight of a molecule expressed in terms of the atomic mass unit, amu.

No-observed-effect level (NOEL). The highest dose of a chemical that does not produce an adverse effect in a person.

Potency factor. The probability of risk produced by a lifetime exposure of an average daily dose of 1.0 mg/kg-day.

Principal organic hazardous constituents (POHC). One or more designated constituents in the effluent upon which the destruction and removal efficiency (DRE) is determined.

Promotion. The stage in cancer development where the cell no longer recognizes growth constraints but continues to grow forming tumors.

Rad. The absorption of 100 ergs of energy as a result of radioactive radiation per gram of substance.

Radioactivity. The spontaneous breakup of the nucleus of an atom.

Reactivity. The instability of a substance due to its speed of reaction.

Relative biological effect (RBE). The ratio of the energy absorbed from a gamma radiation to the energy absorbed from a given radiation that produces the same biological effect as the gamma radiation.

Risk assessment. The act of determining the probability of getting undesirable health effect from exposure to hazards.

Risk characterization. The overall integration of the results of risk assessment.

Risk exposure assessment. The determination of the circumstances of exposure to a given chemical.

Roentgen. The dose of gamma or x radiation that produces 1 esu of electricity in 1 cubic centimeter of air at standard temperature and pressure.

Roentgen equivalent man (rem). The product of the dose in rads and RBE.

Superfund. The other name for the CERCLA (Comprehensive Environmental Response, Compensation, and Liability Act).

Toxic substance. A poisonous substance.

Transporter. The person or facility that receives the hazardous waste from the generator and transports the waste to a TSD facility.

Transuranic waste. Radioactive waste that contains isotopes higher than uranium.

Trial burn. A treatability study that determines whether or not incineration is suitable for destroying a hazardous waste.

PROBLEMS

13-1. A sample of waste is analyzed using the TCLP toxicity test. It contains the following concentrations: lindane, 0.3 mg/L; toxaphene, 0.2 mg/L; and 2,4-D, 15 mg/L. If all the other toxics are absent, is the waste a hazardous waste?

13-2. Repeat Problem 13-1 using the EP toxicity test.

13-3. A waste from an industrial process contains arsenic 4.0 mg/L, mercury 0.02 mg/L, and barium 110 mg/L. Is the waste hazardous?

13-4. What is 2,4-D?

13-5. What is the difference between the EP and TCLP tests?

13-6. What are the masses of the plutonium and uranium atoms expressed in terms of grams and pounds?

13-7. Aside from its radioactivity, why is a waste containing plutonium dangerous?

13-8. On the controversy over the international inspection of North Korea's nuclear facility, the United States believes that the reason for North Korea's objection to the inspection is that one of its facilities is processing plutonium. Why the controversy over plutonium?

13-9. Why can France use plutonium in their reactors whereas underdeveloped countries cannot?

13-10. Global warming is an important issue in today's events, and scientists believe that CO_2 is an important contributing factor. Plutonium can replace fossil fuels for energy production. Are you in favor of using this material for the purpose of alleviating global warming? Why?

13-11. What is the half-life of plutonium?

13-12. What is done to avoid runaway reactions in nuclear reactors?

13-13. What is the mass of one atom of tritium?

13-14. U-238 decays releasing the α particle. What is the atom produced?

13-15. In the decay chain of U-238, what is the stable end product?

13-16. Describe the three levels of radioactive wastes.

13-17. What are the elements $^{40}X_{18}$ and $^{238}X_{92}$?

13-18. A laboratory is contaminated by I-131 at a concentration of 10 times the tolerance level. How long must the laboratory be isolated before it can be occupied again? Solve this problem if the contaminant is I-129.

13-19. Sr-90 decays to yttrium by the release of a β particle. What is the atomic mass number and the atomic number of yttrium?

13-20. The radiation from Co-60 is shielded using concrete. For a thickness of 110 cm, the percent transmission of radiation is 0.001%. The attenuation of radiation with distance has been found to be proportional to the radiation. Calculate the constant of proportionality.

13-21. The rad of radiation A is 10 rads and that of radiation B is 2 rads. If A has an RBE of 1 and B has an RBE of 7, how much more damaging is B compared to A?

13-22. From the result of Problem 13-20, what thickness of concrete will shield the radiation to reduce it by 99.6%?

13-23. Toluene is fed into an incinerator at the rate of 500 kg/h. If the rate at the stack of toluene is 60 g/h, what is the DRE?

13-24. If 95% of the toluene in Problem 13-23 is converted to CO_2 and H_2O and the rest to CO and H_2O, what is the combustion efficiency?

13-25. A trial burn at 1000°C for a hazardous waste mixture designates the following as the POHCs: chlorobenzene, toluene, and xylene. The waste feed and stack emissions are shown below. The stack gas rate of flow is 400 m^3/min at dry standard conditions. The permit requires the following to be met: DRE = 99.99%, HCl = no greater than the larger of 1.8 kg/h or 1.0% of HCl in stack gas, particulates = 180 mg/m^3 at dry standard conditions corrected to 7% O$_2$. Is the incinerator in compliance? The mole fraction of O$_2$ in the stack gas is 14%. Assuming that the stack gas temperature is 900°C, calculate the flue gas composition.

Compound	Inlet (kg/h)	Outlet (kg/h)
Chlorobenzene (C$_6$H$_5$Cl)	153	0.010
Toluene (C$_7$H$_8$)	432	0.037
Xylene (C$_8$H$_{10}$)	435	0.070
HCl	—	1.2
Particulates	—	3.65

13-26. Methylene chloride in an aqueous solution of 6000 mg/L is being incinerated at a rate of 50 L/min. Assuming that all chlorine is converted to HCl, what is the rate of production of HCl?

13-27. Assume that the acid in Problem 13-26 is to be completely scrubbed using lime. Calculate the amount of lime needed.

13-28. In equation (13-9), $P_c = P_m[(21 - X)/(21 - Y)]$, is $21 - Y$ the actual mole fraction of oxygen used in combustion? Why? How about $21 - X$?

13-29. The city of Annapolis, which has a population of 33,178, disinfects its water supply using chlorine. If 100 μg/L of chloroform is being produced (**a**) estimate the maximum lifetime risk of an Annapolitan associated with the drinking of this water. (**b**) What extra cancer deaths would be expected in Annapolis per year from drinking its water?

13-30. The maximum contaminant level in drinking water for trichloroethylene has been set at 5 μg/L. What is the lifetime risk of getting cancer for a person consuming this drinking water?

13-31. The drinking water well of a home near a landfill has been contaminated with trichloroethylene at a concentration of 11 ppb. The owner switches to a public water supply, which also contains chloroform, another suspected carcinogen. If the concentration of chloroform is 55 ppb, has the owner made the right decision?

13-32. An 80-kg man has been working 4 days a week for 30 years at 10 hours per day. He has been inhaling all these years at his workplace an average of 1.0 m^3/h of air contaminated with 0.15 mg/m^3 of trichloroethylene. If the absorption factor for the chemical is 90%, what is the probability that this man will have cancer in his lifetime?

13-33. A person smokes 15 cigarettes per day. One cigarette contains approximately 30 μg of benzo[a]pyrene. If the person weighs 80 kg and has been smoking for the past 30 years, what is his probability of getting cancer in his lifetime?

13-34. Repeat Problem 13-33 for a person weighing 130 lb.

13-35. The chronic water quality standard for DDT in the state of Maryland is 0.001 μg/L. Assume that a risk exposure assessment determines a correction factor to the dose–response data of

0.60. Assuming that the water quality standard is met and that, in fact, the concentration of DDT in the ambient water is 0.001 μg/L, what is the risk of a person dying of cancer due to DDT in Maryland?

13-36. One gram per day of dioxin is leaking into a 40,000-m^3 pond. Assuming complete mixing, what would be the steady state in the pond?

13-37. A factory discharges DDT into a stream, resulting in an instream concentration of 0.09 μg/L just below the outfall. At a distance 100 miles downstream, a town uses this stream as its source of water. Calculate the lifetime risk of getting cancer from drinking this water. Assume that the velocity of the stream is 1.0 mph.

13-38. Repeat Problem 13-36, assuming that dioxin is conservative.

13-39. What is a rem?

13-40. The cancer risk for exposure to cosmic radiation is believed to be one lethal cancer per 8000 person-rem. Cosmic radiation at mean sea level is approximately 40 mrem/yr. What is the lifetime probability of getting cancer for a person living near the ocean?

13-41. Repeat Problem 13-40 for persons living 1 mile above mean sea level. At this altitude the cosmic radiation is approximately 65 mrem/yr.

BIBLIOGRAPHY

DAVIS, M. L., and CORNWELL, D. A. (1991). *Introduction to Environmental Engineering.* McGraw-Hill, New York.

MASTERS, G. M. (1991). *Introduction to Environmental Engineering and Science.* Prentice Hall, Englewood Cliffs, N.J.

USEPA (1988). *Nucleus News.* American Nuclear Society.

USEPA (1989). IRIS database.

USEPA (1989). *Superfund Public Health Evaluation Manual.* Office of Emergency and Remedial Response, Washington, D.C.

VESILIND, P. A., J. J. PEIRCE, and R. F. WEINER (1988). *Environmental Engineering.* Butterworth, Boston.

WENTZ, C. A. (1989). *Hazardous Waste Management.* McGraw-Hill, New York.

CHAPTER 14

Noise Pollution and Control

Noise, an unwanted sound, is a phenomenon that has plagued us from the day we were born. Like wastewater, waste air, and solid waste, it is a pollutant. Noise slows down productivity and can be a cause of complaints from workers subjected to noisy environments. It annoys and hurts people both psychologically and physiologically. In this chapter we discuss noise pollution and its transmission, noise criteria, and control.

SOUND PRESSURE, POWER, AND INTENSITY

There are three important parameters that must be learned to characterize or, otherwise, study noise or sound: pressure, power, and intensity. These are discussed next.

Sound Pressure

As sound is propagated through the air, compression and rarefaction bands are formed. Compression and rarefaction are disturbances over existing conditions. Pressures are high in the compression band and are lower in the rarefaction band. What accounts for the propagation of sound is the differential of these pressures above and below existing atmospheric pressure. Without these differentials, no sound can be transmitted, as the ear will perceive no change.

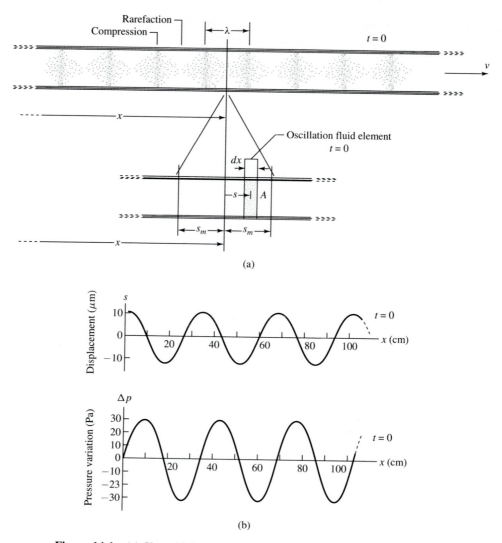

Figure 14-1   (a) Sinusoidal wave sent through a tube. (b) Displacement and pressure waves.

Figure 14-1a shows a sinusoidal air wave propagated through the tube, indicating the compression and rarefaction bands. Call the pressure differential Δp. From physics, the pressure in a traveling sound wave is

$$\Delta p = \Delta p_m \sin(kx - \omega t) \tag{14-1}$$

where Δp_m is the maximum pressure differential, k the angular wave number (the number of radians per unit of displacement x), and ω the angular frequency (the number of radians per unit time t). The plot of equation (14-1) is shown in Figure 14-1b.

In the figure, a differential strip of traveling air wave of thickness dx is isolated and depicted to be at a displacement s from the tip of x. This differential element oscillates to an amplitude of s_m that coincides with the middle of the rarefaction band (at minimum Δp). This means that, when s is at its maximum (s_m), the pressure differential is a minimum, indicating that if Δp uses the sine function, the displacement wave must use the cosine function. Hence let the displacement wave be

$$s = s_m \cos(kx - \omega t) \tag{14-2}$$

The plot of s is shown in Figure 14-1b.

From physics or fluid mechanics, the definition of the bulk modulus of elasticity of a gas is contained in the equation

$$\Delta p = -\beta \frac{\Delta V}{V} \tag{14-3}$$

where β is the bulk modulus of elasticity, V the volume of parcel of air, and ΔV the change in volume due to compression. V is equal to $A \Delta x$ and ΔV is equal to $\Delta(A \Delta x)$ equals $A\Delta(\Delta x) = A \Delta s$, where A is the cross-sectional area of the parcel of air through which sound propagates. Substituting in equation (14-3) and taking limits yields

$$\Delta p = \lim_{\Delta x \to 0} -\beta \frac{\Delta s}{\Delta x} = -\beta \frac{\partial s}{\partial x} \tag{14-4}$$

Partially differentiating equation (14-2) with respect to x and substituting the result in equation (14-4), we have

$$\Delta p = \beta k s_m \sin(kx - \omega t) = \Delta p_m \sin(kx - \omega t) \tag{14-5}$$

From equation (14-5),

$$\Delta p_m = (\beta k)s_m \tag{14-6}$$

From physics, the velocity of propagation of sound v is

$$v = \sqrt{\frac{\beta}{\rho}} \tag{14-7}$$

where ρ is the mass density of air. Therefore, the diferential pressure is

$$\Delta p_m = (v^2 \rho k)s_m = (v \rho \omega)s_m \tag{14-8}$$

Sound Power

The energy of an oscillating sound wave is composed of kinetic and potential energies. When a spring is compressed, it posseses potential energy, and, when it is released against an object placed on it, the energy will be transferred to the object as kinetic energy. Neglecting the effect of friction, the resulting kinetic energy will be equal to the original potential energy. The air propagating the sound wave or noise is like a spring. In its compressed state, it also possesses potential energy. When it expands toward the rarefaction state, the compressed state potential energy will be released not to another object, however, but to the air itself, as kinetic energy. The frictional energy of expansion may be neglected, and the kinetic energy is also therefore equal to the original potential energy. Hence, for any given parcel of air, the energy content of the traveling sound wave is equal to its potential energy plus kinetic energy equals twice its kinetic energy.

The kinetic energy dK of a differential mass of air dm propagating the sound wave is

$$dK = \frac{1}{2}dm v_s^2 \tag{14-9}$$

where v_s is the velocity of oscillation of the air mass—not the velocity of sound, which is v. v_s may be obtained by partially differentiating equation (14-2) with respect to t and the result substituted for the v_s in equation (14-9),

$$dK = \frac{1}{2}(\rho A\, dx)(-\omega s_m)^2 \sin^2(kx - \omega t) \tag{14-10}$$

where $\rho A\, dx$ is dm.

The total energy dE is twice dK. Also, power is the rate of change of E with respect to time, dE/dt. Hence, from equation (14-10) and the fact that dx/dt is v,

$$\frac{dE}{dt} = \frac{dK}{dt} = \rho A v \omega^2 s_m^2 \sin^2(kx - \omega t) \tag{14-11}$$

Taking the average over a whole number of wavelengths, the average power, $\overline{dE/dt}$, is

$$\overline{\frac{dE}{dt}} = P_w = \rho A v \omega^2 s_m^2\, \overline{\sin^2(kx - \omega t)} = \frac{1}{2}\rho A v \omega^2 s_m^2 \tag{14-12}$$

where $\overline{\sin^2(kx - \omega t)} = \frac{1}{2}$ for a whole number of wavelengths.

Sound Intensity

The intensity I of a sound wave is defined as the average rate at which power is transmitted per unit cross-sectional area in the direction of travel. Hence from equation (14-12), I is

$$I = \frac{\frac{1}{2}\rho A v \omega^2 s_m^2}{A} = \frac{1}{2}\rho v \omega^2 s_m^2 \tag{14-13}$$

From equation (14-8), equation (14-13) may also be written as

$$I = \frac{\Delta p_m^2}{2\rho v} \tag{14-14}$$

$\Delta p_m/\sqrt{2}$ is called the *root mean square* (rms) pressure of the sound wave, Δp_{rms}. Hence, in terms of the rms pressure, equation (14-14) becomes

$$I = \frac{\Delta p_{\text{rms}}^2}{\rho v} \tag{14-15}$$

Example 14-1

The maximum differential pressure Δp_m that the ear can tolerate in loud sounds is 28 N/m^2. What is the amplitude s_m for such a sound in air at a frequency of 1000 Hz? Assume β and ρ equal to 20.6 psi and 1.23 kg/m^3 for air, respectively. What are the rms pressure and intensity?

Solution

$$14.696 \text{ psi} = 101,330 \text{ N/m}^2$$

$$\beta = 20.6 \text{ psi} = \frac{101,330}{14.696}(20.6) = 142,038.51 \text{ N/m}^2$$

$$v = \sqrt{\frac{\beta}{\rho}} = \sqrt{\frac{142,038.71}{1.23}} = 339 \text{ m/s}$$

$$s_m = \frac{\Delta p_m}{v\rho\omega} = \frac{28}{339.82(1.23)(1000)(2\pi)} = 1.07(10^{-5}) \text{ m} \quad \textbf{Answer}$$

$$\Delta p_{\text{rms}} = \frac{\Delta p_m}{\sqrt{2}} = \frac{28}{\sqrt{2}} = 19.80 \text{ N/m}^2 \quad \textbf{Answer}$$

$$I = \frac{\Delta p_{\text{rms}}^2}{\rho v} = \frac{19.80^2}{1.23(339.82)} = 0.94 \frac{\text{N}^2/\text{m}^4}{(\text{kg/m}^3)(\text{m/s})}$$

$$= 0.94 \frac{\text{N-m}}{\text{m}^2\text{-s}} = 0.94 \frac{\text{J/s}}{\text{m}^2} = 0.94 \text{ W/m}^2 \quad \textbf{Answer}$$

Example 14-2

Repeat Example 14-1 for the pressure amplitude of $2.8(10^{-5})$ Pa of the faintest detectable sound at 1000 Hz.

Solution

$$\beta = 20.6 \text{ psi} = 142,038.51 \text{ N/m}^2$$

$$v = 339.82 \text{ m/s}$$

$$s_m = \frac{\Delta p_m}{v\rho\omega} = \frac{2.8(10^{-5})}{339.82(1.23)(1000)(2\pi)} = 1.07(10^{-11}) \text{ m} \quad \textbf{Answer}$$

$$\Delta p_{\text{rms}} = \frac{\Delta p_m}{\sqrt{2}} = \frac{2.8(10^{-5})}{\sqrt{2}} = 1.98(10^{-5}) \text{ N/m}^2 \quad \textbf{Answer}$$

$$I = \frac{\Delta p_{\text{rms}}^2}{\rho v} = \frac{[1.98(10^{-5})]^2}{1.23(339.82)} = 9.38(10^{-13}) \text{ W/m}^2 \quad \textbf{Answer}$$

MEASURES OF NOISE

Table 14-1 shows sound powers and pressures coming from various sources. Noise measurements are not normally reported, however, in terms of the units used in the table but in terms of decibels. The decibel is logarithmic.

TABLE 14-1 SOUND POWERS AND PRESSURES COMING
FROM VARIOUS SOURCES

Sound power (W)	Sound pressure (Pa)	Source
—	0.060	City street corner
—	0.020	Conversational speech
—	0.0060	Typical office
—	6.00	Near elevated train
—	2.00	Bottling plant
—	0.60	Full symphony
—	0.20	Inside automobile
—	0.0020	Living room
—	0.00060	Bedroom at night
100	—	75-piece orchestra
1	—	Auto on a highway
0.001	—	Voice shouting

In defining decibel, arbitrary reference values are chosen, depending on the type of decibel to be defined. For example, for power, the reference value is 10^{-12} W. For pressure and intensity, the references are, respectively, $2(10^{-5})$ Pa and 10^{-12} W/m². In terms of general definition, the decibel is

$$\text{decibel (dB)} = 10\log\frac{x}{y} \qquad (14\text{-}16)$$

where x is any one of pressure, power, and intensity and y relates to the respective reference values.

Using the respective decibels, the word *level* is now used. For example, for power, the decibel is the sound power *level* L_w, and for pressure and intensity, the decibels are sound pressure *level* L_p and sound intensity *level* L, respectively. Using equation (14-16), the respective definitions of the appropriate decibels follow.

$$L_w(\text{dB}) = 10\log\frac{P_w}{10^{-12}} \qquad \text{sound power level} \qquad (14\text{-}17)$$

$$L_I(\text{dB}) = 10\log\frac{I}{10^{-12}} \qquad \text{sound intensity level} \qquad (14\text{-}18)$$

L_w and L_I are the sound power and sound intensity levels, respectively.

In the definition of the sound pressure level, the rms value, Δp_{rms}, is used. Since the intensity I is proportional to the square of pressure, to be consistent with equation (14-18), the definition for sound pressure level, L_p, should have the rms pressure and the reference pressure being squared, respectively. Thus

$$L_p \ (\text{dB}) = 10\log\frac{\Delta p_{\text{rms}}^2}{[2(10^{-5})]^2} = 10\log\frac{\Delta p_{\text{rms}}^2}{4.0(10^{-10})} \qquad \text{sound pressure level} \qquad (14\text{-}19)$$

Relationships Among Intensity, Pressure, and Power Levels

Substituting equation (14-15) in equation (14-18) yields

$$L_I = 10 \log \frac{\Delta p_{rms}^2 / \rho v}{10^{-12}} \qquad (14\text{-}20)$$

At $25°C$, $\rho_{air} = 1.184$ kg/m^3, $\beta_{air} = 20.6$ psi $= 142{,}038.51$ N/m^2. Therefore, $v = \sqrt{\beta/\rho} = \sqrt{142{,}038.71/1.184} = 346.4$ m/s. Substituting the values of ρ and v in equation (14-20) gives

$$L_I = 10 \log \frac{\Delta p_{rms}^2}{4.1(10^{-10})} \qquad (14\text{-}21)$$

Although equation (14-21) holds at $25°C$, from equation (14-19), $L_I = L_p$, approximately.

By definition, $P_w = IA$, where A is the area of air that is receiving noise normal to the direction of propagation. Therefore,

$$L_w = 10 \log \frac{P_w}{10^{-12}} = 10 \log \frac{IA}{10^{-12}} = 10 \log \frac{I}{10^{-12}} + 10 \log A = L_I + 10 \log A \qquad (14\text{-}22)$$

Example 14-3

The sound power from a voice shouting is 0.001 W. What is the sound power level? What are the sound intensity, the sound intensity level, the sound pressure, and the sound pressure level at a distance 6 m from the source?

Solution

$$L_w = 10 \log \frac{P_w}{10^{-12}} = 10 \log \frac{0.001}{10^{-12}} = 90 \text{ dB} \quad \textbf{Answer}$$

Assume that the sound radiates from the source in all directions. Hence the area of propagation at 6 m distance is the surface area of a sphere $= 4\pi r^2 = 4\pi(6^2) = 452.39$ m^2. Then

$$I = \frac{0.001}{452.39} = 2.2(10^{-6}) \text{ W/m}^2 \quad \textbf{Answer}$$

$$L_I = 10 \log \frac{2.2(10^{-6})}{10^{-12}} = 63.42 \text{ dB} \quad \textbf{Answer}$$

$$L_I \approx L_p \qquad L_p = 63.42 \text{ dB} \quad \textbf{Answer}$$

$$L_p = 10 \log \frac{\Delta p_{rms}^2}{4.0(10^{-10})} = 63.42$$

$$\Delta p_{rms} = 0.02966 \text{ N/m}^2 \quad \textbf{Answer}$$

HOW DO WE HEAR

Figure 14-2 shows the anatomy of the ear. As shown, the ear is composed of the outer, middle, and inner ears. The pinna, auditory canal, and eardrum or tympanic membrane

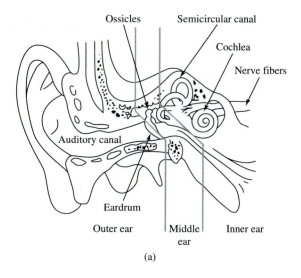

(a)

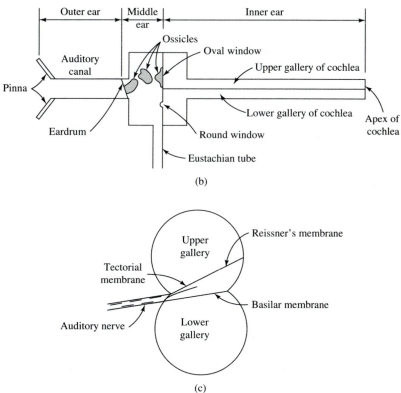

(b)

(c)

Figure 14-2   Ear anatomy. (P. F. Cunniff (1977). *Environmental Noise Pollution*. Reprinted by permission of John Wiley and Sons, Inc.)

are parts of the *outer ear*. Sound transmitted through the air or other medium reaches the *pinna* and continues through the *auditory canal* and strikes the *tympanic membrane*. The membrane is thus set into vibration. The *middle ear*, an air-filled chamber, is composed of three tiny bones called *ossicles*. The ossicles are the *malleous* (hammer), *incus* (anvil), and *stapes* (stirrup). The vibration of the tympanic membrane is transmitted to the malleous, incus, and stapes in the middle ear.

The semicircular canals and the cochlea are parts of the *inner ear*, which contains a fluid called the *perilymph*. The *semicircular canal* provides the sense of balance, and the *cochlea* contains the organ of *Corti*, which is the sense organ for hearing. At the entrance to the inner ear is an element called the *oval window*. Figure 14-2b shows the cochlea being straightened out, and the bottom shows its cross section. The cross section shows a membrane called *Reissner's membrane*, the *basilar membrane*, and the *tectoral membrane*. The cochlea is divided by the basilar membrane into the *upper* and *lower galleries*. Between the tectoral and basilar membranes are some 23,500 minute hair cells.

The vibrations from the stapes are transmitted to the inner ear through the oval window, generating waves in the perilymph fluid. The waves proceed from the oval window, along the upper gallery, around the cochleal apex, and through the lower gallery, terminating at the *round window*. The waves cause displacement of the basilar membrane, which, in turn, sways the tiny hair cells to and fro, producing a straining effect on the cells. The straining produces a piezoelectric effect, and the electricity being produced is transmitted to the brain through the auditory nerve, producing the perception of hearing. The corresponding electric potential produced is called the *cochleal potential*.

As long as sound is not transmitted as a sudden impact, the ear has the capability to protect itself from intense noise by a reflex action called the *acoustic reflex*. This reflex is the stiffening of the tympanic membrane by the action of the two small muscles in the inner ear called the *tensor tympani* and the *stapedius*. The ear can detect sound pressures as low as $2(10^{-5})$ Pa and, before experiencing pain, to as high as 200 Pa.

LOUDNESS

The sound levels of power, intensity, and pressure are physical effects that the ear receives; it is not, however, what the person perceives. What is being perceived or heard depends on how the brain interprets or perceives the sound levels the ear receives. *Loudness* is the brain's perception of the magnitude of the sound levels. If the levels are high, the brain perceives the sound as loud, and if the levels are low, the brain perceives them as not loud. The higher levels of sound produce more vibration in the ear mechanisms, and vice versa; the brain's interpretation of loudness depends on the magnitude of this vibration.

Since what is actually heard is not the decibel level but the interpretation of it, other units are used for loudness: the phon and the sone. The normal, everyday sound being heard is a conglomeration of several frequencies. The audible frequency range is from 16 to 20,000 Hz. Sound below 16 Hz is called *infrasound*, while sound above 20,000 Hz is called *ultrasound*. *Phon* is the loudness of a tone that is numerically equal to the corresponding sound pressure level when heard at 1000 Hz. Thus a sound pressure level at 40 dB is 40 phons when heard at the 1000-Hz frequency. The decibel level for a given loudness may be different at other frequencies. For example, a loudness of 40 phons may have 50, 60, 80, or any other decibel level when heard at other frequencies. Figure 14-3 shows a chart of equal phon or loudness contour.

The phon unit is not additive. For example, assume that there are two sources, each heard separately at 40 phons. If the two sound sources are heard simultaneously, the combined loudness will not be equal to 80 phons. Acoustic workers therefore devised a unit of measurement such that the loudness of two or more sources can simply be added. Take the two 40-phon sources and sound them simultaneously. Stop the sounding and take only one of the two and adjust its tone until it is heard equal to the two sources sounded simultaneously. Using the proper unit of measurement, the resulting value for the adjusted source can be made equal to twice that of either of the two 40-phon sources. Hence if the first 40-phon source is one and the second is also one, the sum is equal to two. Let this be the value of the adjusted 40-phon source. Hence 1 sone for the first 40-phon source plus 1 sone for the second 40-phon source equals 2 sones for the one adjusted to equal the value of the two simultaneously operated 40-phon sources. This new unit, the *sone*, is now defined as the loudness of 40 phons made arbitrarily equal to 1. Using S for the number of sones and P for the number of phons, this definition in equation form is

$$S = 2^{(P-40)/10} \tag{14-23}$$

From this formula, when $P = 40$ phons, $S = 1$ sone, and when $P = 50$ phons, $S = 2$ sones. Hence 40 phons $+$ 40 phons $= 50$ phons, not 80 phons. In terms of sones: 1 sone $+$ 1 sone $= 2$ sones, and the process is additive. The plot of equation (14-23) is also shown in Figure 14-3.

Example 14-4

Calculate the loudness in sones for a simultaneous exposure of the following tones: 30 dB at 350 Hz, 60 dB at 1800 Hz, 70 dB at 3000 Hz, and 45 dB at 4000 Hz. Also calculate the overall number of phons.

Solution   Use the equal-loudness contour of Figure 14-3 to determine the phon levels.

Frequency (Hz)	Sound pressure level (dB)	Loudness level (phons)	Loudness level[a] (sones)
350	30	30	0.5
1800	60	62	4.59
3000	70	79	14.93
4000	45	52	2.3
		Σ	22.32 sones **Answer**

[a]Calculated using $S = 2^{(P-40)/10}$.

$$22.32 = 2^{(P-40)/10} \qquad P = 84.8 \text{ phons} \quad \textbf{Answer}$$

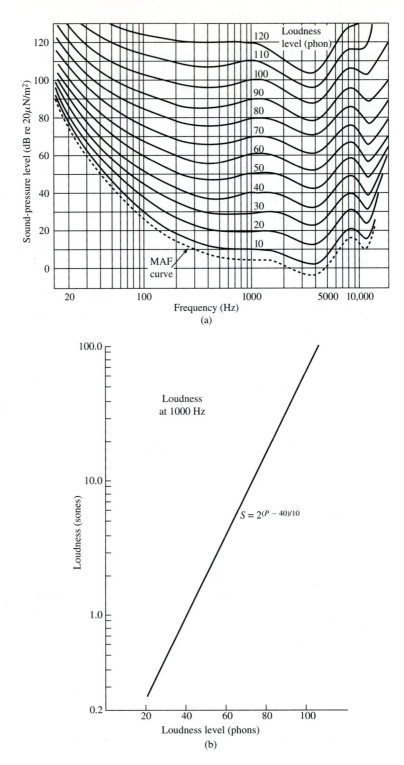

Figure 14-3 (a) Equal-loudness contours. (b) Phon and sone relationship.

MEASUREMENT OF NOISE

Noise is measured by means of a *sound level meter*. The meter is positioned in a desired location with no obstruction from the sound source and the reading is taken. Figure 14-4a is a drawing of a sound level meter attached to a calibrator. A *calibrator* is one that produces a reference sound and when attached to the microphone calibrates the meter. The vertical dashed line in the figure locates the microphone and the point where the calibrator is attached to the meter.

The basic parts of most sound level meters include a microphone, amplifiers, weighting networks, and a display reading in decibels. Figure 14-4b shows the schematic diagram of the basic parts of a sound level meter. Sound pressure coming from a source reaches the microphone, where a pressure transducer such as a diaphragm mimics the pressure pattern of the source and converts the signal to a small current of electricity. The electrical signal thus produced is then amplified in a preamplifier.

Sound pressure that the meter receives is not the same as what the ear would perceive. For this reason an electronic circuitry called a *weighting network* is built into the meter so as to produce a readout that closely resembles a human response. A series of three internationally accepted weighting scales has been adopted. These are the A, B, and C weighting networks. The *A network* approximates human response to low-intensity sounds, the *B network* approximates the response to medium-intensity sounds, and the *C network* approximates the human response to high-intensity sounds. Figure 14-4c shows the plot for the three networks.

The network circuitry electronically subtracts the actual sound pressure level of a particular frequency that reaches the microphone according to in which network the meter is set. For example, using the A network, at a sound frequency of 250 Hz, the actual pressure level reaching the meter is subtracted by 9. The difference is what is shown in the readout. Depending on whether the network used is A, B, or C, readouts are in dBA, dBB, or dBC, respectively. As shown in the figure, at low frequencies, the A network subtracts the incoming signal severely; the C network, less; and the B network, in between. Since the readings are not actually the true sound pressure levels, readouts from the A, B, and C networks are called *sound levels*, not sound pressure levels. This is the reason that the meters are called *sound level meters* and not *sound pressure level meters*.

The A network is more commonly used. Another network, the *D-weighting network*, has also been recommended as closely resembling the human response to noise at airports.

Frequency Band Analysis

Theoretically, the sound pressure level of each frequency can be measured to determine the distribution of levels versus the frequency of a particular sound. This distribution is called the *sound spectrum*. However, this undertaking is hardly warranted and impractical. Therefore, a convention has been set to divide the frequency spectrum into bands. One of these is the *octave band*. A band of one octave is a range of frequencies whose ratio

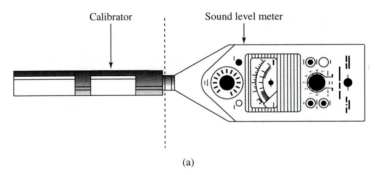

(a)

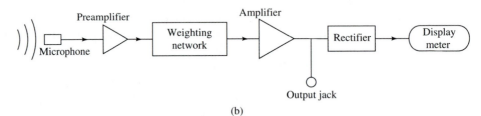

(b)

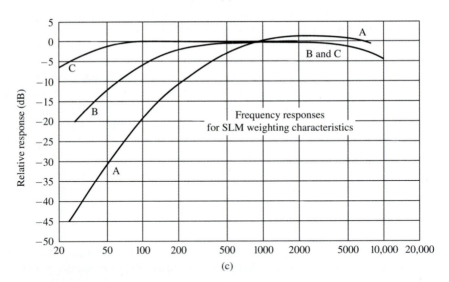

(c)

Figure 14-4 (a) Sound level meter. (b) Schematic drawing of basic parts of sound level meter. (c) Plot of weighting networks. (P. F. Cunniff (1977). *Environmental Noise Pollution.* Reprinted by permission. John Wiley and Sons, Inc.)

of the upper to the lower frequency in the band is 2:1. Thus, calling γ_u and γ_ℓ the upper and lower frequencies, respectively,

$$\gamma_u = 2\gamma_\ell \tag{14-24}$$

The procedure of measuring pressure levels using band analysis is to set the meter at the geometric mean frequency and take the reading. The geometric mean is the average logarithm of the upper and lower frequencies. Letting f_c represent the geometric mean yields

$$\log \gamma_c = \frac{1}{2}(\log \gamma_\ell + \log \gamma_u) = \log \sqrt{\gamma_\ell \gamma_u} \qquad \gamma_c = \sqrt{\gamma_\ell \gamma_u} \qquad (14\text{-}25)$$

To obtain a more accurate characterization of the sound spectrum, the octave band may be subdivided into smaller bands. Hence the band may be divided in two, producing the *1/2 octave band analysis*, or the band may be divided in three, producing the *1/3 band analysis*. In measurement, the geometric mean of these smaller bands is also calculated to read against sound pressure levels or sound levels, as the case may be. Since the lower frequency is always a constant percentage of the upper frequency, the octave, 1/2 octave, 1/3 octave analyses, and so on, are called *constant-percentage frequency analyses*. To differentiate from the 1/2 and 1/3 octave bands, the regular octave band analysis is designated as the *1/1 octave band analysis*.

The sound spectrum may also be analyzed using a constant increment of frequency. This type of analysis is called *constant-frequency analysis*. For example, for a constant-frequency analysis at an interval of 10 Hz, if the sound spectrum ranges from 20 to 10,000 Hz, there will be $(10{,}000 - 20)/10 = 998$ data points to be taken.

The standard set of octave bands used in instrumentation is tabulated in Table 14-2.

TABLE 14-2 STANDARD OCTAVE BANDS

Octave band (Hz)	Center frequency (Hz)
22–44	31.5
44–88	63
88–177	125
177–355	250
355–710	500
710–1,420	1,000
1,420–2,840	2,000
2,840–5,680	4,000
5,680–11,360	8,000
11,360–22,720	16,000

Example 14-5

The upper and lower frequencies of an octave band are 11,360 and 22,720 Hz, respectively. (a) What is the center band frequency? If the lower frequency is 11,360 Hz, what are the upper frequencies of the corresponding (b) 1/2 and (c) 1/3 octaves?

Solution

(a) $\qquad\qquad \gamma_c = \sqrt{11{,}360(22{,}720)} = 16{,}065 \text{ Hz}$ **Answer**

(b)   Let γ_ℓ = lower frequency

γ_{u1} = first upper frequency

γ_u = second and last upper frequencies

x = factor

$$\gamma_{u1} = x\gamma_\ell \qquad \gamma_u = x\gamma_{u1} = x(x\gamma_\ell) = x^2\gamma_\ell = 2\gamma_\ell \qquad x = \sqrt{2}$$

$$\gamma_{u1} = \sqrt{2}(11{,}360) = 16{,}065 \text{ Hz} \quad \textbf{Answer}$$

$$\gamma_u = 2\gamma_\ell = 2(11{,}360) = 22{,}720 \text{ Hz} \quad \textbf{Answer}$$

(c) Let γ_ℓ = lower frequency

γ_{u1} = first upper frequency

γ_{u2} = second upper frequency

γ_u = third and last upper frequencies

x = factor

$$\gamma_{u1} = x\gamma_\ell \qquad \gamma_{u2} = x\gamma_{u1} = x(x\gamma_\ell) = x^2\gamma_\ell$$

$$\gamma_{u3} = x\gamma_{u2} = x(x^2\gamma_\ell) = x^3\gamma_\ell \qquad x = \sqrt[3]{2}$$

$$\gamma_{u1} = \sqrt[3]{2^2}(11{,}360) = 14{,}313 \text{ Hz} \quad \textbf{Answer}$$

$$\gamma_{u2} = \left(\sqrt[3]{2}\right)^2 (11{,}360) = 18{,}033 \text{ Hz} \quad \textbf{Answer}$$

$$\gamma_u = 2\gamma_\ell = 2(11{,}360) = 22{,}720 \text{ Hz} \quad \textbf{Answer}$$

Decibel Addition

Decibel readings of a sound spectrum need to be added and converted to a single reading. This single reading constitutes the overall characterization of the sound or noise. Let there be the following sound pressure level readings: $L_{p1}, L_{p2}, \ldots,$ and L_{pn}. From the definition of L_p, the following expression for the rms differential pressure $\Delta p_{\text{rms}1}$

corresponding to L_{p1} can be obtained:

$$\Delta p_{\text{rms}1} = \sqrt{4.0(10^{-10})10^{L_{p1}/10}} \tag{14-26}$$

The corresponding equations for $L_{p2}, \ldots,$ and L_{pn} are, respectively,

$$\Delta p_{\text{rms}2} = \sqrt{4.0(10^{-10})10^{L_{p2}/10}} \tag{14-27}$$

$$\vdots$$

$$\Delta p_{\text{rms}n} = \sqrt{4.0(10^{-10})10^{L_{pn}/10}} \tag{14-28}$$

Adding the equations above, the total rms pressure, $\Delta p_{\text{rms}T}$, is

$$\Delta p_{\text{rms}T} = \sqrt{4.0(10^{-10})} \sum_{i=1}^{i=n} \sqrt{10^{L_{pi}/10}} \tag{14-29}$$

Hence the total decibel level, L_{pT} is

$$L_{pT} = 10\log\frac{\Delta p^2_{\text{rms}T}}{4.0(10^{-10})} = 10\log\frac{\left[\sqrt{4.0(10^{-10})}\sum_{i=1}^{i=n}\sqrt{10^{L_{pi}/10}}\right]^2}{4.0(10^{-10})}$$

$$\tag{14-30}$$

$$= 10\log\left(\sum_{i=1}^{i=n}\sqrt{10^{L_{pi}/10}}\right)^2$$

By analogy, the total decibel level corresponding to L_w and L_I, respectively, are

$$L_{wT} = 10\log\sum_{i=1}^{i=n} 10^{L_{wi}/10} \tag{14-31}$$

$$L_{IT} = 10\log\sum_{i=1}^{i=n} 10^{L_{Ii}/10} \tag{14-32}$$

Note that the arguments of the logarithms are not squared and the square root of the L values are not taken since, unlike Δp_{rms}, P_w and I are not squared in the definition of decibels.

Example 14-6

The noise spectrum of a cutter equipment at a 5.0-ft distance is analyzed using the 1/1 octave band analysis, producing the results below. (a) What are the sound pressure level L_{pT} and the sound level L_{pAT} generated by the equipment? (b) What is the pressure generated at the given distance? (c) What are the corresponding power and intensity levels?

Center frequency (Hz)	L_p (dB)	A correction[a]	dBA
63	62	−26	36
125	71	−16	55
250	72	−9	63
500	77	−3	74
1000	92	−0	92
2000	79	+1	80
4000	77	+1	78
8000	63	−1	62

[a]From Figure 14-6.

Solution

(a) $L_{pT} = 10 \log \left(\sum\limits_{i=1}^{i=n} \sqrt{10^{L_{pi}/10}} \right)^2$

$= 10 \log \left(\sqrt{10^{62/10}} + \sqrt{10^{71/10}} + \sqrt{10^{72/10}} + \sqrt{10^{77/10}} + \sqrt{10^{92/10}} \right.$

$\left. + \sqrt{10^{79/10}} + \sqrt{10^{77/10}} + \sqrt{10^{63/10}} \right)^2 = 97.28 \text{ dB} \quad \textbf{Answer}$

$L_{pAT} = 10 \log \left(\sqrt{10^{36/10}} + \sqrt{10^{55/10}} + \sqrt{10^{63/10}} + \sqrt{10^{74/10}} + \sqrt{10^{92/10}} \right.$

$\left. + \sqrt{10^{80/10}} + \sqrt{10^{78/10}} + \sqrt{10^{62/10}} \right)^2 = 96.40 \text{ dBA} \quad \textbf{Answer}$

(b) $\Delta p_{\text{rms}} = \sqrt{4.0(10^{-10})10^{L_p/10}} = \sqrt{4.0(10^{-10})10^{97.28/10}}$

$= 1.46 \text{ Pa} \quad \textbf{Answer}$

(c) $L_I = 10 \log \dfrac{\Delta p_{\text{rms}}^2}{4.0(10^{-10})} = 10 \log \dfrac{1.46^2}{4.0(10^{-10})} = 97.27 \text{ dB} \approx L_p \quad \textbf{Answer}$

Assume that the sound radiates in all directions.

$L_w = L_I + 10 \log A = 97.27 + 10 \log 4\pi \left(\dfrac{5.0}{3.281} \right)^2 = 111.92 \text{ dB} \quad \textbf{Answer}$

Miscellaneous Measures of Noise

As a result of the diversity of practice, several units of measure of noise have evolved that are applicable only to specific areas of concern. Thus community noise measurements have evolved differently from industrial noise measurements. Airport noise measurements have also evolved on their own. Units of measurements coming from one area can have no relationship with those coming from other areas.

Also, in discussions so far, sound levels have been expressed as functions of frequencies. This is not necessarily the case, however. Instrument readings may be calibrated to indicate directly and automatically the sum of the decibels of the various frequencies that make up the noise. Thus a 120-dBA noise from a jet airplane may not be a reading for a particular frequency (such as 63 or 125 Hz) but for the whole range of frequencies composing the noise.

Community noise. In the area of community noise, the following measures have been used to assess noise problems: statistical measure L_N, equivalent sound level L_{eq}, day–night average sound level L_{dn}, and noise pollution level L_{NP}. The *statistical measure* is a probability distribution analysis where a sound level equaled or exceeded so much percent of the time is calculated, such as the L_{10} and L_{20}. The values of N, in these cases, are equal to 10 and 20 and are the sound levels exceeded 10% and 20% of the time, respectively. The *equivalent sound level L_{eq}* is an A-weighted average measure over a given time period. The *day–night average sound level L_{dn}* is an L_{eq} A-weighted sound level during a 24-h period ($L_{eq(24)}$) with 10 dBA added to the $L_{eq(1)}$'s during the time from 10 P.M. to 7 A.M. Another noise measure related to L_{dn} is the *community noise equivalent level* (CNEL). In CNEL, 5 dBA is added during the hours of 7 P.M. to 10 P.M. and 10 dBA is added during the hours of 10 P.M. to 7 A.M. over a 24-h period. Finally, the *noise pollution level L_{NP}* is the L_{eq} value that has a probability of exceedance of 0.50%.

The formulas for the measures of noise above are written below.

$$L_{eq} = 10 \log \left(\sum_{i=1}^{n} f_i \sqrt{10^{L_{pAi}/10}} \right)^2 \tag{14-33}$$

where L_{pAi} is the dBA level and f_i is the fraction of time that L_{pAi} is in progress.

$$L_{dn} = 10 \log \left(\sum_{i=7 \text{ AM}}^{i=10 \text{ PM}} f_i \sqrt{10^{L_{eq(1@i)}/10}} + \sum_{i=10 \text{ PM}}^{i=7 \text{ AM}} f_i \sqrt{10^{(L_{eq(1@i)}+10)/10}} \right)^2 \tag{14-34}$$

$$L_{NP} = L_{eq(\infty)} + 2.56\sigma \, (NP\text{dB}) \tag{14-35}$$

where 2.56 is the critical ratio corresponding to 0.50% exceedance (obtained from statistical tables). *Critical ratio* is the ratio of a particular value to the standard deviation. $L_{eq(\infty)}$ is the mean of the L_{eq}'s of a sufficiently long A-weighted sample of noise in dBA. Note that unit of L_{NP} used is NPdB. From statistics, the standard deviation σ is defined as

$$\sigma = \sqrt{\frac{\sum [L_{eqi} - L_{eq(\infty)}]^2}{N-1}} \tag{14-36}$$

where N is the number of L_{eqi} sample values.

Example 14-7

Traffic noise data are shown in the table below. Compute (a) L_{90} and (b) L_{eq}.

Time (s)	dBA
10	71
20	75
30	70
40	78
50	80
60	84
70	76
80	74
90	75
100	74

Solution (a)

Time (s)	dBA	Serial no. (m)	dBA	Percent of time equaled or exceeded $\left(\frac{m}{N+1}\right)$ (100)
10	71	1	84	9.1
20	75	2	80	18.2
30	70	3	78	27.3
40	78	4	76	36.4
50	80	5	75	45.5
60	84	6	75	54.5
70	76	7	74	63.6
80	74	8	74	72.7
90	75	9	71	81.8
100	74	10	70	90.9

$$
\begin{array}{ccc}
71 & \Rightarrow & 81.8 \\
x & \Rightarrow & 90 \\
70 & \Rightarrow & 90.9
\end{array}
\qquad
\frac{x - 71}{70 - 71} = \frac{90 - 81.8}{90.9 - 81.8}
$$

$$
x = L_{90} = 71.1 \text{dBA} \quad \textbf{Answer}
$$

(b) $\quad L_{eq} = 10 \log \left(\sum_{i=1}^{n} f_i \sqrt{10^{L_{pAi}/10}} \right)^2 \qquad f = \frac{10}{100}$

$$
= 10 \log \left[\frac{10}{100} \sqrt{(10^{71/10})} + \frac{10}{100} \sqrt{(10^{75/10})} + \frac{10}{100} \sqrt{(10^{70/10})} \right.
$$

$$
+ \frac{10}{100} \sqrt{(10^{78/10})} + \frac{10}{100} \sqrt{(10^{80/10})} + \frac{10}{100} \sqrt{(10^{84/10})}
$$

$$+ \frac{10}{100}\sqrt{(10^{76/10})} + \frac{10}{100}\sqrt{(10^{74/10})} + \frac{10}{100}\sqrt{(10^{75/10})}$$

$$+ \frac{10}{100}\sqrt{(10^{74/10})} \Bigg]^2 = 76.67 \quad \textbf{Answer}$$

Example 14-8

The hourly L_{eq}'s ($L_{eq(1)}$'s) in a certain neighborhood are shown in the table below. Compute (a) L_{dn} and (b) L_{NP}, assuming that the 24 h of record is sufficient to define $L_{eq(\infty)}$.

Time	$L_{eq(1)}$
24:00–1:00	41
1:00–2:00	31
2:00–3:00	32
3:00–4:00	30
4:00–5:00	30
5:00–6:00	30
6:00–7:00	38
7:00–8:00	51
8:00–9:00	62
9:00–10:00	70
10:00–11:00	60
11:00–12:00	60
12:00–13:00	59
13:00–14:00	61
14:00–15:00	62
15:00–16:00	62
16:00–17:00	60
17:00–18:00	70
18:00–19:00	59
19:00–20:00	61
20:00–21:00	52
21:00–22:00	51
22:00–23:00	40
23:00–24:00	40

Solution

(a) L_{dn}

$$= 10\log\left(\sum_{i=7\text{ AM}}^{i=10\text{ PM}} f_i\sqrt{10^{L_{eq(1@i)}/10}} + \sum_{i=10\text{ PM}}^{i=7\text{ AM}} f_i\sqrt{10^{(L_{eq(1@i)}+10)/10}}\right)^2$$

$$f_i = \frac{1}{24}$$

Time	$L_{eq(1)}$	Adjusted $L_{eq(1)}$	$10^{\text{adjusted } L_{eq(1)}/10}$	$\sqrt{10^{\text{adjusted } L_{eq(1)}/10}}$
24:00–1:00	41	51	125,892.54	354.81
1:00–2:00	31	41	12,589.25	112.20
2:00–3:00	32	42	15,848.93	125.89
3:00–4:00	30	40	10,000	100
4:00–5:00	30	40	10,000	100
5:00–6:00	30	40	10,000	100
6:00–7:00	38	48	63,095.73	251.19
7:00–8:00	51	51	125,892.54	354.81
8:00–9:00	62	62	1,584,893.20	1,258.93
9:00–10:00	70	70	10,000,000	3,162.28
10:00–11:00	60	60	1,000,000	1,000
11:00–12:00	60	60	1,000,000	1,000
12:00–13:00	59	59	794,328.24	891.25
13:00–14:00	61	61	1,258,925.40	1,122.02
14:00–15:00	62	62	1,584,893.20	1,258.93
15:00–16:00	62	62	1,584,893.20	1,258.93
16:00–17:00	60	60	1,000,000	1,000
17:00–18:00	70	70	10,000,000	3,162.28
18:00–19:00	59	59	794,328.24	891.25
19:00–20:00	61	61	1,258,925.40	1,122.02
20:00–21:00	52	52	158,489.32	398.11
21:00–22:00	51	51	125,892.54	354.81
22:00–23:00	40	50	100,000	316.23
23:00–24:00	40	50	100,000	316.23

$$\sum \quad 20,012.17$$

$$L_{dn} = 10 \log \left[\frac{1}{24}(20,012.17) \right]^2 = 58.42 \text{ dBA} \quad \textbf{Answer}$$

(b)
$$L_{NP} = L_{eq(\infty)} + 2.56\sigma \,(\text{NPdB})$$

$$\sigma = \sqrt{\frac{\sum \left[L_{eqi} - L_{eq(\infty)} \right]^2}{N-1}}$$

Time	$L_{eq(i)} - L_{eq(\infty)}$ [a]	$[L_{eq(i)} - L_{eq(\infty)}]^2$	$L_{eq(i)}$
24:00–1:00	41	−9.5	90.25
1:00–2:00	31	−19.5	380.25
2:00–3:00	32	−18.5	342.25
3:00–4:00	30	−20.5	420.25
4:00–5:00	30	−20.5	420.25
5:00–6:00	30	−20.5	420.25

(Continued)

Time	$L_{eq(i)} - L_{eq(\infty)}$[a]	$[L_{eq(i)} - L_{eq(\infty)}]^2$	$L_{eq(i)}$
6:00–7:00	38	−12.5	156.25
7:00–8:00	51	0.5	0.25
8:00–9:00	62	11.5	132.25
9:00–10:00	70	19.5	380.25
10:00–11:00	60	9.5	90.25
11:00–12:00	60	9.5	90.25
12:00–13:00	59	8.5	72.25
13:00–14:00	61	10.5	110.25
14:00–15:00	62	11.5	132.25
15:00–16:00	62	11.5	132.25
16:00–17:00	60	9.5	90.25
17:00–18:00	70	19.5	380.25
18:00–19:00	59	8.5	72.25
19:00–20:00	61	10.5	110.25
20:00–21:00	52	1.5	2.25
21:00–22:00	51	0.5	0.25
22:00–23:00	40	−10.5	110.25
23:00–24:00	40	−10.5	110.25
$\sum$	1,212		4,246

[a] $L_{eq(\infty)} = \frac{1212}{24} = 50.5$ dBA.

$$\sigma = \sqrt{\frac{\sum[L_{eqi} - L_{eq(\infty)}]^2}{N-1}} = \sqrt{\frac{4246}{24-1}} = 13.59$$

$$L_{NP} = 50.5 + 2.56(13.59) = 82.29 (NP\text{dB}) \quad \textbf{Answer}$$

Airport noise. The following measures have been used to evaluate airport noise: perceived noise level, composite noise rating, and noise exposure forecast. *Perceived noise level* (PNL) is a calculated single value based on known levels of people's annoyance from jet flyover noise. The *composite noise rating* (CNR) is a contour map of equal PNLs in and around the vicinity of an airport. The *noise exposure forecast* (NEF) is a contour map similar to CNR but using different definitions, as will be shown later.

For the PNL mode of measure, the chart shown in Figure 14-5a was developed. This is a chart of equal noise that people perceived in jet flyover noises measured in noys. Note that if a given sound pressure level from the ordinate of Figure 14-5 at the 1000-Hz frequency is substituted for the P in equation (14-23), a value is obtained that is equal to the number of noys. Hence at the 1000-Hz frequency level, noys and sones are numerically equal. In addition, the relationship on the right-hand side of Figure 14-5a has the same form as the relationship between phon and sone [equation (14-23)]. Hence using PNdB instead of P, the relationship between PNdB and N (the number of noys) is

$$N = 2^{(\text{PNdB}-40)/10} \tag{14-37}$$

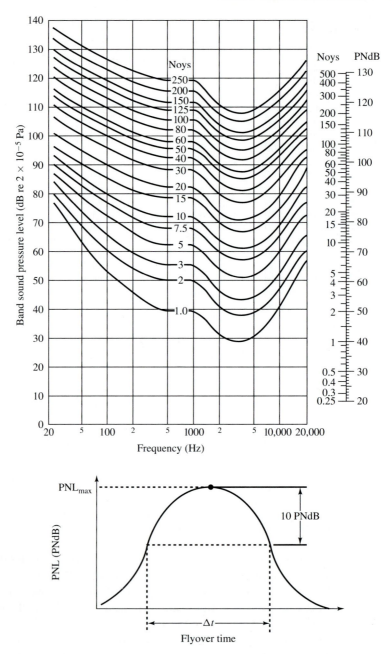

Figure 14-5   (a) Equal noise contours. (b) Measuring the 10-PNdB Δt.

By comparing equations (14-23) and (14-37), this analogy may be made: Phon is to sone as PNdB (perceived noise decibel) is to noy. Note that because of this relationship, noy and sone are almost numerically equal to each other. PNdB is the unit for PNL.

To calculate PNL, octave band data are, first, converted to noys N, using Figure 14-5. After conversion, the effective noy value N_t is calculated by one of the formulas below. N_t is then converted to PNdB by the use of the relationship shown on the right-hand side of Figure 14-5a or by equation (14-37). Note that in this conversion, N of equation (14-37) is the N_t of the following equations:

$$N_t = 0.3 \sum N_i + 0.7 N_{max} \quad \text{(for 1/1 octave band)} \tag{14-38}$$

$$N_t = 0.15 \sum N_i + 0.85 N_{max} \quad \text{(for 1/3 octave band)} \tag{14-39}$$

N_{max} is the maximum noy obtained in the conversion of the octave band data to noy.

Example 14-9

The following data are associated with a jet aircraft at a given location on the ground. Calculate the PNL at the point.

1/1 Octave band (Hz)	L_p (dB)	N_i (noys)
63	91	20
125	103	50
250	105	80
500	100	65
1000	90	33
2000	88	48
4000	81	34
8000	70	14
	$\sum$	344

Solution   The third column has been obtained using Figure 14-5.

$$N_t = 0.3 \sum N_i + 0.7 N_{max} \quad \text{(for 1/1 octave band)}$$

$$= 0.3(344) + 0.7(80) = 159.2 \text{ noys}$$

Note: 80 is the maximum noy in the third column.

From Figure 14-5, PNL = 114 PNdB.   **Answer**

As mentioned before, CNR is simply a contour of equal noise. From CNR, a new method, the noise exposure forecast (NEF), evolved. NEF is also a contour of equal noise but using a different set of definitions. The first is the single event *effective perceived noise level* (EPNL). The EPNL for type of aircraft i flying over path j as perceived at a point on the ground is defined as

$$\text{EPNL}_{ij} = \text{PNL}_{ij\max} + D_{ij} + F \qquad \text{(PNdB)} \tag{14-40}$$

where $\text{PNL}_{\max}$ is the maximum PNL during the flyover of the aircraft; $D = 10\log(\Delta t/20)$, where Δt is the time interval in seconds during which the noise level is within 10 PNdB of $\text{PNL}_{\max}$; and $F \approx 3$ dB. A sketch for finding Δt is shown in Figure 14-5b.

Now, as perceived at a point on the ground, the noise exposure forecast NEF_{ij} for the type of aircraft i flying over a path j for a number of daytime flights n_{Dij} and for a number of nighttime flights n_{Nij} is given empirically by

$$\text{NEF}_{ij} = \text{EPNL}_{ij} + 10\log\left(\frac{n_{Dij}}{20} + \frac{n_{Nij}}{1.2}\right) - 75 \qquad \text{(PNdB)} \tag{14-41}$$

Summing for all aircraft types and possible flight paths, the NEF at a particular point of interest on the ground is (using the method of decibel addition)

$$\text{NEF} = 10\log\left(\sum_i \sum_j 10^{\text{NEF}_{ij}/10}\right) \tag{14-42}$$

An empirical equation to convert from NEF to L_{dn} is

$$L_{dn} = (\text{NEF} + 35) \pm 3 \qquad \text{dBA} \tag{14-43}$$

Example 14-10

Since hundreds of flights can occur in a given airport for several types of aircraft and flight paths, the use of equation (14-42) can become lengthy and tedious. Hence a computer program should be used. However, to illustrate the use of the equations, assume only two aircraft types, two flight paths, two daytime flights, and one nighttime flight at path 1 and one daytime flight and two nighttime flights at path 2 for aircraft types 1 and 2. At a given point on the ground, assume that the $\text{PNL}_{\max}$'s for aircraft types 1 and 2 are 114 and 117 PNdB, respectively, over flight path 1, and 116 and 120, respectively over flight path 2. Also assume that the Δt for both types of aircraft in all paths is 25 s. Find the NEF at the given ground location. Calculate the corresponding L_{dn}.

Solution

$$\text{EPNL}_{ij} = \text{PNL}_{ij\max} + D_{ij} + 3 \qquad \text{(PNdB)}$$

$$\text{EPNL}_{11} = 114 + 10\log\frac{25}{20} + 3 = 117.97 \qquad \text{PNdB}$$

$$\text{EPNL}_{12} = 116 + 10\log\frac{25}{20} + 3 = 119.97 \qquad \text{PNdB}$$

$$\text{EPNL}_{21} = 117 + 10\log\frac{25}{20} + 3 = 120.97 \qquad \text{PNdB}$$

$$\text{EPNL}_{22} = 120 + 10\log\frac{25}{20} + 3 = 123.97 \qquad \text{PNdB}$$

$$\text{NEF}_{ij} = \text{EPNL}_{ij} + 10\log\left(\frac{n_{Dij}}{20} + \frac{n_{Nij}}{1.2}\right) - 75$$

$$n_{D_{11}} = 2 \qquad n_{D_{12}} = 1 \qquad n_{N_{11}} = 1 \qquad n_{N_{12}} = 2$$

$$n_{D_{21}} = 2 \qquad n_{D_{22}} = 1 \qquad n_{N_{21}} = 1 \qquad n_{N_{22}} = 2$$

$$\text{NEF}_{11} = 117.97 + 10\log\left(\frac{2}{20} + \frac{1}{1.2}\right) - 75 = 42.67$$

$$\text{NEF}_{12} = 119.97 + 10\log\left(\frac{1}{20} + \frac{2}{1.2}\right) - 75 = 47.32$$

$$\text{NEF}_{21} = 120.97 + 10\log\left(\frac{2}{20} + \frac{1}{1.2}\right) - 75 = 45.67$$

$$\text{NEF}_{22} = 123.97 + 10\log\left(\frac{1}{20} + \frac{2}{1.2}\right) - 75 = 51.32$$

$$\text{NEF} = 10\log\left(\sum_i \sum_j 10^{\text{NEF}_{ij}/10}\right)$$

$$= 10\log\left(10^{42.67/10} + 10^{47.32/10} + 10^{45.67/10} + 10^{51.32/10}\right) = 53.89 \quad \textbf{Answer}$$

$$L_{dn} = (53.89 + 35) \pm 3 = 88.89 \pm 3 \text{ dBA} \quad \textbf{Answer}$$

Industrial noise. In the area of industrial noise, the following measures have been used: noise exposure rating NER; noise dose, D; noise criteria, NC; preferred noise criteria, PNC; speech interference level, SIL; and preferred speech interference level, PSIL. *Noise exposure rating* is the measure of noise that prorates the actual time of exposure of an employee to the total allowable exposure time for a given noise level. The allowable exposure times are recommended by the Committee on Hearing, Bioacoustics, and Biomechanics (CHABA). The *noise dose* is similar to the NER except that the allowable exposure times are recommended by the Occupational Safety and Health Act (OSHA) and not by CHABA. *Noise criteria* are the measure of noise used to quantify noise levels in interior spaces. The *preferred noise criteria* measure is an improvement over the NC. The *speech interference level* is defined as the average of the sound pressure levels at the octave bands at 600 to 1200, 1200 to 2400, and 2400 to 4800 Hz. The *preferred speech interference level* is the average of the sound pressure levels at the 500-, 1000-, and 2000-Hz octave bands. Approximate PSIL values may be obtained by adding 3 dB to the SIL values.

The exposure times recommended by CHABA are shown in Table 14-3, and those by OSHA are shown in Table 14-4. In conjunction with these tables, the equations defining NER and D are, respectively,

$$\text{NER} = \sum_i \frac{C_i}{T_i} \tag{14-44}$$

$$D = \sum_i \frac{C_i}{T_i} \tag{14-45}$$

TABLE 14-3 ALLOWABLE EXPOSURE TIMES FOR A GIVEN SOUND LEVEL (dBA) AS RECOMMENDED BY CHABA

Cumulative exposure	Number of noise exposures per 8-h workday						
	1	3	7	15	35	75	150 or more
8 h	90						
6 h	91	92	93	94	94	94	94
4 h	93	94	95	96	98	99	100
2 h	96	98	100	103	106	109	112
1 h	99	102	105	109	114	115	
30 min	102	106	110	114	115		
15 min	105	110	115				
8 min	108	115					
4 min	111						

TABLE 14-4 ALLOWABLE NOISE EXPOSURE TIMES (OSHA)

Sound level (dBA)	Time (h)	Sound level (dBA)	Time (h)
85	16.00	86	13.93
87	12.13	88	10.57
89	9.18	90	8.00
91	6.97	92	6.07
93	5.28	94	4.60
95	4.00	96	3.48
97	3.03	98	2.83
99	2.25	100	2.00
101	1.73	102	1.52
103	1.32	104	1.15
105	1.00	106	0.87
107	0.77	108	0.67
109	0.57	110	0.50
111	0.43	112	0.38
113	0.33	114	0.28
115	0.25		

where C is the total exposure time at a given steady-state noise level, and T is the allowable exposure time at the corresponding noise level. The T of equation (14-44) is obtained from Table 14-3 and that of equation (14-45) is obtained from Table 14-4. For the recommendation to be met, NER and D must be less than 1.

Example 14-11

The noise levels in the chiller and compressor areas are 92 dBA and 93 dBA, respectively. If the employee is exposed three times at 2 h per exposure in the chiller area and 2 h in the compressor area determine if the CHABA and OSHA recommendations are met.

Solution

$$NER = \sum_i \frac{C_i}{T_i}$$

$$D = \sum_i \frac{C_i}{T_i}$$

From Table 14-3, for 92 dBA at three exposure times, allowable exposure is 6 h; for the 93 dBA and 1 exposure time, the allowable exposure is 4 h.

$$NER = \frac{3(2)}{6} + \frac{2}{4} = 1.5, \qquad \text{recommendation not met.} \quad \textbf{Answer}$$

From Table 14-4, for 92 dBA allowable exposure is 6.07 h; for the 93 dBA the allowable exposure is 5.28 h.

$$D = \frac{3(2)}{6.07} + \frac{2}{5.28} = 1.37, \quad \text{recommendation not met.} \quad \textbf{Answer}$$

The NC and PNC curves are shown in Figure 14-6. To determine the NC or PNC level around an area, the octave band sound pressure levels are plotted on the ordinate and the band frequencies are plotted on the abscissa to obtain the respective NCs or PNCs. The highest value represents the NC or PNC level, as the case may be.

Example 14-12

The data below were obtained from a cutter after a silencer was installed. What are the corresponding NC and PNC levels? What are the PSIL value and the approximate SIL value?

1/1 Octave band (Hz)	L_p (dB)	NC	PNC
63	63	38	45
125	70	58	62
250	73	>65	>65
500	76	>65	>65
1000	79	>65	>65
2000	80	>65	>65
4000	76	>65	>65
8000	63	64	>65

The corresponding NC and PNC values are both > 65.   **Answer**

$$PSIL = \frac{76 + 79 + 80}{3} = 78.33 \quad \textbf{Answer}$$

$$SIL = 78.33 - 3 = 75.33 \quad \textbf{Answer}$$

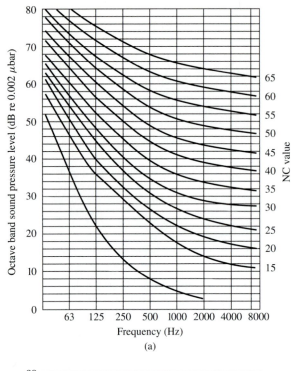

(a)

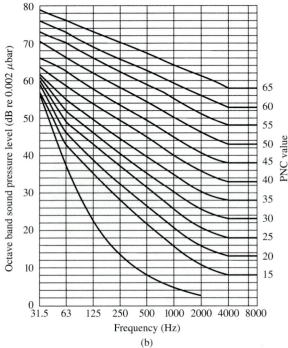

(b)

Figure 14-6 (a) NC curves. (b) PNC curves.

OUTDOOR NOISE PROPAGATION

The sound intensity I at a distance r from a point noise source radiating uniformly in all directions in the surrounding is $P_w/4\pi r^2$, where $4\pi r^2$ is the spherical area receiving the noise. If the source is not a point source, or if the source is not radiating uniformly in all directions, a factor Q, called the directivity, may be factored in. Hence, in general, I at a distance r from the source may be written as

$$I = Q\frac{P_w}{4\pi r^2} \tag{14-46}$$

where $4\pi r^2$ is now a spherical area equivalent to the area receiving the noise. The area receiving the noise may also be designated generally as the area S, whether spherical or nonspherical. With power P_w, the general intensity at the distance r of S is

$$I = \frac{P_w}{S} \tag{14-47}$$

Dividing both sides of equation (14-45) by 10^{-12} W/m^2, taking the "decibel" logarithm, performing the necessary algebra, and using the definition of the respective decibels, the following equation is obtained:

$$L_p \approx L_I = L_w + 10\log\frac{Q}{r^2} - 11 \tag{14-48}$$

This equation is the general representation of the *inverse square law*.

From equation (14-45) and $I = P_w/S$,

$$Q = \frac{4\pi r^2}{S} \tag{14-49}$$

Example 14-13

Determine the value of Q for the following placement of a point source: (a) radiating uniformly in all directions, (b) in a corner, and (c) on a flat surface.

Solution   Let r be the distance.

$$Q = \frac{4\pi r^2}{S}$$

(a) $\qquad\qquad S = 4\pi r^2; \quad Q = \frac{4\pi r^2}{4\pi r^2} = 1$ **Answer**

(b) $\qquad\qquad S = \frac{1}{8}(4\pi r^2); \quad Q = \frac{4\pi r^2}{\frac{1}{8}(4\pi r^2)} = 8$ **Answer**

(c) $\qquad\qquad S = \frac{1}{2}(4\pi r^2); \quad Q = \frac{4\pi r^2}{\frac{1}{2}(4\pi r^2)} = 2$ **Answer**

Example 14-14

The sound power level and the sound pressure level at a distance of 50 ft from one side of a cooling tower have been reported as shown below. Calculate the sound pressure level and the sound level at a distance of 100 ft from the tower.

	Octave band center frequency (Hz)							
Levels	63	125	250	500	1000	2000	4000	8000
L_w (dB)	98	97	92	92	88	83	79	73
L_p (dB)	64	61	58	59	54	50	47	41

Solution

$$L_p \approx L_I = L_w + 10\log\frac{Q}{r^2} - 11$$

$$Q = r^2(10^{L_p - L_w + 11/10})$$

	Octave band center frequency (Hz)							
	63	125	250	500	1000	2000	4000	8000
L_w	98	97	92	92	88	83	79	73
$L_{p(50)}$	64	61	58	59	54	50	47	41
Q	12.53	7.91	12.53	15.77	12.53	15.77	19.86	19.86
$L_{p(100)}$	57.98	54.98	51.98	52.98	47.98	43.98	40.98	34.98
Corrections	−26	−16	−9	−3	0	+1	+1	−1
$L_{pA(100)}$	31.98	38.98	42.98	49.98	47.98	44.98	41.98	33.98

$$L_{pT} = 10\log\left(\sum_{i=1}^{i=n}\sqrt{10^{L_{pi}/10}}\right)^2$$

$$L_{pT(100,\ \text{sound pressure level})} = 10\log\left(\sqrt{10^{57.98/10}} + \sqrt{10^{54.98/10}} + \sqrt{10^{51.98/10}}\right.$$

$$+ \sqrt{10^{52.98/10}} + \sqrt{10^{47.98/10}} + \sqrt{10^{43.98/10}}$$

$$\left.+ \sqrt{10^{40.98/10}} + \sqrt{10^{34.98/10}}\right)^2 = 68.86\ \text{dB} \quad \textbf{Answer}$$

$$L_{pAT(100,\ \text{sound level})} = 10\log\left(\sqrt{10^{31.98/10}} + \sqrt{10^{38.98/10}} + \sqrt{10^{42.98/10}}\right.$$

$$+ \sqrt{10^{49.98/10}} + \sqrt{10^{47.98/10}} + \sqrt{10^{44.98/10}}$$

$$\left.+ \sqrt{10^{41.98/10}} + \sqrt{10^{33.98/10}}\right)^2 = 61.5\ \text{dBA} \quad \textbf{Answer}$$

Attenuating Factors

Equation (14-48) does not apply to the near field, which is the area in the vicinity of the noise source. The area beyond a distance of approximately one wavelength away from the source is called the *far field*. At far-field distances, the equation applies. At these distances, noise will encounter several attenuating factors. These factors can be the attenuation due to molecular absorption by atmospheric air molecules (A_1); the attenuation due to the change in the value of ρv when barometric pressure and temperature change appreciably from normal conditions (A_2); attenuation due to rain, sleet, snow, or fog (A_3); attenuation by barriers (A_4); attenuation by grasses, shrubbery, and trees (A_5); and attenuation owing to wind and temperature gradients (A_6). Including these attenuating factors, equation (14-48) becomes

$$L_p \approx L_I = L_w + 10 \log \frac{Q}{r^2} - 11 - A \qquad (14\text{-}50)$$

where $A = A_1 + A_2 + A_3 + A_4 + A_5 + A_6$.

In the frequency range 125 to 12,500 Hz for temperatures between -10 and $30°C$ and relative humidities between 10 and 90%, the Society of Automotive Engineers[1] derived A_1 as

$$A_1 = \frac{7.4(10^{-8})\gamma^2 r/\phi}{1 + (\beta)(\Delta T)(\gamma)} \qquad \text{dB} \qquad (14\text{-}51)$$

where γ is the center band frequency in hertz, r the distance between source and receiver in meters, ϕ the relative humidity (%), $\beta = 4(10^{-6})$, $\Delta T = T - 20$ in $°C$, and T is the temperature in $°C$.

The equivalence of L_I and L_p is based on the value of $(346.4)(1.184) = 410 \text{ kg/m}^2\text{-s}$ for ρv at the standard temperature and atmosphere of $25°C$ and 1 atm, respectively; however, for values of ρv other than 410, the equivalence fails.

The equations for L_I and L_p derived before are, respectively,

$$L_I = 10 \log \frac{\Delta p_{rms}^2/\rho v}{10^{-12}} \qquad (14\text{-}52)$$

$$L_p = 10 \log \frac{\Delta p_{rms}^2}{4.0(10^{-12})} \qquad (14\text{-}53)$$

Equation (14-52) may be manipulated,

$$L_I = 10 \log \frac{\Delta p_{rms}^2/\rho v}{10^{-12}} \left[\frac{4.1(10^{-10})}{4.1(10^{-10})} \right] \qquad (14\text{-}54)$$

Substituting equation (14-53) into equation (14-54) and simplifying gives

$$L_I = L_p - 10 \log \frac{\rho v}{410} \qquad (14\text{-}55)$$

---

[1] Society of Automotive Engineers (1964). *Standard Values of Atmosphereic Absorption as a Function of Temperature and Humidity for Use in Evaluating Aircraft Flyover Noise (ARP886)*. Society of Automotive Engineers, New York.

The ρv in the numerator represents the correction to L_p if L_I were to be equal to L_p. The correction A_2 is therefore

$$A_2 = 10 \log \frac{\rho v}{410} \qquad (14\text{-}56)$$

A_3 is on the order of 0.5 dB per 1000 m in fogs and is, in general, considered to be zero. Attenuation due to grasses, shrubberies, and trees (A_5) is not easy to generalize. It ranges from 0 to 30 dB per 100 m.

Referring to Figure 14-7, the attenuation due to solid thin barriers A_4 is found as follows: In consistent units, let A be the distance from the noise source to the top of the barrier, B the distance from the noise receiver situated on the other side of the barrier to the top of the barrier, and r the distance between the noise source and noise receiver. From these measurements, a parameter called the Fresnel number N may then be calculated:

$$N = \frac{2(A + B - r)}{\lambda} \qquad (14\text{-}57)$$

where λ is the wavelength. With the value of N known, A_4 may be obtained from Table 14-5. Note that in the table there are two types of noise sources: point and line sources. The moving highway vehicle traffic is an example of a line source.

Attenuation due to wind and temperature gradient (atmospheric stability) is gaged separately as to whether the noise receiver is downwind or upwind from the source. For downwind receivers, A_6 may be estimated from Table 14-6, where γ is the frequency and r is the distance between the source and the receiver. For a receiver upwind from the source, there is a distance X_0 where wind and temperature effects start to deflect the sound waves. For distances less than X_0, wave deflection is negligible, while for distances greater than X_0, deflection is appreciable. Hence X_0 is one of the parameters determined in order to calculate A_6. The space beyond the X_0 limit is called the *shadow zone*. Table 14-7 shows some values of X_0 for source frequencies ranging from 300 to 5000 Hz and for various other environmental conditions, such as night or day, wind speeds, sky conditions, and atmospheric temperature profiles. Finally, Table 14-8 gives the correction A_6 (for upwind receivers) for various values of r/X_0.

The propagation of noise in air involves such a very small perturbation of pressure that the process may be considered frictionless, involving no heat exchange. Thus the process is isentropic. From fluid mechanics, under isentropic conditions, the velocity of sound v in equation (14-56) is

$$v = \sqrt{kRT} \qquad (14\text{-}58)$$

where k is the ratio of specific heats at constant volume to constant pressure, R the gas constant, and T the temperature. For air, $R = 1716$ ft-lb/slug-$^\circ R$ and, for a temperature range of $-40^\circ C$ to $80^\circ C$, $k = 1.4$. Using these values, we have

$$v = 49.01\sqrt{T} \qquad (14\text{-}59)$$

where T is in degrees Rankine.

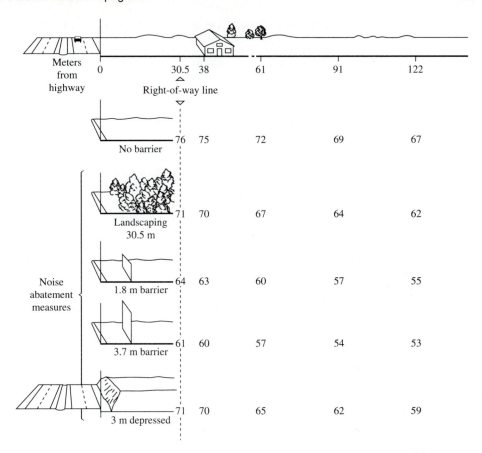

Figure 14-7 Outdoor noise attenuation adjacent to a highway using barriers. (Units of measure in L_{10}, dBA; a four-lane highway at traffic level of 5000 vehicles per hour at 85 km/h and 5% trucks). (P.F. Cunniff (1977). *Environmental Noise Pollution*. Reprinted by permission of John Wiley and Sons, Inc.)

TABLE 14-5 VALUES OF A_4 (dB) FOR CORRESPONDING FRESNEL NUMBERS, N

N	A_4 Point	A_4 Line	N	A_4 Point	A_4 Line	N	A_4 Point	A_4 Line
0.1	8	7	0.2	9	7	0.3	9	8
0.4	10	8	0.5	11	9	0.6	11	9
0.7	12	10	0	13	10	0.9	3	10
1.0	13	11	2	15	12	3	17	14
4	19	15	5	20	16	6	21	16
7	22	17	8	22	18	9	23	19
10	24	19	20	24	21	30	24	22

TABLE 14-6   VALUES OF A_6 FOR VALUES OF γr IN A
DOWNSTREAM NOISE RECEIVER[a]

γr (fps)	A_6(dB)	γr(fps)	A_6 (dB)
$4(10^4)$–$4(10^5)$	0	$6(10^5)$	2
$8(10^5)$	3	10^6	5
$2(10^6)$	9	$4(10^6)$	10
$6(10^6)$	13	$8(10^6)$	14
10^7	15	$2(10^7)$	17

[a]Wind speed: < 40–50 km/h > 3–5 km/h; source–receiver
distance: < 82 km; source height: 3–5 m; receiver height: 1.5 m.

TABLE 14-7   VALUES OF X_0 UNDER VARIOUS ENVIRONMENTAL CONDITIONS

Time		Sky		Temperature profile			Wind speed (m/s)	X_0 (m)
Day	Night	Clear	Overcast	Lapse	Neutral	Inversion		
	×	×				×	1–10	600
×			×		×		1–10	130
×		×		×			1–10	80

TABLE 14-8 A_6 FOR UPWIND
NOISE RECEIVERS[a]

r/X_0	A_6	r/X_0	A_6
0.5–1.0	0	1.5	11
2	21	3	29
4	34	8	42

[a]Source height: 3–5 m; receiver height:
2 m; wind speed: < 25–30 km/h $>$
3–5 km/h; source–receiver distance: $<$
82 km.

Example 14-15

In Example 14-14, what is the sound pressure level at the 100-ft distance given the following additional conditions: A concrete barrier, 30 ft high, is erected between the tower and the 100-ft distance at a point 60 ft from the tower. The tower is 15 ft high, and the receiver point is 6 ft high with the ground assumed horizontal. The atmospheric pressure and temperature are, respectively, 14.7 psi and 60°F. The relative humidity is 80%. Assume that the receiver is downwind from the source. Neglect A_3 and A_5.

Solution

$$A_1 = \frac{7.4(10^{-8})(\gamma^2 r/\phi)}{1 + (\beta)\Delta T)(\gamma)} \qquad \text{dB}$$

$$60°F = (60–32)\left(\frac{5}{9}\right) = 15.56°C = 60 + 460 = 520°R$$

$$A_1 = \frac{7.4(10^{-8})\gamma^2(100/3.281)/80}{1 + (4)(10^{-6})(15.56 - 20)(\gamma)} = \frac{2.82(10^{-8})\gamma^2}{1 - 1.78(10^{-5})\gamma} \quad \text{dB}$$

$$A_2 = 10\log\frac{\rho v}{410}$$

$$p = \rho RT \quad \rho = \frac{p}{RT} = \frac{14.7(144)}{1716(520)} = 0.0024\frac{\text{slug}}{\text{ft}^3}\left(\frac{14.59 \text{ kg}}{\text{slug}}\right)\left(\frac{3.281^3 \text{ ft}^3}{1^3 \text{ m}^3}\right) = 1.237\frac{\text{kg}}{\text{m}^3}$$

$$v = 49.01\sqrt{T} = 49.01\sqrt{520} = 1117.6\frac{\text{ft}}{\text{s}}\left(\frac{1 \text{ m}}{3.281 \text{ ft}}\right) = 340.63 \text{ m/s}$$

$$A_2 = 10\log\frac{(1.237)(340.63)}{410} = 0.12 \text{ dB}$$

$$N = \frac{2(A + B - r)}{\lambda}$$

$$A = \sqrt{60^2 + (30 - 15)^2} = 61.85 \text{ ft}$$

$$B = \sqrt{40^2 + (30 - 6)^2} = 46.65 \text{ ft}$$

$$r = \sqrt{100^2 + (15 - 6)^2} = 100.40 \text{ ft}$$

$$\lambda = \frac{v}{\gamma}$$

$$N = \frac{2(61.85 + 46.65 - 100.40)(\gamma)}{340.63} = 0.048\gamma$$

From Example 14-14, the given data at various frequencies as well as the corrections are as follows:

γ (Hz)	Octave band center frequency (Hz)								
	63	125	250	500	1000	2000	4000	8000	
$L_{p(100)}$	57.98	54.98	51.98	52.98	47.98	43.98	40.98	34.98	[a]
A_1 (10^{-3} dB)	0.11	0.44	1.77	7.11	28.71	117.0	486.0	2104	
A_2 (dB)	−0.12	−0.12	−0.12	−0.12	−0.12	−0.12	−0.12	−0.12	
N	3.02	6.0	12.0	24.0	48.0	96.0	192.0	384.0	
A_4 (dB)	17.2	21	24	24	24	24	24	24	[b]
γr (10^5 fps)	0.06	0.125	0.25	0.5	1.0	2.0	4.0	8.0	
A_6 (dB)	0	0	0	0	0	0	0	3.2	[c]
$L_{p(100)}$ (dB)	41	34	28	29	24	20	17	8	[d]

[a] The data on this line are from Example 14-4.
[b] The data on this line are from Table 14-5, with N known.
[c] The data on this line are from Table 14-6, with γr known.
[d] The data on this line are corrected L_p's.

$$L_{pT\,(100,\ \text{sound pressure level})} = 10\log\left(\sqrt{10^{40.55/10}} + \sqrt{10^{33.86/10}} + \sqrt{10^{27.86/10}}\right.$$

$$+ \sqrt{10^{28.85/10}} + \sqrt{10^{23.83/10}} + \sqrt{10^{19.74/10}}$$

$$\left. + \sqrt{10^{16.37/10}} + \sqrt{10^{5.56/10}}\right)^2 = 47.41\ \text{dB}\quad\textbf{Answer}$$

Note: In this problem, the only effective attenuating factor is the barrier.

INDOOR NOISE PROPAGATION

The propagation of noise in an indoor space depends on the acoustic properties of the walls, ceiling, floors, the materials present inside the space, and the reverberation effect as the sound waves bounce from one surface to the other. It is intuitive that propagation would somehow follow the inverse square law. In fact, the law is used, but in modified form.

L_p had been derived as

$$L_p = L_w + 10\log\frac{Q}{r^2} - 11 \tag{14-60}$$

This equation may also be written as

$$L_p = L_w + 10\log\frac{Q}{4\pi r^2} \tag{14-61}$$

But $Q = 4\pi r^2/S$. Substituting in equation (14-61), we have

$$L_p = L_w + 10\log\frac{1}{S} \tag{14-62}$$

As applied in outdoor noise transmission, S is the area that receives the sound coming from the source. To apply equation (14-62) for indoor transmission, S must be defined in a convenient form. This definition should incorporate indoor receiving parameters. Define S arbitrarily as

$$\frac{1}{S} = \frac{1}{4\pi r^2} + \frac{4}{R} \tag{14-63}$$

where R is called the *room constant*. The expression $4\pi r^2$ is the equivalent spherical area. As shown, S incorporates the indoor parameter r (distance from noise source to a given point) and room constant R. The term containing R expresses the reverberating effect as sound bounces back and forth inside the room. Expressed in terms of total surface area receiving the sound (S_T), R is empirically

$$R = \frac{S_T\bar{\alpha}}{1 - \bar{\alpha}} \tag{14-64}$$

The *sound absorption coefficient* $\bar{\alpha}$ is the fraction of the incident noise that is absorbed by the surface, and $1 - \bar{\alpha}$ is the fraction reflected. $\bar{\alpha}$ is given by

$$\bar{\alpha} = \frac{\sum \alpha_i S_i}{S_T} \tag{14-65}$$

where S is the area with absorption coefficient α and i is the index for the various surfaces. Table 14-9 shows some values of α.

TABLE 14-9 ABSORPTION COEFFICIENT α FOR SOME BUILDING MATERIALS

Material	Octave band center frequency (Hz)					
	125	250	500	1000	2000	4000
Marble or glazed tile	0.01	0.01	0.01	0.01	0.02	0.02
Plaster, gypsum or lime, smooth finish on tile or brick	0.01	0.02	0.02	0.03	0.04	0.05
Plaster, gypsum or lime						
Rough finish on lath	0.13	0.10	0.05	0.05	0.04	0.03
Smooth finish	0.15	0.10	0.05	0.04	0.04	0.03
Plywood paneling, 3/8 in. thick	0.30	0.20	0.20	0.09	0.10	0.11
Brick, unglazed	0.02	0.02	0.03	0.04	0.05	0.06
Brick, unglazed, painted	0.01	0.01	0.02	0.02	0.02	0.02
Carpet, heavy						
On concrete	0.01	0.05	0.15	0.40	0.60	0.70
On 40-oz hairfelt or foam rubber	0.07	0.25	0.60	0.70	0.72	0.74
With impermeable latex backing on 40-oz hairfelt or foam rubber	0.07	0.30	0.40	0.35	0.50	0.65
Concrete block						
Coarse	0.35	0.45	0.30	0.30	0.40	0.25
Painted	0.10	0.05	0.06	0.07	0.09	0.08
Glass						
Large panes of heavy plate glass	0.15	0.05	0.04	0.03	0.02	0.02
Ordinary window glass	0.40	0.30	0.20	0.15	0.07	0.04
Gypsum board, $\frac{1}{2}$ in. nailed to 2 × 4's, 16 in. o.c.	0.30	0.10	0.05	0.04	0.07	0.09
Floors						
Concrete or terrazzo	0.01	0.01	0.02	0.02	0.02	0.02
Asphalt, rubber, or tile on concrete	0.02	0.03	0.03	0.03	0.03	0.02
Wood	0.15	0.10	0.10	0.06	0.06	0.07

L_p is now written as

$$L_p = L_w + 10\log\left(\frac{1}{4\pi r^2} + \frac{4}{R}\right) \tag{14-66}$$

This is the equation for indoor space sound propagation.

Noise Transmission Through Ducts

In heating and cooling systems, fans or blowers convey the air through ducts. Depending on the elements installed, these duct systems may serve to attenuate the sound power generated by the fans, thus lowering the noise power that exits at duct outlets. Or they may exacerbate the noise, increasing the noise power at duct outlets. Table 14-10 shows some values of attenuation for various duct sizes, Table 14-11 shows attenuation in elbows,

TABLE 14-10 APPROXIMATE ATTENUATION (dB/FT) IN BARE[a] RECTANGULAR METAL DUCTS

Duct[b]	Octave band center frequency (Hz)			
	63	125	250	500 and above
Large (72 in. × 72 in.)	0.1	0.1	0.03	0.01
Medium (24 in. × 24 in.)	0.2	0.2	0.1	0.05
Small (6 in. × 6 in.)	0.3	0.3	0.1	0.1

[a]For thermally insulated ducts, double the attenuations.
[b]Dimensions in parentheses are nominal sizes.

TABLE 14-11 APPROXIMATE ATTENUATION (dB) IN ELBOWS (30–90°)

Elbow type	Diameter or width of duct (in.)	Octave band center frequency (Hz)							
		63	125	250	500	1000	2000	4000	8000
Square	5	0	0	0	1	6	8	6	4
	10	0	0	1	6	8	6	4	4
	20	0	1	5	7	5	3	3	3
	40	1	6	8	6	4	4	3	3
Round	5–10	0	0	0	0	1	1	2	3
	11–20	0	0	0	1	1	2	2	3
	21–40	0	0	1	1	2	2	3	3
	41–80	0	1	1	2	2	3	3	3

and Table 14-12 shows sound production in turning vanes at an airflow of 2000 ft/m. Table 14-12 also shows corrections for velocities other than 2000 ft/m. Turning vanes are installed at duct elbows to control heating and ventilating requirements but unfortunately, also serve as additional sound sources.

TABLE 14-12 SOUND POWER PRODUCED BY TURNING VANES AT 2000 FT/MIN

Duct size (ft square)	Octave band center frequency (Hz)							
	63	125	250	500	1000	2000	4000	8000
2	77	71	65	61	56	52	45	38
0.5	57	56	55	53	50	44	38	31
0.25	51	51	51	50	46	42	35	28
0.1	62	61	59	56	52	48	41	33

	Velocity (ft/m)							
	800	1000	1200	1500	2000	2500	3000	4000
Corrections	−24	−18	−12	−8	0	+6	+9	+15

Example 14-16

A factory room measuring 70 ft long by 20 ft wide by 20 ft high is to be heated using a fan with the following sound power level characteristics: 63 Hz, 98 dB; 125 Hz, 88 dB; 250 Hz, 89 dB; 500 Hz, 87 dB; 1,000 Hz, 85 dB; 2,000 Hz, 85 dB; 4,000 Hz, 83 dB; and 8,000 Hz, 71 dB. The main duct, 24 in. by 24 in., has a length of 50 ft with a 90-degree elbow of 24 in. by 24 in. at its terminus. This elbow is used to connect the main to a branch or a continuation duct at a 90-degree turn with the main. The branch duct measures 24 in. by 24 in. by 60 ft long and is located at the ceiling running lengthwise with the 70-ft dimension of the room, perpendicular to and at the middle of the 20-ft dimension. The branch duct has 6 outlets at the ceiling discharging 500 cfm each. The main duct, itself, conveys 3,000 cfm of air. The distances of the outlets to a point in the middle of the room are: outlets 1 and 6 = 32 ft, outlets 2 and 5 = 24 ft, and outlets 3 and 4 = 17 ft. (a) Calculate the sound power level at each duct outlet, and (b) calculate the sound pressure level at the point in the middle of the room. The following are additional information: ceiling (70 ft by 20 ft): acoustical tile, $\alpha(125 \text{ Hz}) = 0.47$, $\alpha(250 \text{ Hz}) = 0.80$, $\alpha(500 \text{ Hz}) = 0.95$, $\alpha(1000 \text{ Hz}) = 1.0$, $\alpha(2000 \text{ Hz}) = 0.9$, $\alpha(4000 \text{ Hz}) = 0.85$, $\alpha(8000 \text{ Hz}) = 1.0$; walls (two 20 ft by 20 ft and two 20 ft by 70 ft): painted concrete block; and floor (70 ft by 20 ft): linoleum.

Solution   (a) Power level at each duct outlet, $L_{w\text{out}}$

	Octave band center frequency (Hz)							
	63	125	250	500	1000	2000	4000	8000
Sound power level of fan (dB)	98	88	89	87	85	85	83	71
Attenuation by 24 in. by 24 in. main duct (Table 14-11, 50 ft)	10	10	5	2.5	2.5	2.5	2.5	2.5
Attenuation by 24 in. by 24 in. elbow (Table 14-11)	0	1	5	7	5	3	3	3
Branch duct (Table 14-10, 60 ft/2)	6	6	3	1.5	1.5	1.5	1.5	1.5
Total attenuation (dB)	16	17	13	11	9	7	7	7
Corrected sound power level of fan (dB)	82*	71	76	76	76	78	76	64
Turning vane power level at 2000 fpm (Table 14-12)	77	71	65	61	56	52	45	38
Correction at 750 fpm	−24	−24	−24	−24	−24	−24	−24	−24
Corrected vane power level (dB)	53**	47	41	37	32	28	21	14
Combined power level at branch	82***	71	76	76	76	78	76	64

$$v = \frac{3000}{(24/12)(24/12)} = 750 \text{ fpm}$$

$$L_{wT} = 10 \log \sum_{i=1}^{i=n} 10^{L_{wi}/10} = 10 \left[\log \left\{ 10^{82^*/10} + 10^{53^{**}/10} \right\} \right] = 82^{***}$$

Let P_{wT} = total power at duct and $P_{w\text{out}}$ = power at each outlet; hence, $P_{w\text{out}} = \frac{1}{6}(P_{wT})$.

$$L_{wT} = 10 \log \sum_{i=1}^{i=n} 10^{L_{wi}/10}$$

$$L_{wT} = 10\log[10^{82/10} + 10^{71/10} + 10^{76/10} + 10^{76/10} + 10^{76/10} +$$
$$10^{78/10} + 10^{76/10} + 10^{64/10}] = 86 \text{ dB}$$
$$10\log P_{w\text{out}} = 10\log\frac{1}{6} + 10\log P_{wT}$$

$$L_{w\text{out}} = 10\log\frac{1}{6} + L_{wT} = -7.78 + 86 = 78 \text{ dB for summation for all frequencies} \quad \textbf{Answer}$$

(b) $$L_p = L_w + 10\log\left(\frac{1}{4\pi r^2} + \frac{4}{R}\right)$$

$$R = \frac{S_T\bar{\alpha}}{1-\bar{\alpha}}$$

$$\bar{\alpha} = \left(\sum \alpha_i S_i\right)/S_T$$

	Octave band center frequency (Hz)							
	63	125	250	500	1000	2000	4000	8000
Combined power level at branch duct outlets L_w (dB)	82	71	76	76	76	78	76	64
$L_{w\text{out}}$ for each outlet $L_{w\text{out}} = (10\log\frac{1}{6} + L_w)$	74	63	68	68	68	70	68	56
α, ceiling $\quad S_1 = 20(70) = 1400 \text{ ft}^2$	—	0.47	0.8	0.95	1.0	0.9	0.85	1
α, walls (Table 14-9) $\quad S_2 = 2(20)(20) + 2(20)(70) = 3600 \text{ ft}^2$		0.1	0.05	0.06	0.07	0.09	0.08	—
α, floor (Table 14-9) $\quad S_3 = 1400 \text{ ft}^2$	—	0.02	0.03	0.03	0.03	0.03	0.02	—
$S_T = S_1 + S_2 + S_3 = 1400 + 3600 + 1400 = 6400$								
$\bar{\alpha} = \left(\sum \alpha_i S_i\right)/S_T$	—	0.16	0.21	0.25	0.26	0.25	0.24	—
$R = \dfrac{S_T\bar{\alpha}}{1-\bar{\alpha}}$	1219*	1219	1701	2133	2249	2133	2021	2021*
$L_p = L_{w\text{out}} + 10\log\left(\dfrac{1}{4\pi r^2} + \dfrac{4}{R}\right)$								
L_p: For $r = 17$ ft, left	50*	39	42	41	41	43	42	30
For $r = 24$ ft, left	49*	38	42	41	41	43	41	29
For $r = 32$ ft, left	49*	38	42	41	41	43	41	29
For $r = 17$ ft, right	50	39	42	41	41	43	42	30
For $r = 24$ ft, right	49	38	42	41	41	43	41	29
For $r = 32$ ft, right	49	38	42	41	41	43	41	29

(Continued)

	Octave band center frequency (Hz)							
	63	125	250	500	1000	2000	4000	8000
L_{pT}:	65**	54	58	57	57	59	57	45

$$L_{pT} = 10 \log \left(\sum_{i=1}^{i=n} \sqrt{10^{L_{pi}/10}} \right)^2$$

$$= 10 \log \left(2 \left\{ \sqrt{10^{50^*/10}} + \sqrt{10^{49^*/10}} + \sqrt{10^{49^*/10}} \right\} \right)^2 = 65**$$

*Assumed

$$L_{p(\text{middle of room})} = 10 \log \left(\sqrt{10^{65/10}} + \sqrt{10^{54/10}} + \sqrt{10^{58/10}} + \sqrt{10^{57/10}} \right.$$

$$\left. + \sqrt{10^{57/10}} + \sqrt{10^{59/10}} + \sqrt{10^{57/10}} + \sqrt{10^{45/10}} \right)^2$$

$$= 75 \text{ dB} \quad \textbf{Answer}$$

Noise Transmission Through Partitions

There are generally two types of partitions: the open barrier type, where sound waves can bend around the open upper part of the structure by diffraction, and the "closed" barrier type, where the sound waves cannot bend around the structure by diffraction (since the structure is closed). An example of the closed type is a bedroom wall where the ceiling and wall are tight (i.e., no opening). Transmission across open barriers has already been discussed.

If the incident noise intensity on the partition is I_i and the transmitted noise intensity is I_t, the *intensity transmission loss* (TL) in decibels is $10 \log(I_i / 10^{-12}) - 10 \log(I_t/10^{-12})$. This may also be written as

$$\text{TL} = 10 \log \frac{I_i}{I_t} \tag{14-67}$$

The *transmission coefficient* is defined as the fraction of the incident noise that is transmitted through the partition or wall, $\tau = I_t/I_i$. Substituting τ into equation (14-67), we have

$$\text{TL} = 10 \log \frac{1}{\tau} \tag{14-68}$$

A high-density medium does not vibrate easily. Since noise is transmitted through vibration, if the medium has a high density, transmission will not be efficient; transmission loss is proportional to density. Also, because of shorter wavelengths, noise at higher frequencies is more easily absorbed and attenuated. Hence transmission loss is also proportional to frequency. Kinsler and Fey[2] showed that in terms of the area density

[2]L. E. Kinsler and A. R. Frey (1962). *Fundamentals of Acoustics*. Wiley, New York.

of the partition ρ_ℓ and frequency γ, the transmission loss, TL, is given empirically by

$$TL = 20 \log \rho_\ell \gamma - 47.4 \qquad (14\text{-}69)$$

where ρ_ℓ is in kg/m^2 and γ is in hertz. Equation (14-69) is called the *mass law* and applies only to rigid barrier with air on either side.

Example 14-17

A brick wall is 0.41 m thick and has an area density of 590 kg/m^2. If the frequency of the incident sound is 500 Hz, calculate TL and τ.

Solution

$$TL = 20 \log \rho_\ell \gamma - 47.4 = 20 \log 590(500) - 47.4 = 61.7 \text{ dB} \quad \textbf{Answer}$$

$$TL = 10 \log \frac{1}{\tau} \qquad 61.7 = 10 \log \frac{1}{\tau} \qquad \tau = 0.0000007 \quad \textbf{Answer}$$

NOISE CONTROL CRITERIA

In this section we present various noise control criteria that have been used in practice. Table 14-13 shows noise criteria in terms of NC (noise criteria) levels for various activities, while Table 14-14 shows speech interference levels for various degrees of communication effectiveness. The U.S. Federal Highway Administration (FHA) developed the standards shown in Table 14-15 for new construction. These levels are those that would be expected to yield no complaints.

TABLE 14-13   ACCEPTABLE NC LEVELS FOR VARIOUS ACTIVITIES

Activity	Suggested range of noise criteria, NC
Acceptable working conditions with minimum speech interference:	
Industrial areas, garages, laundries	NC-45 to NC-55
Moderately fair listening conditions:	
Lobbies, cafeterias, drafting rooms, business machine areas	NC-40 to NC-45
Fair listening conditions required:	
Large offices, restaurants, retail shops and stores, etc.	NC-25 to NC-40
Good listening conditions required:	
Private offices, school classrooms, small conference rooms, libraries, television listening	NC-20 to NC-25
Very good listening conditions required:	
Auditorium, theaters	NC-15 to NC-20
Large meeting and conference rooms	NC-15 to NC-20
Excellent listening conditions required:	
Concert halls, recording studios, etc.	NC-15 to NC-20
Sleeping, resting, and relaxing:	
Suburban and rural: homes, apartments, hotels, hospitals, etc.	NC-15 to NC-20
Urban: homes, apartments, hotels, hospitals, etc.	NC-15 to NC-20

TABLE 14-14 SPEECH INTERFERENCE LEVELS FOR VARIOUS DEGREES OF CONVERSATION EFFECTIVENESS

Telephone communications:

PSIL (dB)	Telephone intelligibility
> 865	Very poor
< 865	Good to very poor

Face-to-face communications:

Vocal effort	PSIL (dB)	Distance from speaker to listener (ft)	Communications intelligibility
Maximum vocal effort	100	1	Difficult
	100 and above	1	Impossible
Shout	90	1	Possible
Very loud voice	70	1	Possible
	60	5	Possible
	50	10	Possible
	40	15	Possible

TABLE 14-15 FHA NOISE STANDARDS FOR HIGHWAYS

Land use category	Design level (dBA) L_{eq}	L_{10}	Description of land use category
A	57	60	Tracts of lands in which serenity and quiet are of extraordinary significance and serve an important public need, and where the preservation of those qualities is essential if the area is to continue to serve its intended purpose. Such areas could include amphitheaters, particular parks or portions of parks, or open spaces that are dedicated or recognized by appropriate local officials for activities requiring special qualities of serenity and quiet.
B	67	70	Residences, motels, hotels, public meeting rooms, schools, churches, libraries, hospitals, picnic areas, recreational areas, playgrounds, active sports, areas, and parks.
C	72	75	Developed lands, properties, or activities not included in categories A and B above.
D	Unlimited	Unlimited	Undeveloped lands.
E	52	55	Public meeting rooms, schools, churches, libraries, hospitals, and similar buildings.

The U.S. Department of Housing and Urban Development (HUD) has also set out guideline criteria for noise exposure at residential sites. These are shown in Table 14-16. Table 14-17 is a land use compatibility table for various NEF values. The NEF zones stated in Table 14-16 are the NEF zones mentioned in Table 14-17.

TABLE 14-16 HUD NOISE ASSESSMENT CRITERIA FOR RESIDENTIAL CONSTRUCTION

General external exposures	Assessment
I. Exceeds 89 dBA 60 min per 24 h Exceeds 75 dBA 8 h per 24 h NEF Zone C (airport environs) > 888 PNdB (exterior)	Unacceptable
II. Exceeds 65 dBA 8 h per 24 h Loud repetitive sounds on site NEF Zone B (airport environs) 74 to 88 PNdB (exterior)	Normally acceptable
III. Does not exceed 65 dBA more than 8 h per 24 h 62 to 74 PNdB (exterior)	Normally acceptable
IV. Does not exceed 45 dBA more than 30 min per 24 h NEF Zone A (airport environs) < 862 PNdB (exterior)	Acceptable

TABLE 14-17 LAND USE COMPATIBILITY FOR VARIOUS NEF VALUES

Land use	Zone A < 830 NEF	Zone B 30–40 NEF	Zone C > 840 NEF
Residential	Yes	[a]	No
Hotels, motels, offices, public buildings	Yes	Yes[b]	No
Schools, hospitals, churches, indoor theaters, auditoriums	Yes[b]	No	No
Commercial, industrial	Yes	Yes	[b]
Outdoor amphitheaters, theaters	Yes[a, c]	No	No
Outdoor recreational (nonspectator)	Yes	Yes	Yes

[a]Case history experience indicates that people in private residences may complain, perhaps vigorously. Concerted group action is possible. New single-dwelling construction should generally be avoided. For apartment construction, note b applies.

[b]An analysis of building noise requirements should be made and needed noise control features should be included in the design.

[c]A detailed noise analysis should be undertaken by qualified personnel for all indoor music auditoriums and all outdoor theaters.

The USEPA has published noise criteria levels deemed necessary to protect the health and welfare of U.S. citizens with an adequate margin of safety. These are shown in Table 14-18.

Finally, the U.S. General Services Administration (GSA) requires that for equipment used on government contracts, the noise levels at the site should not exceed the limits shown in Table 14-19.

TABLE 14-18 NOISE CRITERIA LEVELS TO PROTECT PUBLIC HEALTH AND WELFARE

Effect	Level	Area
Hearing loss	$L_{eq}(24) \leq 70$ dB	All areas
Outdoor activity	$L_{dn} \leq 55$ dB	Outdoor in residential areas and farms, and other outdoor areas where people spend widely varying amounts of time and other places in which quiet is a basis for use.
	$L_{eq}(24) \leq 55$ dB	Outdoor areas where people spend limited amounts of time, such as schoolyards and playgrounds, etc.
Indoor activity	$L_{dn} \leq 45$ dB	Indoor residential areas
Interference and annoyance	$L_{eq}(24) \leq 45$ dB	Other indoor areas with human activities, such as schools

TABLE 14-19 GSA CONSTRUCTION NOISE REQUIREMENTS

Equipment	Sound level (dBA) at 50 ft	Equipment	Sound level (dBA) at 50 ft
Earthmoving		Materials handling	
Front loader	75	Concrete mixer	75
Backhoe	75	Concrete pump	75
Dozer	75	Crane	75
Tractor	75	Derrick	75
Scraper	80	Stationary	
Grader	75	Pump	75
Truck	75	Generator	75
Paver	80	Compressor	75
Impact		Other	
Pile driver	95	Saw	75
Jack hammer	75	Vibrator	75
Rock drill	80		
Pneumatic drill	80		

NOISE CONTROL

There are four general methods of controlling noise: enclosing the noise source, enclosing the noise receiver, putting a barrier between the noise source and the receiver, and controlling the noise generator.

Noise is transmitted by vibration. Hence the property of the enclosure must be such that it should not vibrate when a sound wave hits its surface; otherwise, the enclosure itself can become a noise source. Since vibration is inversely related to the mass of the material, in the use of enclosures, the effectiveness of control is therefore a function of the mass of the enclosure. By the mass law, the ideal enclosure is the heavy enclosure (materials of high density). Table 14-20 shows densities ρ of some common materials of construction.

TABLE 14-20 DENSITIES OF SOME COMMON MATERIALS
OF CONSTRUCTION

Material	ρ (lb/ft^3)
Glass	155
Lead, antimonial (hard)	700
Plexiglas or Lucite	68
Steel	478
Plywood $\left(\frac{1}{4} \text{ to } 1\frac{1}{4} \text{ in.}\right)$	40
Concrete, dense poured	148
Masonry block	
Hollow cinder (nominal 6 in. thick)	49
Hollow cinder $\frac{5}{8}$ in., sand plaster each	58
side (nominal 6 in. thick)	
Solid dense concrete	110
(nominal 4 in. thick)	
Hollow dense concrete (nominal 6 in. thick)	71
Hollow dense concrete, sand-filled voids	105
(nominal 6 in. thick)	
Aluminum	168
Lead, chemical or tellurium	698
Plaster, solid, on metal or gypsum lath	105
Gypsum board ($\frac{1}{2}$ to 2 in.)	40
Wood waste materials bonded	45
with plastic, 5 lb/ft^2	
Concrete (clinker) slab, plastered	99
on both sides, 2 in. thick	
Fir timber	40
Brick	130

Example 14-18

To control the noise coming from a tertiary crusher, the equipment installation is enclosed with masonry blocks of solid dense concrete 4 in. thick. The following is the sound power level characteristics of the crusher at the inside wall of the enclosure: 63 Hz, 98 dB; 125 Hz, 88 dB; 250 Hz, 89 dB; 500 Hz, 87 dB; 1000 Hz, 85 dB; 2000 Hz, 85 dB; 4000 Hz, 83 dB; and 8000 Hz, 71 dB. What vocal effort would be exerted to have a face-to-face conversation immediately outside the enclosure?

Solution

$$TL = 20 \log \rho_\ell \gamma - 47.4$$

From Table 14-20,

$$\rho_\ell = \frac{110}{1/(4/12)} = 36.67 \frac{\text{lb}}{\text{ft}^2} = 36.67(0.454)\left[\frac{1}{(1/3.281)^2}\right] = 179.22 \frac{\text{kg}}{\text{m}^2}$$

Since PSIL is computed only on the basis of the 500-, 1000-, and 2000-Hz octave bands, calculate the TL on the basis of these frequencies only.

$$TL_{500} = 20 \log(179.22)(500) - 47.4 = 51.65 \text{ dB}; \quad \text{dB transmitted} = 87 - 51.65 = 35.35$$

$$TL_{1,000} = 20\log(179.22)(1,000) - 47.4 = 57.67 \text{ dB; dB transmitted} = 85 - 57.67 = 27.33$$

$$TL_{2,000} = 20\log(179.22)(2,000) - 47.4 = 63.67 \text{ dB; dB transmitted} = 85 - 63.67 = 21.33$$

$$\text{PSIL} = \frac{35.35 + 27.33 + 21.33}{3} = 28.00 \text{ dB}$$

From Table 14-14, vocal effort is normal.   **Answer**

Putting a barrier between the source and the receiver is a control used in highways as shown in Figure 14-7. It is worth noting that landscaping is not a particularly good barrier. However, it serves a psychological purpose, since landscaping blocks the sight of the highway.

Control of noise at points of generation may be done with the use of mufflers or silencers and isolation of noise source by vibration control. There are three basic silencers: absorptive silencers, reactive expansion chambers, and diffusers. In the *absorption silencer*, an acoustic material is lined directly on the interior of the duct. The duct may be straight or may have some bends, or the duct may be expanded into a plenum lined with the acoustic material. The acoustic material absorbs the noise, thus attenuating it. The absorption silencer is a *dissipative muffler* (i.e., it dissipates the noise by absorbing it). The *reactive expansion chamber or muffler* has no lining of absorptive materials but attenuates noise by reflecting the sound waves so as to cancel the waves of the incoming noise source. This process is called *destructive interference*. Reactive mufflers are the types found in trucks and automobiles.

High-velocity mass air impinging on stationary air or solid objects produces noise due to the turbulence created. *Diffusers* attenuate the noise by reducing this velocity. The source flow is *diffused out* into a multitude of tiny flows having lower velocities using some appropriate mechanism. The diffuser is an *exhaust muffler*, since it attenuates noise by installing it at the end of a duct or pipe. Schematics of the various types of silencers are shown in Figure 14-8a.

The attenuation in power level in a lined duct may be estimated using the Sabine formula,[3]

$$A = 12.6\frac{P}{S}\alpha^{1.4} \tag{14-70}$$

where A is the attenuation per lineal foot in dB/ft, P the duct perimeter in inches, S the cross-sectional area of the duct in in^2, and α the sound absorption coefficient of the lining. This formula holds only under the following conditions: $\gamma = 250$ to 2000 Hz and $\alpha = 0.2$ to 0.4.

The power level attenuation characteristics of a lined $90°$-duct bend and a lined $180°$-duct bend are shown in Figure 14-9. Referring to Figure 14-8b, the power level attenuation resulting from an acoustically treated air plenum is given empirically by

$$A = 10\log\frac{1}{S_e(\cos\theta/2\pi d^2) + (1 - \alpha)/\alpha S_w} \tag{14-71}$$

[3]A. Thumann and R. K. Miller (1986). *Fundamentals of Noise Control Engineering.* Prentice Hall, Englewood Cliffs, N.J., pp. 225, 228, 233.

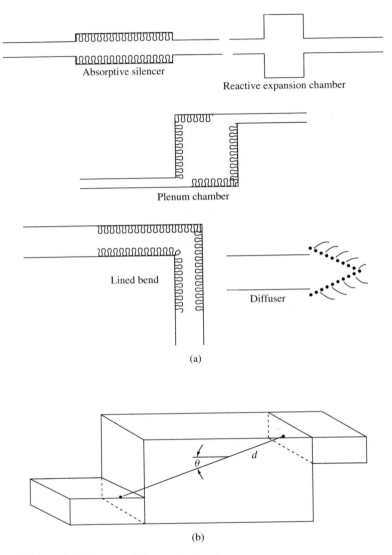

(a)

(b)

Figure 14-8   (a) Basic types of silencers. (b) Fan plenum.

where A is the power level attenuation in dB, S_e the plenum exit area in ft², θ and d are defined in the figure, α is the absorption coefficient, and S_w is the total plenum surface area in ft².

Example 14-19

A factory room measuring 70 ft long by 20 ft wide by 20 ft high is to be heated using a fan with the following sound power level characteristics: 63 Hz, 98 dB; 125 Hz, 88 dB; 250 Hz, 89 dB; 500 Hz, 87 dB; 1000 Hz, 85 dB; 2000 Hz, 85 dB; 4000 Hz, 83 dB; and

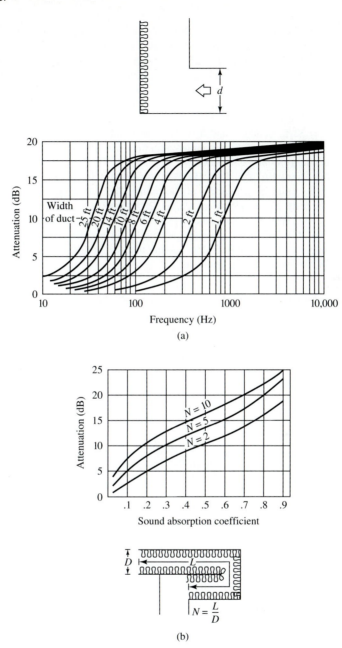

Figure 14-9 (a) Attenuation in lined 90°-duct bends. (b) Attenuation in 180°-duct bend.

8000 Hz, 71 dB. The main duct, 24 in. by 24 in., has a length of 50 ft with an elbow of 24 in. by 24 in. The elbow is used to connect the main to a branch or a continuation duct at a 90-degree turn with the main. The branch duct measures 24 in. by 24 in. by 60 ft long and is located at the ceiling running lengthwise with the 70-ft dimension of the room, perpendicular to and at the middle of the 20-ft dimension. The branch duct has six outlets at the ceiling discharging 500 cfm each. The main duct conveys 3000 cfm of air. The distances of the outlets to a point in the middle of the room are: outlets 1 and 6 = 32 ft, outlets 2 and 5 = 24 ft, and outlets 3 and 4 = 17 ft. (a) Calculate the sound power level at each duct outlet assuming that the duct is lined with an absorptive material having α values as follows: 63 Hz, 0.5; 125 Hz, 0.5; 250 Hz, 0.7; 500 Hz, 0.7; 1000 Hz, 0.7; 2000 Hz, 0.7; 4000 Hz, 0.7; and 8000, 0.6. (b) Calculate the sound pressure level at the point in the middle of the room. The following is additional information: ceiling (70 ft by 20 ft): acoustical tile, $\alpha(125\ \text{Hz}) = 0.47$, $\alpha(250\ \text{Hz}) = 0.80$, $\alpha(500\ \text{Hz}) = 0.95$, $\alpha(1000\ \text{Hz}) = 1.0$, $\alpha(2000\ \text{Hz}) = 0.9$, $\alpha(4000\ \text{Hz}) = 0.85$, walls (two 20 ft by 20 ft and two 20 ft by 70 ft): painted concrete block; and floor (70 ft by 20 ft): linoleum. In the absence of other information, assume that the Sabine formula applies to all the values of α and frequency in this example.

Solution   (a) Power level at each duct outlet, $L_{w\text{out}}$

$$A = 12.6 \frac{P}{S} \alpha^{1.4}$$

Let P_{wT} = total power at duct and $P_{w\text{out}}$ = power at each outlet; hence

$$P_{w\text{out}} = \frac{500}{3000} (P_{wT})$$

$$L_{wT} = 10 \log \sum_{i=1}^{i=n} 10^{L_{wi}/10}$$

$$= 10 \left[\log \left(10^{51.6/10} \right) + \left(10^{45.5/10} \right) + \left(10^{39.5/10} \right) + \left(10^{35.5/10} \right) + \left(10^{30.5/10} \right) \right.$$

$$\left. + \left(10^{26.5/10} \right) + \log \left(10^{19.5/10} \right) + \left(10^{12.5/10} \right) \right] = 52.89 \text{ dB}$$

$$10 \log P_{w\text{out}} = 10 \log \frac{500}{3000} + 10 \log P_{wT}$$

$$L_{w\text{out}} = 10 \log \frac{500}{3000} + L_{wT} = -7.78 + 52.89$$

$$= 45.11 \text{ dB for summation for all frequencies}\quad \textbf{Answer}$$

(b)
$$L_p = L_w + 10 \log \left(\frac{1}{4\pi r^2} + \frac{4}{R} \right)$$

	Octave band center frequency (Hz)							
	63	125	250	500	1000	2000	4000	8000
Sound power level of fan (dB)	98	88	89	87	85	85	83	71
α	0.5	0.5	0.7	0.7	0.7	0.7	0.7	0.6
Attenuation by 24 × 24 in. main duct	39.8ᵃ	39.8	63.73	63.73	63.73	63.73	63.73	51.36
Attenuation by 24 × 24 in. elbow (Figure 14-9)	1	1.5	3	12	17.5	18	18.5	18.5
Branch duct, 60 ft/2 $[A = 12.6(P/S)\alpha^{1.4}]$	23.87	23.87	38.24	38.24	38.24	38.24	38.24	30.81
Total attenuation (dB)	64.67	65.17	105	110	119	120	120	101
Corrected sound power level of fan (dB)	33.33	22.83	—	—	—	—	—	—
Turning vane power level at 2000 fpm	77	71	65	61	56	52	45	38
Correction at 750 fpm (see calculation of velocity belowᵃ)	-25.5	-25.5	-25.5	-25.5	-25.5	-25.5	-25.5	-25.5
Corrected vane power level (dB)	51.5	45.5	39.5	35.5	30.5	26.5	19.5	12.5
Combined power level at branch duct (dB)	51.6	45.5	39.5	35.5	30.5	26.5	19.5	12.5

$$A = 12.6 \frac{P}{S} \alpha^{1.4} = 12.6 \frac{24(4)}{24(24)} (0.5)^{1.4} = 0.796 \text{ dB/ft} \Rightarrow 0.796(50) = 39.8^a$$

ᵃ $v = \dfrac{3,000}{(24/12)(24/12)} = 750$ fpm.

$$R = \frac{S_T \bar{\alpha}}{1 - \bar{\alpha}}$$

$$\bar{\alpha} = \frac{\sum \alpha_i S_i}{S_T}$$

		Octave band center frequency (Hz)						
	63	125	250	500	1000	2000	4000	8000
Combined power level at branch duct L_w (dB)	51.6	45.5	39.5	35.5	30.5	26.5	19.5	12.5
L_{wout} for each outlet $\left(L_{wout} = 10 \log \frac{500}{3000} + L_w\right)$	43.82	37.72	32.72	27.72	22.72	18.72	11.72	4.72
α, ceiling $[20(70) = 1400 \text{ ft}^2]$	—	0.47	0.8	0.95	1.0	0.9	0.85	1
α, walls [Table 14-9, $2(20)(20) + 2(20)(70) = 3600 \text{ ft}^2$]		0.1	0.05	0.06	0.07	0.09	0.08	—
α, floor (Table 14-9, 1400 ft^2)	—	0.02	0.03	0.03	0.03	0.03	0.02	0
$S_T = 1400 + 3600 + 1400 = 6400 \text{ ft}^2$								
$\bar{\alpha} = \dfrac{\sum \alpha_i S_i}{S_T}$	—	0.16	0.21	0.25	0.26	0.25	0.24	—
$R = \dfrac{S_T \bar{\alpha}}{1 - \bar{\alpha}}$	1,219[a]	1,219	1,701	2,133	2,249	2,133	2,021	2,021[a]
$L_p = L_{wout} + 10 \log \left(\dfrac{1}{4\pi r^2} + \dfrac{4}{R}\right)$								
For $r = 17$ ft	19.33	13.23	6.91	1.04	—	—	—	—
For $r = 24$ ft	19.16	13.06	6.68	0.76	—	—	—	—
For $r = 32$ ft	19.08	12.88	6.57	0.63	—	—	—	—
$L_{pT} = 10 \log \left(\sum\limits_{i=1}^{i=n} \sqrt{10^{L_{pi}/10}}\right)^2$	28.89	22.60	16.26	8.38	—	—	—	—

[a] Assumed.

$$L_p = 10 \log \left(2\left(\sqrt{10^{28.89/10}}\right) + 2\left(\sqrt{10^{22.60/10}}\right) + 2\left(\sqrt{10^{16.26/10}}\right) + 2\left(\sqrt{10^{8.38/10}}\right)\right)^2$$

$$= 34.06 \text{ dB} \quad \textbf{Answer}$$

Example 14-20

The fan in a household discharges air into several rooms in the house through a fan plenum. Referring to Figure 14-8, $d = 10$ ft, $\theta = 50°$, $S_w = 350$ ft^2, and $S_e = 25$ ft^2. The absorption coefficients, α, of the glass lining used are as follows: 125 Hz, 0.6; 250 Hz, 0.8; 500 Hz, 0.8; 1000 Hz, 0.9; 2000 Hz, 0.8; 4000 Hz, 0.75. Compute the attenuation corresponding to each octave band.

Solution

$$A = 10 \log \frac{1}{S_e (\cos \theta / 2\pi d^2) + (1 - \alpha)/\alpha S_w}$$

	Octave band center frequency (Hz)					
	125	250	500	1000	2000	4000
α	0.6	0.8	0.8	0.9	0.8	0.75
A (dB)	15.6	15.8	15.8	15.9	15.8	15.7

Answer

Vibration Control

Noise is produced by vibration, and if a noise source is connected to an adjacent structure, the structure may also vibrate, making it a resultant noise source. For example, a machinery with a driving cycle of 10 to 60 Hz and rigidly attached to a foundation may produce a very annoying environment by transmitting its vibration to the foundation. One way to contain the noise is isolating the noise source through *vibration control*. As illustrated in Figure 14-10, vibration isolators may be of the spring type, rubber mount, and so on.

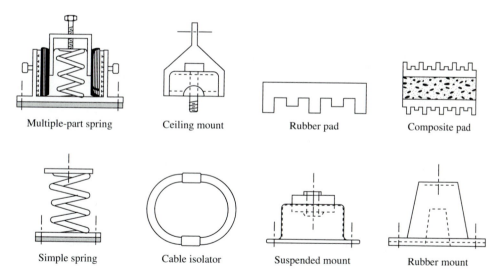

Multiple-part spring   Ceiling mount   Rubber pad   Composite pad

Simple spring   Cable isolator   Suspended mount   Rubber mount

Figure 14-10   Classification of vibration isolators.

Figure 14-11a is used to model the vibration of a noise source. The spring with a spring constant k represents any isolator. The force $F = F_0 \sin \omega t$ is the perturbing force that causes the machine, represented by the block, to be displaced by a distance x from

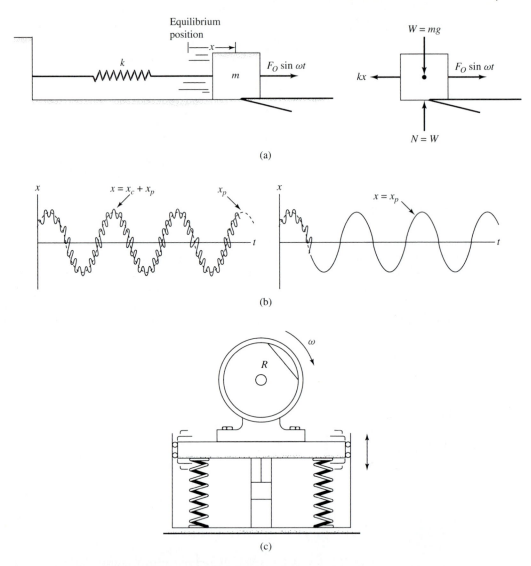

Figure 14-11 Schematics for derivation of vibration control equation.

the equilibrium point. The perturbing force can be the eccentric force that results when a machine is not balanced. The arrow with the half arrowhead at the bottom represents the friction force acting to oppose the perturbing force; it is a *damping force*. W is the weight of the machine, m its mass, g the acceleration due to gravity, and N the normal reactive force. The middle drawing represents the sine waves that result from the action of the perturbing force. The bottom is an example installation.

The friction or the damping force may be modeled as proportional to velocity (i.e., $c\dot{x}$, where c is the damping coefficient and $\dot{x}$ is the velocity). Applying Newton's second law in the horizontal direction and rearranging gives

$$m\ddot{x} + c\dot{x} + kx = F_0 \sin \omega t \tag{14-72}$$

where $\ddot{x}$ is the acceleration. The perturbing force, $F_0 \sin \omega t$, is also called the *forcing function*.

The differential equation corresponding to a system without the perturbing force is a homogeneous equation obtained from equation (14-72) by equating the forcing function to zero. The solution to this homogeneous equation is the complementary solution of equation (14-72), whose *auxiliary equation* in λ is

$$m\lambda^2 + c\lambda + k = 0 \tag{14-73}$$

By the quadratic formula,

$$\lambda = -\frac{c}{2m} \pm \sqrt{\left(\frac{c}{2m}\right)^2 - \frac{k}{m}} \tag{14-74}$$

The value of c that makes the expression inside the radical (the radicand) equal to zero is called the *critical damping coefficient* c_c. From equating the radicand to zero, this coefficient is

$$c_c = 2m\sqrt{\frac{k}{m}} \tag{14-75}$$

When c is greater than c_c, the system is said to be *overdamped; damped*; and when it is less than c_c, the system is said to be *underdamped*.

In equation (14-72), since the forcing function is harmonic (sinusoidal), the system response to this force must also be harmonic. This response is the particular solution x_p since it is a response "particular" to the forcing function. The most general x_p may be written as

$$x_p = A \sin \omega t + B \cos \omega t \tag{14-76}$$

By letting $A = C \cos \phi$ and $B = C \sin \phi$, where C and ϕ are new constants to be determined in place of A and B,

$$x_p = C \sin(\omega t + \phi) \tag{14-77}$$

Since the forcing function is given in terms of ωt and the response function (x_p) is given in terms of $\omega t + \phi$, ϕ represents the phase difference (phase lag) between the applied force and the resulting vibration of the system. x_p may be substituted into equation (14-72), from which, after performing the necessary algebra, the expressions for C and ϕ may be obtained:

$$C = \frac{F_0/k}{\sqrt{[1 - (\omega^2/(k/m))]^2 + 4(c/c_c)^2[\omega^2/(k/m)]}} \tag{14-78}$$

$$\phi = \tan^{-1} \frac{2(c/c_c)\left(\omega/\sqrt{k/m}\right)}{1 - \omega^2/(k/m)} \tag{14-79}$$

Figure 14-11b shows a small curve that wiggles about the large sine curve. In the absence of the forcing function, the large sine curve will be absent and the small curve will then wiggle only about the abscissa. Since the complementary solution x_c is the condition when the forcing function is zero, the small curve wiggling about the abscissa is the graph of the complementary solution. Putting in the forcing function will produce the particular solution x_p. Hence the left-hand side of Figure 14-11b represents the graph of both the complementary and particular solutions of equation (14-72).

The complementary solution is the response of the system when it is allowed to respond without a forcing function. This situation is obtained when the system is originally displaced by an amount x from equilibrium and then released. Upon release, the system will bounce back and forth, but without the forcing function. Since the forcing function is absent, the complementary solution is transitory and will die out with time. Hence the right-hand side of Figure 14-11b is the response of the system after some time when the complementary solution has "disappeared." Since the complementary solution is transitory, it will not be considered; the response of the system will simply be written in terms of its particular solution only. Thus

$$ x = \frac{F_0/k}{\sqrt{[1 - \omega^2/(k/m)]^2 + 4(c/c_c)^2[\omega^2/(k/m)]}} \sin\left[\omega t - \tan^{-1}\frac{2(c/c_c)\left(\omega/\sqrt{k/m}\right)}{1 - \omega^2/(k/m)}\right] $$

$$(14\text{-}80)$$

Let the system be forced by the amplitude F_0 only. Using this as the forcing function in equation (14-72) and going through the same process of simplification as was done with the perturbing force $F_0 \sin \omega t$, the response obtained is

$$ x = x_p = \frac{F_0}{k} \tag{14-81} $$

Since this response is the response of a nonvarying forcing function, it is a *static* response, as opposed to the response portrayed by equation (14-80), which is a *dynamic* response. The ratio of the amplitude of the dynamic response to that of the static response is called the *magnification factor*, MF, obtained from the equations above:

$$ \text{MF} = \frac{1}{\sqrt{[1 - \omega^2/(k/m)]^2 + 4(c/c_c)^2[\omega^2/(k/m)]}} \tag{14-82} $$

MF represents the number of times the response of noncyclical applied force is multiplied over when the force is made a cyclical force. The magnification factor is also called *transmissibility ratio,* TR.

Equation (14-82) shows MF or TR as a function of c/c_c. This ratio is called the *damping factor*. Table 14-21 shows some values of this factor.

Example 14-21

A motor weighing 70 N runs at 3600 rpm. (a) What is the stiffness, k, of the spring isolator to be provided if the desired TR is to be 0.15? (b) What is the phase difference between the applied force and the resulting vibration?

TABLE 14-21   ISOLATOR DAMPING FACTORS

Isolator type	Damping factor, c/c_0
Suspended isolator mount	0.5
Rubber isolator pad	0.15
Composite isolator pad	0.3
Glass fiber block pad	0.15
Simple spring isolator	0.15
Cable isolator	0.25
Multiple part spring	0.2
Rubber isolator mount	0.1
Ceiling isolator mount	0.15

Solution

(a)
$$TR = \frac{1}{\sqrt{[1 - \omega^2/(k/m)]^2 + 4(c/c_c)^2[\omega^2/(k/m)]}}$$

From Table 14-21, use $c/c_0 = 0.15$.

$$\omega = \frac{3600(2\pi)}{60} = 376.99 \text{ rad/s} \qquad m = \frac{70}{9.81} = 7.14 \text{ kg}$$

$$TR = 0.15 = \frac{1}{\sqrt{\left(1 - \frac{376.99^2}{k/7.14}\right)^2 + 4(0.15)^2\left(\frac{376.99^2}{k/7.14}\right)}} \qquad k = 133,262.06 \text{ N/m} \quad \textbf{Answer}$$

(b)
$$\phi = \tan^{-1}\frac{2(c/c_c)(\omega/\sqrt{k/m})}{1 - \omega^2/(k/m)}$$

$$= \tan^{-1}\frac{2(0.15)\left(376.99/\sqrt{133,262.06/7.14}\right)}{1 - 376.99^2/(133,262.06/7.14)} = -7.13°$$

$$\text{or} \quad 180 - 7.13 = 172.87° \quad \textbf{Answer}$$

GLOSSARY

Absorption silencer. A type of silencer where acoustic materials are lined in the interior of a duct to absorb the noise.

Acoustic reflex. A reflex action of the ear designed to protect it from intense noise.

Amplitude. The highest extent of oscillation of the particle in a wave.

A network. The weighting network that resembles human response at low frequencies.

Angular frequency. The number of radians per unit frequency.

Angular wave number. The number of radians per wave displacement.

Basilar membrane. Membrane that divides the cochlea into the upper and lower galleries.

Bel. A unit of expressing ratios of numbers in logarithm.

B network. The weighting network that resembles human response at medium frequencies.

C network. Weighting network that resembles human response at high frequencies.

Cochleal potential. The electric potential developed by the hair cells and transmitted to the brain by the auditory nerve.

Community noise equivalent level (CNEL). An L_{eq} over a 24-h period with 5 dBA added during the hours of 7 P.M. to 10 P.M. and 10 dBA during the hours of 10 P.M. to 7 A.M..

Composite noise rating (CNR). A contour of equal PNL (in PNdB) in and around the vicinity of an airport.

Constant frequency analysis. A frequency band analysis where the sound spectrum is analyzed at a constant increment of frequencies.

Constant percentage frequency analysis. A frequency band analysis where the lower limit frequency in the band is always a constant percentage of the upper limit frequency.

Critical damping coefficient. Value of the damping coefficient when the expression inside the radical of the auxiliary variable of the vibration control equation is equal to zero.

Critically damped system. A vibrating system whose damping coefficient is equal to the critical damping coefficient.

Damping coefficient. A coefficient that expresses the damping force as a linear function of the velocity.

Damping factor. The ratio of the damping coefficient of a vibrating system to its critical damping coefficient.

Damping force. An opposing force; a frictional force.

Day–night average sound level. An L_{eq} over a 24-h period with 10 dBA added to the $L_{eq(1)}$'s during the time from 10 P.M. to 7 A.M.

Decibel. Ten times a bel.

Destructive interference. The attenuation of noise by canceling the incoming noise waves using the reflected noise waves.

Diffuser. A device for attenuating noise by reducing the velocity.

Directivity. A factor that converts from a spherical sound intensity to nonspherical sound intensity.

Dissipative muffler. A silencer that attenuates noise by absorbing it.

D network. A weighting network that resembles human response at airports.

Dynamic response. The response of a vibrating system to a varying perturbing force.

Effective perceived noise level (EPNL). A perceived noise level of measurement that adds additional corrective parameters to the PNL.

Equivalent sound level. A single A-weighted equivalent measure of sound level corresponding to readings taken over a given period of time.

Exhaust muffler. A noise attenuating unit installed at the end of a pipe.

Forcing function. Anything that starts a process into motion.

Frequency. The number of repetitions or cycles per unit time.

Frequency band analysis. The procedure of measuring pressure levels by setting the sound level meter at the geometric center frequency of the band.

Hair cells. Contained between the tectoral and basilar membranes; serve as transducers to convert pressure energy to electrical.

Hertz. One cycle per second.

Infrasound. Sound transmitted below 16 Hz.

Longitudinal waves. Waves in which the particles oscillate in the direction of wave motion.

Loudness. The brain's perception of the magnitude of sound pressures.

Magnification factor (MF). The ratio of the amplitude of the dynamic response of a vibrating system to the amplitude of its static response.

Middle ear. An air-filled chamber of the ear composed of the ossicles: malleous (hammer), incus (anvil), and the stapes (stirrup).

Noise. Unwanted sound.

Noise criteria (NC). A measure of noise developed in a form of a chart to quantify noise levels in interior spaces.

Noise dose. A measure of noise that prorates the actual time of exposure of a person to the total allowable exposure time for a given noise level recommended by OSHA.

Noise exposure forecast (NEF). An improved version of the CNR by utilizing EPNL instead of the PNL in mapping the contours.

Noise exposure rating (NER). A measure of noise that prorates the actual time of exposure or exposures of a person to the total allowable exposure time for a given noise level recommended by CHABA.

Noise pollution level. The L_{eq} equaled or exceeded 0.50% of the time.

Octave band. A band of frequencies where the ratio of the upper limit frequency to the lower limit frequency of the band is 2:1.

1/2 octave band analysis. A frequency band analysis where the octave band is divided into two portions.

1/3 octave band analysis. A frequency band analysis where the octave band is divided into three portions.

Organ of Corti. Contained in the cochlea of the inner ear, it is the sense organ for hearing.

Outer ear. The pinna, auditory canal, and the eardrum (tympanic membrane) of the ear.

Overdamped system. A vibrating system whose damping coefficient is greater than the critical damping coefficient.

Perceived noise level (PNL). A calculated single value of noise based on a known level of people's annoyance from jet flyover noise.

Period. Time per repetition or cycle.

Phon. Loudness of a tone which when heard at 1000 Hz is numerically equal to the corresponding sound pressure level.

Preferred noise criteria (PNC). An improvement over NC.

Preferred speech interference level (PSIL). The arithmetic mean of the sound pressure levels at the 500-, 1000-, and 2000-Hz frequencies.

Rarefaction. In the transmission of sound by longitudinal wave vibration, that portion of the wave where the air is thin.

Reactive expansion muffler. A muffler that attenuates noise by destructive interference.

Semicircular canal. Part of the inner ear that provides the sense of balance.

Shadow zone. The space upwind from a sound source where wind velocity and temperature deflect the sound waves.

Simple harmonic motion. Motion in which acceleration is proportional to position; the motion of a swinging pendulum is such a motion.

Sone. The loudness 40 phons, arbitrarily set equal to 1.

Sound. The auditory sensation produced by the vibration of air, water, and so on.

Sound absorption coefficient. The fraction of incident noise that is absorbed by a surface.

Sound intensity level. Sound intensity expressed in decibels.

Sound level. Sound pressure level measured by the weighting networks A, B, C, and D.

Sound power level. Sound power expressed in decibels.

Sound pressure level. Sound pressure expressed in decibels.

Speech interference level (SIL). The arithmetic mean of the sound pressure levels at the octave bands at 600 to 1200, 1200 to 2400, and 2400 to 4800 Hz.

Standing wave. A wave that does not move.

Static response. The response of a vibrating system to a nonvarying perturbing force.

Transmissibility ratio. The magnification factor.

Transmission coefficient. The fraction of incident noise that is transmitted through a partition or wall.

Transverse waves. Waves in which the particles oscillate perpendicular to the direction of wave motion.

Ultrasound. Sound transmitted above 20,000 Hz.

Underdamped system. A vibrating system whose damping coefficient is less than the critical damping coefficient.

Vibration control. A method of reducing noise by isolating the noise source so as to prevent the propagation of vibration.

Wavelength. The length of displacement of a single wave in a waveform.

Wave number. The reciprocal of wavelength.

Weighting network. The electronic circuitry built into a sound level meter so as to produce a readout that closely resembles a human response.

SYMBOLS

A	attenuation
c	damping coefficient
CNEL	community noise equivalent level
CNR	composite noise rating
dBA	decibel level on the A weighting scale
D	noise dose
EPNL	effective PNL
F	amplitude of forcing function
k	angular wave number; spring constant
L_{dn}	day–night average sound level
L_{eq}	equivalent sound level

L_I	sound intensity level
L_N	statistical measure of noise level
L_{NP}	noise pollution level
L_p	sound pressure level
L_w	sound power level
MF	magnification factor
N	number of sample values; Fresnel number
N_t	effective noy value
NC	noise criteria
NEF	noise exposure forecast
NER	noise exposure rating
NPdB	decibel level on the L_{NP} measure
P	phon
P_w	sound power
Pa	pascal
PNC	preferred NC
PNdB	sound pressure decibel level in the PNL measure
PNL	perceived noise level
PSIL	preferred SIL
Q	directivity
r	distance of noise receiver from noise source
R	gas constant; room constant
s	perturbation displacement of sound about equilibrium (s_m, amplitude)
S	sone; area receiving noise
S_T	total surface area receiving noise
SIL	speech interference level
T	period
TL	transmission loss
TR	transmissibility ratio
v	velocity of sound
v_s	velocity of oscillation of air mass
W	watt
X_0	distance from sound source where temperature and wind start to deflect sound waves
α	absorption coefficient
β	bulk modulus of elasticity
γ	frequency
Δp	sound pressure above or below atmospheric or simply pressure
Δp_{rms}	mean square pressure
κ	wave number
λ	wavelength
ρ	density of medium for sound transmission

ρ_ℓ	areal density of partition
ϕ	relative humidity
ω	angular frequency

PROBLEMS

14-1. Using numerical values show that $\overline{\cos^2(kx - \omega t)} = \frac{1}{2}$.

14-2. Using numerical values show that $\overline{\sin^2(kx - \omega t)} = \frac{1}{2}$.

14-3. The maximum differential pressure, Δp_m, that the ear can tolerate in loud sounds is 28 N/m^2. What is the amplitude, s_m, for such a sound in air at a frequency of 10,000 Hz? Assume β and ρ equal to 20.6 psi and 1.23 kg/m^3 (for air), respectively. What are the rms pressure, intensity, and wavelength?

14-4. What is the difference between sound pressure level and sound level?

14-5. Repeat Problem 14-3 for the pressure amplitude of $2.8(10^{-5})$ Pa of the faintest detectable sound at 1000 Hz.

14-6. What is the sum of 68 dBA and 75 dBA?

14-7. The sound power from an automobile on a highway is 1 W. What is the sound power level? What are the sound intensity, the sound intensity level, the sound pressure, and the sound pressure level at a distance of 6 m from the automobile?

14-8. Using the weighting networks A, B, and C, the following sound levels were measured, respectively: 95 dBA, 95 dBB, and 95 dBC. What is the approximate frequency of the noise?

14-9. Calculate the loudness in sones for a simultaneous exposure for the following tones: 50 dB at 350 Hz, 50 dB at 1800 Hz, 50 dB at 3000 Hz, and 50 dB at 4000 Hz.

14-10. Compute the mean of Problem 14-6.

14-11. The upper and lower frequencies of an octave band are 177 and 355 Hz, respectively. (**a**) What is the center band frequency? If the lower frequency is 11,360 Hz, what are the upper frequencies of the corresponding (**b**) 1/2 and (**c**) 1/3 octaves?

14-12. A sound level of 95 dBA is measured for 5 min followed by a dBA of 60 for a period of 60 min. Calculate the equivalent sound level for the whole duration of 65 min.

14-13. The noise spectrum of a cutter equipment at a 5.0-ft distance is analyzed using the 1/1 octave band analysis, producing the results below. What are the sound pressure level and the sound level, L_{pCT}, in dBC generated by the equipment?

Center frequency (Hz)	L_p
63	62
125	71
250	72
500	77
1000	92
2000	79
4000	77
8000	63

14-14. Find the mean of L_p in Problem 14-13. Compare it to the arithmetic mean.

14-15. Traffic noise data are shown in the following table. Compute (a) L_{10} and (b) L_{50}.

Time (s)	dBA
10	71
20	75
30	70
40	78
50	80
60	84
70	76
80	74
90	75
100	74

14-16. The sound pressure level 55 m downwind from a compressor is 95 dB. If the wind speed is 5 m/s, the temperature is 25°C, the relative humidity is 60%, and the barometric pressure is 100.33 kPa, determine the sound pressure level upstream from the compressor.

14-17. The ambient air temperature is 25°C. If a building is 7.0 m high, express its height in terms of the wavelength of a 500-Hz sound.

14-18. The bihourly L_{eq}'s, $L_{eq(2)}$'s, in a certain neighborhood are shown in the following table. Compute (a) L_{dn} and (b) L_{NP} assuming the 24 h of record is sufficient to define $L_{eq(\infty)}$.

Time	$L_{eq(1)}$
24:00–1:00	41
2:00–3:00	32
4:00–5:00	30
6:00–7:00	38
8:00–9:00	62
10:00–11:00	60
12:00–13:00	59
14:00–15:00	62
16:00–17:00	60
18:00–19:00	59
20:00–21:00	52
22:00–23:00	40

14-19. A shopping mall is proposed to be constructed in a very quiet neighborhood. The noise measurements shown below are then taken in a nearby shopping mall similar to the one proposed. Based on the L_{eq} values, should complaints from the neighborhood be anticipated?

Time	Sound level (dBA)
5:00–7:00	43
7:00–9:00	56
9:00–11:00	64
11:00–13:00	70
13:00–15:00	69
15:00–17:00	55

14-20. The draft fans of the boilers of a 650-MW power plant produce 90 dB at 32.0 Hz at a distance of 190.0 m on a calm day. Determine the dB at a distance of 500 m if the temperature is 25°C, the relative humidity is 70%, and the barometric pressure is 29.92 in. Hg.

14-21. On the stage at an opera house, the following readings were obtained while a singer was performing: 110 dBA, 111 dBB, and 112 dBC. What is the approximate frequency of the singer's voice?

14-22. The following data are associated with a jet aircraft. Calculate the PNL.

1/1 Octave band (Hz)	L_p (dB)	N_i (noys)
125	103	50
500	100	65
1000	90	33
4000	81	34
8000	70	14

14-23. The sound pressure level of a full symphony and in a living room are, respectively, 0.632 Pa and 0.002 Pa. Calculate the respectively sound levels.

14-24. Assume two aircraft types, two flight paths, two daytime flights, and no nighttime flight at path 1 and one daytime flight and one nighttime flight at path 2 for aircraft types 1 and 2. Assume the PNL_{max} values for aircraft types 1 and 2 to be 114 PNdB and 117 PNdB, respectively, over flight path 1 and 116 PNdB and 120 PNdB, respectively, over flight path 2. Also assume the Δt for both types of aircraft in all paths as 25 s. Find the NEFs at flight paths 1 and 2. Calculate the corresponding L_{dn}.

14-25. The basement of a professor's house is leaking at the joint between the basement slab and the wall. He therefore cracked the joints to create a space into which hydraulic cement can be poured to seal the crack. If he uses an air hammer with a dB value of 100, to what L_{eq} will he be subjected if he works 10 h a day?

14-26. The noise levels in the chiller and compressor areas are 92 dBA and 93 dBA, respectively. If the employee is exposed two times at 2 h per exposure in the chiller area and 1 h in the compressor area, determine if the CHABA and OSHA recommendations are met.

14-27. Convert the noise levels in Problem 14-26 to pressure levels.

14-28. Dogs can hear sound pressures as low as $2(10^{-6})$ N/m^2. What is the equivalent sound level? How acute is dog hearing compared to that of humans if the threshold of human hearing is $2.8(10^{-5})$ Pa?

14-29. The data below were obtained from a cutter after a silencer was installed. What are the corresponding NC and PNC levels? What are the PSIL value and approximate SIL value?

1/1 Octave band (Hz)	L_p (dB)
63	53
125	60
250	63
500	66
1000	69
2000	70
4000	66
8000	53

14-30. A machine produces 80 dBA at 100 Hz. What is the corresponding dBC?

14-31. A source is put at the bottom of a cylinder of length ℓ and diameter R. Determine the value of Q.

14-32. The sound power level and the sound pressure level at one side of a cooling tower at a distance of 50 ft have been reported by a manufacturer as shown below. Calculate the sound pressure level and the sound level at a distance of 40 ft from the tower.

Levels	Octave band center frequency (Hz)					
	125	250	1000	2000	4000	8000
L_w (dB)	97	92	88	83	79	73
L_p (dB)	61	58	59	54	50	47

14-33. In Problem 14-32, what is the sound pressure level at the 40-ft distance given the following additional conditions? A concrete barrier 30 ft high is erected between the tower and the 40-ft distance at a point 20 ft from the tower. The tower is 15 ft high and the receiver point is 6 ft high, with the ground assumed horizontal. The atmospheric pressure and temperature are, respectively, 14.7 psi and 60°F. The relative humidity is 80%. Assume that the receiver is downwind from the source. Neglect A_3 and A_5.

14-34. Obtain a sound level meter from your instructor. Using this meter, measure the sound levels in your neighborhood and compare with the reading you obtain on the highway.

14-35. Using the same meter as in Problem 14-34, take the reading as an automobile passes you on your neighborhood street. Compare the reading when no automobile is passing by.

14-36. A factory room measuring 70 ft long by 20 ft wide by 20 ft high is to be heated using a fan with the following sound power level characteristics: 63 Hz, 98 dB; 125 Hz, 88 dB; 250 Hz, 89 dB; 500 Hz, 87 dB; 1000 Hz, 85 dB; 2000 Hz, 85 dB; 4000 Hz, 83 dB; and 8000 Hz, 71 dB. The main duct, 24 in. by 24 in., has a length of 50 ft with a 90° elbow of 24 in. by 24 in. at its terminus. This elbow is used to connect the main to a branch or a continuation duct at a 90° turn with the main. The branch duct measures 24 in. by 24 in. by 60 ft long and is located at the ceiling, running lengthwise with the 70-ft dimension of the room, perpendicular to and at the middle of the 20-ft dimension. The branch duct has six outlets at the ceiling, each discharging an equal amount of air. The main duct itself conveys 15,000 cfm of air. The distances of the outlets to a point in the middle of the room are: outlets 1 and 6 = 32 ft, outlets 2 and 5 = 24 ft, and outlets 3 and 4 = 17 ft. Calculate (**a**) the sound power level at each duct outlet and (**b**) the sound pressure level at the point in the middle of the room. Following is additional information: ceiling (70 ft by 20 ft): acoustical tile, $\alpha(125 \text{ Hz}) = 0.47$, $\alpha(250 \text{ Hz}) = 0.80$, $\alpha(500 \text{ Hz}) = 0.95$, $\alpha(1000 \text{ Hz}) = 1.0$, $\alpha(2000 \text{ Hz}) = 0.9$, $\alpha(4000 \text{ Hz}) = 0.85$, $\alpha(8000 \text{ Hz}) = 1.0$; walls (two 20 ft by 20 ft and two 20 ft by 70 ft): painted concrete block; and floor (70 ft by 20 ft): linoleum.

14-37. A brick wall is 0.41 m thick and has an area density of 590 kg/m^2. If the frequency of the incident sound is 1000 Hz, calculate TL and τ.

14-38. To control the noise coming from a tertiary crusher, the equipment installation is enclosed with $1\frac{1}{4}$-in. plywood. The following are the sound power level characteristics of the crusher at the inside wall of the enclosure: 63 Hz, 98 dB; 125 Hz, 88 dB; 250 Hz, 89 dB; 500 Hz, 87 dB; 1000 Hz, 85 dB; 2000 Hz, 85 dB; 4000 Hz, 83 dB; and 8000 Hz, 71 dB. What vocal effort would be exerted to have a face-to-face conversation immediately outside the enclosure?

14-39. The fan in a household discharges air into several rooms in the house through a fan plenum. Refer to Figure 14-10: $d = 50$ ft, $\theta = 50°$, $S_w = 350$ ft^2, and $S_e = 25$ ft^2. The absorption coefficients, α, of the glass lining used are as follows: 125 Hz, 0.6; 250 Hz, 0.8; 500 Hz, 0.8; 1000 Hz, 0.9; 2000 Hz, 0.8; 4000 Hz, 0.75. Compute the attenuation corresponding to each octave band. Compare the results when $d = 10$ ft.

14-40. A motor weighing 70 N runs at 1200 rpm. (**a**) What is the stiffness, k, of the spring isolator to be provided if the desired TR is to be 0.15? (**b**) What is the phase difference between the applied force and the resulting vibration? Compare result when rpm = 3600.

BIBLIOGRAPHY

BROCH, J. T. (1971). *Acoustic Noise Measurement.* B&K Instruments, Inc.

CUNNIFF, P. F. (1977). *Environmental Noise Pollution.* Wiley, New York.

DAVIS, M. L., and D. A. CORNWELL (1991). *Environmental Engineering.* McGraw-Hill, New York.

HARVEY, S. G. and C. C. OLIVER (1972). "Classification and Performance of Commercial Vibration Isolators." *Proceedings of Inter-Noise 72*, pp. 133–137.

HIBBELER, R. C. (1992). *Engineering Mechanics, Dynamics.* Macmillan, New York.

KINSLER, L. E., and A. R. FREY (1962). *Fundamentals of Acoustics.* Wiley, New York.

PETERSON, A. P. G., and E. E. GROSS (1972). *Handbook of Noise Measurement,* General Radio Company.

SCHULTZ, T. J. (1968). "Noise-Criterion Curves for Use with the USASI Preferred Frequencies." *Journal of the Acoustical Society of America,* 43, pp. 637–638.

SOCIETY OF AUTOMOTIVE ENGINEERS (1964). *Standard Values of Atmospheric Absorption as a Function of Temperature and Humidity for Use in Evaluating Aircraft Flyover Noise (ARP886).* Society of Automotive Engineers, New York.

THUMANN, A., and R. K. MILLER (1986). *Fundamentals of Noise Control Engineering.* Prentice Hall, Englewood Cliffs, N.J.

VESILIND, P. A., J. J. PEIRCE, and R. F. WEINER (1988). *Environmental Engineering.* Butterworth, Boston.

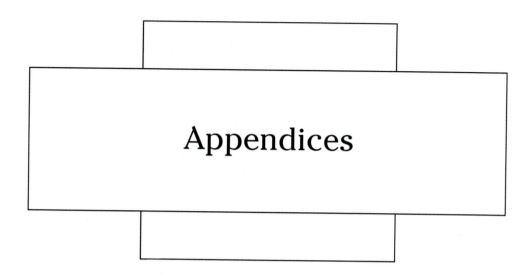

Appendices

APPENDIX 1: ATOMIC WEIGHTS OF THE ELEMENTS BASED ON C-12[a]

Name	Symbol	Atomic number	Atomic weight	Name	Symbol	Atomic number	Atomic weight
Actinium	Ac	89	227.0	Cerium	Ce	58	140.1
Aluminum	Al	13	27.0	Cesium	Cs	55	132.9
Americium	Am	95	(243)	Chlorine	Cl	17	35.5
Antimony	Sb	51	121.8	Chromium	Cr	24	52.0
Argon	Ar	18	39.9	Cobalt	Co	27	58.9
Arsenic	As	33	74.9	Copper	Cu	29	63.5
Astatine	At	85	(210)	Curium	Cm	96	(247)
Barium	Ba	56	137.3	Dysprosium	Dy	66	162.5
Berkelium	Bk	97	(247)	Einsteinium	Es	99	(252)
Beryllium	Be	4	9.012	Erbium	Er	68	167.3
Bismuth	Bi	83	209.0	Europium	Eu	63	152.0
Boron	B	5	10.8	Fermium	Fm	100	(257)
Bromine	Br	35	79.9	Fluorine	F	9	19.0
Cadmium	Cd	48	112.4	Francium	Fr	87	(223)
Calcium	Ca	20	40.1	Gadolinium	Gd	64	157.3
Californium	Cf	98	(251)	Gallium	Ga	31	69.7
Carbon	C	6	12.0	Germanium	Ge	32	72.6

[a]Values in parentheses (isotope atomic masses of longest half-life) are used for radioactive elements whose atomic weights cannot be quoted precisely without knowledge of the origin of the elements.

APPENDIX 1: (CONTINUED)

Name	Symbol	Atomic number	Atomic weight	Name	Symbol	Atomic number	Atomic weight
Gold	Au	79	197.0	Radium	Ra	88	226.0
Hafnium	Hf	72	178.5	Radon	Ra	86	(222)
Helium	He	2	4.0	Rhenium	Re	75	186.2
Holmium	Ho	67	164.9	Rhodium	Rh	45	102.9
Hydrogen	H	1	1.0	Rubidium	Rb	37	85.5
Iodine	I	53	126.9	Ruthenium	Ru	44	101.1
Indium	In	49	114.8	Samarium	Sm	62	150.4
Iridium	Ir	77	192.2	Scandium	Sc	21	45.0
Iron	Fe	26	55.8	Selenium	Se	34	79.0
Krypton	Kr	36	83.8	Silicon	Si	14	28.1
Lanthanum	La	57	138.9	Silver	Ag	47	108.0
Lawrencium	Lr	103	(260)	Sodium	Na	11	23.0
Lead	Pb	82	207.2	Strontium	Sr	38	87.6
Lithium	Li	3	6.9	Sulfur	S	16	32.1
Lutetium	Lu	71	175.9	Tantalum	Ta	73	181.0
Magnesium	Mg	12	24.3	Technetium	Tc	43	(98)
Manganese	Mn	25	54.9	Tellurium	Te	52	127.6
Mendelivium	Md	101	(258)	Terbium	Tb	65	159.0
Mercury	Hg	80	201.0	Thallium	Tl	81	204.4
Molybdenum	Mo	42	95.9	Thorium	Th	90	232.0
Neodymium	Nd	60	144.2	Thulium	Tm	69	169.0
Neon	Ne	10	20.2	Tin	Sn	50	118.7
Neptunium	Np	93	237.0	Titanium	Ti	22	47.9
Nickel	Ni	28	58.7	Tungsten	W	74	183.9
Niobium	Nb	41	92.9	Unnilennium	Une	109	(266)
Nitrogen	N	7	14.0	Unnilhexium	Unh	106	(263)
Nobelium	No	102	(259)	Unniloctium	Uno	108	(265)
Osmium	Os	76	190.2	Unnilpentium	Unp	105	(262)
Oxygen	O	8	16.0	Unnilquadium	Unq	104	(261)
Palladium	Pd	46	106.4	Unnilseptium	Uns	107	(262)
Phosphorus	P	15	31.0	Uranium	U	92	238.0
Platinum	Pt	78	195.1	Vanadium	V	23	50.9
Plutonium	Pu	94	(244)	Xenon	Xe	54	131.3
Polonium	Po	84	(209)	Ytterbium	Yb	70	173.0
Potassium	K	19	39.1	Yttrium	Y	39	88.9
Praseodymium	Pr	59	140.9	Zinc	Zn	30	65.4
Promethium	Pm	61	(145)	Zirconium	Zr	40	91.2
Protactinium	Pa	91	231.0				

APPENDIX 2: SATURATION VALUES OF DISSOLVED OXYGEN EXPOSED TO SATURATED ATMOSPHERE AT ONE ATM PRESSURE

Temperature °C	Chloride concentration (mg/L)					Saturated H_2O vapor pressure (kPa)
	0	5,000	10,000	15,000	20,000	
0	14.6	13.8	13.0	12.0	11.3	0.61
1	14.2	13.4	12.6	11.8	11.0	0.66
2	13.8	13.1	12.3	11.5	10.8	0.71
3	13.5	12.7	12.0	11.2	10.5	0.76
4	13.1	12.4	11.7	11.0	10.3	0.81
5	12.8	12.1	11.4	10.7	10.0	0.87
6	12.5	11.8	11.1	10.5	9.8	0.93
7	12.2	11.5	10.9	10.2	9.6	1.00
8	11.9	11.2	10.6	10.0	9.4	1.07
9	11.6	11.0	10.4	9.8	9.2	1.15
10	11.3	10.7	10.1	9.6	9.0	1.23
11	11.1	10.5	9.9	9.4	8.8	1.31
12	10.8	10.3	9.7	9.2	8.6	1.40
13	10.6	10.1	9.5	9.0	8.5	1.49
14	10.4	9.9	9.3	8.8	8.3	1.60
15	10.2	9.7	9.1	8.6	8.1	1.70
16	10.0	9.5	9.0	8.5	8.0	1.82
17	9.7	9.3	8.8	8.3	7.8	1.94
18	9.5	9.1	8.6	8.2	7.7	2.06
19	9.4	8.9	8.5	8.0	7.6	2.20
20	9.2	8.7	8.3	7.9	7.4	2.34
21	9.0	8.6	8.1	7.7	7.3	2.49
22	8.8	8.4	8.0	7.6	7.1	2.64
23	8.7	8.3	7.9	7.4	7.0	2.81
24	8.5	8.1	7.7	7.3	6.9	2.98
25	8.4	8.0	7.6	7.2	6.7	3.17
26	8.2	7.8	7.4	7.0	6.6	3.36
27	8.1	7.7	7.3	6.9	6.5	3.56
28	7.9	7.5	7.1	6.8	6.4	3.78
29	7.8	7.4	7.0	6.6	6.3	4.01
30	7.6	7.3	6.9	6.5	6.1	4.24

APPENDIX 3: SDWA ACRONYMS

BAT	Best available technology
BTGA	Best technology generally available
CWS	Community water systems
DWEL	Drinking water equivalence level
EMSL	USEPA Environmental Monitoring and Support Laboratory (Cincinnati)
GAC	Granular activated carbon
IOC	Inorganic chemical
IPDWR	Interim primary drinking water regulation
LOAEL	Lowest observed adverse effect level
LOQ	Limit of quantitation
MCL	Maximum contaminant level
MCLG	Maximum contaminant level goal
MDL	Method detection limit
NOAEL	No observed adverse effect level
NPDWR	National primary drinking water regulation
NTNCWS	Nontransient noncommunity water system
PAC	Powdered activated carbon
PHS	Public Health Service
POET	Point-of-entry technology
POUT	Point-of-use technology
PQL	Practical quantitation level
PTA	Packed tower aeration
RFD	Reference dose
RIA	Regulatory impact analysis
RMCL	Recommended maximum contaminant level
RPDWR	Revised primary drinking water regulation
RSC	Relative source contribution
SDWA	Safe Drinking Water Act
SMCL	Secondary maximum contaminant level
SNARL	Suggested no adverse response level
SOC	Synthetic organic chemical
TNCWS	Transient noncommunity water system
UIC	Underground injection control
URTH	Unreasonable risk to health
VOC	Volatile organic chemical

APPENDIX 4: SAMPLE DRINKING WATER VOC VALUES

Contaminant	MCL (mg/L)	BAT
Benzene	0.005–0	PTA or GAC
Carbon tetrachloride	0.005–0	PTA or GAC
p-Dichlorobenzene	0.075	PTA or GAC
1,2-Dichloroethane	0.005–0	PTA or GAC
1,1-Dichloroethylene	0.007	PTA or GAC
1,1,1-Trichloromethane	0.20	PTA or GAC
Trichloroethylene	0.005–0	PTA or GAC
Vinyl chloride	0.002–0	PTA
Bromobenzene	—	—
Bromodichloromethane	—	—
Bromoform	—	—
Bromomethane	—	—
Chlorobenzene	—	—
Chlorodibromomethane	—	—
Chloroethane	—	—
Chloroform	—	—
Chloromethane	—	—
o-Chlorotoluene	—	—
p-Chlorotoluene	—	—
Dibromomethane	—	—
m-Dichlorobenzene	—	—
o-Dichlorobenzene	—	—
1,1-Dichloroethane	—	—
cis-1,2-Dichloroethylene	—	—
trans-1,2-Dichloroethylene	—	—
Dichloromethane	—	—
1,2-Dichloropropane	—	—
1,3-Dichloropropane	—	—
2,2,-Dichloropropane	—	—
1,1-Dichloropropene	—	—
1,3-Dichloropropene	—	—
Ethylbenzene	—	—
Styrene	—	—
1,1,1,2-Tetrachloroethane	—	—
1,1,2,2-Tetrachloroethane	—	—
Tetrachloroethylene	—	—
Toluene	—	—
1,1,2-Trichloroethane	—	—
1,2,3-Trichloropropane	—	—
m-Xylene	—	—
o-Xylene	—	—
p-Xylene	—	—

APPENDIX 5: SAMPLE DRINKING WATER SOC AND IOC VALUES

Contaminant	MCL (mg/L)	Contaminant	MCL (mg/L)
SOC			
Acrylamide	treatment technique	PCBs (as decachloro-biphenyls)	0.0005
Alachlor	0.002	Pentachlorophenol	0.2
Aldicarb	0.01	Styrene	0.005
Aldicarb sulfone	0.04	Tetrachloroethylene	0.005
Aldicarb sulfoxide	0.01	Toluene	2.0
Atrazine	0.003	Toxaphene	0.005
Carbofuran	0.04	2,4,5-TP (Silvex)	0.05
Chlordane	0.002	Xylene	10
2,4-D	0.07		
Dibromochloropropane	0.0002	**IOC**	
o-Dichlorobenzene	0.6	Arsenic	0.05
cis-1,2-Dichloroethylene	0.07	Asbestos	7 million
trans-1,2-Dichloroethylene	0.1		fibers/L
1,2-Dichloropropane	0.005		(longer than
Epichlorohydrin	treatment technique	Barium	10 μm) 5.0
Ethylbenzene	0.7	Cadmium	0.005
Ethylene dibromide (EDB)	0.00005	Chromium	0.1
Heptachlor	0.0004	Mercury	0.002
Heptachlor epoxide	0.0002	Nitrate	10 as N
Lindane	0.0002	Nitrite	1 as N
Methoxychlor	0.04	Selenium	0.05
Monochlorobenzene	0.1	Silver	0.05

APPENDIX 6: DISINFECTANT CHEMICALS AND BY-PRODUCTS

Disinfectants
  Ammonia   Chlorine   Hypochlorite ion
  Chloramines   Chlorine dioxide   Hypochlorous
  Chlorate   Chlorite

Disinfectant by-products
  Haloacetic acids
    Dibromoacetic   Monobromoacetic   Trichloroacetic
    Dichloroacetic   Monochloroacetic

  Haloacetonitriles
    Bromochloroacetonitrile Dichloroacetonitrile
    Dibromoacetonitrile   Trichloroacetonitrile

  Haloketones
    1,1-Dichloropropanone   1,1,1-Trichloropropanone

  Trihalomethanes
    Bromodichloromethane   Chloroform
    Bromoform   Dibromochloromethane

  Chlorophenols
    2-Chlorophenol   2,4-Dichlorophenol   2,4,6-Trichlorophenol

  Aldehydes   3-Chloro-4-(dichloro-   Iodate
  Bromate   methyl)-5-hydroxy-2(5H)- Iodide
  Bromide   furanone [MX] N-Organochloramines
  Bromodichloromethane   Chloropicrin   Ozone by-products
  Chloral hydrate   Cyanogen chloride

APPENDIX 7: MICROBIOLOGICAL AND RADIONUCLIDE CONTAMINANTS AND SECONDARY MCL VALUES FOR A NUMBER OF SUBSTANCES

Microbiological contaminants
    Total coliforms   *Giadia lamblia*   *Legionella*
    Viruses   *Cryptosporidium*

Radionuclide contaminants
    Gross alpha particle   Natural uranium   Beta particle
    Photon radioactivity   Radium 226 and 228   Radon

Secondary MCLs

Contaminant	SMCL (mg/L)
Chloride	250
Color	15 color units
Copper	1
Corrosivity	Noncorrosive
Foaming agents	0.5
Hydrogen sulfide	0.05
Iron	0.3
Manganese	0.05
Odor	3 TON
pH	6.5–8.5
Sulfate	250
Total dissolved solids	500
Zinc	5

APPENDIX 8: SOME PRIMARY DRINKING WATER CRITERIA

Contaminants	Concentration (mg/L, unless noted otherwise)
Arsenic	0.05
Barium	1.00
Cadmium	0.010
Chromium	0.05
Lead	0.05
Mercury	0.002
Nitrate as N	10.00
Selenium	0.01
Silver	0.05
Chlorinated hydrocarbon	
Endrin (1,2,3,4,10,10-hexachloro-6,7-epoxy-1,4,4a,5,6,7,8,8a-octo-hydro-1,4-endo,endo-5,8-dimetha-nonaphthalene)	0.0002
Lindane (1,2,3,4,5,6-hexachloro-cyclohexane, γ isomer)	0.004
Methoxychlor [1,1,1-trichloro-2,2-bis(p-methoxyphenyl)ethane]	0.1
Toxaphene ($C_{10}H_{10}Cl_8$—technical chlorinated camphene, 67-69% chlorine)	0.005
Chlorophenoxys	
2,4-D (2,4-dichlorophenoxyacetic acid)	0.1
2,4,5-TP (2,4,5-trichloro-phenoxypropionic acid)	0.01
Microbiological contaminants	Based on average of two consecutive days: 5 TU. *Membrane filter technique:* not to exceed 1/100 mL on monthly basis; not to exceed 4/100 mL of coliforms in more than one sample in fewer than 20 samples per month on an individual sample basis; not to exceed 4/100 mL of coliforms in more than 5% of sample in more than 20 samples per month on an individual basis. *Fermentation tube method:* 1. *10-mL standard portions.* Coliforms shall not be present in more than 10% of the portions on monthly basis; coliforms shall not be present in three or more portions in more than one sample in fewer than 20 samples per month on an individual basis; coliforms shall not be present in three or more portions in more than 5% of samples in more than 20 samples per month on an individual basis. 2. *100-mL standard portions.* Coliforms shall not be present in more than 60% of the portions on monthly basis; coliforms shall not be present in five portions in more than one sample in fewer than 20 samples per month on an individual basis; coliforms shall not be present in five portions in more than 20% of samples in more than 20 samples per month on an individual basis.
Turbidity	Based on monthly average: 1 TU or up to 5 TU if the water supplier can demonstrate that the higher turbidity does not interfere with disinfection.

APPENDIX 9: SOME SECONDARY DRINKING WATER CRITERIA

Contaminants	Concentration (mg/L, unless noted otherwise)
Chloride	250
Color	15 CU (color units)
Copper	1
Corrosivity	Noncorrosive
Foaming agents	0.5
Hydrogen sulfide	0.05
Iron	0.3
Manganese	0.05
Odor	< 3 TON
pH	6.5–8.5
Sulfate	250
Total dissolved solids	500
Zinc	5

APPENDIX 10: PHYSICAL CONSTANTS

Physical constant	Value
Acceleration due to gravity, g	9.80665 m/s^2 and 32.174 ft/s^2 at sea level
Atomic mass unit, amu	$\frac{1}{12}$ the mass of ^{12}C $= 1.6605402(10^{-27})$ kg
Avogadro's number	$6.0221367(10^{23})$ per g mol
Boltzmann constant	$1.380658(10^{-23})$ J/K
Charge-to-mass ratio for electrons	$1.75881962(10^{11})$ C/kg
Electrical permittivity constant	$8.85(10^{-12})$ C/V-m
Electron charge	$1.60217733(10^{-19})$ C
Electron rest mass	$9.109390(10^{-31})$ kg
Faraday constant	$9.6485309(10^4)$ C per equivalent
g mol	22.4 L ideal gas at STP of 0°C and 1 atm
lb mole	359 ft^3 ideal gas at STP of °C and 1 atm
Neutron rest mass	$1.6749286(10^{-27})$ kg
Planck constant	$6.6260755(10^{-34})$ J-s
Proton rest mass	$1.6726231(10^{-27})$ kg
Speed of light in vacuum	$2.99792458(10^8)$ m/s
Standard room temperature	25°C
Standard temperature and pressure (STP)	1 atm and 0°C
Universal gas constant	$8.205784(10^{-2})$ L-atm-K/g mol, 8.314510 J/g mol-K, 1.987 cal/g mol-K, 82.05 atm-cm^3/g mol-K, $4.968(10^4)$ lb$_m$-ft^2/(lb mol)-°R 49,720 ft-lb$_f$/slug-^{0}R

APPENDIX 11: CONVERSION FACTORS

From	To
Å, angstrom	10^{-8} cm
A	C/s
acre	43,560 ft^2, 0.00156 mi^2, 0.405 ha
atm	101.325 kN/m^2, 14.696 lb_f/in^2, 101.325 kPa, 1.013 bar, 29.92 in. Hg and 760 mmHg at 0°C, 760 torr, 33.936 ft H_2O (60°F), 2116.2 lb_f/ft^2
barrel	42 gal
Btu	252.2 cal, 778.2 ft-lb_f
bu/ha	0.4047 bu/acre
C	$6.2(10^{18})$ electrons, 1 A-s
cal	4.1868 J
cal/g mol	1.8 Btu/lb mol
cm	0.3937 in.
cm^3	1 mL
cP	0.000672 lb_m/s-ft, 10^{-3} Pa-s
dyn	10^{-5} N
erg	1 dyn-cm, 10^{-7} J
esu	$1.59(10^{-19})$ A-s, $1.59(10^{-19})$ C
ft	0.305 m
ft^3	7.481 gal, 28.32 L, 0.0283 m^3
°F	[1.8(°C) + 32]
g	0.0353 oz, 0.0022 lb_m
g/m^3	8.3454 lb_m/Mgal
gal	3.785 L
gpm	0.227 m^3/h
grain	$6.480(10^{-2})$ g
grain/ft^3	2.29 g/m^3
ha	2.4711 acre, 10^4 m^2
hp	746 W; 33,000 ft-lb_f/min; 2545 Btu/h; 0.0738 boiler hp
hp-h	$1.98(10^6)$ ft-lb_f
Hz	1 cycle/s
in.	2.54 cm
J	$2.7778(10^{-7})$ kWh, 0.7376 ft-lb_f, 1.0 W-s, 1.0 N-m, 0.2388 cal, 10^7 ergs
kg	2.2046 lb_m
kg/ha	0.8922 lb_m/acre
kg/ha-day	0.8922 lb_m/acre-day
kg/kWh	1.6440 lb_m/hp-h
kg/m^3	8345.4 lb_m/Mgal
kg/m^2-day	0.2048 lb_m/ft^2-day
kg/m^3-day	62.428 $lb_m/10^3$ ft^3-day
kJ	0.9478 Btu
kJ/kg	0.4303 Btu/lb_m
km	0.6214 mile
kPa	0.0099 atm
kW	0.9478 Btu/s, 1.3410 hp
kWh	3412 Btu, 1.341 hp-h
kW/m^3	5.0763 hp/10^2 gal
$kW/10^3$ m^3	0.0380 hp/10^3 ft^3
K	°C + 273.16

APPENDIX 11: (CONTINUED)

From	To
lb_f	$[lb_m/(g_c\ lb_m/slug)]g$ ($g_c = 32.174$; g is the gravitational acceleration)
lb_m	453.6 g, 7000 grains, 16 oz
long ton	2240 lb_m
L	0.2642 gal, 0.0353 ft^3, 33.8150 oz, 1.057 quarts, $10^{-3}\ m^3$
L/m^2-min	35.3420 gal/ft^2-day
L/m^2	$2.4542(10^{-2})\ gal/ft^2$
m	3.2808 ft, 39.3701 in., 1.0936 yd
m/s^2	3.2808 ft/s^2, 39.3701 $in./s^2$
m^2	$2.471(10^{-4})$ acre
m^3	35.314 ft^3, $8.1071(10^{-4})$ acre-ft, 1000 L, 264.1720 gal
$m^3/10^3\ m^3$	133.6806 ft^3/Mgal
m^3/m-day	80.5196 gal/ft^2-day; 10.7639 ft^3/ft-min
m^3/m^2-day	24.5424 gal/ft^2-day
m^3/m^2-h	589.0173 gal/ft^2-day
m^3/m^3	0.1337 ft^3/gal
mile	5280 ft, 1609 km
Mg	1.1023 short ton, 0.9842 long ton, 1.0 tonne
MJ	0.3725 hp-h
N	0.2248 lb_f, 1 kg-m/s^2
N/m^2	1 Pa
Pa	1 N/m^2, $1.4504(10^{-4})\ lb_f/in^2$
poise	1 g/cm-s, 0.1 Pa-s
$°R$	$°F + 459.49$
short ton	2000 lb_m, 0.907 tonne
stoke	cm^2/s
ton of refrigeration	200 Btu/min
W	0.7376 ft-lb_f/s, 1 J/s, 10^7 ergs/s
$W/m^2-°C$	0.1763 Btu/ft^2-$°F$-h
μm	10^{-6} m, 10^{-3} mm

APPENDIX 12: ENVIRONMENTAL TRAGEDIES

Year	Place	Pollution problem	Deaths, diseases, catastrophe, remarks
1854	Broad Street (pump), London, England	Water	Asiatic cholera epidemic
1872	Lausen, Switzerland	Water	Typhoid fever epidemic
1892	Hamburg, Germany	Water	Cholera epidemic
1923	Chicago, Illinois	Water	Typhoid fever epidemic
1928	Olean, New York	Water	Typhoid fever epidemic
1929	Ogden, Utah	Water	Typhoid fever epidemic
1930	Meuse Valley, Belgium	Air	63 deaths; 6000 became ill
1933	Chicago, Illinois	Water	Amoebic dysentery epidemic
1948	Donora, Pennsylvania	Air	21 deaths; 6000 became ill
1948	London, England	Air	700–800 deaths
1952	London, England	Air	4000 deaths
1956	London, England	Air	1000 deaths
1957	London, England	Air	700–800 deaths
1959	London, England	Air	200–250 deaths
1962	London, England	Air	700 deaths
1963	London, England	Air	700 deaths
1963	New York, New York	Air	200–400 deaths
1972	Stringfellow acid pits, Glen Avon, California	Water, hazardous waste	Groundwater pollution discovered in 1972
1975	Hopewell, Virginia	Water, hazardous waste	Kepone discharge to James River; plant closed in 1975
1975	Back River, Baltimore, Maryland	Water	Fish kills
1975	Waldwick, New Jersey	Water, hazardous waste	Biocraft Laboratories leaked the solvent butanol, acetone, and methylene to the ground; discovered in 1975
1976	Seveso, Italy	Air, hazardous waste	Dioxin cloud over Seveso
1978	Love Canal, Niagara Falls, N.Y.	Hazardous waste	In 1976, puddles of hazardous wastes surfaced at backyards of residents; in 1978, EPA evacuated 237 families; people built on top of chemical dump
1978	France	Oil spill	Called the Amoco Cadiz spill, 68 million gallons of oil was dumped off the coast of France
1979	General Electric site, Oakland, California	Water, hazardous waste	PCB contamination; complaint filed by a GE employee in 1979
1984	Bhopal, India		Over 2000 people died by an accidental release of methyl isocyanate

APPENDIX 12: (CONTINUED)

Year	Place	Pollution problem	Deaths, diseases, catastrophe, remarks
1985	Stratosphere over Antarctica	Air	Announced in 1985 that a big hole over Antarctica the size of the United States has been created due to ozone depletion mostly by CFCs
1986	Chernobyl, Russia	Nuclear	Meltdown of a nuclear reactor at Chernobyl; contamination spread over much of Europe
1989	Valdez, Alaska	Oil spill	The incident called *Exxon Valdez*; the tanker Blight Reef spilled 11 million gallons of crude into Prince William Sound
1991	Sacramento River, California	Hazardous chemicals	Pesticide, metam sodium, spilled from a Southern Pacific train into the 30-mile stretch of stream threatening Shasta Lake, a source of water for California
1991	Luzon, Philippines	Dust, mud flows	Mt. Pinatubo erupted, closing an important U.S. military base
1991	Kuwait, Middle East	Air pollution, spill	The Gulf war; the Iraqi army exploded oil storage tanks, causing massive oil spills into the Persian Gulf; they also set fire to the Kuwaiti oil wells, creating a large air pollution problem
1993	Luzon, Philippines	Dust, mud flows	The Mayon volcano erupted

APPENDIX 13: ENVIRONMENTAL LEGISLATIONS IN THE UNITED STATES

Year	Title	Environmental concern					
		Air	Noise	Water	Solid	Hazardous	Drinking water
1850	City of Boston ordinance		†				
1881	Chicago and Cincinnati antismoke ordinances	†					
1899	Rivers and Harbors Act			†	†		
1912	Public Health Service Act			†			†
1924	Oil Pollution Act			†			
1942	Walsh–Healey Public Contracts Act		†				
1946	Atomic Energy Act					†	
1948	Water Pollution Control Act			†			
1954	Atomic Energy Act					†	
1955	Air Pollution Control Act	†					
1956	Federal Water Pollution Control Act (FWPCA)			†			
1960	Motor Vehicle Exhaust Act	†					
1962	Federal Aid Highways Act Amendments		†				
1963	Clean Air Act	†					
1965	Motor Vehicle Air Pollution Control Act	†					
1965	Water Quality Act			†			
1965	Solid Waste Disposal Act				†		
1966	Department of Transportation Act		†				
1967	Federal Air Quality Act	†					
1968	Radiation and Control for Health and Safety Act					†	
1970	National Environmental Policy Act	†	†	†	†	†	
1970	Resource Recovery Act				†		
1970	Occupational Safety and Health Act		†				
1970	Noise Pollution and Abatement Act		†				
1970	Federal Aid Highways Act		†				
1970	Clean Air Act Amendments	†					
1972	Federal Insecticide, Fungicide, and Rodenticide Act					†	
1972	FWPCA Amendments			†			

APPENDIX 13: (CONTINUED)

		Environmental concern					
Year	Title	Air	Noise	Water	Solid	Hazardous	Drinking water
1972	Marine Protection, Research, and Sanctuaries Act			†			
1974	Safe Drinking Water Act						†
1974	Energy Supply and Environmental Coordination Act	†					
1975	Hazardous Materials Transportation Act					†	
1976	Resource Conservation and Recovery Act				†	†	
1976	Toxic Substances Control Act					†	
1977	Energy Organization Act					†	
1977	Clean Water Act			†			
1977	Clean Air Act Amendments	†					
1980	Comprehensive Environmental Response, Compensation, and Liability Act (Superfund)				†	†	
1980	Acid Precipitation Act	†					
1980	Low-Level Waste Policy Act					†	
1981	Municipal Waste Treatment Construction Grants Amendments			†			
1982	Nuclear Waste Policy Act					†	
1984	Hazardous and Solid Waste Amendments				†	†	
1985	Low-Level Radioactive Waste Policy Act Amendments				†		
1986	Superfund Amendments and Reauthorization Act			†	†		
1986	Radon Gas and Indoor Quality Research Act	†					
1986	Safe Drinking Water Act Amendments						†
1987	Nuclear Waste Policy Act Amendments					†	
1990	Clean Air Act Amendments	†					

APPENDIX 14: USEPA OFFICES

Region	States included in region
I	Maine, New Hampshire, Massachusetts, Vermont, Connecticut, Rhode Island
II	New York, New Jersey
III	Pennsylvania, West Virginia, Virginia, Delaware, Maryland, District of Columbia
IV	Kentucky, Tennessee, North Carolina, South Carolina, Mississippi, Alabama, Georgia, Florida, Virgin Islands
V	Minnesota, Wisconsin, Illinois, Michigan, Indiana, Ohio
VI	New Mexico, Texas, Oklahoma, Arkansas, Louisiana
VII	Nebraska, Kansas, Iowa, Missouri
VIII	Montana, North Dakota, South Dakota, Wyoming, Utah, Colorado
IX	California, Nevada, Arizona, Hawaii, Pacific Territories
X	Washington, Oregon, Idaho, Alaska

APPENDIX 15: ENVIRONMENTAL ORGANIZATIONS

Air and Waste Management Association
P.O. Box 2861
Pittsburgh, PA 15230

American Cetacean Society
P.O. Box 2638
San Pedro, CA 90731

American Forestry Association
1516 P Street, NW
Washington, DC 20005

American Littoral Society
Sandy Hook
Highlands, NJ 07732

American Rivers
801 Pennsylvania Avenue, SE, #303
Washington, DC 20003-2167

American Wilderness Alliance
6700 East Arapahoe, Suite 114
Englewood, CO 80112

Americans for the Environment
1400 16th Street, NW
Washington, DC 20036

Arctic Institute of North America
University of Calgary
2500 University Drive
Calgary, Alberta, Canada T2N 1N4

Atlantic Center for the Environment
39 South Main Street
Ipswich, MA 01938

Canadian Coalition on Acid Rain
112 St. Clair Avenue W, Suite 401
Toronto, Ontario
Canada M4V 2Y3

Canadian Nature Federation
453 Sussex Drive
Canada K1N 6Z4
Ottawa, Ontario

Center for Marine Conservation
1235 DeSales Street, NW
Washington, DC 20036

Citizens Clearinghouse for Hazardous Waste
P.O. Box 926
Arlington, VA 22216

The Conservation Foundation
1250 24th Street, NW
Washington, DC 20037

The Cousteau Society
930 West 21st Street
Norfolk, VA 23517

Defenders of Wildlife
1244 19th Street, NW
Washington, DC 20036

APPENDIX 15: (CONTINUED)

Earth Island Institute
300 Broadway, Suite 28
San Francisco, CA 94133-3312

Environmental Action
1525 New Hampshire Avenue, NW
Washington, DC 20036

Environmental Defense Fund
257 Park Avenue South
New York, NY 10010

Friends of the Earth
Washington, DC 20003
218 D Street, SE

Global Greenhouse Network
1130 17th Street, NW
Washington, DC 20036

Greenpeace, USA
1436 U Street, NW
Washington, DC 20009

Inform, Inc.
381 Park Avenue South
New York, NY 10016

International Research Expeditions
140 University Drive
Menlo Park, CA 94025

Land Stewardship Program
14758 Ostlund Trail N
Marine on St. Croix, MN 55047

Land Trust Alliance
1017 Duke Street
Alexandria, VA 22314

League of Conservation Voters
1150 Connecticut Avenue, NW, Suite 201
Washington, DC 20036

Marine Education and Research Ltd.
17 Hartington Park
Bristol, Great Britain BS6 7ES

National Audubon Society
950 Third Avenue
New York, NY 10022

National Parks and Conservation Association
1015 31st Street, NW
Washington, DC 20007

National Toxics Campaign
37 Temple Place, 4th Floor
Boston, MA 02111

National Wildlife Federation
1412 16th Street, NW
Washington, DC 20036-2266

Natural Resources Defense Council
40 West 20th Street
New York, NY 10011

Nature Conservancy
1815 North Lynn Street
Arlington, VA 22209

Pacific Whale Foundation
101 North Kihei Road, Suite 21
Kihei, Maui, HI 96753

Pollution Probe Foundation
12 Madison Avenue
Toronto, Ontario
Canada M5R 2S1

Public Citizen
P.O. Box 19404
Washington, DC 20036

Rainforest Action Network
301 Broadway, Suite A
San Francisco, CA 94133

Rainforest Alliance
270 Lafayette Street, Suite 5B
New York, NY 10012

Renew America
1400 16th Street, NW, Suite 710
Washington, DC 20036

Sierra Club
730 Polk Street
San Francisco, CA 94109

Student Conservation Association, Inc.
P.O. Box 550
Charlestown, NH 03630

APPENDIX 15: (CONTINUED)

Trust for Public Land
116 New Montgomery, 4th Floor
San Francisco, CA 94105

Water Environment Federation
601 Wythe Street
Alexandria, VA 22314

Wildlife Conservation International
New York Zoological Society
New York, NY 10460-9973

World Society for the Protection of Animals
29 Perkins Street, P.O. Box 190
Boston, MA 02130

World Wildlife Fund
1250 24th Street, NW
Washington, DC 20037

Worldwatch Institute
1776 Massachusetts Avenue, NW
Washington, DC 20036

Zero Population Growth
1400 16th Street, NW
Washington, DC 20036

APPENDIX 16: RELATIVE ROUGHNESS OF VARIOUS KINDS OF NEW PIPES

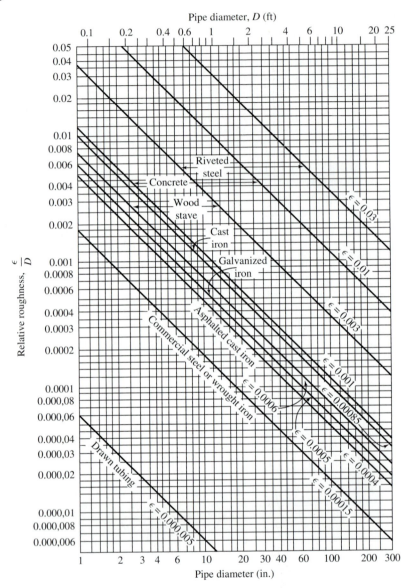

APPENDIX 17: MOODY DIAGRAM

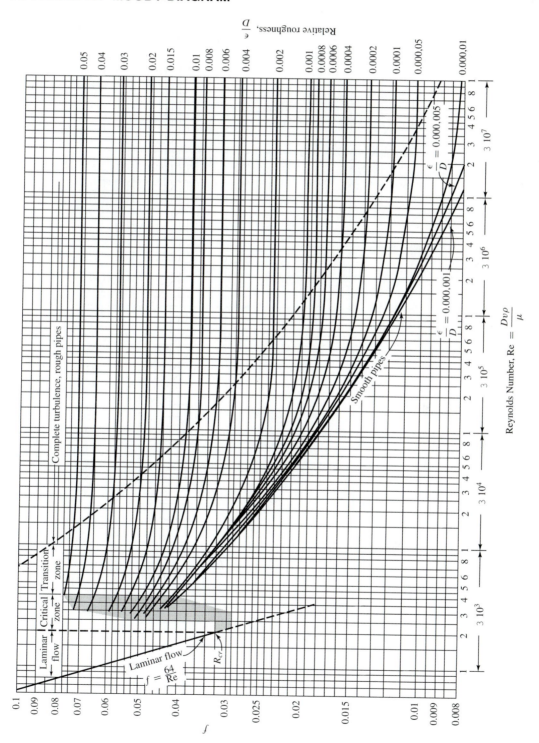

APPENDIX 18: DENSITY AND DYNAMIC VISCOSITY OF WATER

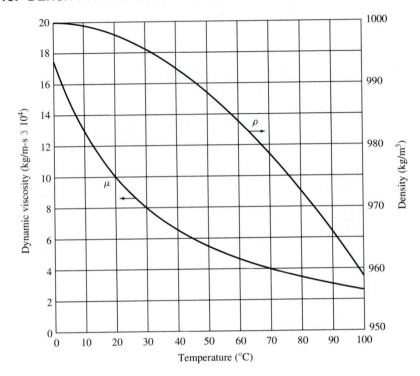

APPENDIX 19: DENSITY AND DYNAMIC VISCOSITY OF AIR

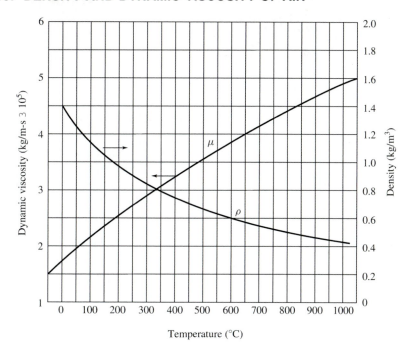

Temperature (°C)

APPENDIX 20: GAUSS–GREEN DIVERGENCE THEOREM

The Gauss–Green divergence theorem converts the area integral to volume integral, and vice versa. Let $\mathbf{S}$ be any vector and $\hat{n}$ be the unit vector normal to area dA, where A is the area surrounding the domain of volume V. Form the volume integral $\int_V \frac{\partial S_3}{\partial z} dV$, where S_3 is the scalar component of $\mathbf{S}$ on the z-axis. Hence,

$$\int_V \frac{\partial S_3}{\partial z} dV = \int_V \frac{\partial S_3}{\partial z} dx\, dy\, dz = \int_A \left\{ \int_{z_1}^{z_2} \frac{\partial S_3}{\partial z} dz \right\} dx\, dy = \int_A \left\{ \int_{z_1}^{z_2} \frac{\partial S_3}{\partial z} dz \right\} [\hat{n} \bullet \hat{n}_3]\, dA$$

where $\hat{n}_3$ is the unit vector in the positive z-direction.

$$\int_A \left\{ \int_{z_1}^{z_2} \frac{\partial S_3}{\partial z} dz \right\} [\hat{n} \bullet \hat{n}_3]\, dA = \int_{A_2} \{S_3(x, y, z_2)\} [\hat{n} \bullet \hat{n}_3]\, dA_2$$

$$- \int_{A_1} \{S_3(x, y, z_1)\} [(-\hat{n} \bullet \hat{n}_3)]\, dA_1$$

A_1 and A_2 are the bounding surface areas in the negative and positive z-directions, respectively. Hence $\hat{n} \bullet \hat{n}_3$ is negative on A_1 (since on $A_1 \hat{n}$ is pointing toward the negative z while $\hat{n}_3$ is pointing toward the positive z). $\hat{n} \bullet \hat{n}_3$ is positive on A_2 (by parallel reasoning).

Therefore,

$$\int_{V} \frac{\partial S_3}{\partial z} dV = \int_{A_2} \{S_3(x, y, z_2)\} [\hat{n} \bullet \hat{n}_3] dA_2 + \int_{A_1} \{S_3(x, y, z_1)\} (\hat{n} \bullet \hat{n}_3) dA_1$$

$$= \int_{A} S_3 [\hat{n} \bullet \hat{n}_3] dA \qquad (20\text{-}1)$$

In a similar manner,

$$\int_{V} \frac{\partial S_1}{\partial x} dV = \int_{A} S_1 [\hat{n} \bullet \hat{n}_1] dA \qquad (20\text{-}2)$$

$$\int_{V} \frac{\partial S_2}{\partial y} dV = \int_{A} S_2 [\hat{n} \bullet \hat{n}_2] dA \qquad (20\text{-}3)$$

Adding equations (20-1), (20-2), and (20-3) produces the *Gauss–Green divergence theorem:*

$$\int_{V} \frac{\partial S_1}{\partial x} dV + \int_{V} \frac{\partial S_2}{\partial y} dV + \int_{V} \frac{\partial S_3}{\partial z} dV = \int_{A} \hat{n} \bullet [S_1 \hat{n}_1 + S_2 \hat{n}_2 + S_3 \hat{n}_3] dS \qquad (20\text{-}4)$$

$$\int_{V} (\nabla \bullet \mathbf{S}) \, dV = \int_{A} (\hat{n} \bullet \mathbf{S}) \, dA \qquad (20\text{-}5)$$

APPENDIX 21: REYNOLDS TRANSPORT THEOREM

Two general methods are used in describing fluid flow: Eulerian and Lagrangian. In the *Eulerian method*, values of properties of the fluid (property is an observable quality of a fluid) are observed at several points in space and time. The *Lagrangian method*, on the other hand, involves tagging each particle of fluid and observing the behavior of the desired property as the fluid moves, independent of spatial location. For example, if velocity is the desired property, this velocity is measured as the particle moves without regard to location in space.

It is possible to convert one method of characterization to the other using the Reynolds transport theorem attributed to Osborne Reynolds. In other words, the theorem converts the Eulerian method of description of fluid flow to the Lagrangian method, and vice versa.

To derive the Reynolds transport theorem, invent two terms: control volume and control mass (Figure 21-1). The solid closed curve on the left with the hatching slanting upward to the right is the *control volume*. The other dashed closed curve with the hatching slanting upward to the left is the *control mass*, that part of the universe composed of masses identified for analysis; as such, if a boundary is used to enclose the masses, no other masses are allowed to pass through this boundary. Control mass is also called a *closed system* or simply *a system,* since no mass is allowed to cross the boundary. The control volume, on the other hand, is a specific volume in space where masses are allowed to come and go, implying that its boundary is permeable to the "traffic" of masses crisscrossing through it.

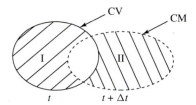

Figure 21-1 Derivation of the Reynolds transport theorem.

Imagine the control mass and the control volume coinciding at an initial time t; that is, the boundaries of both the control volume and the control mass coincide at this initial time. From this position, let them move at different speeds but with the control mass moving faster. After some time Δt later, the control mass and the control volume take different positions; this is the configuration portrayed in Figure 21-1.

Let Φ represent the value of any property of the system. The change of this property when the system goes from t to $t + \Delta t$ is

$$\Delta \Phi_M = \Phi_{M,t+\Delta t} - \Phi_{M,t} \tag{21-1}$$

where Δ is a symbol for change and M refers to control mass. At the initial time, since control mass and control volume coincide, the Φ of the the control mass $\Phi_{M,t}$, is equal to the Φ of the masses inside the control volume $\Phi_{V,t}$. V refers to the control volume. At time Δt later, $\Phi_{M,t+\Delta t}$ is given by

$$\Phi_{M,t+\Delta t} = \Phi_{V,t+\Delta t} - \Phi_{I,t+\Delta t} + \Phi_{II,t+\Delta t} \tag{21-2}$$

where $\Phi_{M,t+\Delta t}$ is the value of Φ for the control mass at $t + \Delta t$. $\Phi_{V,t+\Delta t}$, $\Phi_{I,t+\Delta t}$, and $\Phi_{II,t+\Delta t}$ are, respectively, the values of Φ of the masses in the control volume, space I, and space II at time $t + \Delta t$. Substituting equation (21-2) in equation (21-1), with $\Phi_{V,t}$ equal to $\Phi_{M,t}$, produces

$$\Delta \Phi_M = \Phi_{V,t+\Delta t} - \Phi_{I,t+\Delta t} + \Phi_{II,t+\Delta t} - \Phi_{V,t} \tag{21-3}$$

Rearranging equation (21-3) and dividing all terms by Δt produces

$$\frac{\Delta \Phi_M}{\Delta t} = \frac{\Phi_{V,t+\Delta t} - \Phi_{V,t}}{\Delta t} - \frac{\Phi_{I,t+\Delta t}}{\Delta t} + \frac{\Phi_{II,t+\Delta t}}{\Delta t} \tag{21-4}$$

The function Φ on the left-hand side of equation (21-4) is a function of only one independent variable, t. Hence, in the limit, the term will produce a total derivative as follows:

$$\lim_{\Delta t \to 0} \frac{\Delta \Phi_M}{\Delta t} = \frac{D\Phi_M}{Dt} = \frac{D \int_M \phi \rho \, dV}{Dt} \tag{21-5}$$

where ϕ is the value of the property Φ per unit mass of the system and ρ is the mass density of the system.

For the first term on the right-hand side equation (21-4), Φ is a function of both space (V) and time. Therefore, in the limit, the result will be a partial derivative and

$$\lim_{\Delta t \to 0} \frac{\Phi_{V,t+\Delta t} - \Phi_{V,t}}{\Delta t} = \frac{\partial \int_V \phi \rho \, dV}{\partial t} \tag{21-6}$$

The second term on the right-hand side of equation (21-4) is simply a rate of inflow of the property Φ across the boundary. Similarly, the last term is simply a rate of outflow of this property across the boundary in the opposite side. The volume rates of inflow or outflow are the product of velocities and cross-sectional areas of flow, and the mass rates of inflow and outflow are, accordingly, the product of the mass density and the respective volume rates. The last two terms on the right-hand side of equation (21-4) are, then, in the limit

$$\lim_{\Delta t \to 0} \left(-\frac{\Phi_{I,t+\Delta t}}{\Delta t} + \frac{\Phi_{II,t+\Delta t}}{\Delta t} \right) = \oint_A \phi \rho \mathbf{v} \bullet \hat{n} \, dA \qquad (21\text{-}7)$$

where $\oint_A$ indicates that the terms inside the parentheses on the left-hand side are summed over the area A that surrounds the control volume Ψ. $\mathbf{v}$ and $\hat{n}$ are the velocity and unit normal vectors, respectively. The bullet $\bullet$ is the symbol for the dot product. Equation (21-7) states that volume rate is negative for inflow and positive for outflow. Substituting equations (21-5), (21-6), and (21-7) into equation (21-4), we obtain

$$\frac{D \int_M \phi \rho \, d\Psi}{Dt} = \frac{\partial \int_\Psi \phi \rho \, d\Psi}{\partial t} + \int_A \phi \rho \mathbf{v} \bullet \hat{n} \, dA \qquad (21\text{-}8)$$

The left-hand side of equation (21-8) is a derivative of the property in the control mass. Hence, in a control mass, the derivative is a derivative with respect to time and not of space. This is the derivative that would be observed on a given fluid property, irrespective of where the fluid is in space. This is the Lagrangian method of describing the property and the derivative is called the *Lagrangian derivative.*

The right-hand side of the equation shows the derivatives of the property in the control volume. The derivatives express fluid property as a function of time and space. The function with respect to time is the first term; the function with respect to space is the second term, where the property ϕ is convected by the velocity, $\mathbf{v}$. Since convection involves moving through space, the value of the function must vary as the convection proceeds through space. This method of description is what the observer from a distance is capable of seeing. Standing at a distance, fluid properties as they vary from point to point may be observed (convective derivative); in addition, standing at the same distance, the same fluid properties as they vary with time may also be observed (local derivative). This mode of observation or description is the Eulerian method, and the entire right-hand side of the equation may be called the *Eulerian derivative.*

Equation (21-8) portrays the equivalence of the Lagrangian and the Eulerian views: the Reynolds transport theorem. The left-hand side of the equation is also called *material, substantive,* or *comoving derivative.* The partial derivative on the right-hand side is called *local derivative,* and the second term on this side is called *convective derivative. Since the theorem was derived by shrinking Δt to zero, it must be strongly stressed that the boundaries of both the control mass and the control volume coincide in the application of this theorem.*

APPENDIX 22: LITERATURE ON WAT25 AND ITS INSTRUCTIONS

WAT25 programs the following equations (see Chapter 5 for meaning and derivations). The program can be used to model BOD demanding discharges to both free-flowing and tidal receiving waters.

$$L_c = L_{c0}e^{q_{c-}x} + \frac{G_{Lc}}{k_c'}(1 - e^{q_{c-}x}) \qquad\qquad x \geq 0$$

$$L_c = L_{c0}e^{q_{c+}x} + \frac{G_{Lc}}{k_c'}(1 - e^{q_{c+}x}) \qquad\qquad x \leq 0$$

$$q_{c+} = \frac{u + \sqrt{u^2 + 4E_{xx}k_c'}}{2E_{xx}} \quad \text{and} \quad q_{c-} = \frac{u - \sqrt{u^2 + 4E_{xx}k_c'}}{2E_{xx}}$$

$$L_n = L_{n0}e^{q_{n-}x} + \frac{G_{Ln}}{k_n}(1 - e^{q_{n-}x}) \qquad\qquad x \geq 0$$

$$L_n = L_{n0}e^{q_{n+}x} + \frac{G_{Ln}}{k_n}(1 - e^{q_{n+}x}) \qquad\qquad x \leq 0$$

$$q_{n+} = \frac{u + \sqrt{u^2 + 4E_{xx}k_n}}{2E_{xx}} \quad \text{and} \quad q_{n-} = \frac{u - \sqrt{u^2 + 4E_{xx}k_n}}{2E_{xx}}$$

$$c_4 = c_{40}e^{q_{\mu-}x} \qquad\qquad x \geq 0$$

$$c_4 = c_{40}e^{q_{\mu+}x} \qquad\qquad x \leq 0$$

$$q_{\mu-} = \frac{-u - \sqrt{u^2 + 4E_{xx}(\mu_p - kk)}}{2E_{xx}} \quad \text{and} \quad q_{\mu+} = \frac{-u + \sqrt{u^2 + 4E_{xx}(\mu_p - kk)}}{2E_{xx}}$$

$$D = D_0 e^{q_{0-}x} - \frac{1.24 a_{pc}c_{40}(\mu_p - k_d)}{k_2 - E_{xx}q_{\mu-}^2 + uq_{\mu-}}(e^{q_{\mu-}x} - e^{q_{0-}x})$$

$$+ \frac{L_{c0}k_c}{k_2 - E_{xx}q_{c-}^2 + uq_{c-}}(e^{q_{c-}x} - e^{q_{0-}x}) + \frac{L_{n0}k_n}{k_2 - E_{xx}q_{n-}^2 + uq_{n-}}(e^{q_{n-}x} - e^{q_{0-}x})$$

$$+ G_{Lc}\left[\frac{1 - e^{q_{0-}x}}{k_2} - \frac{e^{q_{c-}x} - e^{q_{0-}x}}{(k_2 - E_{xx}q_{c-}^2 + uq_{c-})}\right] + G_{Ln}\left[\frac{1 - e^{q_{0-}x}}{k_2} - \frac{e^{q_{n-}x} - e^{q_{0-}x}}{(k_2 - E_{xx}q_{n-}^2 + uq_{n-})}\right]$$

$$- \frac{G_O}{k_2}(1 - e^{q_{0-}x}) \qquad\qquad x \geq 0$$

$$q_{0-} = \frac{u - \sqrt{u^2 + 4E_{xx}k_2}}{2E_{xx}}$$

$$D = D_0 e^{q_{0+}x} - \frac{1.24 a_{pc}c_{40}(\mu_p - k_d)}{k_2 - E_{xx}q_{\mu+}^2 + uq_\mu^+}(e^{q_{\mu+}x} - e^{q_{0+}x})$$

$$+ \frac{L_{c0}k_c}{k_2 - E_{xx}q_{c+}^2 + uq_{c+}}(e^{q_{c+}x} - e^{q_{0+}x}) + \frac{L_{n0}k_n}{k_2 - E_{xx}q_{n+}^2 + uq_{n+}}(e^{q_{n+}x} - e^{q_{0+}x})$$

$$+ G_{Lc} \left[\frac{1 - e^{q_{0+}x}}{k_2} - \frac{e^{q_{c+}x} - e^{q_{0+}x}}{(k_2 - E_{xx}q_{c+}^2 + uq_{c-})} \right] + G_{L_n} \left[\frac{1 - e^{q_{0+}x}}{k_2} - \frac{e^{q_{n+}x} - e^{q_{0+}x}}{(k_2 - E_{xx}q_{n+}^2 + uq_{n+})} \right]$$

$$- \frac{G_O}{k_2}(1 - e^{q_{0+}x}) \hspace{8cm} x \le 0$$

$$q_{0+} = \frac{u + \sqrt{u^2 + 4E_{xx}k_2}}{2E_{xx}}$$

$$k_T = k_{20}\theta^{T-20}$$

$$c_{d0} = \frac{c_{db}(Q_b) + c_{dtrib}(Q_{trib1})}{Q_b + Q_{trib}}$$

Running WAT v. 2.5

WAT version 2.5 (or WAT25 or simply WAT) requires the following:

- An IBM PC or compatible computer (80x86 based). A computer with a math coprocessor is recommended.
- A floppy drive and a floppy diskette for data storage.
- A hard drive for placing WAT25.EXE.
- MS-DOS or PC-DOS version 3.1 or later.
- At least 400K of conventional memory available for the program.

Installing WAT to hard disk. As the computer is turned on, the first screen to appear is the root directory. At this directory make the subdirectory WAT25 by typing the command shown at the first bullet below (note that capital letters are to be typed exactly as shown). After creating the subdirectory, move to this subdirectory by typing the command at the next bullet below. Now, you are in the subdirectory WAT25. Insert the diskette provided into a floppy drive (such as A, B, D, and so on). In the command at the third bullet below, the diskette is assumed to be inserted into drive A; if the drive is other than A, substitute the letter of that drive for the A below. Type the command at the third bullet; this copies the model WAT25.EXE into the subdirectory WAT25 of the hard drive. Now, the model is completely installed.

- `MKDIR WAT25`
- `CD WAT25`
- `COPY A:\WAT25.EXE`

Practice run. Practice running the model by using the example provided. There are two example runs provided with the diskette: Examples 5-11 and 5-12. These examples are contained in the datafiles PROJECT.DAT, 00000005.STA, and 00000006.STA. While in the subdirectory WAT25, copy these files by following the commands below (again, the floppy diskette is assumed to be inserted into drive A).

- COPY A:\PROJECT.DAT
- COPY A:\00000005.STA
- COPY A:\00000006.STA

The WAT v. 2.5 program is totally contained in the single DOS executable file WAT25.EXE. Running the program is simply typing "WAT25" (without the quotes) from the DOS command line (DOS prompt).

The datafiles mentioned above were created by WAT. There are two general places where these files may be written to: the hard disk where WAT25.EXE is located and any floppy disk. Writing to the hard disk is not recommended, as this will clutter this disk. It is recommended that a nonprotected floppy be used for datafile storage. *In this practice run, however, the datafile for Examples 5-11 and 5-12 have already been copied to the hard disk (PROJECT.DAT, 00000005.STA, and 00000006.STA). Hence, use the hard disk for this practice. The command to initiate the practice run is*

    WAT25

The command for using other disks for datafile storage would require specifying the disk drive. For example, using A the command would be

    WAT25 A:\

Henceforth, follow the instructions on the screen. After the project has been analyzed, choose "Print Analysis Report" from the "Control Menu" to print the results. Again, just follow the on-screen instructions; WAT is very easy to run.

Generic instructions for running WAT. Practically, the experiences learned from the practice run above are all that are necessary to navigate through WAT25. However, there are still others that are needed to be learned relating to the display.

A predetermined set of display colors has been selected for WAT to be used on systems that have color displays. However, these colors may not look very pleasing or some characters may appear invisible if a monochrome display is being used or a color display is not available. WAT will have to be told if the display is color, monochrome, or LCD. To do so, use the "/B", "/M", or "/L" switch on the command line. The "/B" switch tells WAT to use the black-and-white color palette. This switch would be used if WAT is running with a color graphics adapter (such as a CGA, EGA, or VGA adapter) but is connected to a black-and-white or monochrome monitor. The "/M" switch tells WAT to use the monochrome display adapter (MDA) color palette. This switch would be used, obviously, if WAT is being run on a system using the MDA display adapter. The "/L" switch tells WAT to use the liquid crystal display (LCD) color palette. This switch would be used if WAT is being run on a laptop computer or any other system using an LCD display. For example, if WAT is running on a system with a VGA display adapter but is connected to a monochrome monitor, it would be specified on the command line like this (continuing with the previous example):

```
WAT25 A:\ /B
```

WAT v. 2.5 Source Code and Compiling

Included with WAT25.EXE is the WAT v. 2.5 BASIC and assembly language source code. The source code is composed of eight BASIC modules and three assembly language modules. To compile the BASIC source code, Microsoft QuickBASIC v. 4.5 or later is required. To assemble the assembly language source code, Borland International's Turbo Assembler v. 1.0 or later is required (it may also assemble under Microsoft Macro Assembler v. 5.0 or later with little or no modification). However, assembled versions of the assembly source code (QPRINT.ASM, SAVEAREA.ASM, and RESTAREA.ASM) have already been included. They are placed in the WAT25.LIB library file included with the source code.

Due to the size of the program, *compiling it to memory* in the QuickBASIC environment (QB.EXE) is not possible (unless the extended version of QuickBASIC, QBX.EXE, included with Microsoft BASCOM v. 7.1 Professional Development System is used). Therefore, it can only be compiled *to disk* using the QuickBASIC command line compiler (BC.EXE).

To compile the WAT v. 2.5 program, simply follow these steps:

- Compile all eight modules (WAT25.BAS, WAT25ANA.BAS, WAT25ERR.BAS, WAT25EVT.BAS, WAT25PN1.BAS, WAT25PN2.BAS, WAT25SUP.BAS, and WAT25UTL.BAS) using BC.EXE. Be sure to use the following command line switches: /O, /E, /X, /V, /W, /T, and /C: 512. For example, to compile the WAT25.BAS module, it would be entered on the command line like this:

```
BC WAT25.BAS /O /E /X /V /W /T /C: 512;
```

- After compilation of all eight modules is complete, eight corresponding object files will result. Using the Microsoft Overlay Linker (LINK.EXE, included with the Microsoft QuickBASIC package), link all eight object files together, along with the WAT25.LIB library file. Be sure to use the /EX and /NOE command line switches when invoking LINK. The command to use would be something like this:

```
LINK /EX /NOE WAT25 + WAT25ANA + WAT25ERR + WAT25EVT +
WAT25PN1 + WAT25PN2 + WAT25SUP + WAT25UTL,,,WAT25.LIB
```

The result of the linking process is WAT25.EXE.

Index